HANDBOOK OF MARINE CRAFT HYDRODYNAMICS AND MOTION CONTROL

HANDBOOK OF MARINE CRAFT HYDRODYNAMICS AND MOTION CONTROL

Second Edition

Thor I. Fossen

Norwegian University of Science and Technology
Trondheim, Norway

WILEY

This second edition first published 2021

Edition History
John Wiley & Sons, Ltd. (1e 2011)

Registered Offices
John Wiley & Sons, Inc., 111 River Street, Hoboken, NJ 07030, USA
John Wiley & Sons Ltd, The Atrium, Southern Gate, Chichester, West Sussex, PO19 8SQ, UK

Editorial Office
The Atrium, Southern Gate, Chichester, West Sussex, PO19 8SQ, UK

For details of our global editorial offices, customer services, and more information about Wiley products visit us at www .wiley.com.

Wiley also publishes its books in a variety of electronic formats and by print-on-demand. Some content that appears in standard print versions of this book may not be available in other formats.

Library of Congress Cataloging-in-Publication Data

Names: Fossen, Thor I., author.
Title: Handbook of marine craft hydrodynamics and motion control / Thor I. Fossen, Norwegian University of Science and Technology, Trondheim, Norway.
Description: Second edition. | Hoboken, NJ : Wiley, 2021. | Includes bibliographical references and index.
Identifiers: LCCN 2021007896 (print) | LCCN 2021007897 (ebook) | ISBN 9781119575054 (cloth) | ISBN 9781119575047 (adobe pdf) | ISBN 9781119575030 (epub)
Subjects: LCSH: Ships–Hydrodynamics. | Stability of ships. | Motion control devices. | Automatic pilot (Ships) | Steering-gear. | Ships–Electronic equipment.
Classification: LCC VM156 .F67 2021 (print) | LCC VM156 (ebook) | DDC 623.8/1–dc23
LC record available at https://lccn.loc.gov/2021007896
LC ebook record available at https://lccn.loc.gov/2021007897

Cover Design: Wiley
Cover Image: © MR.Cole_Photographer/Moment/Getty Images

Set in 10/12pt TimesNewRomanMTStd by SPi Global, Chennai, India

C9781119575054_180321

This book is dedicated to my parents Gerd Kristine and Ole Johan Fossen, and my family Heidi, Sindre and Lone Moa who have always been there for me.

Thor I. Fossen

Contents

Part Three Appendices

About the Author

Professor Thor I. Fossen received an MSc degree in Marine Technology in 1987 and a PhD in Engineering Cybernetics in 1991, both from the Norwegian University of Science and Technology (NTNU). Fossen's academic background, besides cybernetics, is computer science, robotics, cybersecurity, aerospace engineering, marine technology and inertial navigation systems. He studied flight control as a Fulbright Scholar at the University of Washington, Seattle. He was appointed professor of guidance, navigation and control at NTNU at age 30. He has been visiting professor at University of California, San Diego (UCSD), University of California, Santa Barbara (UCSB), Aalborg University, Denmark and the Technical University of Denmark (DTU). Professor Fossen has been elected to the Norwegian Academy of Technological Sciences (1998) and elevated to IEEE Fellow (2016). He is one of the founders of the company Marine Cybernetics (2002), which was acquired by DNV GL in 2014. He is also one of the founders of the company SCOUT Drone Inspection (2017).

Professor Fossen is the author of the Wiley textbooks:

Fossen, T. I (2021). *Handbook of Marine Craft Hydrodynamics and Motion Control*, 2nd ed. John Wiley & Sons, Ltd. Chichester, UK

Fossen, T. I (2011). *Handbook of Marine Craft Hydrodynamics and Motion Control*, 1st edition. John Wiley & Sons, Ltd. Chichester, UK

Fossen, T. I (1994). *Guidance and Control of Ocean Vehicles*, John Wiley & Sons, Ltd. Chichester, UK

and co-author of the editorials:

Fossen, T. I, K. Y. Pettersen and H. Nijmeijer (2017). *Sensing and Control for Autonomous Vehicles*, Springer Verlag

Fossen, T. I and H. Nijmeijer (2012). *Parametric Resonance in Dynamical Systems*, Springer Verlag

Nijmeijer, H. and T. I. Fossen (1999). *New Directions in Nonlinear Observer Design*, Springer Verlag.

Professor Fossen has been instrumental in the development of several industrial autopilot, path-following and dynamic positioning (DP) systems. He also has experience in state estimators for marine craft and automotive systems as well as strapdown GNSS/INS navigation systems for ships and aerial vehicles. He received the Automatica Prize Paper Award in 2002 for a concept for weather optimal positioning control of marine craft. In 2008 he received the Arch T. Colwell Merit Award at the SAE 2008 World Congress.

He is currently professor of guidance, navigation and control in the Department of Engineering Cybernetics, NTNU.

Preface

The main motivation for writing the second edition of this book was to include new results on guidance, navigation and control (GNC) and hydrodynamic modeling of marine craft. In addition, the book has been reorganized into 16 chapters to improve readability.

The Wiley book from 1994 was the first attempt to bring hydrodynamic modeling and control system design into a unified notation for modeling, simulation and control. My first book also contains state-of-the-art linear and nonlinear design methods for ships and underwater vehicles up to 1994. In the period 1994–2011 a great deal of work was done on nonlinear control of marine craft. This work resulted in many useful results and lecture notes, which have been collected and included in the *Handbook of Marine Craft Hydrodynamics and Motion Control*. The first edition was published in 2011 and it was used as the main textbook in my course on "Guidance, Navigation and Control of Vehicles" at the Norwegian University of Science and Technology (NTNU) until 2020. Then it was replaced by this edition, which contains many new chapters and sections, general improvements, recent results and MATLAB scripts, which can be downloaded from a GitHub repository. The new chapters of the second edition cover autopilot models, models for underwater vehicles, control forces and moments, model-based navigation systems, Kalman filtering and inertial navigation systems.

Accompanying MATLAB Software

In 2019 the Marine Systems Simulator (MSS), which is an open source MATLAB toolbox, was migrated to GitHub in order to improve version control and support incremental software updates. The MATLAB software accompanying the book can be downloaded from the repository:

MSS toolbox: https://github.com/cybergalactic/MSS

The second edition of the book is fully compliant with the new features of the MSS toolbox and the book contains many examples using the MSS toolbox.

Preview of the Book

Part I of the book covers maneuvering and seakeeping theory and it is explained in detail how the equations of motion can be derived for both cases using both frequency- and time-domain formulations. This includes transformations from the frequency to the time domain and the explanation of fluid-memory effects. Great effort has been made in the development of kinematic equations for effective representation of the equations of motion in seakeeping, body, inertial and geographical coordinates. This is very confusing in the existing literature on hydrodynamics and the need to explain this properly motivated me to find a unifying notation for marine and mechanical systems. This was done in the period 2002–2010 and it is inspired by the elegant formulation used in robotics where systems are represented in a matrix-vector notation. The unified notation dates back to my PhD thesis, which was published in 1991.

Part II of the book covers guidance systems, navigation systems, state estimators and control of marine craft. This second part of the book focuses on state-of-the-art methods for feedback control such as PID control design for linear and nonlinear systems as well as control allocation methods. A chapter with more advanced topics, such as optimal control theory, backstepping, feedback linearization and sliding-mode control, is included for the advanced reader. Case studies and applications are treated at the end of each chapter. The control systems based on PID and optimal control theory are designed with a complexity similar to those used in many industrial systems. The more advanced methods using nonlinear theory are included so the user can compare linear and nonlinear design techniques before a final implementation is made. Many references to existing systems are included so control system vendors can easily find articles describing state-of-the art design methods for marine craft.

Acknowledgments

Most of the results in the book have been developed at NTNU in close cooperation with a large number of my former doctoral students. Our joint efforts have resulted in several patents, industrial implementations and spin-off companies.

The results on maneuvering and seakeeping are joint work with *Dr Tristan Perez*, who visited NTNU as a researcher in 2004–2007. The work with Dr Perez has resulted in several joint publications and I am grateful to him for numerous interesting discussions on hydrodynamic modeling and control.

I am particular grateful to *Professor Tor Arne Johansen* with whom I have co-authored a large number of my publications. Many of our joint results are included in the second edition of the book. *Dr Morten Breivik* and *Associate Professor Anastasios Lekkas* have contributed with many important results on guidance systems. *Professor Mogens Blanke* was instrumental in the development of the the sections on maneuvering, roll damping and propulsion theory. *Associate Professor Torleiv H. Bryne* should be thanked for important contributions on ship control and inertial navigation systems, while *Bjarne Stenberg* should be thanked for creating many of the graphical illustrations. Finally, *Stewart Clark* should be thanked for his assistance with the English language. The book project has been sponsored by the Norwegian Research Council through the author's affiliation as key scientist at the Center of Ships and Ocean Structures (2002–2012) and co-director of the Center of Autonomous Marine Operations and Systems since 2012.

Thor I. Fossen

List of Tables

Part One

Marine Craft Hydrodynamics

De Navium Motu Contra Aquas

1

Introduction to Part I

The subject of this book is *motion control and hydrodynamics of marine craft*. The term marine craft includes ships, high-speed craft, semi-submersibles, floating rigs, submarines, remotely operated and autonomous underwater vehicles, torpedoes, and other propelled and powered structures, for instance a floating air field. Offshore operations involve the use of many marine craft, as shown in Figure 1.1. *Vehicles* that do not travel on land (ocean and flight vehicles) are usually called craft, such as watercraft, sailcraft, aircraft, hovercraft and spacecraft. The term vessel can be defined as follows:

> *Vessel*: "hollow structure made to float upon the water for purposes of transportation and navigation; especially, one that is larger than a rowboat."

The words *vessel, ship* and *boat* are often used interchangeably. In *Encyclopedia Britannica*, a ship and a boat are distinguished by their size through the following definition:

> *Ship*: "any large floating vessel capable of crossing open waters, as opposed to a boat, which is generally a smaller craft. The term formerly was applied to sailing vessels having three or more masts; in modern times it usually denotes a vessel of more than 500 tons of displacement. Submersible ships are generally called boats regardless of their size."

Similar definitions are given for submerged vehicles:

> *Submarine*: "any naval vessel that is capable of propelling itself beneath the water as well as on the water's surface. This is a unique capability among warships, and submarines are quite different in design and appearance from surface ships."
> *Underwater vehicle*: "small vehicle that is capable of propelling itself beneath the water surface as well as on the water's surface. This includes unmanned underwater vehicles (UUV), remotely operated vehicles (ROV), autonomous underwater vehicles (AUV) and underwater robotic vehicles (URV). Underwater vehicles are used both commercially and by the navy."

Handbook of Marine Craft Hydrodynamics and Motion Control, Second Edition. Thor I. Fossen.

Figure 1.1 Marine craft in operation. Source: illustration by B. Stenberg.

From a hydrodynamic point of view, marine craft can be classified according to their maximum operating speed. For this purpose it is common to use the *Froude number*

$$F_{\mathrm{n}} := \frac{U}{\sqrt{gL}} \tag{1.1}$$

where U is the craft speed, L is the overall submerged length of the craft and g is the acceleration of gravity. The pressure carrying the craft can be divided into *hydrostatic* and *hydrodynamic* pressure. The corresponding forces are:

- Buoyancy force due to the hydrostatic pressure (proportional to the displacement of the ship).
- Hydrodynamic force due to the hydrodynamic pressure (approximately proportional to the square of the relative speed to the water).

For a marine craft sailing at constant speed U, the following classifications can be made (Faltinsen 2005):

Displacement vessels ($F_n < 0.4$): The buoyancy force (restoring terms) dominates relative to the hydrodynamic forces (added mass and damping).

Semi-displacement vessel (0.4–$0.5 < F_n < 1.0$–1.2): The buoyancy force is not dominant at the maximum operating speed for a high-speed submerged hull type of craft.

Planing vessel ($F_n > 1.0$–1.2): The hydrodynamic force mainly carries the weight. There will be strong flow separation and the aerodynamic lift and drag forces start playing a role.

Figure 1.2 Displacement vessel.

In this book only displacement vessels are covered; see Figure 1.2.

The Froude number has influence on the hydrodynamic analysis. For displacement vessels, the waves radiated by different parts of the hull do not influence other parts of the hull. For semi-displacement vessels, waves generated at the bow influence the hydrodynamic pressure along the hull towards the stern. These characteristics give rise to different modeling hypotheses, which lead to different hydrodynamic theories.

For displacement ships it is widely accepted that two- and three-dimensional potential theory programs are used to compute the potential coefficients and wave loads; see Section 5.1. For semi-displacement vessels and planing vessels it is important to include the lift and drag forces in the computations (Faltinsen 2005).

Degrees of Freedom and Motion of a Marine Craft

In maneuvering, a marine craft experiences motion in six degrees of freedom (DOFs). The DOFs are the set of independent displacements and rotations that specify completely the displaced position and orientation of the craft. The motion in the horizontal plane is referred to as *surge* (longitudinal motion, usually superimposed on the steady propulsive motion) and *sway* (sideways motion). *Yaw* (rotation about the vertical axis) describes the heading of the craft. The remaining three DOFs are *roll* (rotation about the longitudinal axis), *pitch*(rotation about the transverse axis) and *heave* (vertical motion); see Figure 1.3.

Roll motion is probably the most influential DOF with regards to human performance, since it produces the highest accelerations and, hence, is the principal villain in seasickness. Similarly, pitching and heaving feel uncomfortable to people. When designing ship autopilots, yaw is the primary mode for feedback control. Stationkeeping of a marine craft implies stabilization of the surge, sway and yaw motions.

When designing feedback control systems for marine craft, reduced-order models are often used since most craft do not have actuation in all DOFs. This is usually done by decoupling the motions of the craft according to:

1-DOF models can be used to design forward speed controllers (*surge*), heading autopilots (*yaw*) and roll-damping systems (*roll*).

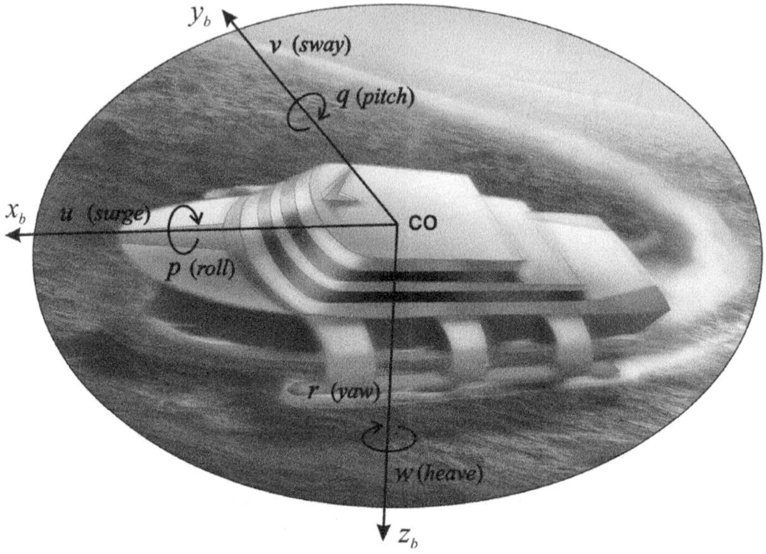

Figure 1.3 Motion in six degrees of freedom (DOFs).

3-DOF models are usually:

- Horizontal-plane models (*surge, sway* and *yaw*) for ships, semi-submersibles and underwater vehicles that are used in dynamic positioning systems, trajectory-tracking control systems and path-following systems. For slender bodies such as submarines, it is also common to assume that the motions can be decoupled into *longitudinal* and *lateral* motions.
- Longitudinal models (*surge, heave* and *pitch*) for forward speed, diving and pitch control.
- Lateral models (*sway, roll* and *yaw*) for turning and heading control.

4-DOF models (*surge, sway, roll* and *yaw)* are usually formed by adding the roll equation to the 3-DOF horizontal-plane model. These models are used in maneuvering situations where it is important to include the rolling motion, usually in order to reduce roll by active control of fins, rudders or stabilizing liquid tanks.

6-DOF models *(surge, sway, heave, roll, pitch* and *yaw*) are fully coupled equations of motion used for simulation and prediction of coupled vehicle motions. These models can also be used in advanced control systems for underwater vehicles that are actuated in all DOFs.

1.1 Classification of Models

The models in this book can be used for prediction, real-time simulation, decision-support systems, situational awareness as well as controller-observer design. The complexity and

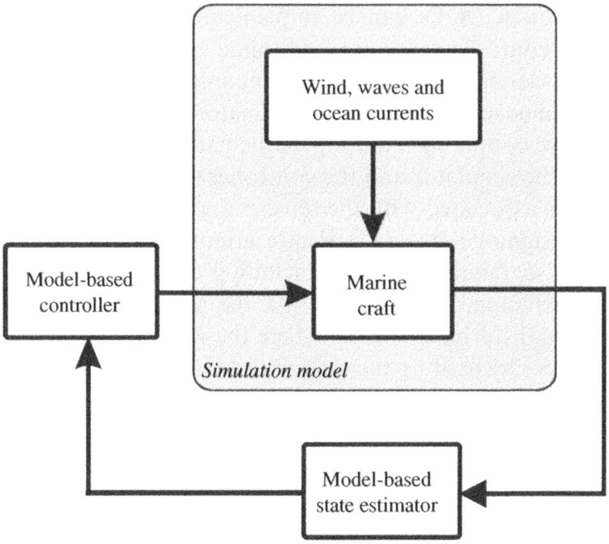

Figure 1.4 Models used in guidance, navigation and control systems.

number of differential equations needed for the various purposes will vary. Consequently, one can distinguish between three types of models (see Figure 1.4):

Simulation model: This model is the most accurate description of a system, for instance a 6-DOF *high-fidelity model* for simulation of coupled motions in the time domain. It includes the marine craft dynamics, propulsion system, measurement system and the environmental forces due to wind, waves and ocean currents. It also includes other features not used for control and observer design that have a direct impact on model accuracy. The simulation model should be able to reconstruct the time responses of the real system and it should also be possible to trigger failure modes to simulate events such as accidents and erroneous signals. Simulation models where the fluid-memory effects are included due to frequency-dependent added mass and potential damping typically consist of 50–200 ordinary differential equations (ODEs) while a maneuvering model can be represented in 6 DOFs with 12 ODEs for generalized position and velocity. In addition, some states are needed to describe the environmental forces and actuators, but still the number of states will be less than 50 for a marine craft.

Control design model: The motion control system is usually designed using a reduced-order or simplified version of the simulation model. In its simplest form, this model is used to compute a set of constant gains for a proportional, integral, derivative (PID) controller. More sophisticated control systems such as model-based control systems use a dynamic model to generate feedforward and feedback signals. The number of ODEs used in conventional model-based ship control systems is usually less than 20. A PID controller typically requires two states: one for the integrator and one for the low-pass filter used to limit noise amplification. Consequently, setpoint regulation in 6 DOFs can be implemented by using 12 ODEs. However, trajectory-tracking controllers require additional states for feedforward as well as filtering so higher-order control laws are not uncommon.

State estimator design model: Stochastic state estimators (Kalman filters) and deterministic state observers are both designed using mathematical models that are different from the models used in the simulator and the controller since the purpose is to capture the additional dynamics associated with the sensors and navigation system as well as disturbances. When designing the state estimator attention is given to accurate modeling of measurement noise, failure situations including dead-reckoning capabilities, filtering and motion prediction. For marine craft, the model-based state estimator often includes a stochastic disturbance model where the goal is to estimate wave, wind and current-induced forces by treating these as colored noise. For marine craft the number of ODEs in the state estimator will typically be 15–20 for a dynamic positioning (DP) system while a basic heading autopilot is implemented with less than five states.

1.2 The Classical Models in Naval Architecture

The motions of a marine craft exposed to wind, waves and ocean currents takes place in 6 DOFs. The equations of motion can be derived using the Newton–Euler or Lagrange equations. The equations of motion are used to simulate ships, high-speed craft, underwater vehicles and floating structures operating under or on the water surface, as shown in Figure 1.5. In Section 3.3 it is shown that a rigid body with constant mass m and center of

Figure 1.5 Ship and semi-submersibles operating offshore. Source: illustration by B. Stenberg.

gravity (x_g, y_g, z_g) relative to a fixed point on the hull can be described by the following coupled differential equations

$$m[\dot{u} - vr + wq - x_g(q^2 + r^2) + y_g(pq - \dot{r}) + z_g(pr + \dot{q})] = X \tag{1.2}$$

$$m[\dot{v} - wp + ur - y_g(r^2 + p^2) + z_g(qr - \dot{p}) + x_g(qp + \dot{r})] = Y \tag{1.3}$$

$$m[\dot{w} - uq + vp - z_g(p^2 + q^2) + x_g(rp - \dot{q}) + y_g(rq + \dot{p})] = Z \tag{1.4}$$

$$\begin{aligned} I_x\dot{p} + (I_z - I_y)qr - (\dot{r} + pq)I_{xz} + (r^2 - q^2)I_{yz} + (pr - \dot{q})I_{xy} \\ +m[y_g(\dot{w} - uq + vp) - z_g(\dot{v} - wp + ur)] = K \end{aligned} \tag{1.5}$$

$$\begin{aligned} I_y\dot{q} + (I_x - I_z)rp - (\dot{p} + qr)I_{xy} + (p^2 - r^2)I_{zx} + (qp - \dot{r})I_{yz} \\ +m[z_g(\dot{u} - vr + wq) - x_g(\dot{w} - uq + vp)] = M \end{aligned} \tag{1.6}$$

$$\begin{aligned} I_z\dot{r} + (I_y - I_x)pq - (\dot{q} + rp)I_{yz} + (q^2 - p^2)I_{xy} + (rq - \dot{p})I_{zx} \\ +m[x_g(\dot{v} - wp + ur) - y_g(\dot{u} - vr + wq)] = N \end{aligned} \tag{1.7}$$

where X, Y, Z, K, M and N denote the external forces and moments. This model is the basis for time-domain simulation of marine craft. The external forces and moments acting on a marine craft are usually modeled by using:

Maneuvering theory: The study of a ship moving at constant positive speed U in calm water within the framework of maneuvering theory is based on the assumption that the maneuvering (hydrodynamic) coefficients are *frequency independent* (no wave excitation). The maneuvering model will in its simplest representation be linear while nonlinear representations can be derived using methods such as cross-flow drag, quadratic damping or Taylor-series expansions; see Chapter 6.

Seakeeping theory: The motions of ships at zero or constant speed in waves can be analyzed using seakeeping theory where the hydrodynamic coefficients and wave forces are computed as a function of the wave excitation frequency using the hull geometry and mass distribution. The seakeeping models are usually derived within a linear framework (Chapter 5) while the extension to nonlinear theory is an important field of research.

For underwater vehicles operating below the wave-affected zone, the wave excitation frequency will not affect the hydrodynamic mass and damping coefficients. Consequently, it is common to model underwater vehicles with constant hydrodynamic coefficients similar to a maneuvering ship.

1.2.1 Maneuvering Theory

Maneuvering theory assumes that the ship is moving in restricted calm water, that is in sheltered waters or in a harbor. Hence, the maneuvering model is derived for a marine craft moving at positive speed U under a zero-frequency wave excitation assumption such that added mass and damping can be represented by using constant hydrodynamic derivatives corresponding to the zero-frequency values of added mass and damping. The zero-frequency assumption is only valid for *surge, sway* and *yaw*, which can be explained by considering a PD-controlled ship in these modes behaving like three mass–damper-spring systems. It is observed that ships will oscillate at 100–150 s under closed-loop control. For 150 s this corresponds to a natural frequency

$$\omega_{\mathrm{n}} = \frac{2\pi}{T} \approx 0.04 \text{ rad s}^{-1} \tag{1.8}$$

which gives confidence in choosing the added mass and damping coefficients at the zero frequency.

The natural frequencies in *heave, roll* and *pitch* are much higher so it is not recommended to use the zero-frequency potential coefficients in these modes. For instance, a ship with a roll period of 10 s will have a natural frequency of 0.628 rad s^{-1} which clearly violates the zero-frequency assumption. This means that hydrodynamic added mass and potential damping should be evaluated at a frequency of 0.628 rad s^{-1} in roll if a pure rolling motion is considered. As a consequence of this, it is common to formulate the ship maneuvering model (1.2)–(1.7) as a coupled *surge–sway–yaw* model and thus neglect heave, roll and pitch motions. In other words

$$m(\dot{u} - vr - x_g r^2 - y_g \dot{r}) = X \tag{1.9}$$

$$m(\dot{v} + ur - y_g r^2 + x_g \dot{r}) = Y \tag{1.10}$$

$$I_z \dot{r} + m(x_g(\dot{v} + ur) - y_g(\dot{u} - vr)) = N. \tag{1.11}$$

It will be shown in the subsequent chapters that the rigid-body kinetics (1.9)–(1.11) can be expressed in matrix-vector form according to (Fossen 1994)

$$\boldsymbol{M}_{\mathrm{RB}}\dot{\boldsymbol{\nu}} + \boldsymbol{C}_{\mathrm{RB}}(\boldsymbol{\nu})\boldsymbol{\nu} = \boldsymbol{\tau}_{\mathrm{RB}} \tag{1.12}$$

$$\boldsymbol{\tau}_{\mathrm{RB}} = \underbrace{\boldsymbol{\tau}_{\mathrm{hyd}} + \boldsymbol{\tau}_{\mathrm{hs}}}_{\substack{\text{hydrodynamic and}\\ \text{hydrostatic forces}}} + \underbrace{\boldsymbol{\tau}_{\mathrm{wind}} + \boldsymbol{\tau}_{\mathrm{wave}}}_{\text{environmental forces}} + \boldsymbol{\tau}_{\mathrm{control}} \tag{1.13}$$

where $\boldsymbol{M}_{\mathrm{RB}}$ is the rigid-body inertia matrix, $\boldsymbol{C}_{\mathrm{RB}}(\boldsymbol{\nu})$ is a matrix of rigid-body Coriolis and centripetal forces and $\boldsymbol{\tau}_{\mathrm{RB}}$ is a vector of generalized forces.

The generalized velocity is conveniently expressed as

$$\boldsymbol{\nu} = [u,\ v,\ w,\ p,\ q,\ r]^{\mathrm{T}} \tag{1.14}$$

where the first three components (u, v, w) are the linear velocities in surge, sway and heave and (p, q, r) are the angular velocities in roll, pitch and yaw. The generalized force acting on the craft is a vector

$$\boldsymbol{\tau}_i = [X_i, \ Y_i, \ Z_i, \ K_i, \ M_i, \ N_i]^{\top}, \quad i \in \{\text{hyd, hs, wind, wave, control}\} \tag{1.15}$$

where the subscripts stand for

- Hydrodynamic added mass and damping
- Hydrostatic forces (spring stiffness)
- Wind forces
- Wave forces (first and second order)
- Control and propulsion forces.

This model is motivated by Newton's second law, $F = ma$, where F represents force, m is the mass and a is the acceleration. The Coriolis and centripetal term is due to the rotation of the body-fixed reference frame with respect to the inertial reference frame. The model (1.12) is used in most textbooks on hydrodynamics and the generalized forces $\boldsymbol{\tau}_i$ can be represented by linear or nonlinear theory.

Linearized Models

In the linear 6-DOF case there will be a total of 36 mass and 36 damping elements proportional to velocity and acceleration. In addition to this, there will be restoring forces, propulsion forces and environmental forces. If the generalized hydrodynamic force $\boldsymbol{\tau}_{\text{hyd}}$ is written in component form using the SNAME (1950) notation, the linear added mass and damping forces become

$$\begin{aligned} X_1 &= X_u u + X_v v + X_w w + X_p p + X_q q + X_r r \\ &\quad + X_{\dot{u}} \dot{u} + X_{\dot{v}} \dot{v} + X_{\dot{w}} \dot{w} + X_{\dot{p}} \dot{p} + X_{\dot{q}} \dot{q} + X_{\dot{r}} \dot{r} \end{aligned} \tag{1.16}$$

$$\vdots$$

$$\begin{aligned} N_1 &= N_u u + N_v v + N_w w + N_p p + N_q q + N_r r \\ &\quad + N_{\dot{u}} \dot{u} + N_{\dot{v}} \dot{v} + N_{\dot{w}} \dot{w} + N_{\dot{p}} \dot{p} + N_{\dot{q}} \dot{q} + N_{\dot{r}} \dot{r} \end{aligned} \tag{1.17}$$

where $X_u, X_v, \ldots, N_r$ are the linear damping coefficients and $X_{\dot{u}}, X_{\dot{v}}, \ldots, N_{\dot{r}}$ represent hydrodynamic added mass.

Nonlinear Models

Application of nonlinear theory implies that many elements must be included in addition to the 36 linear elements. This is usually done by one of the following methods:

- *Truncated Taylor-series expansions* using *odd terms* (first and third order) which are fitted to experimental data, for instance (Abkowitz 1964)

$$X_1 = X_{\dot{u}} \dot{u} + X_u u + X_{uuu} u^3 + X_{\dot{v}} \dot{v} + X_v v + X_{vvv} v^3 + \cdots \tag{1.18}$$

$$\vdots$$

$$N_1 = N_{\dot{u}}\dot{u} + N_u u + N_{uuu}u^3 + N_{\dot{v}}\dot{v} + N_v v + N_{vvv}v^3 + \cdots . \quad (1.19)$$

In this approach added mass is assumed to be linear and damping is modeled by a third-order odd function. Alternatively, *second-order modulus terms* can be used (Fedyaevsky and Sobolev 1963), for instance

$$X_1 = X_{\dot{u}}\dot{u} + X_u u + X_{|u|u}|u|u + X_{\dot{v}}\dot{v} + X_v v + X_{|v|v}|v|v + \cdots \quad (1.20)$$

$$\vdots$$

$$N_1 = N_{\dot{u}}\dot{u} + N_u u + N_{|u|u}|u|u + N_{\dot{v}}\dot{v} + N_v v + N_{|v|v}|v|v + \cdots . \quad (1.21)$$

This is motivated by the square-law damping terms in fluid dynamics and aerodynamics. When applying Taylor-series expansions in model-based control design, the system (1.12) becomes relatively complicated due to the large number of hydrodynamic coefficients on the right-hand side needed to represent the hydrodynamic forces. This approach is quite common when deriving maneuvering models and many of the coefficients are difficult to determine with sufficient accuracy since the model can be overparametrized. Taylor-series expansions are frequently used in commercial planar motion mechanism (PMM) tests where the purpose is to derive the maneuvering coefficients experimentally.

- *First principles* where hydrodynamic effects such as lift and drag are modeled using well established models. This results in physically sound Lagrangian models that preserve energy properties. Models based on first principles usually require a much smaller number of parameters than models based on third order Taylor-series expansions.

1.2.2 Seakeeping Theory

As explained above, maneuvering refers to the study of ship motion in the absence of wave excitation (calm water). Seakeeping, on the other hand, is the study of motion when there is wave excitation and the craft keeps its heading ψ and its speed U constant (which includes the case of zero speed). This introduces a dissipative force (Cummins 1962) known as *fluid-memory effects*. Although both areas are concerned with the same issues, study of motion, stability and control, the separation allows different assumptions to be made that simplify the study in each case. Seakeeping analysis is used in capability analysis and operability calculations to obtain operability diagrams according to the adopted criteria.

The seakeeping theory is formulated using seakeeping axes $\{s\}$ where the state vector $\boldsymbol{\xi} = [\xi_1, \xi_2, \xi_3, \xi_4, \xi_5, \xi_6]^{\mathsf{T}}$ represents perturbations with respect to a fixed equilibrium state; see Figure 1.6. These perturbations can be related to motions in the body frame $\{b\}$ and North-East-Down frame $\{n\}$ by using kinematic transformations; see Section 5.2. The governing model is formulated in the time domain using the *Cummins equation* in the following form (see Section 5.4):

$$(\boldsymbol{M}_{\mathrm{RB}} + \boldsymbol{A}(\infty))\ddot{\boldsymbol{\xi}} + \boldsymbol{B}_{\mathrm{total}}(\infty)\dot{\boldsymbol{\xi}} + \int_0^t \boldsymbol{K}(t-\tau)\dot{\boldsymbol{\xi}}(\tau)\mathrm{d}\tau + \boldsymbol{C}\boldsymbol{\xi} = \boldsymbol{\tau}_{\mathrm{wind}} + \boldsymbol{\tau}_{\mathrm{wave}} + \delta\boldsymbol{\tau} \quad (1.22)$$

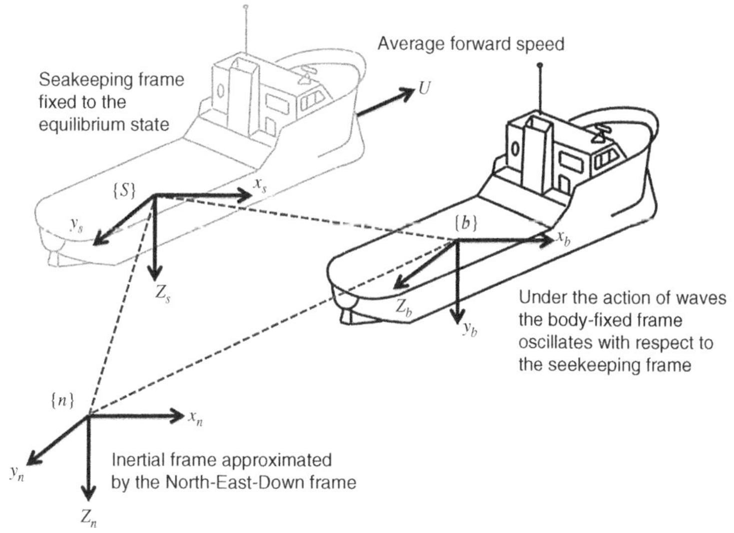

Figure 1.6 Coordinate systems used in seakeeping analysis.

where $\delta\boldsymbol{\tau}$ is the perturbed control input due to propulsion and control surfaces, $\boldsymbol{A}(\infty)$ is the infinite-frequency added mass matrix, $\boldsymbol{B}_{\text{total}}(\infty) = \boldsymbol{B}(\infty) + \boldsymbol{B}_V(\infty)$ is the infinite-frequency damping matrix containing potential and viscous damping terms, $\boldsymbol{C}$ is the spring stiffness matrix and $\boldsymbol{K}(t)$ is a time-varying matrix of *retardation functions* given by

$$\boldsymbol{K}(t) = \frac{2}{\pi}\int_0^{\infty} [\boldsymbol{B}_{\text{total}}(\omega) - \boldsymbol{B}_{\text{total}}(\infty)]\cos(\omega t)\mathrm{d}\omega. \tag{1.23}$$

Equation (1.22) can be transformed to the frequency domain by using the Laplace transformation. Hence, application of $\mathcal{L}\{\ddot{\xi}(t)\} = s^2\xi(s)$ and $\mathcal{L}\{\dot{\xi}\} = s\xi(s)$ together with $s = \mathrm{j}\omega$ gives

$$(-\omega^2[\boldsymbol{M}_{\text{RB}} + \boldsymbol{A}(\omega)] - \mathrm{j}\omega\boldsymbol{B}_{\text{total}}(\omega) + \boldsymbol{C})\xi(\mathrm{j}\omega) = \boldsymbol{\tau}_{\text{wind}}(\mathrm{j}\omega) + \boldsymbol{\tau}_{\text{wave}}(\mathrm{j}\omega) + \delta\boldsymbol{\tau}(\mathrm{j}\omega). \tag{1.24}$$

Naval architects often write the seakeeping model as a *pseudo-differential equation*

$$(\boldsymbol{M}_{\text{RB}} + \boldsymbol{A}(\omega))\ddot{\xi} + \boldsymbol{B}_{\text{total}}(\omega)\dot{\xi} + \boldsymbol{C}\xi = \boldsymbol{\tau}_{\text{wind}} + \boldsymbol{\tau}_{\text{wave}} + \delta\boldsymbol{\tau} \tag{1.25}$$

mixing time and frequency. Unfortunately this is deeply rooted in the literature of hydrodynamics even though it is not correct to mix time and frequency in one single equation. Consequently, it is recommended to use the time- and frequency-domain representations (1.22) and (1.24).

Computer simulations are done under the assumptions of linear theory and harmonic motions such that the resulting response is linear in the time domain. This approach dates back to Cummins (1962) and the necessary derivations are described in Chapter 5.

1.2.3 Unified Theory

A unified theory for maneuvering and seakeeping is useful since it allows for time-domain simulation of a marine craft in a seaway. This is usually done by using the seakeeping representation (1.25) as described in Chapter 5. The next step is to assume linear superposition such that wave-induced forces can be added for different speeds U and sea states. A similar assumption is used to add nonlinear damping and restoring forces so that the resulting model is a unified nonlinear model combining the most important terms from both maneuvering and seakeeping. Care must be taken with respect to "double counting." This refers to the problem that hydrodynamic effects can be modeled twice when merging the results from two theories. The procedure is described in details by Fossen (2005).

1.3 Fossen's Robot-inspired Model for Marine Craft

In order to exploit the physical properties of the maneuvering and seakeeping models, the equations of motion are represented in a vectorial setting that dates back to Fossen (1991). The matrix-vector model is expressed in $\{b\}$ and $\{n\}$ so appropriate kinematic transformations between the reference frames $\{b\}$, $\{n\}$ and $\{s\}$ must be derived. This is done in Chapters 2 and 5. The matrix-vector model is well suited for computer implementation and control systems design.

Component Form

The classical model (1.2)–(1.7) is often combined with expressions such as (1.16)–(1.17) or (1.18)–(1.21) to describe the hydrodynamic forces. This often results in complicated models with hundreds of elements. In most textbooks the resulting equations of motion are in component form. The following introduces a compact notation using matrices and vectors that will simplify the representation of the equations of motion considerably.

Matrix-vector Representation

In order to exploit the physical properties of the models, the equations of motion are represented in a matrix-vector setting. It is often beneficial to exploit physical system properties to reduce the number of coefficients needed for control. This is the main motivation for developing a vectorial representation of the equations of motion. In Fossen (1991) the robot model (Craig 1989; Sciavicco and Siciliano 1996)

$$\boldsymbol{M}(\boldsymbol{q})\ddot{\boldsymbol{q}} + \boldsymbol{C}(\boldsymbol{q}, \dot{\boldsymbol{q}})\dot{\boldsymbol{q}} = \boldsymbol{\tau} \tag{1.26}$$

was used as motivation to derive a compact marine craft model in 6 DOFs using a vectorial setting. In the robot model $\boldsymbol{q}$ is a vector of joint angles, $\boldsymbol{\tau}$ is the torque, while $\boldsymbol{M}$ and $\boldsymbol{C}$ denote the system inertia and Coriolis matrices, respectively. It is found that similar quantities can be identified for marine craft and aircraft. In Fossen (1991) a complete 6-DOF vectorial setting for marine craft was derived based on these ideas. These results were further refined by Sagatun and Fossen (1991), Fossen (1994) and Fossen and Fjellstad (1995). The 6-DOF models considered in this book use the following representation

$$\boldsymbol{M}\dot{\boldsymbol{\nu}} + \boldsymbol{C}(\boldsymbol{\nu})\boldsymbol{\nu} + \boldsymbol{D}(\boldsymbol{\nu})\boldsymbol{\nu} + \boldsymbol{g}(\boldsymbol{\eta}) + \boldsymbol{g}_0 = \boldsymbol{\tau} + \boldsymbol{\tau}_{\text{wind}} + \boldsymbol{\tau}_{\text{wave}} \tag{1.27}$$

where

$$\boldsymbol{\eta} = [x^n,\ y^n,\ z^n,\ \phi,\ \theta,\ \psi]^{\top} \tag{1.28}$$

$$\boldsymbol{\nu} = [u,\ v,\ w,\ p,\ q,\ r]^{\top} \tag{1.29}$$

are vectors of velocities and position/Euler angles, respectively. In fact $\boldsymbol{\nu}$ and $\boldsymbol{\eta}$ are generalized velocities and positions used to describe motions in 6 DOFs. Similarly, $\boldsymbol{\tau}$ is a vector of forces and moments or the generalized forces in 6 DOFs. The model matrices $\boldsymbol{M}$, $\boldsymbol{C}(\boldsymbol{\nu})$ and $\boldsymbol{D}(\boldsymbol{\nu})$ denote inertia, Coriolis and damping, respectively, while $\boldsymbol{g}(\boldsymbol{\eta})$ is a vector of generalized gravitational and buoyancy forces. Static restoring forces and moments due to ballast systems and water tanks are collected in the optionally term $\boldsymbol{g}_0$.

Component Form Versus the Matrix-vector Representation

When designing control systems, there are clear advantages using the vectorial model (1.27) instead of (1.12)–(1.13) and the component forms of the Taylor-series expansions (1.18)–(1.21). The main reasons are that system properties such as symmetry, skew-symmetry and positiveness of matrices can be incorporated into the stability analysis. In addition, these properties are related to passivity of the hydrodynamic and rigid-body models (Berge and Fossen 2000). The system properties represent physical properties of the system, which should be exploited when designing controllers and observers for marine craft. As a consequence, Equation (1.27) is chosen as the foundation for this textbook and the previous book *Guidance and Control of Ocean Vehicles* (Fossen 1994). Equation (1.27) has also been adopted by the international community as a "standard model" for marine control systems design while the "classical model" (1.12)–(1.13) is mostly used in hydrodynamic modeling where isolated effects often are studied in more detail.

It should be noted that the classical model with hydrodynamic forces in component form and the vectorial model (1.27) are equivalent. Therefore it is possible to combine the best of both approaches, that is hydrodynamic component-based modeling and control design models based on vectors and matrices. However, it is much easier to construct multiple-input multiple-output (MIMO) controllers and observers when using the vectorial representation, since the model properties and model reduction follow from the basic matrix properties. This also applies to system analysis since there are many tools for MIMO systems. Finally, it should be pointed out that the vectorial models are beneficial from a computational point of view and in order to perform algebraic manipulations. Readability is also significantly improved thanks to the compact notation.

2

Kinematics

The study of *dynamics* can be divided into two parts: *kinematics*, which treats only geometrical aspects of motion, and *kinetics*, which is the analysis of the forces causing the motion. In this chapter kinematics with application to local and terrestrial navigation is discussed. Kinetics is dealt with in Chapters 3–10.

The interested reader is advised to consult Britting (1971), Maybeck (1979), Forssell (1991), Lin (1992), Hofmann-Wellenhof *et al.* (1994), Parkinson and Spilker (1995), Titterton and Weston (1997) and Farrell (2008) for a discussion of navigation kinematics and kinematics in general. The development of the kinematic equations of motion are also found in Kane *et al.* (1983) and Hughes (1986). Both of these references use spacecraft systems for illustration. An alternative derivation of the Euler angle representation in the context of ship steering is given by Abkowitz (1964). More recent discussions of unit quaternions are found in Chou (1992) and Solà (2016). An analogy to robot manipulators is given by Craig (1989) or Sciavicco and Siciliano (1996), while a more detailed discussion of kinematics is found in Goldstein (1980), and Egeland and Gravdahl (2002).

6-DOF Marine Craft Equations of Motion

The overall goal of Chapters 2–10 is to show that the marine craft equations of motion can be written in the following *matrix-vector form* (Fossen 1991)

$$\dot{\boldsymbol{\eta}} = \boldsymbol{J}_{\Theta}(\boldsymbol{\eta})\boldsymbol{\nu} \tag{2.1}$$

$$\boldsymbol{M}\dot{\boldsymbol{\nu}} + \boldsymbol{C}(\boldsymbol{\nu})\boldsymbol{\nu} + \boldsymbol{D}(\boldsymbol{\nu})\boldsymbol{\nu} + \boldsymbol{g}(\boldsymbol{\eta}) + \boldsymbol{g}_0 = \boldsymbol{\tau} + \boldsymbol{\tau}_{\text{wind}} + \boldsymbol{\tau}_{\text{wave}} \tag{2.2}$$

where the different matrices and vectors and their properties will be defined in the forthcoming sections. This model representation is used as a foundation for model-based control design and stability analysis in Part II.

Handbook of Marine Craft Hydrodynamics and Motion Control, Second Edition. Thor I. Fossen.

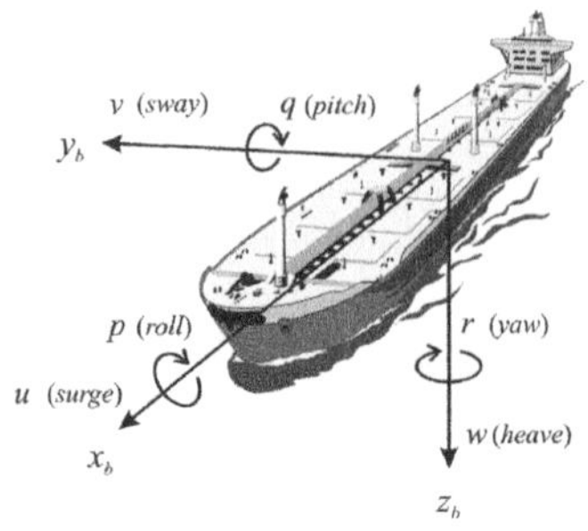

Figure 2.1 The six body-fixed velocities u, v, w, p, q and r and their interpretation in the body-fixed reference frame $x_b y_b z_b$.

Table 2.1 The notation of SNAME (1950) for marine craft. Source: Based on SNAME (1950). The Society of Naval Architects and Marine Engineers. Nomenclature for Treating the Motion of a Submerged Body Through a Fluid. In: Technical and Research Bulletin No. 1–5.

		BODY		NED
DOF		Forces and moments	Linear and angular velocities	Positions and Euler angles
1	Motions in the x_b-direction (surge)	X	u	x^n
2	Motions in the y_b-direction (sway)	Y	v	y^n
3	Motions in the z_b-direction (heave)	Z	w	z^n
4	Rotation about the x_b-axis (roll)	K	p	ϕ
5	Rotation about the y_b-axis (pitch)	M	q	θ
6	Rotation about the z_b-axis (yaw)	N	r	ψ

Motion Variables

For marine craft moving in six *degrees of freedom* (DOFs), six independent coordinates are necessary to determine the position and orientation. The position of the craft is usually chosen as the North–East–Down (NED) position, while orientation is defined by the Euler angles (roll, pitch and yaw angles). The Euler angles describe the orientation the body-fixed reference frame (BODY) with respect to NED.

For marine craft, the six different motion components in the BODY frame are conveniently defined as *surge, sway, heave, roll, pitch* and *yaw* (see Figure 2.1 and Table 2.1).

2.1 Kinematic Preliminaries

2.1.1 Reference Frames

When analyzing the motion of marine craft in 6 DOFs, it is convenient to define two Earth-centered coordinate frames as indicated in Figure 2.2. In addition several geographic reference frames are needed.

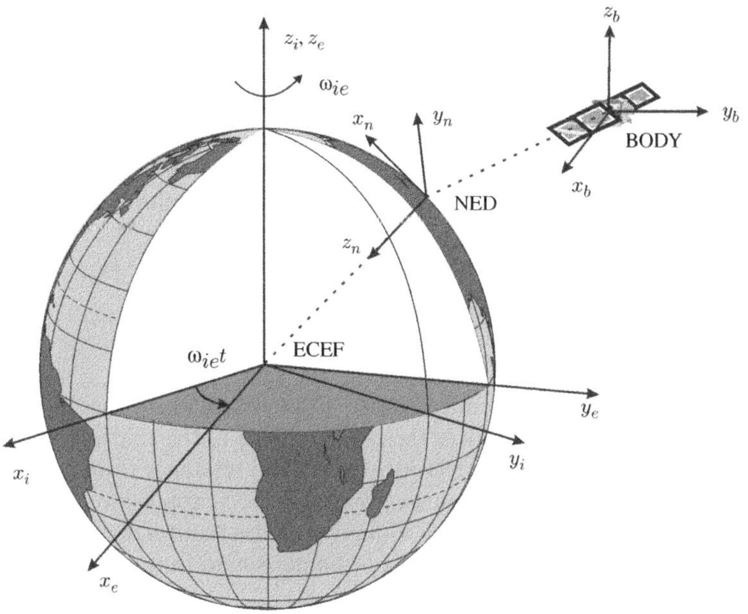

Figure 2.2 The Earth-centered Earth-fixed (ECEF) frame $x_e y_e z_e$ is rotating with angular rate ω_{ie} with respect to an Earth-centered inertial (ECI) frame $x_i y_i z_i$ fixed in space.

Earth-centered Reference Frames

ECI: The Earth-centered inertial (ECI) frame $\{i\} = (x_i,\ y_i,\ z_i)$ is an inertial frame for terrestrial navigation, that is a non-accelerating reference frame in which Newton's laws of motion apply. This is exploited when designing inertial navigation systems. The origin of $\{i\}$ is located at the center o_i of the Earth with axes as shown in Figure 2.2.

ECEF: The Earth-centered Earth-fixed (ECEF) reference frame $\{e\} = (x_e\ , y_e,\ z_e)$ has its origin o_e fixed to the center of the Earth but the axes rotate relative to the inertial frame ECI, which is fixed in space. Moreover

- x_e – axis in the equatorial plane pointing towards the zero/prime meridian; same longitude as the Greenwich observatory
- y_e – axis in the equatorial plane completing the right-hand frame
- z_e – axis pointing along the Earth's rotational axis.

The Earth rotation is $\omega_{ie} = 7.2921 \times 10^{-5}$ rad s^{-1} and the Earth's rotational vector expressed in $\{e\}$ is $\boldsymbol{\omega}^e_{ie} = [0,\ 0,\ \omega_{ie}]^{\top}$. For marine craft moving at relatively low speed, the Earth rotation can be neglected and hence $\{e\}$ can be considered to be inertial. Drifting ships, however, should not neglect the Earth rotation. The coordinate system $\{e\}$ is usually used for global navigation applications, for instance to describe the motion and location of ships in transit between different continents.

Geographic Reference Frames (Tangent Planes)

Geographical reference frames are usually chosen as tangent planes on the surface of the Earth. Tangent planes can be used for both local and terrestrial navigation.

- *Terrestrial navigation:* The tangent plane on the surface of the Earth moves with the craft and its location is specified by time-varying longitude-latitude pairs (l, μ). The tangent frame is usually rotated such that its axes points in the NED directions.
- *Local navigation:*The tangent plane is fixed at constant values (l_0, μ_0) and the position is computed with respect to the local coordinate origin. The axes of the tangent plane are usually chosen to coincide with the NED axes. This is also referred to as "flat-Earth navigation" and the position is accurate to a smaller geographical area, typically 10 × 10 km.

NED: The *North-East-Down* coordinate system is denoted $\{n\} = (x_n, y_n, z_n)$ where

- x_n – axis points towards true *North*
- y_n – axis points towards *East*
- z_n – axis points *downwards* normal to the Earth's surface.

The coordinate origin o_n is defined relative to the Earth's reference ellipsoid (World Geodetic System 1984), usually as the tangent plane to the ellipsoid. An alternative right-handed variant is the East-North-Up (ENU) reference frame. Both the NED and ENU reference frames are commonly used in GNC applications. The location of $\{n\}$ relative to $\{e\}$ is determined by using two angles l and μ denoting the *longitude* and *latitude*, respectively. This is the coordinate system we refer to in our everyday life and it is used for local navigation.

Body-fixed Reference Frames

BODY: The body-fixed reference frame $\{b\} = (x_b, y_b, z_b)$ with origin o_b is a moving coordinate frame that is fixed to the craft. The position and orientation of the craft are described relative to the inertial reference frame (approximated by $\{e\}$ or $\{n\}$) while the linear and angular velocities of the craft should be expressed in the body-fixed coordinate system. The origin o_b is usually chosen to coincide with a point midships in the waterline. For marine craft, the body axes x_b, y_b and z_b are chosen to coincide with the *principal axes of inertia*, and they are usually defined as (see Figure 2.3):

- x_b – longitudinal axis (directed from aft to fore)
- y_b – transversal axis (directed to starboard)
- z_b – normal axis (directed from top to bottom)

FLOW: Flow axes are used to align the x-axis with the craft's velocity vector such that lift is perpendicular to the relative flow and drag is parallel; see Section 2.5. The transformation from FLOW to BODY axes is defined by two principal rotations where the rotation angles are the angle of attack α and the sideslip angle β. The main purpose of the flow axes is to simplify the computations of lift and drag forces. Other applications are line-of-sight (LOS) path-following control systems, which must take into account that α and β may be time varying due to ocean currents.

SEAKEEPING: The seakeeping reference frame $\{s\} = (x_s, y_s, z_s)$ is not fixed to the craft; see Section 5.2. It represents the equilibrium state of a marine craft moving in waves. Hence, in the absence of wave excitation, the $\{s\}$-frame origin o_s coincides with the location of the $\{b\}$-frame origin o_b, which is a fixed point in the craft. Under the action of the waves, the hull is disturbed from its equilibrium and the point o_s oscillates, with respect to its equilibrium position.

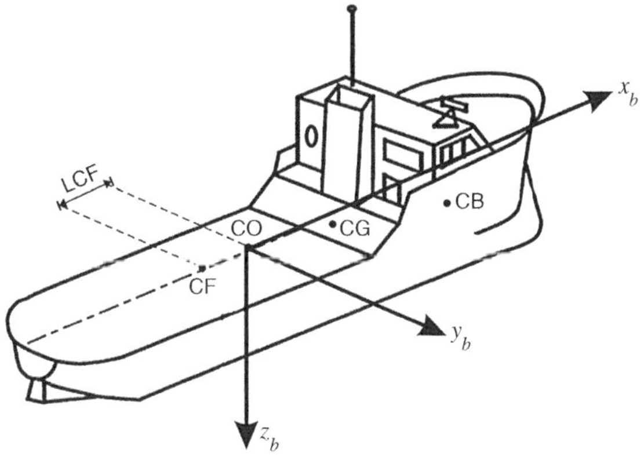

Figure 2.3 Body-fixed reference points. The CO has a constant location while the other reference points can be allowed to have time-varying coordinates with respect to the CO.

2.1.2 *Body-fixed Reference Points*

When designing guidance, navigation and control (GNC) systems for marine craft several reference points are needed to express the equations of motion. The most important reference point is

CO: Coordinate origin o_b of the body-fixed frame $\{b\}$. CO also represents the coordinate origin of the guidance, navigation and motion control systems.

The main idea is that the CO should be body fixed with location specified by the operator, while other reference points can be allowed to have time-varying coordinates with respect to the CO. This is the case for the following reference points, which are all time varying (see Figure 2.3).

CG: Center of gravity located at $\boldsymbol{r}^b_{bg} = [x_g, y_g, z_g]^\top$ relative to the CO.

CB: Center of buoyancy located at $\boldsymbol{r}^b_{bb} = [x_b, y_b, z_b]^\top$ relative to the CO.

CF: Center of flotation located at $\boldsymbol{r}^b_{bf} = [x_f, y_f, z_f]^\top$ relative to the CO.

In general, the locations of the CG, CB and CF depend on loading conditions, fuel burning rate, ballast tank levels, weather conditions and vessel motions. This is in contrast with the CO, which has a constant location.

The CF is the centroid of the waterplane area A_{wp} in calm water. A craft will roll and pitch about the x_f and y_f axes through the CF if the displacement is constant (small-angle assumption). Consequently, this point can be used to compute the pitch and roll periods for moderate sea states. If the CO is located midships in the waterline, $\boldsymbol{r}^b_{\mathrm{bf}} = [\mathrm{LCF},\ 0,\ 0]^\top$ where LCF is the longitudinal center of flotation.

Note that the eigenvalues of the 6-DOF linear model are independent of the reference points (CF, CB, CG and CO) but decoupled equations in heave, roll and pitch will produce incorrect results if they are formulated in a point different from the CF (see Section 4.3).

2.1.3 Generalized Coordinates

For a marine craft not subject to any motion constraints, the number of independent (generalized) coordinates will be equal to the DOFs. The term generalized coordinates refers to the parameters that describe the configuration of the craft relative to some reference configuration. For marine craft, the generalized position is chosen as

$$\boldsymbol{\eta} = [x^n,\ y^n,\ z^n,\ \phi,\ \theta,\ \psi]^\top. \tag{2.3}$$

The generalized velocities are the time derivatives of the generalized coordinates of the system. In other words

$$\dot{\boldsymbol{\eta}} = [\dot{x}^n,\ \dot{y}^n,\ \dot{z}^n,\ \dot{\phi},\ \dot{\theta},\ \dot{\psi}]^\top. \tag{2.4}$$

These quantities are all expressed in the NED frame. However, it is advantageous to express the velocities of the craft in the BODY frame when deriving the equations of motion. The BODY frame velocity vector is denoted as

$$\boldsymbol{\nu} = [u,\ v,\ w,\ p,\ q,\ r]^\top. \tag{2.5}$$

We will use the notation $\vec{u}$ to refer to a *coordinate free vector*, that is a *directed line segment*. When a vector is described relative to a coordinate system $\{n\}$, the following notation will be used

$$\vec{u} = u^n_1\vec{n}_1 + u^n_2\vec{n}_2 + u^n_3\vec{n}_3 \tag{2.6}$$

where $\vec{n}_i$ $(i = 1, 2, 3)$ are the unit vectors that define $\{n\}$, u^n_i are the measures of $\vec{u}$ along $\vec{n}_i$ and $u^n_i\,\vec{n}_i$ are the components of $\vec{u}$ in $\{n\}$. We will also use the *coordinate form* $\boldsymbol{u}^n$ of $\vec{u}$ expressed in $\{n\}$ which is represented by a *column vector* in $\mathbb{R}^3$ mathematically equivalent to

$$\boldsymbol{u}^n = [u^n_1,\ u^n_2,\ u^n_3]^\top. \tag{2.7}$$

For marine craft the coordinate origin o_{b} will be denoted CO and the following notation will be adopted for vectors expressed in the reference frames $\{b\}$, $\{e\}$ and $\{n\}$:

$\boldsymbol{p}^e_{\mathrm{eb}}$	position of the CO relative to o_{e} expressed in $\{e\}$
$\boldsymbol{p}^n_{\mathrm{nb}}$	position of the CO relative to o_{n} expressed in $\{n\}$
$\boldsymbol{v}^e_{\mathrm{nb}}$	linear velocity of the CO relative to o_{n} expressed in $\{e\}$
$\boldsymbol{v}^n_{\mathrm{nb}}$	linear velocity of the CO relative to o_{n} expressed in $\{n\}$
$\boldsymbol{\omega}^b_{\mathrm{nb}}$	angular velocity of $\{b\}$with respect to $\{n\}$ expressed in $\{b\}$
$\boldsymbol{f}^n_{\mathrm{b}}$	force with line of action through the point CO expressed in $\{n\}$
$\boldsymbol{m}^n_{\mathrm{b}}$	moment about the point CO expressed in $\{n\}$
$\boldsymbol{\Theta}_{\mathrm{nb}}$	Euler angles from $\{b\}$ to $\{n\}$
$\boldsymbol{q}^n_b$	Unit quaternion from $\{b\}$ to $\{n\}$.

The different quantities in Table 2.1, as defined by SNAME (1950), can now be conveniently expressed in a vectorial setting according to

ECEF position	$\boldsymbol{p}^e_{eb} = \begin{bmatrix} x^e \\ y^e \\ z^e \end{bmatrix} \in \mathbb{R}^3$	Longitude and latitude	$\boldsymbol{\Theta}_{en} = \begin{bmatrix} l \\ \mu \end{bmatrix} \in \mathbb{T}^2$
NED position	$\boldsymbol{p}^n_{nb} = \begin{bmatrix} x^n \\ y^n \\ z^n \end{bmatrix} \in \mathbb{R}^3$	Attitude (Euler angles)	$\boldsymbol{\Theta}_{nb} = \begin{bmatrix} \phi \\ \theta \\ \psi \end{bmatrix} \in \mathbb{T}^3$
Body-fixed linear velocity	$\boldsymbol{v}^b_{nb} = \begin{bmatrix} u \\ v \\ w \end{bmatrix} \in \mathbb{R}^3$	Body-fixed angular velocity	$\boldsymbol{\omega}^b_{nb} = \begin{bmatrix} p \\ q \\ r \end{bmatrix} \in \mathbb{R}^3$
Body-fixed force	$\boldsymbol{f}^b_b = \begin{bmatrix} X \\ Y \\ Z \end{bmatrix} \in \mathbb{R}^3$	Body-fixed moment	$\boldsymbol{m}^b_b = \begin{bmatrix} K \\ M \\ N \end{bmatrix} \in \mathbb{R}^3$

where $\mathbb{R}^3$ is the *Euclidean space* of dimension three and $\mathbb{T}^3 = \mathbb{S}^1 \times \mathbb{S}^1 \times \mathbb{S}^1 \subset \mathbb{R}^3$ is the three-dimensional torus, defined as the Cartesian product of three circles. The range of the Euler angles is defined modulo 2π radians implying that a valid range could be $[-\pi, \pi)$.

The general motion of a marine craft in 6 DOFs with the CO as coordinate origin is described by the following vectors for generalized position, velocity and force

$$\boldsymbol{\eta} = \begin{bmatrix} \boldsymbol{p}^n_{nb} \\ \boldsymbol{\Theta}_{nb} \end{bmatrix}, \quad \boldsymbol{\nu} = \begin{bmatrix} \boldsymbol{v}^b_{nb} \\ \boldsymbol{\omega}^b_{nb} \end{bmatrix}, \quad \boldsymbol{\tau} = \begin{bmatrix} \boldsymbol{f}^b_b \\ \boldsymbol{m}^b_b \end{bmatrix} \tag{2.8}$$

where $\boldsymbol{\eta} \in \mathbb{R}^3 \times \mathbb{T}^3$ denotes the position and orientation vector, $\boldsymbol{\nu} \in \mathbb{R}^6$ is the linear and angular velocity vector expressed in $\{b\}$ and $\boldsymbol{\tau} \in \mathbb{R}^6$ is used to express the forces and moments acting on the craft in $\{b\}$.

When designing control systems for marine craft operating in a local geographical region, the position vector $\boldsymbol{p}^n_{nb} \in \mathbb{R}^3$ is the preferred choice. For terrestrial navigation, however, it is convenient to express the position of the CO relative to the ECEF coordinate system, that is $\boldsymbol{p}^e_{eb} \in \mathbb{R}^3$. The orientation of the marine craft with respect to NED will be represented by means of the Euler angles $\boldsymbol{\Theta}_{nb} \in \mathbb{T}^3$ alternatively the unit quaternion $\boldsymbol{q}^n_b \in \mathbb{Q}$ where $\mathbb{Q} = \mathbb{S}^3 \subset \mathbb{R}^4$ is the set of unit quaternions. In the next sections, the kinematic equations relating the BODY, NED and ECEF reference frames will be presented.

2.2 Transformations Between BODY and NED

The rotation matrix $\boldsymbol{R}$ between two frames $\{a\}$ and $\{b\}$ is denoted as $\boldsymbol{R}^a_b$, and it is an element in SO(3), that is the special orthogonal group of order 3,

$$SO(3) := \{\boldsymbol{R} | \boldsymbol{R} \in \mathbb{R}^{3\times 3}, \quad \boldsymbol{R} \text{ is orthogonal and } \det(\boldsymbol{R}) = 1\}. \tag{2.9}$$

The group SO(3) is a subset of all orthogonal matrices of order 3, that is SO(3) ⊂ O(3) where O(3) is defined as

$$O(3) := \{\boldsymbol{R} | \boldsymbol{R} \in \mathbb{R}^{3\times3}, \quad \boldsymbol{R}\boldsymbol{R}^{\top} = \boldsymbol{R}^{\top}\boldsymbol{R} = \boldsymbol{I}_3\}. \tag{2.10}$$

Rotation matrices are useful when deriving the kinematic equations of motion for a marine craft. As a consequence of (2.9) and (2.10), the following properties can be stated.

Property 2.1 (Rotation Matrix)
A rotation matrix $\boldsymbol{R} \in SO(3)$ satisfies

$$\boldsymbol{R}\boldsymbol{R}^{\top} = \boldsymbol{R}^{\top}\boldsymbol{R} = \boldsymbol{I}_3, \qquad \det(\boldsymbol{R}) = 1$$

which implies that $\boldsymbol{R}$ is orthogonal. Consequently, the inverse rotation matrix is given by $\boldsymbol{R}^{-1} = \boldsymbol{R}^{\top}$.

In this book, the following notation is adopted when transforming a vector from one coordinate frame to another

$$\boldsymbol{v}^{\text{to}} = \boldsymbol{R}^{\text{to}}_{\text{from}} \boldsymbol{v}^{\text{from}} \tag{2.11}$$

In other words, the vector $\boldsymbol{v}^{\text{from}} \in \mathbb{R}^3$ can be transformed to a new reference frame by applying the rotation matrix $\boldsymbol{R}^{\text{to}}_{\text{from}}$. The result is the vector $\boldsymbol{v}^{\text{to}} \in \mathbb{R}^3$.

A frequently used rotation matrix in GNC applications is the rotation matrix $\boldsymbol{R}^n_b$ from $\{b\}$ to $\{n\}$. When deriving the expression for $\boldsymbol{R}^n_b$ we will make use of the following matrix properties:

Definition 2.1 (Skew-Symmetry of a Matrix)
A matrix $\boldsymbol{S} \in SS(n)$, that is the set of skew-symmetric matrices of order n, is said to be skew-symmetric if

$$\boldsymbol{S} = -\boldsymbol{S}^{\top}.$$

This implies that the off-diagonal elements of $\boldsymbol{S}$ satisfy $s_{ij} = -s_{ji}$ for $i \neq j$ while the diagonal elements are zero.

Definition 2.2 (Cross-Product Operator)
The vector cross product × is defined by

$$\boldsymbol{\lambda} \times \boldsymbol{a} := \boldsymbol{S}(\boldsymbol{\lambda})\boldsymbol{a} \tag{2.12}$$

where $\boldsymbol{S} \in SS(3)$ is

$$\boldsymbol{S}(\boldsymbol{\lambda}) = -\boldsymbol{S}^{\top}(\boldsymbol{\lambda}) = \begin{bmatrix} 0 & -\lambda_3 & \lambda_2 \\ \lambda_3 & 0 & -\lambda_1 \\ -\lambda_2 & \lambda_1 & 0 \end{bmatrix}, \quad \boldsymbol{\lambda} = \begin{bmatrix} \lambda_1 \\ \lambda_2 \\ \lambda_3 \end{bmatrix}. \tag{2.13}$$

The inverse operator is denoted vex(·), *such that*

$$\boldsymbol{\lambda} = \text{vex}(\boldsymbol{S}(\boldsymbol{\lambda})). \tag{2.14}$$

MATLAB

The cross-product operator is included in the MSS toolbox as `Smtrx.m`. Hence, the cross-product $\boldsymbol{b} = \boldsymbol{S}(\boldsymbol{\lambda})\boldsymbol{a}$ can be computed as

```
S = Smtrx(lambda)
b = S * a
```

The inverse operation is

```
lambda = vex(S)
```

Definition 2.3 (Simple Rotation)
The motion of a rigid body or reference frame $\{b\}$ relative to a rigid body or reference frame $\{a\}$ is called a simple rotation of $\{b\}$ in $\{a\}$ if there exists a line L, called an axis of rotation, whose orientation relative to both $\{a\}$ and $\{b\}$ remains unaltered throughout the motion.

Based on this definition, Euler stated the following theorem for rotation of two rigid bodies or reference frames (Euler 1776).

Theorem 2.1 (Euler's Theorem on Rotation)
Every change in the relative orientation of two rigid bodies or reference frames $\{a\}$ and $\{b\}$ can be produced by means of a simple rotation of $\{b\}$ in $\{a\}$.

Let $\boldsymbol{v}^b_{\text{nb}}$ be a vector fixed in BODY and $\boldsymbol{v}^n_{\text{nb}}$ be a vector expressed in NED. Consequently, the vector $\boldsymbol{v}^n_{\text{nb}}$ can be expressed in terms of the vector $\boldsymbol{v}^b_{\text{nb}}$, the unit vector $\boldsymbol{\lambda} = [\lambda_1, \lambda_2, \lambda_3]^\top$, $\|\boldsymbol{\lambda}\| = 1$, parallel to the axis of rotation and β the angle NED is rotated. This rotation is described mathematically by (Hughes 1986, Kane *et al.* 1983)

$$\boldsymbol{v}^n_{\text{nb}} = \boldsymbol{R}^n_b \boldsymbol{v}^b_{\text{nb}} \quad \text{where} \quad \boldsymbol{R}^n_b := \boldsymbol{R}_{\lambda,\beta} \tag{2.15}$$

and $\boldsymbol{R}_{\lambda,\beta}$ is the rotation matrix corresponding to a rotation β about the $\boldsymbol{\lambda}$ axis given by

$$\boldsymbol{R}_{\lambda,\beta} = \boldsymbol{I}_3 + \sin(\beta)\boldsymbol{S}(\boldsymbol{\lambda}) + (1 - \cos(\beta))\boldsymbol{S}^2(\boldsymbol{\lambda}). \tag{2.16}$$

The skew-symmetric matrix $\boldsymbol{S}(\boldsymbol{\lambda})$ is defined according to Definition 2.2. Consequently, $\boldsymbol{S}^2(\boldsymbol{\lambda}) = \boldsymbol{\lambda}\boldsymbol{\lambda}^\top - \boldsymbol{I}_3$ since $\boldsymbol{\lambda}$ is a unit vector.

Expanding (2.16) yields the following expressions for the nine rotation matrix elements

$$R_{11} = (1 - \cos(\beta))\lambda_1^2 + \cos(\beta) \tag{2.17}$$

$$R_{22} = (1 - \cos(\beta))\lambda_2^2 + \cos(\beta) \tag{2.18}$$

$$R_{33} = (1 - \cos(\beta))\lambda_3^2 + \cos(\beta) \tag{2.19}$$

$$R_{12} = (1 - \cos(\beta))\lambda_1\lambda_2 - \lambda_3 \sin(\beta) \tag{2.20}$$

$$R_{21} = (1 - \cos(\beta))\lambda_2\lambda_1 + \lambda_3 \sin(\beta) \tag{2.21}$$

$$R_{23} = (1 - \cos(\beta))\lambda_2\lambda_3 - \lambda_1 \sin(\beta) \tag{2.22}$$

$$R_{32} = (1 - \cos(\beta))\lambda_3\lambda_2 + \lambda_1 \sin(\beta) \tag{2.23}$$

$$R_{31} = (1 - \cos(\beta))\lambda_3\lambda_1 - \lambda_2 \sin(\beta) \tag{2.24}$$

$$R_{13} = (1 - \cos(\beta))\lambda_1\lambda_3 + \lambda_2 \sin(\beta) \tag{2.25}$$

2.2.1 Euler Angle Transformation

The Euler angles, roll (ϕ), pitch (θ) and yaw (ψ), can now be used to express the body-fixed velocity vector $\boldsymbol{v}_{\text{nb}}^{b}$ in the NED reference frame. Let $\boldsymbol{R}(\boldsymbol{\Theta}_{\text{nb}}) : \mathbb{T}^3 \rightarrow SO(3)$ denote the Euler angle rotation matrix with argument $\boldsymbol{\Theta}_{\text{nb}} = [\phi,\ \theta,\ \psi]^{\top}$. Hence,

$$\boldsymbol{v}_{\text{nb}}^{n} = \boldsymbol{R}_b^n \boldsymbol{v}_{\text{nb}}^{b} \tag{2.26}$$

where the rotation matrix from $\{b\}$ to $\{n\}$ is

$$\boldsymbol{R}_b^n := \boldsymbol{R}(\boldsymbol{\Theta}_{\text{nb}}). \tag{2.27}$$

Principal Rotations

The principal rotation matrices (one-axis rotations) can be obtained by setting the unit vector to $\boldsymbol{\lambda} = [1,\ 0,\ 0]^{\top}$, $\boldsymbol{\lambda} = [0,\ 1,\ 0]^{\top}$ and $\boldsymbol{\lambda} = [0,\ 0,\ 1]^{\top}$ for the x, y and z axes, and $\beta = \phi$, $\beta = \theta$ and $\beta = \psi$, respectively, in the formula for $\boldsymbol{R}_{\lambda,\beta}$ given by (2.16). This yields

$$\boldsymbol{R}_{x,\phi} = \begin{bmatrix} 1 & 0 & 0 \\ 0 & c\phi & -s\phi \\ 0 & s\phi & c\phi \end{bmatrix}, \quad \boldsymbol{R}_{y,\theta} = \begin{bmatrix} c\theta & 0 & s\theta \\ 0 & 1 & 0 \\ -s\theta & 0 & c\theta \end{bmatrix}, \quad \boldsymbol{R}_{z,\psi} = \begin{bmatrix} c\psi & -s\psi & 0 \\ s\psi & c\psi & 0 \\ 0 & 0 & 1 \end{bmatrix} \tag{2.28}$$

where $s \cdot = \sin(\cdot)$ and $c \cdot = \cos(\cdot)$.

Linear Velocity Transformation

It is customary to describe $\boldsymbol{R}_b^n = \boldsymbol{R}(\boldsymbol{\Theta}_{\text{nb}})$ by three *principal* rotations about the z, y and x axes (zyx convention). Note that the order in which these rotations is carried out is not arbitrary. In GNC applications it is common to use the zyx convention to describe the rotation *from* $\{b\}$ *to* $\{n\}$ specified in terms of the Euler angles ϕ, θ and ψ. This rotation sequence is mathematically equivalent to

$$\boldsymbol{R}_b^n := \boldsymbol{R}_{z,\psi}\boldsymbol{R}_{y,\theta}\boldsymbol{R}_{x,\phi} \tag{2.29}$$

and the inverse transformation is then written

$$\boldsymbol{R}_n^b = (\boldsymbol{R}_b^n)^\top = \boldsymbol{R}_{x,\phi}^\top\boldsymbol{R}_{y,\theta}^\top\boldsymbol{R}_{z,\psi}^\top \tag{2.30}$$

where we have used the result of Property 2.1. This can also be verified by studying Figure 2.4.

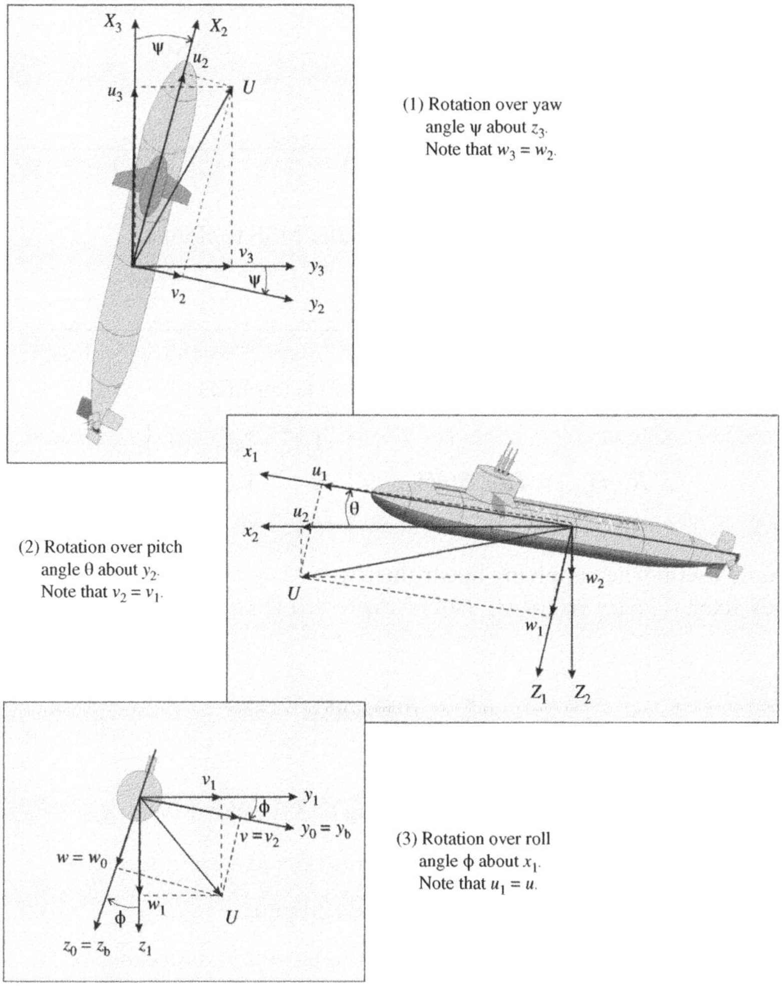

Figure 2.4 Euler angle rotation sequence (zyx rotation sequence). The submarine is rotated from $\{n\}$ to $\{b\}$ by using three principal rotations.

Let $x_3y_3z_3$ be the coordinate system obtained by translating the NED coordinate system $x_\mathrm{n}y_\mathrm{n}z_\mathrm{n}$ parallel to itself until its origin coincides with the origin of the body-fixed coordinate system. The coordinate system $x_3y_3z_3$ is rotated a *yaw* angle ψ about the z_3 axis. This yields the coordinate system $x_2y_2z_2$. The coordinate system $x_2y_2z_2$ is rotated a *pitch* angle θ about the y_2 axis. This yields the coordinate system $x_1y_1z_1$. Finally, the coordinate system $x_1y_1z_1$ is rotated a *roll* angle ϕ about the x_1 axis. This yields the body-fixed coordinate system $x_\mathrm{b}y_\mathrm{b}z_\mathrm{b}$.

Expanding (2.29) finally yields

$$\boldsymbol{R}(\boldsymbol{\Theta}_{\mathrm{nb}}) = \begin{bmatrix} \mathrm{c}\psi\mathrm{c}\theta & -\mathrm{s}\psi\mathrm{c}\phi + \mathrm{c}\psi\mathrm{s}\theta\mathrm{s}\phi & \mathrm{s}\psi\mathrm{s}\phi + \mathrm{c}\psi\mathrm{c}\phi\mathrm{s}\theta \\ \mathrm{s}\psi\mathrm{c}\theta & \mathrm{c}\psi\mathrm{c}\phi + \mathrm{s}\phi\mathrm{s}\theta\mathrm{s}\psi & -\mathrm{c}\psi\mathrm{s}\phi + \mathrm{s}\theta\mathrm{s}\psi\mathrm{c}\phi \\ -\mathrm{s}\theta & \mathrm{c}\theta\mathrm{s}\phi & \mathrm{c}\theta\mathrm{c}\phi \end{bmatrix}. \tag{2.31}$$

MATLAB

The rotation matrix $\boldsymbol{R}(\boldsymbol{\Theta}_{\mathrm{nb}})$ is implemented in the MSS toolbox as

```
R = Rzyx(phi, theta, psi)
```

For small angles $\delta\phi$, $\delta\theta$ and $\delta\psi$ the expression (2.31) simplifies to

$$\boldsymbol{R}(\delta\boldsymbol{\Theta}_{\mathrm{nb}}) \approx \boldsymbol{I}_3 + \boldsymbol{S}(\delta\boldsymbol{\Theta}_{\mathrm{nb}}) = \begin{bmatrix} 1 & -\delta\psi & \delta\theta \\ \delta\psi & 1 & -\delta\phi \\ -\delta\theta & \delta\phi & 1 \end{bmatrix} \tag{2.32}$$

which is quite useful when applying linear theory.

The body-fixed velocity vector $\boldsymbol{v}^b_{\mathrm{nb}}$ can be expressed in $\{n\}$ as

$$\dot{\boldsymbol{p}}^n_{\mathrm{nb}} = \boldsymbol{R}(\boldsymbol{\Theta}_{\mathrm{nb}})\boldsymbol{v}^b_{\mathrm{nb}} \tag{2.33}$$

where $\dot{\boldsymbol{p}}^n_{\mathrm{nb}}$ is the NED velocity vector. Expanding (2.33) yields

$$\begin{aligned} \dot{x}^{\mathrm{n}} &= u\cos(\psi)\cos(\theta) + v(\cos(\psi)\sin(\theta)\sin(\phi) - \sin(\psi)\cos(\phi)) \\ &\quad + w(\sin(\psi)\sin(\phi) + \cos(\psi)\cos(\phi)\sin(\theta)) && (2.34) \\ \dot{y}^{\mathrm{n}} &= u\sin(\psi)\cos(\theta) + v(\cos(\psi)\cos(\phi) + \sin(\phi)\sin(\theta)\sin(\psi)) \\ &\quad + w(\sin(\theta)\sin(\psi)\cos(\phi) - \cos(\psi)\sin(\phi)) && (2.35) \\ \dot{z}^{\mathrm{n}} &= -u\sin(\theta) + v\cos(\theta)\sin(\phi) + w\cos(\theta)\cos(\phi). && (2.36) \end{aligned}$$

The inverse velocity transformation is obtained by Definition 2.1 as

$$\boldsymbol{v}^b_{\text{nb}} = \boldsymbol{R}^{-1}(\boldsymbol{\Theta}_{\text{nb}})\dot{\boldsymbol{p}}^n_{\text{nb}} = \boldsymbol{R}^\top(\boldsymbol{\Theta}_{\text{nb}})\dot{\boldsymbol{p}}^n_{\text{nb}}. \tag{2.37}$$

Example 2.1 (Numerical Computation of Position Trajectory)
The flight path or position trajectory $\boldsymbol{p}^n_{\text{nb}}$ of the craft relative to the NED reference frame is found by numerical integration of (2.33), for instance by using Euler's method

$$\boldsymbol{p}^n_{\text{nb}}[k+1] = \boldsymbol{p}^n_{\text{nb}}[k] + h\boldsymbol{R}(\boldsymbol{\Theta}_{\text{nb}}[k])\boldsymbol{v}^b_{\text{nb}}[k] \tag{2.38}$$

where $h > 0$ is the sampling time and k is the sample index.

Angular Velocity Transformation

The body-fixed angular velocity vector $\boldsymbol{\omega}^b_{\text{nb}} = [p,\ q,\ r]^\top$ and the Euler rate vector $\dot{\boldsymbol{\Theta}}_{\text{nb}} = [\dot{\phi},\ \dot{\theta},\ \dot{\psi}]^\top$ are related through a transformation matrix $\boldsymbol{T}(\boldsymbol{\Theta}_{\text{nb}})$ according to

$$\dot{\boldsymbol{\Theta}}_{\text{nb}} = \boldsymbol{T}(\boldsymbol{\Theta}_{\text{nb}})\boldsymbol{\omega}^b_{\text{nb}}. \tag{2.39}$$

It should be noted that the angular body velocity vector $\boldsymbol{\omega}^b_{\text{nb}} = [p, q, r]^\top$ cannot be integrated directly to obtain actual angular coordinates. This is due to the fact that $\int_0^t \boldsymbol{\omega}^b_{\text{nb}}(\tau)\text{d}\tau$ does not have any immediate physical interpretation; however, the vector $\boldsymbol{\Theta}_{\text{nb}} = [\phi,\ \theta,\ \psi]^\top$ does represent proper generalized coordinates. The transformation matrix $\boldsymbol{T}(\boldsymbol{\Theta}_{nb})$ can be derived in several ways, for instance

$$\boldsymbol{\omega}^b_{\text{nb}} = \begin{bmatrix} \dot{\phi} \\ 0 \\ 0 \end{bmatrix} + \boldsymbol{R}^\top_{x,\phi}\begin{bmatrix} 0 \\ \dot{\theta} \\ 0 \end{bmatrix} + \boldsymbol{R}^\top_{x,\phi}\boldsymbol{R}^\top_{y,\theta}\begin{bmatrix} 0 \\ 0 \\ \dot{\psi} \end{bmatrix} := \boldsymbol{T}^{-1}(\boldsymbol{\Theta}_{\text{nb}})\dot{\boldsymbol{\Theta}}_{\text{nb}}. \tag{2.40}$$

This relationship is verified by inspection of Figure 2.4. Expanding (2.40) yields

$$\boldsymbol{T}^{-1}(\boldsymbol{\Theta}_{\text{nb}}) = \begin{bmatrix} 1 & 0 & -\text{s}\theta \\ 0 & \text{c}\phi & \text{c}\theta\text{s}\phi \\ 0 & -\text{s}\phi & \text{c}\theta\text{c}\phi \end{bmatrix} \quad \Rightarrow \quad \boldsymbol{T}(\boldsymbol{\Theta}_{\text{nb}}) = \begin{bmatrix} 1 & \text{s}\phi\text{t}\theta & \text{c}\phi\text{t}\theta \\ 0 & \text{c}\phi & -\text{s}\phi \\ 0 & \text{s}\phi/\text{c}\theta & \text{c}\phi/\text{c}\theta \end{bmatrix} \tag{2.41}$$

where $\text{s}\,\cdot = \sin(\cdot)$, $\text{c}\,\cdot = \cos(\cdot)$ and $\text{t}\,\cdot = \tan(\cdot)$. Expanding (2.39) yields the Euler angle attitude equations in component form

$$\dot{\phi} = p + q\sin(\phi)\tan(\theta) + r\cos(\phi)\tan(\theta) \tag{2.42}$$

$$\dot{\theta} = q\cos(\phi) - r\sin(\phi) \tag{2.43}$$

$$\dot{\psi} = q\frac{\sin(\phi)}{\cos(\theta)} + r\frac{\cos(\phi)}{\cos(\theta)}, \qquad \theta \neq \pm 90^\circ. \tag{2.44}$$

Note that $\boldsymbol{T}(\boldsymbol{\Theta}_{\text{nb}})$ is undefined for a pitch angle of $\theta = \pm 90^\circ$ and that $\boldsymbol{T}(\boldsymbol{\Theta}_{\text{nb}})$ does not satisfy Property 2.1. Consequently, $\boldsymbol{T}^{-1}(\boldsymbol{\Theta}_{\text{nb}}) \neq \boldsymbol{T}^{\top}(\boldsymbol{\Theta}_{\text{nb}})$. For surface vessels this is not a problem whereas both underwater vehicles and aircraft may operate close to this singularity. In this case, the kinematic equations can be described by two Euler angle representations with different singularities and the singular point can be avoided by switching between them. Another possibility is to use the unit quaternion representation; see Section 2.2.2.

MATLAB

The transformation matrix $\boldsymbol{T}(\boldsymbol{\Theta}_{\text{nb}})$ is implemented in the MSS toolbox as

```
T = Tzyx(phi, theta)
```

For small angles $\delta\phi, \delta\theta$ and $\delta\psi$ the transformation matrix $\boldsymbol{T}(\boldsymbol{\Theta}_{\text{nb}})$ simplifies to

$$\boldsymbol{T}(\delta\boldsymbol{\Theta}_{\text{nb}}) \approx \begin{bmatrix} 1 & 0 & \delta\theta \\ 0 & 1 & -\delta\phi \\ 0 & \delta\phi & 1 \end{bmatrix}. \tag{2.45}$$

The differential equation for the rotation matrix is given by Theorem 2.2.

Theorem 2.2 (Rotation Matrix Differential Equation)
The differential equation for the rotation matrix between the BODY and NED reference frames is

$$\dot{\boldsymbol{R}}_b^n = \boldsymbol{R}_b^n \boldsymbol{S}(\boldsymbol{\omega}_{\text{nb}}^b) \tag{2.46}$$

where

$$\boldsymbol{S}(\boldsymbol{\omega}_{\text{nb}}^b) = \begin{bmatrix} 0 & -r & q \\ r & 0 & -p \\ -q & p & 0 \end{bmatrix}. \tag{2.47}$$

This can be written in component form as nine differential equations

$$\begin{bmatrix} \dot{R}_{11} & \dot{R}_{12} & \dot{R}_{13} \\ \dot{R}_{21} & \dot{R}_{22} & \dot{R}_{23} \\ \dot{R}_{31} & \dot{R}_{23} & \dot{R}_{33} \end{bmatrix} = \begin{bmatrix} R_{12}r - R_{13}q & -R_{11}r + R_{13}p & R_{11}q - R_{12}p \\ R_{22}r - R_{23}q & -R_{21}r + R_{23}p & R_{21}q - R_{22}p \\ R_{23}r - R_{33}q & -R_{31}r + R_{33}p & R_{31}q - R_{23}p \end{bmatrix}. \tag{2.48}$$

Proof. *For a small time increment* Δt *the rotation matrix* $\boldsymbol{R}_b^n$ *satisfies*

$$\boldsymbol{R}_b^n(t+\Delta t) \approx \boldsymbol{R}_b^n(t)\boldsymbol{R}_b^n(\Delta t) \tag{2.49}$$

since $\sin(\Delta t) \approx \Delta t$ *and* $\cos(\Delta t) \approx 1$. *Assume that after time* $t + \Delta t$ *there has been an infinitesimal increment* $\Delta\beta$ *in the rotation angle. From (2.16) we have*

$$\begin{aligned}\boldsymbol{R}_b^n(\Delta t) &= \boldsymbol{I}_3 + \sin(\Delta\beta)\, \boldsymbol{S}(\boldsymbol{\lambda}) + (1-\cos(\Delta\beta))\, \boldsymbol{S}^2(\boldsymbol{\lambda}) \\ &\approx \boldsymbol{I}_3 + \Delta\beta\, \boldsymbol{S}(\boldsymbol{\lambda}).\end{aligned} \tag{2.50}$$

From (2.49), it follows that

$$\boldsymbol{R}_b^n(t+\Delta t) = \boldsymbol{R}_b^n(t)\,(\boldsymbol{I}_3 + \Delta\beta\, \boldsymbol{S}(\boldsymbol{\lambda})). \tag{2.51}$$

Defining the vector $\Delta\boldsymbol{\beta}^b := \Delta\beta\boldsymbol{\lambda}$, *the time derivative of* $\boldsymbol{R}_b^n$ *is found as*

$$\begin{aligned}\dot{\boldsymbol{R}}_b^n(t) &= \lim_{\Delta t\to 0} \frac{\boldsymbol{R}_b^n(t+\Delta t) - \boldsymbol{R}_b^n(t)}{\Delta t} \\ &= \lim_{\Delta t\to 0} \frac{\boldsymbol{R}_b^n(t)\,\Delta\beta\, \boldsymbol{S}(\boldsymbol{\lambda})}{\Delta t} \\ &= \lim_{\Delta t\to 0} \frac{\boldsymbol{R}_b^n(t)\boldsymbol{S}(\Delta\boldsymbol{\beta}^b)}{\Delta t} \\ &= \boldsymbol{R}_b^n(t)\boldsymbol{S}(\boldsymbol{\omega}_{nb}^b)\end{aligned} \tag{2.52}$$

where $\boldsymbol{\omega}_{nb}^b = \lim_{\Delta t\to 0}(\Delta\boldsymbol{\beta}^b/\Delta t)$.

6-DOF Kinematic Equations

Summarizing the results from this section, the 6-DOF kinematic equations can be expressed in matrix-vector form as

$$\dot{\boldsymbol{\eta}} = \boldsymbol{J}_\Theta(\boldsymbol{\eta})\boldsymbol{\nu}$$
$$\Updownarrow$$
$$\begin{bmatrix}\dot{\boldsymbol{p}}_{nb}^n \\ \dot{\boldsymbol{\Theta}}_{nb}\end{bmatrix} = \begin{bmatrix}\boldsymbol{R}(\boldsymbol{\Theta}_{nb}) & \boldsymbol{0}_{3\times 3} \\ \boldsymbol{0}_{3\times 3} & \boldsymbol{T}(\boldsymbol{\Theta}_{nb})\end{bmatrix}\begin{bmatrix}\boldsymbol{v}_{nb}^b \\ \boldsymbol{\omega}_{nb}^b\end{bmatrix} \tag{2.53}$$

where $\boldsymbol{\eta} \in \mathbb{R}^3 \times \mathbb{T}^3$ and $\boldsymbol{\nu} \in \mathbb{R}^6$.

MATLAB

The transformation matrix $\boldsymbol{J}_\Theta(\boldsymbol{\eta})$ and its diagonal elements $\boldsymbol{J}_{11}(\boldsymbol{\eta}) = \boldsymbol{R}(\boldsymbol{\Theta}_{nb})$ and $\boldsymbol{J}_{22}(\boldsymbol{\eta}) = \boldsymbol{T}(\boldsymbol{\Theta}_{nb})$ can be computed by using the MSS toolbox command

```
[J,J11,J22] = eulerang(phi, theta, psi)
```

The differential equations are then found by

```
p_dot      = J11  * v
theta_dot  = J22  * w_nb
```

Alternatively, (2.53) can be written in component form as

$$\dot{x}^{\mathrm{n}} = u\cos(\psi)\cos(\theta) + v(\cos(\psi)\sin(\theta)\sin(\phi) - \sin(\psi)\cos(\phi)) \\ + w(\sin(\psi)\sin(\phi) + \cos(\psi)\cos(\phi)\sin(\theta)) \tag{2.54}$$

$$\dot{y}^{\mathrm{n}} = u\sin(\psi)\cos(\theta) + v(\cos(\psi)\cos(\phi) + \sin(\phi)\sin(\theta)\sin(\psi)) \\ + w(\sin(\theta)\sin(\psi)\cos(\phi) - \cos(\psi)\sin(\phi)) \tag{2.55}$$

$$\dot{z}^{\mathrm{n}} = -u\sin(\theta) + v\cos(\theta)\sin(\phi) + w\cos(\theta)\cos(\phi) \tag{2.56}$$

$$\dot{\phi} = p + q\sin(\phi)\tan(\theta) + r\cos(\phi)\tan(\theta) \tag{2.57}$$

$$\dot{\theta} = q\cos(\phi) - r\sin(\phi) \tag{2.58}$$

$$\dot{\psi} = q\frac{\sin(\phi)}{\cos(\theta)} + r\frac{\cos(\phi)}{\cos(\theta)}, \qquad \theta \neq \pm 90^{\circ}. \tag{2.59}$$

3-DOF Model for Surface Vessels

A frequently used simplification of (2.53) is the 3-DOF (*surge, sway* and *yaw*) representation for surface vessels. This is based on the assumption that ϕ and θ are small during normal operation of ships and floating structures. Consequently, the matrices $\boldsymbol{R}(\boldsymbol{\Theta}_{\mathrm{nb}}) = \boldsymbol{R}_{z,\psi}\boldsymbol{R}_{y,\theta}\boldsymbol{R}_{x,\phi} \approx \boldsymbol{R}_{z,\psi}$ and $\boldsymbol{T}(\boldsymbol{\Theta}_{\mathrm{nb}}) \approx \boldsymbol{I}_3$. Neglecting the elements corresponding to heave, roll and pitch finally yields

$$\dot{\boldsymbol{\eta}} = \boldsymbol{R}(\psi)\boldsymbol{\nu} \tag{2.60}$$

where $\boldsymbol{R}(\psi) := \boldsymbol{R}_{z,\psi}$ with $\boldsymbol{\nu} = [u,\ v,\ r]^{\mathrm{T}}$ and $\boldsymbol{\eta} = [x^n,\ y^n,\ \psi]^{\mathrm{T}}$.

2.2.2 *Unit Quaternions*

An alternative to the Euler angle representation is a four-parameter method based on *unit quaternions* also known as *Euler parameters* (Solà 2016). The main motivation for using four parameters is to avoid the representation singularity of the Euler angles.

A quaternion $\boldsymbol{q}$ is defined as a complex number with one real part η and three imaginary parts given by the vector

$$\boldsymbol{\varepsilon} = [\varepsilon_1,\ \varepsilon_2,\ \varepsilon_3]^{\mathrm{T}}. \tag{2.61}$$

A unit quaternion satisfies $\boldsymbol{q}^\top \boldsymbol{q} = 1$. The set Q of unit quaternions is therefore defined as

$$Q := \{\boldsymbol{q}|\boldsymbol{q}^\top\boldsymbol{q} = 1,\ \boldsymbol{q} = [\eta,\ \boldsymbol{\varepsilon}^\top]^\top, \quad \eta \in \mathbb{R} \text{ and } \boldsymbol{\varepsilon} \in \mathbb{R}^3\}. \tag{2.62}$$

The motion of the body-fixed reference frame relative to the inertial frame will now be expressed in terms of unit quaternions.

From (2.16) it is seen that

$$\boldsymbol{R}_{\beta,\boldsymbol{\lambda}} = \boldsymbol{I}_3 + \sin(\beta)\boldsymbol{S}(\boldsymbol{\lambda}) + (1 - \cos(\beta))\boldsymbol{S}^2(\boldsymbol{\lambda}). \tag{2.63}$$

The real and imaginary parts of the unit quaternions are defined as (Chou 1992)

$$\eta := \cos\left(\frac{\beta}{2}\right) \tag{2.64}$$

$$\boldsymbol{\varepsilon} = [\varepsilon_1,\ \varepsilon_2,\ \varepsilon_3]^\top := \boldsymbol{\lambda}\sin\left(\frac{\beta}{2}\right) \tag{2.65}$$

where $\boldsymbol{\lambda} = [\lambda_1,\ \lambda_2,\ \lambda_3]^\top$ is a unit vector satisfying

$$\boldsymbol{\lambda} = \pm\frac{\boldsymbol{\varepsilon}}{\sqrt{\boldsymbol{\varepsilon}^\top\boldsymbol{\varepsilon}}} \quad \text{if} \quad \sqrt{\boldsymbol{\varepsilon}^\top\boldsymbol{\varepsilon}} \neq 0. \tag{2.66}$$

Consequently, the unit quaternions can be expressed in the form

$$\boldsymbol{q} = \begin{bmatrix} \eta \\ \varepsilon_1 \\ \varepsilon_2 \\ \varepsilon_3 \end{bmatrix} = \begin{bmatrix} \cos\left(\frac{\beta}{2}\right) \\ \boldsymbol{\lambda}\ \sin\left(\frac{\beta}{2}\right) \end{bmatrix} \in Q, \qquad 0 \leq \beta \leq 2\pi. \tag{2.67}$$

This parametrization implies that the unit quaternions satisfy the constraint $\boldsymbol{q}^\top\boldsymbol{q} = 1$, which can be expanded to yield

$$\eta^2 + \varepsilon_1^2 + \varepsilon_2^2 + \varepsilon_3^2 = 1. \tag{2.68}$$

The product of two unit quaternions (*Hamiltonian product*) is

$$\boldsymbol{q}_1 \otimes \boldsymbol{q}_2 = \begin{bmatrix} \eta_{q_1}\eta_{q_2} - \boldsymbol{\varepsilon}_{q_1}^\top\boldsymbol{\varepsilon}_{q_2} \\ \eta_{q_1}\boldsymbol{\varepsilon}_{q_2} + \eta_{q_2}\boldsymbol{\varepsilon}_{q_1} + \boldsymbol{S}(\boldsymbol{\varepsilon}_{q_1})\boldsymbol{\varepsilon}_{q_2} \end{bmatrix}. \tag{2.69}$$

Note that for quaternion operations $\boldsymbol{q}_1 \otimes \boldsymbol{q}_2 \neq \boldsymbol{q}_2 \otimes \boldsymbol{q}_1$ but $\boldsymbol{q}_1 + \boldsymbol{q}_2 = \boldsymbol{q}_2 + \boldsymbol{q}_1$.

Unit Quaternion Rotation Matrix from BODY to NED

From (2.63) with (2.64) and (2.65), the following rotation matrix for the unit quaternion is obtained

$$\boldsymbol{R}(\boldsymbol{q}_b^n) := \boldsymbol{I}_3 + 2\eta \boldsymbol{S}(\boldsymbol{\varepsilon}) + 2\boldsymbol{S}^2(\boldsymbol{\varepsilon}). \tag{2.70}$$

Linear Velocity Transformation

The transformation relating the linear velocity vector in an inertial reference frame to a velocity in the body-fixed reference frame can now be expressed as

$$\dot{\boldsymbol{p}}^n_{\rm nb} = \boldsymbol{R}(\boldsymbol{q}_b^n)\boldsymbol{v}^b_{\rm nb} \tag{2.71}$$

where

$$\boldsymbol{R}(\boldsymbol{q}_b^n) = \begin{bmatrix} 1-2(\varepsilon_2^2+\varepsilon_3^2) & 2(\varepsilon_1\varepsilon_2-\varepsilon_3\eta) & 2(\varepsilon_1\varepsilon_3+\varepsilon_2\eta) \\ 2(\varepsilon_1\varepsilon_2+\varepsilon_3\eta) & 1-2(\varepsilon_1^2+\varepsilon_3^2) & 2(\varepsilon_2\varepsilon_3-\varepsilon_1\eta) \\ 2(\varepsilon_1\varepsilon_3-\varepsilon_2\eta) & 2(\varepsilon_2\varepsilon_3+\varepsilon_1\eta) & 1-2(\varepsilon_1^2+\varepsilon_2^2) \end{bmatrix}. \tag{2.72}$$

Expanding (2.71) yields

$$\dot{x}^{\rm n} = u(1-2\varepsilon_2^2-2\varepsilon_3^2) + 2v(\varepsilon_1\varepsilon_2-\varepsilon_3\eta) + 2w(\varepsilon_1\varepsilon_3+\varepsilon_2\eta) \tag{2.73}$$

$$\dot{y}^{\rm n} = 2u(\varepsilon_1\varepsilon_2+\varepsilon_3\eta) + v(1-2\varepsilon_1^2-2\varepsilon_3^2) + 2w(\varepsilon_2\varepsilon_3-\varepsilon_1\eta) \tag{2.74}$$

$$\dot{z}^{\rm n} = 2u(\varepsilon_1\varepsilon_3-\varepsilon_2\eta) + 2v(\varepsilon_2\varepsilon_3+\varepsilon_1\eta) + w(1-2\varepsilon_1^2-2\varepsilon_2^2). \tag{2.75}$$

As for the Euler angle representation, Property 2.1 implies that the inverse transformation matrix satisfies $\boldsymbol{R}^{-1}(\boldsymbol{q}_b^n) = \boldsymbol{R}^\top(\boldsymbol{q}_b^n)$.

MATLAB

The unit quaternion rotation matrix is easily computed by using the MSS toolbox commands

```
q = [eta, eps1, eps2, eps3]'
R = Rquat(q)
```

Note that $(\boldsymbol{q}_b^n)^\top \boldsymbol{q}_b^n = 1$ must be true for `Rquat(q)` to return a solution. One way to ensure this is to use the transformation

```
q = euler2q(phi, theta, psi)
```

transforming the three Euler angles ϕ, θ and ψ to the unit quaternion vector $\boldsymbol{q}_b^n$; see Section 2.2.3 for details.

Angular Velocity Transformation

The angular velocity transformation can be derived by substituting the expressions for R_{ij} from (2.72) into the differential equation $\dot{\boldsymbol{R}}_b^n = \boldsymbol{R}_b^n \boldsymbol{S}(\boldsymbol{\omega}_{\text{nb}}^b)$; see Theorem 2.2. Some calculations yield

$$\dot{\boldsymbol{q}}_b^n = \frac{1}{2}\boldsymbol{q}_b^n \otimes \begin{bmatrix} 0 \\ \boldsymbol{\omega}_{\text{nb}}^b \end{bmatrix}. \tag{2.76}$$

An alternative formulation is the vector representation (Kane *et al.* 1983)

$$\begin{aligned} \dot{\boldsymbol{q}}_b^n &= \frac{1}{2}\begin{bmatrix} -\boldsymbol{\varepsilon}^\top \\ \eta \boldsymbol{I}_3 + \boldsymbol{S}(\boldsymbol{\varepsilon}) \end{bmatrix} \boldsymbol{\omega}_{\text{nb}}^b \\ &:= \boldsymbol{T}(\boldsymbol{q}_b^n)\boldsymbol{\omega}_{\text{nb}}^b \end{aligned} \tag{2.77}$$

where

$$\boldsymbol{T}(\boldsymbol{q}_b^n) = \frac{1}{2}\begin{bmatrix} -\varepsilon_1 & -\varepsilon_2 & -\varepsilon_3 \\ \eta & -\varepsilon_3 & \varepsilon_2 \\ \varepsilon_3 & \eta & -\varepsilon_1 \\ -\varepsilon_2 & \varepsilon_1 & \eta \end{bmatrix}, \qquad \boldsymbol{T}(\boldsymbol{q}_b^n)\boldsymbol{T}^\top(\boldsymbol{q}_b^n) = \frac{1}{4}\boldsymbol{I}_3. \tag{2.78}$$

Consequently,

$$\dot{\eta} = -\frac{1}{2}(\varepsilon_1 p + \varepsilon_2 q + \varepsilon_3 r) \tag{2.79}$$

$$\dot{\varepsilon}_1 = \frac{1}{2}(\eta p - \varepsilon_3 q + \varepsilon_2 r) \tag{2.80}$$

$$\dot{\varepsilon}_2 = \frac{1}{2}(\varepsilon_3 p + \eta q - \varepsilon_1 r) \tag{2.81}$$

$$\dot{\varepsilon}_3 = \frac{1}{2}(-\varepsilon_2 p + \varepsilon_1 q + \eta r). \tag{2.82}$$

MATLAB

The transformation matrix $\boldsymbol{T}(\boldsymbol{q}_b^n)$ is implemented in the MSS toolbox as

```
T = Tquat(q)
```

6-DOF kinematic equations

The 6-DOF kinematic equations parametrized by unit quaternions result in seven differential equations with $\boldsymbol{\eta} = [x^n, y^n, z^n, \eta, \varepsilon_1, \varepsilon_2, \varepsilon_3]^\top$ as state vector (recall that only six differential equations are needed when using the Euler angle representation). The additional differential equation is needed because of the unity constraint of the Euler parameters. Consequently, unit quaternions do not represent generalized coordinates. The resulting differential equations are

$$\dot{\boldsymbol{\eta}} = \boldsymbol{J}_q(\boldsymbol{\eta})\boldsymbol{v}$$
$$\Updownarrow$$
$$\begin{bmatrix} \dot{\boldsymbol{p}}_{\mathrm{nb}}^n \\ \dot{\boldsymbol{q}}_b^n \end{bmatrix} = \begin{bmatrix} \boldsymbol{R}(\boldsymbol{q}_b^n) & \boldsymbol{0}_{3\times3} \\ \boldsymbol{0}_{4\times3} & \boldsymbol{T}(\boldsymbol{q}_b^n) \end{bmatrix} \begin{bmatrix} \boldsymbol{v}_{\mathrm{nb}}^b \\ \boldsymbol{\omega}_{\mathrm{nb}}^b \end{bmatrix} \tag{2.83}$$

where $\boldsymbol{\eta} \in \mathbb{R}^3 \times \mathbb{Q}$ and $\boldsymbol{v} \in \mathbb{R}^6$, and $\boldsymbol{J}_q(\boldsymbol{\eta}) \in \mathbb{R}^{7\times6}$ is a non-quadratic transformation matrix. Equation (2.83) in component form is given by (2.73)–(2.75) and (2.79)–(2.82).

MATLAB

The transformation matrix $\boldsymbol{J}_q(\boldsymbol{\eta})$ and its elements $\boldsymbol{J}_{11} = \boldsymbol{R}(\boldsymbol{q}_b^n)$ and $\boldsymbol{J}_{22} = \boldsymbol{T}(\boldsymbol{q}_b^n)$ can be computed directly in the MSS toolbox by using the following commands

```
q = [eta, eps1, eps2, eps3]'
[J,J11,J22] = quatern(q)
```

The corresponding differential equations are

```
p_dot = J11 * v
q_dot = J22 * w_nb
```

Implementation Considerations: Unit Quaternion Normalization

When integrating (2.77), a normalization procedure is necessary to ensure that the constraint

$$(\boldsymbol{q}_b^n)^\top \boldsymbol{q}_b^n = \eta^2 + \varepsilon_1^2 + \varepsilon_2^2 + \varepsilon_3^2 = 1 \tag{2.84}$$

is satisfied in the presence of measurement noise and numerical round-off errors. For this purpose, the following discrete-time algorithm can be applied.

Algorithm 2.1 (Discrete-Time Unit Quaternion Normalization)

1. *$k = 0$. Compute the initial value of $\boldsymbol{q}_b^n[0]$.*
2. *Propagate the unit quaternion using Euler's method*

$$\boldsymbol{q}_b^n[k+1] = \boldsymbol{q}_b^n[k] + h\boldsymbol{T}(\boldsymbol{q}_b^n[k])\boldsymbol{\omega}_{\text{nb}}^b[k] \tag{2.85}$$

where h is the sampling time and k is the sampling index.

3. *Normalization*

$$\boldsymbol{q}_b^n[k+1] = \frac{\boldsymbol{q}_b^n[k+1]}{\|\boldsymbol{q}_b^n[k+1]\|} = \frac{\boldsymbol{q}_b^n[k+1]}{\sqrt{\boldsymbol{q}_b^n[k+1]^\top\ \boldsymbol{q}_b^n[k+1]}}. \tag{2.86}$$

4. *Let $k = k + 1$ and return to Step 2.*

A continuous-time algorithm for unit quaternion normalization can be implemented by noting that

$$\frac{\mathrm{d}}{\mathrm{d}t}((\boldsymbol{q}_b^n)^\top \boldsymbol{q}_b^n) = 2(\boldsymbol{q}_b^n)^\top \boldsymbol{T}(\boldsymbol{q}_b^n)\boldsymbol{\omega}_{\text{nb}}^b = 0. \tag{2.87}$$

This shows that if $\boldsymbol{q}_b^n$ is initialized as a unit vector, then it will remain a unit vector. Since integration of the quaternion vector $\boldsymbol{q}_b^n$ from the differential equation (2.77) will introduce numerical errors that will cause the length of $\boldsymbol{q}_b^n$ to deviate from unity, a nonlinear feedback or normalization term is suggested. This can be achieved by replacing the kinematic differential equation (2.77) with

$$\dot{\boldsymbol{q}}_b^n = \boldsymbol{T}(\boldsymbol{q}_b^n)\boldsymbol{\omega}_{\text{nb}}^b + \frac{\gamma}{2}(1 - (\boldsymbol{q}_b^n)^\top \boldsymbol{q}_b^n)\boldsymbol{q}_b^n \tag{2.88}$$

where $\gamma \geq 0$ (typically 100) is a design parameter reflecting the convergence rate of the normalization. This results in

$$\begin{aligned}\frac{\mathrm{d}}{\mathrm{d}t}((\boldsymbol{q}_b^n)^\top \boldsymbol{q}_b^n) &= \underbrace{2(\boldsymbol{q}_b^n)^\top \boldsymbol{T}(\boldsymbol{q}_b^n)\boldsymbol{\omega}_{\text{nb}}^b}_{0 \text{ since } \boldsymbol{q}_b^n(0) \text{ is a unit vector}} + \gamma(1 - (\boldsymbol{q}_b^n)^\top \boldsymbol{q}_b^n)\ (\boldsymbol{q}_b^n)^\top \boldsymbol{q}_b^n \\ &= \gamma(1 - (\boldsymbol{q}_b^n)^\top \boldsymbol{q}_b^n)(\boldsymbol{q}_{\text{b}}^{\text{n}})^\top \boldsymbol{q}_b^n.\end{aligned} \tag{2.89}$$

A change of coordinates $x = 1 - (\boldsymbol{q}_b^n)^\top \boldsymbol{q}_b^n$ and $\dot{x} = -\mathrm{d}/\mathrm{d}t\ ((\boldsymbol{q}_b^n)^\top \boldsymbol{q}_b^n)$ yields

$$\dot{x} = -\gamma x(1 - x) \tag{2.90}$$

Linearization about $x = 0$ results in $\dot{x} = -\gamma x$. Consequently, the normalization algorithm converges with a time constant $T = \gamma^{-1}$.

2.2.3 Unit Quaternion from Euler Angles

If the Euler angles $\boldsymbol{\Theta}_{\rm nb} = [\phi,\ \theta,\ \psi]^{\top}$ are known the corresponding unit quaternion can be computed using the following transformation for the *zyx* rotation sequence (NASA 2013)

$$\boldsymbol{q}_b^n = \begin{bmatrix} \cos\left(\frac{1}{2}\psi\right)\cos\left(\frac{1}{2}\theta\right)\cos\left(\frac{1}{2}\phi\right)+\sin\left(\frac{1}{2}\psi\right)\sin\left(\frac{1}{2}\theta\right)\sin\left(\frac{1}{2}\phi\right) \\ \cos\left(\frac{1}{2}\psi\right)\cos\left(\frac{1}{2}\theta\right)\sin\left(\frac{1}{2}\phi\right)-\sin\left(\frac{1}{2}\psi\right)\sin\left(\frac{1}{2}\theta\right)\cos\left(\frac{1}{2}\phi\right) \\ \sin\left(\frac{1}{2}\psi\right)\cos\left(\frac{1}{2}\theta\right)\sin\left(\frac{1}{2}\phi\right)+\cos\left(\frac{1}{2}\psi\right)\sin\left(\frac{1}{2}\theta\right)\cos\left(\frac{1}{2}\phi\right) \\ \sin\left(\frac{1}{2}\psi\right)\cos\left(\frac{1}{2}\theta\right)\cos\left(\frac{1}{2}\phi\right)-\cos\left(\frac{1}{2}\psi\right)\sin\left(\theta\right)\sin\left(\frac{1}{2}\phi\right) \end{bmatrix}. \tag{2.91}$$

Formula (2.91) is implemented in the MSS toolbox as `euler2q.m`. This formula can also be used to compute the initial values of the Euler parameters corresponding to step 1 of Algorithm 2.1.

Example 2.2 (Unit Quaternion from Euler Angles)
Consider a marine craft with attitude $\phi = 10.0°, \theta = -20.0°$ and $\psi = 30.0°$. The unit quaternion is computed in MATLAB by using the commands

```
phi = 10*(pi/180); psi = 30*(pi/180); theta = -20*(pi/180)
q = euler2q(phi, theta, psi)
q =
    0.9437
    0.1277
   -0.1449
    0.2685
>> norm(q)               % normalization test
ans =
      1
```

2.2.4 Euler Angles from a Unit Quaternion

The relationship between the Euler angles ϕ, θ and ψ (*zyx* convention) and the unit quaternion can be established by requiring that the rotation matrices of the two kinematic representations are equal

$$\boldsymbol{R}(\boldsymbol{\Theta}_{\rm nb}) := \boldsymbol{R}(\boldsymbol{q}_b^n). \tag{2.92}$$

Let the elements of $\boldsymbol{R}(\boldsymbol{q}_b^n)$ be denoted by R_{ij} where the superscripts i and j denote the ith row and jth column. Writing expression (2.92) in component form yields a system of nine equations with three unknowns (ϕ, θ and ψ) given by

$$\begin{bmatrix} c\psi c\theta & -s\psi c\phi + c\psi s\theta s\phi & s\psi s\phi + c\psi c\phi s\theta \\ s\psi c\theta & c\psi c\phi + s\phi s\theta s\psi & -c\psi s\phi + s\theta s\psi c\phi \\ -s\theta & c\theta s\phi & c\theta c\phi \end{bmatrix} = \begin{bmatrix} R_{11} & R_{12} & R_{13} \\ R_{21} & R_{22} & R_{23} \\ R_{31} & R_{32} & R_{33} \end{bmatrix}. \tag{2.93}$$

One solution to (2.93) is

$$\phi = \text{atan2}(R_{32}, R_{33}) \tag{2.94}$$

$$\theta = -\text{asin}(R_{31}), \qquad \theta \neq \pm 90^\circ \tag{2.95}$$

$$\psi = \text{atan2}(R_{21}, R_{11}) \tag{2.96}$$

which gives

$$\phi = \text{atan2}\,(2(\varepsilon_2\varepsilon_3 + \varepsilon_1\eta),\ 1 - 2(\varepsilon_1^2 + \varepsilon_2^2)) \tag{2.97}$$

$$\theta = -\text{asin}\,(2(\varepsilon_1\varepsilon_3 - \varepsilon_2\eta)) \tag{2.98}$$

$$\psi = \text{atan2}\,(2(\varepsilon_1\varepsilon_2 + \varepsilon_3\eta),\ 1 - 2(\varepsilon_2^2 + \varepsilon_3^2)). \tag{2.99}$$

Here atan2(y, x) is the four-quadrant inverse tangent confining the result to $(-\pi, \pi]$. Precautions must be taken against computational errors in the vicinity of $\theta = \pm 90^\circ$. The MSS toolbox script `[phi,theta,psi] = q2euler(q)` is based on Formulas (2.97)–(2.99). A singularity test is included in order to avoid numerical problems for $\theta = \pm 90^\circ$.

Example 2.3 (Euler Angles from a Unit Quaternion)
Consider the marine vessel in Example 2.2 where the Euler angles where converted into a unit quaternion. The inverse transformation `q2euler(q)` *results in*

```
q =[0.9437, 0.1277, -0.1449, 0.2685]'
[phi,theta,psi] = q2euler(q/norm(q))
phi =
    0.1746
theta =
   -0.3491
psi =
    0.5235
```

corresponding to $\phi = 10.0^\circ, \theta = -20.0^\circ$ *and* $\psi = 30.0^\circ$.

2.3 Transformations Between ECEF and NED

Wide-area or terrestrial guidance and navigation implies that the position should be related to the Earth center instead of a local frame on the Earth's surface. This section describes the kinematic transformation between ECEF and NED parametrized in terms of *longitude, latitude* and *height*. The NED reference frame can be represented as a tangent plane on the Earth's surface moving with the craft or an Earth-fixed tangent plane with constant longitude and latitude used for local navigation.

2.3.1 Longitude and Latitude Rotation Matrix

The transformation between the ECEF and NED velocity vectors is

$$\dot{\boldsymbol{p}}^e_{\rm eb} = \boldsymbol{R}^e_n \dot{\boldsymbol{p}}^n_{\rm eb} = \boldsymbol{R}^e_n \boldsymbol{R}^n_b \boldsymbol{v}^b_{\rm eb} \tag{2.100}$$

where $\boldsymbol{R}^e_n = \boldsymbol{R}(\boldsymbol{\Theta}_{\rm en})$ and $\boldsymbol{R}^n_b = \boldsymbol{R}(\boldsymbol{\Theta}_{\rm nb})$. Let $\boldsymbol{\Theta}_{\rm en} = [l,\ \mu]^\top$ denote the vector formed by the longitude and latitude pair $(l,\ \mu)$ shown in Figure 2.5. The rotation matrix $\boldsymbol{R}^e_n$: $\mathbb{T}^2 \rightarrow$ SO(3) between ECEF and NED is found by performing two principal rotations: (1) a rotation l about the z axis and (2) a rotation $(-\mu - \pi/2)$ about the y axis. This gives

$$\begin{aligned}\boldsymbol{R}^e_n &:= \boldsymbol{R}_{z,l}\boldsymbol{R}_{y,-\mu-\frac{\pi}{2}} \\ &= \begin{bmatrix} \cos(l) & -\sin(l) & 0 \\ \sin(l) & \cos(l) & 0 \\ 0 & 0 & 1 \end{bmatrix} \begin{bmatrix} \cos(-\mu-\frac{\pi}{2}) & 0 & \sin(-\mu-\frac{\pi}{2}) \\ 0 & 1 & 0 \\ -\sin(-\mu-\frac{\pi}{2}) & 0 & \cos(-\mu-\frac{\pi}{2}) \end{bmatrix}.\end{aligned} \tag{2.101}$$

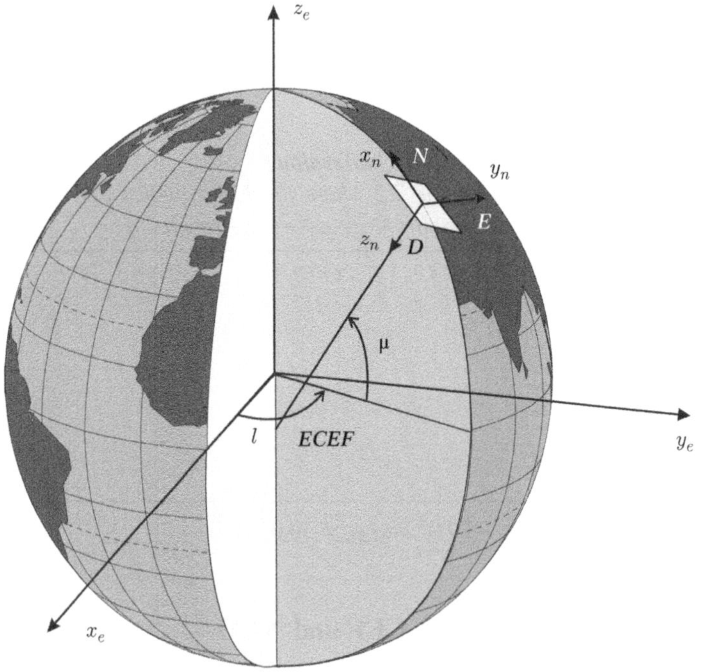

Figure 2.5 Definitions of longitude l and latitude μ and the NED reference frame on the surface of the Earth. The D axis is orthogonal to the Earth's surface.

Using the trigonometric formulae $\cos(-\mu - \pi/2) = -\sin(\mu)$ and $\sin(-\mu - \pi/2) = -\cos(\mu)$ finally yields

$$R_n^e = R(\Theta_{en}) = \begin{bmatrix} -\cos(l)\sin(\mu) & -\sin(l) & -\cos(l)\cos(\mu) \\ -\sin(l)\sin(\mu) & \cos(l) & -\sin(l)\cos(\mu) \\ \cos(\mu) & 0 & -\sin(\mu) \end{bmatrix}. \tag{2.102}$$

Consequently, the ECEF positions $p_{eb}^e = [x^e, y^e, z^e]^\top$ can be found by numerical integration of (2.100).

MATLAB

The rotation matrix R_n^e is computed using the MSS toolbox command

```
R = Rll(l, mu)
```

2.3.2 *Longitude, Latitude and Height from ECEF Coordinates*

The measurements of global navigation satellite systems such as BeiDou, Galileo, GLONASS and Navstar GPS are given in the Cartesian ECEF frame, but these are measurements that do not make much sense to a human operator. The presentation of terrestrial position data $p_{eb}^e = [x^e, y^e, z^e]^\top$ is therefore made in terms of the ellipsoidal parameters longitude l, latitude μ and height h.

The reference ellipsoid used for satellite navigation systems, the WGS-84 ellipsoid, is found by rotating an ellipse around the polar axis (World Geodetic System 1984). Because of symmetry about the polar axis, it is only necessary to look at the meridian plane (latitude equations). The origin of the ellipsoid coincides with the mass center of the Earth. The most important parameters of the WGS-84 ellipsoid are listed in Table 2.2.

In Figure 2.6, μ is the *geodetic* latitude, h is the ellipsoidal height and N is the radius of curvature in the prime vertical. N is calculated by

$$N = \frac{r_e^2}{\sqrt{r_e^2 \cos^2(\mu) + r_p^2 \sin^2(\mu)}} \tag{2.103}$$

where the equatorial and polar earth radii, r_e and r_p, are the semi-axes of the ellipsoid.

Table 2.2 WGS-84 parameters.

Parameters	Comments
$r_e = 6\ 378\ 137.0$ m	Equatorial radius of ellipsoid (semi-major axis)
$r_p = 6\ 356\ 752.314\ 245$ m	Polar axis radius of ellipsoid (semi-minor axis)
$\omega_{ie} = 7.292115 \times 10^{-5}$ rad s^{-1}	Angular velocity of the Earth
$e = 0.0818$	Eccentricity of ellipsoid

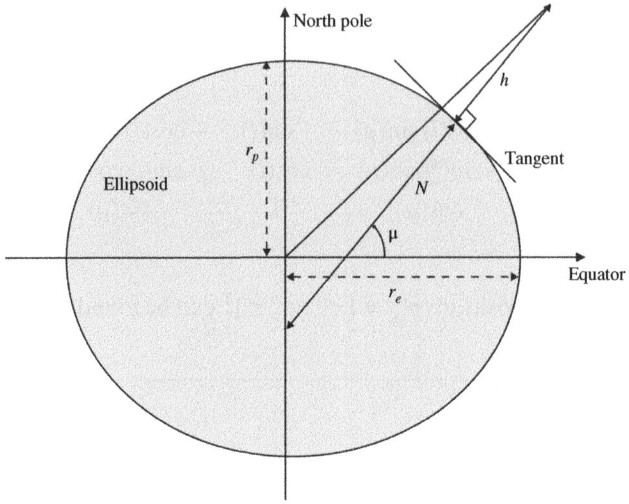

Figure 2.6 Definitions of the ellipsoidal parameters.

Longitude l is given by

$$l = \text{atan}\left(\frac{y^e}{x^e}\right) \tag{2.104}$$

while latitude μ and height h are implicitly computed by

$$\tan(\mu) = \frac{z^e}{p}\left(1 - e^2 \frac{N}{N+h}\right)^{-1} \tag{2.105}$$

$$h = \frac{p}{\cos(\mu)} - N \tag{2.106}$$

where e is the eccentricity of the Earth given by

$$e = \sqrt{1 - \left(\frac{r_\text{p}}{r_\text{e}}\right)^2} \tag{2.107}$$

Note that (2.105)–(2.106) are implicit equations which can be solved iteratively by using Algorithm 2.2 below (Hofmann-Wellenhof *et al.* 1994).

Algorithm 2.2 (Transformation of (x^e, y^e, z^e) to (l, μ, h))

1. *Compute* $p = \sqrt{(x^e)^2 + (y^e)^2}$.
2. *Compute the approximate value* μ_0 *from*

$$\tan(\mu_0) = \frac{z^e}{p}(1 - e^2)^{-1}.$$

3. *Compute an approximate value* N *from*

$$N = \frac{r_e^2}{\sqrt{r_e^2 \cos^2(\mu_0) + r_p^2 \sin^2(\mu_0)}}.$$

4. *Compute the ellipsoidal height by*

$$h = \frac{p}{\cos(\mu_0)} - N.$$

5. *Compute an improved value for the latitude by*

$$\tan(\mu) = \frac{z^e}{p}\left(1 - e^2 \frac{N}{N + h}\right)^{-1}.$$

6. *Check for another iteration step: if* $|\mu - \mu_0| < \varepsilon$, *where* ε *is a small number, then the iteration is complete. Otherwise set* $\mu_0 = \mu$ *and continue with step 3.*

MATLAB

Algorithm 2.2 is programmed in the MSS toolbox as a function

```
[l,mu,h] = ecef2llh(x, y, z)
```

Several other algorithms can be used for this purpose; see Farrell (2008) and references therein. An approximate solution can also be found in Hofmann-Wellenhof *et al.* (1994) and an exact explicit solution is given by Zhu (1993).

Height Transformation

The WGS-84 ellipsoid is a global ellipsoid, which is only an approximation of the mean sea level of the Earth. It can deviate from the real mean sea level by as much as 100 m at certain locations. The Earth's geoid, on the other hand, is defined physically and its center is coincident with the center of the Earth. It is an equipotential surface so that it

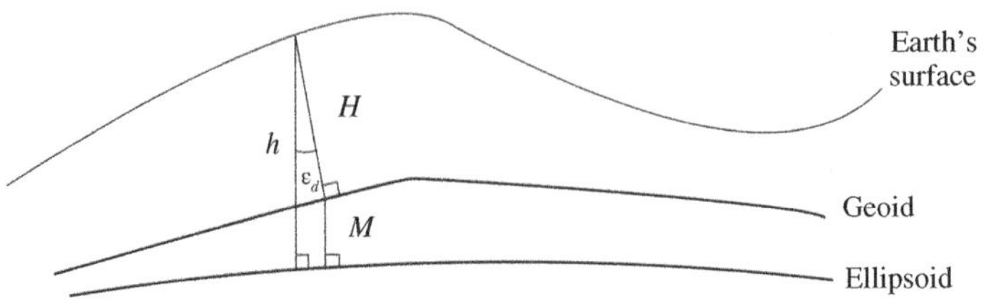

Figure 2.7 Illustration of ellipsoidal and orthonometric heights h and H where ε_d is the deflection of gravity and M is the geoidal height (undulation).

has the same gravitational magnitude all over the surface, and the gravity vector is always perpendicular to the geoid.

The geoid is the surface chosen as a zero level reference. The ellipsoidal height h in Figure 2.7 must therefore be transformed to the *orthometric* height H in Figure 2.7 through the relation

$$h \approx H + M$$

where M is called the *geoidal* height. The angle ε_d is small enough for the above approximation to be sufficiently accurate for all practical purposes. The angle ε_d is known as the deflection of the vertical, and does not exceed 30 arcsec in most of the world. In fact the largest deflection encountered over the entire earth is in the order of 1 arcmin (Britting 1971). The geoidal height M is found through a *datum* transformation (Hofmann-Wellenhof *et al.* 1994).

2.3.3 ECEF Coordinates from Longitude, Latitude and Height

The transformation from $\boldsymbol{\Theta}_\text{en} = [l,\ \mu]^\top$ for given heights h to $\boldsymbol{p}^e_\text{eb} = [x^e,\ y^e,\ z^e]^\top$ is given by (Heiskanen and Moritz 1967)

$$\begin{bmatrix} x^e \\ y^e \\ z^e \end{bmatrix} = \begin{bmatrix} (N+h)\cos(\mu)\cos(l) \\ (N+h)\cos(\mu)\sin(l) \\ \left(\frac{r_\text{p}^2}{r_\text{e}^2}N + h\right)\sin(\mu) \end{bmatrix}. \tag{2.108}$$

For a ship h is the vertical distance from the sea level to the coordinate origin of $\{b\}$.

MATLAB

The transformation from $\boldsymbol{\Theta}_\text{en} = [l,\ \mu]^\top$ to $\boldsymbol{p}^e_\text{eb} = [x^e,\ y^e,\ z^e]^\top$, Equation (2.108), is programmed in the MSS toolbox function

```
[x,y,z] = llh2ecef(l, mu, h)
```

Assume that $l = 10.3°$, $\mu = 63.0°$ and $h = 0$ m. Hence, the ECEF coordinates are computed to be

$$\begin{bmatrix} x^e \\ y^e \\ z^e \end{bmatrix} = \begin{bmatrix} 2\,856\,552 \text{ m} \\ 519\,123 \text{ m} \\ 5\,659\,978 \text{ m} \end{bmatrix}$$

using the MSS MATLAB command

```
[x,y,z] = llh2ecef(10.3*(pi/180), 63.0*(pi/180),0)
```

2.4 Transformations between ECEF and Flat-Earth Coordinates

For local flat-Earth navigation it can be assumed that the NED tangent plane is fixed on the surface of the Earth. Assume that the NED tangent plane is located at l_0 and μ_0 such that

$$\boldsymbol{R}_n^e = \boldsymbol{R}(\boldsymbol{\theta}_{\mathrm{en}}) = \text{constant} \tag{2.109}$$

where $\boldsymbol{\theta}_{\mathrm{en}} = [l_0,\ \mu_0]^\top$. The ECEF coordinates satisfy the differential equation

$$\dot{\boldsymbol{p}}_{\mathrm{eb}}^e = \boldsymbol{R}_n^e \boldsymbol{R}_b^n \boldsymbol{v}_{\mathrm{eb}}^b. \tag{2.110}$$

When designing dynamic positioning (DP) systems for offshore vessels this is a good approximation. However, when designing global waypoint-tracking control systems for ships, "flat Earth" is not a good approximation since $(l,\ \mu)$ will vary largely for ships in transit between the different continents.

2.4.1 *Longitude, Latitude and Height from Flat-Earth Coordinates*

Assume that a flat-Earth coordinate system is defined as the tangent plane to the WGS-84 ellipsoid. The coordinate origin is located at $(l_0,\ \mu_0)$ with reference height h_{ref} in meters above the surface of the Earth. The NED positions with respect to the coordinate origin are denoted as $(x^n,\ y^n,\ z^n)$ with z^n positive downwards.

The Earth radius of curvature in the prime vertical R_{N} and the radius of curvature in the meridian R_{M} are (Farrell 2008)

$$R_{\mathrm{N}} = \frac{r_{\mathrm{e}}}{\sqrt{1 - e^2 \sin^2(\mu_0)}} \tag{2.111}$$

$$R_{\mathrm{M}} = R_{\mathrm{N}} \frac{1 - e^2}{\sqrt{1 - e^2 \sin^2(\mu_0)}} \tag{2.112}$$

where $r_e = 6\ 378\ 137$ m is the semi-minor axis (equatorial radius) and $e = 0.0818$ is the Earth eccentricity. It follows that

$$\Delta l = y^n \, \text{atan2}(1, R_M \cos(\mu_0)) \tag{2.113}$$

$$\Delta \mu = x^n \, \text{atan2}(1, R_N) \tag{2.114}$$

where atan2(y, x) is the four-quadrant inverse tangent confining the result to $(-\pi, \ \pi]$. The triplet $(l, \ \mu, \ h)$ is computed as

$$l = \text{ssa}(l_0 + \Delta l) \tag{2.115}$$

$$\mu = \text{ssa}(\mu_0 + \Delta \mu) \tag{2.116}$$

$$h = h_{\text{ref}} - z^n \tag{2.117}$$

where ssa$(\cdot)$ is the *smallest signed angle* confining the argument to the interval $[-\pi, \pi)$.

MATLAB

The ssa$(\cdot)$ function is implemented in the MSS toolbox as

```
angle = ssa(angle)          % angle [rad]
angle = ssa(angle, type)    % optional argument for [rad] or [deg]
```

Here `type` can be chosen as "`rad`" or "`deg`".

MATLAB

The triplet (l, μ, h) corresponding to the NED coordinates $(x^n, \ y^n, \ z^n)$ in a flat-Earth coordinate origin located at $(l_0, \ \mu_0)$ with reference height h_{ref} in meters above the surface of the Earth is computed as

```
[l,mu,h] = flat2llh(x, y, z, l_0, mu_0, h_ref)
```

2.4.2 Flat-Earth Coordinates from Longitude, Latitude and Height

The NED positions $(x^n, \ y^n, \ z^n)$ with respect to a flat-Earth coordinate system with origin $(l_0, \ \mu_0)$ and reference height h_{ref} are given by (Farrell 2008)

$$x^n = \frac{\Delta \mu}{\text{atan2}(1, R_M)} \tag{2.118}$$

$$y^n = \frac{\Delta l}{\text{atan2}(1, R_N \cos(\mu_0))} \tag{2.119}$$

$$z^n = h_{\text{ref}} - h \tag{2.120}$$

where

$$\Delta l := l - l_0 \tag{2.121}$$

$$\Delta \mu := \mu - \mu_0. \tag{2.122}$$

MATLAB

For a marine craft operating on the sea surface ($h = 0$) or submerged ($h < 0$) with longitude and latitude (l, μ) and a local coordinate origin at (l_0, μ_0, h_{ref}), the NED positions can be computed using the MSS toolbox command

```
[x,y,z] = llh2flat(l, mu, h, l_0, mu_0, h_ref)
```

2.5 Transformations Between BODY and FLOW

The FLOW frame $\{f\} = (x_{\mathrm{f}}, y_{\mathrm{f}}, z_{\mathrm{f}})$ is a body-fixed reference frame used to express the hydrodynamic forces and moments acting on a marine craft. The FLOW axes are found by rotating the BODY axes such that the resulting x_{f} axis is parallel to the freestream flow. Therefore, in FLOW axes, the x_{f} axis points directly into the relative flow while the z_{f} axis remains in the reference plane, but rotates so that it remains perpendicular to the x_{f} axis. The y_{f} axis completes the right-handed system. The transformation is outlined in Section 2.5.3. We will first define the heading, course and crab angles and relate this to the more general definitions of *sideslip angle* and *angle of attack*. Finally, we present the rotation matrix from $\{b\}$ to $\{f\}$ as a function of the sideslip angle and angle of attack.

2.5.1 Definitions of Heading, Course and Crab Angles

The relationship between the angular variables *course, heading* and *sideslip* is important for maneuvering of a marine craft in the horizontal plane. The terms course and heading are used interchangeably in much of the literature on guidance, navigation and control of marine craft, and this leads to confusion. Consequently, definitions utilizing a consistent symbolic notation will now be established. The BODY and NED references frame are denoted $\{b\} = (x_{\mathrm{b}}, y_{\mathrm{b}}, z_{\mathrm{b}})$ and $\{n\} = (x_{\mathrm{n}}, y_{\mathrm{n}}, z_{\mathrm{n}})$, respectively.

Definition 2.4 (Yaw or Heading Angle ψ)
The angle ψ from the x_{n} axis (true North) to the x_{b} axis of the craft, positive rotation about the z_{n} axis by the right-hand screw convention (see Figure 2.8).

The heading angle is usually measured by using a magnetic compass, gyrocompass or two GNSS receivers; see Gade (2016) for a discussion on methods. The heading angle is well defined for zero speed such that it is possible to design a *heading autopilot* to maintain constant heading during stationkeeping and transit. However, during transit it is common to use a *course autopilot* for path following.

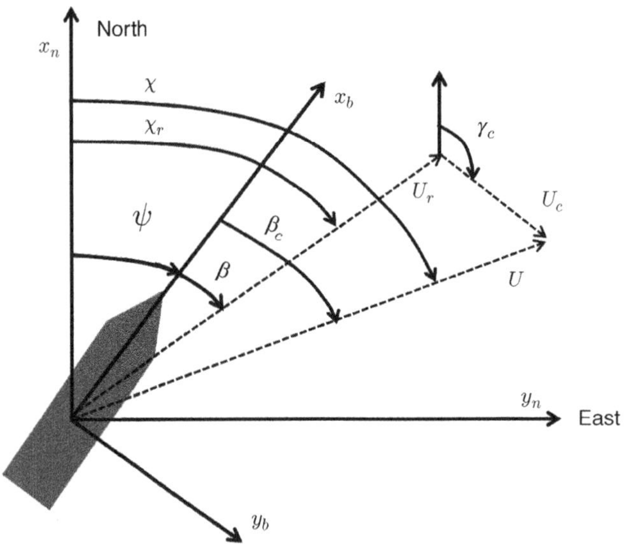

Figure 2.8 Ocean current triangle. Horizontal plane.

Definition 2.5 (Course Angle χ)
The angle χ from the x_n axis (true North) to the velocity vector of the craft, positive rotation about the z_n axis by the right-hand screw convention (see Figure 2.8).

Notice that the course angle is only defined for positive speed. For surface vessels both the course and speed over ground (COG and SOG) can be measured by GNSS. For underwater vehicles it is common to use hydroacoustic reference systems.

From (2.36) it follows that the horizontal motion of a marine craft can be described by

$$\dot{x}^{\rm n} = u\cos(\psi) - v\sin(\psi) \tag{2.123}$$

$$\dot{y}^{\rm n} = u\sin(\psi) + v\cos(\psi). \tag{2.124}$$

These equations can be expressed in *amplitude-phase* form by

$$\dot{x}^{\rm n} = U\cos(\psi + \beta_{\rm c}) := U\cos(\chi) \tag{2.125}$$

$$\dot{y}^{\rm n} = U\sin(\psi + \beta_{\rm c}) := U\sin(\chi) \tag{2.126}$$

where the course angle is defined as

$$\chi := \psi + \beta_{\rm c}. \tag{2.127}$$

Further, the amplitude U and phase variable $\beta_{\rm c}$,

$$U = \sqrt{u^2 + v^2} \tag{2.128}$$

$$\beta_{\rm c} = \text{atan}\left(\frac{v}{u}\right) \tag{2.129}$$

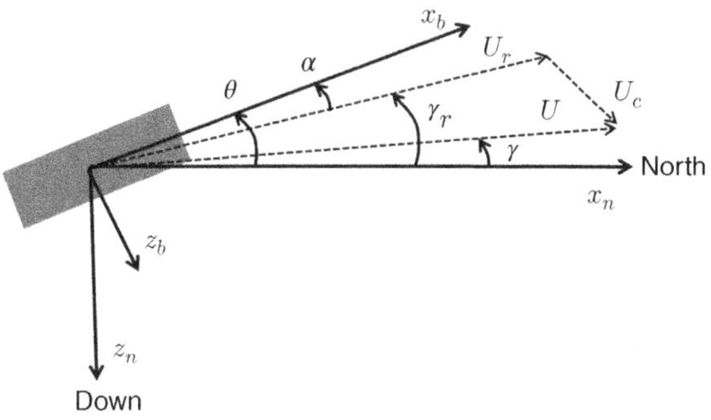

Figure 2.9 Ocean current triangle. Vertical plane.

are equal to the speed in the horizontal plane and the crab angle, respectively. The relationships between the angular variables are shown in Figure 2.8.

Definition 2.6 (Crab Angle β_c)
The angle β_c from the x_b axis to the velocity vector of the craft, positive rotation about the z_b axis by the right-hand screw convention.

$$\beta_c = \text{atan}\left(\frac{v}{u}\right) = \sin^{-1}\left(\frac{v}{U}\right) \overset{\beta_c \text{ small}}{\Rightarrow} \beta_c \approx \frac{v}{U}. \tag{2.130}$$

2.5.2 *Definitions of Angle of Attack and Sideslip Angle*

A marine craft is exposed to ocean currents. Let (u_c, v_c, w_c) denote the velocity components of the current. Hence, the current speed is

$$U_c = \sqrt{u_c^2 + v_c^2 + w_c^2} \tag{2.131}$$

and the *relative speed* is

$$U_r = \sqrt{u_r^2 + v_r^2 + w_r^2} \tag{2.132}$$

where

$$u_r = u - u_c \tag{2.133}$$

$$v_r = v - v_c \tag{2.134}$$

$$w_r = w - w_c. \tag{2.135}$$

It is possible to generalize the results of Section 2.5.1 to incorporate the effects of ocean currents. Figures 2.8 and 2.9 show the angle of attack α and sideslip angle β for a marine craft. The mathematical definitions are given below.

Definition 2.7 (Angle of Attack α)
The angle α from the relative *velocity vector to the x_b axis of the craft, positive rotation about the y_b axis by the right-hand screw convention.*

$$\alpha = \tan^{-1}\left(\frac{w_r}{u_r}\right) \overset{\alpha \text{ small}}{\Rightarrow} \alpha \approx \frac{w_r}{u_r}. \tag{2.136}$$

Definition 2.8 (Sideslip Angle β)
The angle β from the x_b axis to the relative *velocity vector of the craft, positive rotation about the z_b axis by the right-hand screw convention.*

$$\beta = \sin^{-1}\left(\frac{v_r}{U_r}\right) \overset{\beta \text{ small}}{\Rightarrow} \beta \approx \frac{v_r}{U_r}. \tag{2.137}$$

These relationships are easily verified from Figures 2.8 and 2.9 where it is observed that the angle of attack and sideslip angle satisfy

$$\gamma_r = \theta - \alpha \tag{2.138}$$

$$\chi_r = \psi + \beta. \tag{2.139}$$

Here γ_r and χ_r are recognized as the relative flight-path and course angles.

Remark 2.1
In SNAME (1950) and Lewis (1989) the sideslip angle for marine craft is defined according to

$$\beta_{SNAME} := -\beta.$$

Note that the sideslip angle definition in this section follows the sign convention used by the aircraft community, for instance as in Nelson (1998) and Stevens and Lewis (1992). This definition is more intuitive from a guidance point of view, as shown in Figure 2.8.

Time differentiation of β under the assumption that β is small and $U_r > 0$ is constant gives

$$\dot{v}_r = U_r \cos(\beta)\dot{\beta}. \tag{2.140}$$

Consequently, the differential equation for β becomes

$$\dot{\beta} = \frac{1}{U_r \cos(\beta)} \dot{v}_r. \tag{2.141}$$

This relationship is exploited when designing path-following control systems (see Section 12.6).

Example 2.4 (Sideslip Angle: No Ocean Currents)
Consider a ship moving at $U = 10$ m s^{-1} with constant turning rate $r = 0.1$ rad s^{-1} under the assumption of no ocean currents. The steady-state sway velocity during turning is $v =$ 0.8 m s^{-1}. For this case the crab and sideslip angles are equal

$$\beta_c = \beta = \sin^{-1}\left(\frac{v}{U}\right) = \sin^{-1}\left(\frac{0.8}{10}\right) \approx 4.6^\circ. \tag{2.142}$$

Consequently, the heading and course angles satisfy

$$\chi = \psi + 4.6^\circ. \tag{2.143}$$

Example 2.5 (Sideslip Angle: Stationkeeping)
Consider a marine craft at rest and exposed for an ocean current $u_c = v_c = 0.5$ m s^{-1}. Since $u = v = 0$ it follows that $U_r = \sqrt{u_c^2 + v_c^2}$ and

$$\beta_c = 0 \tag{2.144}$$

$$\beta = \sin^{-1}\left(\frac{-v_c}{U_r}\right) \approx -45.0^\circ. \tag{2.145}$$

Example 2.6 (Sideslip Angle: Straight-Line Motion)
Consider a marine craft moving at $u = 10$ m s^{-1} on a straight line under the assumption of zero sway velocity ($v = 0$). For an ocean current $u_c = v_c = 0.5$ m s^{-1} it follows that $U_r = \sqrt{(u - u_c)^2 + v_c^2}$ and

$$\beta_c = 0 \tag{2.146}$$

$$\beta = \sin^{-1}\left(\frac{-v_c}{U_r}\right) \approx -3.0^\circ. \tag{2.147}$$

2.5.3 Flow-axes Rotation Matrix

The transformation from FLOW to BODY axes is defined by two principal rotations. First, the flow axes are rotated by a *negative* sideslip angle $-\beta$ about the z_b axis and the new coordinate system is called *stability axes*. Second, the stability axes are rotated by a *positive* angle α about the new y_b axis. This angle α is called the *angle of attack*.

The names *stability* and *wind axes* are commonly used in aerodynamics to model lift and drag forces, which both are nonlinear functions of α, β and U_r. This convention has been adopted by the marine community and SNAME to describe lift and drag forces on submerged vehicles (SNAME 1950). For marine craft, *wind axes* correspond to *flow axes*.

The main reason for the FLOW axis system is that it is more convenient for calculation of the aerodynamic and hydrodynamic forces. For instance, lift is, by definition, perpendicular to the relative flow, while drag is parallel. With FLOW axes, both lift and drag resolve into a force that is parallel to one of the axes.

Assume that the lift and drag coefficients $C_L(\alpha)$ and $C_D(\alpha)$ are known for a marine craft such that the forces in FLOW axes become

$$F_{\text{drag}} = \frac{1}{2}\rho U_r^2 S C_D(\alpha) \tag{2.148}$$

$$F_{\text{lift}} = \frac{1}{2}\rho U_r^2 S C_L(\alpha) \tag{2.149}$$

where ρ is the density of water and S is a reference area usually chosen as the area of the craft as if it where projected onto the ground below it. Since lift is perpendicular to the relative flow and drag is parallel, the longitudinal forces expressed in BODY axes become

$$\begin{bmatrix} X \\ Z \end{bmatrix} = \begin{bmatrix} \cos(\alpha) & -\sin(\alpha) \\ \sin(\alpha) & \cos(\alpha) \end{bmatrix} \begin{bmatrix} -F_{\text{drag}} \\ -F_{\text{lift}} \end{bmatrix} \tag{2.150}$$

For surface ships it is common to assume that $\alpha = 0$ such that $X = -F_{\text{drag}}$ and $Z = -F_{\text{lift}}$, while submerged vehicles operate at nonzero angles of attack.

Stability and flow axes are also used in path following. For instance, a ship equipped with a single rudder and a main propeller can follow a path even though only two controls are available by simply steering the vessel to the path using the rudder. The speed is controlled by an independent propeller feedback loop (Fossen *et al.* 2003). This means that we control the $x^n y^n$ coordinates and yaw angle ψ of the ship (3 DOF). When doing this, it is optimal to have a zero sideslip angle when there are no ocean currents, wave and wind loads. If the environmental forces are nonzero, it is optimal to have a nonzero sideslip angle, as shown in Figure 2.10. This is referred to as weathervaning. However, it is extremely difficult to track the desired path given by x^n and y^n, and at the same time maintain a constant heading angle ψ unless three controls are available for feedback since this is an underactuated control problem.

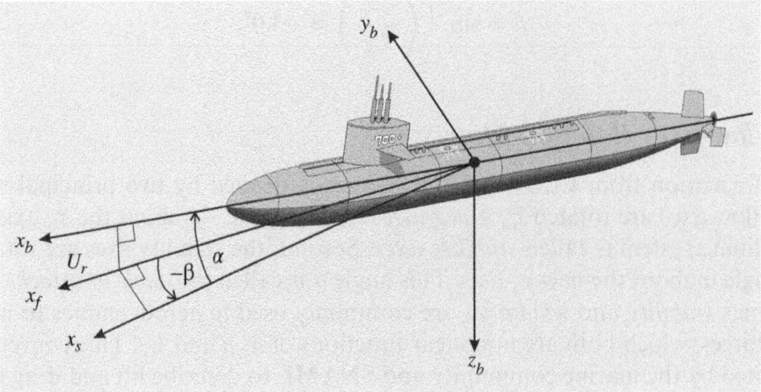

Figure 2.10 Illustration of stability (subscript s) and flow axes (subscript f) in terms of the angle of attack α, sideslip angle β and relative speed U_r.

The transformations between BODY, STABILITY and FLOW axes (subscripts b, s and f) can be mathematically expressed as

$$\boldsymbol{v}_{\mathrm{r}}^{s} = \boldsymbol{R}_{y,\alpha}\boldsymbol{v}_{\mathrm{r}}^{b} \tag{2.151}$$

$$\boldsymbol{v}_{\mathrm{r}}^{f} = \boldsymbol{R}_{z,-\beta}\boldsymbol{v}_{\mathrm{r}}^{s} \tag{2.152}$$

where the subscript r denotes the relative velocities and

$$\boldsymbol{R}_{y,\alpha} = \begin{bmatrix} \cos(\alpha) & 0 & \sin(\alpha) \\ 0 & 1 & 0 \\ -\sin(\alpha) & 0 & \cos(\alpha) \end{bmatrix} \tag{2.153}$$

$$\boldsymbol{R}_{z,-\beta} = \boldsymbol{R}_{z,\beta}^{\top} = \begin{bmatrix} \cos(\beta) & \sin(\beta) & 0 \\ -\sin(\beta) & \cos(\beta) & 0 \\ 0 & 0 & 1 \end{bmatrix} \tag{2.154}$$

The transformation matrix from BODY to FLOW axes then becomes

$$\boldsymbol{R}_{b}^{f} = \boldsymbol{R}_{z,-\beta}\boldsymbol{R}_{y,\alpha} = \begin{bmatrix} \cos(\beta)\cos(\alpha) & \sin(\beta) & \cos(\beta)\sin(\alpha) \\ -\sin(\beta)\cos(\alpha) & \cos(\beta) & -\sin(\beta)\sin(\alpha) \\ -\sin(\alpha) & 0 & \cos(\alpha) \end{bmatrix} \tag{2.155}$$

The velocity transformation $\boldsymbol{v}_{\mathrm{r}}^{f} = \boldsymbol{R}_{b}^{f}\boldsymbol{v}_{\mathrm{r}}^{b}$ can now be rewritten as

$$\boldsymbol{v}_{\mathrm{r}}^{b} = (\boldsymbol{R}_{b}^{f})^{\top}\boldsymbol{v}_{\mathrm{r}}^{f} \tag{2.156}$$

$$\Updownarrow$$

$$\begin{bmatrix} u_{\mathrm{r}} \\ v_{\mathrm{r}} \\ w_{\mathrm{r}} \end{bmatrix} = \boldsymbol{R}_{y,\alpha}^{\top}\boldsymbol{R}_{z,-\beta}^{\top}\begin{bmatrix} U_{\mathrm{r}} \\ 0 \\ 0 \end{bmatrix} \tag{2.157}$$

Writing this expression in component form finally yields

$$u_{\mathrm{r}} = U_{\mathrm{r}}\cos(\alpha)\cos(\beta) \tag{2.158}$$

$$v_{\mathrm{r}} = U_{\mathrm{r}}\sin(\beta) \tag{2.159}$$

$$w_{\mathrm{r}} = U_{\mathrm{r}}\sin(\alpha)\cos(\beta) \tag{2.160}$$

From these expressions it is easy to verify that

$$\alpha = \tan^{-1}\left(\frac{w_r}{u_r}\right) \tag{2.161}$$

$$\beta = \sin^{-1}\left(\frac{v_r}{U_r}\right) \tag{2.162}$$

For small angles of α and β (linear theory), the following approximations are obtained

$$u_r \approx U_r, \qquad v_r \approx \beta U_r, \qquad w_r \approx \alpha U_r \tag{2.163}$$

3

Rigid-body Kinetics

In order to derive the marine craft equations of motion, it is necessary to study the motion of rigid bodies, hydrodynamics and hydrostatics. The overall goal of Chapter is to show that the rigid-body kinetics can be expressed in matrix-vector form according to (Fossen 1991)

$$\boldsymbol{M}_{\mathrm{RB}}\dot{\boldsymbol{\nu}} + \boldsymbol{C}_{\mathrm{RB}}(\boldsymbol{\nu})\boldsymbol{\nu} = \boldsymbol{\tau}_{\mathrm{RB}} \tag{3.1}$$

where $\boldsymbol{M}_{\mathrm{RB}}$ is the rigid-body mass matrix, $\boldsymbol{C}_{\mathrm{RB}}(\boldsymbol{\nu})$ is the rigid-body Coriolis and centripetal matrix due to the rotation of the body-fixed frame $\{b\}$ about the approximative inertial frame $\{n\}$, $\boldsymbol{\nu} = [u, v, w, p, q, r]^{\top}$ is the generalized velocity vector expressed in $\{b\}$ and $\boldsymbol{\tau}_{\mathrm{RB}} = [X, Y, Z, K, M, N]^{\top}$ is a generalized vector of external forces and moments expressed in $\{b\}$.

The rigid-body equations of motion will be derived using the *Newton–Euler formulation* and *vectorial mechanics*. In this context it is convenient to define the vectors without reference to a coordinate frame (*coordinate-free vectors*). The velocity of the CO with respect to $\{n\}$ is a vector $\vec{v}_{\mathrm{nb}}$ that is defined by its magnitude and the direction. The vector $\vec{v}_{\mathrm{nb}}$ expressed in the inertial reference frame is denoted as $\boldsymbol{v}^{i}_{\mathrm{nb}}$, which is also referred to as a *coordinate vector* (see Section 2.1.1). The distance vector $\vec{r}_{\mathrm{bg}}$ from the CO to the CG is decomposed in $\{b\}$ and denoted $\boldsymbol{r}^{b}_{\mathrm{bg}}$.

The equations of motion will be represented in two body-fixed reference points:

CO – coordinate origin o_{b} of $\{b\}$
CG – center of gravity relative to the CO, located at $\boldsymbol{r}^{b}_{\mathrm{bg}} = [x_{\mathrm{g}}, y_{\mathrm{g}}, z_{\mathrm{g}}]^{\top}$.

These points coincide if the vector $\boldsymbol{r}^{b}_{\mathrm{bg}} = \boldsymbol{0}$ (see Figure 3.1). The point CO is usually specified by the control engineer and it is the reference point used to design the guidance, navigation and control systems. For marine craft, it is common to locate this point on

Handbook of Marine Craft Hydrodynamics and Motion Control, Second Edition. Thor I. Fossen.

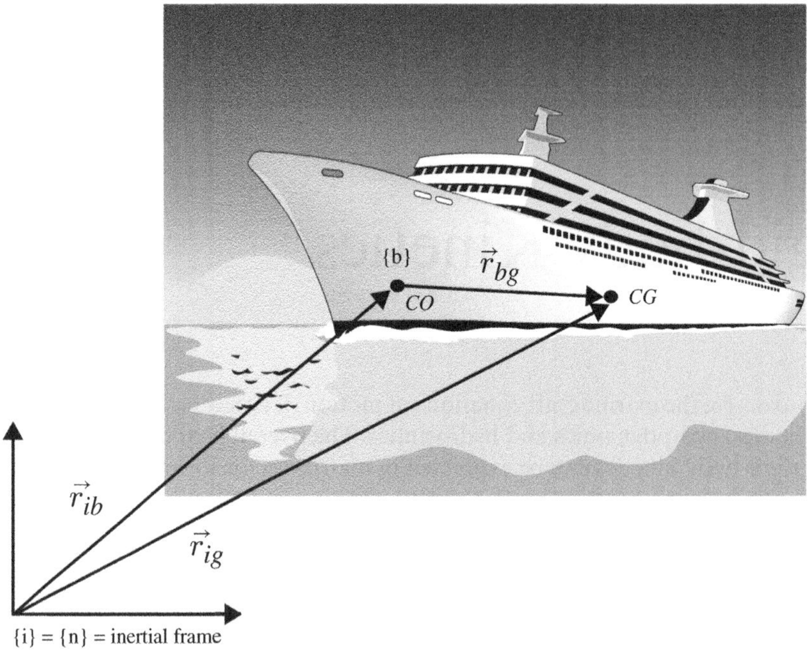

Figure 3.1 Definition of the distance vectors from $\{i\}$ to the CO and the CG.

the centerline midships. It is advantageous to use a fixed reference point CO for control design since the CG will depend on the load condition (see Section 4.3).

3.1 Newton–Euler Equations of Motion about the CG

The *Newton–Euler formulation* is based on *Newton's second law*, which relates mass m, acceleration $\dot{\vec{v}}_{\text{ig}}$ and force $\vec{f}_{\text{g}}$ according to

$$m\dot{\vec{v}}_{\text{ig}} = \vec{f}_{\text{g}} \tag{3.2}$$

where $\vec{v}_{\text{ig}}$ is the velocity of the CG with respect to the *inertial frame* $\{i\}$.

If no force is acting ($\vec{f}_{\text{g}} = \vec{0}$), then the rigid body is moving with constant speed ($\vec{v}_{\text{ig}} =$ constant) or the body is at rest ($\vec{v}_{\text{ig}} = \vec{0}$) – a result known as *Newton's first law*. Newton's laws were published in 1687 by Isaac Newton (1643–1727) in *Philosophia Naturalis Principia Mathematica*.

Euler's First and Second Axioms

Leonhard Euler (1707–1783) showed in his *Novi Commentarii Academiae Scientarium Imperialis Petropolitane* that Newton's second law can be expressed in terms of conservation of both linear momentum $\vec{p}_g$ and angular momentum $\vec{h}_g$. These results are known as *Euler's first and second axioms*, respectively

$$\frac{{}^i\mathrm{d}}{\mathrm{d}t}\vec{p}_g = \vec{f}_g \qquad \vec{p}_g = m\vec{v}_{ig} \tag{3.3}$$

$$\frac{{}^i\mathrm{d}}{\mathrm{d}t}\vec{h}_g = \vec{m}_g \qquad \vec{h}_g = I_g\,\vec{\omega}_{ig} \tag{3.4}$$

where $\vec{f}_g$ and $\vec{m}_g$ are the forces and moments acting on the body's CG and I_g is the inertia dyadic about the body's CG (to be defined later). For a rigid body, the angular velocities of the CO and the CG relative to $\{i\}$ are equal

$$\vec{\omega}_{ig} = \vec{\omega}_{ib} \tag{3.5}$$

Time differentiation in the inertial frame $\{i\}$ is denoted by ${}^i\mathrm{d}/\mathrm{d}t$. The application of these equations is often referred to as *vectorial mechanics* since both conservation laws are expressed in terms of vectors.

When deriving the equations of motion it will be assumed: (1) that the craft is rigid and (2) that the NED frame $\{n\}$ is inertial; see Section 2.1.1. The first assumption eliminates the consideration of forces acting between individual elements of mass while the second eliminates forces due to the Earth's motion relative to a star-fixed inertial reference system. Consequently,

$$\vec{v}_{ig} \approx \vec{v}_{ng} \tag{3.6}$$

$$\vec{\omega}_{ig} = \vec{\omega}_{ib} \approx \vec{\omega}_{nb}. \tag{3.7}$$

Time differentiation of a vector $\vec{a}$ in a rotating reference frame $\{b\}$ satisfies

$$\frac{{}^i\mathrm{d}}{\mathrm{d}t}\vec{a} = \frac{{}^b\mathrm{d}}{\mathrm{d}t}\vec{a} + \vec{\omega}_{ib} \times \vec{a} \tag{3.8}$$

where time differentiation in $\{b\}$ is denoted as

$$\dot{\vec{a}} := \frac{{}^b\mathrm{d}}{\mathrm{d}t}\vec{a}. \tag{3.9}$$

For guidance and navigation applications in space it is usual to use a star-fixed reference frame or a reference frame rotating with the Earth. Marine craft are, on the other hand, usually related to $\{n\}$. This is a good assumption since the forces on marine craft due to the Earth's rotation

$$\omega_{ie} = 7.2921 \times 10^{-5}\,\mathrm{rad/s} \tag{3.10}$$

are quite small compared to the hydrodynamic forces. The Earth's rotation should, however, not be neglected in global navigation applications or if the equations of motion of a drifting ship are analyzed.

3.1.1 Translational Motion About the CG

From Figure 3.1 it follows that

$$\vec{r}_{\text{ig}} = \vec{r}_{\text{ib}} + \vec{r}_{\text{bg}} \tag{3.11}$$

where $\vec{r}_{\text{bg}}$ is the distance vector from the CO to the CG. The velocities of the CG and CO are

$$\vec{v}_{\text{ig}} := \frac{{}^{i}\mathrm{d}}{\mathrm{d}t}\vec{r}_{\text{ig}} \tag{3.12}$$

$$\vec{v}_{\text{ib}} := \frac{{}^{i}\mathrm{d}}{\mathrm{d}t}\vec{r}_{\text{ib}}. \tag{3.13}$$

Further, the assumption that $\{n\}$ is inertial implies that (3.11) can be rewritten as

$$\vec{r}_{\text{ng}} = \vec{r}_{\text{nb}} + \vec{r}_{\text{bg}}. \tag{3.14}$$

Time differentiation of $\vec{r}_{\text{ng}}$ using (3.8) gives

$$\vec{v}_{\text{ng}} = \vec{v}_{\text{nb}} + \left(\frac{{}^{b}\mathrm{d}}{\mathrm{d}t}\vec{r}_{\text{bg}} + \vec{\omega}_{\text{nb}} \times \vec{r}_{\text{bg}}\right) \tag{3.15}$$

For a rigid body, the CG satisfies

$$\frac{{}^{b}\mathrm{d}}{\mathrm{d}t}\vec{r}_{\text{bg}} = \vec{0} \tag{3.16}$$

such that

$$\vec{v}_{\text{ng}} = \vec{v}_{\text{nb}} + \vec{\omega}_{\text{nb}} \times \vec{r}_{\text{bg}}. \tag{3.17}$$

From Euler's first axiom (3.3) it follows that

$$\begin{aligned}
\vec{f}_{\text{g}} &= \frac{{}^{i}\mathrm{d}}{\mathrm{d}t}(m\vec{v}_{\text{ig}}) \\
&= \frac{{}^{i}\mathrm{d}}{\mathrm{d}t}(m\vec{v}_{\text{ng}}) \\
&= \frac{{}^{b}\mathrm{d}}{\mathrm{d}t}(m\vec{v}_{\text{ng}}) + m\vec{\omega}_{\text{nb}} \times \vec{v}_{\text{ng}} \\
&= m(\dot{\vec{v}}_{\text{ng}} + \vec{\omega}_{\text{nb}} \times \vec{v}_{\text{ng}}).
\end{aligned} \tag{3.18}$$

Finally, the vectors can be expressed in $\{b\}$ such that the translational motion about the CG becomes

$$m(\dot{\boldsymbol{v}}^{b}_{\text{ng}} + \boldsymbol{S}(\boldsymbol{\omega}^{b}_{\text{nb}})\boldsymbol{v}^{b}_{\text{ng}}) = \boldsymbol{f}^{b}_{\text{g}} \tag{3.19}$$

where the cross-product is written in matrix form using the skew-symmetric matrix (2.13), that is $\boldsymbol{S}(\boldsymbol{\omega}^b_{\text{nb}})\boldsymbol{v}^b_{\text{ng}} = \boldsymbol{\omega}^b_{\text{nb}} \times \boldsymbol{v}^b_{\text{ng}}$.

3.1.2 *Rotational Motion About the CG*

The rotational dynamics (attitude dynamics) is derived from Euler's second axiom (3.4)

$$\begin{aligned}\vec{m}_{\text{g}} &= \frac{{}^i\text{d}}{\text{d}t}(I_{\text{g}}\,\vec{\omega}_{\text{ib}})\\ &= \frac{{}^i\text{d}}{\text{d}t}(I_{\text{g}}\,\vec{\omega}_{\text{nb}})\\ &= \frac{{}^b\text{d}}{\text{d}t}(I_{\text{g}}\,\vec{\omega}_{\text{nb}}) + \vec{\omega}_{\text{nb}} \times (I_{\text{g}}\,\vec{\omega}_{\text{nb}})\\ &= I_{\text{g}}\,\dot{\vec{\omega}}_{\text{nb}} - (I_{\text{g}}\,\vec{\omega}_{\text{nb}}) \times \vec{\omega}_{\text{nb}}.\end{aligned} \tag{3.20}$$

From this it follows that

$$\boldsymbol{I}^{\text{b}}_{\text{g}}\dot{\boldsymbol{\omega}}^b_{\text{nb}} - \boldsymbol{S}(\boldsymbol{I}^{\text{b}}_{\text{g}}\boldsymbol{\omega}^b_{\text{nb}})\,\boldsymbol{\omega}^b_{\text{nb}} = \boldsymbol{m}^b_{\text{g}} \tag{3.21}$$

since $\boldsymbol{S}(\boldsymbol{I}^{\text{b}}_{\text{g}}\boldsymbol{\omega}^b_{\text{nb}})\boldsymbol{\omega}^b_{\text{nb}} = (\boldsymbol{I}^{\text{b}}_{\text{g}}\boldsymbol{\omega}^b_{\text{nb}}) \times \boldsymbol{\omega}^b_{\text{nb}}$. This expression is also referred to as *Euler's equations*.

Definition 3.1 (Inertia Dyadic)
The inertia dyadic $\boldsymbol{I}^{\text{b}}_{\text{g}} \in \mathbb{R}^{3\times3}$ *about the* CG *is defined as*

$$\boldsymbol{I}^{\text{b}}_{\text{g}} := \begin{bmatrix} I^{CG}_x & -I^{CG}_{xy} & -I^{CG}_{xz}\\ -I^{CG}_{yx} & I^{CG}_y & -I^{CG}_{yz}\\ -I^{CG}_{zx} & -I^{CG}_{zy} & I^{CG}_z \end{bmatrix}, \qquad \boldsymbol{I}^{\text{b}}_{\text{g}} = (\boldsymbol{I}^{\text{b}}_{\text{g}})^{\top} > 0 \tag{3.22}$$

where I^{CG}_x, I^{CG}_y *and* I^{CG}_z *are the* moments of inertia *about the* $x_{\text{b}}, y_{\text{b}}$ *and* z_{b} *axes, and* $I^{CG}_{xy} = I^{CG}_{yx}, I^{CG}_{xz} = I^{CG}_{zx}$ *and* $I^{CG}_{yz} = I^{CG}_{zy}$ *are the* products of inertia *defined as*

$$\begin{aligned} I^{CG}_x &= \textstyle\int_V (y^2 + z^2)\,\rho_{\text{m}}\text{d}V; & I^{CG}_{xy} &= \textstyle\int_V xy\,\rho_{\text{m}}\text{d}V = \int_V yx\,\rho_{\text{m}}\text{d}V = I^{CG}_{yx}\\ I^{CG}_y &= \textstyle\int_V (x^2 + z^2)\,\rho_{\text{m}}\text{d}V; & I^{CG}_{xz} &= \textstyle\int_V xz\,\rho_{\text{m}}\text{d}V = \int_V zx\,\rho_{\text{m}}\text{d}V = I^{CG}_{zx}\\ I^{CG}_z &= \textstyle\int_V (x^2 + y^2)\,\rho_{\text{m}}\text{d}V; & I^{CG}_{yz} &= \textstyle\int_V yz\,\rho_{\text{m}}\text{d}V = \int_V zy\,\rho_{\text{m}}\text{d}V = I^{CG}_{zy} \end{aligned}$$

where ρ_{m} *is the mass density of the rigid body.*

3.1.3 Equations of Motion About the CG

The Newton–Euler equations (3.19) and (3.21) can be represented in matrix-vector form according to

$$\boldsymbol{M}_{\mathrm{RB}}^{\mathrm{CG}} \begin{bmatrix} \dot{\boldsymbol{v}}_{\mathrm{ng}}^{b} \\ \dot{\boldsymbol{\omega}}_{\mathrm{nb}}^{b} \end{bmatrix} + \boldsymbol{C}_{\mathrm{RB}}^{\mathrm{CG}} \begin{bmatrix} \boldsymbol{v}_{\mathrm{ng}}^{b} \\ \boldsymbol{\omega}_{\mathrm{nb}}^{b} \end{bmatrix} = \begin{bmatrix} \boldsymbol{f}_{\mathrm{g}}^{b} \\ \boldsymbol{m}_{\mathrm{g}}^{b} \end{bmatrix} \tag{3.23}$$

$$\Updownarrow$$

$$\underbrace{\begin{bmatrix} m\boldsymbol{I}_3 & \boldsymbol{0}_{3\times3} \\ \boldsymbol{0}_{3\times3} & \boldsymbol{I}_{\mathrm{g}}^{\mathrm{b}} \end{bmatrix}}_{\boldsymbol{M}_{\mathrm{RB}}^{\mathrm{CG}}} \begin{bmatrix} \dot{\boldsymbol{v}}_{\mathrm{ng}}^{b} \\ \dot{\boldsymbol{\omega}}_{\mathrm{nb}}^{b} \end{bmatrix} + \underbrace{\begin{bmatrix} m\boldsymbol{S}(\boldsymbol{\omega}_{\mathrm{nb}}^{b}) & \boldsymbol{0}_{3\times3} \\ \boldsymbol{0}_{3\times3} & -\boldsymbol{S}(\boldsymbol{I}_{\mathrm{g}}^{\mathrm{b}}\boldsymbol{\omega}_{\mathrm{nb}}^{b}) \end{bmatrix}}_{\boldsymbol{C}_{\mathrm{RB}}^{\mathrm{CG}}} \begin{bmatrix} \boldsymbol{v}_{\mathrm{ng}}^{b} \\ \boldsymbol{\omega}_{\mathrm{nb}}^{b} \end{bmatrix} = \begin{bmatrix} \boldsymbol{f}_{\mathrm{g}}^{b} \\ \boldsymbol{m}_{\mathrm{g}}^{b} \end{bmatrix}. \tag{3.24}$$

3.2 Newton–Euler Equations of Motion About the CO

For marine craft it is desirable to derive the equations of motion for an arbitrary origin CO to take advantage of the craft's geometric properties. Since the hydrodynamic forces and moments often are computed in the CO, Newton's laws will be formulated about the CO as well.

In order to do this, we will start with the equations of motion about the CG and transform these expressions to the CO using a coordinate transformation. The needed coordinate transformation is derived from (3.17). In other words

$$\begin{aligned} \boldsymbol{v}_{\mathrm{ng}}^{b} &= \boldsymbol{v}_{\mathrm{nb}}^{b} + \boldsymbol{\omega}_{\mathrm{nb}}^{b} \times \boldsymbol{r}_{\mathrm{bg}}^{b} \\ &= \boldsymbol{v}_{\mathrm{nb}}^{b} - \boldsymbol{r}_{\mathrm{bg}}^{b} \times \boldsymbol{\omega}_{\mathrm{nb}}^{b} \\ &= \boldsymbol{v}_{\mathrm{nb}}^{b} + \boldsymbol{S}^{\top}(\boldsymbol{r}_{\mathrm{bg}}^{b})\boldsymbol{\omega}_{\mathrm{nb}}^{b}. \end{aligned} \tag{3.25}$$

From this it follows that

$$\begin{bmatrix} \boldsymbol{v}_{\mathrm{ng}}^{b} \\ \boldsymbol{\omega}_{\mathrm{nb}}^{b} \end{bmatrix} = \boldsymbol{H}(\boldsymbol{r}_{\mathrm{bg}}^{b}) \begin{bmatrix} \boldsymbol{v}_{\mathrm{nb}}^{b} \\ \boldsymbol{\omega}_{\mathrm{nb}}^{b} \end{bmatrix} \tag{3.26}$$

where $\boldsymbol{r}_{\mathrm{bg}}^{b} = [x_{\mathrm{g}}, y_{\mathrm{g}}, z_{\mathrm{g}}]^{\top}$ and $\boldsymbol{H}(\boldsymbol{r}_{\mathrm{bg}}^{b}) \in \mathbb{R}^{6\times6}$ is the transformation matrix

$$\boldsymbol{H}(\boldsymbol{r}_{\mathrm{bg}}^{b}) := \begin{bmatrix} \boldsymbol{I}_3 & \boldsymbol{S}^{\top}(\boldsymbol{r}_{\mathrm{bg}}^{b}) \\ \boldsymbol{0}_{3\times3} & \boldsymbol{I}_3 \end{bmatrix}, \qquad \boldsymbol{H}^{\top}(\boldsymbol{r}_{\mathrm{bg}}^{b}) = \begin{bmatrix} \boldsymbol{I}_3 & \boldsymbol{0}_{3\times3} \\ \boldsymbol{S}(\boldsymbol{r}_{\mathrm{bg}}^{b}) & \boldsymbol{I}_3 \end{bmatrix}. \tag{3.27}$$

Note that angular velocity is unchanged during this transformation. The next step is to transform (3.23) from the CG to the CO using (3.26). This gives

$$\boldsymbol{H}^{\top}(\boldsymbol{r}_{\mathrm{bg}}^{b})\boldsymbol{M}_{\mathrm{RB}}^{\mathrm{CG}}\boldsymbol{H}(\boldsymbol{r}_{\mathrm{bg}}^{b}) \begin{bmatrix} \dot{\boldsymbol{v}}_{\mathrm{nb}}^{b} \\ \dot{\boldsymbol{\omega}}_{\mathrm{nb}}^{b} \end{bmatrix} + \boldsymbol{H}^{\top}(\boldsymbol{r}_{\mathrm{bg}}^{b})\boldsymbol{C}_{\mathrm{RB}}^{\mathrm{CG}}\boldsymbol{H}(\boldsymbol{r}_{\mathrm{bg}}^{b}) \begin{bmatrix} \boldsymbol{v}_{\mathrm{nb}}^{b} \\ \boldsymbol{\omega}_{\mathrm{nb}}^{b} \end{bmatrix} = \boldsymbol{H}^{\top}(\boldsymbol{r}_{\mathrm{bg}}^{b}) \begin{bmatrix} \boldsymbol{f}_{\mathrm{g}}^{b} \\ \boldsymbol{m}_{\mathrm{g}}^{b} \end{bmatrix}. \tag{3.28}$$

We now define two new matrices in the CO according to (see Appendix C)

$$\boldsymbol{M}_{\mathrm{RB}} := \boldsymbol{H}^{\top}(\boldsymbol{r}^{b}_{\mathrm{bg}})\boldsymbol{M}^{\mathrm{CG}}_{\mathrm{RB}}\boldsymbol{H}(\boldsymbol{r}^{b}_{\mathrm{bg}}) \tag{3.29}$$

$$\boldsymbol{C}_{\mathrm{RB}} := \boldsymbol{H}^{\top}(\boldsymbol{r}^{b}_{\mathrm{bg}})\boldsymbol{C}^{\mathrm{CG}}_{\mathrm{RB}}\boldsymbol{H}(\boldsymbol{r}^{b}_{\mathrm{bg}}). \tag{3.30}$$

Expanding these expressions gives

$$\boldsymbol{M}_{\mathrm{RB}} = \begin{bmatrix} m\boldsymbol{I}_3 & -m\boldsymbol{S}(\boldsymbol{r}^{b}_{\mathrm{bg}}) \\ m\boldsymbol{S}(\boldsymbol{r}^{b}_{\mathrm{bg}}) & \boldsymbol{I}^{\mathrm{b}}_{\mathrm{g}} - m\boldsymbol{S}^2(\boldsymbol{r}^{b}_{\mathrm{bg}}) \end{bmatrix} \tag{3.31}$$

$$\boldsymbol{C}_{\mathrm{RB}} = \begin{bmatrix} m\boldsymbol{S}(\boldsymbol{\omega}^{b}_{\mathrm{nb}}) & -m\boldsymbol{S}(\boldsymbol{\omega}^{b}_{\mathrm{nb}})\boldsymbol{S}(\boldsymbol{r}^{b}_{\mathrm{bg}}) \\ m\boldsymbol{S}(\boldsymbol{r}^{b}_{\mathrm{bg}})\boldsymbol{S}(\boldsymbol{\omega}^{b}_{\mathrm{nb}}) & -m\boldsymbol{S}(\boldsymbol{r}^{b}_{\mathrm{bg}})\boldsymbol{S}(\boldsymbol{\omega}^{b}_{\mathrm{nb}})\boldsymbol{S}(\boldsymbol{r}^{b}_{\mathrm{bg}}) - \boldsymbol{S}(\boldsymbol{I}^{\mathrm{b}}_{\mathrm{g}}\boldsymbol{\omega}^{b}_{\mathrm{nb}}) \end{bmatrix} \tag{3.32}$$

where we have exploited that $\boldsymbol{S}^{\top}(\boldsymbol{r}^{b}_{\mathrm{bg}}) = -\boldsymbol{S}(\boldsymbol{r}^{b}_{\mathrm{bg}})$.

3.2.1 Translational Motion About the CO

From the first row in (3.28) with matrices (3.31) and (3.32) it is seen that the translational motion about the CO satisfies

$$m[\dot{\boldsymbol{v}}^{b}_{\mathrm{nb}} - \boldsymbol{S}(\boldsymbol{r}^{b}_{\mathrm{bg}})\dot{\boldsymbol{\omega}}^{b}_{\mathrm{nb}} + \boldsymbol{S}(\boldsymbol{\omega}^{b}_{\mathrm{nb}})\boldsymbol{v}^{b}_{\mathrm{nb}} - \boldsymbol{S}(\boldsymbol{\omega}^{b}_{\mathrm{nb}})\boldsymbol{S}(\boldsymbol{r}^{b}_{\mathrm{bg}})\boldsymbol{\omega}^{b}_{\mathrm{nb}}] = \boldsymbol{f}^{b}_{\mathrm{g}}. \tag{3.33}$$

Since the translational motion is independent of the attack point of the external force $\boldsymbol{f}^{b}_{\mathrm{g}} = \boldsymbol{f}^{b}_{\mathrm{b}}$ it follows that

$$m[\dot{\boldsymbol{v}}^{b}_{\mathrm{nb}} + \boldsymbol{S}(\dot{\boldsymbol{\omega}}^{b}_{\mathrm{nb}})\boldsymbol{r}^{b}_{\mathrm{bg}} + \boldsymbol{S}(\boldsymbol{\omega}^{b}_{\mathrm{nb}})\boldsymbol{v}^{b}_{\mathrm{nb}} + \boldsymbol{S}^2(\boldsymbol{\omega}^{b}_{\mathrm{nb}})\boldsymbol{r}^{b}_{\mathrm{bg}}] = \boldsymbol{f}^{b}_{\mathrm{b}} \tag{3.34}$$

where we have exploited the fact that $\boldsymbol{S}^{\top}(\boldsymbol{a})\boldsymbol{b} = -\boldsymbol{S}(\boldsymbol{a})\boldsymbol{b} - \boldsymbol{S}(\boldsymbol{b})\boldsymbol{a}$. An alternative representation of (3.34) using vector cross-products is

$$m[\dot{\boldsymbol{v}}^{b}_{\mathrm{nb}} + \dot{\boldsymbol{\omega}}^{b}_{\mathrm{nb}} \times \boldsymbol{r}^{b}_{\mathrm{bg}} + \boldsymbol{\omega}^{b}_{\mathrm{nb}} \times \boldsymbol{v}^{b}_{\mathrm{nb}} + \boldsymbol{\omega}^{b}_{\mathrm{nb}} \times (\boldsymbol{\omega}^{b}_{\mathrm{nb}} \times \boldsymbol{r}^{b}_{\mathrm{bg}})] = \boldsymbol{f}^{b}_{\mathrm{b}}. \tag{3.35}$$

3.2.2 Rotational Motion About the CO

In order to express the rotational motion (attitude dynamics) about the CO we will make use of the parallel-axis theorem that transforms the inertia dyadic to an arbitrarily point.

Theorem 3.1 (Huygens–Steiner's Parallel-axis Theorem)
The inertia dyadic $\boldsymbol{I}_{\mathrm{b}}^{\mathrm{b}} = (\boldsymbol{I}_{\mathrm{b}}^{\mathrm{b}})^{\top} \in \mathbb{R}^{3\times 3}$ *about the origin CO is*

$$\begin{aligned}\boldsymbol{I}_{\mathrm{b}}^{\mathrm{b}} &= \boldsymbol{I}_{\mathrm{g}}^{\mathrm{b}} - m\boldsymbol{S}^2(\boldsymbol{r}_{\mathrm{bg}}^{b}) \\ &= \boldsymbol{I}_{\mathrm{g}}^{\mathrm{b}} + m((\boldsymbol{r}_{\mathrm{bg}}^{b})^{\top}\boldsymbol{r}_{\mathrm{bg}}^{b}\,\boldsymbol{I}_3 - \boldsymbol{r}_{\mathrm{bg}}^{b}(\boldsymbol{r}_{\mathrm{bg}}^{b})^{\top})\end{aligned} \tag{3.36}$$

where $\boldsymbol{I}_{\mathrm{g}}^{\mathrm{b}} = (\boldsymbol{I}_{\mathrm{g}}^{\mathrm{b}})^{\top} \in \mathbb{R}^{3\times 3}$ *is the inertia dyadic about the CG. Expanding (3.36) gives*

$$\boldsymbol{I}_{\mathrm{b}}^{\mathrm{b}} = \boldsymbol{I}_{\mathrm{g}}^{\mathrm{b}} + m\begin{bmatrix} y^2+z^2 & -xy & -xz \\ -yx & x^2+z^2 & -yz \\ -zx & -zy & x^2+y^2 \end{bmatrix}. \tag{3.37}$$

The $M_{\mathrm{RB}}^{\{22\}}$ element in (3.31) is recognized as the parallel-axis theorem (3.36) while the $C_{\mathrm{RB}}^{\{22\}}$ element in (3.32) can be rewritten using the Jacobi identity

$$\boldsymbol{a}\times(\boldsymbol{b}\times\boldsymbol{c}) + \boldsymbol{b}\times(\boldsymbol{c}\times\boldsymbol{a}) + \boldsymbol{c}\times(\boldsymbol{a}\times\boldsymbol{b}) = \boldsymbol{0}. \tag{3.38}$$

Choosing $\boldsymbol{c} = \boldsymbol{a}\times\boldsymbol{b}$ gives

$$\begin{aligned}\boldsymbol{a}\times(\boldsymbol{b}\times(\boldsymbol{a}\times\boldsymbol{b})) &= -\boldsymbol{b}\times((\boldsymbol{a}\times\boldsymbol{b})\times\boldsymbol{a}) \\ &= ((\boldsymbol{a}\times\boldsymbol{b})\times\boldsymbol{a})\times\boldsymbol{b} \\ &= -(\boldsymbol{a}\times(\boldsymbol{a}\times\boldsymbol{b}))\times\boldsymbol{b}\end{aligned} \tag{3.39}$$

which is equivalent to

$$\boldsymbol{S}(\boldsymbol{a})\boldsymbol{S}(\boldsymbol{b})\boldsymbol{S}(\boldsymbol{a})\boldsymbol{b} = -\boldsymbol{S}(\boldsymbol{S}^2(\boldsymbol{a})\boldsymbol{b})\boldsymbol{b}. \tag{3.40}$$

Let $\boldsymbol{a} = \boldsymbol{r}_{\mathrm{bg}}^{b}$ and $\boldsymbol{b} = \boldsymbol{\omega}_{\mathrm{nb}}^{b}$ such that

$$-\boldsymbol{S}(\boldsymbol{r}_{\mathrm{bg}}^{b})\boldsymbol{S}(\boldsymbol{\omega}_{\mathrm{nb}}^{b})\boldsymbol{S}(\boldsymbol{r}_{\mathrm{bg}}^{b})\boldsymbol{\omega}_{\mathrm{nb}}^{b} = \boldsymbol{S}(\boldsymbol{S}^2(\boldsymbol{r}_{\mathrm{bg}}^{b})\boldsymbol{\omega}_{\mathrm{nb}}^{b})\boldsymbol{\omega}_{\mathrm{nb}}^{b}. \tag{3.41}$$

The $C_{\mathrm{RB}}^{\{22\}}$ element in (3.32) multiplied with $\boldsymbol{\omega}_{\mathrm{nb}}^{b}$ is equivalent to

$$\begin{aligned}&-[m\boldsymbol{S}(\boldsymbol{r}_{\mathrm{bg}}^{b})\boldsymbol{S}(\boldsymbol{\omega}_{\mathrm{nb}}^{b})\boldsymbol{S}(\boldsymbol{r}_{\mathrm{bg}}^{b}) + \boldsymbol{S}(\boldsymbol{I}_{\mathrm{g}}^{\mathrm{b}}\boldsymbol{\omega}_{\mathrm{nb}}^{b})]\boldsymbol{\omega}_{\mathrm{nb}}^{b} \\ &\quad = m\boldsymbol{S}(\boldsymbol{S}^2(\boldsymbol{r}_{\mathrm{bg}}^{b})\boldsymbol{\omega}_{\mathrm{nb}}^{b})\boldsymbol{\omega}_{\mathrm{nb}}^{b} - \boldsymbol{S}(\boldsymbol{I}_{\mathrm{g}}^{\mathrm{b}}\boldsymbol{\omega}_{\mathrm{nb}}^{b})\boldsymbol{\omega}_{\mathrm{nb}}^{b} \\ &\quad = m\boldsymbol{S}(\boldsymbol{S}^2(\boldsymbol{r}_{\mathrm{bg}}^{b})\boldsymbol{\omega}_{\mathrm{nb}}^{b})\boldsymbol{\omega}_{\mathrm{nb}}^{b} - \boldsymbol{S}([\boldsymbol{I}_{\mathrm{b}}^{\mathrm{b}} + m\boldsymbol{S}^2(\boldsymbol{r}_{\mathrm{bg}}^{b})]\boldsymbol{\omega}_{\mathrm{nb}}^{b})\boldsymbol{\omega}_{\mathrm{nb}}^{b} \\ &\quad = -\boldsymbol{S}(\boldsymbol{I}_{\mathrm{b}}^{\mathrm{b}}\boldsymbol{\omega}_{\mathrm{nb}}^{b})\boldsymbol{\omega}_{\mathrm{nb}}^{b}.\end{aligned} \tag{3.42}$$

Expanding the last row in (3.28) finally gives the formula for the rotational motion about the CO

$$\boldsymbol{I}_{\mathrm{b}}^{\mathrm{b}}\dot{\boldsymbol{\omega}}_{\mathrm{nb}}^{b} + \boldsymbol{S}(\boldsymbol{\omega}_{\mathrm{nb}}^{b})\boldsymbol{I}_{\mathrm{b}}^{\mathrm{b}}\boldsymbol{\omega}_{\mathrm{nb}}^{b} + m\boldsymbol{S}(\boldsymbol{r}_{\mathrm{bg}}^{b})\dot{\boldsymbol{v}}_{\mathrm{nb}}^{b} + m\boldsymbol{S}(\boldsymbol{r}_{\mathrm{bg}}^{b})\boldsymbol{S}(\boldsymbol{\omega}_{\mathrm{nb}}^{b})\boldsymbol{v}_{\mathrm{nb}}^{b} = \boldsymbol{m}_{\mathrm{b}}^{b} \tag{3.43}$$

where the moment about the CO is

$$\begin{aligned}\boldsymbol{m}_{\mathrm{b}}^{b} &= \boldsymbol{m}_{\mathrm{g}}^{b} + \boldsymbol{r}_{\mathrm{bg}}^{b} \times \boldsymbol{f}_{\mathrm{g}}^{b} \\ &= \boldsymbol{m}_{\mathrm{g}}^{b} + \boldsymbol{S}(\boldsymbol{r}_{\mathrm{bg}}^{b})\boldsymbol{f}_{\mathrm{g}}^{b}.\end{aligned} \tag{3.44}$$

An alternative representation of (3.43) using vector cross-products is

$$\boldsymbol{I}_{\mathrm{b}}^{\mathrm{b}}\dot{\boldsymbol{\omega}}_{\mathrm{nb}}^{b} + \boldsymbol{\omega}_{\mathrm{nb}}^{b} \times \boldsymbol{I}_{\mathrm{b}}^{\mathrm{b}}\boldsymbol{\omega}_{\mathrm{nb}}^{b} + m\boldsymbol{r}_{\mathrm{bg}}^{b} \times (\dot{\boldsymbol{v}}_{\mathrm{nb}}^{b} + \boldsymbol{\omega}_{\mathrm{nb}}^{b} \times \boldsymbol{v}_{\mathrm{nb}}^{b}) = \boldsymbol{m}_{\mathrm{b}}^{b}. \tag{3.45}$$

3.3 Rigid-body Equations of Motion

In the previous sections it was shown how the rigid-body kinetics can be derived by applying *Newtonian* mechanics. In this section, useful properties of the equations of motion are discussed and it is also demonstrated how these properties considerably simplify the representation of the nonlinear equations of motion.

3.3.1 Nonlinear 6-DOF Rigid-body Equations of Motion

Equations (3.35) and (3.45) are usually written in component form according to the SNAME (1950) notation by defining:

$\boldsymbol{f}_{\mathrm{b}}^{b}$	$= [X,\ Y,\ Z]^{\top}$	force through the CO expressed in $\{b\}$
$\boldsymbol{m}_{\mathrm{b}}^{b}$	$= [K,\ M,\ N]^{\top}$	moment about the CO expressed in $\{b\}$
$\boldsymbol{v}_{\mathrm{nb}}^{b}$	$= [u,\ v,\ w]^{\top}$	linear velocity of the CO relative to o_n expressed in $\{b\}$
$\boldsymbol{\omega}_{\mathrm{nb}}^{b}$	$= [p,\ q,\ r]^{\top}$	angular velocity of $\{b\}$relative to $\{n\}$ expressed in $\{b\}$
$\boldsymbol{r}_{\mathrm{bg}}^{b}$	$= [x_{\mathrm{g}},\ y_{\mathrm{g}},\ z_{\mathrm{g}}]^{\top}$	vector from the CO to the CG expressed in $\{b\}$.

Applying these definitions to (3.35) and (3.45) give

$$\begin{aligned}
m[\dot{u} - vr + wq - x_{\mathrm{g}}(q^2 + r^2) + y_{\mathrm{g}}(pq - \dot{r}) + z_{\mathrm{g}}(pr + \dot{q})] &= X \\
m[\dot{v} - wp + ur - y_{\mathrm{g}}(r^2 + p^2) + z_{\mathrm{g}}(qr - \dot{p}) + x_{\mathrm{g}}(qp + \dot{r})] &= Y \\
m[\dot{w} - uq + vp - z_{\mathrm{g}}(p^2 + q^2) + x_{\mathrm{g}}(rp - \dot{q}) + y_{\mathrm{g}}(rq + \dot{p})] &= Z \\
I_x\dot{p} + (I_z - I_y)qr - (\dot{r} + pq)I_{xz} + (r^2 - q^2)I_{yz} + (pr - \dot{q})I_{xy} & \\
+m[y_{\mathrm{g}}(\dot{w} - uq + vp) - z_{\mathrm{g}}(\dot{v} - wp + ur)] &= K \\
I_y\dot{q} + (I_x - I_z)rp - (\dot{p} + qr)I_{xy} + (p^2 - r^2)I_{zx} + (qp - \dot{r})I_{yz} & \\
+m[z_{\mathrm{g}}(\dot{u} - vr + wq) - x_{\mathrm{g}}(\dot{w} - uq + vp)] &= M \\
I_z\dot{r} + (I_y - I_x)pq - (\dot{q} + rp)I_{yz} + (q^2 - p^2)I_{xy} + (rq - \dot{p})I_{zx} & \\
+m[x_{\mathrm{g}}(\dot{v} - wp + ur) - y_{\mathrm{g}}(\dot{u} - vr + wq)] &= N.
\end{aligned} \tag{3.46}$$

The first three equations represent the translational motion, while the last three equations represent the rotational motion.

Matrix-vector Representation

The rigid-body kinetics (3.46) can be expressed in matrix-vector form as (Fossen 1991)

$$\boldsymbol{M}_{\mathrm{RB}}\dot{\boldsymbol{\nu}} + \boldsymbol{C}_{\mathrm{RB}}(\boldsymbol{\nu})\boldsymbol{\nu} = \boldsymbol{\tau}_{\mathrm{RB}} \tag{3.47}$$

where $\boldsymbol{\nu} = [u,\ v,\ w,\ p,\ q,\ r]^{\top}$ is the generalized velocity vector expressed in $\{b\}$ and $\boldsymbol{\tau}_{\mathrm{RB}} = [X,\ Y,\ Z,\ K,\ M,\ N]^{\top}$ is a generalized vector of external forces and moments.

Property 3.1 (Rigid-body System Inertia Matrix $\boldsymbol{M}_{\mathrm{RB}}$)
The representation of the rigid-body system inertia matrix $\boldsymbol{M}_{\mathrm{RB}}$ is unique and satisfies

$$\boldsymbol{M}_{\mathrm{RB}} = \boldsymbol{M}_{\mathrm{RB}}^{\top} > 0, \qquad \dot{\boldsymbol{M}}_{\mathrm{RB}} = \boldsymbol{0}_{6\times 6} \tag{3.48}$$

where

$$\begin{aligned}\boldsymbol{M}_{\mathrm{RB}} &= \begin{bmatrix} m\boldsymbol{I}_3 & -m\boldsymbol{S}(\boldsymbol{r}_{bg}^{b}) \\ m\boldsymbol{S}(\boldsymbol{r}_{bg}^{b}) & \boldsymbol{I}_{\mathrm{b}}^{\mathrm{b}} \end{bmatrix} \\ &= \begin{bmatrix} m & 0 & 0 & 0 & mz_{\mathrm{g}} & -my_{\mathrm{g}} \\ 0 & m & 0 & -mz_{\mathrm{g}} & 0 & mx_{\mathrm{g}} \\ 0 & 0 & m & my_{\mathrm{g}} & -mx_{\mathrm{g}} & 0 \\ 0 & -mz_{\mathrm{g}} & my_{\mathrm{g}} & I_x & -I_{xy} & -I_{xz} \\ mz_{\mathrm{g}} & 0 & -mx_{\mathrm{g}} & -I_{yx} & I_y & -I_{yz} \\ -my_{\mathrm{g}} & mx_{\mathrm{g}} & 0 & -I_{zx} & -I_{zy} & I_z \end{bmatrix}.\end{aligned} \tag{3.49}$$

Here, $\boldsymbol{I}_{\mathrm{b}}^{\mathrm{b}} = (\boldsymbol{I}_{\mathrm{b}}^{\mathrm{b}})^{\top} > 0$ is the inertia dyadic according to Definition 3.1 and $\boldsymbol{S}(\boldsymbol{r}_{bg}^{b})$ is a skew-symmetric matrix according to Definition 2.2.

MATLAB

The rigid-body system inertia matrix M_{RB} can be computed in Matlab as

```
r_g = [10 0 1]';        % location of the CG with respect to the CO
R44 = 10;               % radius of gyration in roll
R55 = 20;               % radius of gyration in pitch
R66 = 5;                % radius of gyration in yaw
m = 1000;               % mass
I_g = m * diag([R44^2 R55^2 R66^2]);   % inertia dyadic (CG)

% rigid-body system inertia matrix
S = Smtrx(r_g);
MRB = [ m * eye(3)    -m * S
        m * S         I_g - m * S^2 ]

MRB =
     1000        0        0        0     1000        0
        0     1000        0    -1000        0    10000
        0        0     1000        0   -10000        0
        0    -1000        0   101000        0   -10000
     1000        0   -10000        0   501000        0
        0    10000        0   -10000        0   125000
```

The rigid-body system inertia matrix can also be computed using the command

```
MRB = rbody(m, R44, R55, R66, zeros(3,1), r_g)
```

The matrix C_{RB} in (3.47) represents the Coriolis vector term $\omega^b_{nb} \times v^b_{nb}$ and the centripetal vector term $\omega^b_{nb} \times (\omega^b_{nb} \times r^b_{bg})$. Contrary to the representation of M_{RB}, it is possible to find a large number of representations for the matrix C_{RB}.

Theorem 3.2 (Coriolis–Centripetal Matrix from System Inertia Matrix)
Let M be a 6×6 system inertia matrix defined as

$$M = M^\top = \begin{bmatrix} M_{11} & M_{12} \\ M_{21} & M_{22} \end{bmatrix} > 0 \tag{3.50}$$

where $M_{21} = M_{12}^\top$. Then the Coriolis–centripetal matrix can always be parameterized such that $C(v) = -C^\top(v)$ by choosing (Sagatun and Fossen 1991)

$$C(\nu) = \begin{bmatrix} 0_{3\times3} & -S(M_{11}\nu_1 + M_{12}\nu_2) \\ -S(M_{11}\nu_1 + M_{12}\nu_2) & -S(M_{21}\nu_1 + M_{22}\nu_2) \end{bmatrix} \tag{3.51}$$

where $\nu_1 := v^b_{nb} = [u, \ v, \ w]^\top$, $\nu_2 := \omega^b_{nb} = [p, \ q, \ r]^\top$ *and* S *is the cross-product operator according to Definition 2.2.*

Proof. *The kinetic energy T is written in the quadratic form*

$$T = \frac{1}{2}\nu^\top M \nu, \quad M = M^\top > 0. \tag{3.52}$$

Expanding this expression yields

$$T = \frac{1}{2}(\nu_1^\top M_{11}\nu_1 + \nu_1^\top M_{12}\nu_2 + \nu_2^\top M_{21}\nu_1 + \nu_2^\top M_{22}\nu_2) \tag{3.53}$$

where $M_{12} = M_{21}^\top$ *and* $M_{21} = M_{12}^\top$. *This gives*

$$\frac{\partial T}{\partial \nu_1} = M_{11}\nu_1 + M_{12}\nu_2 \tag{3.54}$$

$$\frac{\partial T}{\partial \nu_2} = M_{21}\nu_1 + M_{22}\nu_2. \tag{3.55}$$

Using Kirchhoff's equation (Kirchhoff 1869)

$$\frac{\mathrm{d}}{\mathrm{d}t}\left(\frac{\partial T}{\partial \nu_1}\right) + S(\nu_2)\frac{\partial T}{\partial \nu_1} = \tau_1 \tag{3.56}$$

$$\frac{\mathrm{d}}{\mathrm{d}t}\left(\frac{\partial T}{\partial \nu_2}\right) + S(\nu_2)\frac{\partial T}{\partial \nu_2} + S(\nu_1)\frac{\partial T}{\partial \nu_1} = \tau_2 \tag{3.57}$$

where S *is the skew-symmetric cross-product operator in Definition 2.2, it is seen that there are some terms dependent on acceleration, that is* $(\mathrm{d}/\mathrm{d}t)(\partial T/\partial \nu_1)$ *and* $(\mathrm{d}/\mathrm{d}t)(\partial T/\partial \nu_2)$. *The remaining terms are due to Coriolis–centripetal forces. Consequently,*

$$C(\nu)\nu := \begin{bmatrix} S(\nu_2)\frac{\partial T}{\partial \nu_1} \\ S(\nu_2)\frac{\partial T}{\partial \nu_2} + S(\nu_1)\frac{\partial T}{\partial \nu_1} \end{bmatrix} = \begin{bmatrix} 0_{3\times3} & -S(\frac{\partial T}{\partial \nu_1}) \\ -S(\frac{\partial T}{\partial \nu_1}) & -S(\frac{\partial T}{\partial \nu_2}) \end{bmatrix} \begin{bmatrix} \nu_1 \\ \nu_2 \end{bmatrix} \tag{3.58}$$

which after substitution of (3.54) and (3.55) gives (3.51); see Sagatun and Fossen (1991) for the original proof of this theorem.

We next state some useful properties of the Coriolis and centripetal matrix $C_{RB}(\nu)$.

Property 3.2 (Rigid-body Coriolis and Centripetal Matrix C_{RB})
According to Theorem 3.2 the rigid-body Coriolis and centripetal matrix $C_{RB}(\nu)$ can always be represented such that $C_{RB}(\nu)$ is skew-symmetric. In other words

$$C_{RB}(\nu) = -C_{RB}^{\top}(\nu), \qquad \forall \nu \in \mathbb{R}^6. \tag{3.59}$$

The skew-symmetric property is very useful when designing a nonlinear motion control system since the quadratic form $\nu^{\top} C_{RB}(\nu)\nu \equiv 0$. This is exploited in energy-based designs where Lyapunov functions play a key role. The same property is also used in nonlinear observer design. There exist several parametrizations that satisfy Property 3.2. Two of them are presented below.

Lagrangian parametrization: Application of Theorem 3.2 with $M = M_{RB}$ yields the following expression

$$C_{RB}(\nu) = \begin{bmatrix} 0_{3\times3} & -mS(\nu_1) - mS(S(\nu_2)r_{bg}^{b}) \\ -mS(\nu_1) - mS(S(\nu_2)r_{bg}^{b}) & mS(S(\nu_1)r_{bg}^{b}) - S(I_b^{b}\nu_2) \end{bmatrix}. \tag{3.60}$$

This is not a unique parametrization. There are other parameterizations giving the same product $C_{RB}(\nu)\nu$. For instance

$$C_{RB}(\nu) = \begin{bmatrix} 0_{3\times3} & -mS(\nu_1) - mS(\nu_2)S(r_{bg}^{b}) \\ -mS(\nu_1) + mS(r_{bg}^{b})S(\nu_2) & -S(I_b^{b}\nu_2) \end{bmatrix}. \tag{3.61}$$

To illustrate the complexity of 6-DOF modeling, the rigid-body Coriolis and centripetal terms in expression (3.60) are expanded to give

$$C_{RB}(\nu) = \left[\begin{matrix} 0 & 0 & 0 \\ 0 & 0 & 0 \\ 0 & 0 & 0 \\ -m(y_g q + z_g r) & m(y_g p + w) & m(z_g p - v) \\ m(x_g q - w) & -m(z_g r + x_g p) & m(z_g q + u) \\ m(x_g r + v) & m(y_g r - u) & -m(x_g p + y_g q) \end{matrix}\right. \tag{3.62}$$
$$\left.\begin{matrix} m(y_g q + z_g r) & -m(x_g q - w) & -m(x_g r + v) \\ -m(y_g p + w) & m(z_g r + x_g p) & -m(y_g r - u) \\ -m(z_g p - v) & -m(z_g q + u) & m(x_g p + y_g q) \\ 0 & -I_{yz}q - I_{xz}p + I_z r & I_{yz}r + I_{xy}p - I_y q \\ I_{yz}q + I_{xz}p - I_z r & 0 & -I_{xz}r - I_{xy}q + I_x p \\ -I_{yz}r - I_{xy}p + I_y q & I_{xz}r + I_{xy}q - I_x p & 0 \end{matrix}\right].$$

Linear velocity-independent parametrization: By using the property $\boldsymbol{S}(\boldsymbol{v}_1)\boldsymbol{v}_2 = -\boldsymbol{S}(\boldsymbol{v}_2)\boldsymbol{v}_1$, it is possible to move the $-m\boldsymbol{S}(\boldsymbol{v}_1)$ term from $\boldsymbol{C}_{\mathrm{RB}}^{\{12\}}$ to $\boldsymbol{C}_{\mathrm{RB}}^{\{11\}}$ in (3.61). Since $\boldsymbol{S}(\boldsymbol{v}_1)\boldsymbol{v}_1 = \boldsymbol{0}$ this gives an expression for $\boldsymbol{C}_{\mathrm{RB}}(\boldsymbol{v})$ that is independent of linear velocity $\boldsymbol{v}_1$ (Fossen and Fjellstad 1995)

$$\boldsymbol{C}_{\mathrm{RB}}(\boldsymbol{v}) = \begin{bmatrix} m\boldsymbol{S}(\boldsymbol{v}_2) & -m\boldsymbol{S}(\boldsymbol{v}_2)\boldsymbol{S}(\boldsymbol{r}_{\mathrm{bg}}^b) \\ m\boldsymbol{S}(\boldsymbol{r}_{\mathrm{bg}}^b)\boldsymbol{S}(\boldsymbol{v}_2) & -\boldsymbol{S}(\boldsymbol{I}_{\mathrm{b}}^{\mathrm{b}}\boldsymbol{v}_2) \end{bmatrix}. \tag{3.63}$$

Note that this expression is similar to (3.32) which was derived using Newton–Euler equations. Formula (3.63) can also be expressed in terms of $\boldsymbol{H}(\boldsymbol{r}_{\mathrm{bg}}^b)$ for non-zero values of $\boldsymbol{r}_{\mathrm{bg}}^b$ (see (C.23) in Appendix C.2). This gives

$$\boldsymbol{C}_{\mathrm{RB}}(\boldsymbol{v}) = \boldsymbol{H}^{\top}(\boldsymbol{r}_{\mathrm{bg}}^b) \begin{bmatrix} 0 & -mr & mq & 0 & 0 & 0 \\ mr & 0 & -mp & 0 & 0 & 0 \\ -mq & mp & 0 & 0 & 0 & 0 \\ 0 & 0 & 0 & 0 & I_z r & -I_y q \\ 0 & 0 & 0 & -I_z r & 0 & I_x p \\ 0 & 0 & 0 & I_y q & -I_x p & 0 \end{bmatrix} \boldsymbol{H}(\boldsymbol{r}_{\mathrm{bg}}^b). \tag{3.64}$$

Remark 3.1
Formulae (3.63) and (3.64) are useful when irrotational ocean currents

$$\boldsymbol{v}_c = [u_{\mathrm{c}},\ v_{\mathrm{c}},\ w_{\mathrm{c}},\ 0,\ 0,\ 0]^{\top} \tag{3.65}$$

enter the equations of motion. The main reason for this is that $\boldsymbol{C}_{RB}(\boldsymbol{v})$ does not depend on linear velocity $\boldsymbol{v}_1$ (uses only angular velocity $\boldsymbol{v}_2$ and lever arm $\boldsymbol{r}_{bg}^b$). According to Property 10.1 in Section 10.3

$$\boldsymbol{M}_{RB}\dot{\boldsymbol{v}} + \boldsymbol{C}_{RB}(\boldsymbol{v})\boldsymbol{v} \equiv \boldsymbol{M}_{RB}\dot{\boldsymbol{v}}_r + \boldsymbol{C}_{RB}(\boldsymbol{v}_r)\boldsymbol{v}_r \tag{3.66}$$

if the relative velocity vector $\boldsymbol{v}_r = \boldsymbol{v} - \boldsymbol{v}_{\mathrm{c}}$. Since the ocean current (3.65) is assumed to be irrotational, Equation (3.66) can be proven using (3.63). The details are outlined in Section 10.3.

MATLAB

The Lagrangian parametrization (Theorem 3.2) is implemented in the Matlab MSS toolbox in the function `m2c.m`, while the linear-velocity independent parametrization (3.63) is implemented in the more generic function `rbody.m`. The following example demonstrates how $\boldsymbol{C}_{\mathrm{RB}}(\boldsymbol{v})$ can be computed numerically

```
r_g = [10 0 1]';           % location of the CG relative to the CO
R44 = 10;                  % radius of gyration in roll
R55 = 20;                  % radius of gyration in pitch
R66 = 5;                   % radius of gyration in yaw
m = 1000;                  % mass
nu = [8 0.5 0.1 0.2 -0.3 0.2]'; % velocity vector

% Method 1: Linear velocity-independent parametrization
nu2 = nu(4:6);
[MRB,CRB] = rbody(m, R44, R55, R66, nu2, r_g)

MRB =
       1000          0          0          0       1000          0
          0       1000          0      -1000          0      10000
          0          0       1000          0     -10000          0
          0      -1000          0     101000          0     -10000
       1000          0     -10000          0     501000          0
          0      10000          0     -10000          0     125000
CRB =
          0       -200       -300        200       3000      -2000
        200          0       -200          0       2200          0
        300        200          0       -200        300       2000
       -200          0        200          0       2800     120000
      -3000      -2200       -300      -2800          0      -2000
       2000          0      -2000    -120000       2000          0

% Method 2: Lagrangian parametrization
CRB = m2c(MRB, nu)

CRB =
          0          0          0          0       3100      -2300
          0          0          0      -3100          0       7700
          0          0          0       2300      -7700          0
          0       3100      -2300          0      28000     143300
      -3100          0       7700     -28000          0      17700
       2300      -7700          0    -143300     -17700          0
```

Even though the numerical values for the two $\boldsymbol{C}_{\mathrm{RB}}(\boldsymbol{\nu})$ matrices are different, they both produce the same product $\boldsymbol{C}_{\mathrm{RB}}(\boldsymbol{\nu})\boldsymbol{\nu}$.

3.3.2 Linearized 6-DOF Rigid-body Equations of Motion

For a marine craft moving at forward speed U, the rigid-body equations of motion (3.47) can be linearized about $\boldsymbol{\nu}_0 = [U,\ 0,\ 0,\ 0,\ 0,\ 0]^{\top}$. This gives

$$\boldsymbol{M}_{\mathrm{RB}}\dot{\boldsymbol{\nu}} + \boldsymbol{C}^{*}_{\mathrm{RB}}\boldsymbol{\nu} = \boldsymbol{\tau}_{\mathrm{RB}} \tag{3.67}$$

where

$$C^*_{\rm RB} = U\boldsymbol{M}_{\rm RB}\boldsymbol{L}, \qquad C^*_{\rm RB} \neq -(C^*_{\rm RB})^\top \tag{3.68}$$

and $\boldsymbol{L}$ is a selection matrix

$$\boldsymbol{L} = \begin{bmatrix} 0 & 0 & 0 & 0 & 0 & 0 \\ 0 & 0 & 0 & 0 & 0 & 1 \\ 0 & 0 & 0 & 0 & -1 & 0 \\ 0 & 0 & 0 & 0 & 0 & 0 \\ 0 & 0 & 0 & 0 & 0 & 0 \\ 0 & 0 & 0 & 0 & 0 & 0 \end{bmatrix}. \tag{3.69}$$

The linearized Coriolis and centripetal forces are recognized as

$$\boldsymbol{f}_{\rm c} = C^*_{\rm RB}\boldsymbol{\nu} = \begin{bmatrix} 0 \\ mUr \\ -mUq \\ -my_{\rm g}Uq - mz_{\rm g}Ur \\ mx_{\rm g}Uq \\ mx_{\rm g}Ur \end{bmatrix} \tag{3.70}$$

MATLAB

The linearized model (3.67) is computed using the following Matlab commands

```
U = 1;
MRB = [ 1000       0       0       0    1000       0
           0    1000       0   -1000       0   10000
           0       0    1000       0  -10000       0
           0   -1000       0  101000       0  -10000
        1000       0  -10000       0  501000       0
           0   10000       0  -10000       0  125000];

L = zeros(6,6); L(2,6) = 1; L(3,5) = -1;
CRB = U * MRB * L

CRB =
         0         0         0         0         0         0
         0         0         0         0         0      1000
         0         0         0         0     -1000         0
         0         0         0         0         0     -1000
         0         0         0         0     10000         0
         0         0         0         0         0     10000
         0         0         0         0         0     10000
```

Note that the skew-symmetric property is destroyed by linearization. In other words, $C^*_{\rm RB} \neq -(C^*_{\rm RB})^\top$

4

Hydrostatics

Archimedes (287–212 BC) derived the basic laws of fluid statics that are the fundamentals of hydrostatics today. In hydrostatic terminology, the gravitational and buoyancy forces are called *restoring forces* and are equivalent to the spring forces in a *mass–damper–spring* system. In the derivation of the restoring forces and moments it will be distinguished between submersibles and surface vessels:

- **Section 4.1:** underwater vehicles (ROVs, AUVs and submarines)
- **Section 4.2:** surface vessels (ships, semi-submersibles, structures and USVs).

For a floating or submerged vessel, the restoring forces are determined by the volume of the displaced fluid, the location of the center of buoyancy (CB), the area of the water-plane and its associated moments. The forthcoming sections show how these quantities determine the heaving, rolling and pitching motions of a marine craft.

4.1 Restoring Forces for Underwater Vehicles

Consider the submarine in Figure 4.1 where the gravitational force $\boldsymbol{f}_{\rm g}^{b}$ acts through the CG defined by the vector $\boldsymbol{r}_{\rm bg}^{b} := [x_{\rm g}, y_{\rm g}, z_{\rm g}]^{\top}$ relative to the CO. Similar, the buoyancy force $\boldsymbol{f}_{\rm b}^{b}$ acts through the CB defined by the vector $\boldsymbol{r}_{\rm bb}^{b} := [x_{\rm b}, y_{\rm b}, z_{\rm b}]^{\top}$ relative to the CO; see Section 2.1.1.

4.1.1 Hydrostatics of Submerged Vehicles

Let m be the mass of the vehicle including water in free floating space, ∇ the volume of fluid displaced by the vehicle, g the acceleration of gravity (positive downwards) and ρ the water density. According to the SNAME (1950) notation, the submerged weight of the body and buoyancy force are

$$W = mg, \qquad B = \rho g \nabla. \tag{4.1}$$

Handbook of Marine Craft Hydrodynamics and Motion Control, Second Edition. Thor I. Fossen.

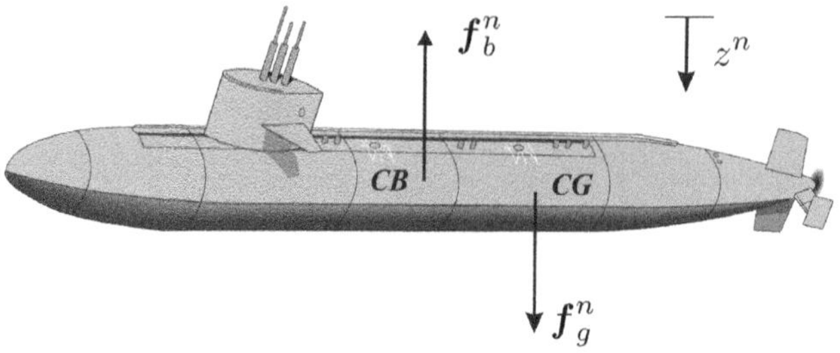

Figure 4.1 Gravitational and buoyancy forces acting on the center of gravity (CG) and center of buoyancy (CB) of a submarine.

These forces act in the vertical plane of $\{n\}$. Consequently,

$$\boldsymbol{f}_{\mathrm{g}}^{n}=\begin{bmatrix}0\\0\\W\end{bmatrix},\qquad \boldsymbol{f}_{\mathrm{b}}^{n}=-\begin{bmatrix}0\\0\\B\end{bmatrix}. \tag{4.2}$$

Note that the z_{n} axis is taken to be positive downwards such that gravity is positive and buoyancy is negative. By applying the results from Section 2.2.1, the weight and buoyancy force can be expressed in $\{b\}$ by

$$\boldsymbol{f}_{\mathrm{g}}^{b}=\boldsymbol{R}^{\top}(\boldsymbol{\Theta}_{\mathrm{nb}})\boldsymbol{f}_{\mathrm{g}}^{n} \tag{4.3}$$

$$\boldsymbol{f}_{\mathrm{b}}^{b}=\boldsymbol{R}^{\top}(\boldsymbol{\Theta}_{\mathrm{nb}})\boldsymbol{f}_{\mathrm{b}}^{n} \tag{4.4}$$

where $\boldsymbol{R}_b^n=\boldsymbol{R}(\boldsymbol{\Theta}_{\mathrm{nb}})$ is the Euler angle coordinate transformation matrix defined in Section 2.2.1. According to (2.2), the sign of the restoring forces and moments $\boldsymbol{f}_i^b$ and $\boldsymbol{m}_i^b=\boldsymbol{r}_{bi}^b\times\boldsymbol{f}_i^b$, $i\in\{g,b\}$, must be changed when moving these terms to the left-hand side of (2.2), corresponding to the vector $\boldsymbol{g}(\boldsymbol{\eta})$. Consequently, the generalized restoring force expressed in $\{b\}$ is

$$\begin{aligned}\boldsymbol{g}(\boldsymbol{\eta})&=-\begin{bmatrix}\boldsymbol{f}_{\mathrm{g}}^{b}+\boldsymbol{f}_{\mathrm{b}}^{b}\\ \boldsymbol{r}_{\mathrm{bg}}^{b}\times\boldsymbol{f}_{\mathrm{g}}^{b}+\boldsymbol{r}_{\mathrm{bb}}^{b}\times\boldsymbol{f}_{\mathrm{b}}^{b}\end{bmatrix}\\ &=-\begin{bmatrix}\boldsymbol{R}^{\top}(\boldsymbol{\Theta}_{\mathrm{nb}})(\boldsymbol{f}_{\mathrm{g}}^{n}+\boldsymbol{f}_{\mathrm{b}}^{n})\\ \boldsymbol{r}_{\mathrm{bg}}^{b}\times\boldsymbol{R}^{\top}(\boldsymbol{\Theta}_{\mathrm{nb}})\boldsymbol{f}_{\mathrm{g}}^{n}+\boldsymbol{r}_{\mathrm{bb}}^{b}\times\boldsymbol{R}^{\top}(\boldsymbol{\Theta}_{\mathrm{nb}})\boldsymbol{f}_{\mathrm{b}}^{n}\end{bmatrix}.\end{aligned} \tag{4.5}$$

Expanding this expression yields

$$\boldsymbol{g}(\boldsymbol{\eta})=\begin{bmatrix}(W-B)\sin(\theta)\\ -(W-B)\cos(\theta)\sin(\phi)\\ -(W-B)\cos(\theta)\cos(\phi)\\ -(y_{\mathrm{g}}W-y_{\mathrm{b}}B)\cos(\theta)\cos(\phi)+(z_{\mathrm{g}}W-z_{\mathrm{b}}B)\cos(\theta)\sin(\phi)\\ (z_{\mathrm{g}}W-z_{\mathrm{b}}B)\sin(\theta)+(x_{\mathrm{g}}W-x_{\mathrm{b}}B)\cos(\theta)\cos(\phi)\\ -(x_{\mathrm{g}}W-x_{\mathrm{b}}B)\cos(\theta)\sin(\phi)-(y_{\mathrm{g}}W-y_{\mathrm{b}}B)\sin(\theta)\end{bmatrix}. \tag{4.6}$$

MATLAB

The generalized restoring force $\boldsymbol{g}(\boldsymbol{\eta})$ for an underwater vehicle can be computed with the CO as coordinate origin by using the MSS toolbox commands:

```
r_g = [0, 0, 0];      % location of the CG relative to the CO
r_b = [0, 0, -10];    % location of the CB relative to the CO
m = 1000;             % mass
g = 9.81;             % acceleration of gravity
W = m * g;            % weight
B = W;                % buoyancy

% pitch and roll angles
theta = 20 * (pi/180); phi = 30 * (pi/180);

% g vector expressed in the CO
g = gvect(W, B, theta, phi, r_g, r_b)

g =
   1.0e+04 *
          0
          0
          0
     4.6092
     3.3552
          0
```

Equation (4.6) is the Euler angle representation of the hydrostatic forces and moments. An alternative representation can be found by applying *unit quaternions*. Then $\boldsymbol{R}(\boldsymbol{q}_{\mathrm{b}}^{\mathrm{n}})$ replaces $\boldsymbol{R}(\boldsymbol{\Theta}_{\mathrm{nb}})$ in (4.3); see Section 2.2.2.

A neutrally buoyant underwater vehicle will satisfy

$$W = B. \tag{4.7}$$

It is convenient to design underwater vehicles with $B > W$ (positive buoyancy) such that the vehicle will surface automatically in the case of an emergency situation, for instance power failure. In this case, the magnitude of B should only be slightly larger than W. If the vehicle is designed such that $B \gg W$, too much control energy is needed to keep the vehicle submerged. Hence, a trade-off between positive buoyancy and controllability must be made.

Example 4.1 (Neutrally Buoyant Underwater Vehicles)
Let the distance from the CB to the CG be defined by the vector

$$[\mathrm{BG}_x,\ \mathrm{BG}_y,\ \mathrm{BG}_z]^{\top} := [x_{\mathrm{g}} - x_{\mathrm{b}},\ y_{\mathrm{g}} - y_{\mathrm{b}},\ z_{\mathrm{g}} - z_{\mathrm{b}}]^{\top}. \tag{4.8}$$

For neutrally buoyant vehicles $W = B$. Hence, (4.6) simplifies to

$$g(\eta) = \begin{bmatrix} 0 \\ 0 \\ 0 \\ -\mathrm{BG}_y W\cos(\theta)\cos(\phi) + \mathrm{BG}_z W\cos(\theta)\sin(\phi) \\ \mathrm{BG}_z W\sin(\theta) + \mathrm{BG}_x W\cos(\theta)\cos(\phi) \\ -\mathrm{BG}_x W\cos(\theta)\sin(\phi) - \mathrm{BG}_y W\sin(\theta) \end{bmatrix}. \tag{4.9}$$

An even simpler representation is obtained for vehicles where the CG and the CB are located on the vertical axis such that $x_b = x_g$ and $y_g = y_b$. This yields

$$g(\eta) = [0,\ 0,\ 0,\ \mathrm{BG}_z W\cos(\theta)\sin(\phi),\ \mathrm{BG}_z W\ \sin(\theta),\ 0]^\top. \tag{4.10}$$

4.2 Restoring Forces for Surface Vessels

Formula (4.6) should only be used for completely submerged vehicles. Static stability considerations due to restoring forces are usually referred to as *metacentric stability* in the hydrostatic literature. A metacentric stable vessel will resist inclinations away from its steady state or equilibrium points in heave, roll and pitch.

For surface vessels, the restoring forces will depend on the vessel's metacentric height, the location of the CG and the CB, as well as the shape and size of the waterplane. Let A_{wp} denote the waterplane area and

GM_T – transverse metacentric height
GM_L – longitudinal metacentric height.

The metacentric height GM_i, where $i \in \{T, L\}$, is the distance between the metacenter M_i and the CG, as shown in Figures 4.2 and 4.3.

Definition 4.1 (Metacenter)
The theoretical point at which an imaginary vertical line through the CB *intersects another imaginary vertical line through a new* CB_1 *created when the body is displaced, or tilted, in the water (see Figure 4.2).*

4.2.1 *Hydrostatics of Floating Vessels*

For a floating vessel at rest, Archimedes stated that buoyancy and weight are in balance

$$mg = \rho g \nabla. \tag{4.11}$$

Assume that the roll and pitch angles are small, and let z^n denote the displacement in heave. Hence, $z^n = 0$ denotes the equilibrium position corresponding to the nominal

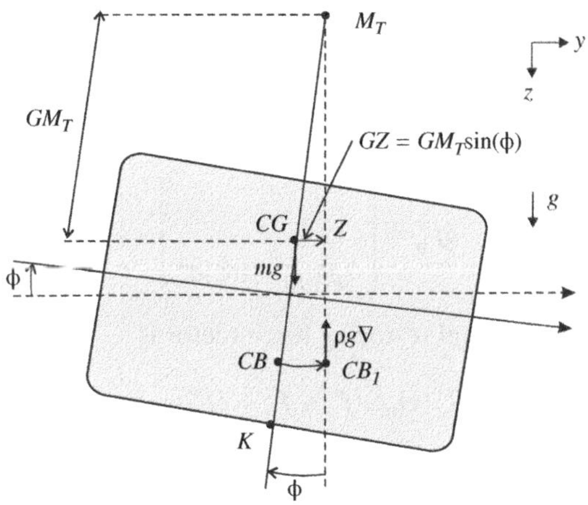

Figure 4.2 Transverse metacentric stability. Notice that $mg = \rho g \nabla$. A similar figure can be drawn to illustrate lateral metacentric stability by simply replacing M_T and ϕ with M_L and θ.

displaced water volume ∇. Moreover, the hydrostatic force in heave will be the difference between the gravitational and the buoyancy forces

$$\begin{aligned} Z_{hs} &= mg - \rho g(\nabla + \delta\nabla(z^n)) \\ &= -\rho g \delta\nabla(z^n) \end{aligned} \tag{4.12}$$

where the change in displaced water $\delta\nabla(z^n)$ is due to variations in vertical position z^n. This is mathematically equivalent to

$$\delta\nabla(z) = \int_0^{z^n} A(\zeta)\mathrm{d}\zeta \tag{4.13}$$

where $A(\zeta)$ is the waterplane area of the vessel as a function of the vertical position. For conventional ships and floating structures, however, it is common to assume that the waterplane area $A(\zeta) \approx A(0) := A_{wp}$ is constant for small perturbations in vertical position. This implies that the hydrostatic restoring force Z_{hs} is linear in z^n such that

$$Z_{hs} \approx -\rho g A_{wp} z^n := Z_z z^n \tag{4.14}$$

where $Z_z = -\rho g A_{wp}$ is recognized as the spring stiffness or restoring coefficient in heave. For small roll and pitch angles, the heave motion can be approximated by a linear mass-damper-spring system

$$\dot{z}^n = w \tag{4.15}$$

$$(m - Z_{\dot{w}})\dot{w} - Z_w w - Z_z z^n = Z_{ext} \tag{4.16}$$

where Z_{ext} is external forcing.

If a floating vessel is forced downwards by an external force such that $z^{\mathrm{n}} > 0$, the buoyancy force becomes larger than the constant gravitational force since the submerged volume ∇ increases by $\delta\nabla$ to $\nabla + \delta\nabla$. Consequently, the perturbed buoyancy force vector due to variations in displaced volume is

$$\delta \boldsymbol{f}_{\mathrm{b}}^{n} = \begin{bmatrix} 0 \\ 0 \\ -\rho g \int_0^{z^{\mathrm{n}}} A(\zeta)\mathrm{d}\zeta \end{bmatrix}. \tag{4.17}$$

From this it follows that the total restoring force vector is

$$\begin{aligned} \boldsymbol{f}_{\mathrm{r}}^{b} &= \boldsymbol{R}^{\top}(\boldsymbol{\Theta}_{\mathrm{nb}})(\boldsymbol{f}_{\mathrm{g}}^{n} + \boldsymbol{f}_{\mathrm{b}}^{n} + \delta \boldsymbol{f}_{\mathrm{b}}^{n}) \\ &= -\rho g \begin{bmatrix} -\sin(\theta) \\ \cos(\theta)\sin(\phi) \\ \cos(\theta)\cos(\phi) \end{bmatrix} \int_0^{z^{\mathrm{n}}} A(\zeta)\mathrm{d}\zeta \end{aligned} \tag{4.18}$$

where we have exploited that $\boldsymbol{f}_{\mathrm{g}}^{n} = -\boldsymbol{f}_{\mathrm{b}}^{n}$.

From Figure 4.2 it is seen that the moment arms in roll and pitch for the force pair $\boldsymbol{f}_{\mathrm{g}}^{n} = -\boldsymbol{f}_{\mathrm{b}}^{n}$ are $\mathrm{GM_T}\sin(\phi)$ and $\mathrm{GM_L}\sin(\theta)$, respectively. Neglecting the moment contribution due to $\delta \boldsymbol{f}_{\mathrm{b}}^{n}$ (only considering $\boldsymbol{f}_{\mathrm{b}}^{n}$), the following moment arm and force vector is obtained

$$\boldsymbol{r}_{\mathrm{GM}}^{b} = \begin{bmatrix} -\mathrm{GM_L}\sin(\theta) \\ \mathrm{GM_T}\sin(\phi) \\ 0 \end{bmatrix} \tag{4.19}$$

$$\begin{aligned} \boldsymbol{f}_{\mathrm{b}}^{b} &= \boldsymbol{R}^{\top}(\boldsymbol{\Theta}_{\mathrm{nb}}) \begin{bmatrix} 0 \\ 0 \\ -\rho g \nabla \end{bmatrix} \\ &= -\rho g \nabla \begin{bmatrix} -\sin(\theta) \\ \cos(\theta)\sin(\phi) \\ \cos(\theta)\cos(\phi) \end{bmatrix}. \end{aligned} \tag{4.20}$$

The restoring moment becomes

$$\begin{aligned} \boldsymbol{m}_{\mathrm{r}}^{b} &= \boldsymbol{r}_{\mathrm{GM}}^{b} \times \boldsymbol{f}_{\mathrm{b}}^{b} \\ &= -\rho g \nabla \begin{bmatrix} \mathrm{GM_T}\sin(\phi)\cos(\theta)\cos(\phi) \\ \mathrm{GM_L}\sin(\theta)\cos(\theta)\cos(\phi) \\ -\mathrm{GM_L}\cos(\theta) + \mathrm{GM_T}\sin(\phi)\sin(\theta) \end{bmatrix}. \end{aligned} \tag{4.21}$$

The assumption that $\boldsymbol{r}_{\mathrm{GM}}^{b} \times \delta \boldsymbol{f}_{\mathrm{b}}^{b} = \boldsymbol{0}$ (no moments due to perturbed heave motions) is a good assumption since this term is small compared to $\boldsymbol{r}_{\mathrm{GM}}^{b} \times \boldsymbol{f}_{\mathrm{b}}^{b}$. Thus,

$$\boldsymbol{g}(\boldsymbol{\eta}) = -\begin{bmatrix} \boldsymbol{f}_{\mathrm{r}}^{b} \\ \boldsymbol{m}_{\mathrm{r}}^{b} \end{bmatrix}. \tag{4.22}$$

Expanding this expression finally gives

$$g(\eta) = \begin{bmatrix} -\rho g(\int_0^{z^n} A(\zeta)d\zeta)\sin(\theta) \\ \rho g(\int_0^{z^n} A(\zeta)d\zeta)\cos(\theta)\sin(\phi) \\ \rho g(\int_0^{z^n} A(\zeta)d\zeta)\cos(\theta)\cos(\phi) \\ \rho g\nabla GM_T \sin(\phi)\cos(\theta)\cos(\phi) \\ \rho g\nabla GM_L \sin(\theta)\cos(\theta)\cos(\phi) \\ \rho g\nabla(-GM_L\cos(\theta)+GM_T)\sin(\phi)\sin(\theta) \end{bmatrix}. \tag{4.23}$$

4.2.2 Linear (Small Angle) Theory for Boxed-shaped Vessels

For surface vessels it is common to use linear theory

$$\boldsymbol{M}\dot{\boldsymbol{v}} + \boldsymbol{N}\boldsymbol{v} + \boldsymbol{G}\boldsymbol{\eta} + \boldsymbol{g}_0 = \boldsymbol{\tau} + \boldsymbol{\tau}_{\text{wind}} + \boldsymbol{\tau}_{\text{wave}} \tag{4.24}$$

implying that (4.23) is approximated by a matrix $\boldsymbol{G}$ of restoring coefficients such that

$$\boldsymbol{g}(\boldsymbol{\eta}) \approx \boldsymbol{G}\boldsymbol{\eta}. \tag{4.25}$$

This is based on the assumption that

$$\int_0^{z^n} A(\zeta)d\zeta \approx A_{wp}z^n$$

and the assumptions that ϕ and θ are small such that

$$\sin(\theta) \approx \theta; \qquad \cos(\theta) \approx 1$$
$$\sin(\phi) \approx \phi; \qquad \cos(\phi) \approx 1.$$

Under these assumptions it follows that (4.23) can be approximated by

$$g(\eta) \approx \begin{bmatrix} -\rho g A_{wp} z^n \theta \\ \rho g A_{wp} z^n \phi \\ \rho g A_{wp} z^n \\ \rho g \nabla GM_T \phi \\ \rho g \nabla GM_L \theta \\ \rho g \nabla(-GM_L + GM_T)\phi\theta \end{bmatrix} \approx \begin{bmatrix} 0 \\ 0 \\ \rho g A_{wp} z^n \\ \rho g \nabla GM_T \phi \\ \rho g \nabla GM_L \theta \\ 0 \end{bmatrix} \tag{4.26}$$

where we have neglected second-order terms. Consequently,

$$\boldsymbol{G}^{CF} = \text{diag}\{0,\ 0,\ \rho g A_{wp},\ \rho g\nabla GM_T,\ \rho g\nabla GM_L,\ 0\}. \tag{4.27}$$

The superscript CF denotes that the matrix is expressed in the center of flotation (see Section 2.1.1). The first moment of areas are zero about the CF. This is mathematically equivalent to

$$\frac{1}{A_{\text{wp}}}\iint_{A_{\text{wp}}} x\,\text{d}A = 0, \qquad \frac{1}{A_{\text{wp}}}\iint_{A_{\text{wp}}} y\,\text{d}A = 0 \tag{4.28}$$

where the integrals are computed about the waterplane centroid, that is the geometric center of A_{wp}. This is the point of rotation for a freely rotating body subject to an applied horizontal moment. The second moment of areas are both positive

$$I_{\text{L}} := \iint_{A_{\text{wp}}} x^2\,\text{d}A, \qquad I_{\text{T}} := \iint_{A_{\text{wp}}} y^2\,\text{d}A. \tag{4.29}$$

For conventional ships the CG and the CB lie on the same vertical line ($x_{\text{b}} = x_{\text{g}}$ and $y_{\text{b}} = y_{\text{g}}$) such that (Newman 1977)

$$\text{GM}_{\text{T}} = \frac{I_{\text{T}}}{\nabla} + z_{\text{g}} - z_{\text{b}} \tag{4.30}$$

$$\text{GM}_{\text{L}} = \frac{I_{\text{L}}}{\nabla} + z_{\text{g}} - z_{\text{b}}. \tag{4.31}$$

It is convenient to represent the equations of motion about the CO usually midships on the centerline with the CG and the CB on the same vertical line. For this case both the first and second moments of area of the waterplane have to be computed about the CO which is located a distance LCF in the x direction. This introduces non-zero coupling terms G_{35} and G_{53} and the formula for $\boldsymbol{G}$ expressed in the CO is modified accordingly

$$\boldsymbol{G} = \boldsymbol{G}^{\top} = \begin{bmatrix} 0 & 0 & 0 & 0 & 0 & 0 \\ 0 & 0 & 0 & 0 & 0 & 0 \\ 0 & 0 & -Z_z & 0 & -Z_\theta & 0 \\ 0 & 0 & 0 & -K_\phi & 0 & 0 \\ 0 & 0 & -M_z & 0 & -M_\theta & 0 \\ 0 & 0 & 0 & 0 & 0 & 0 \end{bmatrix} > 0. \tag{4.32}$$

The numerical value for $\boldsymbol{G}$ can be computed directly from the expression of $\boldsymbol{G}^{\text{CF}}$ by using the transformation (C.16) with distance vector $\boldsymbol{r}^b_{\text{bf}} = [\text{LCF}, 0, 0]^{\top}$ where LCF is the distance from the CO to the CF (negative for conventional ships). This gives

$$
\begin{aligned}
G &= H^{\top}(r^b_{\rm bf})G^{\rm CF}H(r^b_{\rm bf}) \\
&= \begin{bmatrix}
0 & 0 & 0 & 0 & 0 & 0 \\
0 & 0 & 0 & 0 & 0 & 0 \\
0 & 0 & \rho g A_{\rm wp} & 0 & -\rho g A_{\rm wp}\,{\rm LCF} & 0 \\
0 & 0 & 0 & \rho g \nabla {\rm GM_T} & 0 & 0 \\
0 & 0 & -\rho g A_{\rm wp}\,{\rm LCF} & 0 & \rho g (A_{\rm wp}\,{\rm LCF}^2 + \nabla {\rm GM_L}) & 0 \\
0 & 0 & 0 & 0 & 0 & 0
\end{bmatrix}.
\end{aligned}
\tag{4.33}
$$

MATLAB

The 6×6 system spring stiffness matrix G is computed by using the MSS toolbox function `Gmtrx.m` as illustrated by the example below

```
A_wp = 1000;      % waterplane area
nabla = 8000;     % volume displacement
GMT = 1;          % transverse metacentric height
GML = 10;         % longitudinal metacentric heights
LCF = -8;         % location of the CF relative to the CO
r_p = [0 0 0]'; % location of the point CP relative to the CO

% Spring stiffness matrix expressed in the CO
G = Gmtrx(nabla, A_wp, GMT, GML, LCF, r_p)

G =
   1.0e+09 *
         0         0         0         0         0         0
         0         0         0         0         0         0
         0         0    0.0101         0    0.0804         0
         0         0         0    0.0804         0         0
         0         0    0.0804         0    1.4480         0
         0         0         0         0         0         0
```

4.2.3 Computation of Metacenter Heights for Surface Vessels

The metacenter height can be computed by using basic hydrostatics

$$
{\rm GM_T} = {\rm BM_T} - {\rm BG}, \qquad {\rm GM_L} = {\rm BM_L} - {\rm BG}. \tag{4.34}
$$

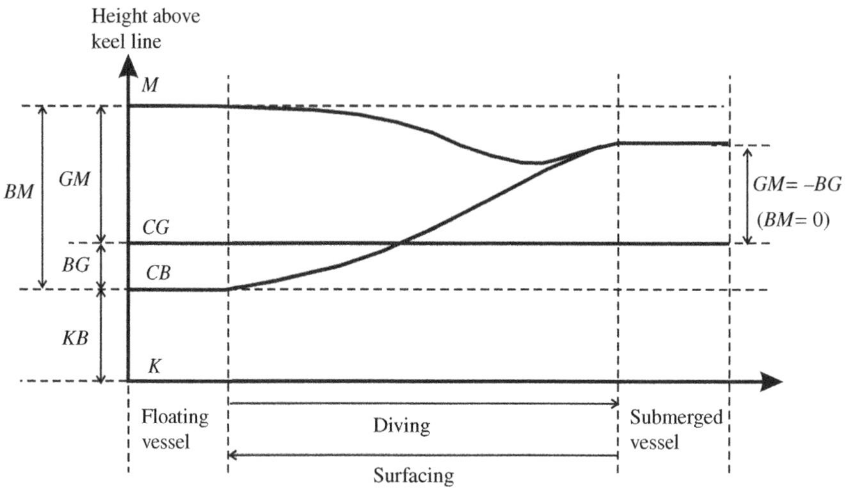

Figure 4.3 Metacenter M, center of gravity CG and center of buoyancy CB for a submerged and a floating vessel. K is the keel line.

This relationship is verified directly from Figure 4.3, where M_T and M_L denote the transverse and longitudinal metacenters (intersections between the vertical lines through the CB and the CB_1 when ϕ and θ approaches zero). The symbol K is used to denote the keel line. For small inclinations ϕ and θ the transverse and longitudinal radii of curvature can be approximated by

$$\mathrm{BM_T} = \frac{I_T}{\nabla}, \qquad \mathrm{BM_L} = \frac{I_L}{\nabla} \tag{4.35}$$

where the second moments of area I_T and I_L are defined by (4.29). For conventional ships an upper bound on these integrals can be found by considering a rectangular waterplane area $A_{wp} = BL$ where B and L are the beam and length of the hull, respectively. This implies that

$$I_T < \frac{1}{12}B^3 L, \qquad I_L < \frac{1}{12}L^3 B. \tag{4.36}$$

At an early stage in the ship design process it is necessary to ensure that the ship will have an adequate $\mathrm{GM_T}$ for all reasonable loading conditions. This requires that KB and $\mathrm{BM_T}$ are estimated. For this purpose, the *Munro-Smith formula* for estimating $\mathrm{BM_T}$ is quite useful (Wilson 2018, pp. 89–90)

$$I_T = \frac{1}{12}B^3 L \frac{6C_w^3}{(1 + C_w)(1 + 2C_w)} \tag{4.37}$$

where $C_w = A_{wp}/(LB)$ is the waterplane area coefficient.

The following method, known as *Morrish's formula*, gives a reasonably accurate estimate of KB (Wilson 2018, pp. 88–89)

$$\mathrm{KB} = \frac{1}{3}\left(\frac{5T}{2} - \frac{\nabla}{A_{\mathrm{wp}}}\right). \tag{4.38}$$

The formulae for KB and $\mathrm{BM_T}$ are very useful when simulating the vessel dynamics at an early stage.

Example 4.2 (Computation of GM Values)
Consider a floating barge with length 100 m *and width* 8 m. *The draft is* 5 m *while the CG is located* 3 m *above the keel line (*KG = 3.0 m*). Since* KB = 2.5 m, *it follows that*

$$\mathrm{BG} = \mathrm{KG} - \mathrm{KB} = 3 - 2.5 = 0.5 \text{ m}. \tag{4.39}$$

Hence,

$$I_{\mathrm{T}} = \frac{1}{12}B^3L = \frac{1}{12}8^3 \times 100 = 4\ 266.7 \text{ m}^4 \tag{4.40}$$

$$I_{\mathrm{L}} = \frac{1}{12}L^3B = \frac{1}{12}100^3 \times 8 = 666\ 666.7 \text{ m}^4. \tag{4.41}$$

The volume displacement is

$$\nabla = 100 \times 8 \times 5 = 4\ 000 \text{ m}^3. \tag{4.42}$$

Consequently,

$$\mathrm{BM_T} = \frac{I_{\mathrm{T}}}{\nabla} = 2.08 \text{ m} \tag{4.43}$$

$$\mathrm{BM_L} = \frac{I_{\mathrm{L}}}{\nabla} = 166.7 \text{ m}. \tag{4.44}$$

Finally,

$$\mathrm{GM_T} = \mathrm{BM_T} - \mathrm{BG} = 2.08 - 0.5 = 1.58 \text{ m} \tag{4.45}$$

$$\mathrm{GM_L} = \mathrm{BM_L} - \mathrm{BG} = 166.7 - 0.5 = 166.2 \text{ m}. \tag{4.46}$$

and

$$\mathrm{KM_T} = \mathrm{KG} + \mathrm{GM_T} = 3 + 1.58 = 4.58 \text{ m} \tag{4.47}$$

$$\mathrm{KM_L} = \mathrm{KG} + \mathrm{GM_L} = 3 + 166.2 = 169.2 \text{ m}. \tag{4.48}$$

Definition 4.2 (Metacenter Stability)
A floating vessel is said to be transverse metacentrically stable if

$$\mathrm{GM_T} \geq \mathrm{GM_{T,min}} > 0 \tag{4.49}$$

and longitudinal metacentrically stable if

$$\mathrm{GM_L} \geq \mathrm{GM_{L,min}} > 0. \tag{4.50}$$

The longitudinal stability requirement (4.50) is easy to satisfy for ships since the pitching motion is quite limited. The transverse requirement, however, is an important design criterion used to predescribe sufficient stability in roll to avoid the craft rolling around. For most ships $\mathrm{GM}_{\mathrm{T,min}} > 0.5$ m while $\mathrm{GM}_{\mathrm{L,min}}$ usually is much larger (more than 100 m).

If the transverse metacentric height GM_{T} is large, the spring is stiff in roll and it is quite uncomfortable for passengers on board the vessel. However, the stability margin and robustness to large transverse waves are good in this case. Consequently, a trade-off between stability and comfort should be made. Another point to consider is that all ships have varying load conditions. This implies that the pitch and roll periods will vary with the loads since GM_{T} varies with the load. This is the topic for the next section.

4.3 Load Conditions and Natural Periods

The chosen load condition or weight distribution will determine the heave, roll and pitch periods of a marine craft. Hydrodynamic codes (see Section 6.2) can be used to compute a linear seakeeping model

$$(\boldsymbol{M}_{RB} + \boldsymbol{A}(\omega))\ddot{\boldsymbol{\xi}} + (\boldsymbol{B}(\omega) + \boldsymbol{B}_{\mathrm{V}}(\omega))\dot{\boldsymbol{\xi}} + \boldsymbol{C}\boldsymbol{\xi} = \boldsymbol{\tau}_{\mathrm{ext}}. \tag{4.51}$$

This is a linear mass–damper-spring system where the seakeeping coordinates $\boldsymbol{\xi} = [\zeta_1, \zeta_2, \zeta_3, \zeta_4, \zeta_5, \zeta_6]^{\mathrm{T}}$ are zero-mean wave-induced perturbations in $\{s\}$ from an equilibrium state defined by a ship moving at constant heading ψ and speed U. In calm water $\dot{\boldsymbol{\xi}} = \boldsymbol{0}$ and $\{s\}$ coincides with $\{b\}$.

The outputs from the hydrodynamic codes are frequency-dependent added mass $\boldsymbol{A}(\omega)$ and potential damping $\boldsymbol{B}(\omega)$. Viscous damping $\boldsymbol{B}_V(\omega)$ is usually added using semi-empirical methods while the restoring matrix $\boldsymbol{C} = \boldsymbol{G}$.

If the decoupled heave, roll and pitch equations are used to compute the natural periods it is important to transform the hydrodynamic data to the CF, which is the point of rotation of a free-floating body. However, in a linear system the natural periods will be independent of the coordinate origin if they are computed using the 6-DOF coupled equations of motion. This is due to the fact that the eigenvalues of a linear system do not change when applying a similarity transformation. However, it is not straightforward to use the linear equations of motion since the potential coefficients depend on the wave frequency. This section presents methods for computation of the natural periods using coupled and decoupled equations in heave, roll and pitch.

4.3.1 Decoupled Computation of Natural Periods

Consider the linear decoupled heave, roll and pitch equations expressed in the CF

$$(m + A_{33}^{\mathrm{CF}}(\omega_3))\ddot{z}^{\mathrm{n}} + (B_{33}^{\mathrm{CF}}(\omega_3) + B_{\mathrm{v},33}^{\mathrm{CF}}(\omega_3))\dot{z}^{\mathrm{n}} + C_{33}^{\mathrm{CF}} z^{\mathrm{n}} = 0 \tag{4.52}$$

$$(I_x^{\mathrm{CF}} + A_{44}^{\mathrm{CF}}(\omega_4))\ddot{\phi} + (B_{44}^{\mathrm{CF}}(\omega_4) + B_{\mathrm{v},44}^{\mathrm{CF}}(\omega_4))\dot{\phi} + C_{44}^{\mathrm{CF}}\phi = 0 \tag{4.53}$$

$$(I_y^{\mathrm{CF}} + A_{55}^{\mathrm{CF}}(\omega_5))\ddot{\theta} + (B_{55}^{\mathrm{CF}}(\omega_5) + B_{\mathrm{v},55}^{\mathrm{CF}}(\omega_5))\dot{\theta} + C_{55}^{\mathrm{CF}}\theta = 0 \tag{4.54}$$

where the potential coefficients A_{ii}^{CF} and B_{ii}^{CF}, viscous damping $B_{\mathrm{v},ii}^{\mathrm{CF}}$, spring stiffness C_{ii}^{CF} ($i = 3,4,5$), and moments of inertia I_x^{CF} and I_y^{CF} are computed in the CF, which is the vessel rotation point for a pure rolling or pitching motion under the assumption of constant volume displacement. In the nonlinear case, the point of rotation as well as the rotation axes will change. If the CF is unknown, a good approximation is to use the midships origin CO and an estimate for LCF. This will only affect the pitching frequency, which is not very sensitive to small translations along the x_b axis. If the natural frequencies are computed in a point far from the CF using the decoupled equations (4.52)–(4.54), the results can be erroneous since the eigenvalues of the decoupled equations depend on the coordinate origin as opposed to the 6-DOF coupled system.

From (4.52)–(4.54) it follows that the natural frequencies and periods of heave, roll and pitch in the CF are given by the implicit equations

$$\omega_3 = \sqrt{\frac{C_{33}^{\mathrm{CF}}}{m + A_{33}^{\mathrm{CF}}(\omega_3)}}, \qquad T_3 = \frac{2\pi}{\omega_3} \tag{4.55}$$

$$\omega_4 = \sqrt{\frac{C_{44}^{\mathrm{CF}}}{I_x^{\mathrm{CF}} + A_{44}^{\mathrm{CF}}(\omega_4)}}, \qquad T_4 = \frac{2\pi}{\omega_4} \tag{4.56}$$

$$\omega_5 = \sqrt{\frac{C_{55}^{\mathrm{CF}}}{I_y^{\mathrm{CF}} + A_{55}^{\mathrm{CF}}(\omega_5)}}, \qquad T_5 = \frac{2\pi}{\omega_5} \tag{4.57}$$

which can be solved in Matlab by using `fsolve.m` (requires the optimization toolbox) or `fzero.m`.

MATLAB

Decoupled analysis for computation of natural periods for the MSS tanker and supply vessel. The hydrodynamic data have been generated using WAMIT and ShipX.

```
%   w_n = natfrequency(vessel,dof,w_0,speed,LCF)
%   vessel = MSS vessel data (computed in the CO)
%   dof = degree of freedom (3,4,5), use -1 for 6-DOF analysis
%   w_0 = initial natural frequency (typical 0.5)
%   speed = index 1,2,3... for hydrodynamic data set
%   LCF = (optionally) x coord. from the CO to the CF (negative)

load tanker;      % WAMIT data file
T_3 = 2 * pi / natfrequency(vessel, 3, 0.5, 1)
T_4 = 2 * pi / natfrequency(vessel, 4, 0.5, 1)
```

```
T_5 = 2 * pi / natfrequency(vessel, 5, 0.5, 1)

T_3 =
     9.6814
T_4 =
    12.5074
T_5 =
     9.0851

load supply;     % shipX data file
T_3 = 2 * pi / natfrequency(vessel, 3, 0.5, 1)
T_4 = 2 * pi / natfrequency(vessel, 4, 0.5, 1)
T_5 = 2 * pi / natfrequency(vessel, 5, 0.5, 1)

T_3 =
     6.3617
T_4 =
    10.8630
T_5 =
     6.0988
```

4.3.2 Computation of Natural Periods in a 6-DOF Coupled System

A 6-DOF coupled analysis of the frequency-dependent data can be done by using modal analysis. The coupled system can be transformed to six decoupled systems and the natural frequencies can be computed for each of them. This involves solving a generalized eigenvalue problem at each frequency.

Let the CO be the coordinate origin of the linear seakeeping model

$$(\boldsymbol{M}_{RB} + \boldsymbol{A}(\omega))\ddot{\boldsymbol{\xi}} + (\boldsymbol{B}(\omega) + \boldsymbol{B}_{\rm V}(\omega) + \boldsymbol{K}_{\rm d})\dot{\boldsymbol{\xi}} + (\boldsymbol{C} + \boldsymbol{K}_{\rm p})\boldsymbol{\xi} = \boldsymbol{0} \tag{4.58}$$

where $\boldsymbol{K}_{\rm p}$ and $\boldsymbol{K}_{\rm d}$ are optional positive definite matrices due to feedback control, $\boldsymbol{A}(\omega)$ and $\boldsymbol{B}(\omega)$ are frequency-dependent added mass and potential damping (see Section 5.3) while $\boldsymbol{B}_{\rm V}(\omega)$ denotes additional viscous damping. The effect of feedback control can be included in the analysis by specifying three PD controllers in surge, sway and yaw using

$$\boldsymbol{K}_p = \text{diag}\{K_{\text{p}_{11}},\ K_{\text{p}_{22}},\ 0,\ 0,\ 0,\ K_{\text{p}_{66}}\} \tag{4.59}$$

$$\boldsymbol{K}_d = \text{diag}\{K_{\text{d}_{11}},\ K_{\text{d}_{22}},\ 0,\ 0,\ 0,\ K_{\text{d}_{66}}\}. \tag{4.60}$$

The gain selection represents a standard feedback control problem for stationkeeping, which can be solved by pole placement or linear-quadratic optimal control theory. Let

$$\boldsymbol{M}(\omega) = \boldsymbol{M}_{RB} + \boldsymbol{A}(\omega) \tag{4.61}$$

$$\boldsymbol{D}(\omega) = \boldsymbol{B}(\omega) + \boldsymbol{B}_{\rm V}(\omega) + \boldsymbol{K}_{\rm d} \tag{4.62}$$

$$\boldsymbol{G} = \boldsymbol{C} + \boldsymbol{K}_{\rm p} \tag{4.63}$$

where $\boldsymbol{M}(\omega) = \boldsymbol{M}^{\top}(\omega) > 0$ and $\boldsymbol{D}(\omega) = \boldsymbol{D}^{\top}(\omega) > 0$ such that

$$\boldsymbol{M}(\omega)\ddot{\xi} + \boldsymbol{D}(\omega)\dot{\xi} + \boldsymbol{G}\xi = \boldsymbol{0}. \tag{4.64}$$

For surface vessels, the restoring matrix takes the following form (see Section 4.2)

$$\boldsymbol{G} = \boldsymbol{G}^{\top} = \begin{bmatrix} K_{\mathrm{p}11} & 0 & 0 & 0 & 0 & 0 \\ 0 & K_{\mathrm{p}22} & 0 & 0 & 0 & 0 \\ 0 & 0 & C_{33} & 0 & C_{35} & 0 \\ 0 & 0 & 0 & C_{44} & 0 & 0 \\ 0 & 0 & C_{53} & 0 & C_{55} & 0 \\ 0 & 0 & 0 & 0 & 0 & K_{\mathrm{p}66} \end{bmatrix}. \tag{4.65}$$

Note that $K_{\mathrm{p}11}$, $K_{\mathrm{p}22}$ and $K_{\mathrm{p}66}$ must be positive to guarantee that $\boldsymbol{G} > 0$. Application of Laplace's transformation $s = \mathrm{j}\omega$ to (4.64) gives

$$(\boldsymbol{G} - \omega^2\boldsymbol{M}(\omega) - \mathrm{j}\omega\boldsymbol{D}(\omega))\xi = \boldsymbol{0}. \tag{4.66}$$

The natural frequencies can be computed for the undamped system $\boldsymbol{D}(\omega) = \boldsymbol{0}$ by solving

$$(\boldsymbol{G} - \omega^2\boldsymbol{M}(\omega))\xi = \boldsymbol{0}. \tag{4.67}$$

The natural frequencies of a marine craft are usually shifted by less than 1.0% when damping is added. Hence, the undamped system (4.67) gives an accurate estimate of the frequencies of oscillation.

Equation (4.67) represents a frequency-dependent *generalized eigenvalue problem*

$$\boldsymbol{G}\boldsymbol{x}_i = \lambda_i\boldsymbol{M}(\omega)\boldsymbol{x}_i \qquad (i = 1, 2, \cdots, 6) \tag{4.68}$$

where $\boldsymbol{x}_i$ is the eigenvector and $\lambda_i = \omega^2$ are the eigenvalues (see Figure 4.4). This is recognized as an algebraic equation

$$|\boldsymbol{G} - \lambda_i\boldsymbol{M}(\omega)| = \boldsymbol{0} \qquad (i = 1, 2, \cdots, 6) \tag{4.69}$$

$$\lambda_i = \omega^2. \tag{4.70}$$

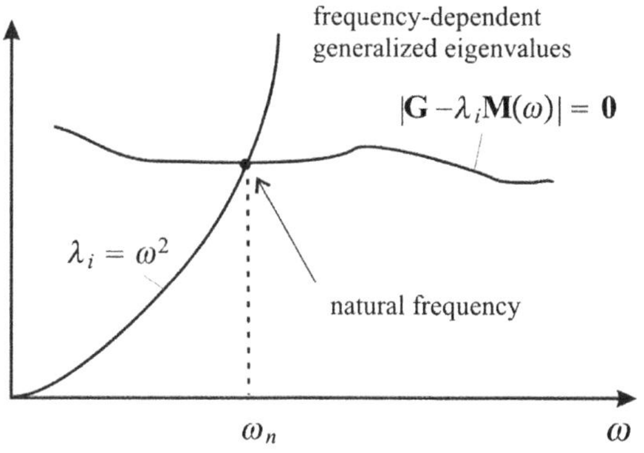

Figure 4.4 Generalized eigenvalue problem.

The characteristic equation of (4.69) is of sixth order

$$\lambda^6 + a_5\lambda^5 + a_4\lambda^4 + a_3\lambda^3 + a_2\lambda^2 + a_1\lambda + a_0 = 0. \tag{4.71}$$

Let the solutions of the eigenvalue problem (4.69) as a function of ω be denoted $\lambda_i^*(\omega)$. Then we can use the *Newton–Raphson method*

$$\omega_{i,k+1} = \omega_{i,k} - \frac{f_i(\omega_{i,k})}{f_i'(\omega_{i,k})} \qquad (i = 1, \ldots, 6, k = 1, \ldots, n) \tag{4.72}$$

where k is the number of iterations, i is the DOF considered and

$$f_i(\omega_{i,k}) = \lambda_i^*(\omega_{i,k}) - \omega_{i,k}^2 \tag{4.73}$$

to satisfy the constraint (4.70). After solving $f_i(\omega_{i,k}) = 0$ for all DOFs to obtain $\omega_{i,n}$, the natural periods in 6 DOFs follow from

$$T_i = \frac{2\pi}{\omega_{i,n}} \tag{4.74}$$

MATLAB

The 6-DOF coupled eigenvalue analysis in Section 4.3.2 is implemented in the MSS toolbox. The natural periods for tanker is computed by using the following commands:

```
dof = -1;           % use -1 for 6-DOF analysis
load tanker;        % load WAMIT tanker data
T = 2 * pi ./ natfrequency(vessel, dof, 0.5, 1)

T =
    9.8261
   12.4543
    8.9536

load supply;        % load ShipX supply ship data
T = 2 * pi ./ natfrequency(vessel, dof, 0.5, 1)

T =
    6.5036
   10.4205
    6.0210
```

Note that that natural periods for the coupled analysis are quite close to the numbers obtained in the decoupled analysis in Section 4.3.1.

4.3.3 Natural Periods as a Function of Load Condition

The roll and pitch periods will depend strongly on the load condition while heave is less affected. From (4.55)–(4.57) it follows that

$$T_3 = 2\pi\sqrt{\frac{m + A_{33}^{\mathrm{CF}}(\omega_3)}{\rho g A_{\mathrm{wp}}}} \tag{4.75}$$

$$T_4 = 2\pi\sqrt{\frac{I_x^{\mathrm{CF}} + A_{44}^{\mathrm{CF}}(\omega_4)}{\rho g \nabla \mathrm{GM_T}}} \tag{4.76}$$

$$T_5 = 2\pi\sqrt{\frac{I_y^{\mathrm{CF}} + A_{55}^{\mathrm{CF}}(\omega_5)}{\rho g \nabla \mathrm{GM_L}}}. \tag{4.77}$$

An estimate of the heave period can be computed by assuming that $A_{33}^{\mathrm{CF}} \approx m = \rho\nabla$ and $\nabla \approx A_{\mathrm{wp}} T$ where T is the draft. This implies that (4.75) can be approximated as

$$T_3 \approx 2\pi\sqrt{\frac{2T}{g}}. \tag{4.78}$$

Formula (4.78) indicates that the natural period in heave mainly depends on the draft T, which again depends on the vessel load condition.

For roll and pitch one can compute the moments of inertia about the CG by using the formulae

$$I_x^{\mathrm{CG}} = mR_{44}^2 \tag{4.79}$$

$$I_y^{\mathrm{CG}} = mR_{55}^2 \tag{4.80}$$

where R_{44} and R_{55} are the radii of gyration with respect to the CG. For offshore vessels $R_{44} \approx 0.35B$ while tankers have $R_{44} \approx 0.37B$. Semi-submersibles have two or more pontoons so $0.40B$ is not uncommon for these vessels. In pitch and yaw it is common to use $R_{55} = R_{66} \approx 0.25L$ for smaller vessels while tankers use $R_{55} = R_{66} \approx 0.27L$.

Assume that $\boldsymbol{r}_{\mathrm{bf}}^{b} = [\mathrm{LCF},\ 0,\ 0]^{\top}$ and $\boldsymbol{r}_{\mathrm{bg}}^{b} = [x_{\mathrm{g}},\ 0,\ z_{\mathrm{g}}]^{\top}$ such that the CO is on the centerline midships. Application of the parallel-axes theorem (3.37) gives

$$I_x^{\mathrm{CF}} = mR_{44}^2 + mz_{\mathrm{g}}^2 := m(R_{44}^{\mathrm{CF}})^2 \tag{4.81}$$

$$I_y^{\mathrm{CF}} = mR_{55}^2 + m((x_{\mathrm{g}} - \mathrm{LCF})^2 + z_{\mathrm{g}}^2) := m(R_{55}^{\mathrm{CF}})^2. \tag{4.82}$$

Assume that the added mass matrix A is computed in the CO. The transformations between the CO and the CF are (see Appendix C.2)

$$A = H^{\top}(r^b_{\mathrm{bf}}) A^{\mathrm{CF}} H(r^b_{\mathrm{bf}}) \tag{4.83}$$

$$A^{\mathrm{CF}} = H^{-\top}(r^b_{\mathrm{bf}}) A H^{-1}(r^b_{\mathrm{bf}}). \tag{4.84}$$

Define κ_4 and κ_5 as the ratios between hydrodynamic added moment of inertia and rigid-body moment of inertia in roll and pitch

$$\kappa_4 := \frac{A^{\mathrm{CF}}_{44}(\omega_4)}{I^{\mathrm{CF}}_x} \tag{4.85}$$

$$\kappa_5 := \frac{A^{\mathrm{CF}}_{55}(\omega_5)}{I^{\mathrm{CF}}_y}. \tag{4.86}$$

Typical values for κ_4 are 0.1–0.3 for ships and 1.0 or more for semi-submersibles. In pitch, κ_5 will be in the range 1.0–2.0. This implies that the roll and pitch periods (4.76)–(4.77) can be computed by

$$T_4 = 2\pi R^{\mathrm{CF}}_{44}\sqrt{\frac{1+\kappa_4}{g\,\mathrm{GM_T}}} \tag{4.87}$$

$$T_5 = 2\pi R^{\mathrm{CF}}_{55}\sqrt{\frac{1+\kappa_5}{g\,\mathrm{GM_L}}}. \tag{4.88}$$

Formulae (4.87)–(4.88) show that the natural periods in roll and pitch mainly depend on $\mathrm{GM_T}$ and $\mathrm{GM_L}$, which again depend on the vessel load condition.

MATLAB

The heave, roll and pitch design periods for the MSS tanker are plotted together with formulae (4.76)–(4.77) and (4.78) in Figure 4.5 using the MSS toolbox commands:

```
load tanker;            % load WAMIT tanker data
loadcond(vessel);       % periods as a function of load condition
```

The roll and pitch periods as a function of $\mathrm{GM_T}$ and $\mathrm{GM_L}$ for a tanker are shown in Figure 4.5. It is seen that T_4 is reduced if $\mathrm{GM_T}$ is increased and vice versa. The same effect is observed for T_5 in pitch. The heave period T_3 is proportional to $\sqrt{T}$ as expected.

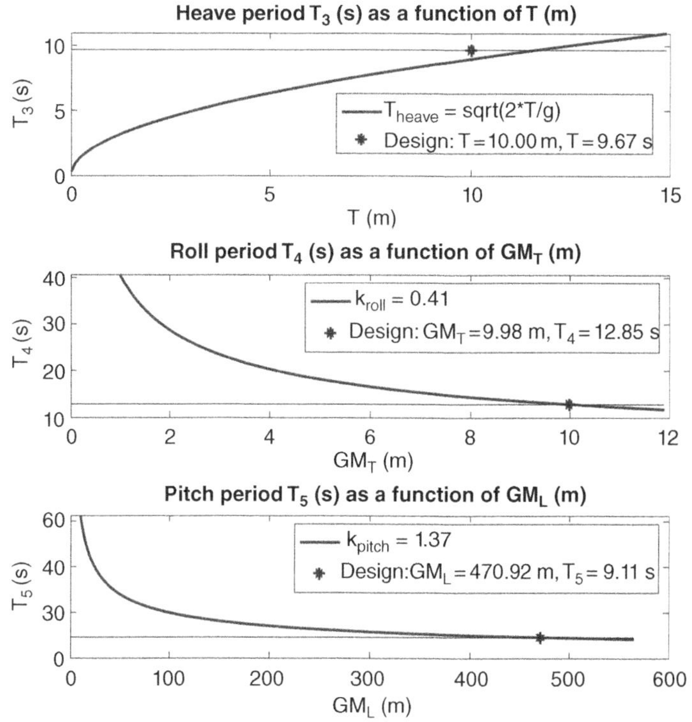

Figure 4.5 Heave, roll and pitch periods of a tanker for varying draft and metacenter heights using formulae (4.76)–(4.77) for roll and pitch. The heave period is approximated by formula (4.78). The design values (asterisks) are computed using the WAMIT data for the operational condition.

4.3.4 *Free-surface Effects*

Many ships are equipped with liquid tanks such as ballast and anti-roll tanks. A partially filled tank is known as a slack tank and in these tanks the liquid can move and endanger the ship's stability. The reduction of metacentric height caused by the liquids in slack tanks is known as the *free-surface effect*. The mass of the liquid and the location of the tanks have no role; it is only the moment of inertia of the surface that affects stability. The effective metacentric height corrected for slack tanks filled with sea water is (Brian 2003)

$$\mathrm{GM}_{\mathrm{T,eff}} = \mathrm{GM}_{\mathrm{T}} - \mathrm{FSC} \tag{4.89}$$

where the *free-surface correction* (FSC) is

$$\mathrm{FSC} = \sum_{r=1}^{N} \frac{\rho}{m} i_r \tag{4.90}$$

and i_r $(r = 1, 2, ..., N)$ are the moments of inertia of the water surfaces. For a rectangular tank with length l_{r} in the x_{b} direction and width b_{r} in the y_{b} direction, the moment of inertia of the surface about an axis through the centroid is

$$i_r = \frac{l_r b_r^3}{12}. \tag{4.91}$$

4.3.5 *Payload Effects*

A reduction in GM_T is observed if a payload with mass m_p is lifted up and suspended at the end of a rope of length h. Then the effective metacentric height is reduced according to

$$GM_{T,eff} = GM_T - h\frac{m_p}{m}. \tag{4.92}$$

Hence, it should be noted that a reduction in GM_T due to slack tanks or lift operations increases the roll period/passenger comfort to the cost of a less stable ship. These effects are also observed in pitch, but pitch is much less affected since $GM_L \gg GM_T$.

4.4 Seakeeping Analysis

In the design of ships and ocean structures, the wave-induced motions are of great importance to the assessment of the comfort and safety of the crew and the passengers. Seakeeping analyses should be performed to estimate seakeeping ability or seaworthiness, that is how well-suited a marine craft is to conditions when underway. In this context the watercraft's response to wave-induced forces is of great importance.

This section presents methods for computation of the heave, roll and pitch responses in regular waves. This also includes resonance analyses, that is the phenomenon that occurs when the frequency of a sinusoidal wave is equal or nearly equal to one of the natural frequencies of the watercraft. This causes the craft to heave, roll and pitch with larger amplitudes than it will experience when the wave-induced force is applied at other frequencies. Resonance can cause a marine craft to roll to very large angles in moderate sea states, leading to cargo damage, loss of containers and, in extreme cases, capsizing of the craft.

4.4.1 *Harmonic Oscillator with Sinusoidal Forcing*

The decoupled heave, roll and pitch motions of a marine craft can be approximated by a second-order mass–damper–spring system

$$m\ddot{x} + d\dot{x} + kx = F\sin(\omega t) \tag{4.93}$$

where F is the amplitude of the sinusoidal driving force and ω is the driving frequency. It is convenient to rewrite (4.93) as

$$\ddot{x} + 2\zeta\omega_n\dot{x} + \omega_n^2 x = \frac{F}{m}\sin(\omega t) \tag{4.94}$$

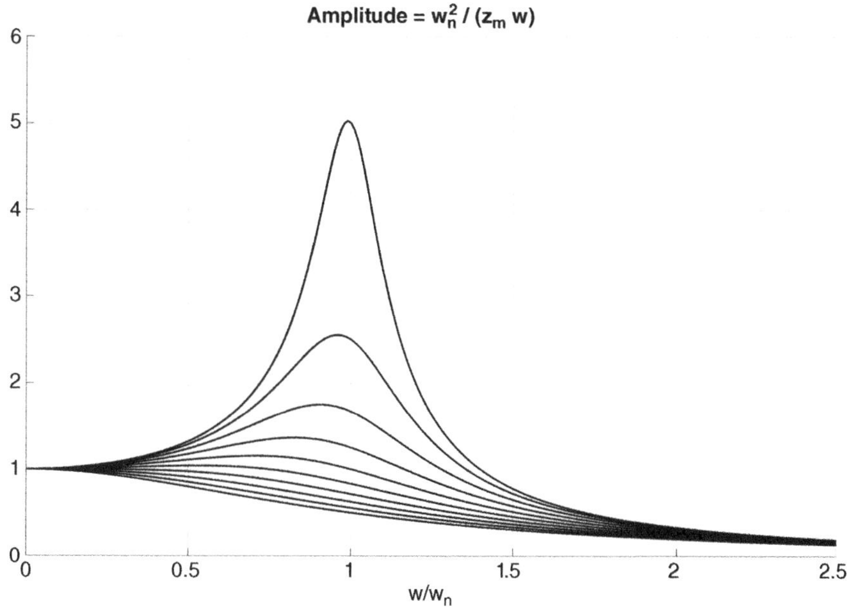

Figure 4.6 Steady-state variation of amplitude $\omega_n^2/(Z_m\omega)$ with relative frequency ω/ω_n for relative damping ratios $\zeta = 0.1, 0.2, ..., 1.0$. The maximum value occurs for the underdamped system $\zeta = 0.1$.

where ζ is the relative damping ratio and ω_n is the natural frequency given by

$$\zeta = \frac{d}{2}\sqrt{\frac{1}{mk}}, \qquad \omega_n = \sqrt{\frac{k}{m}}. \tag{4.95}$$

This is a linear system where the general solution is a sum of a transient solution that depends on the initial conditions, and a steady-state solution that is independent of initial conditions and depends only on the driving amplitude F. The transient solution is a short-lived burst of energy caused by a sudden change of the state. Hence, we only consider the steady-state solution

$$x = \frac{F}{mZ_m\omega}\sin(\omega t + \varepsilon) \tag{4.96}$$

with

$$Z_m = \sqrt{(2\zeta\omega_n)^2 + \frac{1}{\omega^2}(\omega_n^2 - \omega^2)^2} \tag{4.97}$$

$$\varepsilon = \text{atan}\left(\frac{2\zeta\omega_n\omega}{\omega^2 - \omega_n^2}\right). \tag{4.98}$$

Here Z_m is the absolute value of the *impedance* and ε is the *phase* of the oscillation relative to the driving force. Impedance is a measure of how much the craft resists motion when subjected to a sinusoidal force. It represents the ratio of the applied force F to the velocity $\dot{x}$.

Resonant Frequency

The solution (4.96) does not exist for $Z_m = 0$. This is when resonance occur. Solving $Z_m = 0$ from (4.97) for the driving frequency ω gives the resonant frequency

$$\omega_r = \omega_n \sqrt{1 - 2\zeta^2}. \tag{4.99}$$

Since the expression in the square root has to be non-negative, resonance only occurs when $\zeta < 1/\sqrt{2} \approx 0.707$. This corresponds to a significantly underdamped system. For marine craft both roll and pitch are underdamped. Figure 4.6 shows the steady-state variation of the amplitude $\omega_n^2/(Z_m\omega)$ corresponding to $F = m\omega_n^2$ for varying relative damping ratios ζ.

4.4.2 *Steady-state Heave, Roll and Pitch Responses in Regular Waves*

Again, consider the linear decoupled heave, roll and pitch equations (4.52)–(4.54) with harmonic forcing

$$(m + A_{33}^{\mathrm{CF}}(\omega_3))\ddot{z}^{\mathrm{n}} + (B_{33}^{\mathrm{CF}}(\omega_3) + B_{\mathrm{v},33}^{\mathrm{CF}}(\omega_3))\dot{z}^{\mathrm{n}} + C_{33}^{\mathrm{CF}} z^{\mathrm{n}} = F_3 \cos(\omega_{\mathrm{e}} t) \tag{4.100}$$

$$(I_x^{\mathrm{CF}} + A_{44}^{\mathrm{CF}}(\omega_4))\ddot{\phi} + (B_{44}^{\mathrm{CF}}(\omega_4) + B_{\mathrm{v},44}^{\mathrm{CF}}(\omega_4))\dot{\phi} + C_{44}^{\mathrm{CF}} \phi = F_4 \cos(\omega_{\mathrm{e}} t) \tag{4.101}$$

$$(I_y^{\mathrm{CF}} + A_{55}^{\mathrm{CF}}(\omega_5))\ddot{\theta} + (B_{55}^{\mathrm{CF}}(\omega_5) + B_{\mathrm{v},55}^{\mathrm{CF}}(\omega_5))\dot{\theta} + C_{55}^{\mathrm{CF}} \theta = F_5 \sin(\omega_{\mathrm{e}} t) \tag{4.102}$$

where the phase shift of pitch is 90° relative heave and roll. The frequency of encounter is

$$\omega_{\mathrm{e}} = \omega - kU\cos(\beta) \tag{4.103}$$

where β is the encounter angle, that is the angle between the heading angle ψ and the direction γ_{w} of the wave. This implies that head sea corresponds to 180°, see Figure 10.14. The wave frequency ω satisfies the dispersion relation

$$\omega^2 = kg\tanh(kd) \tag{4.104}$$

where k is the wave number and d is the water depth. For large water depths $\omega^2 \approx kg$.

The natural frequencies and relative damping factors are implicit equations which can be derived from (4.95). This gives

$$\omega_3 = \sqrt{\frac{C_{33}^{\mathrm{CF}}}{m + A_{33}^{\mathrm{CF}}(\omega_3)}}, \quad \zeta_3 = \frac{B_{33}^{\mathrm{CF}}(\omega_3) + B_{\mathrm{v},33}^{\mathrm{CF}}(\omega_3)}{2} \sqrt{\frac{1}{(m + A_{33}^{\mathrm{CF}}(\omega_3))\, C_{33}^{\mathrm{CF}}}}$$

$$\omega_4 = \sqrt{\frac{C_{44}^{\mathrm{CF}}}{I_x^{\mathrm{CF}} + A_{44}^{\mathrm{CF}}(\omega_4)}}, \quad \zeta_4 = \frac{B_{44}^{\mathrm{CF}}(\omega_4) + B_{\mathrm{v},44}^{\mathrm{CF}}(\omega_4)}{2} \sqrt{\frac{1}{(I_x^{\mathrm{CF}} + A_{44}^{\mathrm{CF}}(\omega_4))\, C_{44}^{\mathrm{CF}}}}$$

$$\omega_5 = \sqrt{\frac{C_{55}^{\mathrm{CF}}}{I_y^{\mathrm{CF}} + A_{55}^{\mathrm{CF}}(\omega_5)}}, \quad \zeta_5 = \frac{B_{55}^{\mathrm{CF}}(\omega_5) + B_{\mathrm{v},55}^{\mathrm{CF}}(\omega_5)}{2} \sqrt{\frac{1}{(I_y^{\mathrm{CF}} + A_{55}^{\mathrm{CF}}(\omega_5))\, C_{55}^{\mathrm{CF}}}}.$$

The steady-state solutions of (4.52)–(4.54) are obtained from (4.96), and the results are

$$z^{\mathrm{n}} = \frac{F_3}{(m + A_{33}^{\mathrm{CF}}(\omega_3)) Z_{m,3}\, \omega_{\mathrm{e}}} \cos(\omega_{\mathrm{e}} t + \varepsilon_3) \tag{4.105}$$

$$\phi = \frac{F_4}{(I_x^{\mathrm{CF}} + A_{44}^{\mathrm{CF}}(\omega_4)) Z_{m,4}\, \omega_{\mathrm{e}}} \cos(\omega_{\mathrm{e}} t + \varepsilon_4) \tag{4.106}$$

$$\theta = \frac{F_5}{(I_y^{\mathrm{CF}} + A_{55}^{\mathrm{CF}}(\omega_5)) Z_{m,5}\, \omega_{\mathrm{e}}} \sin(\omega_{\mathrm{e}} t + \varepsilon_5) \tag{4.107}$$

where the impedances and phase shifts are

$$Z_{m,3} = \sqrt{(2\zeta_3\omega_3)^2 + \frac{1}{\omega_{\mathrm{e}}^2}(\omega_3^2 - \omega_{\mathrm{e}}^2)^2}, \qquad \varepsilon_3 = \operatorname{atan}\left(\frac{2\zeta_3\omega_3\omega_{\mathrm{e}}}{\omega_{\mathrm{e}}^2 - \omega_3^2}\right) \tag{4.108}$$

$$Z_{m,4} = \sqrt{(2\zeta_4\omega_4)^2 + \frac{1}{\omega_{\mathrm{e}}^2}(\omega_4^2 - \omega_{\mathrm{e}}^2)^2}, \qquad \varepsilon_4 = \operatorname{atan}\left(\frac{2\zeta_4\omega_4\omega_{\mathrm{e}}}{\omega_{\mathrm{e}}^2 - \omega_4^2}\right) \tag{4.109}$$

$$Z_{m,5} = \sqrt{(2\zeta_5\omega_5)^2 + \frac{1}{\omega_{\mathrm{e}}^2}(\omega_5^2 - \omega_{\mathrm{e}}^2)^2}, \qquad \varepsilon_5 = \operatorname{atan}\left(\frac{2\zeta_5\omega_5\omega_{\mathrm{e}}}{\omega_{\mathrm{e}}^2 - \omega_5^2}\right). \tag{4.110}$$

Data from Hydrodynamic Codes

The heave, roll and pitch responses with respect to the CF can be computed using data from hydrodynamic codes. The necessary hydrodynamic terms are the added mass coefficients $A_{ii}^{\mathrm{CF}}(\omega_i)$, potential damping coefficients $B_{ii}^{\mathrm{CF}}(\omega_i)$, viscous damping coefficients

$B^{\mathrm{CF}}_{\mathrm{v},ii}(\omega_i)$ and restoring coefficients C^{CF}_{ii} $(i = 3, 4, 5)$; see Chapter 5. Hydrodynamic codes also compute the wave loads for varying frequencies and encounter angles.

4.4.3 Explicit Formulae for Boxed-shaped Vessels in Regular Waves

A semi-analytical approach can be used to derive frequency response functions for the wave-induced motions of monohull ships.

Jensen *et al.* (2004) have derived closed-form expressions for the heave, roll and pitch responses in regular waves where the required input information for the method is restricted to the main dimensions: length L, breadth B, draught T, block coefficient $C_b = \nabla/(LBT)$ and waterplane area A_w together with speed U and wave encounter angle β. The formulas make it simple to obtain quick estimates of the wave-induced motions and accelerations in the conceptual design phase and to perform a sensitivity study of the variation with main dimensions and operational profile. The method is based on linear strip theory where the coupling terms between heave and pitch are neglected and by assuming a constant sectional added mass equal to the displaced water. Faltinsen (1990, pp. 89–90) presents a similar approach based on strip theory in which explicit formulae for the expressions F_3 and F_5 in (4.105)–(4.107) are derived.

Steady-state Heave and Pitch Responses

Consider the two harmonic oscillators (Jensen *et al.* 2004)

$$\ddot{z}^n + 2\zeta\omega_n\dot{z}^n + \omega_n^2 z^n = \omega_n^2\zeta_a F\cos(\omega_e t) \tag{4.111}$$

$$\ddot{\theta} + 2\zeta\omega_n\dot{\theta} + \omega_n^2\theta = \omega_n^2\zeta_a G\sin(\omega_e t) \tag{4.112}$$

where the wave amplitude ζ_a is the driving term. The oscillators have common relative damping ratio and natural frequency

$$\zeta = \frac{A^2}{B\alpha^3\sqrt{8k^3T}}, \qquad \omega_n = \sqrt{\frac{g}{2T}} \tag{4.113}$$

where A is the sectional hydrodynamic damping

$$A = 2\sin\left(\frac{1}{2}kB\alpha^2\right)\exp(-kT\alpha^2) \tag{4.114}$$

and $\alpha = \omega_e/\omega = 1 - \sqrt{kg}\,U\cos(\beta)$ is the ratio between the frequency of encounter ω_e and the wave frequency ω. Let k_e denote the effective wave number

$$k_e = k|\cos(\beta)| \tag{4.115}$$

and

$$f = \sqrt{(1 - kT)^2 + \left(\frac{A^2}{kB\alpha^3}\right)^2}. \tag{4.116}$$

Then the forcing functions F and G can be expressed as

$$F = \kappa f \frac{\sin(\sigma)}{\sigma} \tag{4.117}$$

$$G = \kappa f \frac{6}{L\sigma}\left(\frac{\sin(\sigma)}{\sigma} - \cos(\sigma)\right) \tag{4.118}$$

where $\sigma = k_e L/2$ and $\kappa = e^{-k_e T}$. The steady-state heave and pitch responses become

$$z^n = \zeta_a \frac{w_n^2}{Z_m w_e} F \cos(w_e t + \varepsilon) \tag{4.119}$$

$$\theta = \zeta_a \frac{w_n^2}{Z_m w_e} G \sin(w_e t + \varepsilon) \tag{4.120}$$

where Z_m and ε are given by (4.97)–(4.98). The heave and pitch responses (4.119)–(4.120) can be plotted in Matlab by using MSS toolbox script `waveresponse345.m`.

Steady-state Roll Response

The roll motion is assumed to be decoupled from the other transverse motions such that the equation of motion for roll in regular waves with unity wave amplitude is (Jensen *et al.* 2004)

$$\left(\frac{T_4}{2\pi}\right)^2 C_{44}\ddot{\phi} + B_{44}\dot{\phi} + C_{44}\phi = M\cos(\omega_e t) \tag{4.121}$$

where

$$C_{44} = \rho\nabla \mathrm{GM_T}. \tag{4.122}$$

The natural period T_4 replaces the mass moment of inertia and the added mass in the equation of motion. Consequently, the variation in the added mass with frequency is neglected. Jensen *et al.* (2004) assumes that the ship can be described as two prismatic beams with the same draft, but different breadths and cross-sectional areas. Under these assumptions a formulae for the 2D sectional inviscid hydrodynamic damping coefficient is derived. An even simpler approach could be to specify the relative damping ratio ζ_4 such that

$$B_{44} = 2\zeta_4 \left(\frac{T_4}{2\pi}\right) C_{44}. \tag{4.123}$$

An estimate of M for varying encounter angles β will then be

$$M = \sqrt{\frac{\rho g^2}{\omega_e} B_{44}} \sin(\beta). \tag{4.124}$$

Finally, the steady-state roll response becomes

$$\phi = \frac{w_4^2}{\rho \nabla \mathrm{GM_T}\, Z_{m,4} w_e} M \cos(w_e t + \varepsilon_4) \tag{4.125}$$

where $Z_{m,4}$ and ε_4 are given by (4.109). The roll response (4.125) can be plotted in Matlab by using the MSS script `waveresponse345.m`.

4.4.4 Case Study: Resonances in the Heave, Roll and Pitch Modes

We will use the MSS supply vessel in the resonance analysis. The wave excitation amplitudes are chosen as

$$F_3 = (m + A_{33}^{\mathrm{CF}}(\omega_3))\omega_3^2 \tag{4.126}$$

$$F_4 = (I_x^{\mathrm{CF}} + A_{44}^{\mathrm{CF}}(\omega_4))\omega_4^2 \tag{4.127}$$

$$F_5 = (I_y^{\mathrm{CF}} + A_{55}^{\mathrm{CF}}(\omega_5))\omega_5^2 \tag{4.128}$$

such that the steady-state amplitudes of (4.105)–(4.107) become $\omega_i^2/(Z_{m,i}\,\omega_e)$ for $i = 3, 4, 5$.

Figure 4.7 shows the amplitudes in heave, roll and pitch for variation in relative frequencies ω_e/ω_i for $i = 3, 4, 5$. The numerical example is included in the MSS Matlab script `ExResonance.m`. As expected, roll is the critical DOF for which the amplitude of the roll angle ϕ is significantly amplified at the resonant frequency $\omega_4 = 0.58$ rad s^{-1} for incoming sinusoidal waves $F_4 \cos(\omega_e t)$ when $\omega_e = \omega_4$. This corresponds to regular waves with a period $T_4 = 10.9$ s, which are likely to happen. The relative damping ratio $\zeta_4 = 0.05$ in roll for the MSS supply vessel is quite low, while $\mathrm{GM_T} = 2.14$ m gives sufficient stability. Hydrodynamic codes usually compute the potential damping terms B_{ii}^{CF} and leaves to the user to add viscous damping $B_{\mathrm{v},ii}^{\mathrm{CF}}$. Hence, the data set should be calibrated by increasing the viscous damping term $B_{\mathrm{v},44}^{\mathrm{CF}}$ if roll is to little damped (Ikeda *et al.* 1976). The positive effect of increasing ζ_4 from 0.05 to 0.10 is clearly seen in the plot. From a physical point of view, roll damping can be increased by adding bilge keels or fins.

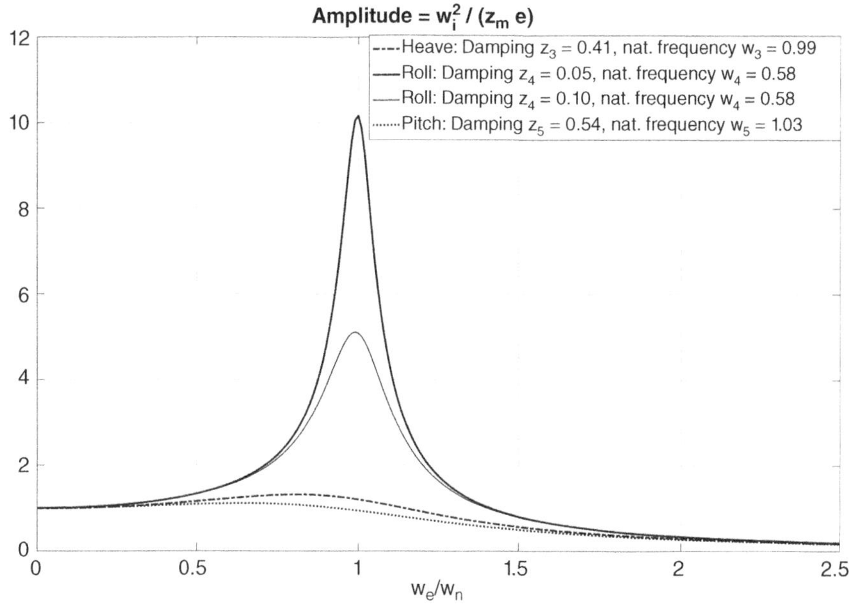

Figure 4.7 Steady-state variation of amplitudes $\omega_i^2/(Z_{m,i}\,\omega_e)$ with relative frequency ω_e/ω_i $(i = 3, 4, 5)$ corresponding to heave, roll and pitch. The numerical analysis is based on the MSS supply vessel. The vessel is underdamped in roll and this gives rise to significant magnification of roll amplitudes when $\omega_e/\omega_4 = 1.0$.

Figure 4.7 shows that the heave and pitch motions are well damped for the supply vessel. The relative damping ratios are $\zeta_3 = 0.41$ and $\zeta_5 = 0.54$. This gives a moderate amplification of the heave and pitch amplitudes.

4.5 Ballast Systems

In addition to the metacentric restoring forces $\boldsymbol{g}(\boldsymbol{\eta})$ described in Section 4.1, the equilibrium point can be changed by pretrimming, for instance by pumping water between the ballast tanks of the vessel. The vessel can only be trimmed in *heave, pitch* and *roll* where restoring forces are present.

Let the equilibrium point be

$$z^n = z^n_{\rm d}, \quad \phi = \phi_{\rm d} \quad \text{and} \quad \theta = \theta_{\rm d}$$

where $z^n_{\rm d}$, $\phi_{\rm d}$ and $\theta_{\rm d}$ are the desired states. The equilibrium states corresponding to these values are found by considering the steady-state solution of

$$\boldsymbol{M}\dot{\boldsymbol{v}} + \boldsymbol{C}(\boldsymbol{v})\boldsymbol{v} + \boldsymbol{D}(\boldsymbol{v})\boldsymbol{v} + \boldsymbol{g}(\boldsymbol{\eta}) + \boldsymbol{g}_0 = \boldsymbol{\tau} + \underbrace{\boldsymbol{\tau}_{\rm wind} + \boldsymbol{\tau}_{\rm wave}}_{\boldsymbol{w}} \tag{4.129}$$

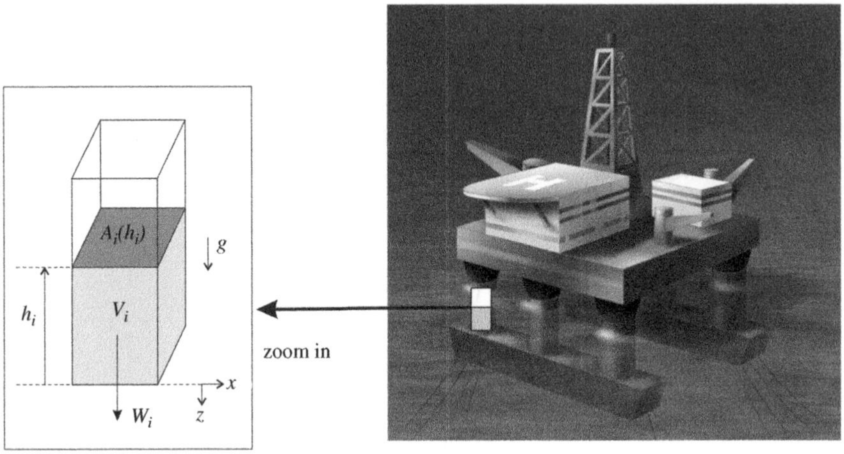

Figure 4.8 Semi-submersible ballast tanks. Source: Illustration by B. Stenberg.

which under assumption of zero acceleration/velocity ($\dot{\boldsymbol{\nu}} = \boldsymbol{\nu} = \boldsymbol{0}$) and no control forces ($\boldsymbol{\tau} = \boldsymbol{0}$) reduces to

$$\boldsymbol{g}(\boldsymbol{\eta}_{\mathrm{d}}) + \boldsymbol{g}_0 = \boldsymbol{w} \tag{4.130}$$

where $\boldsymbol{\eta}_{\mathrm{d}} = [-, -, z_{\mathrm{d}}^n, \phi_{\mathrm{d}}, \theta_{\mathrm{d}}, -]^\top$; that is only three states are used for pretrimming.

The forces and moments $\boldsymbol{g}_0$ due to the ballast tanks are computed using hydrostatic analyses. Consider a marine craft with n ballast tanks of volumes $V_i \leq V_{i,\max}$ for $i = 1, \ldots, n$. For each ballast tank the water volume is

$$V_i(h_i) = \int_0^{h_i} A_i(h)\mathrm{d}h \approx A_i h_i, \qquad A_i(h) = \text{constant} \tag{4.131}$$

where $A_i(h)$ is the area of the ballast tank at height h. Hence, the volume of the water column in each ballast tank can be computed by measuring the water heights h_i. Next, assume that the ballast tanks are located at

$$\boldsymbol{r}_{\mathrm{bi}}^b = [x_i, y_i, z_i]^\top \quad (i = 1, \ldots, n) \tag{4.132}$$

where $\boldsymbol{r}_{\mathrm{bi}}^b$ is the vector from the CO to the geometric center of tank i expressed in $\{b\}$.

The gravitational forces W_i in heave are summed up according to (see Figure 4.8)

$$\begin{aligned} Z_{\mathrm{ballast}} &= \sum_{i=1}^{n} W_i \\ &= \rho g \sum_{i=1}^{n} V_i \end{aligned} \tag{4.133}$$

The moments due to the ballast heave force $\rho g V_i$ are then found from

$$\boldsymbol{m}_i^b = \boldsymbol{r}_{\mathrm{bi}}^b \times \boldsymbol{f}_i^b$$

$$
= \begin{bmatrix} x_i \\ y_i \\ z_i \end{bmatrix} \times \begin{bmatrix} 0 \\ 0 \\ \rho g V_i \end{bmatrix}
$$

$$
= \begin{bmatrix} y_i \rho g V_i \\ -x_i \rho g V_i \\ 0 \end{bmatrix} \tag{4.134}
$$

implying that the roll and pitch moments due to ballast are

$$
K_{\text{ballast}} = \rho g \sum_{i=1}^{n} y_i V_i \tag{4.135}
$$

$$
M_{\text{ballast}} = -\rho g \sum_{i=1}^{n} x_i V_i. \tag{4.136}
$$

The resulting ballast force and moments are

$$
g_0 = \begin{bmatrix} 0 \\ 0 \\ -Z_{\text{ballast}} \\ -K_{\text{ballast}} \\ -M_{\text{ballast}} \\ 0 \end{bmatrix} = \rho g \begin{bmatrix} 0 \\ 0 \\ -\sum_{i=1}^{n} V_i \\ -\sum_{i=1}^{n} y_i V_i \\ \sum_{i=1}^{n} x_i V_i \\ 0 \end{bmatrix}. \tag{4.137}
$$

Metacentric Height Correction

Since ballast tanks are partially filled tanks of liquids, the restoring roll moment will be affected. The formulae for the free-surface correction (4.89)–(4.90) can, however, be applied to correct the transverse metacentric height GM_T in roll.

4.5.1 *Static Conditions for Trim and Heel*

Distribution of water between the ballast tanks can be done manually by pumping water until the desired water levels h_i in each tank are reached or automatically by using feedback control. For manual operation, the steady-state relationships between water levels h_i and the desired pretrimming values z_d^n, ϕ_d and θ_d are needed. Trimming is usually done under the assumptions that ϕ_d and θ_d are small such that linear theory can be applied

$$
\boldsymbol{g}(\boldsymbol{\eta}_d) \approx \boldsymbol{G}\boldsymbol{\eta}_d. \tag{4.138}
$$

Since we are only concerned with the heave, roll and pitch modes it is convenient to use the 3-DOF reduced-order system

$$
\boldsymbol{G}^{\{3,4,5\}} = \begin{bmatrix} -Z_z & 0 & -Z_\theta \\ 0 & -K_\phi & 0 \\ -M_z & 0 & -M_\theta \end{bmatrix}
$$

$$g_0^{\{3,4,5\}} = \rho g \begin{bmatrix} -\sum_{i=1}^n V_i \\ -\sum_{i=1}^n y_i V_i \\ \sum_{i=1}^n x_i V_i \end{bmatrix}$$

$$\eta_d^{\{3,4,5\}} = [z_d^n,\ \phi_d,\ \theta_d]^\top$$

$$w^{\{3,4,5\}} = [w_3,\ w_4,\ w_5]^\top.$$

The key assumption for open-loop pretrimming is that $w^{\{3,4,5\}} = [w_3, w_4, w_5]^\top = \mathbf{0}$, that is no disturbances in heave, roll and pitch. From (4.130) and (4.32) it follows that

$$G^{\{3,4,5\}}\eta_d^{\{3,4,5\}} + g_0^{\{3,4,5\}} = \mathbf{0} \tag{4.139}$$

$$\Updownarrow$$

$$\begin{bmatrix} -Z_z & 0 & -Z_\theta \\ 0 & -K_\phi & 0 \\ -M_z & 0 & -M_\theta \end{bmatrix} \begin{bmatrix} z_d^n \\ \phi_d \\ \theta_d \end{bmatrix} + \rho g \begin{bmatrix} -\sum_{i=1}^n V_i \\ -\sum_{i=1}^n y_i V_i \\ \sum_{i=1}^n x_i V_i \end{bmatrix} = \mathbf{0}.$$

This can be rewritten as

$$Hv = y \tag{4.140}$$

$$\Updownarrow$$

$$\rho g \begin{bmatrix} 1 & \cdots & 1 & 1 \\ y_1 & \cdots & y_{n-1} & y_n \\ -x_1 & \cdots & -x_{n-1} & -x_n \end{bmatrix} \begin{bmatrix} V_1 \\ V_2 \\ \vdots \\ V_n \end{bmatrix} = \begin{bmatrix} -Z_z z_d^n - Z_\theta \theta_d \\ -K_\phi \phi_d \\ -M_z z_d^n - M_\theta \theta_d \end{bmatrix} \tag{4.141}$$

where v is a vector of tank volumes:

$$v = [V_1,\ V_2,\ \dots,\ V_n]^\top \tag{4.142}$$

The tank volumes are computed from (4.140) by using the *Moore–Penrose pseudoinverse*

$$\begin{aligned} v &= H^\dagger y \\ &= H^\top (HH^\top)^{-1} y \end{aligned} \tag{4.143}$$

where it is assumed that $n \geq 3$ and that $HH^\top$ has full rank. Finally, the desired water heights can be computed from

$$V_i(h_i) = \int_o^{h_i} A_i(h)\mathrm{d}h \tag{4.144}$$

$$\Downarrow \quad (A_i(h) = A_i)$$

$$h_i = \frac{V_i}{A_i}. \tag{4.145}$$

Example 4.3 (Semi-submersible Ballast Control)

Consider the semi-submersible shown in Figure 4.9 with four ballast tanks located at $\boldsymbol{r}^b_{b1} = [-x, -y]^\top, \boldsymbol{r}^b_{b2} = [x, -y]^\top, \boldsymbol{r}^b_{b3} = [x, y]^\top$ *and* $\boldsymbol{r}^b_{b4} = [-x, y]^\top$. *In addition, yz symmetry implies that* $Z_\theta = M_z = 0$ *while the diagonal elements in* $\boldsymbol{G}^{\{3,4,5\}}$ *are non-zero. Consequently,*

$$\boldsymbol{H} = \rho g \begin{bmatrix} 1 & 1 & 1 & 1 \\ -y & -y & y & y \\ x & -x & -x & x \end{bmatrix}$$

$$\boldsymbol{y} = \begin{bmatrix} -Z_z z^n_{\mathrm{d}} \\ -K_\phi \phi_{\mathrm{d}} \\ -M_\theta \theta_{\mathrm{d}} \end{bmatrix} = \rho g \begin{bmatrix} A_{\mathrm{wp}} z^n_{\mathrm{d}} \\ \nabla \mathrm{GM_T} \phi_{\mathrm{d}} \\ \nabla \mathrm{GM_L} \theta_{\mathrm{d}} \end{bmatrix}.$$

*The right pseudoinverse of **H** is*

$$\boldsymbol{H}^\dagger = \boldsymbol{H}^\top (\boldsymbol{H}\boldsymbol{H}^\top)^{-1} = \frac{1}{4\rho g} \begin{bmatrix} 1 & -\frac{1}{y} & \frac{1}{x} \\ 1 & -\frac{1}{y} & -\frac{1}{x} \\ 1 & \frac{1}{y} & -\frac{1}{x} \\ 1 & \frac{1}{y} & \frac{1}{x} \end{bmatrix}$$

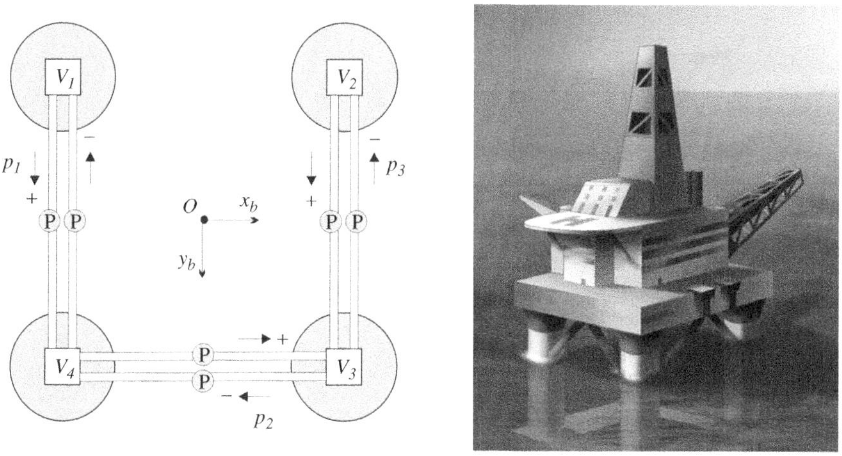

Figure 4.9 Semi-submersible with four ballast tanks. V_i is the water volume in the leg, $i = 1, \ldots, 4$, and p_j is the volume flow for the water pump, $j = 1, \ldots, 3$. Source: Illustration by B. Stenberg.

which finally gives the water volumes V_i corresponding to the desired values z^n_d, ϕ_d and θ_d

$$\boldsymbol{v} = \begin{bmatrix} V_1 \\ V_2 \\ V_3 \\ V_4 \end{bmatrix} = \frac{1}{4\rho g} \begin{bmatrix} 1 & -\frac{1}{y} & \frac{1}{x} \\ 1 & -\frac{1}{y} & -\frac{1}{x} \\ 1 & \frac{1}{y} & -\frac{1}{x} \\ 1 & \frac{1}{y} & \frac{1}{x} \end{bmatrix} \begin{bmatrix} \rho g A_{\mathrm{wp}} z^n_{\mathrm{d}} \\ \rho g \nabla \mathrm{GM_T} \phi_{\mathrm{d}} \\ \rho g \nabla \mathrm{GM_L} \theta_{\mathrm{d}} \end{bmatrix}$$

.

4.5.2 Automatic Ballast Control Systems

In the manual pretrimming case it was assumed that $\boldsymbol{w}^{\{3,4,5\}} = \mathbf{0}$. This assumption can be removed by using feedback from z^n, ϕ and θ. The closed-loop dynamics of a PID-controlled water pump can be described by a first-order model with amplitude saturation

$$T_j \dot{p}_j + p_j = \mathrm{sat}(p_{\mathrm{d}_j}) \tag{4.146}$$

where T_j is a positive time constant, p_j is the volumetric flow rate m^3 s^{-1} produced by the pump, $j = 1, \ldots, m$, and p_{d_j} is the pump setpoint. As shown in Figure 4.9, one separate water pump can be used to pump water in each direction. This implies that the water pump capacity can be different for positive and negative flow directions. In other words

$$\mathrm{sat}(p_{\mathrm{d}_j}) = \begin{cases} p^+_{j,\max} & p_j > p^+_{j,\max} \\ p_{\mathrm{d}_j} & p^-_{j,\max} \le p_{\mathrm{d}_j} \le p^+_{j,\max} \\ p^-_{j,\max} & p_{\mathrm{d}_j} < p^-_{j,\max} \end{cases}. \tag{4.147}$$

The pump time constant T_j is found from a step response, as shown in Figure 4.10. The volume flow $\dot{V}_i$ to tank i is given by linear combinations of flows corresponding to the pumps/pipelines supporting tank i. For the semi-submersible shown in Figure 4.9, we obtain

$$\dot{V}_1 = -p_1 \tag{4.148}$$

$$\dot{V}_2 = -p_3 \tag{4.149}$$

$$\dot{V}_3 = p_2 + p_3 \tag{4.150}$$

$$\dot{V}_4 = p_1 - p_2. \tag{4.151}$$

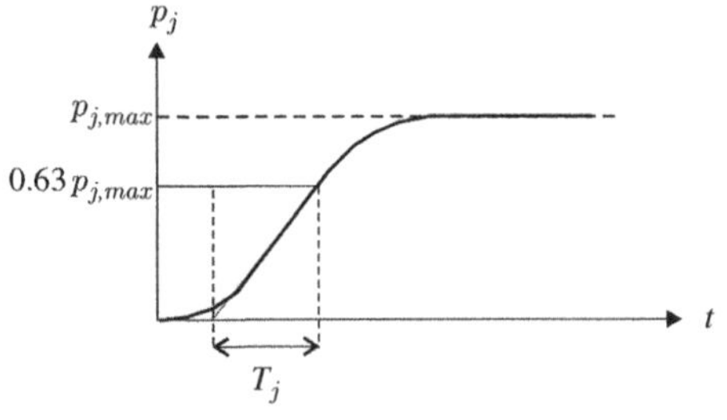

Figure 4.10 The time constant T_j for pump j is found by commanding a step $p_{\mathrm{d}_j} = p_{j,\max}$ as shown in the plot.

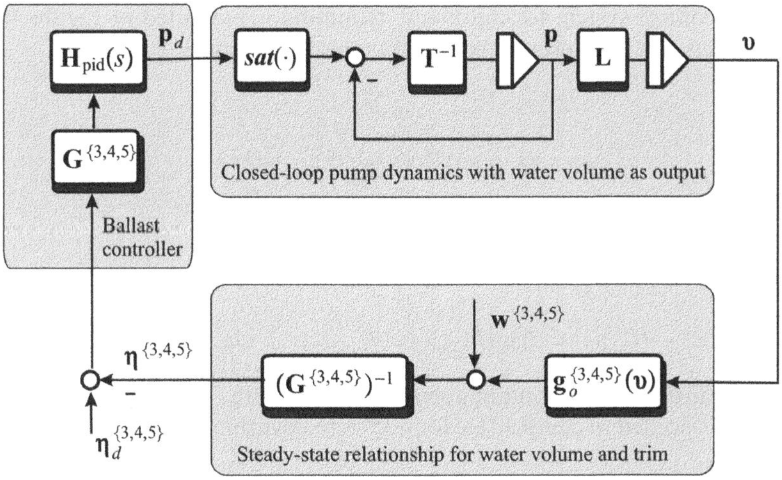

Figure 4.11 Ballast control system using feedback from z^n, ϕ and θ.

More generally, the water flow model can be written

$$\boldsymbol{T}\dot{\boldsymbol{p}} + \boldsymbol{p} = \mathrm{sat}(\boldsymbol{p}_\mathrm{d}) \tag{4.152}$$

$$\dot{\boldsymbol{\upsilon}} = \boldsymbol{L}\boldsymbol{p} \tag{4.153}$$

where $\mathrm{sat}(\boldsymbol{p}_\mathrm{d}) = [\mathrm{sat}(p_{\mathrm{d}_1}), \ldots, \mathrm{sat}(p_{\mathrm{d}_m})]^\top$, $\boldsymbol{p} = [p_1, \ldots, p_m]^\top$ and $\boldsymbol{\upsilon} = [V_1, \ldots, V_n]^\top$ $(m \geq n)$. The mapping from the water volume vector $\boldsymbol{\upsilon}$ to $\boldsymbol{\eta}^{\{3,4,5\}}$ is given by the steady-state condition (see Figure 4.11)

$$\boldsymbol{G}^{\{3,4,5\}}\boldsymbol{\eta}^{\{3,4,5\}} = \boldsymbol{g}_0^{\{3,4,5\}}(\boldsymbol{\upsilon}) + \boldsymbol{w}^{\{3,4,5\}} \tag{4.154}$$

Example 4.4 (Semi-submersible Ballast Control, Continued)
Consider the semi-submersible in Example 4.3. The water flow model corresponding to Figure 4.9 becomes

$$\boldsymbol{\upsilon} = \begin{bmatrix} V_1 \\ V_2 \\ V_3 \\ V_4 \end{bmatrix}, \quad \boldsymbol{p} = \begin{bmatrix} p_1 \\ p_2 \\ p_3 \end{bmatrix}, \quad \boldsymbol{L} = \begin{bmatrix} -1 & 0 & 0 \\ 0 & 0 & -1 \\ 0 & 1 & 1 \\ 1 & -1 & 0 \end{bmatrix} \tag{4.155}$$

reflecting that there are three pumps and four water volumes connected through the configuration matrix ***L***.

A feedback control system for automatic trimming is presented in Figure 4.11. The ballast controllers can be chosen to be of PID type, for instance

$$\boldsymbol{p}_{\mathrm{d}} = \boldsymbol{H}_{\mathrm{pid}}(s)\boldsymbol{G}^{\{3,4,5\}}(\boldsymbol{\eta}_{\mathrm{d}}^{\{3,4,5\}} - \boldsymbol{\eta}^{\{3,4,5\}}) \tag{4.156}$$

where $\boldsymbol{\eta}_{\mathrm{d}}^{\{3,4,5\}} = [z_{\mathrm{d}}^{n}, \phi_{\mathrm{d}}, \theta_{\mathrm{d}}]^{\top}$ and

$$\boldsymbol{H}_{\mathrm{pid}}(s) = \mathrm{diag}\{h_{1,\mathrm{pid}}(s), h_{2,\mathrm{pid}}(s), \ldots, h_{m,\mathrm{pid}}(s)\} \tag{4.157}$$

is a diagonal transfer matrix containing m PID controllers. Note that integral action in the controllers is needed to compensate for non-zero environmental disturbances $\boldsymbol{w}^{\{3,4,5\}}$.

5

Seakeeping Models

The study of ship dynamics has traditionally been covered by two main theories: *maneuvering* and *seakeeping*. Maneuvering refers to the study of ship motion in the absence of wave excitation (calm water). The maneuvering equations of motion are derived in Chapter 6 under the assumption that the hydrodynamic potential coefficients and radiation-induced forces are frequency independent. Seakeeping, on the other hand, refers to the study of motion of marine craft on constant course and speed when there is wave excitation. This includes the trivial case of zero speed. In seakeeping analysis, a dissipative force known as *fluid memory effects* (Cummins 1962) is introduced. Although both areas are concerned with the same issues, study of motion, stability and control, the separation allows us to make different assumptions that simplify the study in each case. A chief distinguishing characteristic of these theories is the use of different coordinates and reference systems to express the equations of motion.

In maneuvering theory, the equations of motion are described relative to $\{b\}$, which is fixed to the marine craft, whereas in seakeeping the motion is described relative to a coordinate system $\{s\}$ fixed to an equilibrium virtual craft that moves at a constant speed and heading corresponding to the average motion of the actual craft. Most hydrodynamic programs compute radiation and wave excitation forces in $\{s\}$.

This chapter presents the seakeeping theory and the classical equation in naval architecture

$$(\boldsymbol{M}_{\rm RB} + \boldsymbol{A}(\omega))\ddot{\boldsymbol{\xi}} + (\boldsymbol{B}(\omega) + \boldsymbol{B}_{\rm V}(\omega))\dot{\boldsymbol{\xi}} + \boldsymbol{C}\boldsymbol{\xi} = \delta\boldsymbol{\tau} + \boldsymbol{\tau}_{\rm wind} + \boldsymbol{\tau}_{\rm wave} \tag{5.1}$$

which is transformed from *equilibrium axes* $\{s\}$ to body-fixed axes $\{b\}$ using the time-domain solution known as the *Cummins equation*. The radiation-induced forces and moment are represented as impulse response functions and state-space models. This is done within a linear framework so viscous damping can be added under the assumption

Handbook of Marine Craft Hydrodynamics and Motion Control, Second Edition. Thor I. Fossen.

of linear superposition. The main results are the $\{b\}$-frame seakeeping equations of motion in the following form

$$\dot{\boldsymbol{\eta}} = \boldsymbol{J}_{\Theta}(\boldsymbol{\eta})\boldsymbol{\nu} \tag{5.2}$$

$$\boldsymbol{M}_{\text{RB}}\dot{\boldsymbol{\nu}} + \boldsymbol{C}_{\text{RB}}^{*}\boldsymbol{\nu} + \boldsymbol{M}_{\text{A}}\dot{\boldsymbol{\nu}}_{\text{r}} + \boldsymbol{C}_{\text{A}}^{*}\boldsymbol{\nu}_{\text{r}} + \boldsymbol{D}\boldsymbol{\nu}_{\text{r}} + \boldsymbol{\mu}_{\text{r}} + \boldsymbol{G}\boldsymbol{\eta} = \boldsymbol{\tau} + \boldsymbol{\tau}_{\text{wind}} + \boldsymbol{\tau}_{\text{wave}} \tag{5.3}$$

where $\boldsymbol{\mu}_{\text{r}}$ is an additional term representing the fluid memory effects. This model is valid in the body-fixed reference frame and describes a maneuvering ship in a seaway. When designing model-based control systems or simulating marine craft motions it is important to have good estimates of the inertia, damping and restoring coefficients. In Chapter 3, formulae for computation of the rigid-body matrices $\boldsymbol{M}_{\text{RB}}$ and $\boldsymbol{C}_{\text{RB}}^{*}$ were given while the restoring and ballast forces, $\boldsymbol{G}\boldsymbol{\eta} + \boldsymbol{g}_0$, were derived in Chapter 4. In this chapter, we will derive formulae for hydrodynamic added mass $\boldsymbol{M}_{\text{A}}$, linear Coriolis–centripetal forces $\boldsymbol{C}_{\text{A}}^{*}$ due to the rotation of the seakeeping reference frame $\{s\}$ about $\{n\}$ and linear potential damping $\boldsymbol{D}_{\text{P}}$. Linear viscous damping $\boldsymbol{D}_{\text{V}}$ will be added manually to obtain a more accurate model.

The terms in (5.3) can be grouped according to

$$\begin{array}{ll}
\text{Inertia forces:} & \boldsymbol{M}_{\text{RB}}\dot{\boldsymbol{\nu}} + \boldsymbol{C}_{\text{RB}}^{*}\boldsymbol{\nu} + \boldsymbol{M}_{\text{A}}\dot{\boldsymbol{\nu}}_{\text{r}} + \boldsymbol{C}_{\text{A}}^{*}\boldsymbol{\nu}_{\text{r}} \\
\text{Damping forces:} & +(\boldsymbol{D}_{\text{P}} + \boldsymbol{D}_{\text{V}})\boldsymbol{\nu}_{\text{r}} + \boldsymbol{\mu}_{\text{r}} \\
\text{Restoring forces:} & +\boldsymbol{G}\boldsymbol{\eta} + \boldsymbol{g}_0 \\
\text{Wind and wave forces:} & = \boldsymbol{\tau}_{\text{wind}} + \boldsymbol{\tau}_{\text{wave}} \\
\text{Propulsion forces:} & +\boldsymbol{\tau}.
\end{array}$$

The matrices $\boldsymbol{M}_{\text{A}}, \boldsymbol{C}_{\text{A}}$ and $\boldsymbol{D}_{\text{P}}$, the fluid memory function $\boldsymbol{\mu}_{\text{r}}$ as well as transfer functions for $\boldsymbol{\tau}_{\text{wave}}$ can be computed using hydrodynamics programs. This requires postprocessing of hydrodynamic data and methods for this are discussed later in this chapter. The environmental forces, $\boldsymbol{\tau}_{\text{wave}}$ and $\boldsymbol{\tau}_{\text{wind}}$, are treated separately in Chapter 10.

Different principles for the computation of the hydrodynamic coefficients can be used. The main tool is *potential theory* where it is assumed that the flow is constant, irrotational and incompressible such that time becomes unimportant. Hence, the discrepancies between real and idealized flow must be compensated by adding dissipative forces, for instance viscous damping.

5.1 Hydrodynamic Concepts and Potential Theory

In order to describe most fluid flow phenomena associated with the waves and the motion of ships in waves, we need to know the velocity of the fluid and the pressure at different locations. The velocity of the fluid at the location $\boldsymbol{x} = [x_1, x_2, x_3]^{\top}$ is given by the *fluid flow velocity vector*

$$\boldsymbol{v}(\boldsymbol{x}, t) = [v_1(\boldsymbol{x}, t), v_2(\boldsymbol{x}, t), v_3(\boldsymbol{x}, t)]^{\top}. \tag{5.4}$$

For the flow velocities involved in ship motion, the fluid can be considered *incompressible*, that is of constant density ρ. Under this assumption, the net volume rate at a volume V enclosed by a closed surface S is

$$\iint_S \boldsymbol{v} \cdot \boldsymbol{n} \, \mathrm{d}s = \iiint_V \mathrm{div}(\boldsymbol{v}) \, \mathrm{d}V = 0. \tag{5.5}$$

Since (5.5) should be valid for all the regions V in the fluid, then by assuming that $\nabla \cdot \boldsymbol{v}$ is continuous we obtain

$$\mathrm{div}(\boldsymbol{v}) = \nabla \cdot \boldsymbol{v} = \frac{\partial v_1}{\partial x} + \frac{\partial v_2}{\partial y} + \frac{\partial v_3}{\partial z} = 0 \tag{5.6}$$

which is the *continuity equation* for incompressible flows.

The conservation of momentum in the flow is described by the *Navier–Stokes equations*; see, for example, Acheson (1990)

$$\rho \left(\frac{\partial \boldsymbol{v}}{\partial t} + \boldsymbol{v} \cdot \nabla \boldsymbol{v} \right) = \rho \, \boldsymbol{F} - \nabla p + \mu \nabla^2 \boldsymbol{v} \tag{5.7}$$

where $\boldsymbol{F} = [0, 0, -g]^{\mathrm{T}}$ are accelerations due to volumetric forces, from which only gravity is considered, $p = p(\boldsymbol{x}, t)$ is the pressure and μ is the viscosity coefficient of the fluid.

To describe the real flow of ships, it is then necessary to solve the Navier–Stokes equations (5.7) together with the continuity equation (5.6). These form a system of nonlinear partial differential equations, which unfortunately do not have analytical solutions and the numerical solutions are still far from being feasible with current computing power.

If viscosity is neglected, the fluid is said to be an *ideal fluid*. This is a common assumption that is made to calculate ship flows because viscosity often matters only in a thin layer close to the ship hull. By disregarding the last term in (5.7), the *Euler equations* of fluid motion are obtained

$$\rho \left(\frac{\partial \boldsymbol{v}}{\partial t} + \boldsymbol{v} \cdot \nabla \boldsymbol{v} \right) = \rho \, \boldsymbol{F} - \nabla p. \tag{5.8}$$

A further simplification of the flow description is obtained by assuming that the flow is *irrotational*

$$\mathrm{curl}(\boldsymbol{v}) = \nabla \times \boldsymbol{v} = 0. \tag{5.9}$$

The term *potential flow* is used to describe irrotational flows of inviscid-incompressible fluids. Under this assumption, there exists a scalar function $\Phi(t, x, y, z)$ called *potential* such that

$$\boldsymbol{v} = \nabla \Phi. \tag{5.10}$$

Hence, if the potential is known the velocities can be calculated as

$$v_1 = \frac{\partial \Phi}{\partial x}, \qquad v_2 = \frac{\partial \Phi}{\partial y}, \qquad v_3 = \frac{\partial \Phi}{\partial z}. \tag{5.11}$$

Using the potential Φ, the continuity equation (5.6) reverts to the *Laplace equation* of the potential

$$\nabla^2 \Phi = \frac{\partial^2 \Phi}{\partial x^2} + \frac{\partial^2 \Phi}{\partial y^2} + \frac{\partial^2 \Phi}{\partial z^2} = 0. \tag{5.12}$$

The potential can then be obtained by solving the Laplace equation (5.12) subject to appropriate boundary conditions, that is by solving a boundary value problem.

The pressure in the fluid can be obtained by integrating the *Euler equation* of fluid motion (5.8). This results in the *Bernoulli equation*

$$\frac{p}{\rho} + \frac{\partial \Phi}{\partial t} + \frac{1}{2}(\nabla \Phi)^2 + gz = C. \tag{5.13}$$

By setting the constant $C = p_0/\rho$, the relative pressure can be computed from

$$p - p_0 = -\rho g z - \rho \frac{\partial \Phi}{\partial t} - \frac{1}{2}\rho(\nabla \Phi)^2 \tag{5.14}$$

For simplicity, the atmospheric pressure p_0 is often considered zero.

To summarize, *potential theory* makes two assumptions:

1. Inviscid fluid (no viscosity)
2. Irrotational flow.

The assumption of irrotational flow leads to the description of the fluid velocity vector as the gradient of a potential function, which has no physical meaning. However, this is a large simplification because the potential is scalar while the velocity is a vector quantity. The potential satisfies the Laplace equation (5.12), which needs to be solved subject to appropriate boundary conditions (on the free surface, sea floor and ship hull). This is another large simplification because the Laplace equation is linear; therefore, superposition holds and the problem can also be solved in the frequency domain, which is the basis of most hydrodynamic programs. Once we have the potential and thus the velocities, the pressure can be computed using Bernoulli's equation. Then, by integrating the pressure over the surface of the hull, the hydrodynamic forces are obtained.

For most problems related to ship motion in waves, potential theory is sufficient to obtain results with appropriate accuracy for engineering purposes. However, because of the simplifying assumptions in some cases we need to complement the results by adding the effects of viscosity. This is important, for example, when considering maneuvering and propeller–rudder–hull interactions. For further discussions on the topics presented in this section, see Newman (1977), Faltinsen (1990), Acheson (1990), Journée and Massie (2001) and Bertram (2004).

5.1.1 Numerical Approaches and Hydrodynamic Codes

In order to evaluate the potentials a boundary value problem needs to be solved. There are different approaches to do this, which lead to different formulations.

Strip Theory (2D Potential Theory)

In some problems, the motion of the fluid can be approximated as two-dimensional. This is characteristic for slender bodies. In this case a good estimate of the hydrodynamic forces can be obtained by applying *strip theory* (Newman 1977; Faltinsen 1990; Journée and Massie 2001). The 2D theory takes into account the fact that variation of the flow in the cross-directional plane is much larger than the variation in the longitudinal direction of the ship. The principle of strip theory involves dividing the submerged part of the craft into a finite number of strips. Hence, 2D hydrodynamic coefficients for added mass can be computed for each strip and then summed over the length of the body to yield the 3D coefficients. The 2D hydrodynamic coefficients can be calculated from boundary element methods or via conformal mapping and analytical expressions. This principle is also used to compute viscous quadratic damping from 2D drag coefficients, as explained in Section 6.4.

Several strip theory programs can be used to compute hydrodynamic added mass $\boldsymbol{M}_{\mathrm{A}}$, potential damping $\boldsymbol{D}_{\mathrm{p}}$ and the hydrostatic matrix $\boldsymbol{G}$. The strip theory programs can be used at both zero speed and forward speed and they calculate frequency-dependent added mass and potential damping coefficients, restoring terms, first- and second-order wave load transfer functions (amplitudes and phases) between the marine craft and the waves for given wave directions and frequencies as well as other hydrodynamic data. Processing of the data is explained later in this chapter.

In this context it will be shown how frequency-dependent added mass and damping can be used to derive the equations of motion where these effects are included as fluid memory effects using retardation functions. In order to compute the retardation functions, asymptotic values for zero and infinite added mass must be used.

Panel Methods (3D Potential Theory)

For potential flows, the integrals over the fluid domain can be transformed to integrals over the boundaries of the fluid domain. This allows the application of panel or boundary element methods to solve the 3D potential theory problem. Panel methods divide the surface of the ship and the surrounding water into discrete elements (panels). On each of these elements, a distribution of sources and sinks is defined that fulfill the Laplace equation. The problem then amounts to finding the strength of these distributions and identifying the potential.

Computer codes based on this approach provide suitable performance for offshore applications at zero-forward speed in either the frequency or time domain. A commercial program WAMIT (2010) has become the de facto industry standard among oil and engineering companies. This program computes frequency-dependent added mass $\boldsymbol{M}_{\mathrm{A}}$, potential damping coefficients $\boldsymbol{D}_{\mathrm{P}}$, restoring terms $\boldsymbol{G}$, and first- and second-order wave load transfer functions (amplitudes and phases) between the marine craft and the waves for given wave directions and frequencies, and much more. One special feature of WAMIT is that the program solves a boundary value problem for zero and infinite added mass. These boundary values are particularly useful when computing the retardation functions describing the fluid memory effects.

Semi-empirical Methods

An alternative and less accurate approach to hydrodynamic programs is to use semi-empirical methods to compute the added mass derivatives; see, for instance, Imlay (1961), Humphreys and Watkinson (1978) and Triantafyllou and Amzallag (1984).

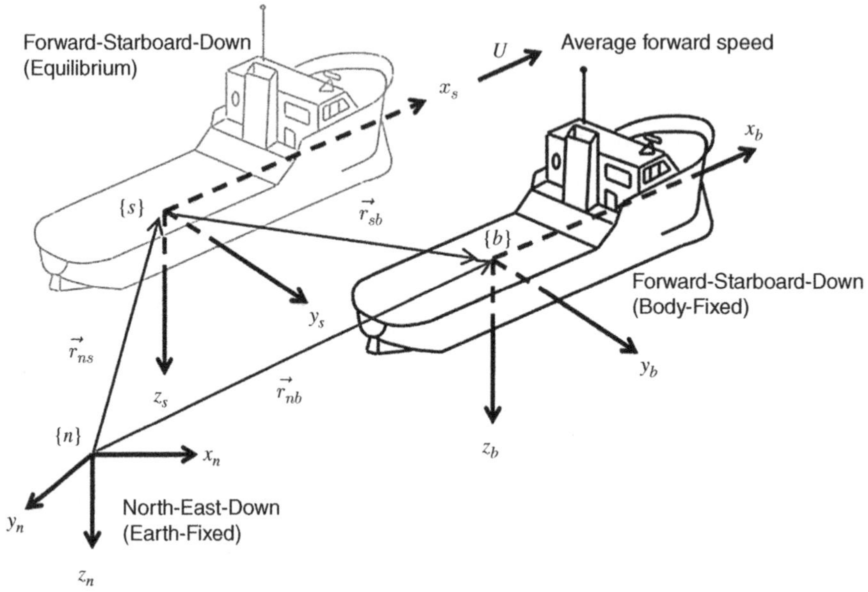

Figure 5.1 The seakeeping coordinate system $\{s\}$ and distance vectors to the reference frames $\{b\}$ and $\{n\}$.

5.2 Seakeeping and Maneuvering Kinematics

This section derives the kinematics needed to transform the equations of motion from the seakeeping reference frame $\{s\}$ to the body-fixed reference frame $\{b\}$ and the NED reference frame $\{n\}$. This is based on Perez and Fossen (2007).

5.2.1 Seakeeping Reference Frame

In seakeeping theory the study of ship motion is performed under the assumption that it can be described as the superposition of an equilibrium state of motion plus perturbations. The equilibrium is determined by a constant heading angle $\bar{\psi}$ and speed U, and the perturbations are zero-mean oscillatory components induced by first-order wave excitations. Note that the case of zero forward speed $U = 0$ is also contemplated as an equilibrium of motion. Due to this, the motion is often described using an equilibrium or seakeeping reference frame.

Seakeeping frame: The seakeeping reference frame $\{s\} = (x_s, y_s, z_s)$ is not fixed to the marine craft; it is fixed to the equilibrium state. Hence, in the absence of wave excitation, the $\{s\}$-frame origin o_s coincides with the location of the $\{b\}$-frame origin o_b (also denoted as CO) which is a fixed point in the ship. Under the action of the waves, the hull is disturbed from its equilibrium and the point o_s oscillates, with respect to its equilibrium position. This is illustrated in Figure 5.1.

The $\{s\}$ frame is considered *inertial* and therefore it is non-accelerating and fixed in orientation with respect to the $\{n\}$ frame (or must vary very slowly). This assumption implies that the $\{s\}$-frame equations of motion are linear. The equilibrium state is defined by a constant heading and speed

$$\boldsymbol{v}^n_{\text{ns}} = [U\cos(\bar{\psi}), U\sin(\bar{\psi}), 0]^\top \tag{5.15}$$

$$\boldsymbol{\omega}^n_{\text{ns}} = [0, 0, 0]^\top \tag{5.16}$$

$$\boldsymbol{\Theta}_{\text{ns}} = [0, 0, \bar{\psi}]^\top \tag{5.17}$$

where $U = \| \boldsymbol{v}^n_{\text{ns}} \|$ is the average forward speed and $\bar{\psi}$ is the equilibrium heading. Hence, the velocity of $\{s\}$ with respect to $\{n\}$ expressed in $\{s\}$ is

$$\boldsymbol{v}^s_{\text{ns}} = \boldsymbol{R}^s_n \boldsymbol{v}^n_{\text{ns}} = [U, 0, 0]^\top \tag{5.18}$$

where $\boldsymbol{R}^s_n = \boldsymbol{R}(\boldsymbol{\Theta}_{\text{ns}})$. The equilibrium heading $\bar{\psi}$ can be computed by averaging the gyro compass measurements ψ over a fixed period (moving horizon) of time.

Seakeeping (Perturbation) Coordinates

The *seakeeping or perturbation coordinates* are defined as (Perez and Fossen 2007)

$$\delta\boldsymbol{\eta} := \begin{bmatrix} \boldsymbol{r}^s_{\text{sb}} \\ \boldsymbol{\Theta}_{\text{sb}} \end{bmatrix} \tag{5.19}$$

$$\delta\boldsymbol{v} := \begin{bmatrix} \boldsymbol{v}^b_{\text{sb}} \\ \boldsymbol{\omega}^b_{\text{sb}} \end{bmatrix}. \tag{5.20}$$

In hydrodynamic textbooks it is common to denote the perturbation coordinates by

$$\boldsymbol{\xi} := \delta\boldsymbol{\eta} \tag{5.21}$$

where $\boldsymbol{\xi} = [\xi_1, \xi_2, \xi_3, \xi_4, \xi_5, \xi_6]^\top$. The first three coordinates (ξ_1, ξ_2, ξ_3) are translational motion perturbations and

$$\boldsymbol{\Theta}_{\text{sb}} = [\xi_4, \xi_5, \xi_6]^\top = [\delta\phi, \delta\theta, \delta\psi]^\top \tag{5.22}$$

are the angular motion perturbations (roll, pitch and yaw).

5.2.2 *Transformation Between BODY and SEAKEEPING*

From the definition of $\{s\}$ and the coordinates $\delta\boldsymbol{\eta}$ and $\delta\boldsymbol{v}$ it follows that

$$\delta\dot{\boldsymbol{\eta}} = \boldsymbol{J}_\Theta(\delta\boldsymbol{\eta})\delta\boldsymbol{v} \tag{5.23}$$

where $\boldsymbol{J}_\Theta(\delta\boldsymbol{\eta})$ is the transformation matrix between $\{b\}$ and $\{s\}$

$$\boldsymbol{J}_\Theta(\delta\boldsymbol{\eta}) = \begin{bmatrix} \boldsymbol{R}(\boldsymbol{\Theta}_{\rm sb}) & \boldsymbol{0}_{3\times3} \\ \boldsymbol{0}_{3\times3} & \boldsymbol{T}(\boldsymbol{\Theta}_{\rm sb}) \end{bmatrix}. \tag{5.24}$$

This expression is similar to the transformation between $\{b\}$ and $\{n\}$. This is an expected result since both $\{n\}$ and $\{s\}$ are assumed inertial while $\{b\}$ rotates about the inertial frame. In addition to position and attitude it is necessary to derive the relationship between the perturbed velocity and acceleration pairs $(\delta\boldsymbol{v}, \delta\dot{\boldsymbol{v}})$ and $(\boldsymbol{v}, \dot{\boldsymbol{v}})$. To obtain these expressions consider the distance vector (see Figure 5.1)

$$\vec{r}_{\rm nb} = \vec{r}_{\rm ns} + \vec{r}_{\rm sb} \tag{5.25}$$

which can be expressed in $\{n\}$ as

$$\boldsymbol{r}^n_{\rm nb} = \boldsymbol{r}^n_{\rm ns} + \boldsymbol{R}^n_s \boldsymbol{r}^s_{\rm sb}. \tag{5.26}$$

Time differentiation gives

$$\dot{\boldsymbol{r}}^n_{\rm nb} = \dot{\boldsymbol{r}}^n_{\rm ns} + \boldsymbol{R}^n_s \dot{\boldsymbol{r}}^s_{\rm sb} + \dot{\boldsymbol{R}}^n_s \boldsymbol{r}^s_{\rm sb} \tag{5.27}$$

where

$$\boldsymbol{R}^n_s = \boldsymbol{R}_{z,\bar{\psi}} = \begin{bmatrix} \cos(\bar{\psi}) & -\sin(\bar{\psi}) & 0 \\ \sin(\bar{\psi}) & \cos(\bar{\psi}) & 0 \\ 0 & 0 & 1 \end{bmatrix}, \qquad \dot{\boldsymbol{R}}^n_s = \boldsymbol{0}. \tag{5.28}$$

Note that the time derivative of $\boldsymbol{R}^n_s$ is zero because $\{s\}$ does not rotate with respect to $\{n\}$. The expression for $\dot{\boldsymbol{r}}^n_{\rm nb}$ can be rewritten as

$$\begin{aligned} \dot{\boldsymbol{r}}^n_{\rm nb} &= \dot{\boldsymbol{r}}^n_{\rm ns} + \boldsymbol{R}^n_b \boldsymbol{R}^b_s \dot{\boldsymbol{r}}^s_{\rm sb} \\ &= \dot{\boldsymbol{r}}^n_{\rm ns} + \boldsymbol{R}^n_b \boldsymbol{v}^b_{\rm sb}. \end{aligned} \tag{5.29}$$

Both sides of (5.29) can be multiplied by $\boldsymbol{R}^b_n$ to obtain

$$\boldsymbol{v}^b_{\rm nb} = \boldsymbol{R}^b_n \boldsymbol{v}^n_{\rm ns} + \boldsymbol{v}^b_{\rm sb}. \tag{5.30}$$

For notational simplicity, the linear and angular velocity vectors are grouped according to

$$\boldsymbol{v} = \begin{bmatrix} \boldsymbol{v}_1 \\ \boldsymbol{v}_2 \end{bmatrix} = \begin{bmatrix} [u, v, w]^\top \\ [p, q, r]^\top \end{bmatrix} \tag{5.31}$$

$$\delta\boldsymbol{v} = \begin{bmatrix} \delta\boldsymbol{v}_1 \\ \delta\boldsymbol{v}_2 \end{bmatrix} = \begin{bmatrix} [\delta u, \delta v, \delta w]^\top \\ [\delta p, \delta q, \delta r]^\top \end{bmatrix}. \tag{5.32}$$

Then it follows from (5.30) that

$$\boldsymbol{v}_1 = \bar{\boldsymbol{v}}_1 + \delta\boldsymbol{v}_1 \tag{5.33}$$

where

$$\bar{\boldsymbol{v}}_1 := \boldsymbol{R}^b_n \begin{bmatrix} U\cos(\bar{\psi}) \\ U\sin(\bar{\psi}) \\ 0 \end{bmatrix} = \boldsymbol{R}^b_s \begin{bmatrix} U \\ 0 \\ 0 \end{bmatrix}. \tag{5.34}$$

To obtain the angular velocity transformation, we make use of

$$\vec{\omega}_{\rm nb} = \vec{\omega}_{\rm ns} + \vec{\omega}_{\rm sb} = \vec{\omega}_{\rm sb} \tag{5.35}$$

since $\vec{\omega}_{\rm ns} = \vec{0}$. In other words, $\{s\}$ does not rotate with respect to $\{n\}$. This leads to

$$\boldsymbol{\omega}^b_{\rm nb} = \boldsymbol{\omega}^b_{\rm sb} \quad \Rightarrow \quad \boldsymbol{v}_2 = \delta \boldsymbol{v}_2. \tag{5.36}$$

The Euler angle transformation matrices $\boldsymbol{R}(\boldsymbol{\Theta}_{\rm sb})$ and $\boldsymbol{T}(\boldsymbol{\Theta}_{\rm sb})$ for $\boldsymbol{\Theta}_{\rm sb} = [\delta\phi, \delta\theta, \delta\psi]^\top$ are similar to those used in Section 2.2. In other words,

$$\boldsymbol{R}(\boldsymbol{\Theta}_{\rm sb}) = \begin{bmatrix} c_{\delta\psi}c_{\delta\theta} & -s_{\delta\psi}c_{\delta\phi} + c_{\delta\psi}s_{\delta\theta}s_{\delta\phi} & s_{\delta\psi}s_{\delta\phi} + c_{\delta\psi}c_{\phi}s_{\delta\theta} \\ s_{\delta\psi}c_{\delta\theta} & c_{\delta\psi}c_{\delta\phi} + s_{\delta\phi}s_{\delta\theta}s_{\delta\psi} & -c_{\delta\psi}s_{\delta\psi} + s_{\delta\theta}s_{\delta\psi}c_{\delta\phi} \\ -s_{\delta\theta} & c_{\delta\theta}s_{\delta\phi} & c_{\delta\theta}c_{\delta\phi} \end{bmatrix} \tag{5.37}$$

$$\boldsymbol{T}(\boldsymbol{\Theta}_{\rm sb}) = \begin{bmatrix} 1 & s_{\delta\phi}t_{\delta\theta} & c_{\delta\phi}t_{\delta\theta} \\ 0 & c_{\delta\phi} & -s_{\delta\phi} \\ 0 & s_{\delta\phi}/c_{\delta\theta} & c_{\delta\phi}/c_{\delta\theta} \end{bmatrix}, \qquad c_{\delta\theta} \neq 0. \tag{5.38}$$

Computing $\bar{\boldsymbol{v}}_1$ under the assumption of small angles gives

$$\begin{aligned} \bar{\boldsymbol{v}}_1 &= \boldsymbol{R}^\top(\boldsymbol{\Theta}_{\rm sb}) \begin{bmatrix} U \\ 0 \\ 0 \end{bmatrix} \\ &= U \begin{bmatrix} c_{\delta\psi}c_{\delta\theta} \\ -s_{\delta\psi}c_{\delta\phi} + c_{\delta\psi}s_{\delta\theta}s_{\delta\phi} \\ s_{\delta\psi}s_{\delta\phi} + c_{\delta\psi}c_{\phi}s_{\delta\theta} \end{bmatrix} \\ &\approx U \begin{bmatrix} 1 \\ -\delta\psi \\ \delta\theta \end{bmatrix}. \end{aligned} \tag{5.39}$$

Finally,

$$\boldsymbol{v} = \bar{\boldsymbol{v}} + \delta\boldsymbol{v} \tag{5.40}$$

where

$$\bar{\boldsymbol{v}} \approx U[1, -\delta\psi, \delta\theta, 0, 0, 0]^\top. \tag{5.41}$$

This can written as

$$\boldsymbol{v} \approx U(\boldsymbol{e}_1 - \boldsymbol{L}\delta\boldsymbol{\eta}) + \delta\boldsymbol{v} \tag{5.42}$$

$$\boldsymbol{e}_1 := \begin{bmatrix} 1 \\ 0 \\ 0 \\ 0 \\ 0 \\ 0 \end{bmatrix}, \qquad \boldsymbol{L} := \begin{bmatrix} 0 & 0 & 0 & 0 & 0 & 0 \\ 0 & 0 & 0 & 0 & 0 & 1 \\ 0 & 0 & 0 & 0 & -1 & 0 \\ 0 & 0 & 0 & 0 & 0 & 0 \\ 0 & 0 & 0 & 0 & 0 & 0 \\ 0 & 0 & 0 & 0 & 0 & 0 \end{bmatrix}. \tag{5.43}$$

Differentiation of (5.42) with respect to time gives

$$\begin{aligned}\dot{\boldsymbol{\nu}} &= -UL\delta\boldsymbol{\nu} + \delta\dot{\boldsymbol{\nu}} \\ &= -UL(\boldsymbol{\nu} - U(\boldsymbol{e}_1 - L\delta\boldsymbol{\eta})) + \delta\dot{\boldsymbol{\nu}} \\ &= -UL\boldsymbol{\nu} + \delta\dot{\boldsymbol{\nu}} \end{aligned} \tag{5.44}$$

since $L\boldsymbol{e}_1 = \boldsymbol{0}$ and $L^2 = \boldsymbol{0}$.

Hence, the linear transformations needed to transform a system from seakeeping coordinates $(\delta\boldsymbol{\eta}, \delta\boldsymbol{\nu})$ to body-fixed coordinates $(\boldsymbol{\eta}, \boldsymbol{\nu})$ are

$$\delta\boldsymbol{\nu} \approx \boldsymbol{\nu} + U(L\delta\boldsymbol{\eta} - \boldsymbol{e}_1) \tag{5.45}$$

$$\delta\dot{\boldsymbol{\nu}} \approx \dot{\boldsymbol{\nu}} + UL\boldsymbol{\nu}. \tag{5.46}$$

The Euler angles are related through the following equation

$$\boldsymbol{\Theta}_{\text{nb}} = \boldsymbol{\Theta}_{\text{ns}} + \boldsymbol{\Theta}_{\text{sb}} \tag{5.47}$$

which gives

$$\begin{bmatrix}\phi \\ \theta \\ \psi\end{bmatrix} = \begin{bmatrix}0 \\ 0 \\ \bar{\psi}\end{bmatrix} + \begin{bmatrix}\delta\phi \\ \delta\theta \\ \delta\psi\end{bmatrix}. \tag{5.48}$$

5.3 The Classical Frequency-domain Model

Frequency-dependent hydrodynamic forces can be determined experimentally or computed using potential theory programs or seakeeping codes. This section describes the transformations needed to obtain what is called the *frequency-domain model* and a method known as *forced oscillations*, which can be used to obtain frequency-dependent added mass and damping experimentally.

The seakeeping equations of motion are considered to be inertial. Hence, the rigid-body kinetics in terms of perturbed coordinates $\delta\boldsymbol{\eta}$ and $\delta\boldsymbol{\nu}$ becomes (see Section 3.3)

$$\delta\dot{\boldsymbol{\eta}} = \boldsymbol{J}_{\Theta}(\delta\boldsymbol{\eta})\delta\boldsymbol{\nu} \tag{5.49}$$

$$\boldsymbol{M}_{\text{RB}}\delta\ddot{\boldsymbol{\eta}} + \boldsymbol{C}_{\text{RB}}(\delta\boldsymbol{\nu})\delta\dot{\boldsymbol{\eta}} = \delta\boldsymbol{\tau}_{\text{RB}}. \tag{5.50}$$

Linear theory suggests that second-order terms can be neglected. Consequently, the rigid-body kinetics in seakeeping coordinates $\boldsymbol{\xi} = \delta\boldsymbol{\eta}$ and $\dot{\boldsymbol{\xi}} = \delta\boldsymbol{\nu}$ is

$$\begin{aligned}\boldsymbol{M}_{\text{RB}}\ddot{\boldsymbol{\xi}} &= \delta\boldsymbol{\tau}_{\text{RB}} \\ &= \boldsymbol{\tau}_{\text{hyd}} + \boldsymbol{\tau}_{\text{hs}} + \boldsymbol{\tau}_{\text{exc}}. \end{aligned} \tag{5.51}$$

The rigid-body kinetics is forced by the term $\delta\boldsymbol{\tau}_{\text{RB}}$ which can be used to model hydrodynamic forces $\boldsymbol{\tau}_{\text{hyd}}$, hydrostatic forces $\boldsymbol{\tau}_{\text{hs}}$ and other external forces $\boldsymbol{\tau}_{\text{exc}}$. Cummins (1962) showed that the radiation-induced hydrodynamic forces in an ideal fluid can be related to frequency-dependent added mass $\boldsymbol{A}(\omega)$ and potential damping $\boldsymbol{B}(\omega)$ according to

$$\boldsymbol{\tau}_{\text{hyd}} = -\bar{\boldsymbol{A}}\ddot{\boldsymbol{\xi}} - \int_0^t \bar{\boldsymbol{K}}(t-\tau)\dot{\boldsymbol{\xi}}(\tau)\mathrm{d}\tau \tag{5.52}$$

where $\bar{\boldsymbol{A}} = \boldsymbol{A}(\infty)$ is the constant infinite-frequency added mass matrix and $\bar{\boldsymbol{K}}(t)$ is a matrix of *retardation functions* given by

$$\bar{\boldsymbol{K}}(t) = \frac{2}{\pi}\int_0^{\infty} \boldsymbol{B}(\omega)\cos(\omega t)\mathrm{d}\omega. \tag{5.53}$$

If linear restoring forces $\boldsymbol{\tau}_{\text{hs}} = -\boldsymbol{C}\boldsymbol{\xi}$ are included in the model, this results in the time-domain model

$$(\boldsymbol{M}_{\text{RB}} + \boldsymbol{A}(\infty))\ddot{\boldsymbol{\xi}} + \int_0^t \bar{\boldsymbol{K}}(t-\tau)\dot{\boldsymbol{\xi}}(\tau)\mathrm{d}\tau + \boldsymbol{C}\boldsymbol{\xi} = \boldsymbol{\tau}_{\text{exc}}. \tag{5.54}$$

This is a vector *integro-differential equation* formulated in the time domain even though the potential coefficients are frequency dependent. In order to understand this, we will consider a floating body forced to oscillate at a given frequency.

5.3.1 Frequency-dependent Hydrodynamic Coefficients

Consider the motions of a floating or submerged body given by

$$\boldsymbol{M}_{\text{RB}}\ddot{\boldsymbol{\xi}} = \boldsymbol{\tau}_{\text{hyd}} + \boldsymbol{\tau}_{\text{hs}} + \boldsymbol{\tau}_{\text{exc}} \tag{5.55}$$

where $\boldsymbol{\tau}_{\text{hyd}}$ and $\boldsymbol{\tau}_{\text{hs}}$ denote the hydrodynamic and hydrostatic forces due to the surrounding water. The vector $\boldsymbol{\tau}_{\text{exc}} = \boldsymbol{f}\cos(\omega t)$ where $\boldsymbol{f} = [f_1, \ldots, f_6]^{\top}$ contains the excitation force amplitudes. In an experimental setup with a restrained scale model, it is then possible to vary the wave excitation frequency ω and the amplitudes f_i for $i = 1, .., 6$ of the excitation force. Hence, by measuring the position and attitude vector $\boldsymbol{\xi}$, the response of the second-order system (5.55) can be fitted to a linear model

$$(\boldsymbol{M}_{\text{RB}} + \boldsymbol{A}(\omega))\ddot{\boldsymbol{\xi}} + \boldsymbol{B}_{\text{total}}(\omega)\dot{\boldsymbol{\xi}} + \boldsymbol{C}\boldsymbol{\xi} = \boldsymbol{\tau}_{\text{exc}} \tag{5.56}$$

for each frequency ω. The total damping matrix is the sum of the potential and viscous damping matrices

$$\boldsymbol{B}_{\text{total}}(\omega) = \boldsymbol{B}(\omega) + \boldsymbol{B}_V(\omega). \tag{5.57}$$

By closer inspection of (5.56), the hydrodynamic and hydrostatic forces are recognized as a frequency-dependent mass–damper–spring system with elements

$$\boldsymbol{\tau}_{\text{hyd}} = \underbrace{-\boldsymbol{A}(\omega)\ddot{\boldsymbol{\xi}} - \boldsymbol{B}(\omega)\dot{\boldsymbol{\xi}}}_{\text{radiation force}} - \underbrace{\boldsymbol{B}_V(\omega)\dot{\boldsymbol{\xi}}}_{\text{viscous damping force}} \tag{5.58}$$

$$\boldsymbol{\tau}_{\text{hs}} = \underbrace{-\boldsymbol{C}\boldsymbol{\xi}}_{\text{restoring force}} . \tag{5.59}$$

The *radiation force* is due to the energy carried away by generated surface waves and it is formed by two components, hydrodynamic inertia forces $\boldsymbol{A}(\omega)\ddot{\boldsymbol{\xi}}$ and potential damping $\boldsymbol{B}(\omega)\dot{\boldsymbol{\xi}}$. The physical interpretations of the matrices are

- $\boldsymbol{A}(\omega)$ added mass matrix
- $\boldsymbol{B}(\omega)$ potential damping matrix
- $\boldsymbol{B}_V(\omega)$ viscous damping matrix
- $\boldsymbol{C}$ spring stiffness or hydrostatic matrix.

If the experiment is repeated for several frequencies ω_i $(i = 1, \ldots N)$, added mass $\boldsymbol{A}(\omega_i)$ and damping $\boldsymbol{B}_{\text{total}}(\omega_i)$ can be computed at different frequencies. Added mass and damping for a conventional ship is plotted as a function of ω in Figures 5.2 and 5.3.

The matrices $\boldsymbol{A}(\omega)$, $\boldsymbol{B}_{\text{total}}(\omega)$ and $\boldsymbol{C}$ in (5.56) represent a *hydrodynamic mass–damper–spring system* that varies with the frequency of the forced oscillation. The added mass matrix $\boldsymbol{A}(\omega)$ should not be understood as additional mass due to a finite amount of water that is dragged with the vessel. A more precise definition is:

Definition 5.1 (Added Mass)
Hydrodynamic added mass can be seen as a virtual mass added to a system because an accelerating or decelerating body must move some volume of the surrounding fluid as it moves through it. Moreover, the object and fluid cannot occupy the same physical space simultaneously.

The model (5.56) is rooted deeply in the literature of hydrodynamics and the abuse of notation of this false time-domain model has been discussed eloquently in the literature. This is in fact not a time-domain model but rather a different way of writing (5.73), which is the frequency response function. The corresponding time-domain model is given by (5.54).

The potential coefficients $\boldsymbol{A}(\omega)$ and $\boldsymbol{B}(\omega)$ are usually computed using a seakeeping program but the frequency response will not be accurate unless viscous damping is included. The viscous matrix $\boldsymbol{B}_V(\omega)$ is an optional matrix that can be used to model viscous damping such as skin friction, surge resistance and viscous roll damping.

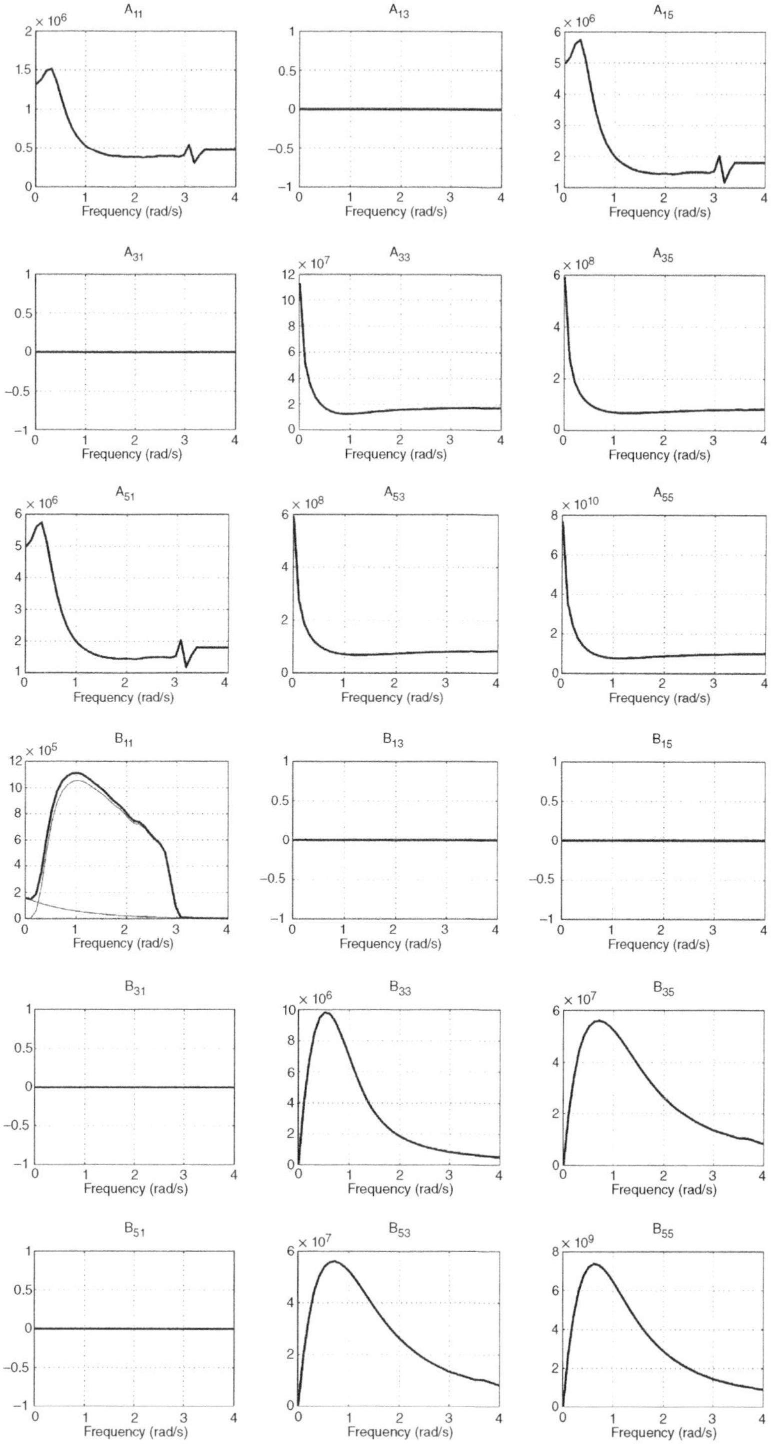

Figure 5.2 Longitudinal added mass and potential damping coefficients as a function of frequency. Exponential decaying viscous damping is included for $B_{11}(\omega)$.

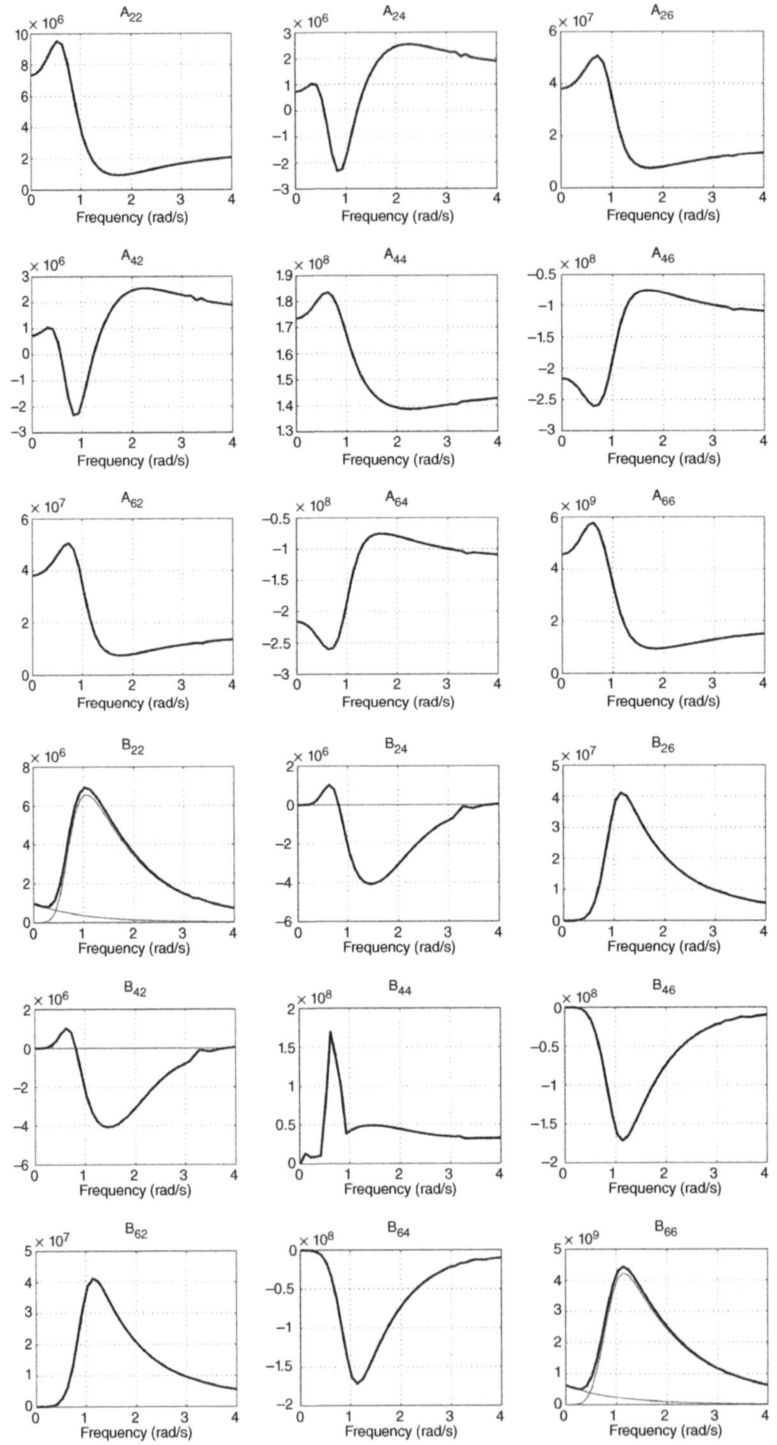

Figure 5.3 Lateral added mass and potential damping coefficients as a function of frequency. Exponential decaying viscous damping is included for $B_{22}(\omega)$ and $B_{66}(\omega)$ while viscous IKEDA damping is included in $B_{44}(\omega)$ (shown as a peak).

The pressure supporting a marine craft can be separated into hydrostatic and hydrodynamic forces. The hydrostatic pressure gives the buoyancy force, which is proportional to the displaced volume. Thus, the *hydrostatic force*, $\boldsymbol{C}\boldsymbol{\xi}$, represents the restoring forces due to gravity and buoyancy that tend to bring the marine craft back to its upright equilibrium position.

The *wave excitation forces*, $\boldsymbol{\tau}_{\text{wave}}$, arise due to changes in pressure due to waves. These have one component that varies linearly with the wave elevation and another that varies nonlinearly. The linear forces are oscillatory forces with a zero mean; these forces are called *first-order wave forces* – Froude–Krylov and diffraction forces. The energy of these forces is distributed at the same frequencies as the wave elevation seen from the moving ship (encounter frequencies). The nonlinear components give rise to non-oscillatory forces – *mean wave drift forces* – and also oscillatory forces, which have energy at frequencies that are both lower and higher than the range of first-order wave forces.

The components at lower frequencies are called second-order *slow wave drift forces*, and together with the mean wave drift and the first-order wave forces constitute the main disturbances for ship motion control. The high-frequency forces are usually of no concern for ship motion control, but can produce oscillation in the structure of the vessel; this effect is known as springing. For further details on wave loads see Faltinsen (1990, 2005).

5.3.2 Viscous Damping

When running seakeeping codes it is important to include an *external viscous damping matrix* $\boldsymbol{B}_{\text{V}}(\omega)$ in order to obtain accurate estimates of the response amplitude operators (RAOs), which are used to compute the motions due to first- and second-order wave forces. The viscous damping coefficients will give additional contributions to the potential damping matrix $\boldsymbol{B}(\omega)$, as shown in the plots for $B_{11}(\omega)$, $B_{22}(\omega)$, $B_{44}(\omega)$ and $B_{66}(\omega)$ in Figures 5.2 and 5.3.

In maneuvering theory, it is standard to use the zero-frequency added mass and damping coefficients in surge, sway and yaw to describe the ship motions. When applying a feedback control system to stabilize the motions in surge, sway and yaw, the natural periods will be in the range of 100–200 s. This implies that the natural frequencies are in the range of 0.03–0.10 rad^{-1} s, which is quite close to the zero wave excitation frequency. Since the potential damping coefficients $B_{11}(\omega)$, $B_{22}(\omega)$ and $B_{66}(\omega)$ are zero for $\omega = 0$, a ship maneuvering model should include viscous damping terms at low frequencies.

Bailey *et al.* (1998) suggest using ramps in surge, sway and yaw to describe the viscous part of the damping forces. However, in the framework of linear theory exponential functions are well suited for this purpose. For instance, $\boldsymbol{B}_{\text{V}}(\omega)$ can be chosen as a diagonal matrix

$$\boldsymbol{B}_V(\omega) = \text{diag}\{\beta_1 \text{e}^{-\alpha\omega} + N_{\text{ITTC}}(A_1), \beta_2 \text{e}^{-\alpha\omega}, \beta_3, \beta_{\text{IKEDA}}(\omega), \beta_5, \beta_6 \text{e}^{-\alpha\omega}\} \tag{5.60}$$

where $\alpha > 0$ is the exponential rate, $\beta_i > 0$ $(i = 1, 2, 6)$ are linear viscous skin friction coefficients, $N_{\text{ITTC}}(A_1)$ is equivalent linear surge resistance depending on the surge velocity amplitude A_1 and $\beta_{\text{IKEDA}}(\omega)$ is frequency-dependent roll damping based on the theory of Ikeda *et al.* (1976). Other models for viscous roll damping can also be used.

One useful property of the exponential function $\beta_i e^{-\alpha\omega}$ is that linear skin friction only affects low-frequency motions. It is also possible to add a frequency-independent linear damper $D_{ii} = \beta_i \dot{\xi}_i$ directly to the equations of motion in the time domain and obtain the same effect as solving the frequency-domain equation with $B_{ii}(\omega) = \beta_i e^{-\alpha\omega}$ (Ross and Fossen 2005).

A rule of thumb for choosing β_i $(i = 1, 2, 6)$ can be to specify the time constants T_i in surge, sway and yaw corresponding to three mass-damper systems where

$$T_1 = \frac{m + A_{11}(0)}{B_{v,11}(0)}, \quad T_2 = \frac{m + A_{22}(0)}{B_{v,22}(0)}, \quad T_6 = \frac{I_z + A_{66}(0)}{B_{v,66}(0)}. \tag{5.61}$$

From this it follows that

$$\beta_1 = \frac{m + A_{11}(0)}{T_1}, \quad \beta_2 = \frac{m + A_{22}(0)}{T_2}, \quad \beta_6 = \frac{I_z + A_{66}(0)}{T_6}. \tag{5.62}$$

Viscous damping β_3 in heave is usually added as a constant value to increase damping at the natural frequency ω_3. The relative damping ratio ζ_3 satisfies

$$2\zeta_3\omega_3 = \frac{B_{33}(\omega_3) + \beta_3}{m + A_{33}(\omega_3)}. \tag{5.63}$$

Consequently, damping can be increased by specifying a percentage increase in damping, e.g. $\beta_3 = pB_{33}(\omega_3)$ where $p > 0$. A similar approach can be used in pitch to determine β_5. In other words

$$2\zeta_5\omega_5 = \frac{B_{55}(\omega_5) + \beta_5}{I_y + A_{55}(\omega_5)}. \tag{5.64}$$

Equivalent Linearization Method and Describing Functions

The surge resistance $N_{\text{ITTC}}(A_1)$ can be found by *equivalent linearization* of the quadratic damping (6.80). Equivalent linearization is a *Fourier-series approximation* where the work done over one period T is the same for the nonlinear and linear terms. This is similar to a sinusoidal-input *describing function* that is frequently used in control engineering. Consider a sinusoidal input

$$u = A\sin(\omega t). \tag{5.65}$$

For static linearities, displaying no dependence upon the derivatives, the describing function for the particular odd polynomial nonlinearity

$$y = c_1 x + c_2 x|x| + c_3 x^3 \tag{5.66}$$

is (Gelb and Vander Velde 1968)

$$N(A) = c_1 + \frac{8A}{3\pi}c_2 + \frac{3A^2}{4}c_3. \tag{5.67}$$

Consequently, the amplitude-dependent linear mapping

$$y = N(A)u \tag{5.68}$$

approximates the nonlinear polynomial (5.66) if the input is a harmonic function. This result is very useful for marine craft since it allows for linear approximation of nonlinear dissipative forces under the assumption of regular waves. For instance, the quadratic damping in surge due to the ITTC surge resistance formulation results in an expression (see Section 6.4.2)

$$\begin{aligned} X &= -X_{|u|u}|u|u \\ &\approx N_{\text{ITTC}}(A_1)u \end{aligned} \tag{5.69}$$

where the surge velocity $u = A_1 \cos(\omega t)$ is assumed to be harmonic. Then it follows from (5.67) that

$$N_{\text{ITTC}}(A_1) = -\frac{8A_1}{3\pi} X_{|u|u}. \tag{5.70}$$

Viscous damping can also be added in sway and yaw using a similar approach. The diagonal terms from the cross-flow drag analysis (see Section 6.4.3) result in similar terms depending on the sway and yaw amplitudes A_2 and A_6. In other words

$$Y = N_{\text{Y, crossflow}}(A_2)v, \qquad N_{\text{Y, crossflow}}(A_2) = -\frac{8A_2}{3\pi} Y_{|v|v} \tag{5.71}$$

$$N = N_{\text{N, crossflow}}(A_6)r, \qquad N_{\text{N, crossflow}}(A_6) = -\frac{8A_6}{3\pi} N_{|r|r}. \tag{5.72}$$

For a ship moving at high speed, the amplitudes A_2 and A_6 will be much smaller than A_1. Hence, it is common to neglect these terms in seakeeping analysis.

5.3.3 *Response Amplitude Operators*

Equation (5.56) can be transformed to the frequency domain by using the Laplace transformation. Hence, application of $\mathcal{L}\{\ddot{\xi}(t)\} = s^2\xi(s)$ and $\mathcal{L}\{\dot{\xi}\} = s\xi(s)$ together with $s = \text{j}\omega$ gives

$$(\boldsymbol{C} - \omega^2[\boldsymbol{M}_{\text{RB}} + \boldsymbol{A}(\omega)] - \text{j}\omega\boldsymbol{B}_{\text{total}}(\omega))\boldsymbol{\xi}(\text{j}\omega) = \boldsymbol{\tau}_{\text{exc}}(\text{j}\omega). \tag{5.73}$$

Assume that $\boldsymbol{\tau}_{\text{exc}}(\text{j}\omega) = F_i\zeta$ for $i = 1, 2, ..., 6$ are harmonic excitation forces proportional to an incoming regular wave $\zeta = \zeta_a e^{\text{j}\omega t}$ where ζ_a is the wave amplitude and F_i denotes the proportional gain. Linear theory implies that $\xi_i = \bar{\xi}_i e^{\text{j}\omega t}$ where $\bar{\xi}_i$ denotes the amplitudes. Consequently,

$$(\boldsymbol{C} - \omega^2[\boldsymbol{M}_{\text{RB}} + \boldsymbol{A}(\omega)] - \text{j}\omega\boldsymbol{B}_{\text{total}}(\omega))\,\bar{\xi}\, e^{\text{j}\omega t} = F_i\zeta_a e^{\text{j}\omega t}. \tag{5.74}$$

Let $\text{RAO}_i(\omega)$ denote the *response amplitude operator* between ζ_a and $\bar{\xi}_i$ for $i = 1, 2, ..., 6$. Hence, the decoupled transfer functions become

$$\text{RAO}_i(\omega) = \frac{\bar{\xi}_i}{\zeta_a} = \frac{F_i}{C_{ii} - \omega^2[M_{RB,ii} + A_{ii}(\omega)] - \text{j}\omega B_{\text{total},ii}(\omega)}. \tag{5.75}$$

Note that $\text{RAO}_i(\omega)$ is a frequency-dependent and complex function. It is common to only consider the absolute value of the response amplitude operator

$$|\text{RAO}_i(\omega)| = \frac{F_i}{\sqrt{(C_{ii} - \omega^2[M_{RB,ii} + A_{ii}(\omega)])^2 + (\omega B_{\text{total},ii}(\omega))^2}}. \tag{5.76}$$

The phase between the wave excitation and the ship motions is

$$\angle\text{RAO}_i(\omega) = -\text{atan}\left(\frac{\omega B_{\text{total},ii}(\omega)}{C_{ii} - \omega^2[M_{RB,ii} + A_{ii}(\omega)]}\right). \tag{5.77}$$

Note the similarity to *Bode* magnitude and phase plots, for which magnitude is logarithmic and given in decibels while phase is plotted in degrees using a common logarithmic frequency axis. The advantage of the logarithmic scale is that asymptotic properties of magnitude and phase are preserved. This is exploited when designing feedback control systems in the frequency domain.

5.4 Time-domain Models including Fluid Memory Effects

The time-domain models are useful both for simulation and control systems design. In particular it is convenient to add nonlinear terms directly in the time domain to describe coupled maneuvers at high speed. Fluid memory effects and wave force terms are kept from the seakeeping theory. Hence, this can be seen as a *unified* approach where seakeeping and maneuvering theory are combined. The basis for the time-domain transformations are the famous papers by Cummins (1962) and Ogilvie (1964), and recent results by Fossen (2005) and Perez and Fossen (2007).

5.4.1 Cummins Equation in SEAKEEPING Coordinates

Cummins (1962) considered the behavior of the fluid and the ship in the time domain *ab initio*. He made the assumption of linearity and considered impulses in the components of motion. This resulted in a boundary value problem in which the potential was separated into two parts: one valid during the duration of the impulses and the other valid after the impulses are extinguished. By expressing the pressure as a function of these potentials and integrating it over the wetted surface of the marine craft, he obtained a vector integro-differential equation, which is known as the *Cummins equation*; see (5.52) in Section 5.3. If we add viscous damping, restoring forces, wave-induced forces and wind forces, the time-domain seakeeping model becomes

$$(\boldsymbol{M}_{\mathrm{RB}} + \bar{\boldsymbol{A}})\ddot{\boldsymbol{\xi}} + \int_{-\infty}^{t} \bar{\boldsymbol{K}}(t-\tau)\dot{\boldsymbol{\xi}}(\tau)\mathrm{d}\tau + \bar{\boldsymbol{C}}\boldsymbol{\xi} = \boldsymbol{\tau}_{\mathrm{wind}} + \boldsymbol{\tau}_{\mathrm{wave}} + \delta\boldsymbol{\tau}. \tag{5.78}$$

In this expression, $\delta\boldsymbol{\tau}$ is the perturbed control input, $\bar{\boldsymbol{A}}$ and $\bar{\boldsymbol{C}}$ are constant matrices to be determined and $\bar{\boldsymbol{K}}(t)$ is a matrix of *retardation functions* given by

$$\bar{\boldsymbol{K}}(t) = \frac{2}{\pi}\int_{0}^{\infty} \boldsymbol{B}_{\mathrm{total}}(\omega)\cos(\omega t)\mathrm{d}\omega. \tag{5.79}$$

Equation (5.78) is a time-domain equation that reveals the structure of the linear equations of motion in $\{s\}$ and it is valid for any excitation, provided the linear assumption is not violated; that is the forces produce small displacements from a state of equilibrium. The terms proportional to the accelerations due to the change in momentum of the fluid have constant coefficients. Moreover, $\bar{\boldsymbol{A}}$ is constant and independent of the frequency of motion as well as forward speed.

Due to the motion of the ship, waves are generated in the free surface. These waves will, in principle, persist at all subsequent times, affecting the motion of the ship. This is known as fluid memory effects, and they are captured by the convolution integral in (5.78). The convolution integral is a function of $\dot{\boldsymbol{\xi}}$ and the retardation functions $\bar{\boldsymbol{K}}(t)$. These functions depend on the hull geometry and the forward speed. This effect appears due to the free surface. For sinusoidal motions, these integrals have components in phase with the motion and 90° out of phase. The latter components contribute to damping, whereas the components in phase with the motion can be added as a frequency-dependent added mass.

The Ogilvie (1964) Transformation

In order to relate the Cummins equation and the matrices $\bar{\boldsymbol{A}}, \bar{\boldsymbol{C}}$ and $\bar{\boldsymbol{K}}$ to the frequency-domain equation, we will rely on a result from Ogilvie (1964). Assume that the floating vessel carries out harmonic oscillations

$$\boldsymbol{\xi} = \bar{\boldsymbol{\xi}}\, e^{\mathrm{j}\omega t} \tag{5.80}$$

with amplitude vector $\bar{\boldsymbol{\xi}}$. Substituting (5.80) into the Cummins equation (5.78) yields

$$-\omega^2[\boldsymbol{M}_{\mathrm{RB}} + \bar{\boldsymbol{A}}]\bar{\boldsymbol{\xi}}\, e^{\mathrm{j}\omega t} + \mathrm{j}\omega \int_{0}^{\infty} \bar{\boldsymbol{K}}(\tau)\bar{\boldsymbol{\xi}}\, e^{\mathrm{j}(\omega t - \omega\tau)}\mathrm{d}\tau + \bar{\boldsymbol{C}}\bar{\boldsymbol{\xi}}\, e^{\mathrm{j}\omega t} = \boldsymbol{\tau}_{\mathrm{wind}} + \boldsymbol{\tau}_{\mathrm{wave}} + \delta\boldsymbol{\tau}$$

where we have replaced τ by $t - \tau$ in the integral. This gives

$$\begin{aligned}&-\omega^2\left\{[\boldsymbol{M}_{\mathrm{RB}} + \bar{\boldsymbol{A}}] - \frac{1}{\omega}\int_{0}^{\infty} \bar{\boldsymbol{K}}(\tau)\sin(\omega\tau)\mathrm{d}\tau\right\}\bar{\boldsymbol{\xi}}\, e^{\mathrm{j}\omega t}\\ &-\mathrm{j}\omega\left\{\int_{0}^{\infty} \bar{\boldsymbol{K}}(\tau)\cos(\omega\tau)\mathrm{d}\tau\right\}\bar{\boldsymbol{\xi}}\, e^{\mathrm{j}\omega t} + \bar{\boldsymbol{C}}\bar{\boldsymbol{\xi}}\, e^{\mathrm{j}\omega t} = \boldsymbol{\tau}_{\mathrm{wind}} + \boldsymbol{\tau}_{\mathrm{wave}} + \delta\boldsymbol{\tau}.\end{aligned} \tag{5.81}$$

The frequency-domain model (5.74) is written

$$-\omega^2[\boldsymbol{M}_{\rm RB} + \boldsymbol{A}(\omega)]\bar{\boldsymbol{\xi}}\, e^{j\omega t} - j\omega \boldsymbol{B}_{\rm total}(\omega)\bar{\boldsymbol{\xi}}\, e^{j\omega t} + \boldsymbol{C}\bar{\boldsymbol{\xi}}\, e^{j\omega t} = \boldsymbol{\tau}_{\rm wind} + \boldsymbol{\tau}_{\rm wave} + \delta\boldsymbol{\tau}. \tag{5.82}$$

By comparing the terms in (5.81) and (5.82), it is seen that

$$\boldsymbol{A}(\omega) = \bar{\boldsymbol{A}} - \frac{1}{\omega}\int_0^\infty \bar{\boldsymbol{K}}(\tau)\sin(\omega\tau)\mathrm{d}\tau \tag{5.83}$$

$$\boldsymbol{B}_{\rm total}(\omega) = \int_0^\infty \bar{\boldsymbol{K}}(\tau)\cos(\omega\tau)\mathrm{d}\tau \tag{5.84}$$

$$\boldsymbol{C} = \bar{\boldsymbol{C}}. \tag{5.85}$$

Equation (5.83) must be valid for all ω. Hence, we choose to evaluate (5.83) at $\omega = \infty$, implying that

$$\bar{\boldsymbol{A}} = \boldsymbol{A}(\infty). \tag{5.86}$$

Equation (5.84) is rewritten using the inverse Fourier transform

$$\bar{\boldsymbol{K}}(t) = \frac{2}{\pi}\int_0^\infty \boldsymbol{B}_{\rm total}(\omega)\cos(\omega t)\mathrm{d}\omega. \tag{5.87}$$

This expression is recognized as a matrix of *retardation functions*. From a numerical point of view is it better to integrate the difference

$$\boldsymbol{K}(t) = \frac{2}{\pi}\int_0^\infty [\boldsymbol{B}_{\rm total}(\omega) - \boldsymbol{B}_{\rm total}(\infty)]\cos(\omega t)\mathrm{d}\omega \tag{5.88}$$

than to use (5.87), since $\boldsymbol{B}_{\rm total}(\omega) - \boldsymbol{B}_{\rm total}(\infty)$ will be exact zero at $\omega = \infty$. Figure 5.4 shows a typical retardation function that converges to zero in 15– 20 s. The tail will oscillate if (5.87) is used instead of (5.88) in the numerical integration.

The relationship between $\bar{\boldsymbol{K}}(t)$ and $\boldsymbol{K}(t)$ follows from

$$\begin{aligned}\bar{\boldsymbol{K}}(t) &= \frac{2}{\pi}\int_0^\infty [\boldsymbol{B}_{\rm total}(\omega) - \boldsymbol{B}_{\rm total}(\infty) + \boldsymbol{B}_{\rm total}(\infty)]\cos(\omega\tau)\mathrm{d}\omega \\ &= \boldsymbol{K}(t) + \frac{2}{\pi}\int_0^\infty \boldsymbol{B}_{\rm total}(\infty)\cos(\omega\tau)\mathrm{d}\omega.\end{aligned} \tag{5.89}$$

Then it follows that

$$\begin{aligned}\int_{-\infty}^t \bar{\boldsymbol{K}}(t-\tau)\dot{\boldsymbol{\xi}}(\tau)\mathrm{d}\tau &= \int_{-\infty}^t \boldsymbol{K}(t-\tau)\dot{\boldsymbol{\xi}}(\tau)\mathrm{d}\tau + \boldsymbol{B}_{\rm total}(\infty)\dot{\boldsymbol{\xi}} \\ &\overset{\rm causal}{=} \int_0^t \boldsymbol{K}(t-\tau)\dot{\boldsymbol{\xi}}(\tau)\mathrm{d}\tau + \boldsymbol{B}_{\rm total}(\infty)\dot{\boldsymbol{\xi}}.\end{aligned} \tag{5.90}$$

We are now ready to state the main result.

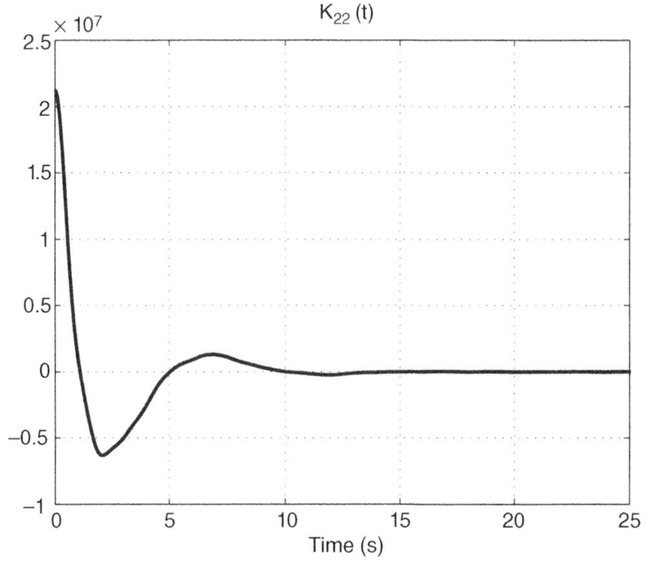

Figure 5.4 Typical plot of the retardation function $K_{22}(t)$ in sway.

Time-domain Seakeeping Equations of Motion

The relationship between the time-domain equation (5.78) and the frequency-domain equation (5.74) is established through (5.83)–(5.85) and (5.90). This gives

$$[\boldsymbol{M}_{\mathrm{RB}} + \boldsymbol{A}(\infty)]\ddot{\boldsymbol{\xi}} + \boldsymbol{B}_{\mathrm{total}}(\infty)\dot{\boldsymbol{\xi}} + \int_0^t \boldsymbol{K}(t-\tau)\dot{\boldsymbol{\xi}}(\tau)\mathrm{d}\tau + \boldsymbol{C}\boldsymbol{\xi} = \boldsymbol{\tau}_{\mathrm{wind}} + \boldsymbol{\tau}_{\mathrm{wave}} + \delta\boldsymbol{\tau} \quad (5.91)$$

where $\boldsymbol{K}(t-\tau)$ is defined by (5.88). The equations of motion (5.91) describe the perturbed motion $\boldsymbol{\xi}$ of a marine craft in 6 DOFs using seakeeping coordinates. We will now transform this result to the rotating frame $\{b\}$.

5.4.2 Linear Time-domain Seakeeping Equations in BODY Coordinates

Two representations in $\{b\}$ are available: one using zero-speed potential coefficients and one using speed-dependent matrices. Motion control systems are usually formulated in $\{b\}$. Consequently, we need to transform the time-domain representation of the Cummins equation (5.91) from $\{s\}$ to $\{b\}$. When transforming the equations of motion to the rotating frame $\{b\}$, *Coriolis and centripetal forces* between $\{s\}$ and $\{b\}$ appear; see Figure 5.5. To illustrate this, consider

$$[\boldsymbol{M}_{\mathrm{RB}} + \boldsymbol{A}(\infty)]\ddot{\boldsymbol{\xi}} + \boldsymbol{B}_{\mathrm{total}}(\infty)\dot{\boldsymbol{\xi}} + \int_0^t \boldsymbol{K}(t-\tau)\dot{\boldsymbol{\xi}}(\tau)\mathrm{d}\tau + \boldsymbol{C}\boldsymbol{\xi} = \boldsymbol{\tau}_{\mathrm{wind}} + \boldsymbol{\tau}_{\mathrm{wave}} + \delta\boldsymbol{\tau} \quad (5.92)$$

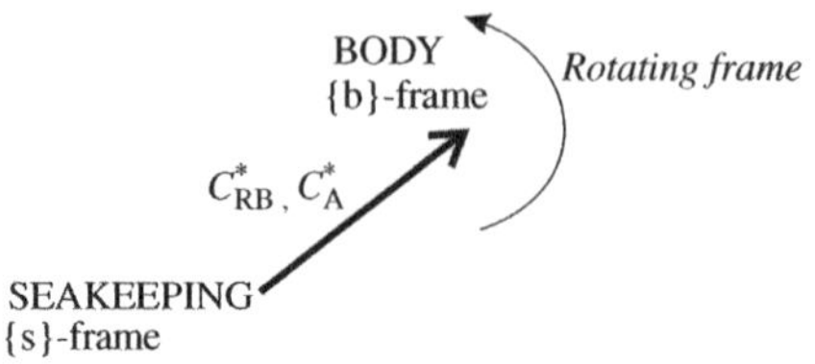

Figure 5.5 Coriolis matrices due to the rotation of the body-fixed frame $\{b\}$ about the inertial frame $\{s\}$.

which can be transformed from $\{s\}$ to $\{b\}$ by using the kinematic transformations (5.21) and (5.45)–(5.46) derived in Section 5.2.2. This gives

$$[\boldsymbol{M}_{\rm RB} + \boldsymbol{A}(\infty)][\dot{\boldsymbol{\nu}} + U\boldsymbol{L}\boldsymbol{\nu}] + \boldsymbol{B}_{\rm total}(\infty)[\boldsymbol{\nu} + U(\boldsymbol{L}\delta\boldsymbol{\eta} - \boldsymbol{e}_1)] + \int_0^t \boldsymbol{K}(t-\tau)\delta\boldsymbol{\nu}(\tau)\mathrm{d}\tau + \boldsymbol{C}\delta\boldsymbol{\eta} = \boldsymbol{\tau}_{\rm wind} + \boldsymbol{\tau}_{\rm wave} + (\boldsymbol{\tau} - \bar{\boldsymbol{\tau}}). \tag{5.93}$$

The steady-state control force $\bar{\boldsymbol{\tau}}$ needed to obtain the forward speed U when $\boldsymbol{\tau}_{\rm wind} = \boldsymbol{\tau}_{\rm wave} = \boldsymbol{0}$ and $\delta\boldsymbol{\eta} = \boldsymbol{0}$ is

$$\bar{\boldsymbol{\tau}} = \boldsymbol{B}_{\rm total}(\infty)U\boldsymbol{e}_1. \tag{5.94}$$

Hence, (5.93) can be rewritten as

$$[\boldsymbol{M}_{\rm RB} + \boldsymbol{A}(\infty)][\dot{\boldsymbol{\nu}} + U\boldsymbol{L}\boldsymbol{\nu}] + \boldsymbol{B}_{\rm total}(\infty)[\boldsymbol{\nu} + U\boldsymbol{L}\delta\boldsymbol{\eta}] + \int_0^t \boldsymbol{K}(t-\tau)\delta\boldsymbol{\nu}(\tau)\mathrm{d}\tau + \boldsymbol{C}\delta\boldsymbol{\eta} = \boldsymbol{\tau}_{\rm wind} + \boldsymbol{\tau}_{\rm wave} + \boldsymbol{\tau}. \tag{5.95}$$

In this expression, the linearized Coriolis–centripetal forces due to rigid-body mass and hydrodynamic added mass are recognized as $\boldsymbol{M}_{\rm RB}U\boldsymbol{L}\boldsymbol{\nu}$ and $\boldsymbol{A}(\infty)U\boldsymbol{L}\boldsymbol{\nu}$, respectively.

When computing the damping and retardation functions, it is common to neglect the influence of $\delta\boldsymbol{\eta}$ on the forward speed such that

$$\delta\boldsymbol{\nu} \approx \boldsymbol{\nu} + U(\boldsymbol{L}\delta\boldsymbol{\eta} - \boldsymbol{e}_1) \approx \boldsymbol{\nu} - U\boldsymbol{e}_1. \tag{5.96}$$

Hence, we can present the linear seakeeping equations expressed in the $\{b\}$ frame.

Linear Equations of Motion Using Zero-speed Potential Coefficients

The kinematic equation between $\{b\}$ and $\{n\}$ is

$$\dot{\boldsymbol{\eta}} = \boldsymbol{J}_\Theta(\boldsymbol{\eta})\boldsymbol{\nu}. \tag{5.97}$$

From (5.95) and (5.96) it follows that

$$\boldsymbol{M}\dot{\boldsymbol{\nu}} + \boldsymbol{C}^*_{\mathrm{RB}}\boldsymbol{\nu} + \boldsymbol{C}^*_{\mathrm{A}}\boldsymbol{\nu} + \boldsymbol{D}\boldsymbol{\nu} + \int_0^t \boldsymbol{K}(t-\tau)[\boldsymbol{\nu}(\tau) - U\boldsymbol{e}_1]\mathrm{d}\tau + \boldsymbol{G}\boldsymbol{\eta} = \boldsymbol{\tau}_{\mathrm{wind}} + \boldsymbol{\tau}_{\mathrm{wave}} + \boldsymbol{\tau} \tag{5.98}$$

where $\boldsymbol{M} = \boldsymbol{M}_{\mathrm{RB}} + \boldsymbol{M}_{\mathrm{A}}$ and

$$\begin{aligned}
\boldsymbol{M}_{\mathrm{A}} &= \boldsymbol{A}(\infty)\\
\boldsymbol{C}^*_{\mathrm{A}} &= U\boldsymbol{A}(\infty)\boldsymbol{L}\\
\boldsymbol{C}^*_{\mathrm{RB}} &= U\boldsymbol{M}_{\mathrm{RB}}\boldsymbol{L}\\
\boldsymbol{D} &= \boldsymbol{B}_{\mathrm{total}}(\infty)\\
\boldsymbol{G} &= \boldsymbol{C}.
\end{aligned}$$

We have here exploited the fact that $\boldsymbol{C}\delta\boldsymbol{\eta} = \boldsymbol{G}\boldsymbol{\eta}$. Notice that $\boldsymbol{C}^*_{\mathrm{RB}}$ and $\boldsymbol{C}^*_{\mathrm{A}}$ are linearized Coriolis and centripetal forces due to the rotation of $\{b\}$ about $\{s\}$. At zero speed, these terms vanish.

Linear Equations of Motion Using Speed-dependent Potential Coefficients

Some potential theory programs compute the potential coefficients as functions of speed and frequency

$$\boldsymbol{A}_U(\omega, U) = \boldsymbol{A}(\omega) + \boldsymbol{\alpha}(\omega, U) \tag{5.99}$$

$$\boldsymbol{B}_U(\omega, U) = \boldsymbol{B}(\omega) + \boldsymbol{\beta}(\omega, U) \tag{5.100}$$

where $\boldsymbol{\alpha}(\omega, U)$ and $\boldsymbol{\beta}(\omega, U)$ denote the speed-dependent terms. For these codes, $\boldsymbol{\beta}(\omega, U)$ can include the matrix $\boldsymbol{C}^*_{\mathrm{A}} = U\boldsymbol{A}(\infty)\boldsymbol{L}$ as well as other effects. A frequently used representation is

$$\boldsymbol{\beta}(\omega, U) = \underbrace{U\boldsymbol{A}(\infty)\boldsymbol{L}}_{\boldsymbol{C}^*_{\mathrm{A}}} + \boldsymbol{B}_{\mathrm{ITTC}}(\omega, U) + \boldsymbol{B}_{\mathrm{IKEDA}}(\omega, U) + \boldsymbol{B}_{\mathrm{transom}}(\omega, U) \tag{5.101}$$

where the subscripts denote linearized ITTC resistance, IKEDA damping and transom stern effects. If the speed-dependent matrices (5.99) and (5.100) are used instead of the zero-speed matrices in (5.98), the equations of motion for each speed, U = constant, take the following form

$$M_U\dot{\nu} + C_{\rm RB}^{*}\nu + D_U\nu + \int_0^t K_U(t-\tau, U)[\nu(\tau) - Ue_1]{\rm d}\tau + G\eta = \tau_{\rm wind} + \tau_{\rm wave} + \tau \tag{5.102}$$

where the matrix $C_{\rm A}^{*}$ is superfluous and

$$M_U = M_{\rm RB} + A_U(\infty, U)$$
$$C_{\rm RB}^{*} = UM_{\rm RB}L$$
$$D_U = B_{{\rm total},U}(\infty, U)$$
$$G = C$$

and

$$K_U(t, U) = \frac{2}{\pi}\int_0^{\infty} [B_{{\rm total},U}(\omega, U) - B_{{\rm total},U}(\infty, U)]\cos(\omega t){\rm d}\omega. \tag{5.103}$$

The speed-dependent equations of motion (5.102) are computed at each speed U = constant while (5.98) is valid for any $U(t)$ provided that $U(t)$ is slowly varying. It is advantageous to use (5.98) since only the zero-speed potential coefficients $A(\omega)$ and $B(\omega)$ are needed in the implementation. This is based on the assumption that the $C_{\rm A}$ matrix is the only element in $B_U(\omega, U)$. In other words, it is assumed that

$$\beta(\omega, U) := C_{\rm A}^{*} = UA(\infty)L. \tag{5.104}$$

When using (5.98) instead of (5.102), it is necessary to add the remaining damping terms directly in the time-domain equations, as explained in Section 5.4.3.

Properties of A, B and K

The following properties are useful when processing the hydrodynamic data:

- Asymptotic values for $\omega = 0$:

$$B(0) = 0.$$

- Asymptotic values for $\omega \rightarrow \infty$:

$$A_U(\infty, U) = 0$$
$$A_U(\infty, U) = A(\infty).$$

These properties can be exploited when computing $\boldsymbol{K}(t)$ numerically since most seakeeping codes only return values on an interval $\omega = [\omega_{\min}, \omega_{\max}]$.

Some useful properties of the retardation functions are:

- Asymptotic value for $t = 0$:

$$\lim_{t \to 0} \boldsymbol{K}(t) \neq \boldsymbol{0} < \infty. \tag{5.105}$$

- Asymptotic value for $t \to \infty$:

$$\lim_{t \to \infty} \boldsymbol{K}(t) = \boldsymbol{0}. \tag{5.106}$$

A plot illustrating the retardation function in sway is shown in Figure 5.4.

5.4.3 Nonlinear Unified Seakeeping and Maneuvering Model with Fluid Memory Effects

Consider the seakeeping model (5.98) based on the potential coefficients

$$\begin{aligned} &\boldsymbol{M}\dot{\boldsymbol{v}} + \boldsymbol{C}^*_{\mathrm{RB}}\boldsymbol{v} + \boldsymbol{C}^*_{\mathrm{A}}\boldsymbol{v} + \boldsymbol{D}\boldsymbol{v} \\ &\quad + \int_0^t \boldsymbol{K}(t-\tau)[\boldsymbol{v}(\tau) - U\boldsymbol{e}_1]\mathrm{d}\tau + \boldsymbol{G}\boldsymbol{\eta} = \boldsymbol{\tau}_{\mathrm{wind}} + \boldsymbol{\tau}_{\mathrm{wave}} + \boldsymbol{\tau}. \end{aligned} \tag{5.107}$$

For this model, the linearized Coriolis and centripetal matrices $\boldsymbol{C}^*_{\mathrm{RB}}$ and $\boldsymbol{C}^*_{\mathrm{A}}$ can be replaced by their nonlinear counterparts $\boldsymbol{C}_{\mathrm{RB}}(\boldsymbol{v})$ and $\boldsymbol{C}_A(\boldsymbol{v})$; see Section 6.3.1. In addition, the nonlinear damping $\boldsymbol{D}(\boldsymbol{v})\boldsymbol{v}$ or maneuvering coefficients can be added directly in the time domain.

Unified Seakeeping and Maneuvering Model

Some authors refer to (5.107) as a *unified model* when nonlinear maneuvering terms are included since it merges the maneuvering and seakeeping theories (see Bailey et al. 1998; Fossen 2005). This gives a *unified* seakeeping and maneuvering model in the following form

$$\begin{aligned} &\boldsymbol{M}_{\mathrm{RB}}\dot{\boldsymbol{v}} + \boldsymbol{C}_{\mathrm{RB}}(\boldsymbol{v})\boldsymbol{v} \\ &\quad + \boldsymbol{M}_{\mathrm{A}}\dot{\boldsymbol{v}}_{\mathrm{r}} + \boldsymbol{C}_{\mathrm{A}}(\boldsymbol{v}_{\mathrm{r}})\boldsymbol{v}_{\mathrm{r}} + \boldsymbol{D}(\boldsymbol{v}_{\mathrm{r}})\boldsymbol{v}_{\mathrm{r}} + \boldsymbol{\mu}_{\mathrm{r}} + \boldsymbol{G}\boldsymbol{\eta} = \boldsymbol{\tau}_{\mathrm{wind}} + \boldsymbol{\tau}_{\mathrm{wave}} + \boldsymbol{\tau} \end{aligned} \tag{5.108}$$

where the velocity vector $\boldsymbol{v}$ in the hydrodynamic terms has been replaced by relative velocity $\boldsymbol{v}_{\mathrm{r}}$ to model the effects of ocean currents. The seakeeping *fluid memory effects* are captured in the term

$$\boldsymbol{\mu}_{\mathrm{r}} := \int_0^t \boldsymbol{K}(t-\tau)\underbrace{[\boldsymbol{v}_{\mathrm{r}}(\tau) - U_{\mathrm{r}}\boldsymbol{e}_1]}_{\delta \boldsymbol{v}_{\mathrm{r}}}\mathrm{d}\tau \tag{5.109}$$

where $\delta \boldsymbol{\nu}_r = (\boldsymbol{\nu} - \boldsymbol{\nu}_c) - U_r \boldsymbol{e}_1$ and $U_r = U - U_c$ is the relative speed. If the currents are neglected the formula for the fluid memory effects simplifies to

$$\boldsymbol{\mu} := \int_0^t \boldsymbol{K}(t-\tau)\underbrace{[\boldsymbol{\nu}(\tau) - U\boldsymbol{e}_1]}_{\delta \boldsymbol{\nu}}\mathrm{d}\tau \tag{5.110}$$

Constant and Irrotational Ocean Currents

The model (5.108) can be simplified if the *ocean currents* are assumed to be *constant* and *irrotational* in $\{n\}$ such that Property 10.1 is satisfied. Following the approach in Section 10.3 this gives

$$\dot{\boldsymbol{\eta}} = \boldsymbol{J}_\Theta(\boldsymbol{\eta})\boldsymbol{\nu} \tag{5.111}$$

$$\boldsymbol{M}\dot{\boldsymbol{\nu}}_r + \boldsymbol{C}(\boldsymbol{\nu}_r)\boldsymbol{\nu}_r + \boldsymbol{D}(\boldsymbol{\nu}_r)\boldsymbol{\nu}_r + \boldsymbol{\mu}_r + \boldsymbol{G}\boldsymbol{\eta} = \boldsymbol{\tau}_{\text{wind}} + \boldsymbol{\tau}_{\text{wave}} + \boldsymbol{\tau} \tag{5.112}$$

where $\boldsymbol{M} = \boldsymbol{M}_{\text{RB}} + \boldsymbol{M}_{\text{A}}$ and $\boldsymbol{C} = \boldsymbol{C}_{\text{RB}} + \boldsymbol{C}_{\text{A}}$.

Example 5.1 (Zero-speed Model for DP with Fluid Memory Effects)
For stationkeeping ($U = 0$ and $r = 0$), the model (5.112) reduces to

$$\dot{\boldsymbol{\eta}} = \boldsymbol{J}_\Theta(\boldsymbol{\eta})\boldsymbol{\nu} \tag{5.113}$$

$$\boldsymbol{M}\dot{\boldsymbol{\nu}}_r + \boldsymbol{D}\boldsymbol{\nu}_r + \boldsymbol{\mu}_r + \boldsymbol{G}\boldsymbol{\eta} = \boldsymbol{\tau}_{\text{wind}} + \boldsymbol{\tau}_{\text{wave}} + \boldsymbol{\tau} \tag{5.114}$$

under the assumptions that $\boldsymbol{C}(\boldsymbol{\nu}_r) = \boldsymbol{0}$ and $\boldsymbol{D}(\boldsymbol{\nu}_r) = \boldsymbol{D}$. This is similar to the result of Fossen and Smogeli (2004).

5.5 Identification of Fluid Memory Effects

Kristiansen and Egeland (2003) and Kristiansen *et al.* (2005) developed a state-space approximation for $\boldsymbol{\mu}$ using realization theory. Other methods such as the impulse response LS fitting can also be used (see Yu and Falnes 1995, 1998). The time-domain methods are usually used in conjunction with model reduction in order to obtain a state-space model of smaller dimension suited for feedback control and time-domain simulation. This often results in a state-space model $(\boldsymbol{A}_r, \boldsymbol{B}_r, \boldsymbol{C}_r, \boldsymbol{D}_r)$ where the $\boldsymbol{D}_r$ matrix is non-zero (Perez and Fossen 2008). This is non-physical since potential damping should not amplify signals at low frequencies. Hence, care must be taken when using time-domain methods. As a consequence of this, frequency-domain identification methods are much more accurate and they have the advantage that a transfer function of correct relative degree can be chosen prior to the identification process. Hence, model reduction in the time domain can be avoided since the estimated transfer function can be converted into a $(\boldsymbol{A}_r, \boldsymbol{B}_r, \boldsymbol{C}_r)$ state-space model exploiting the structural constraint $\boldsymbol{D}_r = \boldsymbol{0}$ directly. A more detailed discussion of the identification methods are found in Perez and Fossen (2008) while practical aspects are reported in Perez and Fossen (2011).

5.5.1 Frequency-domain Identification Using the MSS FDI Toolbox

This section illustrates how the fluid memory effects can be accurately approximated using frequency-domain identification. The main tool for this is the MSS FDI toolbox (Perez and Fossen 2009). When using the frequency-domain approach, the property that the mapping $\delta \boldsymbol{v}_r \rightarrow \boldsymbol{\mu}_r$ has relative degree one is exploited. Hence, the fluid memory effects $\boldsymbol{\mu}_r$ can be approximated by a matrix $\boldsymbol{H}(s)$ containing relative degree one transfer functions (see Figure 5.6)

$$h_{ij}(s) = \frac{p_r s^r + p_{r-1} s^{r-1} + \cdots + p_0}{s^n + q_{n-1} s^{n-1} + \cdots + q_0}, \qquad r = n-1, \quad n \geq 2 \tag{5.115}$$

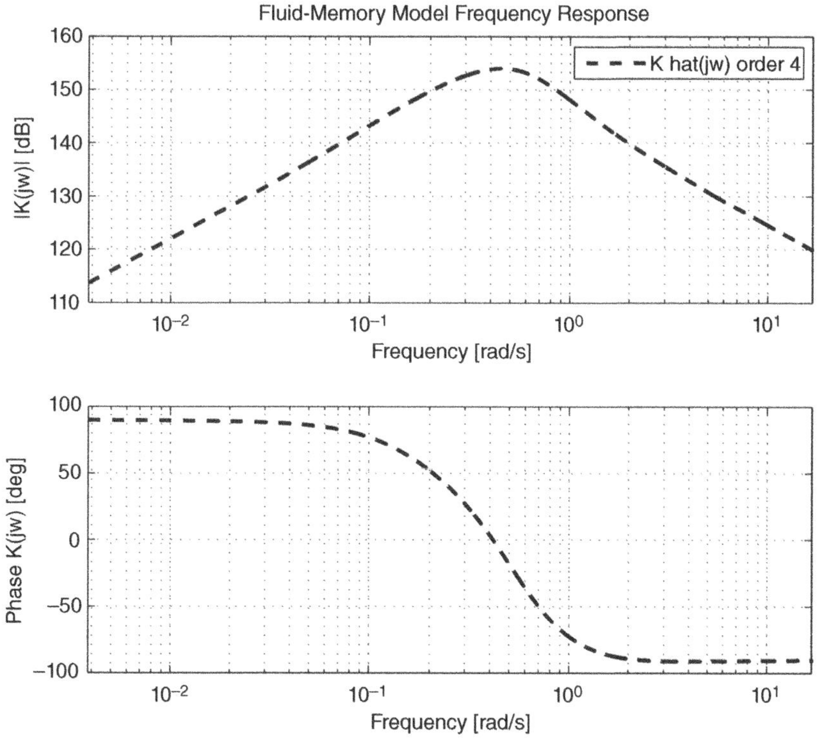

Figure 5.6 Bode plot showing the identified transfer function $h_{33}(s)$ of an FPSO when $A_{33}(\infty)$ is treated as an unknown to be estimated.

such that

$$\boldsymbol{\mu}_{\mathrm{r}} = \boldsymbol{H}(s)\delta\boldsymbol{\nu}_{\mathrm{r}} \tag{5.116}$$

with

$$\boldsymbol{H}(s) = \boldsymbol{C}_{\mathrm{r}}(s\boldsymbol{I} - \boldsymbol{A}_{\mathrm{r}})^{-1}\boldsymbol{B}_{\mathrm{r}}. \tag{5.117}$$

Consequently, the corresponding state-space model is in the form

$$\dot{\boldsymbol{x}} = \boldsymbol{A}_{\mathrm{r}}\boldsymbol{x} + \boldsymbol{B}_{\mathrm{r}}\delta\boldsymbol{\nu}_{\mathrm{r}} \tag{5.118}$$

$$\boldsymbol{\mu}_{\mathrm{r}} = \boldsymbol{C}_{\mathrm{r}}\boldsymbol{x}. \tag{5.119}$$

The states $\boldsymbol{x}$ in (5.119) reflect the fact that once the marine craft changes the momentum of the fluid, this will affect the forces in the future. In other words, the radiation forces at a particular time depend on the history of the velocity of the marine craft up to the present time. The dimension of $\boldsymbol{x}$ and the matrices $\boldsymbol{A}_{\mathrm{r}}$, $\boldsymbol{B}_{\mathrm{r}}$ and $\boldsymbol{C}_{\mathrm{r}}$ depend on the order of the identified transfer functions (usually 2–20).

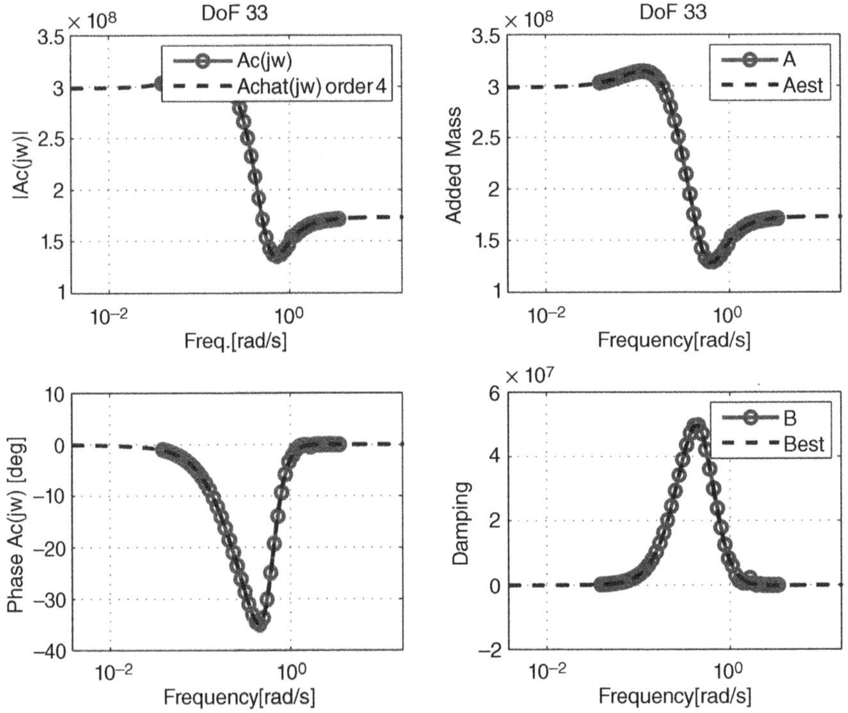

Figure 5.7 FPSO identification results for $h_{33}(s)$ without using the infinite added mass $A_{33}(\infty)$. The left-hand-side plots show the complex coefficient and its estimate while added mass and damping are plotted on the right-hand-side plots.

MATLAB:

The fluid memory transfer function (5.115) can be computed using the MSS toolbox, which includes the FDI toolbox for frequency-domain identification (Perez and Fossen 2009). The toolbox includes two demo files for the cases where infinite added mass is unknown (2D strip theory codes) or computed by the hydrodynamic code, for instance the 3D code by WAMIT.

Example 5.2 (Computation of Fluid Memory Effects)
Consider the FPSO data set in the MSS toolbox and assume that the infinite-frequency added mass matrix is unknown. Hence, we can estimate the fluid transfer function $h_{33}(s)$ by using the following Matlab code

```
load fpso
Dof = [3,3];                    % Use coupling 3-3 heave-heave
Nf = length(vessel.freqs);
w = vessel.freqs(1:Nf-1)';
Ainf = vessel.A(Dof(1),Dof(2),Nf)  % Ainf computed by WAMIT
Ainf =
   1.7283e+08

A = reshape(vessel.A(Dof(1),Dof(2),1:Nf-1),1,length(w))';
B = reshape(vessel.B(Dof(1),Dof(2),1:Nf-1),1,length(w))';
```

A fourth-order transfer function of relative degree 1 is found by using the following options (see Perez and Fossen (2009) for a more detailed explanation).

```
FDIopt.OrdMax = 20;
FDIopt.AinfFlag = 0;
FDIopt.Method = 2;
FDIopt.Iterations = 20;
FDIopt.PlotFlag = 0;
FDIopt.LogLin = 1;
FDIopt.wsFactor = 0.1;
FDIopt.wminFactor = 0.1;
FDIopt.wmaxFactor = 5;

[Krad,Ainf_hat] = FDIRadMod(W,A,0,B,FDIopt,Dof)

Krad =

     1.647e07 s^3 + 2.358e07 s^2 + 2.122e06 s
  ------------------------------------------------
  s^4 + 1.253 s^3 + 0.7452 s^2 + 0.2012 s + 0.01686

Ainf_hat =
  1.7265e+08
```

The state-space model (5.119) is obtained by calling

```
[num,den] = tfdata(Krad,'v');
[A_r,B_r,C_r,D_r] = tf2ss(num,den)

A_r =
  -1.2529    -0.7452    -0.2012    -0.0169
   1.0000          0          0          0
        0     1.0000          0          0
        0          0     1.0000          0
B_r =
   1
   0
   0
   0
C_r =
  1.0e+07 *
   1.6472     2.3582     0.2122          0
D_r =
   0
```

The identified transfer function $h_{33}(s)$ *is plotted in Figure 5.6 while curve fitting of amplitude, phase, added mass and potential damping are shown in Figure 5.7. The estimated transfer function and potential coefficients are matching the experimental data with good accuracy. Notice that the asymptotic behavior satisfies the properties of added mass* $A_{33}(\omega)$ *and potential damping* $B_{33}(\omega)$ *as expected.*

6

Maneuvering Models

In Chapter 5 the 6-DOF seakeeping equations of motion for a ship in a seaway were presented. The seakeeping model is based on linear theory and a potential theory program is used to compute the frequency-dependent hydrodynamic forces for varying wave excitation frequencies. The time-domain representation of the seakeeping model is very useful for accurate prediction of motions and sealoads of floating structures offshore. The seakeeping theory can also be applied to displacement ships moving at constant speed. Seakeeping time-domain models are limited to linear theory since it is necessary to approximate the fluid memory effects by impulse responses or transfer functions.

An alternative to the seakeeping formalism is to use maneuvering theory to describe the motions of marine craft in 3 DOFs (*surge*, *sway* and *yaw*). Sometimes roll is augmented to the horizontal-plane model to describe more accurately the coupled lateral motions, that is *sway–roll–yaw* couplings while *surge* is left decoupled; see Section 6.6. In maneuvering theory, frequency-dependent added mass and potential damping are approximated by constant values and thus it is not necessary to compute the fluid-memory effects. The main results of this chapter are based on the assumption that the hydrodynamic forces and moments can be approximated at one frequency of oscillation such that the fluid-memory effects can be neglected. The result is a nonlinear mass–damper–spring system with constant coefficients.

In the following sections, it is shown that the maneuvering equations of motion can be represented by (Fossen 1991)

$$\boldsymbol{M}\dot{\boldsymbol{\nu}} + \boldsymbol{C}(\boldsymbol{\nu})\boldsymbol{\nu} + \boldsymbol{D}(\boldsymbol{\nu})\boldsymbol{\nu} + \boldsymbol{g}(\boldsymbol{\eta}) + \boldsymbol{g}_0 = \boldsymbol{\tau} + \boldsymbol{\tau}_{\rm wind} + \boldsymbol{\tau}_{\rm wave}. \tag{6.1}$$

In the case of *irrotational ocean currents*, the relative velocity vector

$$\boldsymbol{\nu}_{\rm r} = \boldsymbol{\nu} - \boldsymbol{\nu}_{\rm c}, \qquad \boldsymbol{\nu}_{\rm c} = [u_{\rm c},\ v_{\rm c},\ w_{\rm c},\ 0,\ 0,\ 0]^{\top}$$

Handbook of Marine Craft Hydrodynamics and Motion Control, Second Edition. Thor I. Fossen.

contributes to the hydrodynamic terms such that

$$\underbrace{M_{\mathrm{RB}}\dot{\nu} + C_{\mathrm{RB}}(\nu)\nu}_{\text{rigid-body forces}} + \underbrace{M_{\mathrm{A}}\dot{\nu}_{\mathrm{r}} + C_{\mathrm{A}}(\nu_{\mathrm{r}})\nu_{\mathrm{r}} + D(\nu_{\mathrm{r}})\nu_{\mathrm{r}}}_{\text{hydrodynamic forces}} + \underbrace{g(\eta) + g_0}_{\text{hydrostatic forces}} = \tau + \tau_{\mathrm{wind}} + \tau_{\mathrm{wave}}. \tag{6.2}$$

The model (6.2) can be simplified if the *ocean currents* are assumed to be *constant* and *irrotational* in $\{n\}$ such that (see Section 10.3)

$$\dot{\nu}_{\mathrm{c}} = \begin{bmatrix} -S(\omega_{\mathrm{nb}}^b) & 0_{3\times3} \\ 0_{3\times3} & 0_{3\times3} \end{bmatrix} \nu_{\mathrm{c}}. \tag{6.3}$$

According to Property 10.1, it is then possible to represent the equations of motion by relative velocities only

$$M\dot{\nu}_{\mathrm{r}} + C(\nu_{\mathrm{r}})\nu_{\mathrm{r}} + D(\nu_{\mathrm{r}})\nu_{\mathrm{r}} + g(\eta) + g_0 = \tau + \tau_{\mathrm{wind}} + \tau_{\mathrm{wave}} \tag{6.4}$$

where

$M = M_{\mathrm{RB}} + M_{\mathrm{A}}$ – system inertia matrix (including added mass)
$C(\nu_{\mathrm{r}}) = C_{\mathrm{RB}}(\nu_{\mathrm{r}}) + C_{\mathrm{A}}(\nu_{\mathrm{r}})$ – Coriolis–centripetal matrix (including added mass)
$D(\nu_{\mathrm{r}})$ – damping matrix
$g(\eta)$ – vector of gravitational/buoyancy forces and moments
g_0 – vector used for pretrimming (ballast control)
τ – vector of control inputs
τ_{wind} – vector of generalized wind forces
τ_{wave} – vector of generalized wave-induced forces.

The expressions for M, $C(\nu_{\mathrm{r}})$, $D(\nu_{\mathrm{r}})$, $g(\eta)$ and g_0 are derived in the forthcoming sections while the environmental forces τ_{wind} and τ_{wave} are treated separately in Chapter 10. The maneuvering model presented in this chapter is mainly intended for controller–observer design, prediction and computer simulations in combination with system identification and parameter estimation.

Hydrodynamic programs compute mass, inertia, potential damping and restoring forces while a more detailed treatment of viscous dissipative forces are found in the extensive literature on hydrodynamics; see Faltinsen (1990, 2005), Newman (1977), Sarpkaya (1981) and Triantafyllou and Hover (2002). Other useful references on marine craft modeling are Lewandowski (2004) and Perez (2005).

6.1 Rigid-body Kinetics

Recall from Chapter 3 that the rigid-body kinetics can be expressed as

$$\boldsymbol{M}_{\mathrm{RB}}\dot{\boldsymbol{\nu}} + \boldsymbol{C}_{\mathrm{RB}}(\boldsymbol{\nu})\boldsymbol{\nu} = \boldsymbol{\tau}_{\mathrm{RB}} \tag{6.5}$$

where $\boldsymbol{M}_{\mathrm{RB}} = \boldsymbol{M}_{\mathrm{RB}}^{\top} > 0$ is the rigid-body mass matrix and $\boldsymbol{C}_{\mathrm{RB}}(\boldsymbol{\nu}) = -\boldsymbol{C}_{\mathrm{RB}}^{\top}(\boldsymbol{\nu})$ is the rigid-body Coriolis and centripetal matrix due to the rotation of $\{b\}$ about the approximate inertial frame $\{n\}$. The horizontal motion of a maneuvering ship or semi-submersible is given by the motion components in surge, sway and yaw. Consequently, the state vectors are chosen as $\boldsymbol{\nu} = [u,\ v,\ r]^{\top}$ and $\boldsymbol{\eta} = [x^n,\ y^n,\ \psi]^{\top}$. It is also common to assume that the craft has homogeneous mass distribution and xz-plane symmetry so that

$$I_{xy} = I_{yz} = 0. \tag{6.6}$$

Let the $\{b\}$-frame coordinate origin be set in the centerline of the craft in the point CO, such that $y_g = 0$. Under the previously stated assumptions, the matrices (3.49) and (3.62) associated with the rigid-body kinetics reduce to

$$\boldsymbol{M}_{\mathrm{RB}} = \begin{bmatrix} m & 0 & 0 \\ 0 & m & mx_g \\ 0 & mx_g & I_z \end{bmatrix}, \quad \boldsymbol{C}_{\mathrm{RB}}(\boldsymbol{\nu}) = \begin{bmatrix} 0 & -mr & -mx_g r \\ mr & 0 & 0 \\ mx_g r & 0 & 0 \end{bmatrix}. \tag{6.7}$$

Note that surge is decoupled from sway and yaw in $\boldsymbol{M}_{\mathrm{RB}}$ due to symmetry considerations of the system inertia matrix (see Section 3.3).

The linear approximation to (6.5) about $u = U = \text{constant}$, $v = 0$ and $r = 0$ is

$$\boldsymbol{M}_{\mathrm{RB}}\dot{\boldsymbol{\nu}} + \boldsymbol{C}_{\mathrm{RB}}^{*}\boldsymbol{\nu} = \boldsymbol{\tau}_{\mathrm{RB}} \tag{6.8}$$

where

$$\boldsymbol{C}_{\mathrm{RB}}^{*} = \begin{bmatrix} 0 & 0 & 0 \\ 0 & 0 & mU \\ 0 & 0 & mx_g U \end{bmatrix}. \tag{6.9}$$

6.2 Potential Coefficients

Hydrodynamic potential theory programs can be used to compute the added mass and damping matrices by integrating the pressure of the fluid over the wetted surface of the hull; see Section 5.1. These programs assume that viscous effects can be neglected. Consequently, it is necessary to add viscous forces manually. The programs are also based on the assumptions that first- and second-order wave forces can be linearly superimposed.

The potential coefficients are usually represented as frequency-dependent matrices for 6-DOF motions. The matrices are

- $\boldsymbol{A}(\omega)$ added mass
- $\boldsymbol{B}(\omega)$ potential damping

where ω is the wave excitation frequency of a sinusoidal (regular) wave generated by a wave maker or the ocean. Figure 6.1 illustrates the components in sway.

In seakeeping analysis, the equations of motion (see Chapter 5)

$$(\boldsymbol{M}_{\mathrm{RB}} + \boldsymbol{A}(\omega))\ddot{\boldsymbol{\xi}} + (\boldsymbol{B}(\omega) + \boldsymbol{B}_{\mathrm{V}}(\omega))\dot{\boldsymbol{\xi}} + \boldsymbol{C}\boldsymbol{\xi} = \delta\boldsymbol{\tau} + \boldsymbol{\tau}_{\mathrm{wind}} + \boldsymbol{\tau}_{\mathrm{wave}} \tag{6.10}$$

are perturbations $\boldsymbol{\xi} = [\delta x^n,\ \delta y^n,\ \delta z^n,\ \delta\phi,\ \delta\theta,\ \delta\psi]^{\mathsf{T}}$ about an inertial equilibrium frame.

Equation (6.10) should not be used in computer simulations. As earlier mentioned, (6.10) is not an ordinary differential equation since it combines time and frequency. As stressed in Chapter 5, the time-domain seakeeping model should be represented by the Cummins equation, which is an integro-differential equation (Cummins 1962).

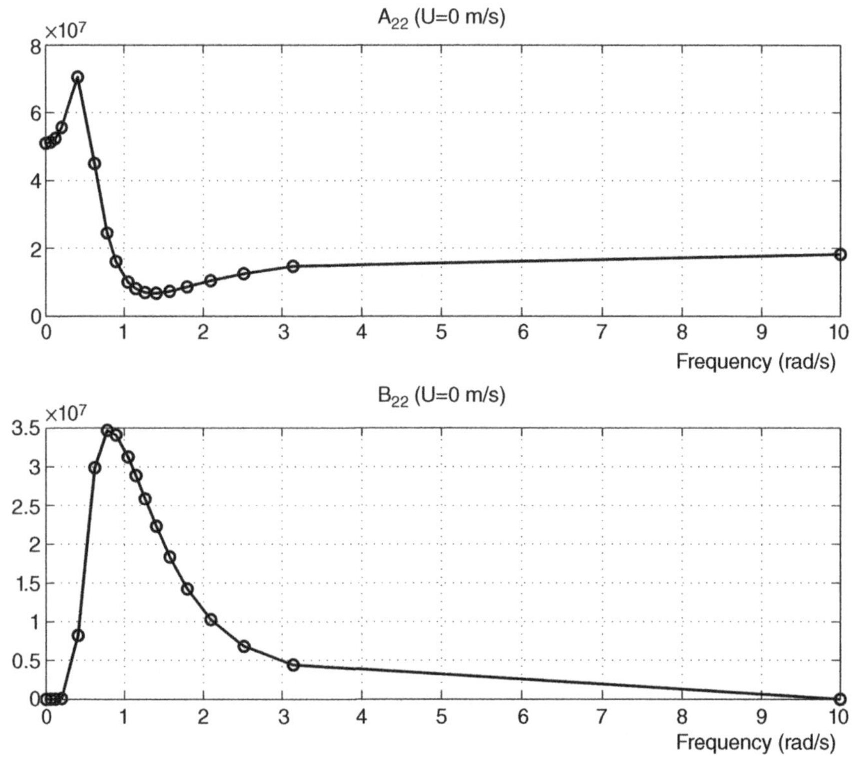

Figure 6.1 Numerical computation of added mass $A_{22}(\omega)$ and potential damping $B_{22}(\omega)$ in sway as a function of wave frequency ω for a large tanker.

For surface vessels it is common to solve the Cummins equation in the time domain under the assumption of linear theory (see Section 5.4)

$$(\boldsymbol{M}_{\mathrm{RB}} + \boldsymbol{M}_{A})\dot{\boldsymbol{v}} + \boldsymbol{C}^{*}_{\mathrm{RB}}\boldsymbol{v} + \boldsymbol{C}^{*}_{\mathrm{A}}\boldsymbol{v} + \boldsymbol{D}\boldsymbol{v} + \boldsymbol{\mu} + \boldsymbol{G}\boldsymbol{\eta} = \boldsymbol{\tau}_{\mathrm{wind}} + \boldsymbol{\tau}_{\mathrm{wave}} + \boldsymbol{\tau} \tag{6.11}$$

where the hydrostatic and hydrodynamic terms are

$$\begin{aligned} \boldsymbol{M}_{\mathrm{A}} &= \boldsymbol{A}(\infty) \\ \boldsymbol{C}^{*}_{\mathrm{A}} &= U\boldsymbol{A}(\infty)\boldsymbol{L} \\ \boldsymbol{D} &= \boldsymbol{B}(\infty) + \boldsymbol{B}_{\mathrm{V}}(\infty) \\ \boldsymbol{G} &= \boldsymbol{C}. \end{aligned}$$

This introduces *fluid-memory effects* $\boldsymbol{\mu}$, which can be interpreted as filtered potential damping forces. These forces are retardation functions that can be approximated by transfer functions and state-space models, as shown in Section 5.4.

6.2.1 Frequency-independent Added Mass and Potential Damping

The classical maneuvering model makes use of the following assumption:

Definition 6.1 (Zero-frequency Models for Surge, Sway and Yaw)
The horizontal motions (surge, sway and yaw) of a marine craft with port-starboard symmetry can be described by a zero-frequency model where

$$\boldsymbol{M}_{\mathrm{A}} = \boldsymbol{A}^{\{1,2,6\}}(0) = \begin{bmatrix} A_{11}(0) & 0 & 0 \\ 0 & A_{22}(0) & A_{26}(0) \\ 0 & A_{62}(0) & A_{66}(0) \end{bmatrix} \tag{6.12}$$

$$\boldsymbol{D}_{\mathrm{p}} = \boldsymbol{B}^{\{1,2,6\}}(0) = \boldsymbol{0} \tag{6.13}$$

are constant matrices.

When applying a feedback control system to stabilize the motions in surge, sway and yaw, the natural periods of the closed-loop system will be in the range 100–200 s. This implies that the natural frequencies are in the range 0.03–0.10 rad^{-1}s, which is quite close to the zero wave excitation frequency. This gives confidence in choosing $\boldsymbol{A}^{\{1,2,6\}}(0)$ and $\boldsymbol{B}^{\{1,2,6\}}(0)$ to approximate the dominating dynamics. Also note that the viscous damping matrix $\boldsymbol{B}_{\mathrm{V}}^{\{1,2,6\}}(\omega)$ will dominate the potential damping matrix $\boldsymbol{B}^{\{1,2,6\}}(\omega)$ at low frequency and that fluid memory effects $\boldsymbol{\mu}$ can be neglected at higher speeds.

Definition 6.1 implies that (6.11) can be approximated at a single frequency $\omega = 0$ in surge, sway and yaw and thus avoid the fluid memory effects ($\boldsymbol{\mu} = \boldsymbol{0}$). This gives

$$(\boldsymbol{M}_{\mathrm{RB}} + \boldsymbol{M}_A)\dot{\boldsymbol{\nu}} + \boldsymbol{C}^*_{\mathrm{RB}}\boldsymbol{\nu} + \boldsymbol{C}^*_{\mathrm{A}}\boldsymbol{\nu} + \boldsymbol{D}\boldsymbol{\nu} = \boldsymbol{\tau}_{\mathrm{wind}} + \boldsymbol{\tau}_{\mathrm{wave}} + \boldsymbol{\tau} \tag{6.14}$$

where

$$\boldsymbol{M}_{\mathrm{A}} = \boldsymbol{A}^{\{1,2,6\}}(0)$$
$$\boldsymbol{C}^*_{\mathrm{A}} = U\boldsymbol{A}^{\{1,2,6\}}(0)\boldsymbol{L}^{\{1,2,6\}}$$
$$\boldsymbol{D} = \boldsymbol{B}_{\mathrm{V}}^{\{1,2,6\}}(0)$$

and

$$\boldsymbol{L}^{\{1,2,6\}} = \begin{bmatrix} 0 & 0 & 0 \\ 0 & 0 & 1 \\ 0 & 0 & 0 \end{bmatrix}. \tag{6.15}$$

The viscous damping matrix $\boldsymbol{B}_{\mathrm{V}}^{\{1,2,6\}}(0)$ can be computed using the methods in Section 5.3.2.

Underwater Vehicles

For vehicles operating at water depths below the wave-affected zone, the hydrodynamic coefficients will be independent of the wave excitation frequency. Consequently,

$$\boldsymbol{A}(\omega) = \text{constant } \forall\omega \tag{6.16}$$

$$\boldsymbol{B}(\omega) = \boldsymbol{0}. \tag{6.17}$$

This means that if a seakeeping code is used to compute the potential coefficients, only one frequency is needed to obtain an estimate of the added mass matrix. In addition, there will be no potential damping. However, viscous damping $\boldsymbol{B}_{\mathrm{V}}(\omega)$ will be present.

6.2.2 Extension to 6-DOF Models

One limitation of Definition 6.1 is that it cannot be applied to *heave, roll* and *pitch*. These modes are second-order mass–damper–spring systems where the dominating frequencies are the natural frequencies. Hence, the constant frequency models in heave, roll and pitch should be formulated at their respective natural frequencies and not at the zero frequency.

The natural frequencies for the decoupled motions in heave, roll and pitch are given by the implicit equations

$$\omega_3 = \sqrt{\frac{C_{33}^{\mathrm{CF}}}{m + A_{33}^{\mathrm{CF}}(\omega_3)}} \tag{6.18}$$

$$\omega_4 = \sqrt{\frac{C_{44}^{\mathrm{CF}}}{I_x^{\mathrm{CF}} + A_{44}^{\mathrm{CF}}(\omega_4)}} \tag{6.19}$$

$$\omega_5 = \sqrt{\frac{C_{55}^{\mathrm{CF}}}{I_y^{\mathrm{CF}} + A_{55}^{\mathrm{CF}}(\omega_5)}} \tag{6.20}$$

where the potential coefficients $A_{ii}^{\mathrm{CF}}(\omega)$ and $B_{ii}^{\mathrm{CF}}(\omega)$ $(i = 3, 4, 5)$, and moments of inertia I_x^{CF} and I_y^{CF} are computed about the CF.

The potential coefficient matrices $\boldsymbol{A}(\omega)$ and $\boldsymbol{B}(\omega)$ can be computed using a hydrodynamic code. If we rely on Definition 6.1 to approximate $\boldsymbol{M}_{\mathrm{A}}$ and $\boldsymbol{D} = \boldsymbol{D}_{\mathrm{P}} + \boldsymbol{D}_{\mathrm{V}}$ in surge, sway and yaw it is necessary to assume that there are no couplings between the surge, heave–roll–pitch and the sway–yaw subsystems. Hence, added mass and potential damping can be approximated by two constant matrices

$$\boldsymbol{M}_{\mathrm{A}} \approx \begin{bmatrix} A_{11}(0) & 0 & & \cdots & & 0 \\ 0 & A_{22}(0) & & & & A_{26}(0) \\ & & \boxed{\begin{matrix} A_{33}(\omega_3) & 0 & 0 \\ 0 & A_{44}(\omega_4) & 0 \\ 0 & 0 & A_{55}(\omega_5) \end{matrix}} & & & \\ & \cdots & & & & \cdots \\ 0 & A_{62}(0) & & \cdots & & A_{66}(0) \end{bmatrix} \tag{6.21}$$

$$\boldsymbol{D}_{\mathrm{P}} \approx \begin{bmatrix} 0 & 0 & & \cdots & & 0 \\ 0 & 0 & & & & 0 \\ & & \boxed{\begin{matrix} B_{33}(\omega_3) & 0 & 0 \\ 0 & B_{44}(\omega_4) & 0 \\ 0 & 0 & B_{55}(\omega_5) \end{matrix}} & & & \\ & \cdots & & & & \cdots \\ 0 & 0 & & \cdots & & 0 \end{bmatrix}. \tag{6.22}$$

The natural frequencies ω_3, ω_4 and ω_5 can be computed using the methods in Sections 4.3.1–4.3.2. The linear viscous damping terms are usually approximated by a diagonal matrix

$$\boldsymbol{D}_{\mathrm{V}} \approx \mathrm{diag}\{B_{11v},\ B_{22v},\ B_{33v},\ B_{44v},\ B_{55v},\ B_{66v}\} \tag{6.23}$$

where the elements B_{iiv} $(i = 1, ..., 6)$ can be computed using the methods in Section 5.3.2.

6.3 Added Mass Forces in a Rotating Coordinate System

In seakeeping theory, the body frame $\{b\}$ rotates about $\{s\}$, which is assumed to be inertial. This results in two linear Coriolis and centripetal matrices $\boldsymbol{C}_{\mathrm{RB}}^*$ and $\boldsymbol{C}_{\mathrm{A}}^*$. In maneuvering theory it is assumed that $\{b\}$ rotates about $\{n\}$ as illustrated in Figure 6.2. This suggests that the linear terms $\boldsymbol{C}_{\mathrm{RB}}^*$ and $\boldsymbol{C}_{\mathrm{A}}^*$ can be replaced by their nonlinear counterparts $\boldsymbol{C}_{\mathrm{RB}}(\boldsymbol{\nu})$ and $\boldsymbol{C}_{\mathrm{A}}(\boldsymbol{\nu})$, which transforms (6.14) to

$$(\boldsymbol{M}_{\mathrm{RB}} + \boldsymbol{M}_A)\dot{\boldsymbol{\nu}} + \boldsymbol{C}_{\mathrm{RB}}(\boldsymbol{\nu})\boldsymbol{\nu} + \boldsymbol{C}_{\mathrm{A}}(\boldsymbol{\nu})\boldsymbol{\nu} + \boldsymbol{D}\boldsymbol{\nu} = \boldsymbol{\tau}_{\mathrm{wind}} + \boldsymbol{\tau}_{\mathrm{wave}} + \boldsymbol{\tau}. \tag{6.24}$$

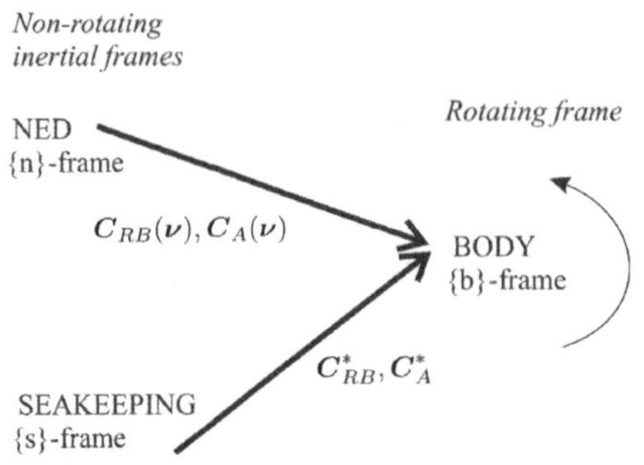

Figure 6.2 Coriolis matrices due to the rotation of the body-fixed frame $\{b\}$ about the approximative inertial frames $\{n\}$ or $\{s\}$.

The nonlinear Coriolis and centripetal matrix $\boldsymbol{C}_{\mathrm{RB}}(\boldsymbol{\nu})$ for a rigid-body is given by (6.7), while $\boldsymbol{C}_{\mathrm{A}}(\boldsymbol{\nu})$ depends on the added mass terms. This is in general a complicated function but it is straightforward to find the formula by using Lagrangian mechanics.

6.3.1 Lagrangian Mechanics

In Section 3.1, it was shown that the rigid-body kinetics of a marine craft can be derived by applying the *Newtonian* formulation. As for the rigid-body kinetics, it is advantageous to separate the added mass forces and moments in terms that belong to the *added mass matrix* $\boldsymbol{M}_{\mathrm{A}}$ and a matrix of hydrodynamic Coriolis and centripetal terms denoted $\boldsymbol{C}_{\mathrm{A}}(\boldsymbol{\nu})$. To derive the expressions for these two matrices, an *energy approach* based on Kirchhoff's equations will now be presented. Detailed discussions of Newtonian and Lagrangian mechanics are found in Goldstein (1980), Hughes (1986), Kane *et al.* (1983), Meirovitch (1990) and Egeland and Gravdahl (2002).

The Lagrangian L is formed by using kinetic energy T and potential energy V, according to

$$L = T - V. \tag{6.25}$$

The *Euler–Lagrange equation* is

$$\frac{\mathrm{d}}{\mathrm{d}t}\left(\frac{\partial L}{\partial \dot{\boldsymbol{\eta}}}\right) - \frac{\partial L}{\partial \boldsymbol{\eta}} = \boldsymbol{J}_{\Theta}^{-\top}(\boldsymbol{\eta})\boldsymbol{\tau} \tag{6.26}$$

which in component form corresponds to a set of six second-order differential equations. From the above formula it is seen that the Lagrangian mechanics describes the system dynamics in terms of energy. Formula (6.26) is valid in any reference frame, inertial and body-fixed, as long as *generalized coordinates* are used.

For a marine craft not subject to any motion constraints, the number of independent (*generalized*) coordinates is equal to the number of DOFs. For a marine craft moving in 6 DOFs the generalized coordinates in $\{n\}$ can be chosen as

$$\boldsymbol{\eta} = [x^n,\ y^n,\ z^{\mathrm{n}},\ \phi,\ \theta,\ \psi]^{\top}. \tag{6.27}$$

It should be noted that the alternative unit quaternion representation

$$\boldsymbol{\eta} = [x^n,\ y^n,\ z^{\mathrm{n}},\ \eta,\ \varepsilon_1,\ \varepsilon_2,\ \varepsilon_3]^{\mathrm{T}} \tag{6.28}$$

cannot be used in a Lagrangian approach since this representation is defined by seven parameters. Hence, these parameters are not *generalized coordinates*. It is not straightforward to formulate the Euler–Lagrange equation of motion in $\{b\}$ since

$$\boldsymbol{\nu} = [u,\ v,\ w,\ p,\ q,\ r]^{\mathrm{T}} \tag{6.29}$$

cannot be integrated to yield a set of generalized coordinates in terms of position and orientation. In fact the integral $\int_0^t \boldsymbol{\nu}(\tau)\mathrm{d}\tau$ has no immediate physical interpretation. Consequently, the Euler-Lagrange equation cannot be directly used to formulate the equations of motion in $\{b\}$. However, this problem is circumvented by applying Kirchhoff's equations of motion, or the so-called *quasi-Lagrangian* approach; see Meirovitch and Kwak (1989) for details.

6.3.2 *Kirchhoff's Equation*

Consider a marine craft with linear velocity $\boldsymbol{\nu}_1 := [u,\ v,\ w]^{\mathrm{T}}$ and angular velocity $\boldsymbol{\nu}_2 := [p,\ q,\ r]^{\mathrm{T}}$ expressed in $\{b\}$. Hence, the force $\boldsymbol{\tau}_1 := [X,\ Y,\ Z]^{\mathrm{T}}$ and moment $\boldsymbol{\tau}_2 := [K,\ M,\ N]^{\mathrm{T}}$ are related to the kinetic energy (Kirchhoff 1869)

$$T = \frac{1}{2}\boldsymbol{\nu}^{\mathrm{T}}\boldsymbol{M}\boldsymbol{\nu} \tag{6.30}$$

by the vector equations

$$\frac{\mathrm{d}}{\mathrm{d}t}\left(\frac{\partial T}{\partial \boldsymbol{\nu}_1}\right) + \boldsymbol{S}(\boldsymbol{\nu}_2)\frac{\partial T}{\partial \boldsymbol{\nu}_1} = \boldsymbol{\tau}_1 \tag{6.31}$$

$$\frac{\mathrm{d}}{\mathrm{d}t}\left(\frac{\partial T}{\partial \boldsymbol{\nu}_2}\right) + \boldsymbol{S}(\boldsymbol{\nu}_2)\frac{\partial T}{\partial \boldsymbol{\nu}_2} + \boldsymbol{S}(\boldsymbol{\nu}_1)\frac{\partial T}{\partial \boldsymbol{\nu}_1} = \boldsymbol{\tau}_2 \tag{6.32}$$

where $\boldsymbol{S}$ is the skew-symmetric cross-product operator in Definition 2.2. *Kirchhoff's equation* will prove to be very useful in the derivation of the expression for added inertia. Note that Kirchhoff's equation do not include the gravitational forces.

6.3.3 *Added Mass and Coriolis–Centripetal Matrices*

The matrix $\boldsymbol{C}_{\mathrm{A}}^{*}$ in (6.11) represents linearized forces due to a rotation of $\{b\}$ about the seakeeping frame $\{s\}$. Instead of using $\{s\}$ as the inertial frame, we will assume that $\{n\}$ is the inertial frame and that $\{b\}$ rotates about $\{n\}$. The nonlinear Coriolis and centripetal matrix $\boldsymbol{C}_{\mathrm{A}}(\boldsymbol{\nu})$ due to a rotation of $\{b\}$ about the inertial frame $\{n\}$ can be derived using an energy formulation based on the constant matrix $\boldsymbol{M}_{\mathrm{A}}$. Since any motion of the marine craft will induce a motion in the otherwise stationary fluid, the fluid must move aside and then close behind the craft in order to let the craft pass through the fluid.

Figure 6.3 Rigid-body and fluid kinetic energy (ocean surrounding the ship). Source: Illustration by B. Stenberg.

As a consequence, the fluid motion possesses kinetic energy that it would lack otherwise (see Figure 6.3). The expression for the fluid kinetic energy T_A is written as a quadratic form (Lamb 1932)

$$T_A = \frac{1}{2}\boldsymbol{v}^{\top}\boldsymbol{M}_A\boldsymbol{v}, \qquad \dot{\boldsymbol{M}}_A = \boldsymbol{0} \tag{6.33}$$

where $\boldsymbol{M}_A = \boldsymbol{M}_A^{\top} \geq 0$ is the 6×6 *system inertia matrix* of added mass terms

$$\boldsymbol{M}_A = -\begin{bmatrix} X_{\dot{u}} & X_{\dot{v}} & X_{\dot{w}} & X_{\dot{p}} & X_{\dot{q}} & X_{\dot{r}} \\ Y_{\dot{u}} & Y_{\dot{v}} & Y_{\dot{w}} & Y_{\dot{p}} & Y_{\dot{q}} & Y_{\dot{r}} \\ Z_{\dot{u}} & Z_{\dot{v}} & Z_{\dot{w}} & Z_{\dot{p}} & Z_{\dot{q}} & Z_{\dot{r}} \\ K_{\dot{u}} & K_{\dot{v}} & K_{\dot{w}} & K_{\dot{p}} & K_{\dot{q}} & K_{\dot{r}} \\ M_{\dot{u}} & M_{\dot{v}} & M_{\dot{w}} & M_{\dot{p}} & M_{\dot{q}} & M_{\dot{r}} \\ N_{\dot{u}} & N_{\dot{v}} & N_{\dot{w}} & N_{\dot{p}} & N_{\dot{q}} & N_{\dot{r}} \end{bmatrix}. \tag{6.34}$$

The notation of SNAME (1950) for the hydrodynamic derivatives is used in this expression; for instance the hydrodynamic added mass force Y along the y axis due to an acceleration $\dot{u}$ in the x direction is written as

$$Y = -Y_{\dot{u}}\dot{u}, \qquad Y_{\dot{u}} := \frac{\partial Y}{\partial \dot{u}}. \tag{6.35}$$

This implies that $\{\boldsymbol{M}_A\}_{21} = -Y_{\dot{u}}$ in the example above.

Property 6.1 (Hydrodynamic System Inertia Matrix $\boldsymbol{M}_A$)
For a rigid body at rest or moving at forward speed $U \geq 0$ in ideal fluid, the hydrodynamic system inertia matrix $\boldsymbol{M}_A$ is positive semidefinite

$$\boldsymbol{M}_A = \boldsymbol{M}_A^{\top} \geq 0.$$

Proof. Newman (1977) has shown this for zero speed. The results extend to forward speed by using the approach presented in Chapter 5.

Remark 6.1
In a real fluid the 36 elements of $\boldsymbol{M}_{\mathrm{A}}$ *may all be distinct but still* $\boldsymbol{M}_{\mathrm{A}} \geq 0$. *Experience has shown that the numerical values of the added mass derivatives in a real fluid are usually in good agreement with those obtained from ideal theory (see Wendel 1956).*

Remark 6.2
If experimental data are used, the inertia matrix can be symmetrized by using

$$\boldsymbol{M}_A = \frac{1}{2}(\hat{\boldsymbol{M}}_A + \hat{\boldsymbol{M}}_A^{\mathrm{T}}) \tag{6.36}$$

where $\hat{\boldsymbol{M}}_A$ *contains the experimentally data.*

Added Mass Forces and Moments

Based on the kinetic energy T_{A} of the fluid, it is straightforward to derive the added mass forces and moments. Substituting (6.33) into (6.31) and (6.32) gives the following expressions for the added mass terms (Imlay 1961)

$$\begin{aligned} X_{\mathrm{A}} = {} & X_{\dot{u}}\dot{u} + X_{\dot{w}}(\dot{w} + uq) + X_{\dot{q}}\dot{q} + Z_{\dot{w}}wq + Z_{\dot{q}}q^2 \\ & + X_{\dot{v}}\dot{v} + X_{\dot{p}}\dot{p} + X_{\dot{r}}\dot{r} - Y_{\dot{v}}vr - Y_{\dot{p}}rp - Y_{\dot{r}}r^2 \\ & - X_{\dot{v}}ur - Y_{\dot{w}}wr \\ & + Y_{\dot{w}}vq + Z_{\dot{p}}pq - (Y_{\dot{q}} - Z_{\dot{r}})qr \end{aligned} \tag{6.37}$$

$$\begin{aligned} Y_{\mathrm{A}} = {} & X_{\dot{v}}\dot{u} + Y_{\dot{w}}\dot{w} + Y_{\dot{q}}\dot{q} \\ & + Y_{\dot{v}}\dot{v} + Y_{\dot{p}}\dot{p} + Y_{\dot{r}}\dot{r} + X_{\dot{v}}vr - Y_{\dot{w}}vp + X_{\dot{r}}r^2 + (X_{\dot{p}} - Z_{\dot{r}})rp - Z_{\dot{p}}p^2 \\ & - X_{\dot{w}}(up - wr) + X_{\dot{u}}ur - Z_{\dot{w}}wp \\ & - Z_{\dot{q}}pq + X_{\dot{q}}qr \end{aligned} \tag{6.38}$$

$$\begin{aligned} Z_{\mathrm{A}} = {} & X_{\dot{w}}(\dot{u} - wq) + Z_{\dot{w}}\dot{w} + Z_{\dot{q}}\dot{q} - X_{\dot{u}}uq - X_{\dot{q}}q^2 \\ & + Y_{\dot{w}}\dot{v} + Z_{\dot{p}}\dot{p} + Z_{\dot{r}}\dot{r} + Y_{\dot{v}}vp + Y_{\dot{r}}rp + Y_{\dot{p}}p^2 \\ & + X_{\dot{v}}up + Y_{\dot{w}}wp \\ & - X_{\dot{v}}vq - (X_{\dot{p}} - Y_{\dot{q}})pq - X_{\dot{r}}qr \end{aligned} \tag{6.39}$$

$$\begin{aligned} K_{\mathrm{A}} = {} & X_{\dot{p}}\dot{u} + Z_{\dot{p}}\dot{w} + K_{\dot{q}}\dot{q} - X_{\dot{v}}wu + X_{\dot{r}}uq - Y_{\dot{w}}w^2 - (Y_{\dot{q}} - Z_{\dot{r}})wq + M_{\dot{r}}q^2 \\ & + Y_{\dot{p}}\dot{v} + K_{\dot{p}}\dot{p} + K_{\dot{r}}\dot{r} + Y_{\dot{w}}v^2 - (Y_{\dot{q}} - Z_{\dot{r}})vr + Z_{\dot{p}}vp - M_{\dot{r}}r^2 - K_{\dot{q}}rp \\ & + X_{\dot{w}}uv - (Y_{\dot{v}} - Z_{\dot{w}})vw - (Y_{\dot{r}} + Z_{\dot{q}})wr - Y_{\dot{p}}wp - X_{\dot{q}}ur \\ & + (Y_{\dot{r}} + Z_{\dot{q}})vq + K_{\dot{r}}pq - (M_{\dot{q}} - N_{\dot{r}})qr \end{aligned} \tag{6.40}$$

$$
\begin{aligned}
M_A = {} & X_{\dot{q}}(\dot{u} + wq) + Z_{\dot{q}}(\dot{w} - uq) + M_{\dot{q}}\dot{q} - X_{\dot{w}}(u^2 - w^2) - (Z_{\dot{w}} - X_{\dot{u}})wu \\
& + Y_{\dot{q}}\dot{v} + K_{\dot{q}}\dot{p} + M_{\dot{r}}\dot{r} + Y_{\dot{p}}vr - Y_{\dot{r}}vp - K_{\dot{r}}(p^2 - r^2) + (K_{\dot{p}} - N_{\dot{r}})rp \\
& - Y_{\dot{w}}uv + X_{\dot{v}}vw - (X_{\dot{r}} + Z_{\dot{p}})(up - wr) + (X_{\dot{p}} - Z_{\dot{r}})(wp + ur) \\
& - M_{\dot{r}}pq + K_{\dot{q}}qr
\end{aligned} \tag{6.41}
$$

$$
\begin{aligned}
N_A = {} & X_{\dot{r}}\dot{u} + Z_{\dot{r}}\dot{w} + M_{\dot{r}}\dot{q} + X_{\dot{v}}u^2 + Y_{\dot{w}}wu - (X_{\dot{p}} - Y_{\dot{q}})uq - Z_{\dot{p}}wq - K_{\dot{q}}q^2 \\
& + Y_{\dot{r}}\dot{v} + K_{\dot{r}}\dot{p} + N_{\dot{r}}\dot{r} - X_{\dot{v}}v^2 - X_{\dot{r}}vr - (X_{\dot{p}} - Y_{\dot{q}})vp + M_{\dot{r}}rp + K_{\dot{q}}p^2 \\
& - (X_{\dot{u}} - Y_{\dot{v}})uv - X_{\dot{w}}vw + (X_{\dot{q}} + Y_{\dot{p}})up + Y_{\dot{r}}ur + Z_{\dot{q}}wp \\
& - (X_{\dot{q}} + Y_{\dot{p}})vq - (K_{\dot{p}} - M_{\dot{q}})pq - K_{\dot{r}}qr.
\end{aligned} \tag{6.42}
$$

Imlay (1961) arranged the equations in four lines with longitudinal components on the first line and lateral components on the second. The third line consists of mixed terms involving u or w as one factor. If one or both of these velocities are large enough to be treated as constants, the third line may be treated as an additional term to the lateral equations of motion. The fourth line contains mixed terms that usually can be neglected as second-order terms.

It should be noted that the off-diagonal elements of $\boldsymbol{M}_A$ will be small compared to the diagonal elements for most practical applications. A more detailed discussion on the different added mass derivatives can be found in Humphreys and Watkinson (1978).

Property 6.2 (Hydrodynamic Coriolis–Centripetal Matrix $\boldsymbol{C}_A(\boldsymbol{\nu})$)
For a rigid body moving through an ideal fluid the hydrodynamic Coriolis and centripetal matrix $\boldsymbol{C}_A(\boldsymbol{\nu})$ can always be parameterized such that it is skew-symmetric

$$
\boldsymbol{C}_A(\boldsymbol{\nu}) = -\boldsymbol{C}_A^\top(\boldsymbol{\nu}), \qquad \forall \boldsymbol{\nu} \in \mathbb{R}^6. \tag{6.43}
$$

One parametrization satisfying (6.43) is

$$
\boldsymbol{C}_A(\boldsymbol{\nu}) = \begin{bmatrix} \boldsymbol{0}_{3\times3} & -\boldsymbol{S}(\boldsymbol{A}_{11}\boldsymbol{\nu}_1 + \boldsymbol{A}_{12}\boldsymbol{\nu}_2) \\ -\boldsymbol{S}(\boldsymbol{A}_{11}\boldsymbol{\nu}_1 + \boldsymbol{A}_{12}\boldsymbol{\nu}_2) & -\boldsymbol{S}(\boldsymbol{A}_{21}\boldsymbol{\nu}_1 + \boldsymbol{A}_{22}\boldsymbol{\nu}_2) \end{bmatrix} \tag{6.44}
$$

where $\boldsymbol{A}_{ij} \in \mathbb{R}^{3\times3}$ is defined as

$$
\boldsymbol{M}_A := \begin{bmatrix} \boldsymbol{A}_{11} & \boldsymbol{A}_{12} \\ \boldsymbol{A}_{21} & \boldsymbol{A}_{22} \end{bmatrix}. \tag{6.45}
$$

Proof. *Substituting $\boldsymbol{M}_A$ for $\boldsymbol{M}$ in (3.51) in Theorem 3.2 directly proves (6.44).*

Formula (6.44) can be written in component form according to

$$C_A(\boldsymbol{\nu}) = \begin{bmatrix} 0 & 0 & 0 & 0 & -a_3 & a_2 \\ 0 & 0 & 0 & a_3 & 0 & -a_1 \\ 0 & 0 & 0 & -a_2 & a_1 & 0 \\ 0 & -a_3 & a_2 & 0 & -b_3 & b_2 \\ a_3 & 0 & -a_1 & b_3 & 0 & -b_1 \\ -a_2 & a_1 & 0 & -b_2 & b_1 & 0 \end{bmatrix} \tag{6.46}$$

where

$$a_1 = X_{\dot{u}}u + X_{\dot{v}}v + X_{\dot{w}}w + X_{\dot{p}}p + X_{\dot{q}}q + X_{\dot{r}}r \tag{6.47}$$

$$a_2 = Y_{\dot{u}}u + Y_{\dot{v}}v + Y_{\dot{w}}w + Y_{\dot{p}}p + Y_{\dot{q}}q + Y_{\dot{r}}r \tag{6.48}$$

$$a_3 = Z_{\dot{u}}u + Z_{\dot{v}}v + Z_{\dot{w}}w + Z_{\dot{p}}p + Z_{\dot{q}}q + Z_{\dot{r}}r \tag{6.49}$$

$$b_1 = K_{\dot{u}}u + K_{\dot{v}}v + K_{\dot{w}}w + K_{\dot{p}}p + K_{\dot{q}}q + K_{\dot{r}}r \tag{6.50}$$

$$b_2 = M_{\dot{u}}u + M_{\dot{v}}v + M_{\dot{w}}w + M_{\dot{p}}p + M_{\dot{q}}q + M_{\dot{r}}r \tag{6.51}$$

$$b_3 = N_{\dot{u}}u + N_{\dot{v}}v + N_{\dot{w}}w + N_{\dot{p}}p + N_{\dot{q}}q + N_{\dot{r}}r. \tag{6.52}$$

Properties 6.2 and 10.1 imply that the marine craft dynamics can be represented in terms of nonlinear Coriolis and centripetal forces using relative velocity

$$\boldsymbol{M}\dot{\boldsymbol{\nu}}_r + \boldsymbol{C}(\boldsymbol{\nu}_r)\boldsymbol{\nu}_r + \boldsymbol{D}(\boldsymbol{\nu}_r)\boldsymbol{\nu}_r + \boldsymbol{g}(\boldsymbol{\eta}) = \boldsymbol{\tau} + \boldsymbol{\tau}_{\text{wind}} + \boldsymbol{\tau}_{\text{wave}} \tag{6.53}$$

where

$$\boldsymbol{M} = \boldsymbol{M}_{\text{RB}} + \boldsymbol{M}_{\text{A}} \tag{6.54}$$

$$\boldsymbol{C}(\boldsymbol{\nu}_r) = \boldsymbol{C}_{\text{RB}}(\boldsymbol{\nu}_r) + \boldsymbol{C}_{\text{A}}(\boldsymbol{\nu}_r) \tag{6.55}$$

while classical seakeeping theory uses linear matrices $\boldsymbol{C}^*_{\text{RB}}$ and $\boldsymbol{C}^*_{\text{A}}$ as explained in Section 6.2.

Example 6.1 (Added Mass for Surface Vessels)
For surface ships such as tankers, cargo ships, supply vessels and cruise-liners it is common to decouple the surge mode from the steering dynamics due to xz plane symmetry. Similar, the heave, pitch, and roll modes are neglected under the assumption that these motion variables are small. Hence, $\boldsymbol{\nu}_r = [u_r, v_r, r]^\top$ *implies that the added mass derivatives for a surface ship are*

$$\boldsymbol{M}_A = \boldsymbol{M}_A^\top = -\begin{bmatrix} X_{\dot{u}} & 0 & 0 \\ 0 & Y_{\dot{v}} & Y_{\dot{r}} \\ 0 & Y_{\dot{r}} & N_{\dot{r}} \end{bmatrix} \qquad (N_{\dot{v}} = Y_{\dot{r}}) \tag{6.56}$$

$$\boldsymbol{C}_A(\boldsymbol{\nu}_r) = -\boldsymbol{C}_A^\top(\boldsymbol{\nu}_r) = \begin{bmatrix} 0 & 0 & Y_{\dot{v}}v_r + Y_{\dot{r}}r \\ 0 & 0 & -X_{\dot{u}}u_r \\ -Y_{\dot{v}}v_r - Y_{\dot{r}}r & X_{\dot{u}}u_r & 0 \end{bmatrix}. \tag{6.57}$$

The Coriolis and centripetal forces corresponding to (6.57) are

$$
\boldsymbol{C}_A(\boldsymbol{\nu}_r)\boldsymbol{\nu}_r = \begin{bmatrix} Y_{\dot{v}} v_r r + Y_{\dot{r}} r^2 \\ -X_{\dot{u}} u_r r \\ \underbrace{(X_{\dot{u}} - Y_{\dot{v}}) u_r v_r}_{\text{Munk moment}} - Y_{\dot{r}} u_r r \end{bmatrix} \tag{6.58}
$$

where the first term in the yaw moment is the nonlinear Munk moment, which is known to have destabilizing effects.

Example 6.2 (Added Mass for Underwater Vehicles)
In general, the motion of an underwater vehicle moving in 6 DOFs at high speed will be highly nonlinear and coupled. However, in many AUV and ROV applications the vehicle will only be allowed to move at low speed. If the vehicle also has three planes of symmetry, this suggests that the contribution from the off-diagonal elements in the matrix $\boldsymbol{M}_A$ can be neglected. Hence, the following simple expressions for the matrices $\boldsymbol{M}_A$ and $\boldsymbol{C}_A$ are obtained

$$
\boldsymbol{M}_A = \boldsymbol{M}_A^\top = -\mathrm{diag}\{X_{\dot{u}}, Y_{\dot{v}}, Z_{\dot{w}}, K_{\dot{p}}, M_{\dot{q}}, N_{\dot{r}}\} \tag{6.59}
$$

$$
\boldsymbol{C}_A(\boldsymbol{\nu}_r) = -\boldsymbol{C}_A^\top(\boldsymbol{\nu}_r) = \begin{bmatrix} 0 & 0 & 0 & 0 & -Z_{\dot{w}} w_r & Y_{\dot{v}} v_r \\ 0 & 0 & 0 & Z_{\dot{w}} w_r & 0 & -X_{\dot{u}} u_r \\ 0 & 0 & 0 & -Y_{\dot{v}} v_r & X_{\dot{u}} u_r & 0 \\ 0 & -Z_{\dot{w}} w_r & Y_{\dot{v}} v_r & 0 & -N_{\dot{r}} r & M_{\dot{q}} q \\ Z_{\dot{w}} w_r & 0 & -X_{\dot{u}} u_r & N_{\dot{r}} r & 0 & -K_{\dot{p}} p \\ -Y_{\dot{v}} v_r & X_{\dot{u}} u_r & 0 & -M_{\dot{q}} q & K_{\dot{p}} p & 0 \end{bmatrix}. \tag{6.60}
$$

The diagonal structure is often used since it is time consuming to determine the off-diagonal elements from experiments as well as theory. In practice, the diagonal approximation is found to be quite good for many applications. This is due to the fact that the off-diagonal elements of a positive inertia matrix will be much smaller than their diagonal counterparts.

6.4 Dissipative Forces

Hydrodynamic damping for marine craft is mainly caused by:

Potential damping: We recall from the beginning of Section 6.2 that *added mass, damping* and *restoring* forces and moments are encountered when a body is forced to oscillate with the wave excitation frequency in the absence of incident waves. The radiation-induced damping term is usually referred to as *linear frequency-dependent potential damping* $\boldsymbol{B}(\omega)$.

Skin friction: Linear frequency-dependent skin friction $\boldsymbol{B}_v(\omega)$ due to laminar boundary layer theory is important when considering the low-frequency motion of marine craft (Faltinsen and Sortland 1987). In addition to linear skin friction, there will be a high-frequency contribution due to a turbulent boundary layer. This is usually referred to as a quadratic or nonlinear skin friction.

Wave drift damping: Wave drift damping can be interpreted as added resistance for surface vessels advancing in waves. This type of damping is derived from second-order wave theory. Wave drift damping is the most important damping contribution to surge for higher sea states. This is due to the fact that the wave drift forces are proportional to the square of the significant wave height H_s. Wave drift damping in sway and yaw is small relative to eddy-making damping (vortex shedding). A rule of thumb is that second-order wave drift forces are less than 1 % of the first-order wave forces when the significant wave height is equal to 1 m and 10 % when the significant wave height is equal to 10 m.

Damping due to vortex shedding: *D'Alambert's paradox* states that no hydrodynamic forces act on a solid moving completely submerged with constant velocity in a non-viscous fluid. In a viscous fluid, frictional forces are present such that the system is not conservative with respect to energy. This is commonly referred to as *interference drag*. It arises due to the shedding of vortex sheets at sharp edges. The viscous damping force due to vortex shedding can be modeled as

$$f(u) = -\frac{1}{2}\rho\ C_D(R_n)\ A|u|u \tag{6.61}$$

where u is the velocity of the craft, A is the projected cross-sectional area under water, $C_D(R_n)$ is the drag coefficient based on the representative area and ρ is the water density. This expression is recognized as one of the terms in *Morison's equation* (see Faltinsen 1990). The drag coefficient $C_D(R_n)$ is a function of the *Reynolds number* $R_n = uD/\nu$ where D is the characteristic length of the body and ν is the kinematic viscosity coefficient ($\nu = 1.56 \times 10^{-6}$ for salt water at 5 °C with salinity 3.5%).

Lifting forces: Hydrodynamic lift forces arise from two physical mechanisms. The first is due to the linear circulation of water around the hull. The second is a nonlinear effect, commonly called cross-flow drag, which acts from a momentum transfer from the body to the fluid. This secondary effect is closely linked to vortex shedding.

The different damping terms contribute to both linear and quadratic damping. However, it is in general difficult to separate these effects. In many cases, it is convenient to write total hydrodynamic damping as

$$\boldsymbol{D}(\boldsymbol{\nu}_r) = \boldsymbol{D} + \boldsymbol{D}_n(\boldsymbol{\nu}_r) \tag{6.62}$$

where $\boldsymbol{D}$ is the *linear damping matrix* due to potential damping and possible skin friction and $\boldsymbol{D}_n(\boldsymbol{\nu}_r)$ is the *nonlinear damping matrix* due to quadratic damping and higher-order terms. Hydrodynamic damping satisfies the following dissipative property.

Property 6.3 (Hydrodynamic Damping Matrix $\boldsymbol{D}(\boldsymbol{\nu}_r)$)
For a rigid body moving through an ideal fluid the hydrodynamic damping matrix

$$\boldsymbol{D}(\boldsymbol{\nu}_r) = \frac{1}{2}(\boldsymbol{D}(\boldsymbol{\nu}_r) + \boldsymbol{D}^\top(\boldsymbol{\nu}_r)) + \frac{1}{2}(\boldsymbol{D}(\boldsymbol{\nu}_r) - \boldsymbol{D}^\top(\boldsymbol{\nu}_r)) \tag{6.63}$$

will be real, non-symmetric and strictly positive

$$\boldsymbol{D}(\boldsymbol{v}_\text{r}) > 0, \quad \forall \boldsymbol{v}_\text{r} \in \mathbb{R}^6 \tag{6.64}$$

or

$$\boldsymbol{x}^\top \boldsymbol{D}(\boldsymbol{v}_\text{r})\boldsymbol{x} = \frac{1}{2}\boldsymbol{x}^\top(\boldsymbol{D}(\boldsymbol{v}_\text{r}) + \boldsymbol{D}^\top(\boldsymbol{v}_\text{r}))\boldsymbol{x} > 0 \;\; \forall \; \boldsymbol{x} \neq \boldsymbol{0}. \tag{6.65}$$

Some of the damping terms can be determined by using well-established methods from the literature and experimental techniques.

6.4.1 Linear Damping

For surface ships such as tankers, cargo ships, supply vessels and cruise-liners it is common to decouple the surge mode from the steering dynamics due to xz plane symmetry. Hence, the linear damping matrix in the CO with decoupled surge dynamics can be written

$$\begin{aligned} \boldsymbol{D} &= \boldsymbol{D}_\text{P} + \boldsymbol{D}_\text{V} \\ &= -\begin{bmatrix} X_u & 0 & 0 & 0 & 0 & 0 \\ 0 & Y_v & 0 & Y_\text{p} & 0 & Y_\text{r} \\ 0 & 0 & Z_w & 0 & Z_q & 0 \\ 0 & K_v & 0 & K_\text{p} & 0 & K_\text{r} \\ 0 & 0 & M_w & 0 & M_q & 0 \\ 0 & N_v & 0 & N_\text{p} & 0 & N_\text{r} \end{bmatrix}. \end{aligned} \tag{6.66}$$

Example 6.3 (Linear Damping for Surface Vessels)
In maneuvering theory, the heave, pitch, and roll modes are neglected under the assumption that these motion variables are small. Hence, the linear damping matrix for a surface vessel reduces to

$$\boldsymbol{D} = -\begin{bmatrix} X_u & 0 & 0 \\ 0 & Y_v & Y_\text{r} \\ 0 & N_v & N_\text{r} \end{bmatrix}. \tag{6.67}$$

The diagonal terms in (6.67) relate to seakeeping theory according to

$$-X_u = B_{11v}(0) \tag{6.68}$$

$$-Y_v = B_{22v}(0) \tag{6.69}$$

$$-N_\text{r} = B_{66v}(0) \tag{6.70}$$

where the expressions for B_{iiv} $(i = 1, 2, 6)$ are given by (5.60).

Example 6.4 (Linear Damping for Underwater Vehicles)
In general, the motion of an underwater vehicle moving in 6 DOFs at high speed will be highly nonlinear and coupled. However, in many AUV and ROV applications the vehicle will only be allowed to move at low speed. If the vehicle also has three planes of symmetry, this suggests that the contribution from the off-diagonal elements in the matrix **D** *can be neglected. Consequently,*

$$\boldsymbol{D} = -\mathrm{diag}\{X_u,\ Y_v,\ Z_w,\ K_p,\ M_q,\ N_r\} \tag{6.71}$$

where X_u, Y_v and N_r are approximated by the zero-frequency formulae (6.68)–(6.70). Recall that the potential damping matrix $\boldsymbol{B}(\omega) = \boldsymbol{0}$ is zero for submerged vehicles operating below the wave-affected zone. For heave, roll and pitch it is common to approximate linear damping as

$$-Z_w = B_{33v}(\omega_3) \tag{6.72}$$

$$-K_p = B_{44v}(\omega_4) \tag{6.73}$$

$$-M_q = B_{55v}(\omega_5) \tag{6.74}$$

where (5.60) can be used to compute the viscous terms. Alternatively, linear damping can be computed by specifying three relative damping ratios ζ_3, ζ_4 and ζ_5 in the formula

$$D_{ii} = 2\zeta_i \omega_i\, M_{ii} \qquad (i = 3, 4, 5). \tag{6.75}$$

Here M_{ii} denotes the sum of rigid-body and hydrodynamic added mass (constant for submerged vehicles) while ω_i are given by (6.18)–(6.20).

The diagonal structure is often used since it is time consuming to determine the off-diagonal elements from experiments as well as theory.

6.4.2 Nonlinear Surge Damping

In surge, the viscous damping for ships may be modeled as (Lewis 1989)

$$X = -\frac{1}{2}\rho S(1 + k)C_f(u_r)|u_r|u_r \tag{6.76}$$

$$C_f(u_r) = \underbrace{\frac{0.075}{(\log_{10}(R_n) - 2)^2}}_{C_F} + C_R \tag{6.77}$$

where ρ is the density of water, S is the wetted surface of the hull,

$$\begin{aligned} u_r &= u - u_c \\ &= u - V_c \cos(\beta_{V_c} - \psi) \end{aligned} \tag{6.78}$$

is the relative surge velocity (see Section 10.3), k is the form factor giving a viscous correction, C_F is the flat plate friction from the ITTC 1957 line and C_R represents *residual friction* due to hull roughness, pressure resistance, wave-making resistance and wave-breaking resistance. For ships in transit k is typically 0.1 whereas this value is much higher in DP, typically $k = 0.25$ (Hoerner 1965). The friction coefficient C_F depends on the *Reynolds number*

$$R_n = \frac{L_{pp}}{\nu}|u_r| \geq 0 \tag{6.79}$$

where $\nu = 1 \times 10^{-6}$ m s^{-2} is the *kinematic viscosity* at 20°C. Note that the Reynolds number is non-negative since it is independent of the flow direction. From a numerical point of view, the denominator of C_F should be replaced with, $(\log_{10}(R_n) - 2)^2 + \epsilon$, where ϵ is a small number to ensure that (6.76) and (6.77) are well defined at $u_r = 0$.

The damping model in surge can also be expressed in terms of the hydrodynamic derivative $X_{|u|u}$ since

$$X = X_{|u|u}|u_r|u_r \tag{6.80}$$

$$X_{|u|u} = -\frac{1}{2}\rho S(1+k)C_f(u_r) < 0. \tag{6.81}$$

Low-speed Calibration

For DP applications (low-speed maneuvering), the ITTC formulae (6.76) and (6.77) give too little damping compared to what is observed in a wind tunnel experiment with a scale model. For a ship at zero speed, the current force is (see Section 6.7.1)

$$X = \frac{1}{2}\rho A_{F_c} C_X(\gamma_c) V_c^2 \tag{6.82}$$

where V_c and γ_c are the current speed and angle of attack, respectively. The frontal projected current area is denoted A_{F_c} while ρ is the density of water. The current coefficient $C_X(\gamma_c)$ is shown in Figure 6.11. Assume that the ship is facing the current such that $\gamma_c = 0$ and $u_c = -V_c$. Formulae (6.80) and (6.81) with $u_r = 0 - u_c = V_c$ can be rewritten as

$$X = -\frac{1}{2}\rho S(1+k)C_f(u_r)V_c^2. \tag{6.83}$$

Comparing Formula (6.83) with (6.82) at $u_r = 0$ gives

$$C_f(0) = -\frac{A_{F_c}}{S(1+k)} C_X(0) \tag{6.84}$$

where $C_X(0) < 0$ as shown in Figure 6.11. Hence, one way to obtain sufficient damping at low speed is to modify the resistance curve $C_f(u_r)$ according to

$$C_f^{new}(u_r) = (1 - \sigma(u_r))C_f(u_r) + \sigma(u_r)\left(-\frac{A_{F_c}}{S(1+k)}C_X(0)\right) \tag{6.85}$$

where $\sigma(u_r)$ is a blending function

$$\sigma(u_r) = e^{-\alpha u_r^2} \tag{6.86}$$

with $\alpha > 0$ (typically 5.0). The modified resistance curve $C_f^{new}(u_r)$ is plotted together with $C_f(u_r)$ in Figure 6.4. Note that the resistance curve is increased at lower velocities due to the contribution of the current coefficient $C_X(0)$. The second plot shows the current force for the original and modified resistance curves. Note that the effect of the current coefficient vanishes at higher speeds thanks to the exponentially decaying weight $e^{-\alpha u_r^2}$.

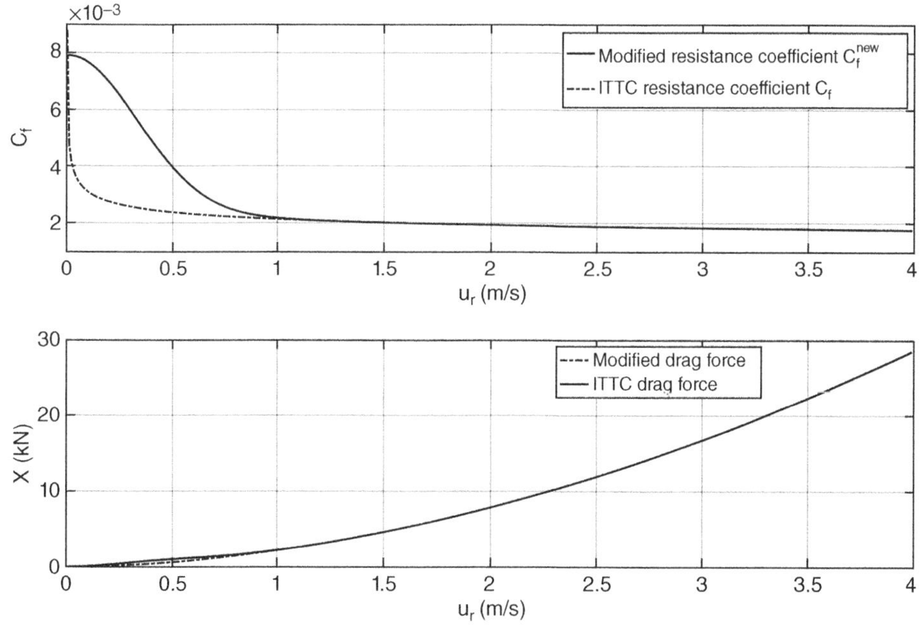

Figure 6.4 Modified resistance curve $C_f^{new}(u_r)$ and $C_f(u_r)$ as a function of u_r when $C_R = 0$ and $\sigma(u_r) = e^{-5u_r^2}$. Here $C_f^{new}(0) = -(A_{F_c}/S(1+k))C_X(0) = 0.008$ where $C_X(0) = -0.16$ is the current coefficient.

6.4.3 Cross-flow Drag Principle

For relative current angles $|\beta_{V_c} - \psi| \gg 0$, where β_{V_c} is the current direction, the cross-flow drag principle may be applied to calculate the nonlinear damping force in sway and the yaw moment (Faltinsen 1990)

$$Y = -\frac{1}{2}\rho \int_{-\frac{L_{pp}}{2}}^{\frac{L_{pp}}{2}} T(x)C_{\mathrm{d}}^{\mathrm{2D}}(x)|v_r + xr|(v_r + xr)\mathrm{d}x \tag{6.87}$$

$$N = -\frac{1}{2}\rho \int_{-\frac{L_{pp}}{2}}^{\frac{L_{pp}}{2}} T(x)C_{\mathrm{d}}^{\mathrm{2D}}(x)x|v_r + xr|(v_r + xr)\mathrm{d}x \tag{6.88}$$

where $C_{\mathrm{d}}^{\mathrm{2D}}(x)$ is the 2D drag coefficient, $T(x)$ is the draft and

$$\begin{aligned} v_r &= v - v_{\mathrm{c}} \\ &= v - V_{\mathrm{c}} \sin(\beta_{\beta_{V_{\mathrm{c}}}} - \psi) \end{aligned} \tag{6.89}$$

is the relative sway velocity (see Section 10.3). This is a strip theory approach where each hull section contributes to the integral. Drag coefficients for different hull forms are found in Hooft (1994). A constant 2D current coefficient can also be estimated using Hoerner's curve (see Figure 6.5).

MATLAB

The 2D drag coefficients $C_{\mathrm{d}}^{\mathrm{2D}}$ can be computed as a function of beam B and length T using Hoerner's curve. This is implemented in the Matlab MSS toolbox as

```
Cd = Hoerner(B,T)
```

A 3D representation of (6.87) and (6.88) eliminating the integrals can be found by curve fitting formula (6.87) and (6.88) to *second-order modulus terms* to obtain a maneuvering model similar to that of Fedyaevsky and Sobolev (1963)

$$Y = Y_{|v|v}|v_r|v_r + Y_{|v|r}|v_r|r + Y_{v|r|}v_r|r| + Y_{|r|r}|r|r \tag{6.90}$$

$$N = N_{|v|v}|v_r|v_r + N_{|v|r}|v_r|r + N_{v|r|}v_r|r| + N_{|r|r}|r|r \tag{6.91}$$

where $Y_{|v|v}, Y_{|v|r}, Y_{v|r|}, Y_{|r|r}|r|r|, N_{|v|v}, N_{|v|r}, N_{v|r|}$, and $N_{|r|r}$ are maneuvering coefficients defined using the SNAME notation. In the next section, this approach will be used to derive maneuvering models in 3 DOFs.

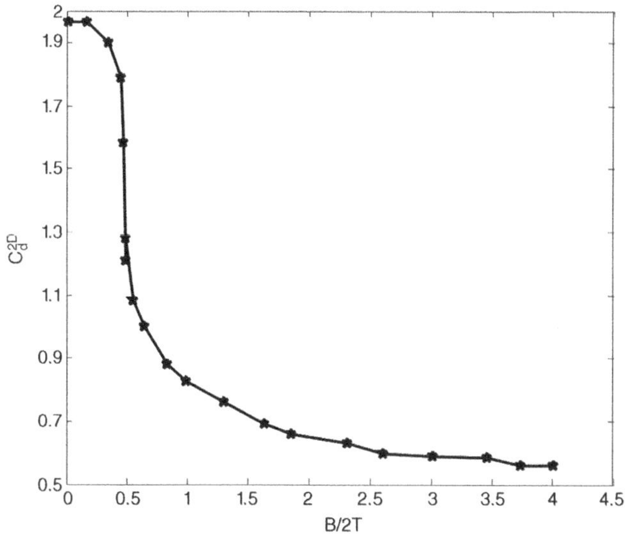

Figure 6.5 2D cross-flow coefficient C_d^{2D} as a function of $B/2T$ (Hoerner 1965).

6.5 Ship Maneuvering Models (3 DOFs)

This section summarizes the linear and nonlinear maneuvering equations using the results in Sections 6.1–6.4.

MATLAB

Several MSS maneuvering models are located in the toolbox catalog `../MSS/VESSELS/`

```
[xdot,U] = container(x,tau)    % Container ship L = 175 m
[xdot,U] = Lcontainer(x,tau)   % Linearized container ship
[xdot,U] = mariner(x,ui,U0)    % Mariner class vessel L = 161 m
[xdot,U] = navalvessel(x,tau)  % Naval vessel L = 51.5 m
[xdot,U] = tanker(x,ui)        % Tanker L = 304.8 m
```

The models can be simulated under feedback control using the user editable scripts

```
Sim<model name>.m
```

6.5.1 Nonlinear Equations of Motion

The matrix-vector representation for a surface vessel with generalized position vector $\boldsymbol{\eta} = [x^n, \ y^n, \ \psi]^\top$ and velocity vector $\boldsymbol{\nu} = [u, \ v, \ r]^\top$ is

$$\dot{\boldsymbol{\eta}} = \boldsymbol{R}(\psi)\boldsymbol{\nu} \tag{6.92}$$

$$\underbrace{\boldsymbol{M}_{\mathrm{RB}}\dot{\boldsymbol{\nu}} + \boldsymbol{C}_{\mathrm{RB}}(\boldsymbol{\nu})\boldsymbol{\nu}}_{\text{rigid-body forces}} + \underbrace{\boldsymbol{M}_{\mathrm{A}}\dot{\boldsymbol{\nu}}_{\mathrm{r}} + \boldsymbol{N}(\boldsymbol{\nu}_{\mathrm{r}})\boldsymbol{\nu}_{\mathrm{r}}}_{\text{hydrodynamic forces}} = \boldsymbol{\tau} + \boldsymbol{\tau}_{\mathrm{wind}} + \boldsymbol{\tau}_{\mathrm{wave}} \tag{6.93}$$

where $\boldsymbol{\nu}_{\mathrm{r}} = \boldsymbol{\nu} - \boldsymbol{\nu}_{\mathrm{c}}$ is the relative velocity vector, $\boldsymbol{\nu}_{\mathrm{c}} = [u_{\mathrm{c}},\ \upsilon_{\mathrm{c}},\ 0]^{\top}$ is the current velocity vector and

$$\boldsymbol{N}(\boldsymbol{\nu}_{\mathrm{r}}) := \boldsymbol{C}_{\mathrm{A}}(\boldsymbol{\nu}_{\mathrm{r}}) + \boldsymbol{D} + \boldsymbol{D}_{\mathrm{n}}(\boldsymbol{\nu}_{\mathrm{r}}). \tag{6.94}$$

Note that the added mass Coriolis and centripetal terms together with hydrodynamic damping terms are collected into the matrix $\boldsymbol{N}(\boldsymbol{\nu}_{\mathrm{r}})$. This is convenient since it is difficult to distinguish terms in $\boldsymbol{C}_{\mathrm{A}}(\boldsymbol{\nu}_{\mathrm{r}})$ with similar terms in $\boldsymbol{D}_{\mathrm{n}}(\boldsymbol{\nu}_{\mathrm{r}})$. Consequently, only the sum of these terms is used in the model in order to avoid overparametrization. The model matrices are

$$\boldsymbol{R}(\psi) = \begin{bmatrix} \cos(\psi) & -\sin(\psi) & 0 \\ \sin(\psi) & \cos(\psi) & 0 \\ 0 & 0 & 1 \end{bmatrix} \tag{6.95}$$

$$\boldsymbol{M}_{\mathrm{RB}} = \begin{bmatrix} m & 0 & 0 \\ 0 & m & mx_g \\ 0 & mx_g & I_z \end{bmatrix} \tag{6.96}$$

$$\boldsymbol{C}_{\mathrm{RB}}(\boldsymbol{\nu}) = \begin{bmatrix} 0 & -mr & -mx_g r \\ mr & 0 & 0 \\ mx_g r & 0 & 0 \end{bmatrix} \tag{6.97}$$

$$\boldsymbol{M}_{\mathrm{A}} = -\begin{bmatrix} X_{\dot{u}} & 0 & 0 \\ 0 & Y_{\dot{\upsilon}} & Y_{\dot{r}} \\ 0 & N_{\dot{\upsilon}} & N_{\dot{r}} \end{bmatrix} = \begin{bmatrix} A_{11}(0) & 0 & 0 \\ 0 & A_{22}(0) & A_{26}(0) \\ 0 & A_{62}(0) & A_{66}(0) \end{bmatrix} \tag{6.98}$$

$$\begin{aligned} \boldsymbol{C}_{\mathrm{A}}(\boldsymbol{\nu}_{\mathrm{r}}) &= \begin{bmatrix} 0 & 0 & Y_{\dot{\upsilon}}\upsilon_r + Y_{\dot{r}}r \\ 0 & 0 & -X_{\dot{u}}u_r \\ -Y_{\dot{\upsilon}}\upsilon_r - Y_{\dot{r}}r & X_{\dot{u}}u_r & 0 \end{bmatrix} \\ &= \begin{bmatrix} 0 & 0 & -A_{22}(0)\upsilon_r - A_{26}(0)r \\ 0 & 0 & A_{11}(0)u_r \\ A_{22}(0)\upsilon_r + A_{26}(0)r & -A_{11}(0)u_r & 0 \end{bmatrix} \end{aligned} \tag{6.99}$$

$$\boldsymbol{D} = -\begin{bmatrix} X_u & 0 & 0 \\ 0 & Y_v & Y_r \\ 0 & N_v & N_r \end{bmatrix} \approx \begin{bmatrix} B_{11v}(0) & 0 & 0 \\ 0 & B_{22v}(0) & 0 \\ 0 & 0 & B_{66v}(0) \end{bmatrix} \tag{6.100}$$

$$\boldsymbol{D}_{\mathrm{n}}(\boldsymbol{\nu}_{\mathrm{r}}) = -\begin{bmatrix} X_{|u|u}|u_r| & 0 & 0 \\ 0 & Y_{|v|v}|v_r| + Y_{|r|v}|r| & Y_{|v|r}|v_r| + Y_{|r|r}|r| \\ 0 & N_{|v|v}|v_r| + N_{|r|v}|r| & N_{|v|r}|v_r| + N_{|r|r}|r| \end{bmatrix}. \tag{6.101}$$

In the case of ocean currents it is possible to express (6.93) using only the relative velocity vector $\boldsymbol{\nu}_{\mathrm{r}}$ and thus avoiding terms in $\boldsymbol{\nu}$. Since $\boldsymbol{C}_{\mathrm{RB}}(\boldsymbol{\nu})$ is parametrized independent of linear velocity it follows from Property 10.1 in Section 10.3 that (6.93) can be rewritten as

$$\underbrace{\boldsymbol{M}\dot{\boldsymbol{\nu}}_{\mathrm{r}} + \boldsymbol{C}(\boldsymbol{\nu}_{\mathrm{r}})\boldsymbol{\nu}_{\mathrm{r}} + \boldsymbol{D}\boldsymbol{\nu}_{\mathrm{r}} + \boldsymbol{D}_{\mathrm{n}}(\boldsymbol{\nu}_{\mathrm{r}})\boldsymbol{\nu}_{\mathrm{r}}}_{\boldsymbol{N}(\boldsymbol{\nu}_{\mathrm{r}})\boldsymbol{\nu}_{\mathrm{r}}} = \boldsymbol{\tau} + \boldsymbol{\tau}_{\mathrm{wind}} + \boldsymbol{\tau}_{\mathrm{wave}} \tag{6.102}$$

where

$$\boldsymbol{M} = \boldsymbol{M}_{\mathrm{A}} + \boldsymbol{M}_{\mathrm{RB}} \tag{6.103}$$

$$\boldsymbol{C}(\boldsymbol{\nu}_{\mathrm{r}}) = \boldsymbol{C}_{\mathrm{A}}(\boldsymbol{\nu}_{\mathrm{r}}) + \boldsymbol{C}_{\mathrm{RB}}(\boldsymbol{\nu}_{\mathrm{r}}). \tag{6.104}$$

In this representation the generalized velocity $\boldsymbol{\nu}_{\mathrm{r}}$ is the only velocity vector while (6.93) uses both $\boldsymbol{\nu}$ and $\boldsymbol{\nu}_{\mathrm{r}}$.

State-space Model Including Ocean Currents

For an ocean current with constant speed V_{c} and direction $\beta_{V_{\mathrm{c}}}$ the components of the ocean current velocity vector $\boldsymbol{\nu}_{\mathrm{c}} = [u_{\mathrm{c}}, v_{\mathrm{c}}, 0]^{\top}$ are computed as (see Section 10.3)

$$u_{\mathrm{c}} = V_{\mathrm{c}} \cos(\beta_{V_{\mathrm{c}}} - \psi) \tag{6.105}$$

$$v_{\mathrm{c}} = V_{\mathrm{c}} \sin(\beta_{V_{\mathrm{c}}} - \psi). \tag{6.106}$$

Time differentiation of u_{c} and v_{c} gives the accelerations

$$\dot{u}_{\mathrm{c}} = rV_{\mathrm{c}} \sin(\beta_{V_{\mathrm{c}}} - \psi) = rv_{\mathrm{c}} \tag{6.107}$$

$$\dot{v}_{\mathrm{c}} = -rV_{\mathrm{c}} \cos(\beta_{V_{\mathrm{c}}} - \psi) = -ru_{\mathrm{c}}. \tag{6.108}$$

The state-space model for (6.104) in terms of generalized position and relative velocity can be summarized as (Fossen 2012)

$$\dot{\boldsymbol{\eta}} = \boldsymbol{R}(\psi)\left(\boldsymbol{\nu}_{\mathrm{r}} + \begin{bmatrix} u_{\mathrm{c}} \\ v_{\mathrm{c}} \\ 0 \end{bmatrix}\right) \tag{6.109}$$

$$\dot{\boldsymbol{\nu}}_{\mathrm{r}} = \boldsymbol{M}^{-1}(\boldsymbol{\tau} + \boldsymbol{\tau}_{\mathrm{wind}} + \boldsymbol{\tau}_{\mathrm{wave}} - \boldsymbol{C}(\boldsymbol{\nu}_{\mathrm{r}})\boldsymbol{\nu}_{\mathrm{r}} - \boldsymbol{D}\boldsymbol{\nu}_{\mathrm{r}} - \boldsymbol{D}_{\mathrm{n}}(\boldsymbol{\nu}_{\mathrm{r}})\boldsymbol{\nu}_{\mathrm{r}}). \tag{6.110}$$

Figure 6.6 Displacement vessel where the horizontal-plane model can be used for DP and maneuvering.

Alternatively, the state-space model can be expressed using absolute velocities

$$\dot{\boldsymbol{\eta}} = \boldsymbol{R}(\psi)\boldsymbol{\nu} \tag{6.111}$$

$$\dot{\boldsymbol{\nu}} = \begin{bmatrix} r v_c \\ -r u_c \\ 0 \end{bmatrix} + \boldsymbol{M}^{-1}(\boldsymbol{\tau} + \boldsymbol{\tau}_{\text{wind}} + \boldsymbol{\tau}_{\text{wave}} - \boldsymbol{C}(\boldsymbol{\nu}_r)\boldsymbol{\nu}_r - \boldsymbol{D}\boldsymbol{\nu}_r - \boldsymbol{D}_n(\boldsymbol{\nu}_r)\boldsymbol{\nu}_r) \tag{6.112}$$

where $\boldsymbol{\nu}_r = \boldsymbol{\nu} - [u_c, \ v_c, \ 0]^\top$.

6.5.2 Nonlinear Maneuvering Model Based on Surge Resistance and Cross-flow Drag

If we use the surge resistance and cross-flow drag models in Section 6.4, the $\boldsymbol{N}(\boldsymbol{\nu}_r)$ matrix in the maneuvering model (6.94) can be redefined according to

$$\boldsymbol{N}(\boldsymbol{\nu}_r)\boldsymbol{\nu}_r := \boldsymbol{C}_A(\boldsymbol{\nu}_r)\boldsymbol{\nu}_r + \boldsymbol{D}\boldsymbol{\nu}_r + \boldsymbol{d}(\boldsymbol{\nu}_r) \tag{6.113}$$

where

$$C_A(\nu_r) = \begin{bmatrix} 0 & 0 & Y_{\dot{v}}v_r + Y_{\dot{r}}r \\ 0 & 0 & -X_{\dot{u}}u_r \\ -Y_{\dot{v}}v_r - Y_{\dot{r}}r & X_{\dot{u}}u_r & 0 \end{bmatrix} \tag{6.114}$$

$$D = \begin{bmatrix} -X_u & 0 & 0 \\ 0 & -Y_v & -Y_r \\ 0 & -N_v & -N_r \end{bmatrix} \tag{6.115}$$

$$d(\nu_r) = \begin{bmatrix} \frac{1}{2}\rho S(1+k)C_f^{\mathrm{new}}(u_r)|u_r|u_r \\ \frac{1}{2}\rho \int_{-\frac{L_{pp}}{2}}^{\frac{L_{pp}}{2}} T(x)C_d^{2D}(x)|v_r + xr|(v_r + xr)\,\mathrm{d}x \\ \frac{1}{2}\rho \int_{-\frac{L_{pp}}{2}}^{\frac{L_{pp}}{2}} T(x)C_d^{2D}(x)x|v_r + xr|(v_r + xr)\,\mathrm{d}x \end{bmatrix}. \tag{6.116}$$

The linear damping matrix D in this expression is important for low-speed maneuvering and stationkeeping while the term $d(\nu_r)$ dominates at higher speeds. Linear damping also guarantees that the velocity converges exponentially to zero.

6.5.3 Nonlinear Maneuvering Model Based on Second-order Modulus Functions

The idea of using second-order *modulus functions* to describe the nonlinear dissipative terms in $N(\nu_r)$ dates back to Fedyaevsky and Sobolev (1963). Within this framework, a simplified form of Norrbin's nonlinear model (Norrbin 1970), which retains the most important terms for steering and propulsion loss assignment, was proposed by Blanke (1981). This model corresponds to fitting the cross-flow drag integrals (6.87) and (6.88) to second-order modulus functions

$$\begin{aligned} N(\nu_r)\nu_r &= C_A(\nu_r)\nu_r + D(\nu_r)\nu_r \\ &= \begin{bmatrix} Y_{\dot{v}}v_r r + Y_{\dot{r}}r^2 \\ -X_{\dot{u}}u_r r \\ (X_{\dot{u}} - Y_{\dot{v}})u_r v_r - Y_{\dot{r}}u_r r \end{bmatrix} \left(\text{alter-natively: } \begin{bmatrix} X_{vr}v_r r + X_{rr}r^2 \\ Y_{ur}u_r r \\ N_{uv}u_r v_r + N_{ur}u_r r \end{bmatrix} \right) \\ &+ \begin{bmatrix} -X_{|u|u}|u_r|u_r \\ -Y_{|v|v}|v_r|v_r - Y_{|v|r}|v_r|r - Y_{v|r|}v_r|r| - Y_{|r|r}|r|r| \\ -N_{|v|v}|v_r|v_r - N_{|v|r}|v_r|r - N_{v|r|}v_r|r| - N_{|r|r}|r|r| \end{bmatrix} \end{aligned}$$

$$= \begin{bmatrix} -X_{|u|u}|u_r|u_r + Y_{\dot{v}}v_r r + Y_{\dot{r}}r^2 \\ -X_{\dot{u}}u_r r - Y_{|v|v}|v_r|v_r - Y_{|v|r}|v_r|r - Y_{v|r|}v_r|r| - Y_{|r|r}|r|r| \\ (X_{\dot{u}} - Y_{\dot{v}})u_r v_r - Y_{\dot{r}}u_r r - N_{|v|v}|v_r|v_r \\ \qquad -N_{|v|r}|v_r|r - N_{v|r|}v_r|r| - N_{|r|r}|r|r| \end{bmatrix}. \tag{6.117}$$

The $\boldsymbol{C}_A$ terms can also be denoted as $X_{vr}v_r r$, $X_{rr}r^2$, $Y_{ur}u_r r$, $N_{uv}u_r v_r$ and $N_{ur}u_r r$. If these terms are experimentally obtained, viscous effects will be included in addition to the potential coefficients $Y_{\dot{v}}$, $X_{\dot{u}}$ and $Y_{\dot{r}}$.

From this expression, we obtain the $\boldsymbol{C}_A(\boldsymbol{v}_r)$ and $\boldsymbol{D}(\boldsymbol{v}_r)$ matrices

$$\boldsymbol{C}_A(\boldsymbol{v}_r) = \begin{bmatrix} 0 & 0 & Y_{\dot{v}}v_r + Y_{\dot{r}}r \\ 0 & 0 & -X_{\dot{u}}u_r \\ -Y_{\dot{v}}v_r - Y_{\dot{r}}r & X_{\dot{u}}u_r & 0 \end{bmatrix} \tag{6.118}$$

$$\boldsymbol{D}(\boldsymbol{v}_r) = \begin{bmatrix} -X_{|u|u}|u_r| & 0 & 0 \\ 0 & -Y_{|v|v}|v_r| - Y_{|r|v}|r| & -Y_{|v|r}|v_r| - Y_{|r|r}|r| \\ 0 & -N_{|v|v}|v_r| - N_{|r|v}|r| & -N_{|v|r}|v_r| - N_{|r|r}|r| \end{bmatrix} \tag{6.119}$$

Recall that $\boldsymbol{D}(\boldsymbol{v}_r) = \boldsymbol{D} + \boldsymbol{D}_n(\boldsymbol{v}_r)$. However, linear potential damping and skin friction $\boldsymbol{D}$ are neglected in (6.117) since the nonlinear quadratic terms $\boldsymbol{D}_n(\boldsymbol{v}_r)$ dominate at higher speeds (see Figure 6.7). This is a good assumption for maneuvering while stationkeeping models should include a non-zero $\boldsymbol{D}$.

Figure 6.7 shows the significance of the linear and quadratic terms for different ship speeds. It is recommended to use different damping models depending on the regime of the control system. In many cases, it is important to include both linear and quadratic damping, since only quadratic damping in the model will cause oscillatory behavior at low speed. The main reason is that linear damping is needed for exponential convergence to zero. For marine craft operating in waves, linear damping will always be present due to potential damping and linear skin friction (Faltinsen and Sortland 1987). For large ships Blanke (1981) suggests simplifying (6.119) according to

$$\boldsymbol{D}_n(\boldsymbol{v}_r) = \begin{bmatrix} -X_{|u|u}|u_r| & 0 & 0 \\ 0 & -Y_{|v|v}|v_r| & -Y_{|v|r}|v_r| \\ 0 & -N_{|v|v}|v_r| & -N_{|v|r}|v_r| \end{bmatrix}. \tag{6.120}$$

This gives

$$\boldsymbol{N}(\boldsymbol{v}_r) = \boldsymbol{C}_A(\boldsymbol{v}_r) + \boldsymbol{D}(\boldsymbol{v}_r)$$

$$= \begin{bmatrix} -X_{|u|u}|u_r| & 0 & Y_{\dot{v}}v_r + Y_{\dot{r}}r \\ 0 & -Y_{|v|v}|v_r| & -X_{\dot{u}}u_r - Y_{|v|r}|v_r| \\ -Y_{\dot{v}}v_r - Y_{\dot{r}}r & X_{\dot{u}}u_r - N_{|v|v}|v_r| & -N_{|v|r}|v_r| \end{bmatrix}. \tag{6.121}$$

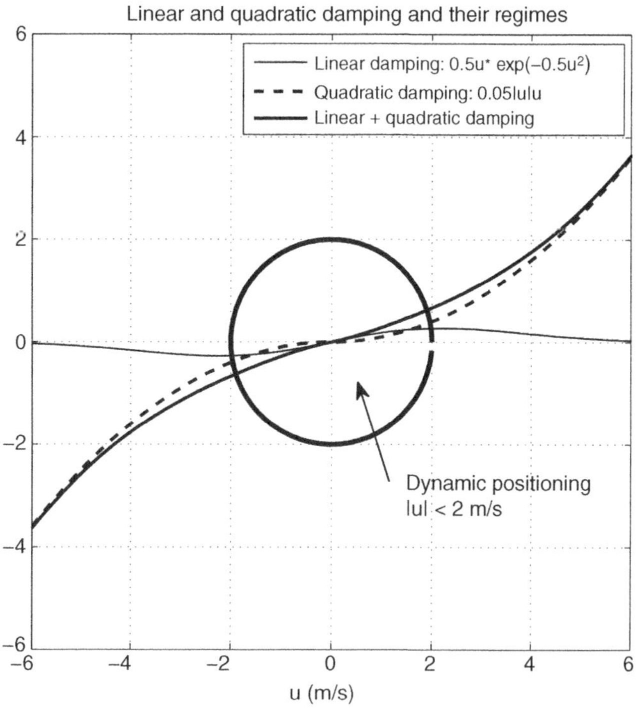

Figure 6.7 Linear and quadratic damping and their speed regimes. Note that the linear part goes to zero for higher speeds.

6.5.4 Nonlinear Maneuvering Model Based on Odd Functions

So far, we have discussed nonlinear maneuvering models based on first principles such as surge resistance and cross-flow drag, which have been approximated by second-order *modulus functions* (see Fedyaevsky and Sobolev 1963; Norrbin 1970).

In many cases a more pragmatic approach is used for curve fitting of experimental data (Clarke 2003). This is typically done by using *Taylor series* of first- and second-order terms (Abkowitz 1964) to describe the nonlinear terms in $\boldsymbol{N}(\boldsymbol{v}_r)$.

The Nonlinear Model of Abkowitz (1964)

One of the standard nonlinear ship models in the literature is that of Abkowitz (1964). Consider the nonlinear rigid-body kinetics

$$\boldsymbol{M}_{\mathrm{RB}}\dot{\boldsymbol{v}} + \boldsymbol{C}_{\mathrm{RB}}(\boldsymbol{v})\boldsymbol{v} = \boldsymbol{\tau}_{\mathrm{RB}} \tag{6.122}$$

with external forces and moment

$$\boldsymbol{\tau}_{\mathrm{RB}} = [X(\boldsymbol{x}),\ \ Y(\boldsymbol{x}),\ \ N(\boldsymbol{x})]^{\top} \tag{6.123}$$

where $\boldsymbol{x} = [u,\ v,\ r,\ \dot{u},\ \dot{v},\ \dot{r},\ \delta]^{\top}$ and δ is the rudder angle. Based on these equations, Abkowitz (1964) proposed a third-order truncated *Taylor-series* expansion of the functions $X(\boldsymbol{x})$, $Y(\boldsymbol{x})$ and $N(\boldsymbol{x})$ at $\boldsymbol{x}_0 = [U,\ 0,\ 0,\ 0,\ 0,\ 0,\ 0]^{\top}$. This gives

$$X(\boldsymbol{x}) \approx X(\boldsymbol{x}_0) + \sum_{i=1}^{n}\left(\left.\frac{\partial X(\boldsymbol{x})}{\partial x_i}\right|_{\boldsymbol{x}_0} \Delta x_i + \frac{1}{2}\left.\frac{\partial^2 X(\boldsymbol{x})}{(\partial x_i)^2}\right|_{\boldsymbol{x}_0} \Delta x_i^2 + \frac{1}{6}\left.\frac{\partial^3 X(\boldsymbol{x})}{(\partial x_i)^3}\right|_{\boldsymbol{x}_0} \Delta x_i^3 \right)$$

$$Y(\boldsymbol{x}) \approx Y(\boldsymbol{x}_0) + \sum_{i=1}^{n}\left(\left.\frac{\partial Y(\boldsymbol{x})}{\partial x_i}\right|_{\boldsymbol{x}_0} \Delta x_i + \frac{1}{2}\left.\frac{\partial^2 Y(\boldsymbol{x})}{(\partial x_i)^2}\right|_{\boldsymbol{x}_0} \Delta x_i^2 + \frac{1}{6}\left.\frac{\partial^3 Y(\boldsymbol{x})}{(\partial x_i)^3}\right|_{\boldsymbol{x}_0} \Delta x_i^3 \right)$$

$$N(\boldsymbol{x}) \approx Z(\boldsymbol{x}_0) + \sum_{i=1}^{n}\left(\left.\frac{\partial N(\boldsymbol{x})}{\partial x_i}\right|_{\boldsymbol{x}_0} \Delta x_i + \frac{1}{2}\left.\frac{\partial^2 N(\boldsymbol{x})}{(\partial x_i)^2}\right|_{\boldsymbol{x}_0} \Delta x_i^2 + \frac{1}{6}\left.\frac{\partial^3 N(\boldsymbol{x})}{(\partial x_i)^3}\right|_{\boldsymbol{x}_0} \Delta x_i^3 \right)$$

where $\Delta\boldsymbol{x} = \boldsymbol{x} - \boldsymbol{x}_0 = [\Delta x_1,\ \Delta x_2,\ \ldots\ ,\Delta x_7]^{\top}$. Unfortunately, a third-order Taylor-series expansion results in a large number of terms. By applying some physical insight, the complexity of these expressions can be reduced. Abkowitz (1964) makes the following assumptions:

1. *Most ship maneuvers can be described by a third-order truncated Taylor series expansion about the steady-state condition $u = u_0$.*
2. *Only first-order acceleration terms are considered.*
3. *Standard port/starboard symmetry simplifications except terms describing the constant force and moment arising from single-screw propellers.*
4. *The coupling between the acceleration and velocity terms is negligible.*

Simulations of standard ship maneuvers show that these assumptions are quite good. Applying these assumptions to the expressions $X(\boldsymbol{x})$, $Y(\boldsymbol{x})$ and $N(\boldsymbol{x})$ yields

$$\begin{aligned} X = {} & X^* + X_{\dot{u}}\dot{u} + X_u\Delta u + X_{uu}\Delta u^2 + X_{uuu}\Delta u^3 + X_{vv}v^2 + X_{rr}r^2 + X_{\delta\delta}\delta^2 \\ & + X_{rv\delta}rv\delta + X_{r\delta}r\delta + X_{v\delta}v\delta + X_{vvu}v^2\Delta u + X_{rru}r^2\Delta u + X_{\delta\delta u}\delta^2\Delta u \\ & + X_{rvu}rvu + X_{r\delta u}r\delta\Delta u + X_{v\delta u}v\delta\Delta u \end{aligned} \tag{6.124}$$

$$\begin{aligned} Y = {} & Y^* + Y_u\Delta u + Y_{uu}\Delta u^2 + Y_r r + Y_v v + Y_{\dot{r}}\dot{r} + Y_{\dot{v}}\dot{v} + Y_\delta\delta + Y_{rrr}r^3 + Y_{vvv}v^3 \\ & + Y_{\delta\delta\delta}\delta^3 + Y_{rr\delta}r^2\delta + Y_{\delta\delta r}\delta^2 r + Y_{rrv}r^2 v + Y_{vvr}v^2 r + Y_{\delta\delta v}\delta^2 v + Y_{vv\delta}v^2\delta \\ & + Y_{\delta vr}\delta vr + Y_{vu}v\Delta u + Y_{vuu}v\Delta u^2 + Y_{ru}r\Delta u + Y_{ruu}r\Delta u^2 + Y_{\delta u}\delta\Delta u \\ & + Y_{\delta uu}\delta\Delta u^2 \end{aligned} \tag{6.125}$$

$$\begin{aligned} N = {} & N^* + N_u\Delta u + N_{uu}\Delta u^2 + N_r r + N_v v + N_{\dot{r}}\dot{r} + N_{\dot{v}}\dot{v} + N_\delta\delta + N_{rrr}r^3 \\ & + N_{vvv}v^3 + N_{\delta\delta\delta}\delta^3 + N_{rr\delta}r^2\delta + N_{\delta\delta r}\delta^2 r + N_{rrv}r^2 v + N_{vvr}v^2 r + N_{\delta\delta v}\delta^2 v \\ & + N_{vv\delta}v^2\delta + N_{\delta vr}\delta vr + N_{vu}v\Delta u + N_{vuu}v\Delta u^2 + N_{ru}r\Delta u + N_{ruu}r\Delta u^2 \\ & + N_{\delta u}\delta\Delta u + N_{\delta uu}\delta\Delta u^2. \end{aligned} \tag{6.126}$$

The hydrodynamic derivatives are defined using the notation

$$F^* = F(\boldsymbol{x}_0), \qquad F_{x_i} = \left.\frac{\partial F(\boldsymbol{x})}{\partial x_i}\right|_{x_0}$$

$$F_{x_i x_j} = \frac{1}{2}\left.\frac{\partial^2 F(\boldsymbol{x})}{\partial x_i \partial x_j}\right|_{x_0}, \qquad F_{x_i x_j x_k} = \frac{1}{6}\left.\frac{\partial^3 F(\boldsymbol{x})}{\partial x_i \partial x_j \partial x_k}\right|_{x_0}$$

where $F \in \{X, \ Y, \ N\}$.

PMM Models

The hydrodynamic coefficients can be experimentally determined by using a planar-motion-mechanism (PMM) system, which is a device for experimentally determining the hydrodynamic derivatives required in the equations of motion. This includes coefficients usually classified into the three categories of static stability, rotary stability and acceleration derivatives. The PMM device is capable of oscillating a ship (or submarine) model while it is being towed in a testing tank. The forces are measured on the scale model and fitted to odd functions based on Taylor-series expansions. The resulting model is usually referred to as the PMM model and this model is scaled up to a full-scale ship by using Froude number similarity. This ensures that the ratio between the inertial and gravitational forces is kept constant.

6.5.5 Linear Maneuvering Model

The linear maneuvering equations in *surge*, *sway* and *yaw* is a special case of the nonlinear model (6.102). Consider

$$\underbrace{(\boldsymbol{M}_{\rm RB} + \boldsymbol{M}_{\rm A})}_{\boldsymbol{M}}\dot{\boldsymbol{v}}_{\rm r} + \underbrace{(\boldsymbol{C}^*_{\rm RB} + \boldsymbol{C}^*_{\rm A} + \boldsymbol{D})}_{\boldsymbol{N}}\boldsymbol{v}_{\rm r} = \boldsymbol{\tau} + \boldsymbol{\tau}_{\rm wind} + \boldsymbol{\tau}_{\rm wave} \tag{6.127}$$

where nonlinear Coriolis, centripetal and damping forces are linearized about the velocities $v = r = 0$ and the cruise speed

$$U = \sqrt{u^2 + v^2} \approx u. \tag{6.128}$$

Linearization is performed under the assumption that the unknown current velocities are negligible, that is $u_{\rm c} = v_{\rm c} = 0$. Thus, the expressions for $\boldsymbol{C}^*_{\rm RB}$, $\boldsymbol{C}^*_{\rm A}$ and $\boldsymbol{D}$ are

$$\boldsymbol{C}^*_{\rm RB} = \begin{bmatrix} 0 & 0 & 0 \\ 0 & 0 & mU \\ 0 & 0 & mx_g U \end{bmatrix} \tag{6.129}$$

$$\boldsymbol{C}_{\mathrm{A}}^{*}=\begin{bmatrix}0 & 0 & 0\\ 0 & 0 & -X_{\dot{u}}U\\ 0 & (X_{\dot{u}}-Y_{\dot{v}})U & -Y_{\dot{r}}U\end{bmatrix} \tag{6.130}$$

$$\boldsymbol{D}=-\begin{bmatrix}X_u & 0 & 0\\ 0 & Y_v & Y_{\mathrm{r}}\\ 0 & N_v & N_{\mathrm{r}}\end{bmatrix}. \tag{6.131}$$

Starboard–port symmetry implies that surge is decoupled from sway and yaw. Equation (6.127) can be expanded as

$$\begin{bmatrix}m-X_{\dot{u}} & 0 & 0\\ 0 & m-Y_{\dot{v}} & mx_g-Y_{\dot{r}}\\ 0 & mx_g-N_{\dot{v}} & I_z-N_{\dot{r}}\end{bmatrix}\begin{bmatrix}\dot{u}_{\mathrm{r}}\\ \dot{v}_{\mathrm{r}}\\ \dot{r}\end{bmatrix} + \begin{bmatrix}-X_u & 0 & 0\\ 0 & -Y_v & (m-X_{\dot{u}})U-Y_{\mathrm{r}}\\ 0 & (X_{\dot{u}}-Y_{\dot{v}})U-N_v & (mx_g-Y_{\dot{r}})U-N_{\mathrm{r}}\end{bmatrix}\begin{bmatrix}u_r\\ v_r\\ r\end{bmatrix}=\begin{bmatrix}\tau_1\\ \tau_2\\ \tau_6\end{bmatrix}. \tag{6.132}$$

For a rudder in the propeller slipstream

$$\tau_1=-X_{\delta\delta}\delta^2+(1-t)T \tag{6.133}$$

$$\tau_2=-Y_{\delta}\delta \tag{6.134}$$

$$\tau_6=-N_{\delta}\delta \tag{6.135}$$

where δ is the rudder angle; see Equations (9.105) and (9.106) in Section 9.5. The thrust T is reduced by a factor $1-t$ due to the extra resistance on the hull caused by the propeller. The thrust deduction number t is typically 0.05–0.2; see Section 9.1.

Linear Forward Speed Model (Surge Subsystem)

The first row in (6.132) is recognized as the linear surge equation

$$(m-X_{\dot{u}})\dot{u}_{\mathrm{r}}-X_u u_r=-X_{\delta\delta}\delta^2+(1-t)T \tag{6.136}$$

where $-X_u>0$ and $X_{\delta\delta}\delta^2>0$ is added resistance due to rudder deflections as defined by (9.104). The linear damping coefficient X_u can be approximated by equivalent linearization; see Section 6.4.2. In other words

$$X_u=\frac{8A_1}{3\pi}X_{|u|u} \tag{6.137}$$

where the relative surge velocity $u_r=A_1\cos(\omega t)$ is assumed to be harmonic.

Linear Maneuvering Model (Sway–Yaw Subsystem)

The second and third rows in (6.132) are the sway–yaw subsystem. This model is also known as the *potential theory representation* (Fossen 1994, Clarke and Horn 1997)

$$\boldsymbol{M}\dot{\boldsymbol{v}}_{\rm r} + \boldsymbol{N}\boldsymbol{v}_{\rm r} = \boldsymbol{b}\delta \tag{6.138}$$

where $\boldsymbol{v}_{\rm r} = [v_r, r]^{\rm T}$ and δ is the rudder angle. Consequently,

$$\boldsymbol{M} = \begin{bmatrix} m - Y_{\dot{v}} & mx_g - Y_{\dot{r}} \\ mx_g - N_{\dot{v}} & I_z - N_{\dot{r}} \end{bmatrix} \tag{6.139}$$

$$\boldsymbol{N} = \begin{bmatrix} -Y_v & (m - X_{\dot{u}})U - Y_{\rm r} \\ (X_{\dot{u}} - Y_{\dot{v}})U - N_v & (mx_g - Y_{\dot{r}})U - N_{\rm r} \end{bmatrix} \tag{6.140}$$

$$\boldsymbol{b} = \begin{bmatrix} -Y_\delta \\ -N_\delta \end{bmatrix}. \tag{6.141}$$

Comment 6.1 *Davidson and Schiff (1946) assumed that the hydrodynamic forces $\boldsymbol{\tau}_{RB}$ are linear in $\dot{\boldsymbol{v}}_{\rm r}$, $\boldsymbol{v}_{\rm r}$ and $\delta_{\rm R}$ (linear strip theory) such that*

$$\boldsymbol{\tau}_{RB} = \underbrace{\begin{bmatrix} Y_{\dot{v}} & Y_{\dot{r}} \\ N_{\dot{v}} & N_{\dot{r}} \end{bmatrix}}_{\boldsymbol{M}_{\rm A}} \dot{\boldsymbol{v}}_{\rm r} + \underbrace{\begin{bmatrix} Y_v & Y_{\rm r} \\ N_v & N_{\rm r} \end{bmatrix}}_{\boldsymbol{D}} \boldsymbol{v}_{\rm r} + \underbrace{\begin{bmatrix} Y_\delta \\ N_\delta \end{bmatrix}}_{\boldsymbol{b}} \delta_{\rm R} \tag{6.142}$$

where $\delta_{\rm R} = -\delta$ is the actual rudder angle. This gives

$$N = \begin{bmatrix} -Y_v & mU - Y_{\rm r} \\ -N_v & mx_g U - N_{\rm r} \end{bmatrix}. \tag{6.143}$$

Note that the Munk moment $(X_{\dot{u}} - Y_{\dot{v}})Uv_r$ is missing in the yaw equation when compared to (6.140). This is a destabilizing moment known from aerodynamics which tries to turn the craft; see Faltinsen (1990, pp. 188–189). Also note that the less important terms $X_{\dot{u}}Ur$ and $Y_{\dot{r}}Ur$ are removed from $\boldsymbol{N}$ when compared to (6.140). All missing terms terms are due to the $\boldsymbol{C}_{\rm A}(\boldsymbol{v}_{\rm r})$ matrix, which is omitted in the linear expression (6.142). Consequently, it is recommended to use (6.140), which includes the terms from the $\boldsymbol{C}_{\rm A}(\boldsymbol{v}_{\rm r})$ matrix.

6.6 Ship Maneuvering Models Including Roll (4 DOFs)

The maneuvering models presented in Section 6.5.1 only describe the horizontal motions (*surge, sway* and *yaw*). These models are intended for the design and simulation of DP systems, heading autopilots, trajectory-tracking and path-following control systems. Many vessels, however, are equipped with actuators that can reduce the rolling motion. This could be anti-rolling tanks, rudders and fin stabilizers. In order to design a control system for roll damping, it is necessary to add the roll equation to the horizontal-plane model. Inclusion of roll means that the restoring moment due to buoyancy and gravity must be included. The resulting model is a 4-DOF maneuvering model that includes roll (*surge, sway, roll* and *yaw*).

The speed equation (6.136) can be decoupled from the sway, roll and yaw modes. The resulting model takes the form

$$\boldsymbol{M}\dot{\boldsymbol{v}}_{\mathrm{r}} + \boldsymbol{N}\boldsymbol{v}_{\mathrm{r}} + \boldsymbol{G}\boldsymbol{\eta} = \boldsymbol{\tau} \tag{6.144}$$

where $\boldsymbol{v}_r = [v_r,\ p,\ r]^{\top}$ and $\boldsymbol{\eta} = [y^n,\ \phi,\ \psi]^{\top}$ are the states while $\boldsymbol{\tau}$ is a vector of control forces. For a ship with homogeneous mass distribution and xz plane symmetry, $I_{xy} = I_{yz} = 0$ and $y_g = 0$.

From the general expressions (3.49) and (6.34) in Sections 3.3 and 6.3.1, respectively, we get (with non-zero I_{xz})

$$\boldsymbol{M} = \begin{bmatrix} m - Y_{\dot{v}} & -mz_g - Y_{\dot{p}} & mx_g - Y_{\dot{r}} \\ -mz_g - K_{\dot{v}} & I_x - K_{\dot{p}} & -I_{xz} - K_{\dot{r}} \\ mx_g - N_{\dot{v}} & -I_{xz} - N_{\dot{p}} & I_z - N_{\dot{r}} \end{bmatrix}. \tag{6.145}$$

The expression for $\boldsymbol{N}$ is obtained by linearization of $\boldsymbol{C}(\boldsymbol{v}_{\mathrm{r}})$ and $\boldsymbol{D}(\boldsymbol{v}_{\mathrm{r}})$ about $u_r = U$ and $v_r = r = 0$ which gives

$$\boldsymbol{N} = \begin{bmatrix} -Y_v & -Y_{\mathrm{p}} & mU - Y_{\mathrm{r}} \\ -K_v & -K_{\mathrm{p}} & -mz_g U - K_{\mathrm{r}} \\ -N_v & -N_{\mathrm{p}} & mx_g U - N_{\mathrm{r}} \end{bmatrix}. \tag{6.146}$$

Recall from Section 4.1 that the linear restoring forces and moments for a surface vessel can be written

$$\boldsymbol{G} = \mathrm{diag}\{0,\ \mathrm{GM_T} W,\ 0\} \tag{6.147}$$

where $W = mg$ is the weight and $\mathrm{GM_T}$ is the transverse metacenter height.

In addition to these equations, the kinematic equations (assuming $q = \theta = 0$)

$$\dot{\phi} = p \tag{6.148}$$

$$\dot{\psi} = r \tag{6.149}$$

must be augmented to the system model. The general kinematic expressions are found in Section 2.2.1.

State-space Model

The linearized model (6.144) together with (6.148) and (6.149) can be written in state-space form by defining the state vector as $\boldsymbol{x} := [v_r,\ p,\ r,\ \phi,\ \psi]^\top$. The elements associated with the matrices $\boldsymbol{A}$ and $\boldsymbol{B}$ are given by

$$\dot{\boldsymbol{x}} = \underbrace{\begin{bmatrix} a_{11} & a_{12} & a_{13} & a_{14} & 0 \\ a_{21} & a_{22} & a_{23} & a_{24} & 0 \\ a_{31} & a_{32} & a_{33} & a_{34} & 0 \\ 0 & 1 & 0 & 0 & 0 \\ 0 & 0 & 1 & 0 & 0 \end{bmatrix}}_{\boldsymbol{A}} \boldsymbol{x} + \underbrace{\begin{bmatrix} b_{11} & b_{12} & \cdots & b_{1r} \\ b_{21} & b_{22} & \cdots & b_{2r} \\ b_{31} & b_{32} & \cdots & b_{3r} \\ 0 & 0 & \cdots & 0 \\ 0 & 0 & \cdots & 0 \end{bmatrix}}_{\boldsymbol{B}} \boldsymbol{u} \tag{6.150}$$

where the elements a_{ij} are found from

$$\begin{bmatrix} a_{11} & a_{12} & a_{13} \\ a_{21} & a_{22} & a_{23} \\ a_{31} & a_{32} & a_{33} \end{bmatrix} = -\boldsymbol{M}^{-1}\boldsymbol{N}, \qquad \begin{bmatrix} * & a_{14} & * \\ * & a_{24} & * \\ * & a_{34} & * \end{bmatrix} = -\boldsymbol{M}^{-1}\boldsymbol{G} \tag{6.151}$$

while the elements b_{ij} depend on what type of actuators are in use. Finally, the roll and yaw outputs are chosen as

$$\phi = \underbrace{[0,\ 0,\ 0,\ 1,\ 0]}_{\boldsymbol{c}_\phi^\top}\boldsymbol{x}, \qquad \psi = \underbrace{[0,\ 0,\ 0,\ 0,\ 1]}_{\boldsymbol{c}_\psi^\top}\boldsymbol{x}. \tag{6.152}$$

Decompositions in Roll and Sway–Yaw Subsystems

To simplify the system for further analysis, the state vector is reorganized such that state variables associated with the steering and roll dynamics are separated. Consequently, (6.150) is rewritten as

$$\begin{bmatrix} \dot{v}_r \\ \dot{r} \\ \dot{\psi} \\ \hline \dot{p} \\ \dot{\phi} \end{bmatrix} = \begin{bmatrix} a_{11} & a_{13} & 0 & a_{12} & a_{14} \\ a_{31} & a_{33} & 0 & a_{32} & a_{34} \\ 0 & 1 & 0 & 0 & 0 \\ \hline a_{21} & a_{23} & 0 & a_{22} & a_{24} \\ 0 & 0 & 0 & 1 & 0 \end{bmatrix} \begin{bmatrix} v_r \\ r \\ \psi \\ \hline p \\ \phi \end{bmatrix} + \begin{bmatrix} b_{11} & b_{12} & \cdots & b_{1r} \\ b_{31} & b_{32} & \cdots & b_{3r} \\ 0 & 0 & \cdots & 0 \\ \hline b_{21} & b_{22} & \cdots & b_{2r} \\ 0 & 0 & \cdots & 0 \end{bmatrix} \boldsymbol{u}. \tag{6.153}$$

Let

$$\begin{bmatrix} \dot{\boldsymbol{x}}_\psi \\ \dot{\boldsymbol{x}}_\phi \end{bmatrix} = \begin{bmatrix} \boldsymbol{A}_{\psi\psi} & \boldsymbol{A}_{\psi\phi} \\ \boldsymbol{A}_{\phi\psi} & \boldsymbol{A}_{\phi\phi} \end{bmatrix} \begin{bmatrix} \boldsymbol{x}_\psi \\ \boldsymbol{x}_\phi \end{bmatrix} + \begin{bmatrix} \boldsymbol{B}_\psi \\ \boldsymbol{B}_\phi \end{bmatrix} \boldsymbol{u} \tag{6.154}$$

where $\boldsymbol{x}_\psi = [v_r,\ r,\ \psi]^\top$ and $\boldsymbol{x}_\phi = [p,\ \phi]^\top$.

If the coupling matrices are small, that is $\boldsymbol{A}_{\psi\phi} = \boldsymbol{A}_{\phi\psi} = \boldsymbol{0}$, the following subsystems

$$\begin{bmatrix} \dot{p} \\ \dot{\phi} \end{bmatrix} = \begin{bmatrix} a_{22} & a_{24} \\ 1 & 0 \end{bmatrix} \begin{bmatrix} p \\ \phi \end{bmatrix} + \begin{bmatrix} b_{21} & b_{22} & \cdots & b_{2r} \\ 0 & 0 & \cdots & 0 \end{bmatrix} \boldsymbol{u} \tag{6.155}$$

and

$$\begin{bmatrix} \dot{v}_r \\ \dot{r} \\ \dot{\psi} \end{bmatrix} = \begin{bmatrix} a_{11} & a_{13} & 0 \\ a_{31} & a_{33} & 0 \\ 0 & 1 & 0 \end{bmatrix} \begin{bmatrix} v_r \\ r \\ \psi \end{bmatrix} + \begin{bmatrix} b_{11} & b_{12} & \cdots & b_{1r} \\ b_{31} & b_{32} & \cdots & b_{3r} \\ 0 & 0 & \cdots & 0 \end{bmatrix} \boldsymbol{u} \tag{6.156}$$

will describe the ship dynamics. The last expression is recognized as the second-order Nomoto model (7.20) with r control inputs.

Transfer Functions for Steering and Rudder-roll Damping

The linearized model (6.150) is useful for frequency analysis of rudder-roll damping (RRD) systems. For simplicity consider a ship with one rudder $u = \delta$ and

$$\boldsymbol{b} = [b_{11},\ b_{21},\ b_{31},\ 0,\ 0]^\top. \tag{6.157}$$

For the state-space model (6.150) the transfer functions

$$\frac{\phi}{\delta}(s) = \boldsymbol{c}_\phi^\top (s\boldsymbol{I}_5 - \boldsymbol{A})^{-1}\boldsymbol{b} \tag{6.158}$$

$$\frac{\psi}{\delta}(s) = \boldsymbol{c}_\psi^\top (s\boldsymbol{I}_5 - \boldsymbol{A})^{-1}\boldsymbol{b} \tag{6.159}$$

take the following forms

$$\frac{\phi}{\delta}(s) = \frac{b_2 s^2 + b_1 s + b_0}{s^4 + a_3 s^3 + a_2 s^2 + a_1 s + a_0} \tag{6.160}$$

$$\frac{\psi}{\delta}(s) = \frac{c_3 s^3 + c_2 s^2 + c_1 s + c_0}{s(s^4 + a_3 s^3 + a_2 s^2 + a_1 s + a_0)}. \tag{6.161}$$

Note that these expressions are based on the full state-space model. The transfer functions can be simplified by model reduction for instance by using `modred.m` in Matlab. A decoupled approach could be to remove the yaw states from the roll equation and vice versa such that

$$\frac{\phi}{\delta}(s) \approx \frac{K_{\text{roll}}\, \omega_{\text{roll}}^2\, (1 + T_5 s)}{(1 + T_4 s)(s^2 + 2\zeta\omega_{\text{roll}} s + \omega_{\text{roll}}^2)} \tag{6.162}$$

$$\frac{\psi}{\delta}(s) \approx \frac{K_{\text{yaw}}\, (1 + T_3 s)}{s(1 + T_1 s)(1 + T_2 s)}. \tag{6.163}$$

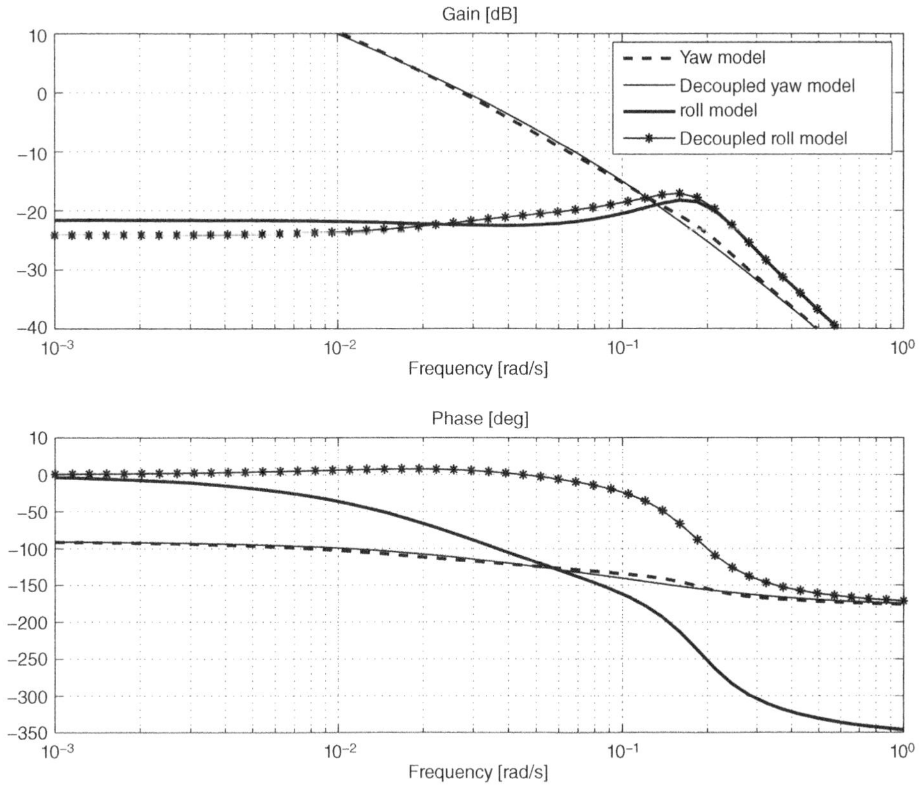

Figure 6.8 Transfer functions for the roll and sway–yaw subsystems corresponding to the Son and Nomoto container ship.

In most cases, this approximation is only rough so care should be taken. In Figure 6.8 it is seen that the phase of the roll transfer function is inaccurate for the reduced-order model.

Also, parametric investigations show that cross-couplings between steering and roll might give robust performance problems of RRD control systems (Blanke and Christensen 1993). This is also documented in Blanke (1996), who has identified the ship parameters for several loading conditions during sea trials with a series of ships. The results clearly reveal changes in the dynamics between the different ships in the series, indicating that there is a robustness problem due to changes in load conditions and rudder shape. Nonlinear effects also give rise to the same problem. Identification of ship steering-roll models are discussed by Blanke and Tiano (1997). The interested reader is also advised to consult Van der Klugt (1987) for a discussion of decoupled linear models for RRD, while nonlinear models are presented in Section 6.6.1.

Example 6.5 (Roll and Sway–Yaw Transfer Functions)
The roll and yaw transfer functions corresponding to the model of Son and Nomoto (1981) are plotted in Figure 6.8 using the MSS toolbox script `ExRRD1.m`*. The plots show both the full state-space model (6.150) and the reduced-order models. The numerical results are further discussed below.*

MATLAB

The model considered is a container ship of length $L = 175$ m and with a displacement volume of 21222 m^3 (Son and Nomoto 1981). The ship is moving at service speed $U = 7.0$ m s^{-1}. The model is based on a third-order Taylor-series expansion (see Section 6.5.4) of the hydrodynamic forces including higher-order restoring terms replacing (6.147). The nonlinear model is included in the MSS toolbox under the file name `container.m` while a linearized version of this model is found in `Lcontainer.m`. Consider the Matlab code

```
U = 7;                           % service speed (m/s)
L = 175;                         % length of ship (m)
T    = diag([ 1 1/L 1/L]);       % scaling, see Appendix D
Tinv = diag([ 1 L L]);

% nondimensional matrices: nu = [v p r]
M = [ 0.01497     0.0003525      -0.0002205
     -0.0002205   0.0000210       0
      0.0003525   0               0.000875    ];

N = [ 0.012035    0.00522     0
     -0.000314    0.0000075   0.0000692
      0.0038436 -0.000213     0.00243      ];

G = [ 0 0.0000704  0
      0 0.0004966  0
      0 0.0001468  0];

b = [-0.002578 0.0000855 0.00126    ]';

% dimensional state-space model: x = [v p r phi psi]
Minv = inv(T*M*Tinv);
A11  = - Minv * (U/L)*T*N*Tinv;
A12  = - Minv * (U/L)^2*T*G*Tinv;
B1   =   Minv * (U^2/L)*T*b;

A = [ A11         A12(:,2:3)
       0   1   0   0   0
       0   0   1   0   0   ]

B = [ B1 ; 0 ; 0  ]

% transfer functions (without yaw integrator)
roll = ss(A(1:4,1:4),B(1:4,1),[0 0 0 1],0)
yaw  = ss(A(1:4,1:4),B(1:4,1),[0 0 1 0],0)
yaw_integrator = tf(1,[1 0]);
```

```
% model reduction
red_yaw  = ss(modred(yaw,[2,4],'del'));   % removing roll states
red_roll = ss(modred(roll,[3],'del'));    % removing yaw state

% display transfer functions
zpk(series(yaw,yaw_integrator)); zpk(roll)
zpk(series(red_yaw,yaw_integrator)); zpk(red_roll)
```

The Matlab script produces the transfer functions

$$\frac{\phi}{\delta}(s) = \frac{0.0032(s-0.036)(s+0.077)}{(s+0.026)(s+0.116)(s^2+0.136s+0.036)} \approx \frac{0.083(1+49.1s)}{(1+31.5s)(s^2+0.134s+0.033)} \tag{6.164}$$

and

$$\frac{\psi}{\delta}(s) = \frac{0.0024(s+0.0436)(s^2+0.162s+0.035)}{s(s+0.0261)(s+0.116)(s^2+0.136s+0.036)} \approx \frac{0.032(1+16.9s)}{s(1+24.0s)(1+9.2s)}. \tag{6.165}$$

For roll we obtain the natural frequency and relative damping ratio

$$\omega_{\text{roll}} = 0.189 \text{ rad s}^{-1} \tag{6.166}$$

$$\zeta = 0.36. \tag{6.167}$$

It is seen that the amplitudes of the roll and yaw models are quite close. However, the reduced-order model in roll does not describe the phase with sufficient accuracy, so stability problems could be an issue when designing a model-based RRD. The main reason for this is that one pole–zero pair is omitted in the reduced-order roll model. Since this is a right-half-plane zero,

$$z = 0.036 \text{ rad s}^{-1} \tag{6.168}$$

the pole–zero pair gives an additional phase lag of $-180°$, as observed in the plot of the full model. This will of course result in serious stability problems when trying to damp the roll motion.

In practice it will be difficult to design an RRD for this system since the controller should reduce the energy at the peak frequency which is much higher than the right-half-plane zero $z = 0.036$ rad s^{-1}. This is a non-minimum phase property that cannot be changed with feedback (recall that only poles and not zeros can be moved using feedback control). The non-minimum phase characteristic is observed as an inverse response in roll when a step input is applied (see Figure 6.9).

The plots in Figure 6.9 are generated by simulating the nonlinear model of Son and Nomoto using `ExRRD3.m` in the MSS toolbox. The nonminimum phase behavior due to

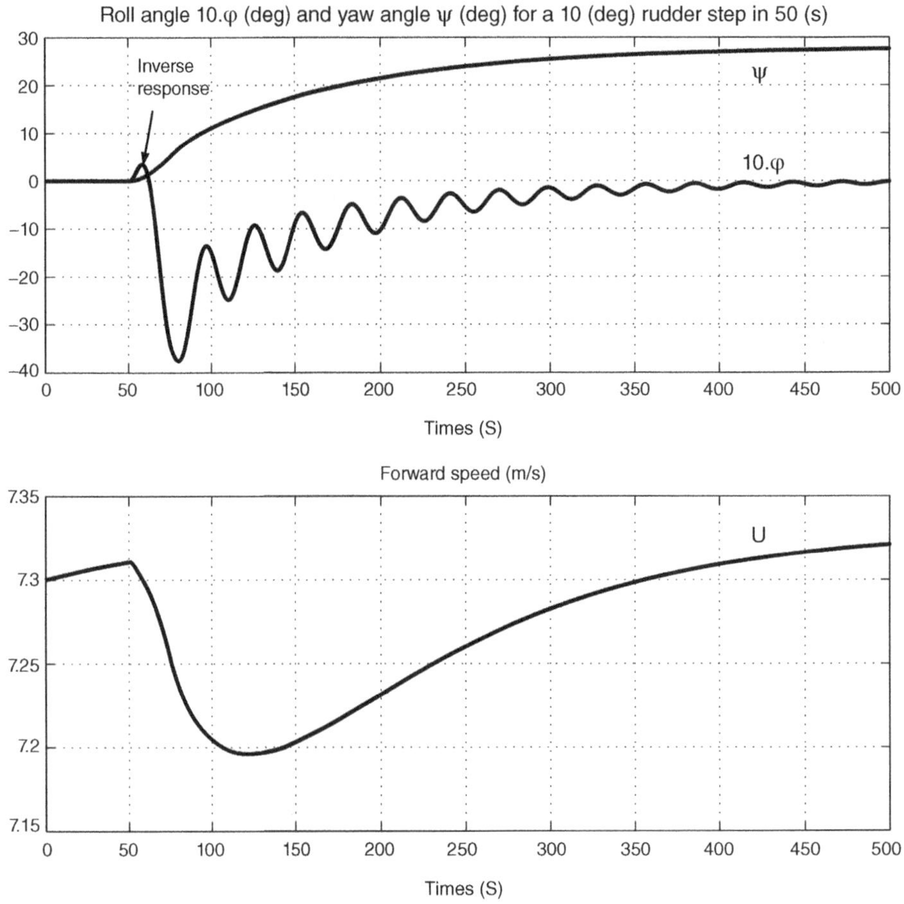

Figure 6.9 Roll angle 10ϕ and yaw angle ψ versus time for a 10° rudder step in 50 s. Notice the inverse response in roll and speed reduction during turning.

the right-half-plane zero is discussed in more detail by Fossen and Lauvdal (1994), where both linear and nonlinear analyses of the models of Son and Nomoto are considered. The nonlinear equivalent to a right-half-plane zero is unstable zero dynamics.

6.6.1 The Nonlinear Model of Son and Nomoto

A nonlinear model including roll for a high-speed container ship has been proposed by Son and Nomoto (1981, 1982)

$$(m+m_x)\dot{u} - (m+m_y)vr = X + \tau_1 \tag{6.169}$$

$$(m+m_y)\dot{v} + (m+m_x)ur + m_y\alpha_y\dot{r} - m_yl_y\dot{p} = Y + \tau_2 \tag{6.170}$$

$$(I_x+J_x)\dot{p} - m_yl_y\dot{v} - m_xl_xur = K - \mathrm{GM_T}W\phi + \tau_4 \tag{6.171}$$

$$(I_z+J_z)\dot{r} + m_y\alpha_y\dot{v} = N - x_gY + \tau_6 \tag{6.172}$$

where $m_x = A_{11}(0)$, $m_y = A_{22}(0)$, $J_x = A_{44}(\omega_4)$ and $J_z = A_{66}(0)$ denote the added mass and added moments of inertia. The control forces and moments are recognized as $\boldsymbol{\tau} = [\tau_1, \tau_2, \tau_4, \tau_6]^\top$. The added mass x coordinates of m_x and m_y are denoted by α_x and α_y, while l_x and l_y are the added mass z coordinates of m_x and m_y, respectively.

The terms on the right-hand side of (6.169)–(6.172) are defined in terms of a third-order Taylor-series expansion where small coefficients are neglected. The remaining terms are

$$X = X(u) + (1-t)T + X_{vr}vr + X_{vv}v^2 + X_{rr}r^2 + X_{\phi\phi}\phi^2 + X_\delta \sin\delta + X_{\text{ext}} \tag{6.173}$$

$$Y = Y_v v + Y_{\text{r}} r + Y_\phi \phi + Y_{\text{p}} p + Y_{vvv}v^3 + Y_{rrr}r^3 + Y_{vvr}v^2 r + Y_{vrr}vr^2 + Y_{vv\phi}v^2\phi + Y_{v\phi\phi}v\phi^2 + Y_{rr\phi}r^2\phi + Y_{r\phi\phi}r\phi^2 + Y_\delta \cos\delta + Y_{\text{ext}} \tag{6.174}$$

$$K = K_v v + K_{\text{r}} r + K_\phi \phi + K_{\text{p}} p + K_{vvv}v^3 + K_{rrr}r^3 + K_{vvr}v^2 r + K_{vrr}vr^2 + K_{vv\phi}v^2\phi + K_{v\phi\phi}v\phi^2 + K_{rr\phi}r^2\phi + K_{r\phi\phi}r\phi^2 + K_\delta \cos\delta + K_{\text{ext}} \tag{6.175}$$

$$N = N_v v + N_{\text{r}} r + N_\phi \phi + N_{\text{p}} p + N_{vvv}v^3 + N_{rrr}r^3 + N_{vvr}v^2 r + N_{vrr}vr^2 + N_{vv\phi}v^2\phi + N_{v\phi\phi}v\phi^2 + N_{rr\phi}r^2\phi + N_{r\phi\phi}r\phi^2 + N_\delta \cos\delta + N_{\text{ext}} \tag{6.176}$$

where $X(u)$ is usually modeled as quadratic drag $X(u) = X_{|u|u}|u|u$ and the subscript ext denotes external forces and moments due to wind, waves and ocean currents.

MATLAB

The nonlinear container ship model is implemented in the MSS toolbox as

```
[xdot,U] = container(x, ui)
```

The linearized model for $U = U_0$ is accessed as

```
[xdot,U] = Lcontainer(x, ui, U0)
```

where

```
x = [u v r x y psi p phi delta]'     % state vector
ui = [delta_c n_c]'                  % rudder and RPM inputs
```

In the linear case only one input, `delta_c`, is used since the forward speed `U0` is constant. For the nonlinear model, propeller rpm, `n_c`, should be positive.

6.6.2 *The Nonlinear Model of Blanke and Christensen*

An alternative model formulation describing the steering and roll motions of ships has been proposed by Blanke and Christensen (1993). This model is written as

$$\boldsymbol{M}\dot{\boldsymbol{v}} + \boldsymbol{C}_{\text{RB}}(\boldsymbol{v})\boldsymbol{v} + \boldsymbol{G}\boldsymbol{\eta} = \boldsymbol{\tau}_{\text{hyd}} + \boldsymbol{\tau}_{\text{wind}} + \boldsymbol{\tau}_{\text{wave}} + \boldsymbol{\tau} \tag{6.177}$$

where $\boldsymbol{v} = [v,\ p,\ r]^\top, \boldsymbol{\tau}_{\text{hyd}} = [Y,\ K,\ N]^\top$ and

$$\boldsymbol{M} = \begin{bmatrix} m - Y_{\dot{v}} & -mz_g - Y_{\dot{p}} & mx_g - Y_{\dot{r}} \\ -mz_g - K_{\dot{v}} & I_x - K_{\dot{p}} & 0 \\ mx_g - N_{\dot{v}} & 0 & I_z - N_{\dot{r}} \end{bmatrix} \tag{6.178}$$

$$\boldsymbol{C}_{\text{RB}}(\boldsymbol{v}) = \begin{bmatrix} 0 & 0 & mu \\ 0 & 0 & 0 \\ -mu & 0 & 0 \end{bmatrix} \tag{6.179}$$

$$\boldsymbol{G} = \begin{bmatrix} 0 & 0 & 0 \\ 0 & \text{GM}_{\text{T}} W & 0 \\ 0 & 0 & 0 \end{bmatrix}. \tag{6.180}$$

The hydrodynamic forces in $\boldsymbol{\tau}_{\text{hyd}}$ include both damping and hydrodynamic Coriolis and centripetal terms

$$\begin{aligned} Y = {} & Y_{|u|v}|u|v + Y_{ur}ur + Y_{v|v|}|v|v + Y_{v|r|}v|r| + Y_{|v|r}|v|r \\ & + Y_{\phi|uv|}\phi|uv| + Y_{\phi|ur|}\phi|ur| + Y_{\phi uu}\phi u^2 + Y_{\text{ext}} \end{aligned} \tag{6.181}$$

$$\begin{aligned} K = {} & K_{|u|v}|u|v + K_{ur}ur + K_{v|v|}|v|v + K_{v|r|}v|r| + K_{|v|r}|v|r \\ & + K_{\phi|uv|}\phi|uv| + K_{\phi|ur|}\phi|ur| + K_{\phi uu}\phi u^2 + K_{|u|p}|u|p \\ & + K_{p|p|}p|p| + K_{\text{p}}p + K_{\phi\phi\phi}\phi^3 + K_{\text{ext}} \end{aligned} \tag{6.182}$$

$$\begin{aligned} N = {} & N_{|u|v}|u|v + N_{|u|r}|u|r + N_{r|r|}|r|r + N_{v|r|}v|r| + N_{|v|r}|v|r \\ & + N_{\phi|uv|}\phi|uv| + N_{\phi|ur|}\phi|ur| + N_{\text{p}}p + N_{|p|p}|p|p + N_{|u|p}|u|p \\ & + N_{\phi u|u|}\phi|u|u| + N_{\text{ext}} \end{aligned} \tag{6.183}$$

where the forces and moments associated with the roll motion are assumed to involve the square terms of the surge speed u^2 and $|u|u$. The terms Y_{ext}, K_{ext} and N_{ext} consist of possible contributions from external disturbances while control inputs such as rudders, propellers and bow thrusters are included in $\boldsymbol{\tau}$.

MATLAB

A nonlinear naval ship model is implemented in the MSS toolbox as

```
[xdot,U] = navalvessel(x, tau)
```

where

```
x = [u v p r phi psi]'                   % state vector
tau = [X_ext Y_ext K_ext N_ext]'   % external forces and moments
```

6.7 Low-Speed Maneuvering Models for Dynamic Positioning (3 DOFs)

Models for dynamic positioning (DP) are derived under the assumption of low speed. The DP models are valid for stationkeeping and low-speed maneuvering up to approximately 2 m s^{-1}, as indicated by the speed regions shown in Figure 6.7. This section presents a nonlinear DP model based on current coefficients and linear exponential damping that can be used for accurate simulation and prediction. In addition to this, a linearized model intended for controller observer design is derived.

Consider the nonlinear maneuvering model (6.93) in *surge, sway* and *yaw*

$$\dot{\boldsymbol{\eta}} = \boldsymbol{R}(\psi)\boldsymbol{\nu} \tag{6.184}$$

$$\boldsymbol{M}\dot{\boldsymbol{\nu}} + \boldsymbol{C}_{\mathrm{RB}}(\boldsymbol{\nu})\boldsymbol{\nu} + \boldsymbol{N}(\boldsymbol{\nu}_{\mathrm{r}})\boldsymbol{\nu}_{\mathrm{r}} = \boldsymbol{\tau} + \boldsymbol{\tau}_{\mathrm{wind}} + \boldsymbol{\tau}_{\mathrm{wave}} \tag{6.185}$$

where $\boldsymbol{\eta} = [x^n,\ y^n,\ \psi]^{\top}$ and $\boldsymbol{\nu} = [u,\ v,\ r]^{\top}$. This implies that the dynamics associated with the motion in heave, roll and pitch are neglected, that is $w = p = q = 0$. The nonlinear terms are due to to Coriolis, centripetal and damping forces

$$\boldsymbol{N}(\boldsymbol{\nu}_{\mathrm{r}})\boldsymbol{\nu}_{\mathrm{r}} := \boldsymbol{C}_{\mathrm{A}}(\boldsymbol{\nu}_{\mathrm{r}})\boldsymbol{\nu}_{\mathrm{r}} + \boldsymbol{D}(\boldsymbol{\nu}_{\mathrm{r}})\boldsymbol{\nu}_{\mathrm{r}}. \tag{6.186}$$

6.7.1 *Current Coefficients*

For low-speed applications, ocean currents and damping can be modeled by three *current coefficients* C_X, C_Y and C_N. These can be experimentally obtained using scale models in wind tunnels. The resulting forces are measured on the model, which is restrained from moving ($U = 0$). The current coefficients can also be related to the surge resistance, cross-flow drag and the Munk moment used in maneuvering theory. For a ship moving at forward speed $U > 0$, quadratic damping will be embedded in the current coefficients if relative speed is used.

In many textbooks and papers, for instance Blendermann (1994), wind and current coefficients are defined in $\{b\}$ relative to the bow using a *counter clockwise rotation* γ_{c} (see Figure 6.10). The current forces on a marine craft at rest ($U = 0$) can be expressed in terms of the area-based current coefficients $C_X(\gamma_{\mathrm{c}})$, $C_Y(\gamma_{\mathrm{c}})$ and $C_N(\gamma_{\mathrm{c}})$ as

$$X_{\mathrm{current}} = \frac{1}{2}\rho A_{\mathrm{Fc}} C_X(\gamma_{\mathrm{c}}) V_{\mathrm{c}}^2 \tag{6.187}$$

$$Y_{\mathrm{current}} = \frac{1}{2}\rho A_{\mathrm{Lc}} C_Y(\gamma_{\mathrm{c}}) V_{\mathrm{c}}^2 \tag{6.188}$$

$$N_{\mathrm{current}} = \frac{1}{2}\rho A_{\mathrm{Lc}} L_{\mathrm{oa}} C_N(\gamma_{\mathrm{c}}) V_{\mathrm{c}}^2 \tag{6.189}$$

where V_{c} is the speed of the ocean current. The frontal and lateral projected currents areas are denoted A_{Fc} and A_{Lc}, respectively, while L_{oa} is the length overall and ρ is the density of water. Typical experimental current coefficients are shown in Figure 6.11.

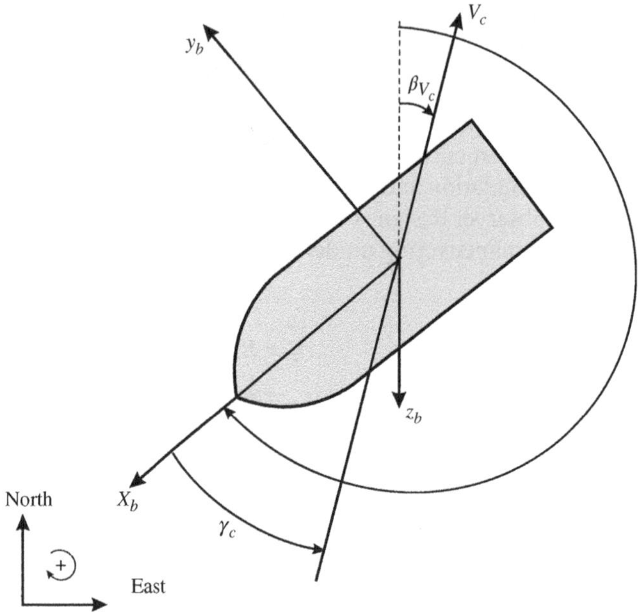

Figure 6.10 Current speed V_c, current direction β_{V_c} and current angle of attack γ_c relative to the bow.

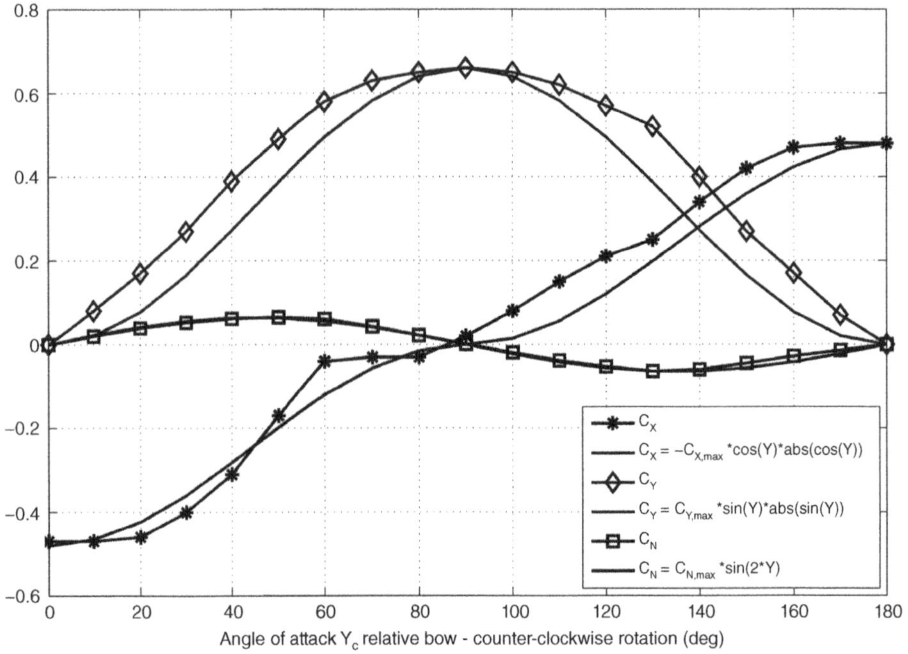

Figure 6.11 Experimental current coefficients C_X, C_Y and C_N for a tanker. Notice that γ_c is a counter clockwise rotation and the angle of attack $\gamma_c = 0°$ for a current in the bow.

Equations (6.187)–(6.189) do not compensate for the ship motions since they only depend on the current speed V_c. For a ship moving at relative forward speed, $U_r > 0$, current forces and quadratic damping in surge and sway should be included as

$$X_{current} = \frac{1}{2}\rho A_{Fc} C_X(\gamma_{rc}) V_{rc}^2 \tag{6.190}$$

$$Y_{current} = \frac{1}{2}\rho A_{Lc} C_Y(\gamma_{rc}) V_{rc}^2 \tag{6.191}$$

$$N_{current} = \frac{1}{2}\rho A_{Lc} L_{oa} C_N(\gamma_{rc}) V_{rc}^2. \tag{6.192}$$

These expressions are functions of the relative speed V_{rc} and angle of attack γ_{rc} given by the equations

$$V_{rc} = \sqrt{u_{rc}^2 + v_{rc}^2} = \sqrt{(u - u_c)^2 + (v - v_c)^2} \tag{6.193}$$

$$\gamma_{rc} = -\text{atan2}(v_{rc}, u_{rc}) \tag{6.194}$$

where

$$u_c = V_c \cos(\beta_{V_c} - \psi) \tag{6.195}$$

$$v_c = V_c \sin(\beta_{V_c} - \psi) \tag{6.196}$$

are the current velocities (see Section 10.3).

Ocean Current Angle of Attack

From Figure 6.10, it is seen that the angles associated with an ocean current in the horizontal plane for a marine craft at rest satisfy

$$\gamma_c = \psi - \beta_{V_c} - \pi \tag{6.197}$$

where β_{V_c} is the direction of the ocean current and γ_c is specified relative to the bow. Hence, the velocity components (6.195) and (6.196) can be written

$$u_c = -V_c \cos(\gamma_c) \tag{6.198}$$

$$v_c = V_c \sin(\gamma_c). \tag{6.199}$$

The magnitude of the ocean current is

$$V_c = \sqrt{u_c^2 + v_c^2}. \tag{6.200}$$

Note that for zero speed the expressions (6.193) and (6.194) become

$$V_{rc} = \sqrt{(u - u_c)^2 + (v - v_c)^2} \overset{u=v=0}{=} V_c \tag{6.201}$$

$$\tan(\gamma_{rc}) = -\frac{v - v_c}{u - v_c} \overset{u=v=0}{=} -\frac{v_c}{u_c} = \tan(\gamma_c). \tag{6.202}$$

This means that the angles γ_{rc} and γ_c as well as the speeds V_{rc} and V_c in general are different for $U > 0$. Consequently, the geometrical relationship (6.197) shown in Figure 6.10 only holds for $U = 0$.

Relationship Between Current Coefficients and Quadratic Drag

The current coefficients can be related to the surge resistance (6.80), and cross-flow drag (6.90) and (6.91) coefficients by assuming low speed such that $u \approx 0$ and $v \approx 0$. This is a good assumption for DP vessels. From (6.198) and (6.199) it follows that the quadratic terms satisfy

$$\begin{aligned} u_r|u_r| &\approx -u_c|-u_c| \\ &= V_c^2 \cos(\gamma_c)|\cos(\gamma_c)| \end{aligned} \tag{6.203}$$

$$\begin{aligned} v_r|v_r| &\approx -v_c|-v_c| \\ &= -V_c^2 \sin(\gamma_c)|\sin(\gamma_c)| \end{aligned} \tag{6.204}$$

$$\begin{aligned} u_r v_r &\approx u_c v_c \\ &= -\frac{1}{2} V_c^2 \sin(2\gamma_c). \end{aligned} \tag{6.205}$$

The next step is to neglect terms in r (no rotations during stationkeeping) in (6.90) and (6.91) and require that C_X, C_Y and C_N in (6.187)–(6.189) satisfy

$$X_{\text{current}} = \frac{1}{2}\rho A_{\text{Fc}} C_X(\gamma_c) V_c^2 := X_{|u|u}|u_r|u_r \tag{6.206}$$

$$Y_{\text{current}} = \frac{1}{2}\rho A_{\text{Lc}} C_Y(\gamma_c) V_c^2 := Y_{|v|v}|v_r|v_r \tag{6.207}$$

$$N_{\text{current}} = \frac{1}{2}\rho A_{\text{Lc}} L_{\text{oa}} C_N(\gamma_c) V_c^2 := N_{|v|v}|v_r|v_r - \underbrace{(X_{\dot{u}} - Y_{\dot{v}})u_r v_r}_{\text{Munk moment}} \tag{6.208}$$

for $u = v = r = 0$. Note that the Munk moment $(Y_{\dot{v}} - X_{\dot{u}})u_r v_r$ in the yaw equation is included in the expression for N_{current} (see Section 6.5.5). The other terms are recognized as diagonal quadratic damping terms in $\boldsymbol{D}(\boldsymbol{\nu}_r)$.

This gives the following analytical expressions for the area-based current coefficients

$$C_X(\gamma_c) = -2\left(\frac{-X_{|u|u}}{\rho A_{Fc}}\right)\cos(\gamma_c)|\cos(\gamma_c)| \tag{6.209}$$

$$C_Y(\gamma_c) = 2\left(\frac{-Y_{|v|v|}}{\rho A_{Lc}}\right)\sin(\gamma_c)|\sin(\gamma_c)| \tag{6.210}$$

$$C_N(\gamma_c) = \frac{2}{\rho A_{Lc} L_{oa}}(-N_{|v|v}\sin(\gamma_c)|\sin(\gamma_c)| + \frac{1}{2}\underbrace{(X_{\dot{u}} - Y_{\dot{v}})}_{A_{22}-A_{11}}\sin(2\gamma_c)). \tag{6.211}$$

These results are similar to Faltinsen (1990, pp. 187–188). The trigonometric functions in (6.209)–(6.211) will be quite close to the shape of the experimental current coefficients shown in Figure 6.11. For tankers, the current coefficients can be computed using the formulae of Leite *et al.* (1998) whereas the ITTC and cross-flow drag principles are commonly used for other hull forms.

6.7.2 Nonlinear DP Model Based on Current Coefficients

The nonlinear DP model based on current coefficients takes the following form

$$\dot{\boldsymbol{\eta}} = \boldsymbol{R}(\psi)\boldsymbol{\nu} \tag{6.212}$$

$$\boldsymbol{M}\dot{\boldsymbol{\nu}} + \boldsymbol{C}_{RB}(\boldsymbol{\nu})\boldsymbol{\nu} + \boldsymbol{D}e^{-\alpha V_{rc}}\,\boldsymbol{\nu}_r + \boldsymbol{d}(V_{rc}, \gamma_{rc}) = \boldsymbol{\tau} + \boldsymbol{\tau}_{wind} + \boldsymbol{\tau}_{wave} \tag{6.213}$$

where

$$\boldsymbol{d}(V_{rc}, \gamma_{rc}) = \begin{bmatrix} -\frac{1}{2}\rho A_{Fc} C_X(\gamma_{rc}) V_{rc}^2 \\ -\frac{1}{2}\rho A_{Lc} C_Y(\gamma_{rc}) V_{rc}^2 \\ -\frac{1}{2}\rho A_{Lc} L_{oa} C_N(\gamma_{rc}) V_{rc}^2 - N_{|r|r} r|r| \end{bmatrix} \tag{6.214}$$

and $-N_{|r|r} > 0$ is an optional quadratic damping coefficient used to counteract the destabilizing Munk moment in yaw since the current coefficients do not include nonlinear damping in yaw. The model also includes an optional linear damping matrix

$$\boldsymbol{D} = \begin{bmatrix} -X_u & 0 & 0 \\ 0 & -Y_v & -Y_r \\ 0 & -N_v & -N_r \end{bmatrix} \tag{6.215}$$

to ensure exponential convergence at low relative speed V_r. This is done by tuning $\alpha > 0$. At higher speeds $V_{rc} \gg 0$ and the nonlinear term $\boldsymbol{d}(V_{rc}, \gamma_{rc})$ dominates over the linear term, which vanishes at higher speeds.

It is also possible to eliminate $\boldsymbol{\nu}$ in (6.213) by using Property 10.1 in Section 10.3. The key assumption is that $\boldsymbol{C}_{\mathrm{RB}}(\boldsymbol{\nu})$ must be parametrized according to (3.63). Hence, it follows that

$$\boldsymbol{M}\dot{\boldsymbol{\nu}}_{\mathrm{r}} + \boldsymbol{C}_{\mathrm{RB}}(\boldsymbol{\nu}_{\mathrm{r}})\boldsymbol{\nu}_{\mathrm{r}} + \boldsymbol{D}e^{-\alpha V_{\mathrm{rc}}}\,\boldsymbol{\nu}_{\mathrm{r}} + \boldsymbol{d}(V_{\mathrm{rc}}, \gamma_{\mathrm{rc}}) = \boldsymbol{\tau} + \boldsymbol{\tau}_{\mathrm{wind}} + \boldsymbol{\tau}_{\mathrm{wave}} \tag{6.216}$$

where $\boldsymbol{\nu}_{\mathrm{r}}$ is the state vector.

6.7.3 Linear Time-varying DP Model

As shown in Section 6.4, linear damping is a good assumption for low-speed applications. Similarly, the quadratic velocity terms given by $\boldsymbol{C}(\boldsymbol{\nu}_{\mathrm{r}})\boldsymbol{\nu}_{\mathrm{r}}$ and $\boldsymbol{d}(V_{\mathrm{rc}}, \gamma_{\mathrm{rc}})$ can be neglected when designing DP control systems if the ocean currents (drift) are properly compensated for by using integral action. One way to do this is to treat the ocean currents as a slowly-varying bias vector $\boldsymbol{b}$ expressed in $\{n\}$. Hence, the relative velocity vector $\boldsymbol{\nu}_{\mathrm{r}}$ is superfluous.

The generalized position $\boldsymbol{\eta} = [x^n, y^n, \psi]^{\top}$ is usually measured using GNSS and a gyrocompass. Since the heading angle $\psi = \psi(t)$ is accurately measured, the kinematic nonlinearity due to the rotation matrix can be removed by assuming that

$$\boldsymbol{R}_{z,\psi}(\psi(t)) := \boldsymbol{R}(t) \tag{6.217}$$

is known for all $t \geq 0$. Consequently, the resulting DP model will be a linear time-varying (LTV) model

$$\dot{\boldsymbol{\eta}} = \boldsymbol{R}(t)\boldsymbol{\nu} \tag{6.218}$$

$$\boldsymbol{M}\dot{\boldsymbol{\nu}} + \boldsymbol{D}\boldsymbol{\nu} = \boldsymbol{R}^{\top}(t)\boldsymbol{b} + \boldsymbol{\tau} + \boldsymbol{\tau}_{\mathrm{wind}} + \boldsymbol{\tau}_{\mathrm{wave}} \tag{6.219}$$

$$\dot{\boldsymbol{b}} = \boldsymbol{0}. \tag{6.220}$$

Note that the currents are assumed constant in $\{n\}$ and therefore transformed to $\{b\}$ by $\boldsymbol{R}^{\top}(t)\boldsymbol{b}$. The generalized control force vector $\boldsymbol{\tau} \in \mathbb{R}^3$ is assumed to be linear in the control inputs $\boldsymbol{u} \in \mathbb{R}^r$ such that

$$\boldsymbol{\tau} = \boldsymbol{B}\boldsymbol{u} \tag{6.221}$$

where $\boldsymbol{B}$ is the input matrix. The resulting state-space model is a nine-state LTV system

$$\dot{\boldsymbol{x}} = \underbrace{\begin{bmatrix} \boldsymbol{0}_{3\times 3} & \boldsymbol{R}(t) & \boldsymbol{0}_{3\times 3} \\ \boldsymbol{0}_{3\times 3} & -\boldsymbol{M}^{-1}\boldsymbol{D} & \boldsymbol{M}^{-1}\boldsymbol{R}^{\top}(t) \\ \boldsymbol{0}_{3\times 3} & \boldsymbol{0}_{3\times 3} & \boldsymbol{0}_{3\times 3} \end{bmatrix}}_{\boldsymbol{A}(t)} \boldsymbol{x} + \underbrace{\begin{bmatrix} \boldsymbol{0}_{3\times r} \\ \boldsymbol{M}^{-1}\boldsymbol{B} \\ \boldsymbol{0}_{3\times r} \end{bmatrix}}_{\boldsymbol{B}} \boldsymbol{u} + \underbrace{\begin{bmatrix} \boldsymbol{0}_{3\times 3} \\ \boldsymbol{M}^{-1} \\ \boldsymbol{0}_{3\times 3} \end{bmatrix}}_{\boldsymbol{E}} \boldsymbol{w} \tag{6.222}$$

$$\boldsymbol{y} = \underbrace{[\boldsymbol{I}_3 \;\; \boldsymbol{0}_{3\times 3} \;\; \boldsymbol{0}_{3\times 3}]}_{\boldsymbol{C}} \boldsymbol{x} + \boldsymbol{\varepsilon} \tag{6.223}$$

where $\boldsymbol{x} = [\boldsymbol{\eta}^{\top}, \ \boldsymbol{\nu}^{\top}, \ \boldsymbol{b}^{\top}]^{\top}$ is the state vector, $\boldsymbol{y} = \boldsymbol{\eta} + \boldsymbol{\varepsilon}$ is the measurement vector and

$$\boldsymbol{w} = \boldsymbol{\tau}_{\text{wind}} + \boldsymbol{\tau}_{\text{wave}}. \tag{6.224}$$

This model is intended for controller–observer design. Since the model is linear, it is possible to use an LTV Kalman filter design to estimate the state vector. Similarly, a linear-quadratic optimal controller can be designed to compute the control inputs.

MATLAB

A linear model of a supply vessel is included in the MSS toolbox as

```
[xdot,U] = supply(x, tau)       % Supply vessel L = 76.2 m
```

where `U` is the speed and

```
u = [x, y, psi, u, v, r]'       % state vector
tau = [tau_1, tau_2, tau_6]     % control inputs
```

7

Autopilot Models for Course and Heading Control

This chapter presents mathematical models of marine craft for *course* and *heading* control. The presented models are state-of-the art models used to design autopilot systems for surface craft, USVs, AUVs and ROVs. In *Encyclopedia Britannica*, an autopilot is defined as follows:

> *Automatic pilot*, "also called autopilot, or autohelmsman, device for controlling an aircraft or other vehicle without constant human intervention".

The earliest automatic pilots could do no more than maintain an aircraft in straight and level flight by controlling *roll, pitch* and *yaw* movements; and they are still used most often to relieve the pilot during routine cruising. Modern automatic pilots can, however, execute complex maneuvers such as 3D attitude and path-following control and they are used for aircraft, marine craft, automotive systems, etc.

It is important to stress the concepts for course and heading control since there are many conceptual misunderstandings regarding the course and heading of a marine craft. The course angle χ of a marine craft is the cardinal direction in which the craft is moving. Hence, the course angle is to be distinguished from the heading angle ψ, which is the compass direction in which the craft's bow or nose is pointed. The difference between the course and heading angles is the crab angle

$$\beta_c = \sin^{-1}(v/U), \qquad U > 0 \tag{7.1}$$

which satisfies (see Section 2.5.2)

$$\chi = \psi + \beta_c. \tag{7.2}$$

Handbook of Marine Craft Hydrodynamics and Motion Control, Second Edition. Thor I. Fossen.

Note that β_c is not defined for $U = 0$ and thus the course angle is not defined for zero speed but, to the contrary, the yaw angle (compass direction) is well-defined for zero and forward speed. This is important to take into account when designing autopilot systems for marine craft and we will distinguish our design philosophy between autopilots for stationkeeping and "flying vehicles".

7.1 Autopilot Models for Course Control

Surface craft are usually equipped with a global navigation satellite system (GNSS) receiver, which measures:

COG – course over ground (χ)
SOG – speed over ground (U).

Underwater vehicles, however, use hydroacoustic reference systems to determine their position, velocity and course since GNSS signals cannot be received under water.

Path-following controllers are usually implemented as two control loops using an inner-loop course autopilot, as shown in Figure 7.1. The path that a marine craft follows over the ground is called COG or *track*. Furthermore, the intended track is referred to as a *route*. For ships and aircraft, routes are typically straight-line segments between waypoints.

7.1.1 State-space Model for Course Control

If the course angle χ is available as a direct measurement it is straightforward to design a PID controller for course control. Alternatively, the course angle can be computed from the north-east positions according to

$$\chi[k] \approx \text{atan2}\left(y^n[k] - y^n[k-1],\ x^n[k] - x^n[k-1]\right). \tag{7.3}$$

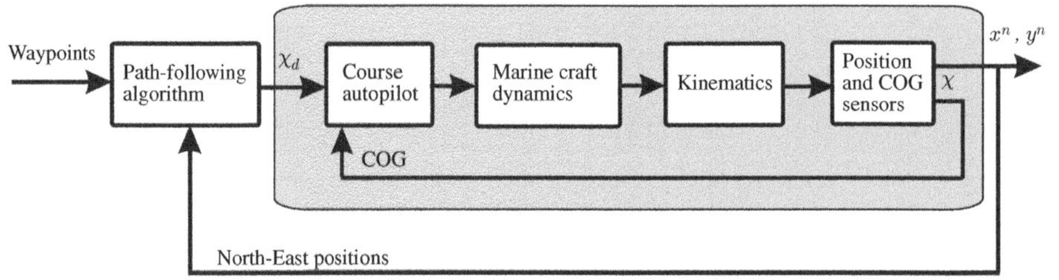

Figure 7.1 Inner-loop course angle autopilot and outer-loop path-following control system.

The PID controller gains are usually tuned using the maneuvering models presented in Section 6.5 to verify the time responses. Let the generalized position and velocity vectors of a surface craft be denoted by $\boldsymbol{\eta} = [x^n, y^n, \psi]^\top$ and $\boldsymbol{\nu} = [u, v, r]^\top$, respectively. From (6.102) we have the equations of relative motion

$$\boldsymbol{M}\dot{\boldsymbol{\nu}}_r + \boldsymbol{C}(\boldsymbol{\nu}_r)\boldsymbol{\nu}_r + \boldsymbol{D}\boldsymbol{\nu}_r + \boldsymbol{D}_n(\boldsymbol{\nu}_r)\boldsymbol{\nu}_r = \boldsymbol{\tau} + \boldsymbol{\tau}_{\text{wind}} + \boldsymbol{\tau}_{\text{wave}}. \tag{7.4}$$

Assume that north-east positions (x^n, y^n), SOG and COG are measured (see Section 13.1), giving a total of four measurements y_i $(i = 1, 2, 3, 4)$. The state-space model corresponding to these measurements is

$$\dot{\boldsymbol{\eta}} = \boldsymbol{R}(\psi)\boldsymbol{\nu} \tag{7.5}$$

$$\dot{\boldsymbol{\nu}} = \begin{bmatrix} rv_c \\ -ru_c \\ 0 \end{bmatrix} + \boldsymbol{M}^{-1}(\boldsymbol{\tau} + \boldsymbol{\tau}_{\text{wind}} + \boldsymbol{\tau}_{\text{wave}} - \boldsymbol{C}(\boldsymbol{\nu}_r)\boldsymbol{\nu}_r - \boldsymbol{D}\boldsymbol{\nu}_r - \boldsymbol{D}_n(\boldsymbol{\nu}_r)\boldsymbol{\nu}_r) \tag{7.6}$$

$$\begin{aligned} y_1 &= x^n + \varepsilon_1 && \text{(north position)} \\ y_2 &= y^n + \varepsilon_2 && \text{(east position)} \\ y_3 &= \sqrt{u^2 + v^2} + \varepsilon_3 && \text{(SOG – speed over ground)} \\ y_4 &= \psi + \sin^{-1}\left(\frac{v}{U}\right) + \varepsilon_4 && \text{(COG – course over ground)} \end{aligned} \tag{7.7}$$

where $\boldsymbol{\nu}_r = \boldsymbol{\nu} - [u_c, v_c, 0]^\top$ and ε_i $(i = 1, 2, 3, 4)$ is Gaussian white noise measurement noise.

The control objective of the course autopilot system in Figure 7.1 is to ensures that the inner loop satisfies

$$\frac{\chi}{\chi_d} \approx 1 \tag{7.8}$$

such that an outer loop path-following algorithm can be used to generate the reference signal χ_d as described in Section 12.4. The control loops can be designed to meet the bandwidth limitations of the two control loops using successive loop closure techniques (see Section 15.2) or nonlinear methods.

7.1.2 Course Angle Transfer Function

For a marine craft moving on a straight line at forward speed U the crab angle $\beta_c = 0$. During turning, the crab angle will be non-zero and act as a disturbance when controlling the course angle. A linear design model for PID control can be found by differentiating the course angle (7.2) with respect to time and approximate the yaw dynamics by a mass–damper system. In other words

$$\dot{\chi} = r + \dot{\beta}_c \tag{7.9}$$

$$(I_z - N_{\dot{r}})\dot{r} - N_r r = \tau_N \tag{7.10}$$

where τ_N is the yaw moment control input and β_c is treated as a disturbance, which need to be canceled by integral action. Consequently, the yaw rate transfer function becomes

$$\frac{r}{\tau_N}(s) = \frac{K}{Ts + 1} \tag{7.11}$$

where $K = 1/(-N_r)$ and $T = (I_z - N_{\dot{r}})/(-N_r)$. Hence,

$$\chi(s) = \frac{K}{s(Ts + 1)}\tau_N(s) + \beta_c(s). \tag{7.12}$$

7.2 Autopilot Models for Heading Control

Ships are usually equipped with a gyrocompass, which is a non-magnetic compass based on a fast-spinning disc. A north-seeking gyro gives a highly accurate measurement of the yaw angle and this is the preferred sensor from the safety point of view (see Section 13.1). Magnetic compasses are not used on-board commercial ships as navigational devices since they are very sensitive to magnetic disturbances. An alternative measurement could be to use two GNSS antennas on the same receiver with a known offset vector to compute the heading angle. This solution is, however, sensitive to ionospheric disturbances, multipath, loss of signals, the number of available satellites, etc.

If the heading angle ψ is available as a direct measurement, it is straightforward to design a PID controller for heading control. The PID controller gains can be computed as a function of the model parameters if the yaw dynamics is known. Models for this are presented in Sections 7.2.1–7.2.3 and heading autopilot design is discussed in Section 15.3.4. Figure 7.2 shows a conventional heading autopilot system.

Model-based heading autopilots for marine craft are usually based on the model of Nomoto *et al.* (1957) for which the sway force τ_Y and yaw moment τ_N are generated by a single rudder with deflection δ (see Section 9.5.1)

$$\tau_Y = -Y_\delta \delta \tag{7.13}$$

$$\tau_N = -N_\delta \delta. \tag{7.14}$$

It is straightforward to modify τ_Y and τ_N to include other control inputs as shown in Chapter 9.

7.2.1 Second-order Nomoto Model

A linear autopilot model for heading control can be derived from the linear maneuvering model (see Section 6.5.5)

$$\boldsymbol{M}\dot{\boldsymbol{\nu}}_r + \boldsymbol{N}\boldsymbol{\nu}_r = \boldsymbol{b}\,\delta \tag{7.15}$$

where $\boldsymbol{v}_\mathrm{r} = [v_r,\ r]^\top$ and

$$\boldsymbol{M} = \begin{bmatrix} m - Y_{\dot{v}} & mx_g - Y_{\dot{r}} \\ mx_g - N_{\dot{v}} & I_z - N_{\dot{r}} \end{bmatrix} \tag{7.16}$$

$$\boldsymbol{N} = \begin{bmatrix} -Y_v & (m - X_{\dot{u}})U - Y_r \\ (X_{\dot{u}} - Y_{\dot{v}})U - N_v & (mx_g - Y_{\dot{r}})U - N_r \end{bmatrix} \tag{7.17}$$

$$\boldsymbol{b} = \begin{bmatrix} -Y_\delta \\ -N_\delta \end{bmatrix}. \tag{7.18}$$

Choosing the yaw rate r as output

$$r = \boldsymbol{c}^\top \boldsymbol{v}_\mathrm{r}, \qquad \boldsymbol{c}^\top = [0,\ 1] \tag{7.19}$$

and application of the *Laplace transformation* yields

$$\frac{r}{\delta}(s) = \frac{K(T_3 s + 1)}{(T_1 s + 1)(T_2 s + 1)}. \tag{7.20}$$

A similar expression is obtained for the sway motion

$$\frac{v_r}{\delta}(s) = \frac{K_v(T_v s + 1)}{(T_1 s + 1)(T_2 s + 1)} \tag{7.21}$$

where K_v and T_v differ from K and T_3 in the yaw equation. Equation (7.20) is known as *Nomoto's second-order model* (Nomoto *et al.* 1957). The time-domain representation of (7.20) is

$$T_1 T_2 \psi^{(3)} + (T_1 + T_2)\ddot{\psi} + \dot{\psi} = K(T_3\dot{\delta} + \delta). \tag{7.22}$$

This model can be normalized using speed U and length L (see Appendix D)

$$\left(\frac{L}{U}\right)^2 T_1'\, T_2'\, \psi^{(3)} + \left(\frac{L}{U}\right)(T_1' + T_2')\, \ddot{\psi} + \dot{\psi} = K'\, T_3'\, \dot{\delta} + \left(\frac{U}{L}\right) K'\, \delta. \tag{7.23}$$

The normalized model can be used as basis for gain-scheduling control by choosing U as scheduling variable. Consequently, the PID controller gains will be functions of the model parameters and the direct measurement U. Parameter adaption based on classical methods will in general be slower than gain scheduling.

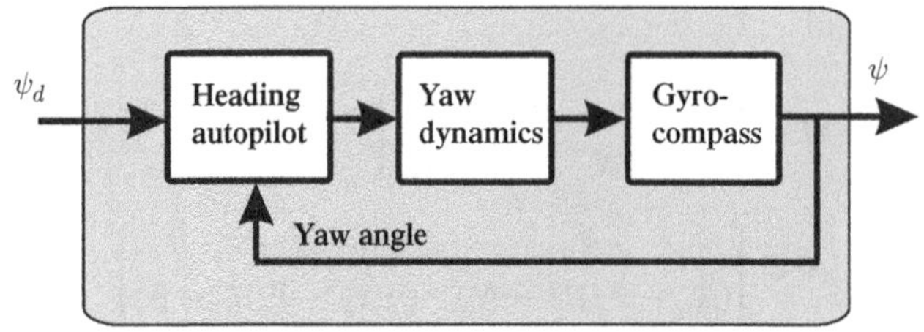

Figure 7.2 Heading autopilot system where the yaw angle is measured by using a gyrocompass.

7.2.2 First-order Nomoto Model

The *first-order Nomoto model* is obtained by defining the *equivalent time constant*

$$T := T_1 + T_2 - T_3. \tag{7.24}$$

Consequently, (7.20) can be approximated as

$$\frac{r}{\delta}(s) = \frac{K}{Ts+1}. \tag{7.25}$$

Using, $\dot{\psi} = r$ yields

$$\frac{\psi}{\delta}(s) = \frac{K}{s(Ts+1)} \tag{7.26}$$

which is the transfer function that is used in most commercial autopilot systems. The time-domain representation is

$$T\ddot{\psi} + \dot{\psi} = K\delta. \tag{7.27}$$

The first-order Nomoto model can be normalized using speed U and length L according to (see Appendix D)

$$\left(\frac{L}{U}\right) T' \ddot{\psi} + \dot{\psi} = \left(\frac{U}{L}\right) K' \delta. \tag{7.28}$$

The normalized model is used to design a gain-scheduled autopilot where U is treated as a time-varying measurement.

The accuracy of the first-order Nomoto model when compared to the second-order model is illustrated in Example 7.1 where a course stable cargo ship and a course unstable oil tanker are studied (see Section 15.1.1).

Table 7.1 Parameters for a cargo ship and a fully loaded oil tanker.

	L (m)	u_0 (m s)$^{-1}$	∇ (dwt)	K (s^{-1})	T_1 (s)	T_2 (s)	T_3 (s)
Cargo ship	161	7.7	16622	0.185	118.0	7.8	18.5
Oil tanker	350	8.1	389100	−0.019	−124.1	16.4	46.0

Example 7.1 (Frequency Responses for Nomoto First- and Second-Order Models)
Consider a Mariner class cargo ship (Chislett and Strøm-Tejsen 1965a) and a fully loaded tanker (Dyne and Trägårdh 1975) given by the parameters in Table 7.1.

MATLAB:

The Bode diagram is generated by using the MSS toolbox commands:

```
T1 = 118; T2 = 7.8; T3 = 18.5; K = 0.185;
nomoto(T1, T2, T3, K)

T1 = -124.1; T2 = 16.4; T3 = 46.0; K = -0.019;
nomoto(T1, T2, T3, K);

function nomoto(T1,T2,T3,K);
% NOMOTO(T1,T2,T3,K) generates the Bode plots for
%               K                        K (1+T3s)
% H1(s) = ------------       H2(s) = -------------------
%             s(1+Ts)                  s(1+T1s)(1+T2s)

T = T1 + T2 - T3;
d1 = [T 1 0]; n1 = K;
d2 = [T1*T2 T1+T2 1 0]; n2 = K*[T3 1];
[mag1,phase1,w] = bode(n1,d1);
[mag2,phase2]   = bode(n2,d2,w);

% shift phase with 360 deg for course-unstable ship
if K <  0
        phase1 = phase1-360;
        phase2 = phase2-360;
end

subplot(211),semilogx(w,20*log10(mag1)),grid;
xlabel('Frequency [rad/s]'),title('Gain [dB]');
hold on,semilogx(w,20*log10(mag2),'--'),hold off;

subplot(212),semilogx(w,phase1),grid;
xlabel('Frequency [rad/s]'),title('Phase [deg]');
hold on,semilogx(w,phase2,'--'),hold off;
```

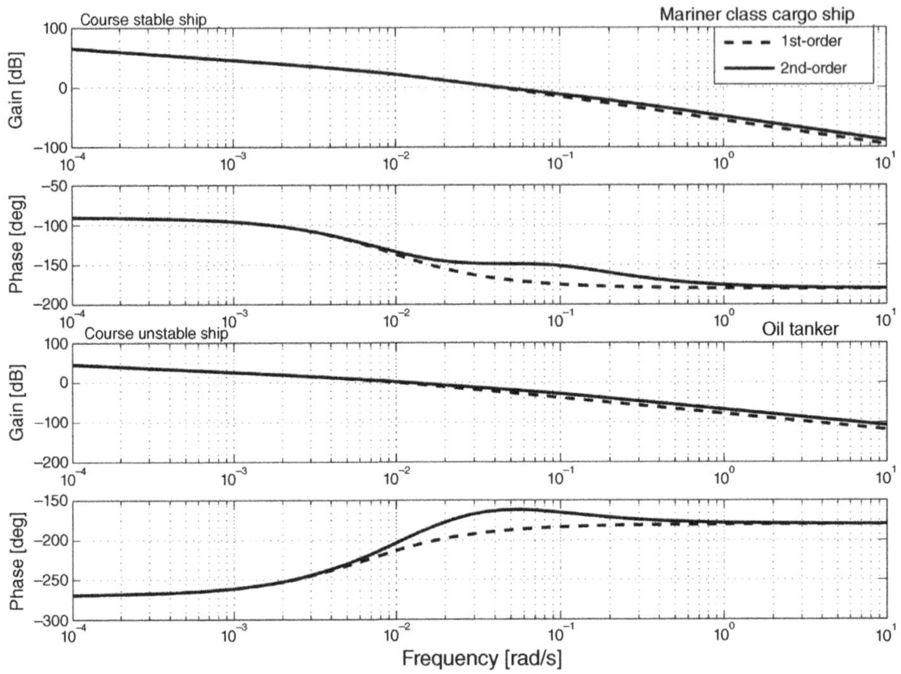

Figure 7.3 First-order and second-order Nomoto transfer functions for a course-stable Mariner class cargo ship and a course-unstable oil tanker.

It is seen from Figure 7.3 that the first-order approximation is quite accurate up to 0.1 rad s^{-1} for the cargo ship and the tanker. A small deviation in the phase around 0.5 rad s^{-1} is observed. This is due to the cancellation of the sway dynamics.

7.2.3 Nonlinear Extensions of Nomoto's Model

The linear Nomoto model can be extended to include nonlinear effects by adding a *static nonlinearity* to describe the *maneuvering characteristics*.

Nonlinear Extension of Nomoto's First-order Model

In Norrbin (1963) the following first-order model was proposed

$$T\dot{r} + H_{\mathrm{N}}(r) = K\delta \tag{7.29}$$

$$H_{\mathrm{N}}(r) = n_3 r^3 + n_2 r^2 + n_1 r + n_0 \tag{7.30}$$

where $H_N(r)$ is a nonlinear function. The linear model (7.27) is a special case of the nonlinear model corresponding to $H_N(r) = r$.

MATLAB:

Two nonlinear Nomoto models based on the data set of Van Amerongen (1982) are located under the MSS toolbox catalog `../MSS/VESSELS/`. The models are called using:

```
% Frigate L = 100 m
[psi_dot,r_dot,delta_dot] = frigate(r,U,delta,delta_c,d_r)

% ROV Sefakkel (training ship) L = 45 m
[psi_dot,r_dot,delta_dot] = ROVzefakkel(r,U,delta,delta_c,d_r)
```

Nonlinear Extension of Nomoto's Second-order Model

Bech and Wagner Smith (1969) proposed a second-order model

$$T_1 T_2 \ddot{r} + (T_1 + T_2)\dot{r} + K H_B(r) = K(\delta + T_3 \dot{\delta}) \tag{7.31}$$

$$H_B(r) = b_3 r^3 + b_2 r^2 + b_1 r + b_0 \tag{7.32}$$

where $H_B(r)$ can be found from Bech's reverse spiral maneuver. The linear equivalent (7.22) is obtained for $H_B(r) = r$.

The linear and nonlinear maneuvering characteristics are shown in Figure 15.12 in Section 15.1.2. They are generated by solving for r as a function of δ using the steady-state solutions of (7.29) and (7.31) corresponding to

$$H_N(r) = K\delta \tag{7.33}$$

$$H_B(r) = \delta. \tag{7.34}$$

The nonlinear maneuvering characteristics can also be generated from full-scale maneuvering tests. For stable ships both the *Bech* and *Dieudonne spiral tests* can be applied, while the Bech spiral is the only one avoiding the hysteresis effect for course-unstable ships; see Section 15.1.2 for details.

For a course-unstable ship, $b_1 < 0$, whereas a course-stable ship satisfies $b_1 > 0$. A single-screw propeller or asymmetry in the hull will cause a non-zero value of b_0. Similarly, symmetry in the hull implies that $b_2 = 0$. Since a constant rudder angle is required to compensate for constant steady-state wind and current forces, the bias term

b_0 could conveniently be treated as an additional rudder off set. This in turn implies that a large number of ships can be described by the polynomial

$$H_B(r) = b_3 r^3 + b_1 r. \tag{7.35}$$

The coefficients b_i for $i = 0, \dots, 3$ are related to those in Norrbin's model n_i by

$$n_i = \frac{b_i}{|b_1|} \tag{7.36}$$

resulting in

$$H_N(r) = n_3 r^3 + n_1 r. \tag{7.37}$$

This implies that $n_1 = 1$ for a course-stable ship and $n_1 = -1$ for a course-unstable ship.

7.2.4 Pivot Point

The pivot point is a useful tool in ship handling and the location of the pivot point in a maneuvering situations is of great importance for the captain of the ship. The pivot point is the point around which the ship is yawing when the ship rotates then at this point there is no transverse velocity. The location of pivot point depends on the instantaneous set of forces acting on the ship and consequently its location will change over time. The *pivot point* in yaw is defined as follows.

Definition 7.1 (Pivot Point)
A ship's pivot point x_p is a point on the centerline measured from the CG at which sway and yaw completely cancel each other (Tzeng 1998a)

$$v_{np} = v_{ng} + x_p r \equiv 0 \tag{7.38}$$

where v_{ng} is the sway velocity of the CG with respect to $\{n\}$. The pivot point will scribe the ship's turning circle and all other points appear to be turning about this point.

It is possible to compute the pivot point for a turning ship at time t by measuring the velocity $v_{ng}(t)$ in the CG and the turning rate $r(t)$. From (7.38) it follows that

$$x_p(t) = -\frac{v_{ng}(t)}{r(t)}, \quad r(t) \neq 0. \tag{7.39}$$

This expression is not defined for a zero yaw rate corresponding to a straight-line motion. This means that the pivot point is located at infinity when moving on a straight line or in a pure sway motion.

It is well known to pilots that the pivot point of a turning ship is located at about 1/5 to 1/4 ship length aft of bow (Tzeng 1998a). The location of the pivot point of a rudder controlled ship is related to the ratio of the *sway-rudder* and *yaw-rudder* gain coefficients. This can be explained by studying the steady-state solution of the linear maneuvering model. Combining (7.20) and (7.21) under the assumption of no currents ($v_c = 0$) gives

$$\frac{v}{r} = \frac{K_v(T_v s + 1)}{K(T_3 s + 1)} \stackrel{s=0}{=} \frac{K_v}{K} \tag{7.40}$$

Consequently, the steady-state location of the pivot point is given by the ratio

$$x_{\mathrm{p,ss}} = -\frac{K_v}{K} \tag{7.41}$$

This expression can also be related to the hydrodynamic derivatives according to

$$x_{\mathrm{p,ss}} = -\frac{N_r Y_\delta - (Y_r - mU)N_\delta}{Y_v N_\delta - N_v Y_\delta}. \tag{7.42}$$

Note that $x_{\mathrm{p,ss}}$ depends on the forward speed U. The non-dimensional form becomes (see Appendix D)

$$\begin{aligned} x'_{\mathrm{p,ss}} &= \frac{x_{\mathrm{p,ss}}}{L_{\mathrm{pp}}} \\ &= -\frac{N'_r Y'_\delta - (Y'_r - m')N'_\delta}{Y'_v N'_\delta - N'_v Y'_\delta}. \end{aligned} \tag{7.43}$$

Example 7.2 (Pivot Point for the Mariner Class Vessel)
Consider the Mariner Class vessel (Chislett and Strøm-Tejsen 1965b) where the non-dimensional linear maneuvering coefficients for $U = 7.175$ m s^{-1} (15 knots) *are given as*

$$\begin{aligned} Y'_v &= -1160 \times 10^{-5} & N'_v &= -264 \times 10^{-5} \\ Y'_r - m' &= -499 \times 10^{-5} & N'_r &= -166 \times 10^{-5} \\ Y'_\delta &= 278 \times 10^{-5} & N'_\delta &= -139 \times 10^{-5}. \end{aligned}$$

The Mariner Class vessel is programmed in the MSS toolbox file `mariner.m`*. The non-dimensional pivot point is computed from (7.43). This gives*

$$x'_{\mathrm{p,ss}} = 0.4923 \quad \Rightarrow \quad x_{\mathrm{p,ss}} = 0.4923\ L_{\mathrm{pp}} \tag{7.44}$$

where $L_{\mathrm{pp}} = 160.93$ m *is the length between the perpendiculars AP and FP. The overall length is* $L_{oa} = 171.8$ m. *The pivot point* $x_{\mathrm{p,\,ss}}$ *is located ahead of the CG. Since the CG is located at* $x_g = -0.023\ L_{\mathrm{pp}}$ *and the bow is approximately* $0.03\ L_{\mathrm{pp}}$ *fore of the FP, the pivot point is* $0.06\ L_{\mathrm{pp}}$ *aft of the bow; see Figure 7.4.*

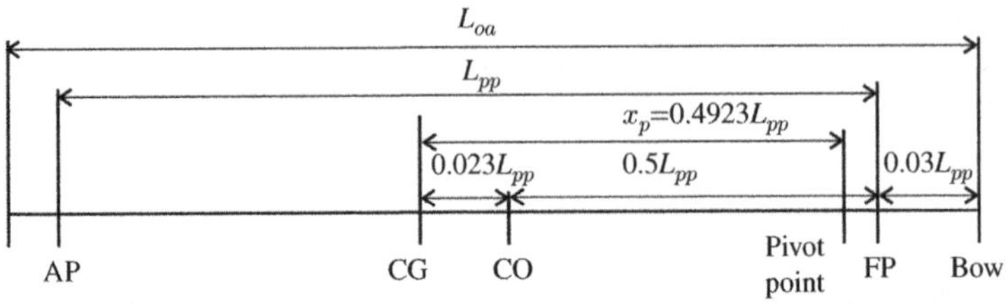

Figure 7.4 Location of the pivot point for the Mariner Class vessel.

8

Models for Underwater Vehicles

This chapter presents models for underwater vehicles. The foundation for the models are the kinematic equations (Chapter 2), rigid-body kinetics (Chapter 3), hydrostatics (Chapter 4) and maneuvering theory (Chapter 6). Results from these chapters are combined to obtain forward speed models, horizontal-plane models for steering, vertical-plane models for diving and 6-DOF coupled models for advanced maneuvers.

8.1 6-DOF Models for AUVs and ROVs

Ship models are usually reduced-order models for control of the horizontal-plane motions (*surge, sway* and *yaw*) in combination with *roll* if roll damping is an issue. In this section we will discuss 6-DOF models, which are useful for underwater vehicles with actuation in all DOFs. Such vehicles can control the position and attitude in 6 DOFs by using thrusters, moving weights, spinning rotors and control surfaces.

A 6-DOF model is usually implemented in a computer to describe all dynamic effects as accurately as possible. This is the high-fidelity *simulation model* shown in Figure 1.4 in Section 1.1. The simulation model should be able to reconstruct the time responses of the physical system. Model-based controllers and state estimators, however, can be designed using reduced-order or simplified models.

When designing feedback control systems it can be advantageous to formulate the equations of motion in both the $\{b\}$ and $\{n\}$ frames in order to exploit physical properties of the model. This section includes nonlinear transformations that can be used to represent the equations of motion in both reference frames.

8.1.1 Equations of Motion Expressed in BODY

Consider the nonlinear equations of motion expressed in the following form

$$\dot{\boldsymbol{\eta}} = \boldsymbol{J}_k(\boldsymbol{\eta})(\boldsymbol{\nu}_r + \boldsymbol{\nu}_c) \tag{8.1}$$

$$\boldsymbol{M}\dot{\boldsymbol{\nu}}_r + \boldsymbol{C}(\boldsymbol{\nu}_r)\boldsymbol{\nu}_r + \boldsymbol{D}(\boldsymbol{\nu}_r)\boldsymbol{\nu}_r + \boldsymbol{g}(\boldsymbol{\eta}) + \boldsymbol{g}_0 = \boldsymbol{\tau} \tag{8.2}$$

Handbook of Marine Craft Hydrodynamics and Motion Control, Second Edition. Thor I. Fossen.

where $\boldsymbol{v}_c = [u_c,\ v_c,\ w_c,\ 0,\ 0,\ 0]^\top$ is the current velocity vector and

$$\boldsymbol{M} = \boldsymbol{M}_{RB} + \boldsymbol{M}_A \tag{8.3}$$

$$\boldsymbol{C}(\boldsymbol{v}_r) = \boldsymbol{C}_{RB}(\boldsymbol{v}_r) + \boldsymbol{C}_A(\boldsymbol{v}_r) \tag{8.4}$$

$$\boldsymbol{D}(\boldsymbol{v}_r) = \boldsymbol{D} + \boldsymbol{D}_n(\boldsymbol{v}_r) \tag{8.5}$$

The expressions for $\boldsymbol{\eta}$ and $\boldsymbol{J}_k(\boldsymbol{\eta})$ depend on the kinematic representation. Two different choices for $\boldsymbol{J}_k(\boldsymbol{\eta})$ will be presented where the subscript $k \in \{\Theta, \mathrm{q}\}$ denotes the Euler angle and unit quaternion representations, respectively. In other words

$$\boldsymbol{J}_\Theta(\boldsymbol{\eta}) = \begin{bmatrix} \boldsymbol{R}(\boldsymbol{\Theta}_{nb}) & \boldsymbol{0}_{3\times3} \\ \boldsymbol{0}_{3\times3} & \boldsymbol{T}(\boldsymbol{\Theta}_{nb}) \end{bmatrix}, \quad \boldsymbol{\eta} = [x^n,\ y^n,\ z^n,\ \phi,\ \theta,\ \psi]^\top \tag{8.6}$$

$$\boldsymbol{J}_q(\boldsymbol{\eta}) = \begin{bmatrix} \boldsymbol{R}(\boldsymbol{q}_b^n) & \boldsymbol{0}_{3\times3} \\ \boldsymbol{0}_{4\times3} & \boldsymbol{T}(\boldsymbol{q}_b^n) \end{bmatrix}, \quad \boldsymbol{\eta} = [x^n,\ y^n,\ z^n,\ \eta,\ \varepsilon_1,\ \varepsilon_2,\ \varepsilon_3]^\top. \tag{8.7}$$

The system inertia matrix $\boldsymbol{M}$ for an underwater vehicle follows the symmetry considerations presented in Section 8.1.4. Recall that for a starboard–port symmetrical underwater vehicle with $y_g = 0$ and $I_{xy} = I_{yz} = 0$,

$$\boldsymbol{M} = \begin{bmatrix} m - X_{\dot{u}} & 0 & -X_{\dot{w}} & 0 & mz_g - X_{\dot{q}} & 0 \\ 0 & m - Y_{\dot{v}} & 0 & -mz_g - Y_{\dot{p}} & 0 & mx_g - Y_{\dot{r}} \\ -X_{\dot{w}} & 0 & m - Z_{\dot{w}} & 0 & -mx_g - Z_{\dot{q}} & 0 \\ 0 & -mz_g - Y_{\dot{p}} & 0 & I_x - K_{\dot{p}} & 0 & -I_{zx} - K_{\dot{r}} \\ mz_g - X_{\dot{q}} & 0 & -mx_g - Z_{\dot{q}} & 0 & I_y - M_{\dot{q}} & 0 \\ 0 & mx_g - Y_{\dot{r}} & 0 & -I_{zx} - K_{\dot{r}} & 0 & I_z - N_{\dot{r}} \end{bmatrix}. \tag{8.8}$$

Consequently, it is straightforward to compute the Coriolis and centripetal matrix $\boldsymbol{C}(\boldsymbol{v}_r)$ using the results in Sections 3.3 and 6.3.3 when the structure of $\boldsymbol{M}$ has been determined. In general, the damping of an underwater vehicle moving in 6 DOFs at high speed will be highly nonlinear and coupled. This could be described mathematically as

$$\boldsymbol{D}_n(\boldsymbol{v}_r)\boldsymbol{v}_r = \begin{bmatrix} |\boldsymbol{v}_r|^\top \boldsymbol{D}_{n1} \boldsymbol{v}_r \\ |\boldsymbol{v}_r|^\top \boldsymbol{D}_{n2} \boldsymbol{v}_r \\ |\boldsymbol{v}_r|^\top \boldsymbol{D}_{n3} \boldsymbol{v}_r \\ |\boldsymbol{v}_r|^\top \boldsymbol{D}_{n4} \boldsymbol{v}_r \\ |\boldsymbol{v}_r|^\top \boldsymbol{D}_{n5} \boldsymbol{v}_r \\ |\boldsymbol{v}_r|^\top \boldsymbol{D}_{n6} \boldsymbol{v}_r \end{bmatrix} \tag{8.9}$$

where $|\boldsymbol{v}_r|^\top = [\ |u_r|,\ |v_r|,\ |w_r|,\ |p|,\ |q|,\ |r|\ ]$ and $\boldsymbol{D}_{ni}(\boldsymbol{v}_r)$ for $i = 1, \ldots, 6$ are 6×6 damping matrices. Nevertheless, one rough approximation could be to use quadratic drag in surge and the cross-flow drag in sway and yaw (see Section 6.4.3). Alternatively, if the vehicle is performing a non-coupled motion, it makes sense to assume a diagonal structure of $\boldsymbol{D}(\boldsymbol{v}_r)$ such that

$$\begin{aligned} \boldsymbol{D}(\boldsymbol{v}_r) = & -\mathrm{diag}\{X_u, Y_v, Z_w, K_p, M_q, N_r\} \\ & -\mathrm{diag}\{X_{|u|u}|u_r|,\ Y_{|v|v}|v_r|,\ Z_{|w|w}|w_r|,\ K_{|p|p}|p|,\ M_{|q|q}|q|,\ N_{|r|r}|r|\}. \end{aligned} \tag{8.10}$$

It is also possible to use the current coefficient representation in Section 6.7.1 to model the damping. This can be done by replacing $\boldsymbol{D}_{\mathrm{n}}(\boldsymbol{v}_{\mathrm{r}})\boldsymbol{v}_{\mathrm{r}}$ with 6-DOF current coefficients (see Section 6.7.1)

$$\boldsymbol{d}(V_{\mathrm{rc}}, \gamma_{\mathrm{rc}}) = -\frac{1}{2}\rho V_{\mathrm{rc}}^2 \begin{bmatrix} A_{\mathrm{F_c}} C_X(\gamma_{\mathrm{rc}}) \\ A_{\mathrm{L_c}} C_Y(\gamma_{\mathrm{rc}}) \\ A_{\mathrm{F_c}} C_Z(\gamma_{\mathrm{rc}}) \\ A_{\mathrm{L_c}} H_{\mathrm{L_c}} C_K(\gamma_{\mathrm{rc}}) \\ A_{\mathrm{F_c}} H_{\mathrm{F_c}} C_M(\gamma_{\mathrm{rc}}) \\ A_{\mathrm{L_c}} L_{oa} C_N(\gamma_{\mathrm{rc}}) \end{bmatrix} \tag{8.11}$$

where C_X, C_Y, C_Z, C_K, C_M and C_N are the current coefficients and $H_{\mathrm{F_c}}$ and $H_{\mathrm{L_c}}$ are the centroids above the water line of the frontal and lateral projected areas $A_{\mathrm{F_c}}$ and $A_{\mathrm{L_c}}$, respectively.

The submerged weight of the body and buoyancy force are (see Section 4.1)

$$\boldsymbol{g}(\boldsymbol{\eta}) = \begin{bmatrix} (W - B)\sin(\theta) \\ -(W - B)\cos(\theta)\sin(\phi) \\ -(W - B)\cos(\theta)\cos(\phi) \\ -(y_{\mathrm{g}} W - y_b B)\cos(\theta)\cos(\phi) + (z_{\mathrm{g}} W - z_b B)\cos(\theta)\sin(\phi) \\ (z_{\mathrm{g}} W - z_b B)\sin(\theta) + (x_{\mathrm{g}} W - x_b B)\cos(\theta)\cos(\phi) \\ -(x_{\mathrm{g}} W - x_b B)\cos(\theta)\sin(\phi) - (y_{\mathrm{g}} W - y_b B)\sin(\theta) \end{bmatrix}. \tag{8.12}$$

8.1.2 Equations of Motion Expressed in NED

For simplicity, assume that $\boldsymbol{v}_{\mathrm{c}} = \boldsymbol{0}$. Hence, the equations of motion (8.2) when transformed to $\{n\}$ take the following form

$$\boldsymbol{M}^*(\boldsymbol{\eta})\ddot{\boldsymbol{\eta}} + \boldsymbol{C}^*(\boldsymbol{v}, \boldsymbol{\eta})\dot{\boldsymbol{\eta}} + \boldsymbol{D}^*(\boldsymbol{v}, \boldsymbol{\eta})\dot{\boldsymbol{\eta}} + \boldsymbol{g}^*(\boldsymbol{\eta}) + \boldsymbol{g}_0^*(\boldsymbol{\eta}) = \boldsymbol{\tau}^* \tag{8.13}$$

where the expressions for $\boldsymbol{M}^*(\boldsymbol{\eta})$, $\boldsymbol{C}^*(\boldsymbol{v}, \boldsymbol{\eta})$, $\boldsymbol{D}^*(\boldsymbol{v}, \boldsymbol{\eta})$, $\boldsymbol{g}^*(\boldsymbol{\eta})$, $\boldsymbol{g}_0^*(\boldsymbol{\eta})$, $\boldsymbol{\tau}^*$ and the associated kinematic transformations depend on how attitude is represented. The Euler angles and unit quaternion choices are outlined below.

1. **Positions and Euler angles:** The Euler angle representation (2.53) is based on the three parameters ϕ, θ and ψ. This gives

$$\boldsymbol{J}_\Theta(\boldsymbol{\eta}) = \begin{bmatrix} \boldsymbol{R}(\boldsymbol{\Theta}_{\mathrm{nb}}) & \boldsymbol{0}_{3\times3} \\ \boldsymbol{0}_{3\times3} & \boldsymbol{T}(\boldsymbol{\Theta}_{\mathrm{nb}}) \end{bmatrix}, \quad \boldsymbol{J}_\Theta^{-1}(\boldsymbol{\eta}) = \begin{bmatrix} \boldsymbol{R}^\top(\boldsymbol{\Theta}_{\mathrm{nb}}) & \boldsymbol{0}_{3\times3} \\ \boldsymbol{0}_{3\times3} & \boldsymbol{T}^{-1}(\boldsymbol{\Theta}_{\mathrm{nb}}) \end{bmatrix} \tag{8.14}$$

   where $\boldsymbol{\eta} := [x^n, y^n, z^n, \phi, \theta, \psi]^\top$. The representation singularity at $\theta = \pm\pi/2$ in the expression for $\boldsymbol{T}_\Theta$ implies that the inverse matrix $\boldsymbol{J}_\Theta^{-1}(\boldsymbol{\eta})$ does not exist at these values. The state transformations are as follows

$$\begin{aligned} \dot{\boldsymbol{\eta}} &= \boldsymbol{J}_\Theta(\boldsymbol{\eta})\boldsymbol{v} & &\iff \boldsymbol{v} = \boldsymbol{J}_\Theta^{-1}(\boldsymbol{\eta})\dot{\boldsymbol{\eta}} \\ \ddot{\boldsymbol{\eta}} &= \boldsymbol{J}_\Theta(\boldsymbol{\eta})\dot{\boldsymbol{v}} + \dot{\boldsymbol{J}}_\Theta(\boldsymbol{\eta})\boldsymbol{v} & &\iff \dot{\boldsymbol{v}} = \boldsymbol{J}_\Theta^{-1}(\boldsymbol{\eta})[\ddot{\boldsymbol{\eta}} - \dot{\boldsymbol{J}}_\Theta(\boldsymbol{\eta})\boldsymbol{J}_\Theta^{-1}(\boldsymbol{\eta})\dot{\boldsymbol{\eta}}] \end{aligned}$$

and

$$\begin{aligned}
M^*(\eta) &= J_\Theta^{-\top}(\eta) M J_\Theta^{-1}(\eta) \\
C^*(\nu, \eta) &= J_\Theta^{-\top}(\eta)[C(\nu) - M J_\Theta^{-1}(\eta)\dot{J}_\Theta(\eta)] J_\Theta^{-1}(\eta) \\
D^*(\nu, \eta) &= J_\Theta^{-\top}(\eta) D(\nu) J_\Theta^{-1}(\eta) \\
g^*(\eta) + g_0^*(\eta) &= J_\Theta^{-\top}(\eta)[g(\eta) + g_0] \\
\tau^* &= J_\Theta^{-\top}(\eta)\tau.
\end{aligned} \tag{8.15}$$

2. **Positions and unit quaternions:** The unit quaternion representation (2.83) avoids the singular points $\theta \neq \pm\pi/2$ by using four parameters $\eta, \varepsilon_1, \varepsilon_2$ and ε_3 to represent attitude

$$J_q(\eta) = \begin{bmatrix} R(q_b^n) & 0_{3\times3} \\ 0_{4\times3} & T(q_b^n) \end{bmatrix}, \qquad J_q^\dagger(\eta) = \begin{bmatrix} R^\top(q_b^n) & 0_{3\times4} \\ 0_{4\times3} & 4\, T^\top(q_b^n) \end{bmatrix}. \tag{8.16}$$

Note that pseudo-inverse $J_q^\dagger(\eta)$ is computed using the left *Moore–Penrose pseudo-inverse* and by exploiting the property $T^\top(q_b^n)T(q_b^n) = 1/4\, I_3$. Hence, the left inverse of $T(q_b^n)$ is

$$\begin{aligned}
T^\dagger(q_b^n) &= \left(T^\top(q_b^n)T(q_b^n)\right)^{-1} T^\top(q_b^n) \\
&= 4\, T^\top(q_b^n).
\end{aligned} \tag{8.17}$$

For this case, $\eta = [x^n,\ y^n,\ z^n,\ \eta,\ \varepsilon_1,\ \varepsilon_2,\ \varepsilon_3]^\top$ and

$$\begin{aligned}
\dot{\eta} &= J_q(\eta)\nu & &\Longleftrightarrow & \nu &= J_q^\dagger(\eta)\dot{\eta} \\
\ddot{\eta} &= J_q(\eta)\dot{\nu} + \dot{J}_q(\eta)\nu & &\Longleftrightarrow & \dot{\nu} &= J_q^\dagger(\eta)[\ddot{\eta} - \dot{J}_q(\eta)J_q^\dagger(\eta)\dot{\eta}]
\end{aligned}$$

and

$$\begin{aligned}
M^*(\eta) &= J_q^\dagger(\eta)^\top M J_q^\dagger(\eta) \\
C^*(\nu, \eta) &= J_q^\dagger(\eta)^\top[C(\nu) - M J_q^\dagger(\eta)\dot{J}_q(\eta)] J_q^\dagger(\eta) \\
D^*(\nu, \eta) &= J_q^\dagger(\eta)^\top D(\nu) J_q^\dagger(\eta) \\
g^*(\eta) + g_0^*(\eta) &= J_q^\dagger(\eta)^\top[g(\eta) + g_0] \\
\tau^* &= J_q^\dagger(\eta)^\top \tau.
\end{aligned} \tag{8.18}$$

8.1.3 Properties of the 6-DOF Model

When designing feedback control systems for underwater vehicles in 6 DOFs, there are clear advantages of using the matrix-vector representations. The main reasons are that system properties such as symmetry, skew-symmetry and positiveness of matrices can be incorporated into the stability analysis.

The following properties hold for the body-fixed vector representation:

Property 8.1 (System Inertia Matrix M)
For a rigid body the system inertia matrix is positive definite and constant, that is

$$\boldsymbol{M} = \boldsymbol{M}^\top > 0, \quad \dot{\boldsymbol{M}} = \boldsymbol{0}.$$

Property 8.2 (Coriolis and Centripetal Matrix C)
For a rigid body moving through an ideal fluid the Coriolis and centripetal matrix $\boldsymbol{C}(\boldsymbol{\nu})$ can always be parameterized such that it is skew-symmetric, that is

$$\boldsymbol{C}(\boldsymbol{\nu}) = -\boldsymbol{C}^\top(\boldsymbol{\nu}), \quad \forall \boldsymbol{\nu} \in \mathbb{R}^6.$$

Note that $\boldsymbol{C}(\boldsymbol{\nu})$ is skew-symmetric since we always can find matrices $\boldsymbol{C}_{\rm RB}(\boldsymbol{\nu})$ and $\boldsymbol{C}_{\rm A}(\boldsymbol{\nu})$ that are skew-symmetric.

For the model expressed in $\{n\}$ it is straightforward to verify that

1. $\boldsymbol{M}^*(\boldsymbol{\eta}) = \boldsymbol{M}^*(\boldsymbol{\eta})^\top > 0, \quad \forall \boldsymbol{\eta}$
2. $\boldsymbol{x}^\top[\dot{\boldsymbol{M}}^*(\boldsymbol{\eta}) - 2\boldsymbol{C}^*(\boldsymbol{\nu}, \boldsymbol{\eta})]\boldsymbol{x} = 0, \quad \forall \boldsymbol{x} \neq \boldsymbol{0}, \forall \boldsymbol{\nu}, \boldsymbol{\eta}$
3. $\boldsymbol{D}^*(\boldsymbol{\nu}, \boldsymbol{\eta}) > 0, \quad \forall \boldsymbol{\nu}, \boldsymbol{\eta}$

since $\boldsymbol{M} = \boldsymbol{M}^\top > 0$ and $\dot{\boldsymbol{M}} = \boldsymbol{0}$. It should be noted that $\boldsymbol{C}^*(\boldsymbol{\nu}, \boldsymbol{\eta})$ will not be skew-symmetric although $\boldsymbol{C}(\boldsymbol{\nu})$ is skew-symmetric.

Example 8.1 (Lyapunov Analysis Exploiting 6-DOF Model Properties)
Consider the following 6-DOF model

$$\dot{\boldsymbol{\eta}} = \boldsymbol{J}_k(\boldsymbol{\eta})\boldsymbol{\nu} \tag{8.19}$$

$$\boldsymbol{M}\dot{\boldsymbol{\nu}} + \boldsymbol{C}(\boldsymbol{\nu})\boldsymbol{\nu} + \boldsymbol{D}(\boldsymbol{\nu})\boldsymbol{\nu} + \boldsymbol{g}(\boldsymbol{\eta}) = \boldsymbol{\tau} \tag{8.20}$$

where $\boldsymbol{J}_k(\boldsymbol{\eta})$ can be represented by $\boldsymbol{J}_\Theta(\boldsymbol{\eta})$ or $\boldsymbol{J}_{\rm q}(\boldsymbol{\eta})$. The obvious Lyapunov function candidate is based on kinetic and potential energies

$$V = \frac{1}{2}\boldsymbol{\nu}^\top \boldsymbol{M}\boldsymbol{\nu} + \frac{1}{2}\boldsymbol{\eta}^\top \boldsymbol{K}_{\rm p}\boldsymbol{\eta} \tag{8.21}$$

where $\boldsymbol{K}_{\rm p} = \boldsymbol{K}_{\rm p}^\top > 0$ is a constant gain matrix. Since $\boldsymbol{M} = \boldsymbol{M}^\top > 0$ and $\dot{\boldsymbol{M}} = \boldsymbol{0}$, it follows that

$$\begin{aligned}\dot{V} &= \boldsymbol{\nu}^\top \boldsymbol{M}\dot{\boldsymbol{\nu}} + \boldsymbol{\eta}^\top \boldsymbol{K}_{\rm p}\dot{\boldsymbol{\eta}} \\ &= \boldsymbol{\nu}^\top \boldsymbol{M}\dot{\boldsymbol{\nu}} + \boldsymbol{\eta}^\top \boldsymbol{K}_{\rm p}\boldsymbol{J}_k(\boldsymbol{\eta})\boldsymbol{\nu} \\ &= \boldsymbol{\nu}^\top(\boldsymbol{M}\dot{\boldsymbol{\nu}} + \boldsymbol{J}_k^\top(\boldsymbol{\eta})\boldsymbol{K}_{\rm p}\boldsymbol{\eta}).\end{aligned} \tag{8.22}$$

Substituting (8.20) into the expression for $\dot{V}$ gives

$$\dot{V} = \boldsymbol{v}^{\top}(\boldsymbol{\tau} - \boldsymbol{C}(\boldsymbol{v})\boldsymbol{v} - \boldsymbol{D}(\boldsymbol{v})\boldsymbol{v} - \boldsymbol{g}(\boldsymbol{\eta}) + \boldsymbol{J}_k^{\top}(\boldsymbol{\eta})\boldsymbol{K}_{\mathrm{p}}\boldsymbol{\eta}) \tag{8.23}$$

Since $\boldsymbol{v}^{\top}\boldsymbol{C}(\boldsymbol{v})\boldsymbol{v} \equiv \boldsymbol{0}$ and $\boldsymbol{v}^{\top}\boldsymbol{D}(\boldsymbol{v})\boldsymbol{v} > 0$, we can choose a PD control law with feedforward according to

$$\boldsymbol{\tau} = \boldsymbol{g}(\boldsymbol{\eta}) - \boldsymbol{K}_{\mathrm{d}}\boldsymbol{v} - \boldsymbol{J}_k^{\top}(\boldsymbol{\eta})\boldsymbol{K}_{\mathrm{p}}\boldsymbol{\eta} \tag{8.24}$$

with $\boldsymbol{K}_{\mathrm{d}} > 0$ such that

$$\dot{V} = -\boldsymbol{v}^{\top}(\boldsymbol{K}_{\mathrm{d}} + \boldsymbol{D}(\boldsymbol{v}))\boldsymbol{v} \leq 0 \tag{8.25}$$

Consequently, GAS follows from Krasowskii–LaSalle's theorem if $\boldsymbol{J}_k(\boldsymbol{\eta})$ is non-singular; see Appendix A.1.

8.1.4 Symmetry Considerations of the System Inertia Matrix

We have seen that the 6-DOF nonlinear equations of motion, in their most general representation, require that a large number of hydrodynamic derivatives are known. From a practical point of view this is an unsatisfactory situation. However, the number of unknown parameters can be drastically reduced by using body-symmetry conditions.

In general

$$\boldsymbol{M} = \begin{bmatrix} m - X_{\dot{u}} & -X_{\dot{v}} & -X_{\dot{w}} & -X_{\dot{p}} & mz_{\mathrm{g}} - X_{\dot{q}} & -my_{\mathrm{g}} - X_{\dot{r}} \\ -X_{\dot{v}} & m - Y_{\dot{v}} & -Y_{\dot{w}} & -mz_{\mathrm{g}} - Y_{\dot{p}} & -Y_{\dot{q}} & mx_{\mathrm{g}} - Y_{\dot{r}} \\ -X_{\dot{w}} & -Y_{\dot{w}} & m - Z_{\dot{w}} & my_{\mathrm{g}} - Z_{\dot{p}} & -mx_{\mathrm{g}} - Z_{\dot{q}} & -Z_{\dot{r}} \\ -X_{\dot{p}} & -mz_{\mathrm{g}} - Y_{\dot{p}} & my_{\mathrm{g}} - Z_{\dot{p}} & I_x - K_{\dot{p}} & -I_{xy} - K_{\dot{q}} & -I_{zx} - K_{\dot{r}} \\ mz_{\mathrm{g}} - X_{\dot{q}} & -Y_{\dot{q}} & -mx_{\mathrm{g}} - Z_{\dot{q}} & -I_{xy} - K_{\dot{q}} & I_y - M_{\dot{q}} & -I_{yz} - M_{\dot{r}} \\ -my_{\mathrm{g}} - X_{\dot{r}} & x_{\mathrm{g}} - Y_{\dot{r}} & -Z_{\dot{r}} & -I_{zx} - K_{\dot{r}} & -I_{yz} - M_{\dot{r}} & I_z - N_{\dot{r}} \end{bmatrix}. \tag{8.26}$$

From the definitions of $\boldsymbol{M}_{\mathrm{A}}$ and $\boldsymbol{M}_{\mathrm{RB}}$ it is straightforward to verify the following cases (notice that $m_{ij} = m_{ji}$)

(i) xy plane of symmetry (bottom/top symmetry)

$$\boldsymbol{M} = \begin{bmatrix} m_{11} & m_{12} & 0 & 0 & 0 & m_{16} \\ m_{21} & m_{22} & 0 & 0 & 0 & m_{26} \\ 0 & 0 & m_{33} & m_{34} & m_{35} & 0 \\ 0 & 0 & m_{43} & m_{44} & m_{45} & 0 \\ 0 & 0 & m_{53} & m_{54} & m_{55} & 0 \\ m_{61} & m_{62} & 0 & 0 & 0 & m_{66} \end{bmatrix}.$$

(ii) xz plane of symmetry (port/starboard symmetry)

$$\boldsymbol{M} = \begin{bmatrix} m_{11} & 0 & m_{13} & 0 & m_{15} & 0 \\ 0 & m_{22} & 0 & m_{24} & 0 & m_{26} \\ m_{31} & 0 & m_{33} & 0 & m_{35} & 0 \\ 0 & m_{42} & 0 & m_{44} & 0 & m_{46} \\ m_{51} & 0 & m_{53} & 0 & m_{55} & 0 \\ 0 & m_{62} & 0 & m_{64} & 0 & m_{66} \end{bmatrix}.$$

(iii) yz plane of symmetry (fore/aft symmetry)

$$\boldsymbol{M} = \begin{bmatrix} m_{11} & 0 & 0 & 0 & m_{15} & m_{16} \\ 0 & m_{22} & m_{23} & m_{24} & 0 & 0 \\ 0 & m_{32} & m_{33} & m_{34} & 0 & 0 \\ 0 & m_{42} & m_{43} & m_{44} & 0 & 0 \\ m_{51} & 0 & 0 & 0 & m_{55} & m_{56} \\ m_{61} & 0 & 0 & 0 & m_{65} & m_{66} \end{bmatrix}.$$

(iv) xz and yz planes of symmetry (port/starboard and fore/aft symmetries)

$$\boldsymbol{M} = \begin{bmatrix} m_{11} & 0 & 0 & 0 & m_{15} & 0 \\ 0 & m_{22} & 0 & m_{24} & 0 & 0 \\ 0 & 0 & m_{33} & 0 & 0 & 0 \\ 0 & m_{42} & 0 & m_{44} & 0 & 0 \\ m_{51} & 0 & 0 & 0 & m_{55} & 0 \\ 0 & 0 & 0 & 0 & 0 & m_{66} \end{bmatrix}.$$

More generally, the resulting inertia matrix for a body with ij and jk planes of symmetry is formed by the intersection $M_{ij \cap jk} = M_{ij} \cap M_{jk}$.

(v) xz, yz and xy planes of symmetry (port/starboard, fore/aft and bottom/top symmetries)

$$\boldsymbol{M} = \mathrm{diag}\{\, m_{11},\ m_{22},\ m_{33},\ m_{44},\ m_{55},\ m_{66} \,\}. \tag{8.27}$$

8.2 Longitudinal and Lateral Models for Submarines

The 6-DOF equations of motion can in many cases be divided into two non-interacting (or lightly interacting) subsystems:

- Longitudinal subsystem: states u_r, w_r, q and θ
- Lateral subsystem: states v_r, p, r, ϕ and ψ.

This decomposition is good for slender symmetrical bodies (large length/width ratio) or so-called "flying vehicles", as shown in Figure 8.1; typical applications are aircraft, missiles and submarines (Gertler and Hagen 1967; Feldman 1979; Tinker 1982). This can also be seen from the expression of the system inertia matrix in the case of starboard–port

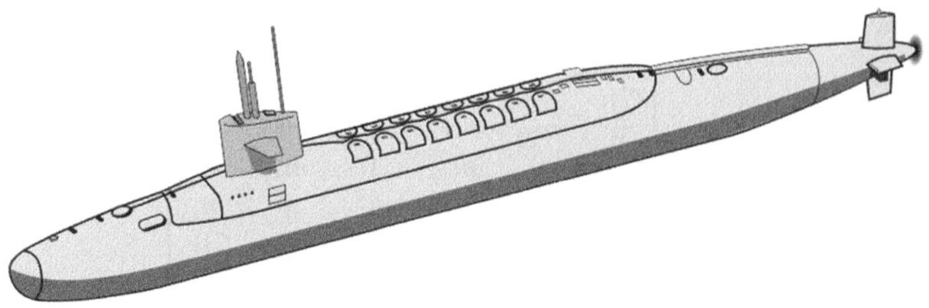

Figure 8.1 Slender body submarine (large length/width ratio).

symmetry (see Section 8.1.4)

$$M = \begin{bmatrix} m_{11} & 0 & m_{13} & 0 & m_{15} & 0 \\ 0 & m_{22} & 0 & m_{24} & 0 & m_{26} \\ m_{31} & 0 & m_{33} & 0 & m_{35} & 0 \\ 0 & m_{42} & 0 & m_{44} & 0 & m_{46} \\ m_{51} & 0 & m_{53} & 0 & m_{55} & 0 \\ 0 & m_{62} & 0 & m_{64} & 0 & m_{66} \end{bmatrix} \tag{8.28}$$

which clearly confirms that the two subsystems

$$M_{\rm long} = \begin{bmatrix} m_{11} & m_{13} & m_{15} \\ m_{31} & m_{33} & m_{35} \\ m_{51} & m_{53} & m_{55} \end{bmatrix}, \quad M_{\rm lat} = \begin{bmatrix} m_{22} & m_{24} & m_{26} \\ m_{42} & m_{44} & m_{46} \\ m_{62} & m_{64} & m_{66} \end{bmatrix} \tag{8.29}$$

are decoupled.

8.2.1 Longitudinal Subsystem

Under the assumption that the lateral states v, p, r, ϕ are small, the longitudinal kinematics for surge, heave and pitch are, see (2.36) and (2.43),

$$\begin{aligned} \begin{bmatrix} \dot{z}^n \\ \dot{\theta} \end{bmatrix} &= \begin{bmatrix} \cos(\theta) & 0 \\ 0 & 1 \end{bmatrix} \begin{bmatrix} w \\ q \end{bmatrix} + \begin{bmatrix} -\sin(\theta) \\ 0 \end{bmatrix} u \\ &\approx \begin{bmatrix} w_{\rm r} - U\theta \\ q \end{bmatrix} + \begin{bmatrix} 1 \\ 0 \end{bmatrix} w_{\rm c} \end{aligned} \tag{8.30}$$

where $u \approx U$, $w_{\rm r} = w - w_{\rm c}$, $\sin(\theta) \approx \theta$ and $\cos(\theta) \approx 1$. For simplicity, it is assumed that higher-order damping terms can be neglected, that is $\boldsymbol{D}_{\rm n}(\boldsymbol{\nu}) = \boldsymbol{0}$. Coriolis is, however, modeled by assuming that $U \gg 0$ and that second-order terms in v, w, p, q and r are small. Hence, linearization of (3.62) about the forward speed U gives

$$\boldsymbol{C}_{\mathrm{RB}}(\boldsymbol{\nu}_{\mathrm{r}})\boldsymbol{\nu}_{\mathrm{r}} \approx \underbrace{\begin{bmatrix} 0 & 0 & 0 \\ 0 & 0 & -mU \\ 0 & 0 & mx_{\mathrm{g}}U \end{bmatrix}}_{\boldsymbol{C}^{*}_{\mathrm{RB}}} \begin{bmatrix} u_{\mathrm{r}} \\ w_{\mathrm{r}} \\ q \end{bmatrix}. \tag{8.31}$$

Note that $\boldsymbol{C}^{*}_{\mathrm{RB}} \neq -(\boldsymbol{C}^{*}_{\mathrm{RB}})^{\top}$ for the decoupled model. Assuming a diagonal $\boldsymbol{M}_{\mathrm{A}}$ as in Example 6.2, the corresponding added mass terms are

$$\boldsymbol{C}_{\mathrm{A}}(\boldsymbol{\nu}_{\mathrm{r}})\boldsymbol{\nu}_{\mathrm{r}} = \begin{bmatrix} -Z_{\dot{w}}w_{\mathrm{r}}q + Y_{\dot{v}}v_{\mathrm{r}}r \\ -Y_{\dot{v}}v_{\mathrm{r}}p + X_{\dot{u}}u_{\mathrm{r}}q \\ (Z_{\dot{w}} - X_{\dot{u}})u_{\mathrm{r}}w_{\mathrm{r}} + (N_{\dot{r}} - K_{\dot{p}})pr \end{bmatrix} \approx \underbrace{\begin{bmatrix} 0 & 0 & 0 \\ 0 & 0 & X_{\dot{u}}U \\ 0 & (Z_{\dot{w}} - X_{\dot{u}})U & 0 \end{bmatrix}}_{\boldsymbol{C}^{*}_{A}} \begin{bmatrix} u_{\mathrm{r}} \\ w_{\mathrm{r}} \\ q \end{bmatrix}. \tag{8.32}$$

From Section 6.5.1 and (4.6) with $W = B$ and $x_{\mathrm{g}} = x_b$, the dynamics becomes

$$\begin{aligned} &\begin{bmatrix} m - X_{\dot{u}} & -X_{\dot{w}} & mz_{\mathrm{g}} - X_{\dot{q}} \\ -X_{\dot{w}} & m - Z_{\dot{w}} & -mx_{\mathrm{g}} - Z_{\dot{q}} \\ mz_{\mathrm{g}} - X_{\dot{q}} & -mx_{\mathrm{g}} - Z_{\dot{q}} & I_y - M_{\dot{q}} \end{bmatrix} \begin{bmatrix} \dot{u}_{\mathrm{r}} \\ \dot{w}_{\mathrm{r}} \\ \dot{q} \end{bmatrix} \\ &+ \begin{bmatrix} -X_u & -X_w & -X_{\mathrm{q}} \\ -Z_u & -Z_w & -Z_{\mathrm{q}} \\ -M_u & -M_w & -M_{\mathrm{q}} \end{bmatrix} \begin{bmatrix} u_{\mathrm{r}} \\ w_{\mathrm{r}} \\ q \end{bmatrix} + \begin{bmatrix} 0 & 0 & 0 \\ 0 & 0 & -(m - X_{\dot{u}})U \\ 0 & (Z_{\dot{w}} - X_{\dot{u}})U & mx_{\mathrm{g}}U \end{bmatrix} \begin{bmatrix} u_{\mathrm{r}} \\ w_{\mathrm{r}} \\ q \end{bmatrix} \\ &+ \begin{bmatrix} 0 \\ 0 \\ \mathrm{BG}_z W \sin(\theta) \end{bmatrix} = \begin{bmatrix} \tau_1 \\ \tau_3 \\ \tau_5 \end{bmatrix}. \end{aligned} \tag{8.33}$$

This model is the basis for forward speed control (state u_{r}) and depth/diving autopilot design (states $w_{\mathrm{r}}, q, \theta$). The speed dynamics can be removed by assuming that the speed controller stabilizes the forward speed such that $u = U = \text{constant}$. Hence, (8.33) reduces to a combined pitch and diving model where the relative surge velocity u_{r} is linearized about the forward speed U to give

$$\begin{aligned} &\begin{bmatrix} m - Z_{\dot{w}} & -mx_{\mathrm{g}} - Z_{\dot{q}} \\ -mx_{\mathrm{g}} - Z_{\dot{q}} & I_y - M_{\dot{q}} \end{bmatrix} \begin{bmatrix} \dot{w}_{\mathrm{r}} \\ \dot{q} \end{bmatrix} + \begin{bmatrix} -Z_w & -Z_{\mathrm{q}} \\ -M_w & -M_{\mathrm{q}} \end{bmatrix} \begin{bmatrix} w_{\mathrm{r}} \\ q \end{bmatrix} \\ &+ \begin{bmatrix} 0 & -(m - X_{\dot{u}})U \\ (Z_{\dot{w}} - X_{\dot{u}})U & mx_{\mathrm{g}}U \end{bmatrix} \begin{bmatrix} w_{\mathrm{r}} \\ q \end{bmatrix} + \begin{bmatrix} 0 \\ \mathrm{BG}_z W \sin(\theta) \end{bmatrix} = \begin{bmatrix} \tau_3 \\ \tau_5 \end{bmatrix}. \end{aligned} \tag{8.34}$$

Consequently, if $\dot{w}_r = w_r = 0$ and θ is small such that $\sin(\theta) \approx \theta$, the linear pitch dynamics becomes

$$(I_y - M_{\dot{q}})\ddot{\theta} - M_q\dot{\theta} + \mathrm{BG}_z W\theta = \tau_5. \tag{8.35}$$

The natural frequency and period are recognized as

$$\omega_5 = \sqrt{\frac{\mathrm{BG}_z W}{I_y - M_{\dot{q}}}}, \qquad T_5 = \frac{2\pi}{\omega_5}. \tag{8.36}$$

8.2.2 *Lateral Subsystem*

Under the assumption that the longitudinal states $u,\ w,\ p,\ r,\ \phi$ and θ are small, the lateral kinematics, see 6.5.1 and (2.44), reduce to

$$\dot{\phi} = p \tag{8.37}$$

$$\dot{\psi} = r. \tag{8.38}$$

Again it is assumed that higher-order velocity terms can be neglected so that nonlinear damping $\boldsymbol{D}_n(\boldsymbol{\nu}_r) = \boldsymbol{0}$ and that the Coriolis terms can be linearized about the constant speed U, see (3.62). Consequently,

$$\boldsymbol{C}_{RB}(\boldsymbol{\nu}_r)\boldsymbol{\nu}_r \approx \underbrace{\begin{bmatrix} 0 & 0 & mU \\ 0 & 0 & 0 \\ 0 & 0 & mx_g U \end{bmatrix}}_{\boldsymbol{C}^*_{RB}} \begin{bmatrix} v_r \\ p \\ r \end{bmatrix}. \tag{8.39}$$

Under the assumption of a diagonal $\boldsymbol{M}_A$ structure as in Example 6.2, the corresponding added mass terms are

$$\boldsymbol{C}_A(\boldsymbol{\nu}_r)\boldsymbol{\nu}_r = \begin{bmatrix} Z_{\dot{w}} w_r p - X_{\dot{u}} u_r r \\ (Y_{\dot{v}} - Z_{\dot{w}}) v_r w_r + (M_{\dot{q}} - N_{\dot{r}}) qr \\ (X_{\dot{u}} - Y_{\dot{v}}) u_r v_r + (K_{\dot{p}} - M_{\dot{q}}) pq \end{bmatrix}$$

$$\approx \underbrace{\begin{bmatrix} 0 & 0 & -X_{\dot{u}} U \\ 0 & 0 & 0 \\ (X_{\dot{u}} - Y_{\dot{v}})U & 0 & 0 \end{bmatrix}}_{\boldsymbol{C}^*_A} \begin{bmatrix} v_r \\ p \\ r \end{bmatrix}. \tag{8.40}$$

Next, assume that $W = B$, $x_g = x_b$ and $y_g = y_b$. Then it follows from Section 6.5.1 that

$$\begin{bmatrix} m - Y_{\dot{v}} & -mz_g - Y_{\dot{p}} & mx_g - Y_{\dot{r}} \\ -mz_g - Y_{\dot{p}} & I_x - K_{\dot{p}} & -I_{zx} - K_{\dot{r}} \\ mx_g - Y_{\dot{r}} & -I_{zx} - K_{\dot{r}} & I_z - N_{\dot{r}} \end{bmatrix} \begin{bmatrix} \dot{v}_r \\ \dot{p} \\ \dot{r} \end{bmatrix} + \begin{bmatrix} -Y_v & -Y_p & -Y_r \\ -K_v & -K_p & -K_r \\ -N_v & -N_p & -N_r \end{bmatrix} \begin{bmatrix} v_r \\ p \\ r \end{bmatrix}$$
$$+ \begin{bmatrix} 0 & 0 & (m - X_{\dot{u}})U \\ 0 & 0 & 0 \\ (X_{\dot{u}} - Y_{\dot{v}})U & 0 & mx_g U \end{bmatrix} \begin{bmatrix} v_r \\ p \\ r \end{bmatrix} + \begin{bmatrix} 0 \\ \mathrm{BG}_z W \sin(\phi) \\ 0 \end{bmatrix} = \begin{bmatrix} \tau_2 \\ \tau_4 \\ \tau_6 \end{bmatrix}. \tag{8.41}$$

For vehicles where $\dot{p}$ and p are small (small roll motions), this reduces to

$$\begin{bmatrix} m - Y_{\dot{v}} & mx_g - Y_{\dot{r}} \\ mx_g - Y_{\dot{r}} & I_z - N_{\dot{r}} \end{bmatrix} \begin{bmatrix} \dot{v}_r \\ \dot{r} \end{bmatrix} + \begin{bmatrix} -Y_v & -Y_r \\ -N_v & -N_r \end{bmatrix} \begin{bmatrix} v_r \\ r \end{bmatrix}$$
$$+ \begin{bmatrix} 0 & (m - X_{\dot{u}})U \\ (X_{\dot{u}} - Y_{\dot{v}})U & mx_g U \end{bmatrix} \begin{bmatrix} v_r \\ r \end{bmatrix} = \begin{bmatrix} \tau_2 \\ \tau_6 \end{bmatrix} \tag{8.42}$$

which is the sway–yaw maneuvering model (see Section 6.5.5). The decoupled linear roll equation under the assumption of a small ϕ is

$$(I_x - K_{\dot{p}})\ddot{\phi} - K_p \dot{\phi} + \mathrm{BG}_z W \phi = \tau_4. \tag{8.43}$$

From this it follows that the natural frequency and period are

$$\omega_4 = \sqrt{\frac{\mathrm{BG}_z W}{I_x - K_{\dot{p}}}}, \qquad T_4 = \frac{2\pi}{\omega_4}. \tag{8.44}$$

8.3 Decoupled Models for "Flying Underwater Vehicles"

For slender symmetrical bodies (large length/width ratio) or so-called "flying underwater vehicles" it is common to decompose the 6-DOF equations of motion into three non-interacting (or lightly interacting) subsystems:

- Forward speed subsystem: state u_r
- Course angle subsystem: states v_r, p, r, ϕ and χ
- Pitch-depth subsystem: states w_r, q, z and θ.

These subsystems are used to design forward speed, course and pitch–depth control systems for underwater vehicles.

8.3.1 Forward Speed Subsystem

For slender bodies moving at forward speed $U \gg 0$ it is standard to assume that the surge dynamics is decoupled from the other motions of the vehicle due to the large length/width ratio. In other words

$$(m - X_{\dot{u}})\dot{u}_{\mathrm{r}} - X_{|u|u}|u_{\mathrm{r}}|u_{\mathrm{r}} = T \tag{8.45}$$

where T is the thrust (see Section 9.1). The linear skin friction term $X_u u_{\mathrm{r}}$ is assumed to be negligible at higher speeds. The transfer function between the thrust T and the relative surge velocity u_{r} can be found by using the equivalent linearization method, see (5.66), to approximate the nonlinear damping term

$$X_{|u|u}|u_{\mathrm{r}}|u_{\mathrm{r}} \approx \frac{8A_1}{3\pi} X_{|u|u} u_{\mathrm{r}} \tag{8.46}$$

where the surge velocity $u_{\mathrm{r}} = A_1 \cos(\omega t)$ is assumed to be harmonic. Hence, the transfer function becomes

$$u(s) = \frac{K_u}{T_u s + 1} T(s) + d_u(s) \tag{8.47}$$

where

$$d_u(s) = \frac{1}{T_u s + 1} u_{\mathrm{c}}(s) \tag{8.48}$$

is a disturbance term due to the ocean currents and

$$K_u = \frac{1}{-\frac{8A_1}{3\pi} X_{|u|u}}, \qquad T_u = \frac{m - X_{\dot{u}}}{-\frac{8A_1}{3\pi} X_{|u|u}}. \tag{8.49}$$

The value for $X_{|u|u}$ is given by (6.81) while $X_{\dot{u}} = -A_{11}(\omega) =$ constant for all ω is the added mass in surge.

8.3.2 Course Angle Subsystem

Consider the linear maneuvering model (see Section 6.5.5)

$$\begin{bmatrix} m - Y_{\dot{v}} & mx_{\mathrm{g}} - Y_{\dot{r}} \\ mx_{\mathrm{g}} - Y_{\dot{r}} & I_z - N_{\dot{r}} \end{bmatrix} \begin{bmatrix} \dot{v}_{\mathrm{r}} \\ \dot{r} \end{bmatrix} + \begin{bmatrix} -Y_v & -Y_r \\ -N_v & -N_r \end{bmatrix} \begin{bmatrix} v_{\mathrm{r}} \\ r \end{bmatrix}$$
$$+ \begin{bmatrix} 0 & (m - X_{\dot{u}})U \\ (X_{\dot{u}} - Y_{\dot{v}})U & mx_{\mathrm{g}} U \end{bmatrix} \begin{bmatrix} v_{\mathrm{r}} \\ r \end{bmatrix} = \begin{bmatrix} \tau_2 \\ \tau_6 \end{bmatrix}. \tag{8.50}$$

For simplicity assume that the control input is a single tail rudder (see Section 9.7). This gives the following state-space model

$$\dot{x} = Ax + bu \tag{8.51}$$

$$\Updownarrow$$

$$\begin{bmatrix} \dot{v}_r \\ \dot{r} \\ \dot{\psi} \end{bmatrix} = \begin{bmatrix} a_{11} & a_{12} & 0 \\ a_{21} & a_{22} & 0 \\ 0 & 1 & 0 \end{bmatrix} \begin{bmatrix} v_r \\ r \\ \psi \end{bmatrix} + \begin{bmatrix} b_1 \\ b_2 \\ 0 \end{bmatrix} \delta_r \tag{8.52}$$

where $x = [v_r, \ r, \ \psi]^\top$ and $u = \delta_r$ is the rudder angle.

The transfer function between the rudder deflection and the heading angle is (see Section 7.2.1)

$$\begin{aligned} \psi(s) &= \frac{K(T_3 s + 1)}{s(T_1 s + 1)(T_2 s + 1)} \delta_r(s) + d_\psi(s) \\ &\approx \frac{K}{s(Ts + 1)} \delta_r(s) + d_\psi(s) \end{aligned} \tag{8.53}$$

where K and $T = T_1 + T_2 - T_3$ are recognized as the Nomoto gain and time constants, and d_ψ is a disturbance term representing unmodeled dynamics and ocean currents. Since, $\chi = \psi + \beta_c$, the transfer function for course control takes the following form

$$\chi(s) = \frac{K}{s(Ts + 1)} \delta_r(s) + d_\chi(s) \tag{8.54}$$

where $d_\chi = d_\psi + \beta_c$.

8.3.3 Pitch–Depth Subsystem

Pitch and depth control of underwater vehicles is usually done by using control surfaces, thrusters, moving masses, spinning rotors and ballast systems. For a neutrally buoyant vehicle, stern rudders δ_e are effective for diving and depth changing maneuvers since they require relatively little control energy compared to thrusters; see Section 9.7. The longitudinal model (8.33) in Section 8.2 can be linearized about the speed U to give

$$\begin{aligned} &\begin{bmatrix} m - Z_{\dot{w}} & -mx_g - Z_{\dot{q}} \\ -mx_g - Z_{\dot{q}} & I_y - M_{\dot{q}} \end{bmatrix} \begin{bmatrix} \dot{w}_r \\ \dot{q} \end{bmatrix} + \begin{bmatrix} -Z_w & -Z_q \\ -M_w & -M_q \end{bmatrix} \begin{bmatrix} w_r \\ q \end{bmatrix} \\ &+ \begin{bmatrix} 0 & -(m - X_{\dot{u}})U \\ (Z_{\dot{w}} - X_{\dot{u}})U & mx_g U \end{bmatrix} \begin{bmatrix} w_r \\ q \end{bmatrix} + \begin{bmatrix} 0 \\ \mathrm{BG}_z W \sin(\theta) \end{bmatrix} = \begin{bmatrix} \tau_3 \\ \tau_5 \end{bmatrix}. \end{aligned} \tag{8.55}$$

A state-space representation of this model is

$$\dot{x} = Ax + bu + \varphi(t) \tag{8.56}$$

$$\Updownarrow$$

$$\begin{bmatrix} \dot{w}_r \\ \dot{q} \\ \dot{\theta} \\ \dot{z}^n \end{bmatrix} = \begin{bmatrix} a_{11} & a_{12} & 0 & 0 \\ a_{21} & a_{22} & a_{23} & 0 \\ 0 & 1 & 0 & 0 \\ 1 & 0 & -U & 0 \end{bmatrix} \begin{bmatrix} w_r \\ q \\ \theta \\ z^n \end{bmatrix} + \begin{bmatrix} b_1 \\ b_2 \\ 0 \\ 0 \end{bmatrix} \delta_e + \begin{bmatrix} 0 \\ 0 \\ 0 \\ 1 \end{bmatrix} w_c \tag{8.57}$$

where $x = [w_r,\ q,\ \theta,\ z^n]^\top$, $u = \delta_e$ is the stern rudder (elevator) and w_c is the vertical current velocity given by an exogenous system $\varphi(t)$. The kinematic equations are based on the approximations (see Section 2.2.1)

$$\dot{\theta} = q\cos(\phi) - r\sin(\phi) \approx q \tag{8.58}$$

$$\dot{z}^n = -u\sin(\theta) + v\cos(\theta)\sin(\phi) + w\cos(\theta)\cos(\phi) \approx (w_r + w_c) - U\theta \tag{8.59}$$

for $v = r = 0$ and small values of θ and ϕ. A further simplification could be to assume that $\dot{w}_r$ and w_r are small during depth-changing maneuvers. This gives the transfer functions

$$z^n(s) = -\frac{U}{s}\theta(s) + d_z(s) \tag{8.60}$$

$$\theta(s) = \frac{a_3}{s^2 + a_1 s + a_2}\delta_e(s) + d_\theta(s) \tag{8.61}$$

where $d_z(s)$ and are $d_\theta(s)$ are disturbance terms due to unmodeled dynamics and ocean currents. Consequently, the transfer function for depth control using stern rudders (elevators) as control inputs becomes

$$z^n(s) = \frac{-a_3 U}{s(s^2 + a_1 s + a_2)}\delta_e(s) + \overline{d}_z(s) \tag{8.62}$$

where $\overline{d}_z(s)$ is the resulting disturbance term.

8.4 Cylinder-Shaped Vehicles and Myring-type Hulls

Figure 8.2 shows a cylinder-shaped AUV. Assume that the CO is located in the center of the AUV and that the CB coincides with the CO such that $r^b_{bb} = [0,\ 0,\ 0]^\top$. Furthermore assume that the CG is located at $r^b_{bg} = [x_g,\ y_g,\ z_g]^\top$. Hence, the AUV equations of motion takes the following form

$$\dot{\eta} = J_k(\eta)(v_r + v_c) \tag{8.63}$$

$$M\dot{v}_r + C(v_r)v_r + Dv_r + d(v_r) + g(\eta) = \tau \tag{8.64}$$

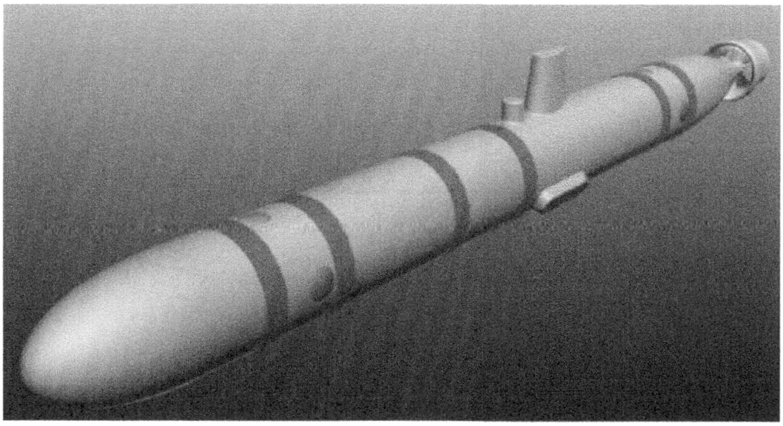

Figure 8.2 Cylinder-shaped AUV. Source: From T.-H., K. Sammut, F. He and S.-K. Lee (2012). Shape Optimization of an Autonomous Underwater Vehicle with a Ducted Propeller using Computational Fluid Dynamics Analysis. International Journal of Naval Architecture and Ocean Engineering 4, 44-56. ©2012, Society of Naval Architects of Korea.

where $\boldsymbol{\eta}$ and $\boldsymbol{J}_k(\boldsymbol{\eta})$ for $k \in \{\Theta, q\}$ can be parametrized using Euler angles or unit quaternions; see Section 8.1.

Cylinder-shaped AUVs such as the *light autonomous underwater vehicle* (LAUV) (Madureira *et al.* 2013) and the *remote environmental monitoring unit* (REMUS) AUV (Hydroid 2019; REMUS 2019) can be represented by the Myring hull profile equations (Myring 1976), which are known to produce minimum drag force for a given fineness ratio (L/D), that is, the ratio of its length L to its maximum diameter D.

8.4.1 Myring-type Hull

The shapes of the nose, cylinder and tail sections of the Myring-type hull are determined by (see Figure 8.3)

$$r(x) = \begin{cases} \frac{1}{2}D\left(1 - \left(\frac{x-L_\mathrm{n}}{L_\mathrm{n}}\right)^2\right)^{1/n}, & 0 \leq x \leq L_\mathrm{n} \\ \frac{D}{2}, & L_\mathrm{n} < x < L_\mathrm{n} + L_\mathrm{c} \\ \frac{1}{2}D - \left(\frac{3D}{2L_\mathrm{t}^2} - \frac{\tan(\alpha_\mathrm{t})}{L_\mathrm{t}}\right)(x - (L_\mathrm{n} + L_\mathrm{c}))^2 & \\ \quad + \left(\frac{D}{L_\mathrm{t}^3} - \frac{\tan(\alpha_\mathrm{t})}{L_\mathrm{t}^2}\right)(x - (L_\mathrm{n} + L_\mathrm{c}))^3, & L_\mathrm{n} + L_\mathrm{c} \leq x \leq L \end{cases} \tag{8.65}$$

where $r(x)$ is the radius at a point along the x axis, L_n is the length of the nose, L_c is the length of body, L_t is the length of the tail section, D is the cylinder diameter, n is an exponential parameter that can be varied to give different body shapes, and α_t is the included angle at the tip of the tail. Hence, the length of the vehicle is $L = L_\mathrm{n} + L_\mathrm{c} + L_\mathrm{t}$.

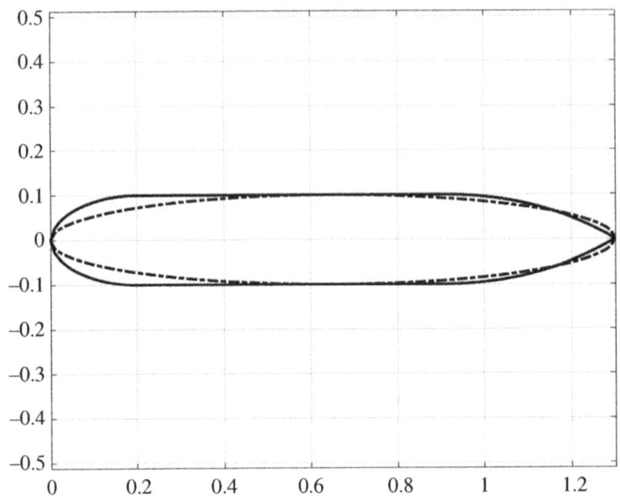

Figure 8.3 Myring-type hull (solid) for $D = 0.2$ m, $L = 1.3$ m, $L_n = 0.2$ m, $L_c = 0.7$ m, $L_t = 0.4$ m, $n = 1.8$, and $\alpha_t = 30^\circ$ approximated by a spheroid (dotted line).

8.4.2 Spheroid Approximation

An *ellipsoid* is a surface that may be obtained from a sphere by deforming it. Consider an ellipsoid totally submerged and with the origin at the center of the ellipsoid

$$\frac{x^2}{a^2} + \frac{y^2}{b^2} + \frac{z^2}{c^2} = 1 \tag{8.66}$$

where a, b and c are the semi-axes, see Figure 8.4. A *prolate spheroid* is obtained by letting $b = c$ and $a > b$. Consequently, the mass of a prolate spheroid is

$$m = \frac{4}{3}\pi\rho abc = \frac{4}{3}\pi\rho ab^2 \tag{8.67}$$

and the moments of inertia are

$$I_x = \frac{1}{5}m(b^2 + c^2) = \frac{2}{5}mb^2 \tag{8.68}$$

$$I_y = \frac{1}{5}m(a^2 + c^2) = \frac{1}{5}m(a^2 + b^2) \tag{8.69}$$

$$I_z = \frac{1}{5}m(a^2 + b^2) = I_y. \tag{8.70}$$

The largest difference between a Myring-type shape and a spheroid lies in the ends of the body. To approximate the hull with a spheroid as shown in Figure 8.3, the method of equivalent ellipsoid is used (Korotkin 2009). This involves choosing the function $r(x)$ such that the volumes of the spheroid and the Myring-type hull are equal

$$\int_0^L \pi r^2(x)\mathrm{d}x = \frac{4}{3}\pi\rho ab^2. \tag{8.71}$$

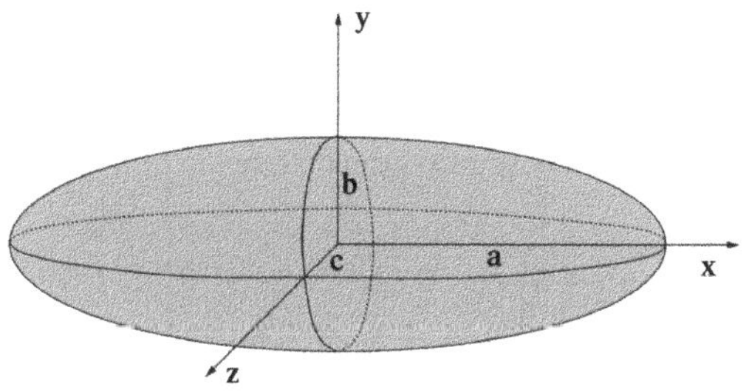

Figure 8.4 Ellipsoid with semi-axes a, b and c.

Consequently, the inertia dyadic can be approximated by the formulae for the spheroid

$$I_g^b = \text{diag}\left\{\frac{2}{5}mb^2,\ \frac{1}{5}m(a^2+b^2),\ \frac{1}{5}m(a^2+b^2)\right\}. \tag{8.72}$$

For non-zero values of $\boldsymbol{r}_{bg}^b$, it follows from (C.22) that the rigid-body inertia matrix is

$$\boldsymbol{M}_{RB} = \boldsymbol{H}^\top(\boldsymbol{r}_{bg}^b)\ \text{diag}\left\{m,\ m,\ m,\ \frac{2}{5}mb^2,\ \frac{1}{5}m(a^2+b^2),\ \frac{1}{5}m(a^2+b^2)\right\}\boldsymbol{H}(\boldsymbol{r}_{bg}^b). \tag{8.73}$$

Fortunately, many of the added mass derivatives contained in the general expressions for added mass are either zero or mutually related when the body has various symmetries. Imlay (1961) gives the following expressions for the diagonal added mass derivatives (cross-coupling terms will be zero when the CO coincides with the CB due to body symmetry about three planes)

$$X_{\dot{u}} = -\frac{\alpha_0}{2-\alpha_0}\ m \tag{8.74}$$

$$Y_{\dot{v}} = Z_{\dot{w}} = -\frac{\beta_0}{2-\beta_0}\ m \tag{8.75}$$

$$K_{\dot{p}} = 0 \tag{8.76}$$

$$N_{\dot{r}} = M_{\dot{q}} = -\frac{1}{5}\ \frac{(b^2-a^2)^2(\alpha_0-\beta_0)}{2(b^2-a^2)+(b^2+a^2)(\beta_0-\alpha_0)}\ m. \tag{8.77}$$

The constants α_0 and β_0 can be calculated as

$$\alpha_0 = \frac{2(1-e^2)}{e^3}\left(\frac{1}{2}\ln\frac{1+e}{1-e} - e\right) \tag{8.78}$$

$$\beta_0 = \frac{1}{e^2} - \frac{1-e^2}{2e^3}\ln\frac{1+e}{1-e} \tag{8.79}$$

where the eccentricity is defined as

$$e := 1 - (b/a)^2. \tag{8.80}$$

An alternative representation of these mass derivatives is presented by Lamb (1932) who defines Lamb's k-factors as

$$k_1 = \frac{\alpha_0}{2 - \alpha_0} \tag{8.81}$$

$$k_2 = \frac{\beta_0}{2 - \beta_0} \tag{8.82}$$

$$k' = \frac{e^4(\beta_0 - \alpha_0)}{(2 - e^2)(2e^2 - (2 - e^2)(\beta_0 - \alpha_0))}. \tag{8.83}$$

Hence, the added system inertia matrix can be written in terms of Lamb's k-factors

$$\begin{aligned} \boldsymbol{M}_A &= -\text{diag}\{X_{\dot{u}},\ Y_{\dot{v}},\ Z_{\dot{w}},\ K_{\dot{p}},\ M_{\dot{q}},\ N_{\dot{r}}\} \\ &= \text{diag}\{mk_1,\ mk_2,\ mk_2,\ 0,\ k'I_y,\ k'I_y\}. \end{aligned} \tag{8.84}$$

The final expression for the system inertia matrix then becomes $\boldsymbol{M} = \boldsymbol{M}_{\text{RB}} + \boldsymbol{M}_A$.

A more general discussion on added mass derivatives for bodies with various symmetries is found in Imlay (1961). Other useful references discussing methods for computation of the added mass derivatives are Humphreys and Watkinson (1978), and Triantafyllou and Amzallag (1984).

The Coriolis and centripetal matrix $\boldsymbol{C}(\boldsymbol{v}_r) = \boldsymbol{C}_{\text{RB}}(\boldsymbol{v}_r) + \boldsymbol{C}_A(\boldsymbol{v}_r)$ is computed using (3.64). Moreover,

$$\boldsymbol{C}_{\text{RB}}(\boldsymbol{v}_r) = \boldsymbol{H}^{\top}(\boldsymbol{r}^b_{bg}) \begin{bmatrix} 0 & -mr & mq & 0 & 0 & 0 \\ mr & 0 & -mp & 0 & 0 & 0 \\ -mq & mp & 0 & 0 & 0 & 0 \\ 0 & 0 & 0 & 0 & I_z r & -I_y q \\ 0 & 0 & 0 & -I_z r & 0 & I_x p \\ 0 & 0 & 0 & I_y q & -I_x p & 0 \end{bmatrix} \boldsymbol{H}(\boldsymbol{r}^b_{bg}) \tag{8.85}$$

where I_x, I_y and I_z are given by (8.68)–(8.70). From Example 6.2 it follows that

$$\boldsymbol{C}_A(\boldsymbol{v}_r) = \begin{bmatrix} 0 & 0 & 0 & 0 & -Z_{\dot{w}}w_r & Y_{\dot{v}}v_r \\ 0 & 0 & 0 & Z_{\dot{w}}w_r & 0 & -X_{\dot{u}}u_r \\ 0 & 0 & 0 & -Y_{\dot{v}}v_r & X_{\dot{u}}u_r & 0 \\ 0 & -Z_{\dot{w}}w_r & Y_{\dot{v}}v_r & 0 & -N_{\dot{r}}r & M_{\dot{q}}q \\ Z_{\dot{w}}w_r & 0 & -X_{\dot{u}}u_r & N_{\dot{r}}r & 0 & -K_{\dot{p}}p \\ -Y_{\dot{v}}v_r & X_{\dot{u}}u_r & 0 & -M_{\dot{q}}q & K_{\dot{p}}p & 0 \end{bmatrix}. \tag{8.86}$$

Assume that the lift and drag coefficients $C_L(\alpha)$ and $C_D(\alpha)$ are known for the AUV such that the forces expressed in the FLOW axes are (see Section 2.5.2)

$$F_{\text{lift}} = \frac{1}{2}\rho U_r^2 S C_L(\alpha) \tag{8.87}$$

$$F_{\text{drag}} = \frac{1}{2}\rho U_r^2 S C_D(\alpha) \tag{8.88}$$

where ρ is the density of water, U_r is the relative speed and S is a reference area usually chosen as the area of the vehicle as if it were projected down onto the ground below it. The lift and drag coefficients can be approximated by linear theory

$$C_L(\alpha) \approx C_{L_0} + C_{L_\alpha}\alpha + C_{L_{\delta_e}}\delta_e \tag{8.89}$$

$$C_D(\alpha) \approx C_{D_0} + C_{D_\alpha}\alpha + C_{D_{\delta_e}}\delta_e \tag{8.90}$$

where $C_{D_0} = C_{L_0} = 0$ for symmetrical vehicles and δ_e is an optional control surface (elevator) for depth-changing maneuvers. Since lift is perpendicular to the relative flow and drag is parallel, the longitudinal forces expressed in the BODY axes are

$$\begin{bmatrix} X \\ Z \end{bmatrix} = \begin{bmatrix} \cos(\alpha) & -\sin(\alpha) \\ \sin(\alpha) & \cos(\alpha) \end{bmatrix} \begin{bmatrix} -F_{\text{drag}} \\ -F_{\text{lift}} \end{bmatrix}. \tag{8.91}$$

The transverse force depends on the sideslip angle according to

$$Y = \frac{1}{2}\rho U_r^2 S C_Y(\beta) \tag{8.92}$$

where

$$C_Y(\beta) \approx C_{Y_0} + C_{Y_\beta}\beta + C_{Y_{\delta_r}}\delta_r. \tag{8.93}$$

Here it can be assumed that $C_{Y_0} = 0$ for symmetrical vehicles. The rudder angle δ_r is an optional control surface for turning maneuvers (see Section 9.7.1).

Linear damping

$$\boldsymbol{D} = -\text{diag}\{X_u,\ Y_v,\ Z_w,\ K_p,\ M_q,\ N_r\} \tag{8.94}$$

is important when operating at low speed, while the nonlinear hydrodynamic terms (8.91) and (8.92) given by

$$\boldsymbol{d}(\boldsymbol{\nu}_r) = \frac{1}{2}\rho U_r^2 S \begin{bmatrix} C_D(\alpha)\cos(\alpha) - C_L(\alpha)\sin(\alpha) \\ C_Y(\beta) \\ C_D(\alpha)\sin(\alpha) - C_L(\alpha)\cos(\alpha) \\ 0 \\ 0 \\ 0 \end{bmatrix} \tag{8.95}$$

dominates at higher speeds. For high-speed vehicles such as cylinder-shaped torpedos it is necessary to include hydrodynamic coefficients for the roll, pitch and yaw moments in the expression for $\boldsymbol{d}(\boldsymbol{v}_r)$. The procedure is similar to the method used to design small unmanned aerial vehicles (UAVs), see Beard and McLain (2012) for details.

Finally, $\boldsymbol{r}^b_{bb} = [0,\ 0,\ 0]^\top$, $\boldsymbol{r}^b_{bg} = [x_g,\ y_g,\ z_g]^\top$ and $W = B$ implies that (8.12) reduces to

$$\boldsymbol{g}(\boldsymbol{\eta}) = \begin{bmatrix} 0 \\ 0 \\ 0 \\ z_g W \cos(\theta)\sin(\phi) - y_g W \cos(\theta)\cos(\phi) \\ z_g W \sin(\theta) + x_g W \cos(\theta)\cos(\phi) \\ x_g W \cos(\theta)\sin(\phi) - y_g W \sin(\theta) \end{bmatrix} \tag{8.96}$$

8.5 Spherical-Shaped Vehicles

Spherical-shaped underwater vehicles are easy to model due to symmetry. An example is the omni-directional intelligent navigator (ODIN) shown in Figure 8.5, which is an AUV that has been developed at the University of Hawaii (Choi *et al.* 2003). Consider the 6-DOF equations of motion

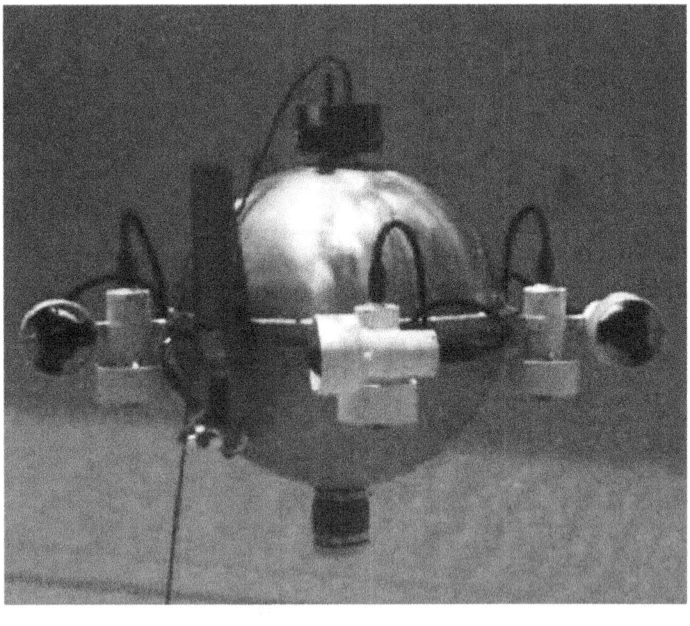

Figure 8.5 The ODIN underwater vehicle. Source: Choi, H. T., A. Hanai, S. K. Choi and J. Yuh (2003). Development of an Underwater Robot, ODIN-III. Proc. of the 2003 IEEURSJ International Conference on Intelligent Robots and Systems Las Vegas, Nevada. October 2003.

$$\dot{\boldsymbol{\eta}} = \boldsymbol{J}_k(\boldsymbol{\eta})(\boldsymbol{\nu}_r + \boldsymbol{\nu}_c) \tag{8.97}$$

$$\boldsymbol{M}\dot{\boldsymbol{\nu}}_r + \boldsymbol{C}(\boldsymbol{\nu}_r)\boldsymbol{\nu}_r + \boldsymbol{D}\boldsymbol{\nu}_r + \boldsymbol{g}(\boldsymbol{\eta}) = \boldsymbol{\tau} \tag{8.98}$$

where $\boldsymbol{\eta}$ and $\boldsymbol{J}_k(\boldsymbol{\eta})$ for $k \in \{\Theta, q\}$ can be parametrized using Euler angles or unit quaternions; see Section 8.1. Assume that the CO is located in the center of the sphere and that the CB coincides with the CO, while the CG is located a distance z_g below the CO. Hence, $\boldsymbol{r}^b_{bb} = [0,\ 0,\ 0]^\top$ and $\boldsymbol{r}^b_{bg} = [0,\ 0,\ z_g]^\top$. Under these assumptions, the rigid-body system inertia matrix (3.49) reduces to

$$\boldsymbol{M}_{\mathrm{RB}} = \begin{bmatrix} m & 0 & 0 & 0 & mz_g & 0 \\ 0 & m & 0 & -mz_g & 0 & 0 \\ 0 & 0 & m & 0 & 0 & 0 \\ 0 & -mz_g & 0 & I_x & 0 & 0 \\ mz_g & 0 & 0 & 0 & I_y & 0 \\ 0 & 0 & 0 & 0 & 0 & I_z \end{bmatrix}. \tag{8.99}$$

The moments of inertia for a sphere with radius R are

$$I_x = I_y = I_z = \frac{2}{5}mR^2. \tag{8.100}$$

Furthermore, the added mass system inertia matrix for a spherical-shaped vehicle is

$$\boldsymbol{M}_A = -\mathrm{diag}\{X_{\dot{u}},\ Y_{\dot{v}},\ Z_{\dot{w}},\ 0,\ 0,\ 0\} \tag{8.101}$$

where the hydrodynamic derivatives are

$$X_{\dot{u}} = Y_{\dot{v}} = Z_{\dot{w}} = -\rho\frac{2}{3}\pi R^2. \tag{8.102}$$

Consequently,

$$\boldsymbol{M} = \boldsymbol{M}_{\mathrm{RB}} + \boldsymbol{M}_A = \begin{bmatrix} m - X_{\dot{u}} & 0 & 0 & 0 & mz_g & 0 \\ 0 & m - Y_{\dot{v}} & 0 & -mz_g & 0 & 0 \\ 0 & 0 & m - Z_{\dot{w}} & 0 & 0 & 0 \\ 0 & -mz_g & 0 & I_x & 0 & 0 \\ mz_g & 0 & 0 & 0 & I_y & 0 \\ 0 & 0 & 0 & 0 & 0 & I_z \end{bmatrix}. \tag{8.103}$$

The Coriolis–centripetal matrix, $\boldsymbol{C}(\boldsymbol{\nu}_r) = \boldsymbol{C}_{\mathrm{RB}}(\boldsymbol{\nu}_r) + \boldsymbol{C}_A(\boldsymbol{\nu}_r)$, consists of two parts where the expression for $\boldsymbol{C}_{\mathrm{RB}}(\boldsymbol{\nu}_r)$ can be derived from (3.63) while $\boldsymbol{C}_{\mathrm{A}}(\boldsymbol{\nu}_r)$ is given by Example 6.2. This gives

$$\boldsymbol{C}_{\mathrm{RB}}(\boldsymbol{\nu}_r) = \begin{bmatrix} 0 & -mr & mq & mz_g r & 0 & 0 \\ mr & 0 & -mp & 0 & mz_g r & 0 \\ -mq & mp & 0 & -mz_g p & -mz_g q & 0 \\ -mz_g r & 0 & mz_g p & 0 & I_z r & -I_y q \\ 0 & -mz_g r & mz_g q & -I_z r & 0 & I_x p \\ 0 & 0 & 0 & I_y q & -I_x p & 0 \end{bmatrix} \tag{8.104}$$

$$\boldsymbol{C}_{\mathrm{A}}(\boldsymbol{\nu}_r) = \rho \frac{2}{3} \pi R^2 \begin{bmatrix} 0 & 0 & 0 & 0 & w_r & -v_r \\ 0 & 0 & 0 & -w_r & 0 & u_r \\ 0 & 0 & 0 & v_r & -u_r & 0 \\ 0 & w_r & -v_r & 0 & 0 & 0 \\ -w_r & 0 & u_r & 0 & 0 & 0 \\ v_r & -u_r & 0 & 0 & 0 & 0 \end{bmatrix}. \tag{8.105}$$

Finally, linear hydrodynamic damping is chosen as

$$\boldsymbol{D} = -\mathrm{diag}\{X_u, \ Y_v, \ Z_w, \ K_{\mathrm{p}}, \ M_{\mathrm{q}}, \ N_r\} \tag{8.106}$$

while for a neutrally buoyant vehicle ($W = B$), Equation (8.12) reduces to

$$\boldsymbol{g}(\boldsymbol{\eta}) = \mathrm{diag}\{0, \ 0, \ 0, \ z_g W \cos(\theta) \sin(\phi), \ z_g W \sin(\theta), \ 0\}. \tag{8.107}$$

9

Control Forces and Moments

This chapter discusses mathematical models for control forces and moments. A surface craft is usually controlled in surge, sway and yaw while cargo ships and passenger ferries also have systems for roll damping. Submerged vehicles, however, are controlled in surge, sway, heave and yaw. Some vehicles also control roll and pitch during hovering and intervention operations. Alternatively, roll and pitch can be left uncontrolled if the vehicle is metacentrically stable.

The forthcoming sections discus the following systems:

- Propellers as thrust devices
- Ship propulsion systems
- USV and underwater vehicle propulsion systems
- Thrusters
- Rudder in propeller slipstream
- Fin stabilizators
- Control surfaces
- Control moment gyroscopes
- Moving mass actuators.

9.1 Propellers as Thrust Devices

The two main types of propellers available for ordinary merchant vessels are fixed-blade or fixed-pitch (FP) propellers and controllable-pitch (CP) propellers. These two types are widely used as prime mover thrust devices and main devices for azimuth and tunnel thrusters.

9.1.1 Fixed-pitch Propeller

A first-order approximation of the propeller thrust T and torque Q can be found from lift force calculations using a quasi-static approach (Lewis 1967, Carlton 1994). Marine craft usually operate with variable forward speed. Therefore the performance of the

Handbook of Marine Craft Hydrodynamics and Motion Control, Second Edition. Thor I. Fossen.

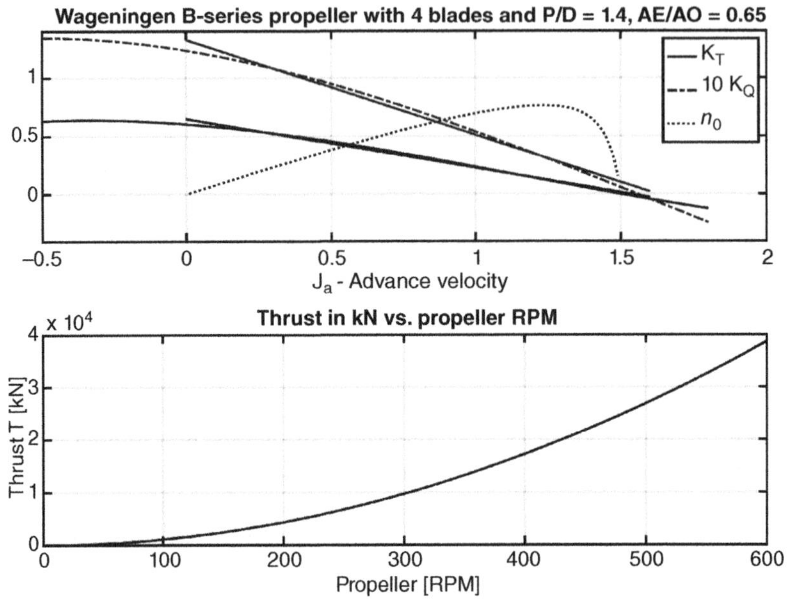

Figure 9.1 Upper plot: thrust K_T, torque K_Q and efficiency η_0 curves (including bilinear approximations) for a propeller. Lower plot: propeller thrust T as a function of propeller speed n (see the MSS toolbox script `ExWageningen.m`).

propeller will be a function of the speed of the water in the wake of the hull (*advance speed* u_a), propeller revolutions n per second (rps) and propeller diameter D (m s^{-1}). The non-dimensional open-water characteristics in Figure 9.1 are defined in terms of the *open-water advance coefficient*

$$J_a = \frac{u_a}{nD} \tag{9.1}$$

The range of J_a values relevant to normal operation is quite narrow. It is only during heavy accelerations and decelerations that the propeller gets exposed to larger parts of the diagram.

The non-dimensional propeller thrust and propeller torque coefficients K_T and K_Q and the thruster open-water efficiency η_0, that is the efficiency in undisturbed water, are given by

$$K_T = \frac{T}{\rho D^4 |n| n}, \qquad K_Q = \frac{Q}{\rho D^5 |n| n}, \qquad \eta_0 = \frac{J_a}{2\pi} \cdot \frac{K_T}{K_Q} \tag{9.2}$$

where ρ (kgm^{-3}) is the water density and T (N) and Q (Nm) are the propeller thrust and torque, respectively. The difference between the ship speed and the average flow velocity over the propeller disc is called the wake. For a ship in transit, it is common to define the relative speed reduction by introducing the advance speed at the propeller, that is the speed of the water going into the propeller, as

$$u_a = (1 - w)\, u \tag{9.3}$$

where w is the wake fraction number (typically 0.1–0.4) and u (m s^{-1}) is the forward speed of the ship. In practice, the wake fraction number can be determined directly from the open-water test results.

Another effect to be considered is the so-called *thrust deduction*. An increase in the flow velocity in the boundary layer behind the ship as a result of the propeller will disturb the pressure balance between the bow and stern. This phenomenon causes extra resistance on the hull which can be described by the thrust deduction number t (typically 0.05–0.2) by modifying the propeller thrust T to $(1 - t)\ T$. The thrust deduction number will strongly depend on the shape of the stern. Hence, the influence of the hull will be described by the hull efficiency

$$\eta_H = \frac{1-t}{1-w}. \tag{9.4}$$

In practice, the ratio between the propeller thrust and torque in open water and behind the stern will differ. This effect can be described by the ratio

$$\eta_B = \frac{J_a}{2\pi}\frac{K_T}{K_{QB}} = \frac{K_Q}{K_{QB}} \tag{9.5}$$

where K_{QB} is the torque coefficient measured for a propeller behind the stern. Let the relative rotative efficiency η_R be defined as the ratio $\eta_R = \eta_B/\eta_0$. Hence the total propeller thrust efficiency can be defined as the product

$$\eta_{tot} = \eta_0 \cdot \eta_M \cdot \eta_H \cdot \eta_B \tag{9.6}$$

where η_M is the mechanical efficiency (typically 0.8–0.9). The open-water test is usually performed by using a towing carriage or a cavitation tunnel. Then force and torque sensors can be applied to measure the propeller force T and torque Q, respectively. Since the speed u_a of the towing carriage or the water stream in the cavitation tunnel also can be measured, K_T, K_Q and η_0 can be calculated from (9.2). This is usually done by applying a nominal (design) value for n.

Bilinear Thrust Model

The propeller thrust and torque follow from (9.2),

$$T = \rho D^4 K_T(J_a)\ |n|n \tag{9.7}$$

$$Q = \rho D^5 K_Q(J_a)\ |n|n. \tag{9.8}$$

Typical curves for K_T and $10 \cdot K_Q$ versus J_a are shown in Figure 9.1 together with the linear approximations

$$T = T_{|n|n}\ |n|n - T_{|n|u_a}\ |n|u_a \tag{9.9}$$

$$Q = Q_{|n|n}\ |n|n - Q_{|n|u_a}\ |n|u_a \tag{9.10}$$

and

$$K_T(J_a) \approx \alpha_1 - \alpha_2\ J_a \tag{9.11}$$

$$K_Q(J_a) \approx \beta_1 - \beta_2\ J_a. \tag{9.12}$$

The four constants α_1, α_2, β_1 and β_2 relate to the coefficients by

$$\begin{array}{ll} T_{|n|n} = \rho D^4 \alpha_1, & Q_{|n|n} = \rho D^5 \beta_1 \\ T_{|n|u_a} = \rho D^3 \alpha_2, & Q_{|n|u_a} = \rho D^4 \beta_2 \end{array}. \tag{9.13}$$

9.1.2 *Controllable-pitch Propeller*

Controllable-pitch (CP) propellers are screw-blade propellers where the blades can be turned under the control of a hydraulic-servo system. CP propellers are used where maneuvering properties need to be improved, where a ship has equipment that requires constant shaft speed, or with most twin-screw ships. Equipment that requires constant shaft speed includes axis generators coupled directly to the shaft via a gear, that is the generator runs with a multiple of the shaft's angular speed, and certain types of trawl drives used in the fishing industry.

For the FP propeller, developed thrust and propeller shaft torque were determined by the bilinear relation with propeller turn rate n and the water velocity u_a at the propeller disc. This is also the case for a CP propeller. Consider

$$T = T_{|n|n}(\theta)\ |n|n - T_{|n|u_a}(\theta)\ |n|u_a \tag{9.14}$$

$$Q = Q_{|n|n}(\theta)\ |n|n - Q_{|n|u_a}(\theta)\ |n|u_a. \tag{9.15}$$

The coefficients $T_{|n|n}(\theta)$, $T_{|n|u_a}(\theta)$, $Q_{|n|n}(\theta)$ and $Q_{|n|u_a}(\theta)$ are complex functions of the pitch angle θ. This is apparent from Figures 9.2 and 9.3 showing K_T and K_Q curves for a CP propeller with various values of relative pitch, between full ahead (100 %) and full

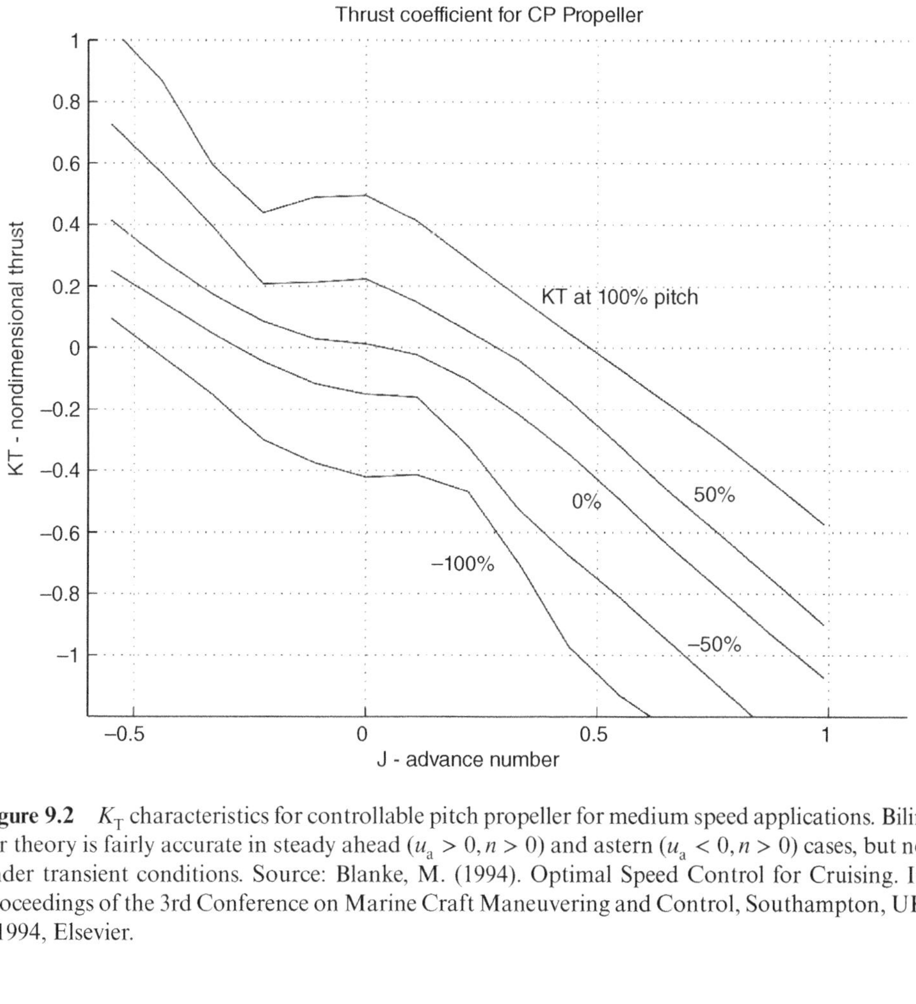

Figure 9.2 K_T characteristics for controllable pitch propeller for medium speed applications. Bilinear theory is fairly accurate in steady ahead ($u_a > 0, n > 0$) and astern ($u_a < 0, n > 0$) cases, but not under transient conditions. Source: Blanke, M. (1994). Optimal Speed Control for Cruising. In: Proceedings of the 3rd Conference on Marine Craft Maneuvering and Control, Southampton, UK. ©1994, Elsevier.

astern (−100 %). On closer inspection, the curves are not too difficult to approximate, and in a simplified analysis we can use one of the following approximations

$$K_T(J_a) \approx (\alpha_1 - \alpha_2 \, J_a) \, \theta \tag{9.16}$$

$$K_T(J_a) \approx (\alpha_1 - \alpha_2 \, J_a) \, |\theta|\theta. \tag{9.17}$$

For the K_T curves in Figure 9.2, application of (9.16) implies that $T_{|n|n}(\theta) = T_{\theta|n|n} \, \theta$ and $T_{|n|u_a}(\theta) = T_{|n|u_a} \, \theta$. Consequently, we obtain the following thrust for the CP propeller

$$T \approx T_{\theta|n|n} \, \theta \, |n|n - T_{\theta|n|u_a} \, \theta|n|u_a \tag{9.18}$$

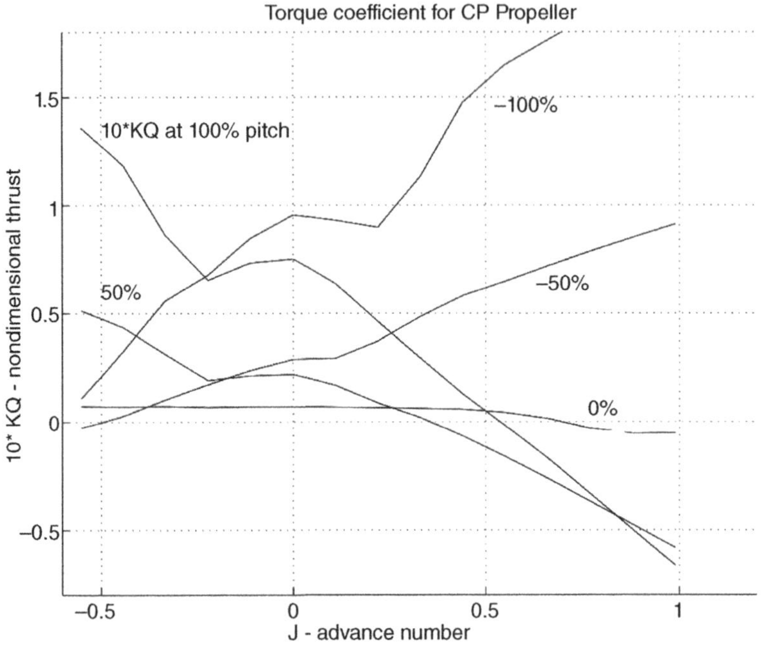

Figure 9.3 K_Q characteristic for controllable pitch propeller for medium speed application. Pitch values from −100% to 100% are shown for positive n. Bilinear theory is seen to be fairly accurate in steady ahead and astern cases but not otherwise. Source: Blanke, M. (1994). Optimal Speed Control for Cruising. In: Proceedings of the 3rd Conference on Marine Craft Maneuvering and Control, Southampton, UK. ©1994, Elsevier.

Bollard Pull

For low-speed applications, the thrust coefficients can be approximated by using *Bollard pull* data, which is the conventional measure of the pulling (or towing) power of a watercraft. It is defined as the force exerted by a vessel under full power, on a shore-mounted bollard through a tow-line. For a CP propeller the propeller rpm can be held constant, while pitch is varied. Consequently, $J_a = 0$ and $n = n_0$ implies that (9.17) takes the following form

$$K_T(0) = \alpha_1 \; |\theta|\theta. \tag{9.19}$$

This gives the thrust

$$T(n_0) = K(n_0) \; |\theta|\theta \tag{9.20}$$

where

$$K(n_0) = \rho D^4 \alpha_1 \; |n_0|n_0. \tag{9.21}$$

The quadratic behavior in θ for dynamic positioning (DP) is confirmed by the experimental data in Fossen *et al.* (1996), where an offshore supply vessel equipped with main propeller and tunnel thrusters are considered. The measured thrust in Bollard pull is

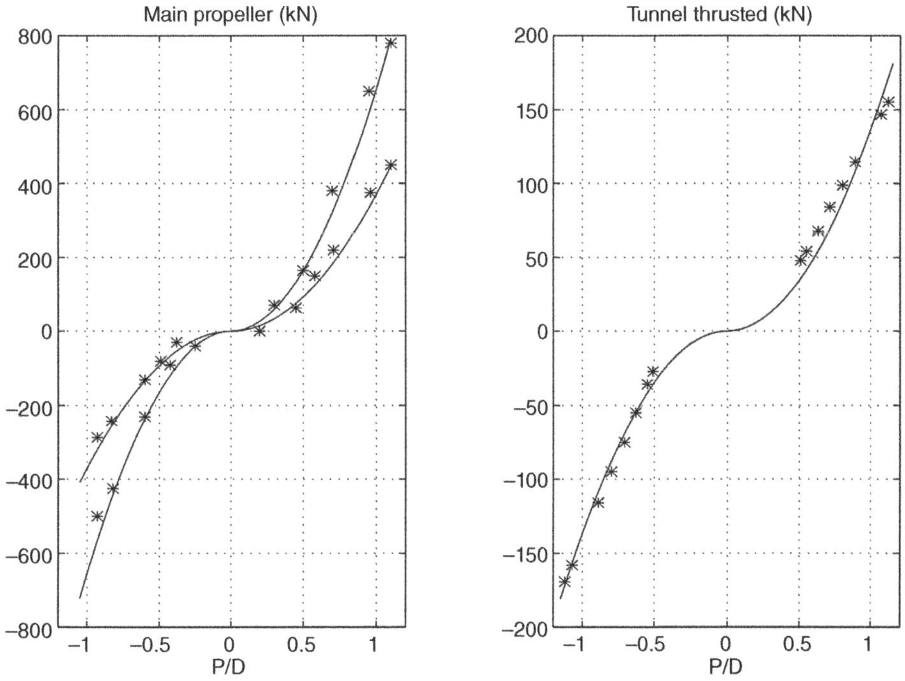

Figure 9.4 Thrust $T(n_0) = K(n_0)|\theta|\theta$ versus pitch θ for a main propeller (left-hand plot) and a tunnel thruster (right-hand plot) operating a constant propeller revolution n_0. The asterisks are experimental measured values and the solid lines are least-square fits to a quadratic model.

shown as asterisks in Figure 9.4 while the solid lines are least-square fits to (9.20). The main propeller operated at $n_0 = 122$ rpm and $n_0 = 160$ rpm, while the tunnel thruster ran at $n_0 = 236$ rpm resulting in

$$\begin{aligned} &\text{Main propeller:} \quad && T(122) = 370\ |\theta|\theta \qquad T(160) = 655\ |\theta|\theta \\ &\text{Tunnel thruster:} \quad && T(236) = 137\ |\theta|\theta. \end{aligned}$$

For DP ships using fixed-speed CP propellers it is common to operate at one or two FP propellers such that only θ is used for active control by the DP system. For ships in transit a constant demand for thrust and power suggests that a fixed-speed CP propeller should be used while low-speed applications such as DP operations require little thrust in good weather, suggesting that a variable-speed FP propeller might be advantageous (see Figure 9.5). Note that the fixed-speed CP propeller also requires power at zero thrust.

MATLAB

The Wageningen B-series of propellers were designed and tested at the Netherlands ship model basin in Wageningen (Barnitsas *et al.* 1981). The open-water characteristics of 120 propellers were tested and fitted to polynomials. The data

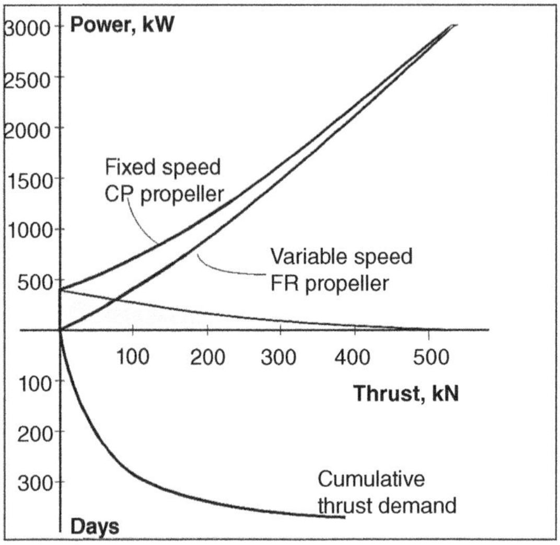

Figure 9.5 Power consumption of fixed-speed CP and variable-speed FP propellers.

set is available in the MSS toolbox as `WageningData.mat`. The propellers can be configured in Matlab using:

```
% see ExWageningen.m
rho = 1025;       % Density of water (kg/m^3)
D = 5;            % Propeller diameter (m)
PD = 1.4;         % pitch/diameter ratio
AEAO = 0.65;      % blade area ratio
z = 4;            % number of propeller blades

% Comput KT and KQ for advance velocites Ja
Ja = -0.8:0.01:1.8;
for i = 1:length(Ja)
    [KT(i), KQ(i)] = wageningen(Ja(i),PD,AEAO,z);
end

% Compute KT and KQ for Ja = 0 (Bollard pull)
[KT_0, KQ_0] = wageningen(0,PD,AEAO,z);

% Compute thrust [N]
n = 0:0.1:10;                           % propeller [RPS]
T = rho * D^4 * KT_0 * n .* abs(n);     % thrust [N]

% Fit KT and KQ data to straight lines
Jdata = 0:0.01:1.6;
for i = 1:length(Jdata)
    [KTdata(i), KQdata(i)] = wageningen(Jdata(i),PD,AEAO,z);
end
```

```
alpha = polyfit(Jdata,KTdata,1)    % KT = alpha(1)*Ja + alpha(2)
beta  = polyfit(Jdata,KQdata,1)    % KQ = beta(1) *Ja + beta(2)
```

Figure 9.1 shows the computed data graphically.

9.2 Ship Propulsion Systems

Marine propulsion is the mechanism or system used to generate thrust to move a ship or boat across water. There exists many different types of marine propulsion systems such as:

- Diesel engines
- Diesel-electric drives
- All-electric propulsion
- Podded propulsion
- Gas turbines
- Steam turbines
- Fuel cells
- Gas fuel
- Biodiesel fuel
- Water jets.

Some of these systems are discussed in the forthcoming sections.

9.2.1 Podded Propulsion Units

A podded propulsion unit consists of an FP propeller mounted on a steerable gondola (pod), which also contains the electric motor driving the propeller (see Figure 9.6). In the traditional azimuth thrusters, the propeller is driven by an electric motor or a diesel engine inside the ship's hull. The propeller is coupled to the prime mover with shafts and gears that allow rotating the propeller about a vertical axis. In a podded unit, an electric motor is mounted inside the propulsion unit and the propeller is connected directly to the motor shaft. The pod's propeller usually faces forward because in this pulling (or tractor) configuration the propeller is more efficient due to operation in undisturbed flow. Because it can rotate about its mount axis, the pod can apply its thrust in any direction similar to an azimuth thruster.

The thrust from a podded propulsion unit is approximated by (see Section 9.1)

$$T = \rho D^4 K_{\mathrm{T}}(J_{\mathrm{a}})|n|n \approx T_{|n|n}\ |n|n - T_{|n|u_{\mathrm{a}}}\ (1-w)\ |n|u_{\mathrm{r}} \tag{9.22}$$

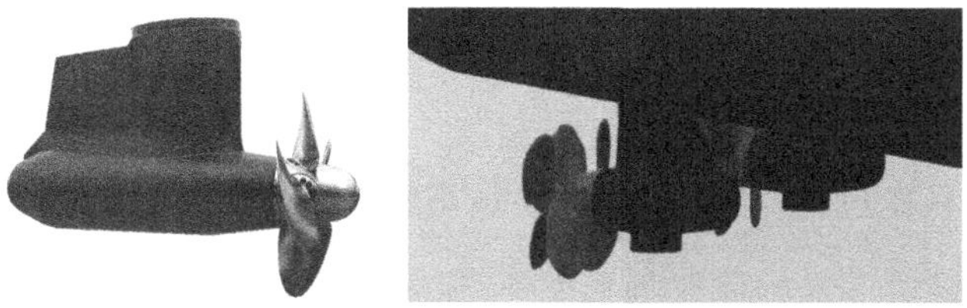

Figure 9.6 Left: podded propulsion unit. Right: two-podded propulsion unit mounted aft of a ship.

where the second term has been modified by using $u_a = (1 - w)u_r$ for a ship in transit. Maneuvering theory implies that (see Section 6.5.1)

$$\boldsymbol{M}\dot{\boldsymbol{\nu}}_r + \boldsymbol{C}(\boldsymbol{\nu}_r)\boldsymbol{\nu}_r + \boldsymbol{D}(\boldsymbol{\nu}_r)\boldsymbol{\nu}_r = \boldsymbol{\tau} + \boldsymbol{\tau}_{\text{wind}} + \boldsymbol{\tau}_{\text{wave}} \tag{9.23}$$

where $\boldsymbol{\tau} = [X,\ Y,\ N]^\top$ is a vector of control inputs. Let α denote the azimuth angle. For the podded propulsion unit (9.22) this gives the following control forces and moment in surge, sway and yaw

$$\boldsymbol{\tau} = (1 - t)\ T_{|n|n}\ |n|n \begin{bmatrix} \cos(\alpha) \\ \sin(\alpha) \\ l_x\ \sin(\alpha) - l_y\ \cos(\alpha) \end{bmatrix} - \boldsymbol{d}_{\text{loss}}(n, \alpha)u_r \tag{9.24}$$

where $(l_x,\ l_y)$ is the location of the pod with respect to the CO and

$$\boldsymbol{d}_{\text{loss}}(n, \alpha) := (1 - t)(1 - w)\ T_{|n|u_a}\ |n| \begin{bmatrix} \cos(\alpha) \\ \sin(\alpha) \\ l_x\ \sin(\alpha) - l_y\ \cos(\alpha) \end{bmatrix}. \tag{9.25}$$

Equation (9.25) is a dissipative propeller loss term, which vanishes for $n = 0$. It is possible to express (9.24) as a function of $\boldsymbol{u} = [u_1,\ u_2]^\top$ where

$$u_1 = (1 - t)\ T_{|n|n}\ |n|n\ \cos(\alpha) \tag{9.26}$$

$$u_2 = (1 - t)\ T_{|n|n}\ |n|n\ \sin(\alpha). \tag{9.27}$$

This gives

$$\boldsymbol{\tau} = \boldsymbol{B}\boldsymbol{u} - \boldsymbol{D}_{\text{loss}}(n, \alpha)\boldsymbol{\nu}_r \tag{9.28}$$

where

$$\boldsymbol{B} = \begin{bmatrix} 1 & 0 \\ 0 & 1 \\ -l_y & l_x \end{bmatrix}, \qquad \boldsymbol{D}_{\text{loss}}(n, \alpha) = \left[\boldsymbol{d}_{\text{loss}}(n, \alpha)\ \ \boldsymbol{0}_{3\times 2} \right]. \tag{9.29}$$

Consequently, the ship model (9.23) takes the following form

$$\boldsymbol{M}\dot{\boldsymbol{\nu}}_r + \boldsymbol{C}(\boldsymbol{\nu}_r)\boldsymbol{\nu}_r + (\boldsymbol{D}(\boldsymbol{\nu}_r) + \boldsymbol{D}_{\text{loss}}(n, \alpha))\boldsymbol{\nu}_r = \boldsymbol{B}\boldsymbol{u} + \boldsymbol{\tau}_{\text{wind}} + \boldsymbol{\tau}_{\text{wave}} \tag{9.30}$$

where $\boldsymbol{D}_{\text{loss}}(n, \alpha)$ is an additional dissipative term due to the podded propulsion unit. It is straightforward to design a control law for (9.30) since the control force $\boldsymbol{Bu}$ is linear in the input $\boldsymbol{u}$. The control inputs $\boldsymbol{u}$ are mapped to physical thruster signals by

$$|n|n = \frac{1}{(1-t)\,T_{|n|n}}\sqrt{u_1^2 + u_2^2} \tag{9.31}$$

$$\alpha = \text{atan2}(u_2, u_1) \tag{9.32}$$

Hence, the propeller rps is computed as $n = \text{sgn}(\tau)\sqrt{|\tau|}$ where $\tau = |n|n$.

9.2.2 *Prime Mover System*

The section describes the rotational shaft dynamics, engine dynamics and governor of a modern ship-speed propulsion system. The dynamics of the prime mover and its control system is tightly coupled to the speed dynamics of the ship via the propeller thrust T and torque Q according to (see Figure 9.7)

$$I_m \dot{n} = Q_m - Q - Q_f \tag{9.33}$$

$$(m - X_{\dot{u}})\,\dot{u}_r = X_{|u|u}\,|u_r|u_r + (1-t)T + d_u \tag{9.34}$$

where

- u_r relative surge velocity (m/s)
- n shaft speed (rad/s)
- I_m inertia of the rotating parts including the propeller and added inertia of the water (kgm^2)
- Q_m produced torque developed by the main engine (Nm)
- Q_f friction torque (Nm)
- d_u unmodeled dynamics and external disturbances (N).

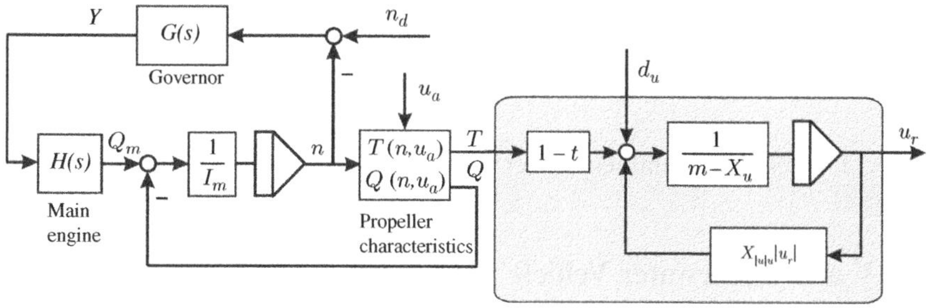

Figure 9.7 Simplified diagram showing the speed-propulsion system.

The transfer function $H(s)$ from the control input $Y(s)$ to the produced torque developed by the ship engine $Q_m(s)$ (see Figure 9.7) can be approximated by a first-order system with time delay

$$H(s) = \frac{Q_{\rm m}}{Y}(s) \approx \frac{K_{\rm m}}{T_{\rm m}+1}{\rm e}^{-\tau s} \tag{9.35}$$

where τ represents the time delay, $K_{\rm m}$ is the gain constant and $T_{\rm m}$ is the time constant. The approximative model can be used to simulate the dynamics of a large number of propulsion systems such as standard diesel engines, diesel-electric ship propulsion systems and all-electric ships. However, the model does not include the dynamics of the power system so it should only be used to test and verify the ship-speed control system.

The control input $Y(s)$ depends on the engine system. For a diesel engine $Y(s)$ will be the fuel pump command, while for an electric propulsion system $Y(s)$ will represent the power needed to generate the motor torque $Q_{\rm m}$. The control objective of the governor $G(s)$ in Figure 9.7 is to ensure that

$$\frac{n}{n_{\rm d}} \approx 1 \tag{9.36}$$

up to the frequency $1/\tau$, which is the bandwidth limit imposed by the engine time delay.

Ship Dynamics

The ship model (9.34) can be used to relate the surge acceleration $\dot{u}$ to the propeller speed n. To see this, the expressions for relative surge velocity and acceleration (see Section 10.3)

$$u_{\rm r} = u - u_{\rm c} = u - V_{\rm c}\cos(\beta_{V_{\rm c}}) \tag{9.37}$$

$$\dot{u}_{\rm r} = \dot{u} - rv_{\rm c} = \dot{u} - rV_{\rm c}\sin(\beta_{V_{\rm c}}) \tag{9.38}$$

must be substituted into (9.34) to obtain the state-space model

$$\begin{aligned}\dot{u} = rV_{\rm c}\sin(\beta_{V_{\rm c}}) + \frac{1}{m - X_{\dot{u}}}[&X_{|u|u}\,|u - V_{\rm c}\cos(\beta_{V_{\rm c}})|\,(u - V_{\rm c}\cos(\beta_{V_{\rm c}})) \\ &+(1-t)(T_{|n|n}(\theta)\,|n|n + T_{|n|V_a}(\theta)\,|n|u_{\rm a}) + d_{\rm u}]\end{aligned} \tag{9.39}$$

where $u_{\rm a} = (1 - w)u_{\rm r}$ is the ambient water velocity.

9.3 USV and Underwater Vehicle Propulsion Systems

Small USVs and underwater vehicles use propellers for propulsion, maneuvering, attitude control and dynamic positioning (see Figure 9.8). A propeller is a mechanical device with

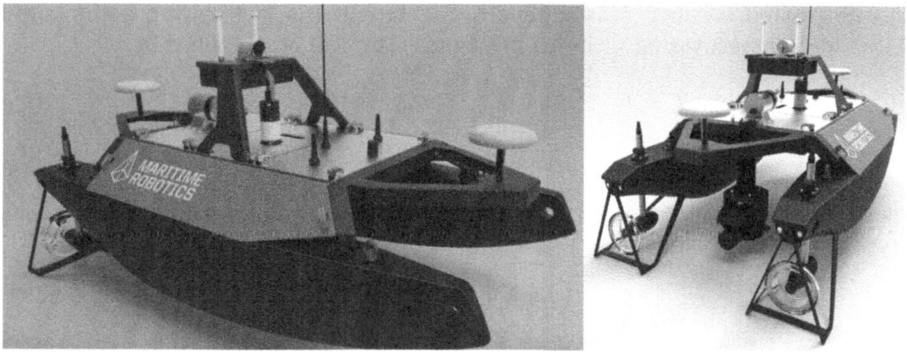

Figure 9.8 The two aft propellers used to control the Maritime Robotics Otter USV. Source: Reproduced with kind permission of www.maritimerobotics.com.

shaped blades that turn on a shaft. It can be used to propel aircraft and ocean vehicles. For both USVs and submerged vehicles it is common to use ducted propellers to increase the efficiency at lower speeds, which allows the designer to consider smaller diameter of the blades. A duct will also protect the propeller from debris. A ducted propeller is known as a *Kort nozzle*. Ludwig Kort was a German fluid dynamicist who in the 1930s tried to reduce canal erosion and by this he discovered that directing the wake of a propeller through a short, stationary nozzle increased the thrust.

The propeller is usually driven by a direct current (DC) motor, which converts direct current electrical energy into mechanical energy. This section describes mathematical models for the propeller shaft and the DC motor dynamics.

9.3.1 Propeller Shaft Speed Models

Yoerger *et al.* (1991) proposed a *one-state model* with quadratic damping for propeller shaft speed. The linearized version of this model is

$$J_{\mathrm{m}}\dot{n} + K_{\mathrm{n}}\, n = Q_{\mathrm{m}} - Q \tag{9.40}$$

where n is the shaft speed and Q_{m} is the control input (shaft torque). For marine craft in transit, only positive values $u_{\mathrm{a}} > 0$ are considered (see Figure 9.9). A large number of marine craft, however, operate in regimes where the ambient water velocity can be

Figure 9.9 Ambient water velocity u_{a} (upstream), axial flow velocity in propeller disc u_{p} and velocity in the wake u_{w} (downstream) for an underwater vehicle moving at velocity u.

both positive and negative. It is common to assume that $u_a = 0$ during stationkeeping and low-speed maneuvering such that (9.9) and (9.10) can be simplified by

$$T = T_{|n|n}\ |n|n - T_{|n|u_a}\ |n|u_a \overset{u_a=0}{=} T_{|n|n}\ |n|n \tag{9.41}$$

$$Q = Q_{|n|n}\ |n|n - Q_{|n|u_a}\ |n|u_a \overset{u_a=0}{=} Q_{|n|n}\ |n|n. \tag{9.42}$$

Healey *et al.* (1995) have modified the model (9.40) to describe overshoots in thrust, which are typical in experimental data. The result is a *two-state model*

$$J_m\dot{n} + K_n n = Q_m - Q \tag{9.43}$$

$$m_f\dot{u}_p + d_f|u_p - u|(u_p - u) = T \tag{9.44}$$

where u is the forward speed of the vehicle and u_p is axial-flow velocity in the propeller disc. This was done by modeling the control volume of water around the propeller as a mass-damper system with parameters m_f and d_f. The mass-damper of the control volume interacts with the vehicle speed dynamics, which is also represented by a mass-damper system. Experimental verifications of the one- and two-state models are presented by Whitcomb and Yoerger (1999).

The signal u_p can be measured by using a laser-Doppler velocimeter (LDV) system, a particle image velocimeter (PIV) system or an acoustic Doppler velocimeter system for instance. These are expensive devices and the alternative is to estimate u_p using a Kalman filter or a nonlinear observer; see Fossen and Blanke (2000) and references therein.

A more general model is the *three-state* propeller shaft speed model (Blanke *et al.* 2000b)

$$J_m\dot{n} + K_n n = Q_m - Q \tag{9.45}$$

$$m_f\dot{u}_p + d_{f_0}u_p + d_f|u_p|(u_p - u_a) = T \tag{9.46}$$

$$(m - X_{\dot{u}})\dot{u}_r - X_u u_r - X_{|u|u}|u_r|u_r = (1 - t)T \tag{9.47}$$

where damping in surge is modeled as the sum of linear laminar skin friction, $-X_u u_r$ and nonlinear quadratic drag, $-X_{|u|u}|u_r|u_r$. Similar, linear damping, $d_{f_0}u_p$, is included in the axial-flow model since quadratic damping $d_f|u_p|u_p$ alone would give an unrealistic response at low speeds. Linear skin friction gives exponential convergence to zero at low speeds. The ambient water velocity u_a in (9.46) can be approximated by the steady-state condition

$$u_a = (1 - w)u_r \tag{9.48}$$

where $0 < w < 1$ is the *wake fraction number* (Lewis 1989).

9.3.2 Motor Armature Current Control

Consider a DC motor

$$L_a\frac{\mathrm{d}}{\mathrm{d}t}i_m = -R_a i_m - K_m n + V_m \tag{9.49}$$

$$Q_m = K_m i_m \tag{9.50}$$

where V_m is the armature voltage, i_m is the armature current and Q_m is the motor torque. In addition, L_a is the armature inductance, R_a is the armature resistance and K_m is the motor torque constant.

Since the electrical time constant $T_a = L_a/R_a$ is small compared to the mechanical time constant, time-scale separation suggests

$$\frac{L_a}{R_a}\frac{d}{dt}i_m \approx 0. \tag{9.51}$$

Hence, the shaft speed dynamics is given by

$$0 = -R_a i_m - K_m n + V_m \tag{9.52}$$

$$J_m \dot{n} = K_m i_m - Q. \tag{9.53}$$

The motor current i_m can be controlled by using a P controller

$$V_m = -K_p(i_m - i_d), \qquad K_p > 0 \tag{9.54}$$

where i_d is the desired motor current. From (9.52) we get

$$(R_a + K_p)i_m = -K_m n + K_p i_d. \tag{9.55}$$

The motor dynamics (9.53) for the current-controlled motor therefore takes the form

$$J_m \dot{n} + \frac{K_m^2}{R_a + K_p} n = \frac{K_m K_p}{R_a + K_p} i_d - Q. \tag{9.56}$$

If a high-gain controller $K_p \gg R_a > 0$ is used, this expression simplifies to

$$J_m \dot{n} = K_m i_d - Q. \tag{9.57}$$

The desired motor torque is chosen as

$$Q_d = K_m i_d \tag{9.58}$$

such that

$$J_m \dot{n} = Q_d - Q. \tag{9.59}$$

Hence, $Q = Q_d$ in steady state.

The current set point is computed by using the thrust to torque ratio

$$Q = D\frac{K_Q(J_a)}{K_T(J_a)}\,T \tag{9.60}$$

which follows directly from (9.7) and (9.8). The vehicle velocity can be controlled such that $u = u_d$ by using thrust T as the control input. Consequently, $T = T_d$ in steady state.

Since $i_d = Q_d / K_m$ we finally arrive at

$$i_d = \frac{D}{K_m} \frac{K_Q(J_a)}{K_T(J_a)} T_d \tag{9.61}$$

where

$$J_a = \frac{u_a}{nD} \approx 0 \tag{9.62}$$

for low-speed maneuvering and stationkeeping applications. The armature-current controller (9.54) and the desired signals are shown in Figure 9.10.

9.3.3 Motor Speed Control

In many cases it is desirable to control the propeller revolutions instead of the armature current. Combining (9.52) and (9.53) gives

$$J_m \dot{n} + \frac{K_m^2}{R_a} n = \frac{K_m}{R_a} V_m - Q. \tag{9.63}$$

This suggests that the motor speed controller can be chosen as

$$V_m = -K_p(n - n_d), \qquad K_p > 0 \tag{9.64}$$

where n_d is the desired propeller revolutions, usually computed by inverting the thrust equation

$$T_d = \rho D^4 K_T(0) |n_d| n_d \tag{9.65}$$

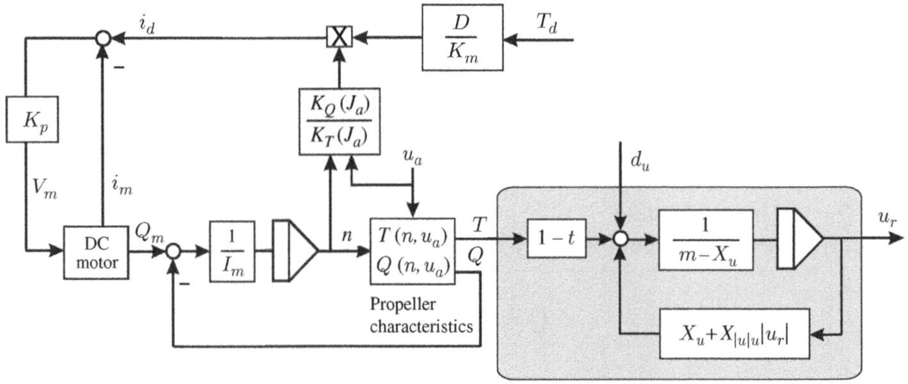

Figure 9.10 Block diagram showing the DC motor controller and propeller.

according to

$$n_d = \text{sgn}(T_d)\sqrt{\frac{|T_d|}{\rho D^4 K_T(0)}} \tag{9.66}$$

where sgn(·) is the signum function.

9.4 Thrusters

A thruster is a device for propelling a marine craft. It is a more complex arrangement than a single propeller. A thruster can consist of several propellers, for instance twin contra-rotating propellers (CRP). In addition, a thruster system can be podded to protect mechanical devices and reduce the resistance (energy consumption) during transit.

There exists different types of thruster systems such as:

- **Tunnel thruster** – propeller installed in a transverse tunnel, producing a transverse force. Both bow and stern thrusters are used to dynamically position a ship or a boat.
- **Externally mounted thrusters** – instead of a tunnel thruster, marine craft may have externally mounted thrusters. They are usually used by craft where it is impossible or undesirable to install a tunnel thruster, due to hull shape or outfitting.
- **Azimuth thruster** – Ship propellers placed in pods that can be rotated on the horizontal plane, making a rudder unnecessary. This is also referred to as *thrust vectoring* since both the magnitude and direction of the force can be controlled.
- **CRP thruster** – an azimuthing thruster equipped with twin contra-rotating propellers.
- **Jet thruster** – a pump arranged to take suction from beneath or close to the keel and to discharge to either side, to develop port or starboard thrust, or in many cases through 360°.

The generalized force in 6 DOFs corresponding to the thrust vector $\boldsymbol{f}_t^b = [F_x, F_y, F_z]^\top$ expressed in $\{b\}$ is

$$\boldsymbol{\tau} = \begin{bmatrix} \boldsymbol{f}_t^b \\ \boldsymbol{r}_{bt}^b \times \boldsymbol{f}_t^b \end{bmatrix} = \begin{bmatrix} F_x \\ F_y \\ F_z \\ l_y F_z - l_z F_y \\ l_z F_x - l_x F_z \\ l_x F_y - l_y F_x \end{bmatrix} \overset{\text{surge, sway and yaw}}{\Longrightarrow} \boldsymbol{\tau} = \begin{bmatrix} F_x \\ F_y \\ l_x F_y - l_y F_x \end{bmatrix} \tag{9.67}$$

where $\boldsymbol{r}_{bt}^b = [l_x, l_y, l_z]^\top$ is a vector of thruster lever arms with respect to the CO.

9.4.1 *Tunnel Thrusters*

Tunnel thrusters are transverse thrusters going through the hull of the craft (see Figure 9.11). The propeller unit is mounted inside a transverse tube and it produces a

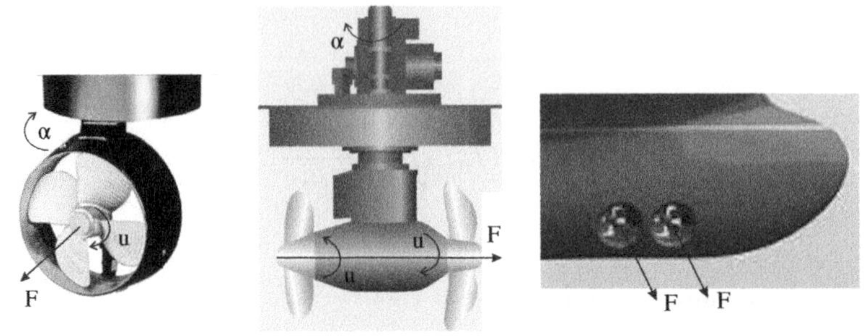

Figure 9.11 Azimuth thruster, twin contra-rotating propeller and tunnel thrusters. The force (thrust) is denoted F and $u = |n|n$ is the squared propeller revolution. For thrust vectoring, α is the direction of the force.

force F_y in the y direction. Tunnel thrusters are only effective at low speed which limits their use to low-speed maneuvering and stationkeeping.

The generalized control force produced by a tunnel thruster is

$$\boldsymbol{\tau} = [0,\ F_y,\ 0, -l_z F_y,\ 0,\ l_x F_y]^\top. \tag{9.68}$$

For tunnel thrusters the advance speed u_a will be small. Hence, $J_a \approx 0$ and the propeller produces a transverse force (see Section 9.1)

$$\text{FP propeller:} \quad F_y = \rho D^4 K_T(0)|n|n \approx T_{|n|n}\ |n|n \tag{9.69}$$

$$\text{CP propeller:} \quad F_y \approx T_{\theta|n|n}\ \theta\ |n|n. \tag{9.70}$$

9.4.2 *Azimuth Thrusters*

Azimuth thrusters can be rotated an angle α about the z axis and produce two force components (F_x, F_y) in the horizontal plane (see Figure 9.11). They are usually mounted under the hull of the craft and the most sophisticated units are retractable. Azimuth thrusters are frequently used in DP systems since they can produce forces in different directions. Hence, this becomes an overactuated control problem that can be optimized with respect to power and possible failure situations. The generalized control force is

$$\boldsymbol{\tau} = [F\cos(\alpha),\ F\sin(\alpha),\ 0, -l_z F\sin(\alpha),\ l_z F\cos(\alpha),\ l_x F\sin(\alpha) - l_y F\cos(\alpha)]^\top. \tag{9.71}$$

Azimuth thrusters used for maneuvering produce a force

$$F = \rho D^4 K_T(J_a)|n|n. \tag{9.72}$$

For DP applications it is common to assume that $J_a = 0$ such that

$$\text{FP propeller:} \quad F = \rho D^4 K_T(0)|n|n \approx T_{|n|n}\ |n|n \tag{9.73}$$

$$\text{CP propeller:} \quad F \approx T_{\theta|n|n}\, \theta\, |n|n. \tag{9.74}$$

For a surface vessel, an azimuth thruster will produce the following control forces and moment in the horizontal plane

$$\boldsymbol{\tau} = \begin{bmatrix} F\cos(\alpha) \\ F\sin(\alpha) \\ l_x\, F\sin(\alpha) - l_y\, F\cos(\alpha) \end{bmatrix}. \tag{9.75}$$

Example 9.1 (Thrust Configuration Matrix for a DP Vessel)
Consider the ship in Figure 9.12, which is equipped with two azimuth thrusters F_1 and F_3, two tunnel thrusters F_2 and F_4, and two main propellers (forces F_5 and F_6). Assume that all thruster systems ($i = 1, 2, ..., 6$) have FP propellers and the thrust is given by (see Section 9.1)

$$F_i = \rho D_i^4 K_{T_i}(0)\ |n_i|n_i := K_i u_i \tag{9.76}$$

where $K_i := \rho D_i^4 K_{T_i}(0)$ and

$$u_i := |n_i|n_i. \tag{9.77}$$

The azimuth forces F_1 and F_3 are decomposed along the x and y axes by defining the following virtual control inputs ($i = 1, 3$)

$$u_{i_x} := u_i \cos(\alpha_i) \tag{9.78}$$

$$u_{i_y} := u_i \sin(\alpha_i) \tag{9.79}$$

such that the forces in the x and y directions become

$$F_{i_x} = K_i u_{i_x} \tag{9.80}$$

$$F_{i_y} = K_i u_{i_y}. \tag{9.81}$$

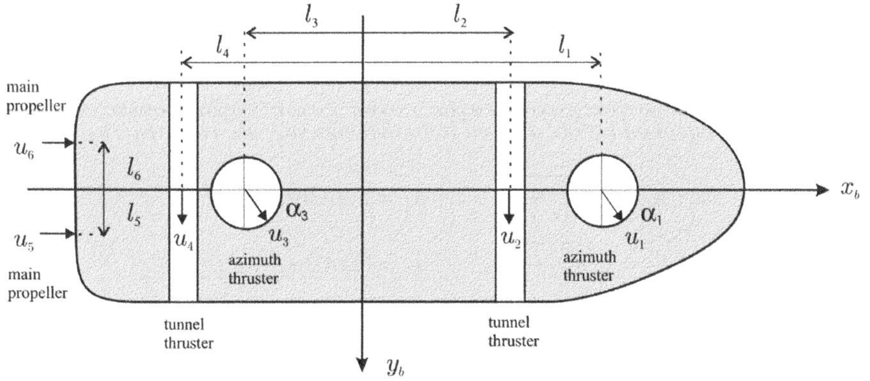

Figure 9.12 Ship equipped with two azimuth thrusters (forces F_1 and F_3), two tunnel thrusters (forces F_2 and F_4) and two main propellers (forces F_5 and F_6).

The generalized force vector for motions in the surge, sway and yaw is expressed in terms of the input matrix $\boldsymbol{B} \in \mathbb{R}^{3\times 8}$ according to

$$\boldsymbol{\tau} = \boldsymbol{B}\boldsymbol{u}_{\mathrm{e}}, \qquad \boldsymbol{B} = \boldsymbol{T}_{\mathrm{e}}\boldsymbol{K}_{\mathrm{e}} \tag{9.82}$$

where $\boldsymbol{u}_{\mathrm{e}} = [u_{1_x}, u_{1_y}, u_2, u_{3_x}, u_{3_y}, u_4, u_5, u_6]^{\top}$ is the extended control input vector, $\boldsymbol{T}_{\mathrm{e}} \in \mathbb{R}^{3\times 8}$ is the extended thrust configuration matrix and $\boldsymbol{K}_{\mathrm{e}} \in \mathbb{R}^{8\times 8}$ is the extended thrust coefficient matrix; see Section 11.2.1. Then it follows that the forces and moment X, Y and N in surge, sway and yaw, respectively for the thruster configuration shown in Figure 9.12 satisfy

$$\boldsymbol{\tau} = \boldsymbol{T}_{\mathrm{e}}\boldsymbol{K}_{\mathrm{e}}\boldsymbol{u}_{\mathrm{e}} \tag{9.83}$$

$$\Updownarrow$$

$$\begin{bmatrix} X \\ Y \\ N \end{bmatrix} = \begin{bmatrix} 1 & 0 & 0 & 1 & 0 & 0 & 1 & 1 \\ 0 & 1 & 1 & 0 & 1 & 1 & 0 & 0 \\ 0 & l_1 & l_2 & 0 & l_3 & l_4 & -l_5 & -l_6 \end{bmatrix}$$

$$\cdot \begin{bmatrix} K_1 & 0 & 0 & 0 & 0 & 0 & 0 & 0 \\ 0 & K_1 & 0 & 0 & 0 & 0 & 0 & 0 \\ 0 & 0 & K_2 & 0 & 0 & 0 & 0 & 0 \\ 0 & 0 & 0 & K_3 & 0 & 0 & 0 & 0 \\ 0 & 0 & 0 & 0 & K_3 & 0 & 0 & 0 \\ 0 & 0 & 0 & 0 & 0 & K_4 & 0 & 0 \\ 0 & 0 & 0 & 0 & 0 & 0 & K_5 & 0 \\ 0 & 0 & 0 & 0 & 0 & 0 & 0 & K_6 \end{bmatrix} \begin{bmatrix} u_{1_x} \\ u_{1_y} \\ u_2 \\ u_{3_x} \\ u_{3_y} \\ u_4 \\ u_5 \\ u_6 \end{bmatrix}. \tag{9.84}$$

If the control input $\boldsymbol{u}_{\mathrm{e}}$ is computed using the pseudoinverse (see Section 11.2)

$$\begin{aligned} \boldsymbol{u}_{\mathrm{e}} &= \boldsymbol{B}^{\dagger}\boldsymbol{\tau} \\ &= \boldsymbol{K}_{\mathrm{e}}^{-1}\boldsymbol{T}_{\mathrm{e}}^{\dagger}\boldsymbol{\tau} \end{aligned} \tag{9.85}$$

the propeller commands (usually rpm*) become ($i = 1, 2, \ldots, 6$)*

$$n_i = \mathrm{sgn}(u_i)\sqrt{|u_i|} \tag{9.86}$$

while the azimuth control inputs are derived from the pairs (u_{1_x}, u_{1_y}) and (u_{3_x}, u_{3_y}) according to

$$u_1 = \sqrt{u_{1_x}^2 + u_{1_y}^2}, \qquad \alpha_1 = \mathrm{atan2}(u_{1y}, u_{1x}) \tag{9.87}$$

$$u_3 = \sqrt{u_{3_x}^2 + u_{3_y}^2}, \qquad \alpha_3 = \mathrm{atan2}(u_{3_y}, u_{3_x}). \tag{9.88}$$

9.5 Rudder in the Propeller Slipstream

Rudders are primary steering devices for merchant ships. Also USVs and underwater vehicles can be equipped with stern rudders as an alternative to use propellers and thrusters

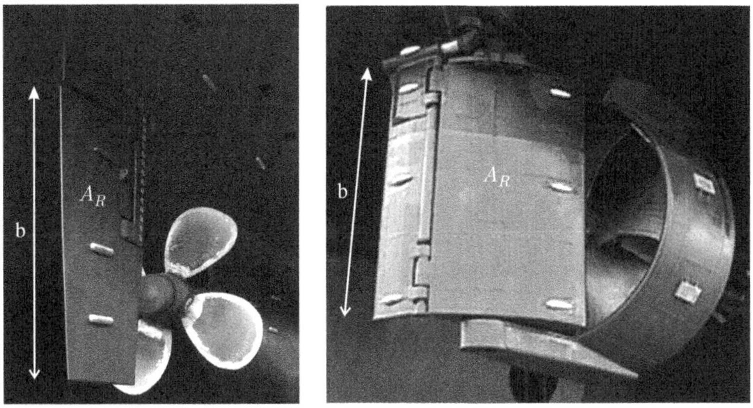

Figure 9.13 Left: Conventional stern rudder for a ship. Right: High-lift flap rudder consisting of two or more sections which move relative to each other as helm is applied. A_R denotes the area of the rudder and b is the rudder height.

for steering. Rudders outside of the propeller slipstream are ineffective at small or zero ship speed (e.g. DP and berthing operations). Insufficient rudder effectiveness at slow ship speed can be temporarily increased by increasing the propeller rpm.

The main purpose of using rudders is to generate forces for course keeping and maneuvering. They can also be used for stationkeeping and emergency stops. Furthermore, rudders affect propeller thrust efficiency and total ship resistance. The performance of rudders depends on the rudder hydrodynamic characteristics, which are affected by the design choices. An overview of rudder design for the last 60 years is found in Liu and Hekkenberg (2017).

Figure 9.13 shows a conventional rudder and a flap rudder. The flap rudder is designed to improve the effective lift, which again improves the maneuverability of the craft. The flap rudder consists of two or more sections which move relative to each other as helm is applied and the angle of the main or driven section moves. Thus the shape of the rudder changes dynamically as the angle of helm is changed. Flap rudders give a much higher lift per rudder angle and a 60–70% higher maximum lift compared to the conventional rudder of same shape, size and area can be expected.

9.5.1 Rudder Forces and Moment

Vertical rudders are usually placed at the craft's stern behind the propellers to produce a transverse force and a steering moment about the craft's CG by deflecting the water flow to a direction of the foil plane. Let A_R denote the area of the rudder and b the rudder height. The aspect ratio is then (see Figure 9.13)

$$\Lambda = \frac{b^2}{A_R}. \tag{9.89}$$

Data for different rudder profiles are found in Bertram (2004) and references therein. The rudder normal force is expressed as (Kijima *et al.* 1990)

$$F_{\mathrm{N}} = \frac{1}{2}\rho U_{\mathrm{R}}^2 A_{\mathrm{R}} C_{\mathrm{N}} \sin(\alpha_{\mathrm{R}}) \tag{9.90}$$

where the resultant rudder inflow speed U_{R} and effective rudder angle α_{R} are

$$U_{\mathrm{R}} = \sqrt{u_{\mathrm{R}}^2 + v_{\mathrm{R}}^2} \tag{9.91}$$

$$\alpha_{\mathrm{R}} = \delta - \tan^{-1}\left(\frac{v_{\mathrm{R}}}{u_{\mathrm{R}}}\right) \approx \delta - \beta_{\mathrm{R}} \tag{9.92}$$

where the ratio $\beta_{\mathrm{R}} \approx v_{\mathrm{R}}/u_{\mathrm{R}}$ is the drift angle at the position of the rudder. This value will be larger than the ship sideslip angle β. The rudder angle δ is defined such that a positive δ gives a positive turning rate. This means that the actual rudder angle

$$\delta_{\mathrm{R}} := -\delta. \tag{9.93}$$

According to Fujii (1960), and Fujii and Tsuda (1961, 1962), the coefficient C_{N} can be estimated as follows

$$C_{\mathrm{N}} = \frac{6.13\Lambda}{\Lambda + 2.25}. \tag{9.94}$$

The inflow speed is in general complicated to compute. For simulation studies and testing of control systems under normal operation of the rudder (small rudder angles δ) it is quite accurate to assume that $\beta_{\mathrm{R}} \approx 0$ such that $U_{\mathrm{R}} \approx u_{\mathrm{R}}$ and $\alpha_{\mathrm{R}} \approx \delta$. The *Maneuvering Modeling Group* (MMG) has proposed a standard method for ship maneuvering prediction with procedures based on captive model tests to compute the hydrodynamic coefficients including the rudder forces (Yasukawa and Yoshimura 2015). In this work the inflow surge velocity is computed as

$$u_{\mathrm{R}} = \varepsilon u(1 - w_{\mathrm{P}})\sqrt{\eta\left(1 + \kappa\left(\sqrt{1 + \frac{8K_{\mathrm{T}}}{\pi J_{\mathrm{a}}^2}} - 1\right)\right)^2 + (1 - \eta)} \tag{9.95}$$

where u is the surge velocity, K_{T} is the thrust coefficient, J_{a} is the advance number, ε is the ratio of wake fraction at rudder position to that at propeller position (typically 1.09) defined by $\varepsilon := (1 - w_{\mathrm{R}})/(1 - w_{\mathrm{P}})$ and κ is an experimental constant (typically 0.5). Let D denote the propeller diameter and b the rudder height. From Yasukawa and Yoshimura (2015) it follows that $\eta \approx D/b$.

The rudder normal force contributes to forces in surge and sway as well as a yaw moment

$$X_{\mathrm{R}} = -(1 - t_{\mathrm{R}})F_{\mathrm{N}} \sin(\delta) \tag{9.96}$$

$$Y_R = -(1 + a_H)F_N \cos(\delta) \tag{9.97}$$

$$N_R = -(x_R + a_H x_H)F_N \cos(\delta) \tag{9.98}$$

where x_R is the longitudinal coordinate of the rudder position (typically $-0.5L_{pp}$). The coefficient for additional drag t_R can be approximated by *Matsumoto's method* (Matsumoto and Suemitsu 1980)

$$1 - t_R = 0.28\ C_B + 0.55 \tag{9.99}$$

where $C_B = \nabla/(LBT)$ is the block coefficient. The rudder force increase factor a_H and the longitudinal coordinate x_H of the additional lateral force are usually determined experimentally. However, Kijima *et al.* (1990) propose to use the curves in Figure 9.14 as a first estimate.

The assumption that $\beta_R = 0$ implies that $\alpha_R \approx \delta$ and $U_R \approx u_R$ such that (9.96)–(9.98) become

$$X_R = -(1 - t_R)\left(\frac{1}{2}\rho U_R^2 A_R C_N\right)\sin^2(\delta) \tag{9.100}$$

$$Y_R = -(1 + a_H)\left(\frac{1}{2}\rho U_R^2 A_R C_N\right)\frac{1}{2}\sin(2\delta) \tag{9.101}$$

$$N_R = -(x_R + a_H x_H)\left(\frac{1}{2}\rho U_R^2 A_R C_N\right)\frac{1}{2}\sin(2\delta). \tag{9.102}$$

Finally, if δ is small, the following formulae are obtained

$$\boldsymbol{\tau}_R = \begin{bmatrix} -\frac{1}{2}(1 - t_R)\rho U_R^2 A_R C_N \sin^2(\delta) \\ -\frac{1}{4}(1 + a_H)\rho U_R^2 A_R C_N \sin(2\delta) \\ -\frac{1}{4}(x_R + a_H x_H)\rho U_R^2 A_R C_N \sin(2\delta) \end{bmatrix} \approx \begin{bmatrix} -X_{\delta\delta}\,\delta^2 \\ -Y_\delta\,\delta \\ -N_\delta\,\delta \end{bmatrix}. \tag{9.103}$$

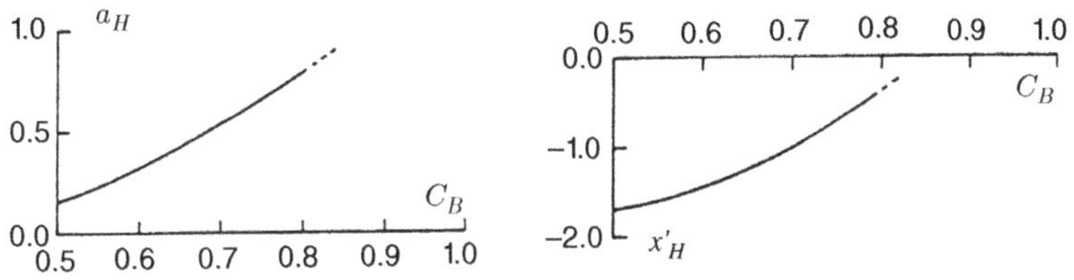

Figure 9.14 The interaction coefficients a_H and $x'_H = x_H/L_{pp}$ as a function of C_B (Kijima *et al.* 1990).

This implies that the rudder coefficients are

$$X_{\delta\delta} = \frac{1}{2}(1 - t_{\mathrm{R}})\rho U_{\mathrm{R}}^2 A_{\mathrm{R}} C_{\mathrm{N}} > 0 \tag{9.104}$$

$$Y_{\delta} = \frac{1}{4}(1 + a_{\mathrm{H}})\rho U_{\mathrm{R}}^2 A_{\mathrm{R}} C_{\mathrm{N}} > 0 \tag{9.105}$$

$$N_{\delta} = \frac{1}{4}(x_{\mathrm{R}} + a_{\mathrm{H}} x_{\mathrm{H}})\rho U_{\mathrm{R}}^2 A_{\mathrm{R}} C_{\mathrm{N}} < 0. \tag{9.106}$$

9.5.2 Steering Machine Dynamics

The steering machine, hydraulic or electric, is controlled by a feedback control system that ensures that the rudder angle δ is close to the commanded rudder angle δ_c. However, the steering machine is a nonlinear system with two important physical limitations: the maximum rudder angle and rudder rate. In computer simulations and when designing autopilots, Van Amerongen (1982) suggests using a simplified representation of the steering machine where the maximum rudder angle δ_{max} and rudder rate $\dot{\delta}_{\mathrm{max}}$ are specified (see Figure 9.15). This simplification is based on the assumption that the steering machine dynamics is much faster than the saturated turning rate commands generated by the autopilot. Generally, the rudder angle and rudder rate limiters in Figure 9.15 will be in the ranges

$$\delta_{\mathrm{max}} = 35 \text{ (deg)}; \qquad 2.3 \text{ (deg/s)} \leq \dot{\delta}_{\mathrm{max}}$$

for commercial ships. The requirement for minimum average rudder rate is specified by the classification societies. Typically, it is required that the rudder can be turned from 35 deg on either side to 30 deg on the other sides respectively within 28 s for the class notations *Tug* or *Offshore service vessel*. For the class notations for *Icebreaker* and *Pusher* the requirements are 15 and 20 s, respectively.

The 28 s requirement corresponds to the lower turning rate limit of 2.3 degs^{-1}. Recently, much faster steering machines have been designed with rudder speeds up to

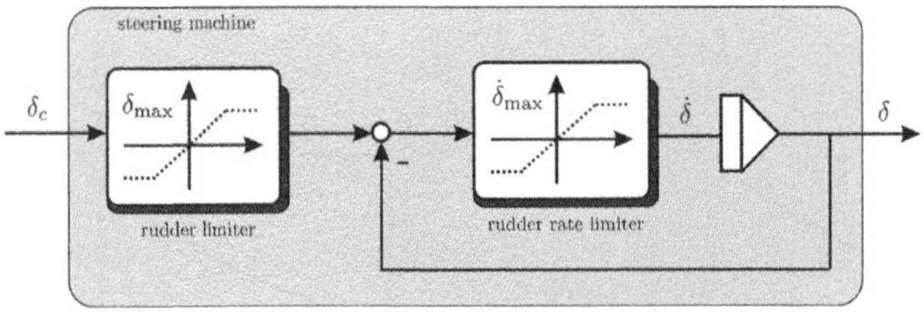

Figure 9.15 Simplified diagram of the rudder control loop. Source: Based on Van Amerongen, J. (1982). Adaptive Steering of Ships – A Model Reference Approach to Improved Maneuvering and Economical Course Keeping. PhD thesis. Delft University of Technology, Netherlands.1982, Delft University Press.

15–20 deg s^{-1}. A rudder speed of 5–20 deg s^{-1} is usually required for a rudder-roll stabilization (RRS) system to work properly.

Another model of the rudder dynamics could be (Rios-Neto and Da Cruz 1985)

$$\dot{\delta} = \begin{cases} \dot{\delta}_{\max}\left(1 - \mathrm{e}^{-\frac{\delta_c - \delta}{\Delta}}\right) & \text{if } \delta_c - \delta \geq 0 \\ -\dot{\delta}_{\max}\left(1 - \mathrm{e}^{-\frac{\delta_c - \delta}{\Delta}}\right) & \text{if } \delta_c - \delta < 0 \end{cases}. \tag{9.107}$$

The parameter Δ will depend on the moment of inertia of the rudder. Typical values will be in the range $3 \leq \Delta \leq 10$.

The limitations of the rudder angle and the rudder speed can be illustrated with the following two simple examples adopted from Van der Klugt (1987).

Example 9.2 (Limitation of the Rudder Angle)

Consider the rudder angle limiter in Figure 9.16 where δ_c is the commanded rudder angle and δ is the actual rudder angle. Let the controller output be given by

$$\delta_c = A \sin(\omega_0 t). \tag{9.108}$$

Figure 9.17 shows the actual rudder angle for three different cases; $A = 3/4\ \delta_{\max}$, $A = \delta_{\max}$ and $A = 4/3\ \delta_{\max}$ where $\delta_{\max} = 30$ deg and $\omega_0 = \pi/10$ rad s^{-1}. It is seen from Figure 9.17 that no extra phase lag is introduced for any of the cases. However, an obvious reduction in amplitude is observed for the saturated case. Consequently, a PID controller will usually suffer from reduced performance but it will be stable.

Example 9.3 (Limitation of the Rudder Rate)

Consider the rudder rate limiter in Figure 9.18 where δ_c is the commanded rudder angle and δ is the actual rudder angle. Let the controller output be given by

$$\delta_c = A \sin(\omega_0 t). \tag{9.109}$$

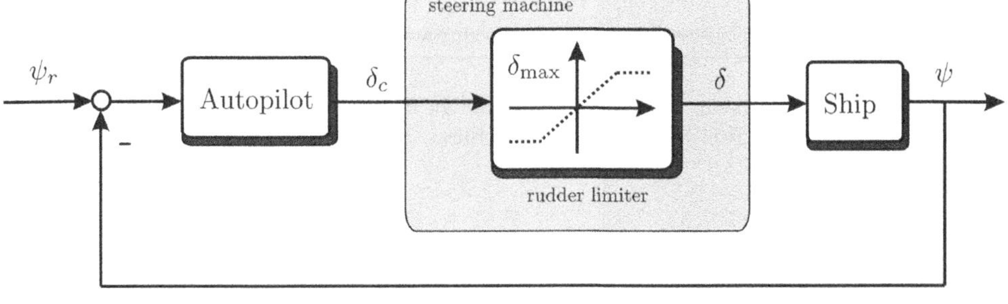

Figure 9.16 Simplified system with rudder limiter. Source: Based on Van der Klugt, P. G. M. (1987). Rudder Roll Stabilization. PhD thesis. Delft University of Technology, Delft, Netherlands.1982, Delft University Press.

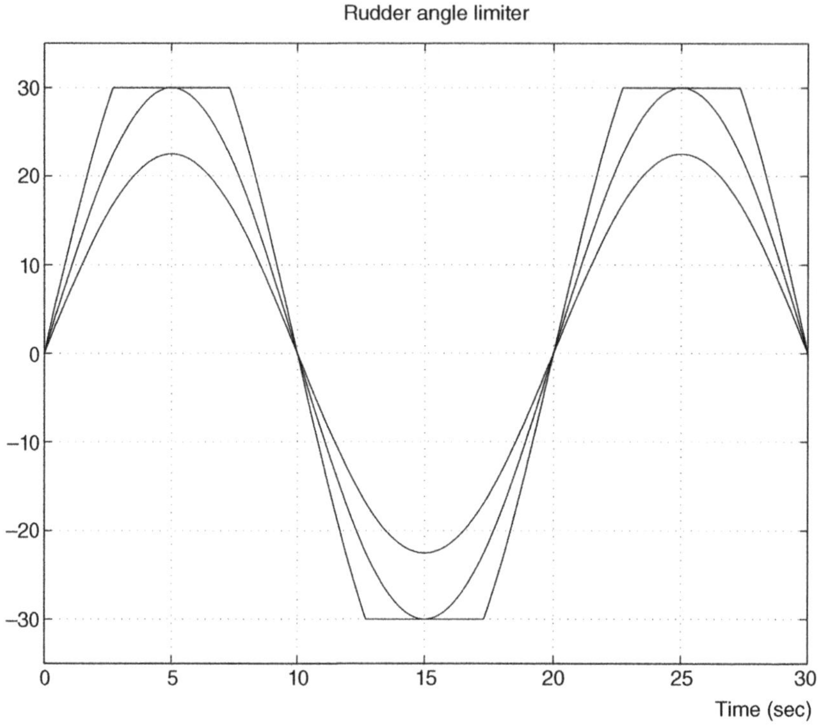

Figure 9.17 Influence of the rudder limiter. Source: Modified from Van der Klugt, P. G. M. (1987). Rudder Roll Stabilization. PhD thesis. Delft University of Technology, Delft, Netherlands.

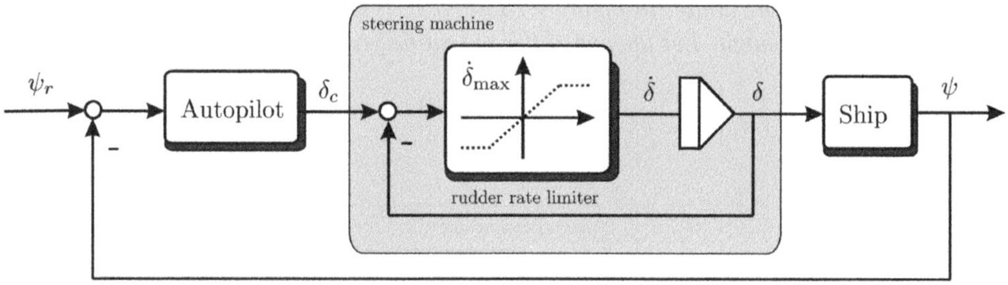

Figure 9.18 Simplified system with rudder rate limiter. Source: Modified from Van der Klugt, P. G. M. (1987). Rudder Roll Stabilization. PhD thesis. Delft University of Technology, Delft, Netherlands.

Figure 9.19 shows the actual and commanded rudder angle for $\dot{\delta}_{\max} = 4$ deg/s, $A = 30$ deg *and* $\omega_0 = \pi/10$ rad s^{-1}. *Besides saturation we now observe that an additional phase lag has been introduced. Reduced phase margins can lead to severe stability problems for the control system. In practice, rudder rate limitations are typical in extreme weather conditions since compensation of high-frequency disturbances require a faster rudder.*

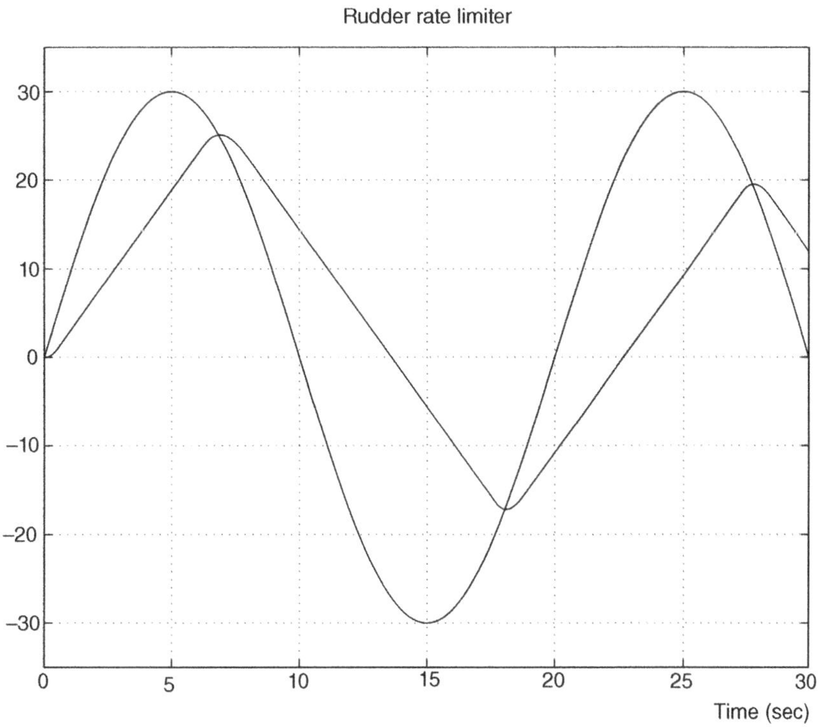

Figure 9.19 Influence of the rudder rate limiter. Source: Modified from Van der Klugt, P. G. M. (1987). Rudder Roll Stabilization. PhD thesis. Delft University of Technology, Delft, Netherlands.

9.6 Fin Stabilizators

Fin stabilizers are primarily used for roll damping (see Figure 9.20). They provide considerable damping if the speed of the ship is not too low. The disadvantage with additional fins is increased hull resistance and high costs associated with the installation, since at least two new hydraulic systems must be installed. Retractable fins are popular, since they are inside the hull when not in use (no additional drag). It should be noted that fins are not effective at low speed and that they cause underwater noise in addition to drag.

Fin stabilizers were patented by John I. Thornycroft in 1889 and there exists a large number of commercial systems today. Modern systems are attractive for roll reduction since they are highly effective, work on a large number of ships, and they are easy to control, even for varying load conditions and actuator configurations. Fins stabilizers are effective at high speed, but at the price of additional drag and added noise. The most economical systems are retractable fins, where additional drag is avoided during normal operation, since fin stabilizers are not needed in moderate weather. Fin stabilizer can also be used in combination with rudder-roll damping systems (see Källström 1981, Roberts and Braham 1990, Roberts 1992, Hickey *et al.* 1997, Hearns *et al.* 2000, Katebi *et al.* 2000, Perez 2005).

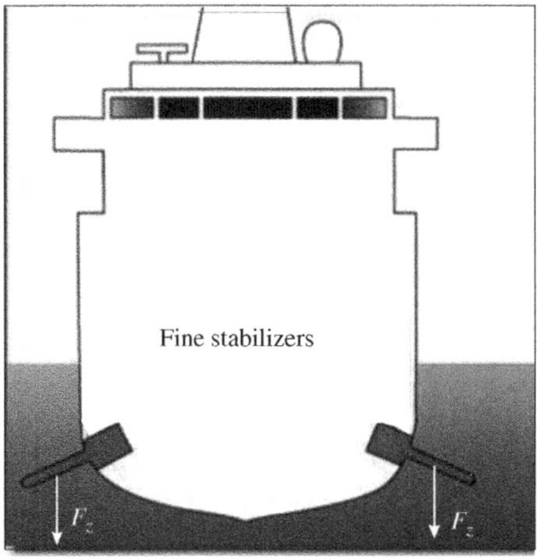

Figure 9.20 Fin stabilized ship where the vertical force $F_z = K_\alpha \alpha_F$ is proportional to the angle α_F for small deflections.

Some obvious benefits of fin stabilizing systems for roll damping are:

- Prevent cargo damage and to increase the effectiveness of the crew by avoiding or reducing seasickness. This is also important from a safety point of view.
- For naval ships and offshore vessels, critical marine operations like landing a helicopter, formation control and underway replenishment benefit from additional roll damping.

9.6.1 Lift and Drag Forces on Fins

The water flowing past the surface of the fin exerts a force on it (see Figure 9.20). Lift is the component of this force that is perpendicular to the oncoming flow direction. It contrasts with the drag force, which is the component of the force parallel to the flow direction. The angle α_F of the fin stabilizers can be adjusted by the control system. Assume that the lift and drag coefficients $C_L(\alpha_F)$ and $C_D(\alpha_F)$ are known such that the forces expressed in FLOW axes are

$$F_{\text{drag}} = \frac{1}{2}\rho U_r^2 A_F C_D(\alpha_F) \tag{9.110}$$

$$F_{\text{lift}} = \frac{1}{2}\rho U_r^2 A_F C_L(\alpha_F) \tag{9.111}$$

where ρ is the density of water, A_F is the fin area and U_r is the relative speed. Since lift is perpendicular to the relative flow and drag is parallel, the longitudinal forces expressed in BODY axes are

$$\begin{bmatrix} F_x \\ F_z \end{bmatrix} = \begin{bmatrix} \cos(\alpha_F) & -\sin(\alpha_F) \\ \sin(\alpha_F) & \cos(\alpha_F) \end{bmatrix} \begin{bmatrix} -F_{\text{drag}} \\ -F_{\text{lift}} \end{bmatrix}. \tag{9.112}$$

The roll moment τ_4 produced by a single fin stabilizer follows from (9.67),

$$\tau_4 = l_y F_z \tag{9.113}$$

where l_y is the location of the fin with respect to the CO.

9.6.2 Roll Moment Produced by Symmetrical Fin Stabilizers

Consider a ship with a pair of symmetrical fin stabilizers (see Figure 9.20). If the two control signals are chosen equal, the roll moment becomes

$$\tau_4 = 2\, l_y F_z. \tag{9.114}$$

Expanding (9.112) gives

$$F_z = -F_{\mathrm{drag}} \sin(\alpha_\mathrm{F}) - F_{\mathrm{lift}} \cos(\alpha_\mathrm{F}). \tag{9.115}$$

The lift and drag coefficients can be approximated by linear theory

$$C_\mathrm{L}(\alpha_\mathrm{F}) \approx C_{\mathrm{L}_0} + C_{\mathrm{L}_\alpha} \alpha_\mathrm{F} \tag{9.116}$$

$$C_\mathrm{D}(\alpha_\mathrm{F}) \approx C_{\mathrm{D}_0} + C_{\mathrm{D}_\alpha} \alpha_\mathrm{F} \tag{9.117}$$

Assume that $C_{\mathrm{L}_0} = C_{\mathrm{D}_0} = 0$ and that α_F is small such that $\sin(\alpha_\mathrm{F}) \approx \alpha_\mathrm{F}$ and $\cos(\alpha_\mathrm{F}) \approx 1$. Then

$$F_z \approx -\frac{1}{2} \rho U_\mathrm{r}^2 A_\mathrm{F} (C_{\mathrm{D}_\alpha} \alpha_\mathrm{F}^2 + C_{\mathrm{L}_\alpha} \alpha_\mathrm{F}). \tag{9.118}$$

Linear theory suggests that the second-order term α_F^2 can be neglected such that

$$F_z \approx -\frac{1}{2} \rho U_\mathrm{r}^2 A_\mathrm{F} C_{\mathrm{L}_\alpha} \alpha_\mathrm{F}. \tag{9.119}$$

Consequently, $\tau_4 = 2\, l_y F_z$ gives the roll moment generated by a pair of fin stabilizers. In other words

$$\tau_4 = -l_y \rho U_\mathrm{r}^2 A_\mathrm{F} C_{\mathrm{L}_\alpha} \alpha_\mathrm{F}. \tag{9.120}$$

9.7 Underwater Vehicle Control Surfaces

Control surfaces are hydrodynamic devices allowing an operator or autopilot to control the velocity and attitude of the vehicle. They are used on underwater vehicles such as AUVs, submarines and torpedoes. The primary control surfaces are (see Figure 9.21):

- **Rudders** – stern vertical surfaces used for turning
- **Dive planes** – bow and stern horizontal surfaces used for depth control.

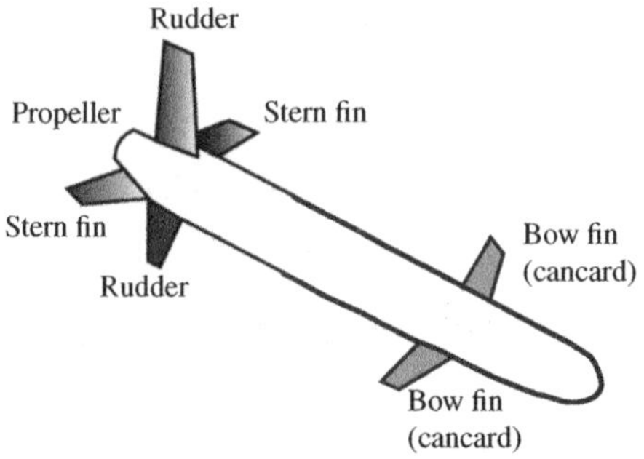

Figure 9.21 Stern: Horizontal fin and vertical rudder. This is also called cross-form rudder. Bow: *canard*, a small fin located in front, for additional stability and lift.

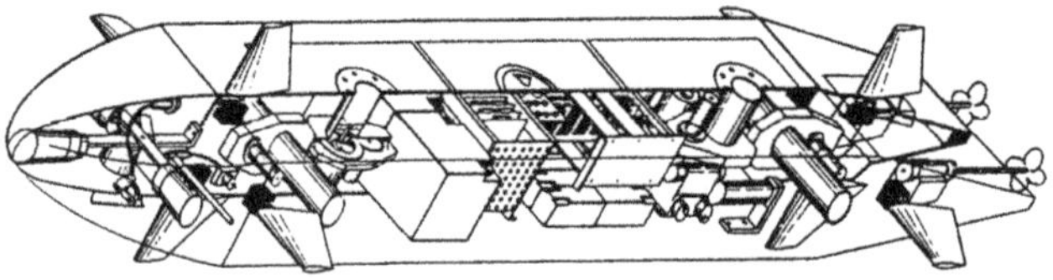

Figure 9.22 Schematic drawing of the NPS II AUV, length $L = 5.3$ m and weight $W = 53.4$ kN, showing the vertical rudder, propellers and dive planes. Source: Healey, A. J. and D. Lienard (1993). Multivariable Sliding Mode Control for Autonomous Diving and Steering of Unmanned Underwater Vehicles. IEEE Journal of Ocean Engineering, OE-18(3), 327-339. ©1993, IEEE.

Figure 9.22 shows the NPS AUV II, which is a "flying vehicle" (Healey and Lienard 1993). The vehicle is controlled by a vertical rudder, two propellers and several dive planes. The 6-DOF mathematical model is included in the MSS toolbox.

MATLAB

```
function [xdot,U] = npsauv(x,ui)
% States: x  = [ u v w p q r x y z phi theta psi ]'
% Inputs: ui = [ delta_r delta_s delta_b delta_bp delta_bs n ]'
% where
%   delta_r  = rudder angle                      (rad)
%   delta_s  = port and starboard stern plane    (rad)
%   delta_b  = top and bottom bow plane          (rad)
%   delta_bp = port bow plane                    (rad)
%   delta_bs = starboard bow plane               (rad)
%   n        = propeller shaft speed             (rpm)
```

9.7.1 Rudder

The rudder is typically mounted aft of the vehicle as shown in Figure 9.21. The rudder can be deflected an angle δ_R, which will force the vehicle to turn. The rudder forces are function of the rudder lift and drag coefficients $C_L(\delta_R)$ and $C_D(\delta_R)$ according to

$$F_{\text{drag}} = \frac{1}{2}\rho U_r^2 A_R C_D(\delta_R) \tag{9.121}$$

$$F_{\text{lift}} = \frac{1}{2}\rho U_r^2 A_R C_L(\delta_R) \tag{9.122}$$

where ρ is the density of water, A_R is the rudder area and U_r is the relative speed. Since lift is perpendicular to the relative flow and drag is parallel, the longitudinal forces expressed in the BODY axes are

$$\begin{bmatrix} X_R \\ Y_R \end{bmatrix} = \begin{bmatrix} \cos(\delta_R) & -\sin(\delta_R) \\ \sin(\delta_R) & \cos(\delta_R) \end{bmatrix} \begin{bmatrix} -F_{\text{drag}} \\ -F_{\text{lift}} \end{bmatrix}. \tag{9.123}$$

The lift and drag coefficients can be approximated by linear theory

$$C_L(\delta_R) \approx C_{L_0} + C_{L_\delta}\delta_R \tag{9.124}$$

$$C_D(\delta_R) \approx C_{D_0} + C_{D_\delta}\delta_R. \tag{9.125}$$

Figure 9.23 confirms that lift and drag can be accurately described by linear theory for small rudder angles (Söding 1999). Since underwater vehicle rudders are streamlined to

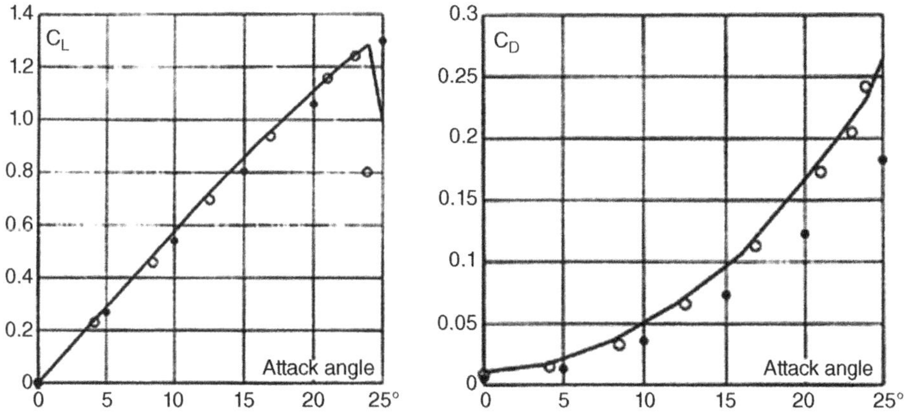

Figure 9.23 Experimental rudder lift and drag curves (circles) compared to Rans calculations (solid lines). The curves verify the linear behavior for small rudder angles. Source: Söding, H. (1999). Limits of Potential Theory in Rudder Flow Predictions In: Twenty- Second Symposium on Naval Hydrodynamics Chapter: 10 Weinblum Lecture, pp. 622-637. The National Academies Press, USA. ©1999, National Academy of Sciences.

produce high lift with minimum drag a standard assumption is $F_{\text{drag}} \approx 0$. Figure 9.23 suggests that $C_L(\delta_R) \approx C_{L_\delta}\delta_R$ such that

$$F_{\text{lift}} \approx \frac{1}{2}\rho U_r^2 A_R C_{L_\delta}\delta_R. \tag{9.126}$$

Assume that δ_R is small such that $\sin(\delta_R) \approx \delta_R$ and $\cos(\delta_R) \approx 1$. Then the rudder forces (9.123) in the x and y directions become

$$X_R = -\frac{1}{2}\rho U_r^2 A_r C_{L_\delta}\delta_R^2 \tag{9.127}$$

$$Y_R = -\frac{1}{2}\rho U_r^2 A_r C_{L_\delta}\delta_R. \tag{9.128}$$

For the AUV in Figure 9.21, the rudder is located on the centerline at $\boldsymbol{r}^b_{bR} = [x_R, 0, z_R]^\top$ with respect to the CO. A twin-rudder system will have two stern rudders located at $\boldsymbol{r}^b_{bR_1} = [x_{R_1}, y_{R_1}, z_{R_1}]^\top$ and $\boldsymbol{r}^b_{bR_2} = [x_{R_2}, y_{R_2}, z_{R_2}]^\top$ where $x_{R_1} = x_{R_2}$, $y_{R_1} = -y_{R_2}$ and $z_{R_1} = z_{R_2}$. From (9.67) it follows that the 6-DOF generalized rudder forces are

$$\boldsymbol{\tau}_R \overset{\substack{\text{single}\\ \text{rudder}}}{=} \begin{bmatrix} X_R \\ Y_R \\ 0 \\ -z_R Y_R \\ z_R X_R \\ x_R Y_R \end{bmatrix}, \quad \boldsymbol{\tau}_R \overset{\substack{\text{twin}\\ \text{rudder}}}{=} \begin{bmatrix} X_{R_1} + X_{R_2} \\ Y_{R_1} + Y_{R_2} \\ 0 \\ -z_{R_1} Y_{R_1} - z_{R_2} Y_{R_2} \\ z_{R_1} X_{R_1} + z_{R_2} X_{R_2} \\ x_{R_1} Y_{R_1} - y_{R_1} X_{R_1} + x_{R_2} Y_{R_2} - y_{R_2} X_{R_2} \end{bmatrix}. \tag{9.129}$$

9.7.2 *Dive Planes*

Dive planes are control surfaces, usually located at the stern of an underwater vehicle. They control the vehicle's pitch angle, and therefore the angle of attack and lift of the dive plane (see Figure 9.21). For aircraft dive planes are called elevators. Additional dive planes can also be located in the front of the vehicle similar to *canard* wings on an aircraft. Cancards increase the lift and pitch response of the vehicle but the downside is additional drag and fuel consumption.

Consider the vehicle in Figure 9.21, which is controller by a stern rudder δ_R, stern dive planes δ_S and bow dive planes δ_B. Assume that the dive planes are streamlined to produce high lift with minimum drag such that drag can be neglected. Hence,

$$F_{\text{S, lift}} = -\frac{1}{2}\rho U_r^2 A_S C_{L_\delta}\delta_S, \qquad F_{\text{B, lift}} = -\frac{1}{2}\rho U_r^2 A_B C_{L_\delta}\delta_B \tag{9.130}$$

where $C_{L_\delta} = \partial C_L / \partial \delta$ is the slope of the lift curve at zero angle, A_S is the stern-plane area and A_B is the bow-plane area. Consequently, rotating the lift force from FLOW to BODY under the assumption that the stern and bow plane angles are small give

$$X_S = -\frac{1}{2}\rho U_r^2 A_S C_{L_\delta} \delta_S^2, \qquad X_B = -\frac{1}{2}\rho U_r^2 A_B C_{L_\delta} \delta_B^2 \tag{9.131}$$

$$Z_S = -\frac{1}{2}\rho U_r^2 A_S C_{L_\delta} \delta_S, \qquad Z_B = -\frac{1}{2}\rho U_r^2 A_B C_{L_\delta} \delta_B. \tag{9.132}$$

Furthermore, assume that the two dive planes are located at $\boldsymbol{r}_{bS}^b = [x_S, y_S, z_S]^\top$ and $\boldsymbol{r}_{bB}^b = [x_B, y_B, z_B]^\top$. From (9.67) it follows that the 6-DOF generalized rudder forces are

$$\boldsymbol{\tau}_S = \begin{bmatrix} X_S \\ 0 \\ Z_S \\ y_S Z_S \\ z_S X_S - x_S Z_S \\ -y_S X_S \end{bmatrix}, \qquad \boldsymbol{\tau}_B = \begin{bmatrix} X_B \\ 0 \\ Z_B \\ y_B Z_B \\ z_B X_B - x_B Z_B \\ -y_B X_B \end{bmatrix}. \tag{9.133}$$

9.8 Control Moment Gyroscope

A control moment gyroscope (CMG) is a device that is used in spacecraft, ship and underwater vehicle attitude control systems. A CMG consists of a spinning rotor and one or more motorized gimbals that tilt the rotor's angular momentum. As the rotor tilts, the changing angular momentum causes a gyroscopic torque that rotates the craft. In this section, we will discuss applications to ships and underwater vehicles including

- Ship roll gyrostabilizers
- CMGs for underwater vehicles.

Unlike rudder and fins, the gyroscope does not rely on the forward speed of the vehicle to generate roll, pitch and yaw stabilizing moments for attitude control. Hence, CMG systems can be used during hovering.

9.8.1 *Ship Roll Gyrostabilizer*

The principle for use of dedicated spinning wheels that generate gyroscopic forces for reducing the roll motion of ships has been known for more than 100 years (Schlick 1904).

Figure 9.24 Two 25-ton roll-stabilizing gyroscopes being installed on the transport USS Henderson during construction in 1917, the first large ship to use gyroscopic stabilization. Source: https://commons.wikimedia.org/wiki/File:Ship_stabilizing_gyroscopes_USS_Henderson_1917.jpg #/media/File:Ship_stabilizing_gyroscopes_USS_Henderson_1917.jpg

The American company Sperry developed a system that addressed the problem of the Schlick gyroscope by using an electric motor and a brake to control the precession of the spinning wheel, see Figure 9.24. However, early designs of the gyrostabilizer were not a success due to their relatively large size and the inability of roll damping control systems in certain sea states and sailing conditions (speed and heading relative to the waves). Today, advances in material technology, mechanical design, electrical drives, and feedback control theory have resulted in feasible systems.

The gyroscopic torque produced by a gyrostabilizer on a ship opposes the roll moment generated by the waves. The principle for the gyroscopic torque is conservation of angular momentum. The wave-pressure forces on the hull induce roll motion and an excitation torque on the gyro that is proportional to the roll rate. This excitation torque changes the angular momentum such that the spinning wheels develop precession motion. The cross product of the spin angular velocity and the precession rate induce a torque that opposes the excitation torque, and thus the roll excitation moment on the vessel. Effective ship installations require rotors having a weight of approximately 3–5% of a vessel's displacement.

A mathematical model for the linear ship roll motion together with an n-spinning-wheel gyrostabilizer was proposed by Perez and Steinmann (2009),

$$(I_x - K_{\dot{p}} + I_g)\dot{p} - K_p p + \mathrm{GM_T} W \phi = \tau_{\mathrm{wave}} + \tau_{\mathrm{wind}} - nK_g \cos(\alpha)\dot{\alpha} \tag{9.134}$$

$$I_g \ddot{\alpha} + D_g \dot{\alpha} + G_g \sin(\alpha) = K_g \cos(\alpha)\, p + \tau_p \tag{9.135}$$

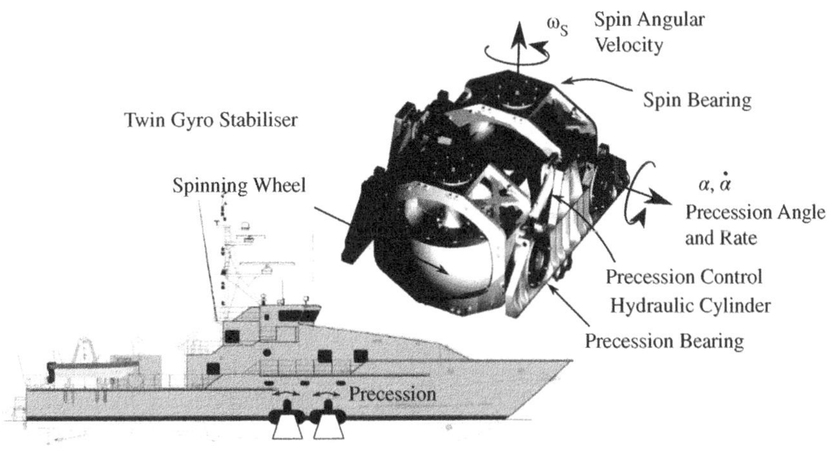

Figure 9.25 The Halcyon's twin gyrostabilizer ($n = 2$). Source: Perez, T. and P. D. Steinmann (2009). Analysis of Ship Roll Gyrostabiliser Control. In: Proceedings of the 8th IFAC International Conference on Manoeuvring and Control of Marine Craft MCMC'09. Guaruja (SP), Brazil. pp. 30–36.

where ϕ is the roll angle, α is the gyro precession angle and

I_x ship moment of inertia
I_g inertia of a single spinning wheel along the precession axis
$K_{\dot{p}}$ added moment of inertia coefficient caused by the ship hull (negative)
K_p linear roll damping coefficient including viscous effects (negative)
K_g spinning angular momentum, that is $K_g = I_{spin}\ \omega_{spin}$
D_g damping coefficient associated with friction in the precession bearings
G_g restoring coefficient associated with the mass distribution of the spinning wheel (pendulum effect)
τ_p precession control torque.

For small precession angles α, the system (9.134)–(9.135) can be approximated by a linear mass–damper–spring system

$$\begin{bmatrix} I_x - K_{\dot{p}} + I_g & 0 \\ 0 & I_g \end{bmatrix} \ddot{\boldsymbol{x}} + \begin{bmatrix} -K_p & nK_g \\ -K_g & D_g \end{bmatrix} \dot{\boldsymbol{x}} + \begin{bmatrix} \mathrm{GM_T}W & 0 \\ 0 & G_g \end{bmatrix} \boldsymbol{x} = \begin{bmatrix} 0 \\ 1 \end{bmatrix} u + \begin{bmatrix} \tau_{wave} + \tau_{wind} \\ 0 \end{bmatrix} \tag{9.136}$$

where $\boldsymbol{x} = [\phi,\ \alpha]^{\top}$ and $u = \tau_p$.

As observed by Elmer A. Sperry, co-inventor with Herman Anschütz-Kaempfe of the gyrocompass, the gyrostabilizer can be used to generate roll rate, which again generates a roll moment (Chalmers 1931). Hence, we could design a precession torque controller such that

$$\dot{\alpha} \approx \kappa p, \qquad \kappa > 0. \tag{9.137}$$

This is clearly a desired control objective since the ship roll dynamics (9.134) will experience additional roll damping for $\kappa > 0$ as seen from

$$(I_x - K_{\dot{p}})\ddot{\phi} + (nK_g\kappa - K_p)\dot{\phi} + \mathrm{GM_T}W\phi = \tau_{\mathrm{wave}} + \tau_{\mathrm{wind}}. \tag{9.138}$$

To achieve the control objective (9.137), a full-state feedback controller using feedback from ϕ, p, α and $\dot{\alpha}$ can be designed. In a simplified decoupled design, feedback from the roll states are avoided. This suggests that the gyrostabilizer can be designed using a PD precession controller

$$\tau_p = -K_{\mathrm{p}}\alpha - K_{\mathrm{d}}\dot{\alpha}. \tag{9.139}$$

The closed-loop transfer function follows from (9.135),

$$\dot{\alpha} = \frac{K_{\mathrm{g}}\ s}{s^2 I_{\mathrm{g}} + (D_{\mathrm{g}} + K_{\mathrm{d}})s + (G_{\mathrm{g}} + K_p)}\ p \approx \kappa p \tag{9.140}$$

where the controller gains should be chosen such that κ has its maximum value for the peak frequency (wave encounter frequency) of the wave loads.

Perez and Steinmann (2009) present a case study of a navy patrol boat, for which the gyro stabilizer is a twin-wheel stabilizer with a total mass equivalent to 3.5% of the vessel displacement. The simulation study shows that there are fundamental limitations associated with the roll reduction for certain frequency ranges, while other frequencies are well damped. This confirms that it is important to tune the controller such that maximum roll damping is achieved for frequencies where the wave loads are at their maximum. In other words, the control system needs to adapt to changes in the sea state in order to avoid roll amplification. Adaptive and constrained control designs are methods, which can overcome these limitations in a practical design.

9.8.2 Control Moment Gyros for Underwater Vehicles

Internal rotors (CMGs) can control three-axis attitude of an underwater vehicle even at zero speed. The CMGs, in combination with thrusters, rudders and dive planes, improve controllability and maneuverability of underwater vehicles in particular at low speed. Applications to underwater robots and vehicles have been reported by Woolsey and Leonard (2002a), Thornton *et al.* (2005, 2007, 2008) and Xua *et al.* (2019).

Angular Momentum of a Spinning Wheel

The angular momentum h of a spinning wheel is

$$h = I_{\text{wheel}}\, \omega_{\text{wheel}} \tag{9.141}$$

where I_{wheel} is the inertia moment and ω_{wheel} is the angular velocity of the spinning wheel. Both quantities can be changed to increase or decrease h. However, an alternative approach is to keep ω_{wheel} = constant such that

$$h = I_{\text{wheel}}\, \omega_{\text{wheel}} = \text{constant} \tag{9.142}$$

and instead use a gimbal motor to change h (see Figure 9.27). Torque-induced gyroscopic precession is the phenomenon in which the axis of a spinning object describes a cone in space when an external torque is applied to it. Hence, by changing the torque of the gimbal motor it is possible to control the angular momentum h of the flywheel. The CMG pyramid configuration in Figure 9.26 is frequently used to produce a 3D torque for attitude control.

CMG Angular Momentum – Pyramid Configuration

A popular CMG configuration is the pyramid shown in Figure 9.26, which will generate three-axis attitude control torques for an underwater vehicle. For this configuration, the skew angle will be $\beta = \cos^{-1}(1/\sqrt{3})$ and the four single gimbal CMGs will have the minimal redundancy needed to control the attitude in three directions (Thornton *et al.* 2007).

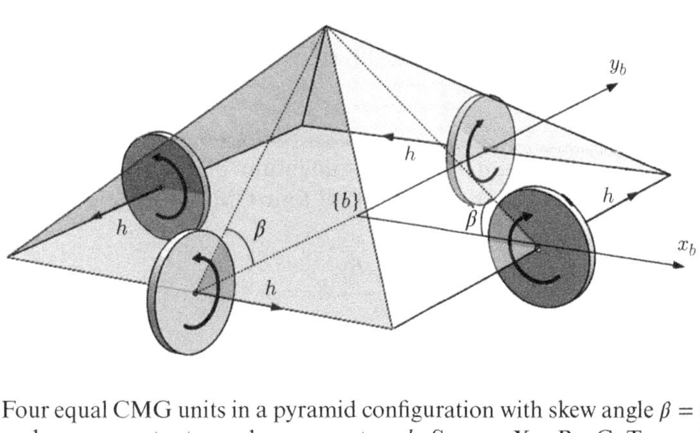

Figure 9.26 Four equal CMG units in a pyramid configuration with skew angle $\beta = \cos^{-1}(1/\sqrt{3})$. Each CMG produces a constant angular momentum h. Source: Xu, R., G. Tanga, L. Hana and D. Xiea (2019). Trajectory Tracking Control for a CMG-Based Underwater Vehicle with Input Saturation in 3D Space. Ocean Engineering 173, 587–598. ©2009, Elsevier.

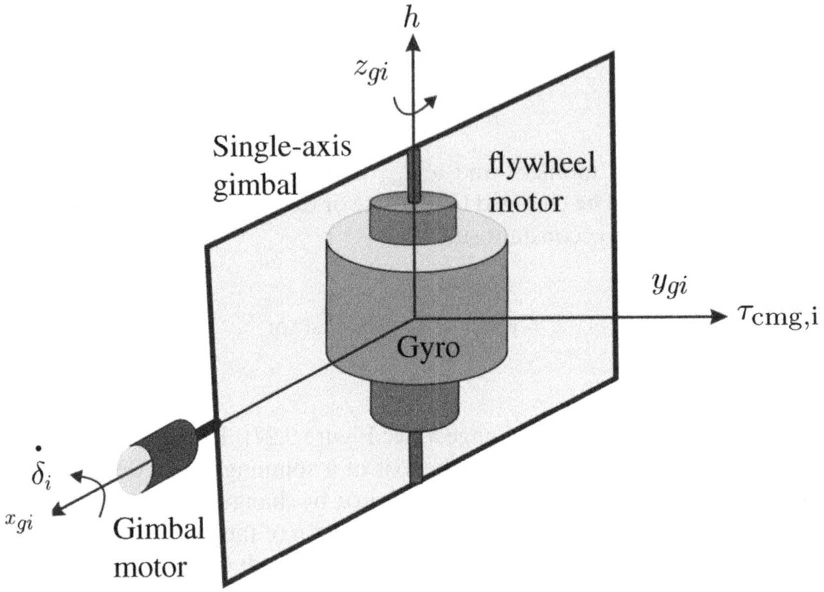

Figure 9.27 Single CMG gimbal unit: the flywheel rotates with constant angular rate and the angular rate $\dot{\delta}_i$ of the ith gimbal change the flywheel's angular momentum h. This produces a gyroscopic torque $\tau_{\mathrm{cmg},i}$.

Consider four equal gyros where δ_i $(i = 1, 2, 3, 4)$ are the gimbal angles. Hence, the angular momentum of the CMG system is (Kurokawa 1997)

$$\boldsymbol{h}^b_{\mathrm{cmg}}(\boldsymbol{\delta}) = h\begin{bmatrix} -\cos(\beta)\sin(\delta_1) - \cos(\delta_2) + \cos(\beta)\sin(\delta_3) + \cos(\delta_4) \\ \cos(\delta_1) - \cos(\beta)\sin(\delta_2) - \cos(\delta_3) + \cos(\beta)\sin(\delta_4) \\ \sin(\beta)(\sin(\delta_1) + \sin(\delta_2) + \sin(\delta_3) + \sin(\delta_4)) \end{bmatrix} \tag{9.143}$$

where h represents the constant angular momentum of each CMG expressed in the gimbal-fixed frame, that is h is the same for all four CMGs (see Figure 9.27). The time derivative of $\boldsymbol{h}^b_{\mathrm{cmg}}$ is

$$\dot{\boldsymbol{h}}^b_{\mathrm{cmg}} = \frac{\partial \boldsymbol{h}^b_{\mathrm{cmg}}(\boldsymbol{\delta})}{\partial \boldsymbol{\delta}}\,\dot{\boldsymbol{\delta}} := h\boldsymbol{A}(\boldsymbol{\delta})\dot{\boldsymbol{\delta}} \tag{9.144}$$

here $\boldsymbol{A}(\boldsymbol{\delta}) \in \mathbb{R}^{3\times 4}$ is the Jacobian matrix

$$\boldsymbol{A}(\boldsymbol{\delta}) = \begin{bmatrix} -\cos(\beta)\cos(\delta_1) & \sin(\delta_2) & \cos(\beta)\cos(\delta_3) & -\sin(\delta_4) \\ -\sin(\delta_1) & -\cos(\beta)\cos(\delta_2) & \sin(\delta_3) & \cos(\beta)\cos(\delta_4) \\ \sin(\beta)\cos(\delta_1) & \sin(\beta)\cos(\delta_2) & \sin(\beta)\cos(\delta_3) & \sin(\beta)\cos(\delta_4) \end{bmatrix}. \tag{9.145}$$

The CMG gimbal angular velocity vector (see Figure 9.27)

$$\dot{\boldsymbol{\delta}} = [\dot{\delta}_1,\ \dot{\delta}_2,\ \dot{\delta}_3,\ \dot{\delta}_4]^{\top} \tag{9.146}$$

is chosen as control input.

AUV Equations of Motion

Consider the cylinder-shaped AUV in Figure 8.2. Assume that the CO is located in the center of the AUV and that the CB coincides with the CO such that $\boldsymbol{r}^b_{\mathrm{bb}} = [0,\ 0,\ 0]^{\top}$. Furthermore, assume that the CG is located at $\boldsymbol{r}^b_{\mathrm{bg}} = [x_{\mathrm{g}},\ y_{\mathrm{g}},\ z_{\mathrm{g}}]^{\top}$ and $W = B = mg$ (neutrally buoyant vehicle). The system inertia matrix $\boldsymbol{M}(\boldsymbol{\delta})$ and the Coriolis and centripetal matrix $\boldsymbol{C}(\boldsymbol{\delta}, \boldsymbol{\nu}_{\mathrm{r}})$ of an CMG-actuated underwater vehicle will both depend on the gimbal angular velocity vector $\boldsymbol{\delta}$. Consequently, the resulting model will be in the following form

$$\dot{\boldsymbol{\eta}} = \boldsymbol{J}_k(\boldsymbol{\eta})(\boldsymbol{\nu}_{\mathrm{r}} + \boldsymbol{\nu}_{\mathrm{c}}) \tag{9.147}$$

$$\boldsymbol{M}(\boldsymbol{\delta})\dot{\boldsymbol{\nu}}_{\mathrm{r}} + \boldsymbol{C}(\boldsymbol{\delta}, \boldsymbol{\nu}_{\mathrm{r}})\boldsymbol{\nu}_{\mathrm{r}} + \boldsymbol{D}\boldsymbol{\nu}_{\mathrm{r}} + \boldsymbol{d}(\boldsymbol{\nu}_{\mathrm{r}}) + \boldsymbol{g}(\boldsymbol{\eta}) = \begin{bmatrix} \boldsymbol{0}_{3\times1} \\ \boldsymbol{\tau}_{\mathrm{cmg}} \end{bmatrix} + \boldsymbol{\tau} \tag{9.148}$$

where $\boldsymbol{\eta}$ and $\boldsymbol{J}_k(\boldsymbol{\eta})$ for $k \in \{\Theta, q\}$ can be parametrized using Euler angles or unit quaternions; see Section 8.1. To derive the expressions for $\boldsymbol{M}(\boldsymbol{\delta})$ and $\boldsymbol{C}(\boldsymbol{\delta}, \boldsymbol{\nu}_{\mathrm{r}})$ a kinetic energy approach will be applied. The kinetic energy of a rigid-body, including the CMGs and hydrodynamic added inertia, is given by (Woolsey and Leonard 2002a)

$$T = \frac{1}{2}[\boldsymbol{\nu}_1^{\top},\ \boldsymbol{\nu}_2^{\top}] \begin{bmatrix} m\boldsymbol{I}_3 + \boldsymbol{A}_{11} & -m\boldsymbol{S}(\boldsymbol{r}^b_{\mathrm{bg}}) + \boldsymbol{A}_{12} \\ m\boldsymbol{S}(\boldsymbol{r}^b_{\mathrm{bg}}) + \boldsymbol{A}_{21} & \boldsymbol{I}^b_{\mathrm{b}} + \boldsymbol{A}_{22} + \boldsymbol{I}^b_{\mathrm{cmg}}(\boldsymbol{\delta}) \end{bmatrix} \begin{bmatrix} \boldsymbol{\nu}_1 \\ \boldsymbol{\nu}_2 \end{bmatrix} \tag{9.149}$$

where $\boldsymbol{\nu}_1 = [u,\ v,\ w]^{\top}$ and $\boldsymbol{\nu}_2 = [p,\ q,\ r]^{\top}$. The inertia dyadic of the rigid body and the CMGs are $\boldsymbol{I}^b_{\mathrm{b}} = \boldsymbol{I}^b_g - m\boldsymbol{S}^2(\boldsymbol{r}^b_{\mathrm{bg}})$ and $\boldsymbol{I}^b_{\mathrm{cmg}}(\boldsymbol{\delta})$, respectively. Hydrodynamic added mass is included by the matrices $\boldsymbol{A}_{11}$, $\boldsymbol{A}_{12} = \boldsymbol{A}_{21}^{\top}$ and $\boldsymbol{A}_{22}$. The translational and rotational dynamics can be derived from *Kirchhoff's equations*, which relates the kinetic energy T to the force vector $\boldsymbol{\tau}_1$ and moment vector $\boldsymbol{\tau}_2$ acting on the body. From Section 6.3.1, we have

$$\frac{\mathrm{d}}{\mathrm{d}t}\left(\frac{\partial T}{\partial \boldsymbol{\nu}_1}\right) + \boldsymbol{S}(\boldsymbol{\nu}_2)\frac{\partial T}{\partial \boldsymbol{\nu}_1} = \boldsymbol{\tau}_1 \tag{9.150}$$

$$\frac{\mathrm{d}}{\mathrm{d}t}\left(\frac{\partial T}{\partial \boldsymbol{\nu}_2}\right) + \boldsymbol{S}(\boldsymbol{\nu}_2)\frac{\partial T}{\partial \boldsymbol{\nu}_2} + \boldsymbol{S}(\boldsymbol{\nu}_1)\frac{\partial T}{\partial \boldsymbol{\nu}_1} = \boldsymbol{\tau}_2. \tag{9.151}$$

Expanding (9.149) gives the following formula for the kinetic energy

$$T = \frac{1}{2}\boldsymbol{\nu}_1^{\top}(m\boldsymbol{I}_3 + \boldsymbol{A}_{11})\boldsymbol{\nu}_1 + \frac{1}{2}\boldsymbol{\nu}_2^{\top}(\boldsymbol{I}^b_{\mathrm{b}} + \boldsymbol{I}^b_{\mathrm{cmg}}(\boldsymbol{\delta}) + \boldsymbol{A}_{22})\boldsymbol{\nu}_2 + \boldsymbol{\nu}_1^{\top}(-m\boldsymbol{S}(\boldsymbol{r}^b_{\mathrm{g}}) + \boldsymbol{A}_{12})\boldsymbol{\nu}_2. \tag{9.152}$$

Hence,

$$\frac{\partial T}{\partial \boldsymbol{v}_1} = (m\boldsymbol{I}_3 + \boldsymbol{A}_{11})\boldsymbol{v}_1 + (-m\boldsymbol{S}(\boldsymbol{r}^b_{\text{bg}}) + \boldsymbol{A}_{12})\boldsymbol{v}_2 \tag{9.153}$$

$$\begin{aligned}\frac{\partial T}{\partial \boldsymbol{v}_2} &= \boldsymbol{A}_{21}\boldsymbol{v}_1 + (\boldsymbol{I}^b_{\text{b}} + \boldsymbol{I}^b_{\text{cmg}}(\boldsymbol{\delta}) + \boldsymbol{A}_{22})\boldsymbol{v}_2 + \boldsymbol{I}^b_{\text{cmg}}(\boldsymbol{\delta})\,\dot{\boldsymbol{\delta}} \\ &= (m\boldsymbol{S}(\boldsymbol{r}^b_{\text{bg}}) + \boldsymbol{A}_{21})\boldsymbol{v}_1 + (\boldsymbol{I}^b_{\text{b}} + \boldsymbol{I}^b_{\text{cmg}}(\boldsymbol{\delta}) + \boldsymbol{A}_{22})\boldsymbol{v}_2 + \boldsymbol{h}^b_{\text{cmg}}(\boldsymbol{\delta})\end{aligned} \tag{9.154}$$

$$\frac{\mathrm{d}}{\mathrm{d}t}\left(\frac{\partial T}{\partial \boldsymbol{v}_1}\right) = (m\boldsymbol{I}_3 + \boldsymbol{A}_{11})\dot{\boldsymbol{v}}_1 + (-m\boldsymbol{S}(\boldsymbol{r}^b_{\text{bg}}) + \boldsymbol{A}_{12})\dot{\boldsymbol{v}}_2 \tag{9.155}$$

$$\begin{aligned}\frac{\mathrm{d}}{\mathrm{d}t}\left(\frac{\partial T}{\partial \boldsymbol{v}_2}\right) &= (m\boldsymbol{S}(\boldsymbol{r}^b_{\text{bg}}) + \boldsymbol{A}_{21})\dot{\boldsymbol{v}}_1 + (\boldsymbol{I}^b_{\text{b}} + \boldsymbol{I}^b_{\text{cmg}}(\boldsymbol{\delta}) + \boldsymbol{A}_{22})\dot{\boldsymbol{v}}_2 \\ &\quad + \dot{\boldsymbol{I}}^b_{\text{cmg}}(\boldsymbol{\delta})\,\boldsymbol{v}_2 + h\boldsymbol{A}(\boldsymbol{\delta})\dot{\boldsymbol{\delta}}\end{aligned} \tag{9.156}$$

where $\boldsymbol{h}^b_{\text{cmg}}(\boldsymbol{\delta}) = \boldsymbol{I}^b_{\text{cmg}}(\boldsymbol{\delta})\dot{\boldsymbol{\delta}}$ and $\dot{\boldsymbol{h}}^b_{\text{cmg}}(\boldsymbol{\delta}) = h\boldsymbol{A}(\boldsymbol{\delta})\dot{\boldsymbol{\delta}}$. The control torque $\boldsymbol{\tau}_{\text{cmg}}$ is applied to the four internal rotors such that

$$\dot{\boldsymbol{h}}^b_{\text{cmg}}(\boldsymbol{\delta}) = h\boldsymbol{A}(\boldsymbol{\delta})\dot{\boldsymbol{\delta}} = -\boldsymbol{\tau}_{\text{cmg}}. \tag{9.157}$$

Hence, it follows that

$$\boldsymbol{M}(\boldsymbol{\delta}) = \begin{bmatrix} m\boldsymbol{I}_3 + \boldsymbol{A}_{11} & -m\boldsymbol{S}(\boldsymbol{r}^b_{\text{bg}}) + \boldsymbol{A}_{12} \\ m\boldsymbol{S}(\boldsymbol{r}^b_{\text{bg}}) + \boldsymbol{A}_{21} & \boldsymbol{I}^b_{\text{b}} + \boldsymbol{A}_{22} \end{bmatrix} + \boldsymbol{M}_{\text{cmg}}(\boldsymbol{\delta}) \tag{9.158}$$

$$\boldsymbol{C}(\boldsymbol{\delta}, \boldsymbol{v}) = \begin{bmatrix} m\boldsymbol{S}(\boldsymbol{v}_2) & -m\boldsymbol{S}(\boldsymbol{v}_2)\boldsymbol{S}(\boldsymbol{r}^b_{\text{bg}}) - \boldsymbol{S}(\boldsymbol{A}_{11}\boldsymbol{v}_1 + \boldsymbol{A}_{12}\boldsymbol{v}_2) \\ m\boldsymbol{S}(\boldsymbol{r}^b_{\text{bg}})\boldsymbol{S}(\boldsymbol{v}_2) - \boldsymbol{S}(\boldsymbol{A}_{11}\boldsymbol{v}_1 + \boldsymbol{A}_{12}\boldsymbol{v}_2) & \boldsymbol{S}(\boldsymbol{I}^b_{\text{b}}\boldsymbol{v}_2) - \boldsymbol{S}(\boldsymbol{A}_{21}\boldsymbol{v}_1 + \boldsymbol{A}_{22}\boldsymbol{v}_2) \end{bmatrix} + \boldsymbol{C}_{\text{cmg}}(\boldsymbol{\delta}) \tag{9.159}$$

where

$$\boldsymbol{M}_{\text{cmg}}(\boldsymbol{\delta}) = \begin{bmatrix} \boldsymbol{0}_{3\times 3} & \boldsymbol{0}_{3\times 3} \\ \boldsymbol{0}_{3\times 3} & \boldsymbol{I}^b_{\text{cmg}}(\boldsymbol{\delta}) \end{bmatrix} \tag{9.160}$$

$$\boldsymbol{C}_{\text{cmg}}(\boldsymbol{\delta}) = \begin{bmatrix} \boldsymbol{0}_{3\times 3} & \boldsymbol{0}_{3\times 3} \\ \boldsymbol{0}_{3\times 3} & \dot{\boldsymbol{I}}^b_{\text{cmg}}(\boldsymbol{\delta}) - \boldsymbol{S}(\boldsymbol{h}^b_{\text{cmg}}(\boldsymbol{\delta})) \end{bmatrix}. \tag{9.161}$$

Singularity Avoidance

When designing the control system it is important to notice that singularities may exist in the Jacobian $A(\delta)$ in (9.157) when transforming the control input $\tau_{\rm cmg}$ to desired angular rates $\dot{\delta}$. At least three single-axis CMGs are necessary for control of vehicle attitude. However, no matter how many CMGs the vehicle uses, gimbal motion can lead to relative orientations that produce no usable output torque along certain directions. The singularities can be monitored by computing

$$\det\left(A(\delta)A^{\top}(\delta)\right). \tag{9.162}$$

A classical solution to the singularity problem is the damped least-squares method in robotics (Nakamura and Hanafusa 1986, Chiaverini 1993). This leads to the solution

$$\dot{\delta} = -\frac{1}{h}A^{\top}(\delta)\,(A(\delta)A^{\top}(\delta) + \lambda^2 I_3)^{-1}\tau_{\rm cmg} \tag{9.163}$$

where $\lambda \geq 0$ is the damping factor. Note that, when $\lambda = 0$, the damped least-squares solution reduces to a regular matrix inversion. Different algorithms for λ are reviewed by Chiaverini *et al.* (1994). A survey of steering laws for CMGs is found in Kurokawa (2007).

Inertia Dyadic for Pyramid-type CMG Systems

The inertia dyadic $I^b_{\rm cmg}(\delta)$ can be modeled as (MacKunis *et al.* 2008)

$$I^b_{\rm cmg}(\delta) = I^b_0 + \sum_{i=1}^{4}\left(R^b_{gi}(\delta_i)I^{gi}_{{\rm cmg},i}R^b_{gi}(\delta_i)^{\top} - m_{\rm cmg}\, S^2(r^b_{bi})\right) \tag{9.164}$$

where the parallel-axis theorem (3.36) has been applied and

I^b_0 = constant	rotational-independent inertia expressed in $\{b\}$
$I^{gi}_{{\rm cmg},i} = {\rm diag}\{I_{x_i}, I_{y_i}, I_{z_i}\}$	inertia of CMG i expressed in the gimbal-fixed frame $\{gi\}$
$R^b_{gi}(\delta_i)$	rotation matrix from $\{gi\}$ to $\{b\}$
$r^b_{bi} = [x_i,\ y_i,\ z_i]^{\top}$	location of CMG number i
$m_{\rm cmg}$	mass of one CMG.

Because the gimbal and rotor are taken to be dynamically balanced, their respective inertia dyadics include no products of inertia. The rotation matrix from the ith gimbal-fixed frame $\{gi\}$ to $\{b\}$ consists of three principal rotations (Sun *et al.* 2010)

$$R^b_{gi}(\delta_i) = \begin{bmatrix} 0 & -1 & 0 \\ 1 & 0 & 0 \\ 0 & 0 & 1 \end{bmatrix}\begin{bmatrix} 1 & 0 & 0 \\ 0 & \cos(\beta) & -\sin(\beta) \\ 0 & \sin(\beta) & \cos(\beta) \end{bmatrix}\begin{bmatrix} \cos(\delta_i) & -\sin(\delta_i) & 0 \\ \sin(\delta_i) & \cos(\delta_i) & 0 \\ 0 & 0 & 1 \end{bmatrix}. \tag{9.165}$$

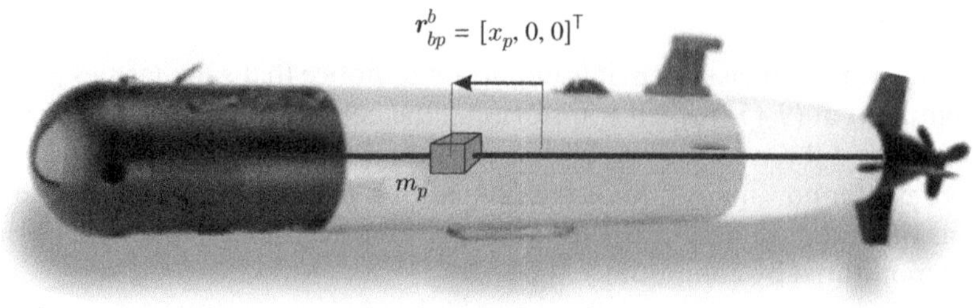

Figure 9.28 Cylinder-shaped AUV with a moving mass m_p that can translate in the x direction.

9.9 Moving Mass Actuators

Internal moving mass actuators as well as internal rotors (CMGs), are promising alternatives to propellers and control surfaces (Woolsey and Leonard 2002b, Li and Shao 2008, Li 2016). Advanced autonomous vehicles such as AUVs, gliders and USVs operate with limited battery power (see Figure 9.28). For these type of vehicles reduced power consumption is important to increase endurance and range. Unlike propeller-driven AUVs and USVs, gliders have fixed wings and tails, and use internal actuators to travel by concatenating a series of upwards and downwards glides. The endurance of underwater gliders is significantly longer than propeller-driven AUVs. Gliders, typically operate at a speed of about 0.3 m s^{-1} compared to 1.5–2.0 m s^{-1} of AUVs and 1.0–10.0 m s^{-1} of small USVs. A detailed review and overview of underwater gliders are found in Graver (2006).

AUV Equations of Motion

Consider an underwater vehicle with constant mass m and a moving point mass m_p (see Figure 9.28). The total mass of the vehicle and the point mass is

$$m_{\text{total}} = m + m_p. \tag{9.166}$$

Hence, the center of gravity is is shifted according to

$$\boldsymbol{r}^b_{bg'} = \frac{m\,\boldsymbol{r}^b_{bg} + m_p\,\boldsymbol{r}^b_{bp}}{m_{\text{total}}} \tag{9.167}$$

where $\boldsymbol{r}^b_{bp} = [x_p,\ y_p,\ z_p]^\top$ is the time-varying location of the point mass with respect to the CO expressed in $\{b\}$. Let $\boldsymbol{v}^b_{bp} = \dot{\boldsymbol{r}}^b_{bp}$ denote the velocity of the point mass. Then the AUV kinetic energy with an internal moving mass becomes (Woolsey and Leonard 2002b)

$$T = \frac{1}{2}[(\boldsymbol{v}^b_{bp})^\top,\ \boldsymbol{\nu}_1^\top,\ \boldsymbol{\nu}_2^\top]\boldsymbol{M}(\boldsymbol{r}^b_{bp})\begin{bmatrix} \boldsymbol{v}^b_{bp} \\ \boldsymbol{\nu}_1 \\ \boldsymbol{\nu}_2 \end{bmatrix} \tag{9.168}$$

where $\boldsymbol{\nu}_1 = [u,\ v,\ w]^\top$, $\boldsymbol{\nu}_2 = [p,\ q,\ r]^\top$ and

$$\boldsymbol{M}(\boldsymbol{r}^b_{\mathrm{bp}}) = \boldsymbol{M}_{\mathrm{RB}}(\boldsymbol{r}^b_{\mathrm{bp}}) + \boldsymbol{M}_{\mathrm{A}}. \tag{9.169}$$

The matrices are

$$\boldsymbol{M}_{\mathrm{RB}}(\boldsymbol{r}^b_{\mathrm{bp}}) = \begin{bmatrix} m_{\mathrm{p}}\boldsymbol{I}_3 & m_{\mathrm{p}}\boldsymbol{I}_3 & -m_{\mathrm{p}}\boldsymbol{S}(\boldsymbol{r}^b_{\mathrm{bp}}) \\ m_{\mathrm{p}}\boldsymbol{I}_3 & (m+m_{\mathrm{p}})\boldsymbol{I}_3 & -m_{\mathrm{p}}\boldsymbol{S}(\boldsymbol{r}^b_{\mathrm{bp}}) - m\boldsymbol{S}(\boldsymbol{r}^b_{\mathrm{bg}}) \\ m_{\mathrm{p}}\boldsymbol{S}(\boldsymbol{r}^b_{\mathrm{bp}}) & m_{\mathrm{p}}\boldsymbol{S}(\boldsymbol{r}^b_{\mathrm{bp}}) + m\boldsymbol{S}(\boldsymbol{r}^b_{\mathrm{bg}}) & \boldsymbol{I}^b_{\mathrm{b}} - m_{\mathrm{p}}\boldsymbol{S}^2(\boldsymbol{r}^b_{\mathrm{bp}}) \end{bmatrix} \tag{9.170}$$

$$\boldsymbol{M}_{\mathrm{A}} = \begin{bmatrix} \boldsymbol{0}_{3\times3} & \boldsymbol{0}_{3\times3} & \boldsymbol{0}_{3\times3} \\ \boldsymbol{0}_{3\times3} & \boldsymbol{A}_{11} & \boldsymbol{A}_{12} \\ \boldsymbol{0}_{3\times3} & \boldsymbol{A}_{21} & \boldsymbol{A}_{22} \end{bmatrix}. \tag{9.171}$$

The linear and angular momentums are obtained as

$$\mathcal{P}^b_{\mathrm{p}} = \frac{\partial T}{\partial \boldsymbol{v}^b_{\mathrm{bp}}}, \qquad \mathcal{P}^b = \frac{\partial T}{\partial \boldsymbol{\nu}_1}, \qquad \mathcal{H}^b = \frac{\partial T}{\partial \boldsymbol{\nu}_2}. \tag{9.172}$$

Consequently,

$$\begin{aligned} \begin{bmatrix} \mathcal{P}^b_{\mathrm{p}} \\ \mathcal{P}^b \\ \mathcal{H}^b \end{bmatrix} &= \boldsymbol{M}(\boldsymbol{r}^b_{\mathrm{bp}}) \begin{bmatrix} \boldsymbol{v}^b_{\mathrm{bp}} \\ \boldsymbol{\nu}_1 \\ \boldsymbol{\nu}_2 \end{bmatrix} \\ &= \begin{bmatrix} m_{\mathrm{p}}\boldsymbol{v}^b_{\mathrm{bp}} + m_{\mathrm{p}}\boldsymbol{\nu}_1 - m_{\mathrm{p}}\boldsymbol{S}(\boldsymbol{r}^b_{\mathrm{bp}})\boldsymbol{\nu}_2 \\ m_{\mathrm{p}}\boldsymbol{v}^b_{\mathrm{bp}} + (\boldsymbol{A}_{11} + (m+m_{\mathrm{p}})\boldsymbol{I}_3)\boldsymbol{\nu}_1 \\ \quad +(\boldsymbol{A}_{12} - m_{\mathrm{p}}\boldsymbol{S}(\boldsymbol{r}^b_{\mathrm{bp}}) - m\boldsymbol{S}(\boldsymbol{r}^b_{\mathrm{bg}}))\boldsymbol{\nu}_2 \\ m_{\mathrm{p}}\boldsymbol{S}(\boldsymbol{r}^b_{\mathrm{bp}})\boldsymbol{v}^b_{\mathrm{bp}} + (\boldsymbol{A}_{21} + m_{\mathrm{p}}\boldsymbol{S}(\boldsymbol{r}^b_{\mathrm{bp}}) + m\boldsymbol{S}(\boldsymbol{r}^b_{\mathrm{bg}}))\boldsymbol{\nu}_1 \\ \quad + (\boldsymbol{A}_{22} + \boldsymbol{I}^b_{\mathrm{b}} - m_{\mathrm{p}}\boldsymbol{S}^2(\boldsymbol{r}^b_{\mathrm{bp}}))\boldsymbol{\nu}_2 \end{bmatrix}. \end{aligned} \tag{9.173}$$

The velocity vectors are computed from the linear and angular momentums by inverting the mass matrix

$$\begin{bmatrix} \boldsymbol{v}^b_{\mathrm{bp}} \\ \boldsymbol{\nu}_1 \\ \boldsymbol{\nu}_2 \end{bmatrix} = \boldsymbol{M}^{-1}(\boldsymbol{r}^b_{\mathrm{bp}}) \begin{bmatrix} \mathcal{P}^b_{\mathrm{p}} \\ \mathcal{P}^b \\ \mathcal{H}^b \end{bmatrix}. \tag{9.174}$$

The kinematic equations for the point mass and the vehicle are

$$\dot{\boldsymbol{r}}^b_{\mathrm{bp}} = \boldsymbol{v}^b_{\mathrm{bp}} \tag{9.175}$$

$$\dot{\boldsymbol{\eta}} = \boldsymbol{J}_\Theta(\boldsymbol{\eta}) \begin{bmatrix} \boldsymbol{\nu}_1 \\ \boldsymbol{\nu}_2 \end{bmatrix}. \tag{9.176}$$

Suppose that the only forces and moments which act on the AUV are due to gravity and buoyancy, and to the internal control force $\boldsymbol{\tau}_{\rm p}$ used to control the position $\boldsymbol{r}^b_{\rm bp}$ of the point mass. Furthermore, assume that the CO is located in the center of the AUV and that the CB coincides with the CO. Consequently, $\boldsymbol{r}^b_{\rm bb} = [0,\ 0,\ 0]^\top$. Then for a neutrally buoyant vehicle the restoring moments (4.5) can be expressed as

$$\boldsymbol{m}^b_{\rm g} = \boldsymbol{S}(\boldsymbol{r}^b_{\rm bg})\boldsymbol{R}^\top(\boldsymbol{\Theta}_{nb})\boldsymbol{f}^n_{\rm g} \tag{9.177}$$

$$\boldsymbol{m}^b_{\rm p} = \boldsymbol{S}(\boldsymbol{r}^b_{\rm bp})\boldsymbol{R}^\top(\boldsymbol{\Theta}_{nb})\boldsymbol{f}^n_{\rm p} \tag{9.178}$$

where $\boldsymbol{f}^n_{\rm g} = [0,\ 0,\ mg]^\top$ and $\boldsymbol{f}^n_{\rm p} = [0,\ 0,\ m_{\rm p}g]^\top$ are the gravitational force of the vehicle and point mass, respectively. Hence, the state-space model in terms of linear and angular moments become (Woolsey and Leonard 2002b)

$$\dot{\boldsymbol{\mathcal{P}}}^b_{\rm p} = -\boldsymbol{S}(\boldsymbol{\nu}_2)\boldsymbol{\mathcal{P}}^b_{\rm p} + \boldsymbol{R}^\top(\boldsymbol{\Theta}_{\rm nb})\boldsymbol{f}^n_{\rm p} + \boldsymbol{\tau}_{\rm p} \tag{9.179}$$

$$\dot{\boldsymbol{\mathcal{P}}}^b = -\boldsymbol{S}(\boldsymbol{\nu}_2)\boldsymbol{\mathcal{P}}^b \tag{9.180}$$

$$\dot{\boldsymbol{\mathcal{H}}}^b = -\boldsymbol{S}(\boldsymbol{\nu}_1)\boldsymbol{\mathcal{P}}^b - \boldsymbol{S}(\boldsymbol{\nu}_2)\boldsymbol{\mathcal{H}}^b + \boldsymbol{S}(\boldsymbol{r}^b_{\rm bp})\boldsymbol{R}^\top(\boldsymbol{\Theta}_{\rm nb})\boldsymbol{f}^n_{\rm p} + \boldsymbol{S}(\boldsymbol{r}^b_{\rm bg})\boldsymbol{R}^\top(\boldsymbol{\Theta}_{\rm nb})\boldsymbol{f}^n_{\rm g}. \tag{9.181}$$

Note the resemblance to Kirchoff's equations (3.56) and (3.57).

Control Strategies

The point mass $m_{\rm p}$ affects the angular momentum $\mathcal{H}^b$ given by (9.181) and thus the vehicle's attitude via the gravitational force $\boldsymbol{f}^n_{\rm p}$. Hence, the control strategy can be to move the point such that attitude is stabilized. The position of the point mass $\boldsymbol{r}^b_{\rm bp}$ is easily controlled by means of the control force $\boldsymbol{\tau}_{\rm p}$ in (9.179).

In Woolsey and Leonard (2002b) a potential-shaping control law is used to stabilize the system. The stability analysis shows that the feedback gain must be chosen sufficient large and the point mass must be big enough to influence the AUV dynamics. These conditions are derived using Lyapunov stability theory since the AUV dynamics with a moving point mass is highly nonlinear. A similar analysis is presented by Li (2016) and Li and Su (2016) who have studied the horizontal-plane motion of a cylinder-shaped AUV. In this work an LQR heading autopilot is designed using an internal moving mass to control the vehicle's heading.

10

Environmental Forces and Moments

Chapters 1–9 present the marine craft equations of motion. In this chapter, complementary models for environmental disturbances are derived. The three external loads on a marine craft are

- Wind
- Waves
- Ocean currents.

The purpose of the chapter is to offer realistic models for simulation, testing and verification of feedback control systems in varying environmental conditions. Complementary textbooks on hydrodynamic loads are Faltinsen (1990), Newman (1977) and Sarpkaya (1981).

Superposition of Wind and Wave Disturbances

For control system design it is common to assume the *principle of superposition* when considering wind and wave disturbances. For most marine control applications this is a good approximation. In general, the environmental forces will be highly nonlinear and both additive and multiplicative to the dynamic equations of motion. An accurate description of the environmental forces and moments are important in vessel simulators that are produced for human operators.

In Chapter 6 it was shown that the nonlinear dynamic equations of motion can be written

$$\boldsymbol{M}\dot{\boldsymbol{\nu}} + \boldsymbol{C}(\boldsymbol{\nu})\boldsymbol{\nu} + \boldsymbol{D}(\boldsymbol{\nu})\boldsymbol{\nu} + \boldsymbol{g}(\boldsymbol{\eta}) + \boldsymbol{g}_0 = \underbrace{\boldsymbol{\tau}_{\text{wind}} + \boldsymbol{\tau}_{\text{wave}}}_{\boldsymbol{w}} + \boldsymbol{\tau}. \tag{10.1}$$

The principle of superposition suggests that the generalized wind- and wave-induced forces are added to the right-hand side of (10.1) by defining

Handbook of Marine Craft Hydrodynamics and Motion Control, Second Edition. Thor I. Fossen.

$$\boldsymbol{w} := \boldsymbol{\tau}_{\text{wind}} + \boldsymbol{\tau}_{\text{wave}} \tag{10.2}$$

where $\boldsymbol{\tau}_{\text{wind}} \in \mathbb{R}^6$ and $\boldsymbol{\tau}_{\text{wave}} \in \mathbb{R}^6$ represent the generalized forces due to wind and waves. Computer-effective models for the simulation of generalized wind and wave forces are presented in Sections 10.1 and 10.2.

Equations of Relative Motion for Simulation of Ocean Currents

The forces on a marine craft due to ocean currents can be implemented by replacing the generalized velocity vector in the hydrodynamic terms with relative velocities

$$\boldsymbol{\nu}_{\text{r}} = \boldsymbol{\nu} - \boldsymbol{\nu}_{\text{c}} \tag{10.3}$$

where $\boldsymbol{\nu}_{\text{c}} \in \mathbb{R}^6$ is the velocity of the ocean current expressed in $\{b\}$. The equations of motion including ocean currents become

$$\underbrace{\boldsymbol{M}_{\text{RB}}\dot{\boldsymbol{\nu}} + \boldsymbol{C}_{\text{RB}}(\boldsymbol{\nu})\boldsymbol{\nu}}_{\text{rigid-body terms}} + \underbrace{\boldsymbol{M}_{\text{A}}\dot{\boldsymbol{\nu}}_r + \boldsymbol{C}_{\text{A}}(\boldsymbol{\nu}_{\text{r}})\boldsymbol{\nu}_{\text{r}} + \boldsymbol{D}(\boldsymbol{\nu}_{\text{r}})\boldsymbol{\nu}_{\text{r}}}_{\text{hydrodynamic terms}} + \underbrace{\boldsymbol{g}(\boldsymbol{\eta}) + \boldsymbol{g}_0}_{\text{hydrostatic terms}} = \boldsymbol{\tau} + \boldsymbol{w}. \tag{10.4}$$

Note that the rigid-body kinetics is independent of the ocean current. A frequently used simplification is to assume that the ocean currents are *irrotational* and *constant* in $\{n\}$. In Section 10.3 it is shown that this assumption implies that (10.4) can be transformed to the equations of relative motions

$$\boldsymbol{M}\dot{\boldsymbol{\nu}}_r + \boldsymbol{C}(\boldsymbol{\nu}_{\text{r}})\boldsymbol{\nu}_r + \boldsymbol{D}(\boldsymbol{\nu}_{\text{r}})\boldsymbol{\nu}_{\text{r}} + \boldsymbol{g}(\boldsymbol{\eta}) + \boldsymbol{g}_0 = \boldsymbol{\tau}_{\text{wind}} + \boldsymbol{\tau}_{\text{wave}} + \boldsymbol{\tau} \tag{10.5}$$

where all mass, Coriolis–centripetal and damping terms are functions of the relative acceleration and velocity vectors only. The matrices $\boldsymbol{M}$ and $\boldsymbol{C}(\boldsymbol{\nu}_{\text{r}})$ in this model become

$$\boldsymbol{M} = \boldsymbol{M}_{\text{RB}} + \boldsymbol{M}_{\text{A}} \tag{10.6}$$

$$\boldsymbol{C}(\boldsymbol{\nu}_{\text{r}}) = \boldsymbol{C}_{\text{RB}}(\boldsymbol{\nu}_{\text{r}}) + \boldsymbol{C}_{\text{A}}(\boldsymbol{\nu}_{\text{r}}). \tag{10.7}$$

In the linear case, Equation (10.5) reduces to

$$\boldsymbol{M}\dot{\boldsymbol{\nu}}_r + \boldsymbol{N}\boldsymbol{\nu}_{\text{r}} + \boldsymbol{G}\boldsymbol{\eta} + \boldsymbol{g}_0 = \boldsymbol{\tau} + \boldsymbol{w}. \tag{10.8}$$

Models for simulation of ocean currents in terms of $\boldsymbol{\nu}_{\text{c}}$ are presented in Section 10.3.

10.1 Wind Forces and Moments

Wind is defined as the movement of air relative to the surface of the Earth. Mathematical models of wind forces and moments are used in motion control systems to improve the performance and robustness of the system in extreme conditions. Models for this are presented in the forthcoming sections.

10.1.1 *Wind Forces and Moments on Marine Craft at Rest*

Let $V_{\rm w}$ and $\gamma_{\rm w}$ denote the wind speed and angle of attack, respectively; see Figure 10.1. The wind forces and moments acting on a surface craft are computed using a similar approach to that of the current coefficients defined in Section 6.7.1. For zero speed it is common to write

$$X_{\rm wind} = q\ C_X(\gamma_{\rm w}) A_{\rm F_w} \tag{10.9}$$

$$Y_{\rm wind} = q\ C_Y(\gamma_{\rm w}) A_{\rm L_w} \tag{10.10}$$

$$Z_{\rm wind} = q\ C_Z(\gamma_{\rm w}) A_{\rm F_w} \tag{10.11}$$

$$K_{\rm wind} = q\ C_K(\gamma_{\rm w}) A_{\rm L_w} H_{\rm L_w} \tag{10.12}$$

$$M_{\rm wind} = q\ C_{\rm M}(\gamma_{\rm w}) A_{\rm F_w} H_{\rm F_w} \tag{10.13}$$

$$N_{\rm wind} = q\ C_N(\gamma_{\rm w}) A_{\rm L_w} L_{\rm oa} \tag{10.14}$$

where $H_{\rm F_w}$ and $H_{\rm L_w}$ are the centroids above the water line of the frontal and lateral projected areas $A_{\rm F_w}$ and $A_{\rm L_w}$, respectively, $L_{\rm oa}$ is the length overall and

$$\gamma_{\rm w} = \psi - \beta_{V_{\rm w}} - \pi \tag{10.15}$$

where $\beta_{V_{\rm w}}$ is the wind direction (see Figure 10.1).

The dynamic pressure of the apparent wind is

$$q = \frac{1}{2}\rho_{\rm a} V_{\rm w}^2 \tag{10.16}$$

where $\rho_{\rm a}$ is the air density given by Table 10.1.

The mean velocity profile satisfies a boundary-layer profile (Bretschneider 1969)

$$V_{\rm w}(h) = V_{10}(h/10)^{\alpha} \tag{10.17}$$

where V_{10} is the wind velocity 10 m above the sea surface, h is the height above the sea surface and $\alpha = 1/7$. The non-dimensional wind coefficients C_X, C_Y, C_Z, C_K, $C_{\rm M}$ and C_N are usually computed using 10 m as reference height. To convert the non-dimensional wind coefficients to a different reference height, the ratio between the dynamic pressures at the two heights are used

$$\frac{\frac{1}{2}\rho_{\rm a} V_{\rm w}(h_1)^2}{\frac{1}{2}\rho_{\rm a} V_{\rm w}(h_2)^2} = \frac{V_{\rm w}(h_1)^2}{V_{\rm w}(h_2)^2} = \frac{[V_{10}(h_1/10)^{\alpha}]^2}{[V_{10}(h_2/10)^{\alpha}]^2} = \left(\frac{h_1}{h_2}\right)^{2\alpha}. \tag{10.18}$$

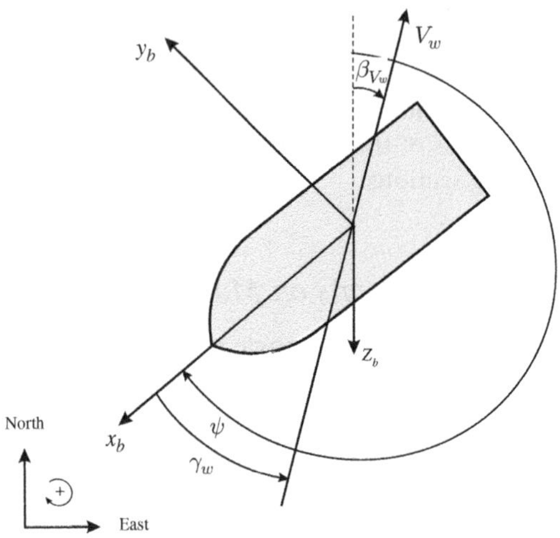

Figure 10.1 Wind speed V_w, wind direction β_{V_w} and wind angle of attack γ_w relative to the bow.

Table 10.1 Air density as a function of temperature.

°C	Air density, ρ_a (kg m^{-3})
−10	1.342
−5	1.317
0	1.292
5	1.269
10	1.247
15	1.225
20	1.204
25	1.184
30	1.165

Consequently, the non-dimensional wind coefficients at height h_1 can be converted to height h_2 by multiplying with

$$\left(\frac{h_1}{h_2}\right)^{2\alpha}. \tag{10.19}$$

For surface ships it is common to assume that $Z_{\text{wind}} = M_{\text{wind}} = 0$ while the roll moment K_{wind} is used for ships and ocean structures where large rolling angles are an issue. For semi-submersibles both K_{wind} and M_{wind} are needed in addition to the horizontal motion components X_{wind}, Y_{wind} and N_{wind}.

The wind speed is usually specified in terms of *Beaufort numbers*, as shown in Table 10.2.

Table 10.2 Definition of Beaufort numbers (Price and Bishop 1974).

Beaufort number	Description of wind	Wind speed (knots)
0	Calm	0–1
1	Light air	2–3
2	Light breeze	4–7
3	Gentle breeze	8–11
4	Moderate breeze	12–16
5	Fresh breeze	17–21
6	Strong breeze	22–27
7	Moderate gale	28–33
8	Fresh gale	34–40
9	Strong gale	41–48
10	Whole gale	49–56
11	Storm	57–65
12	Hurricane	More than 65

Wind Coefficient Approximation for Symmetrical Ships

For ships that are symmetrical with respect to the xz and yz planes, the wind coefficients for horizontal plane motions can be approximated by

$$C_X(\gamma_w) \approx -c_x \cos(\gamma_w) \tag{10.20}$$

$$C_Y(\gamma_w) \approx c_y \sin(\gamma_w) \tag{10.21}$$

$$C_N(\gamma_w) \approx c_n \sin(2\gamma_w) \tag{10.22}$$

which are convenient formulae for computer simulations. Experiments with ships indicate that $c_x \in [0.50,\ 0.90]$, $c_y \in [0.70,\ 0.95]$ and $c_n \in [0.05,\ 0.20]$. However, these values should be used with care.

10.1.2 Wind Forces and Moments on Moving Marine Craft

For a ship moving at a forward speed, (10.9)–(10.14) should be redefined in terms of relative wind speed V_{rw} and angle of attack γ_{rw} according to

$$\boldsymbol{\tau}_{\text{wind}} = \frac{1}{2}\rho_a V_{rw}^2 \begin{bmatrix} C_X(\gamma_{rw})A_{F_w} \\ C_Y(\gamma_{rw})A_{L_w} \\ C_Z(\gamma_{rw})A_{F_w} \\ C_K(\gamma_{rw})A_{L_w}H_{L_w} \\ C_M(\gamma_{rw})A_{F_w}H_{F_w} \\ C_N(\gamma_{rw})A_{L_w}L_{oa} \end{bmatrix} \tag{10.23}$$

with

$$V_{\text{rw}} = \sqrt{u_{\text{rw}}^2 + v_{\text{rw}}^2} \tag{10.24}$$

$$\gamma_{\text{rw}} = -\text{atan2}(v_{\text{rw}}, u_{\text{rw}}). \tag{10.25}$$

The relative velocities are

$$u_{\text{rw}} = u - u_{\text{w}} \tag{10.26}$$

$$v_{\text{rw}} = v - v_{\text{w}} \tag{10.27}$$

while the components of V_{w} in the x and y directions are (see Figure 10.1)

$$u_{\text{w}} = V_{\text{w}} \cos(\beta_{V_{\text{w}}} - \psi) \tag{10.28}$$

$$v_{\text{w}} = V_{\text{w}} \sin(\beta_{V_{\text{w}}} - \psi). \tag{10.29}$$

The wind speed V_{w} and its direction $\beta_{V_{\text{w}}}$ can be measured by an anemometer and a weathervane, respectively. Anemometer is derived from the Greek word *anemos*, meaning wind. Anemometers can be divided into two classes: those that measure the wind's speed and those that measure the wind's pressure. If the pressure is measured, a formula relating pressure with speed must be applied.

The wind measurements should be low-pass filtered since only the mean wind forces and moments can be compensated for by the propulsion system. In fact, since the inertia of the craft is so large, it is unnecessary for the control system to compensate for wind gust. In order to implement wind feedforward compensation for a DP vessel using (10.23), only the wind coefficients C_X, C_Y and C_N are needed. They can be experimentally obtained by using a scale model located in a wind tunnel or computed numerically. Different models for computation of the wind coefficients for varying hull geometries will now be discussed.

10.1.3 Wind Coefficients Based on Helmholtz–Kirchhoff Plate Theory

Blendermann (1994) applies a simple load concept to compute the wind coefficients. This is based on the *Helmholtz–Kirchhoff* plate theory. The wind load functions are parametrized in terms of four primary parameters: longitudinal and transverse resistance CD_{l} and CD_{t}, respectively, the cross-force parameter δ and the rolling moment factor κ. Numerical values for different vessels are given in Table 10.3.

The longitudinal resistance coefficient $\text{CD}_{\text{l}_{\text{AF}}}(\gamma_{\text{w}})$ in Table 10.3 is scaled according to

$$\text{CD}_{\text{l}} = \text{CD}_{\text{l}_{\text{AF}}}(\gamma_{\text{w}}) \frac{A_{\text{F}_{\text{w}}}}{A_{\text{L}_{\text{w}}}} \tag{10.30}$$

where values for two angles $\gamma_{\text{w}} \in \{0, \pi\}$ are given. The value $\text{CD}_{\text{l}_{\text{AF}}}(0)$ corresponds to head wind while $\text{CD}_{\text{l}_{\text{AF}}}(\pi)$ should be used for tail wind. By using these two values in the

Table 10.3 Coefficients of lateral and longitudinal resistance, cross-force and rolling moment (Blendermann 1994).

Type of vessel	CD_t	$CD_{l_{AF}}(0)$	$CD_{l_{AF}}(\pi)$	δ	κ
1. Car carrier	0.95	0.55	0.60	0.80	1.2
2. Cargo vessel, loaded	0.85	0.65	0.55	0.40	1.7
3. Cargo vessel, container on deck	0.85	0.55	0.50	0.40	1.4
4. Container ship, loaded	0.90	0.55	0.55	0.40	1.4
5. Destroyer	0.85	0.60	0.65	0.65	1.1
6. Diving support vessel	0.90	0.60	0.80	0.55	1.7
7. Drilling vessel	1.00	0.70–1.00	0.75–1.10	0.10	1.7
8. Ferry	0.90	0.45	0.50	0.80	1.1
9. Fishing vessel	0.95	0.70	0.70	0.40	1.1
10. Liquefied natural gas tanker	0.70	0.60	0.65	0.50	1.1
11. Offshore supply vessel	0.90	0.55	0.80	0.55	1.2
12. Passenger liner	0.90	0.40	0.40	0.80	1.2
13. Research vessel	0.85	0.55	0.65	0.60	1.4
14. Speed boat	0.90	0.55	0.60	0.60	1.1
15. Tanker, loaded	0.70	0.90	0.55	0.40	3.1
16. Tanker, in ballast	0.70	0.75	0.55	0.40	2.2
17. Tender	0.85	0.55	0.55	0.65	1.1

Source: Blendermann, W. (1994). Parameter Identification of Wind Loads on Ships. Journal of Wind Engineering and Industrial Aerodynamics 51, pp. 339–351. ©2009, Elsevier

regions $|\gamma_w| \leq \pi/2$ and $|\gamma_w| > \pi/2$, respectively, a non-symmetrical wind load function for surge can be computed. In other words, this gives different wind loads for head and tail winds, as shown in Figure 10.2. Alternatively, a symmetrical wind profile is obtained by using $CD_{l_{AF}}(0)$ or the mean of $CD_{l_{AF}}(0)$ and $CD_{l_{AF}}(\pi)$.

Let the mean height of the area A_{L_w} be denoted by

$$H_M = \frac{A_{L_w}}{L_{oa}} \tag{10.31}$$

and let the coordinates $(s_L, s_H) = (s_L, H_{L_w})$ describe the centroid of the transverse project area A_{L_w} with respect to the main section and above the water line. Based on these quantities, Blendermann (1994) gives the following expressions for the wind coefficients

$$C_X(\gamma_w) = \underbrace{-CD_l \frac{A_{L_w}}{A_{F_w}}}_{CD_{lAF}} \frac{\cos(\gamma_w)}{1 - \frac{\delta}{2}\left(1 - \frac{CD_l}{CD_t}\right)\sin^2(2\gamma_w)} \tag{10.32}$$

$$C_Y(\gamma_w) = CD_t \frac{\sin(\gamma_w)}{1 - \frac{\delta}{2}\left(1 - \frac{CD_l}{CD_t}\right)\sin^2(2\gamma_w)} \tag{10.33}$$

$$C_K(\gamma_w) = \kappa\, C_Y(\gamma_w) \tag{10.34}$$

$$C_N(\gamma_w) = \left[\frac{s_L}{L_{oa}} - 0.18\left(\gamma_w - \frac{\pi}{2}\right)\right] C_Y(\gamma_w) \tag{10.35}$$

where the expression for $C_K(\gamma_w)$ has been modified to comply with (10.12). Note that in Blendermann (1994)

$$C_K^{\text{Blendermann}}(\gamma_w) = \frac{s_H}{H_M} C_K(\gamma_w) \tag{10.36}$$

where $s_H = H_{L_w}$. The numerical values for several vessel types are given in Table 10.3.

Consider the research vessel in Table 10.3 with $A_{F_w} = 160.7\ \text{m}^2$, $A_{L_w} = 434.8\ \text{m}^2$, $s_L = 1.48\ \text{m}$, $s_H = 5.10\ \text{m}$, $L_{oa} = 55.0\ \text{m}$, $L_{pp} = 48.0\ \text{m}$ and $B = 12.5\ \text{m}$. For this vessel, the wind coefficients are computed in Matlab according to:

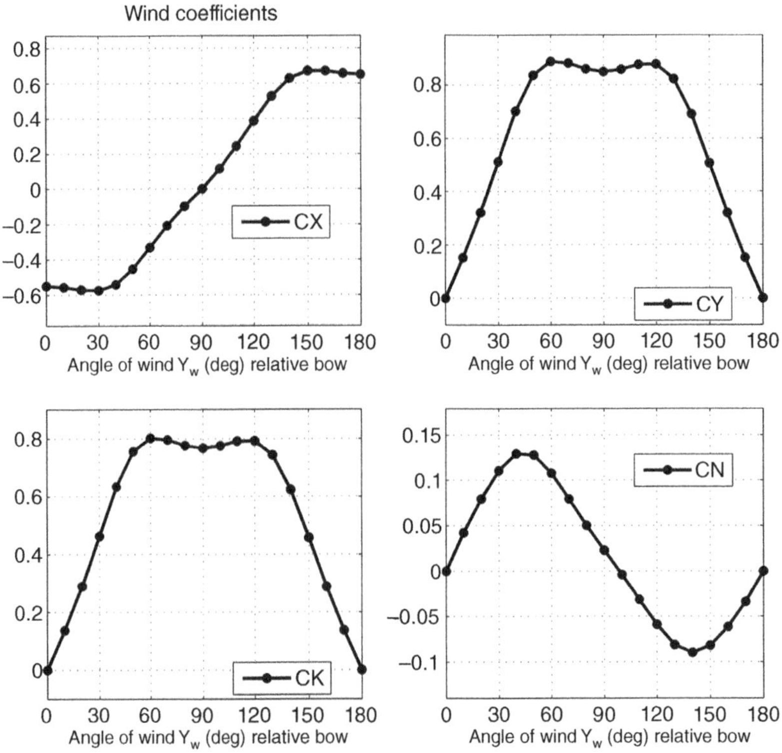

Figure 10.2 Wind coefficients for the research vessel in Table 10.3 (vessel #13). C_X is generated using $CD_{l_{AF}}(0)$ and $CD_{l_{AF}}(\pi)$ for $|\gamma_w| \leq \pi/2$ and $|\gamma_w| > \pi/2$, respectively.

MATLAB

The wind coefficients are plotted in Figure 10.2 using the MSS toolbox example file `ExWindForce.m`. The data sets of Blendermann (1994) are programmed in the Matlab function:

```
[w_wind,CX,CY,CK,CN] = ...
    blendermann94(gamma_r,V_r,AFw,ALw,sH,sL,Loa,vessel_no)
```

This function computes the non-symmetrical version of C_X.

10.1.4 Wind Coefficients for Merchant Ships

Isherwood (1972) derived a set of wind coefficients by using multiple regression techniques to fit experimental data of merchant ships. The wind coefficients are parametrized in terms of the following eight parameters:

L_{oa}	length overall
B	beam
A_{L_w}	lateral projected area
A_{T_w}	transverse projected area
A_{SS}	lateral projected area of superstructure
S	length of perimeter of lateral projection of model excluding water line and slender bodies such as masts and ventilators
C	distance from bow of centroid of lateral projected area
M	number of distinct groups of masts or king posts seen in lateral projection; king posts close against the bridge front are not included.

From regression analyses it was concluded that the measured data were best fitted to the following three equations:

$$C_X = -\left(A_0 + A_1\frac{2A_L}{L^2} + A_2\frac{2A_T}{B^2} + A_3\frac{L}{B} + A_4\frac{S}{L} + A_5\frac{C}{L} + A_6 M\right)$$

$$C_Y = B_0 + B_1\frac{2A_L}{L^2} + B_2\frac{2A_T}{B^2} + B_3\frac{L}{B} + B_4\frac{S}{L} + B_5\frac{C}{L} + B_6\frac{A_{SS}}{A_L}$$

$$C_N = C_0 + C_1\frac{2A_L}{L^2} + C_2\frac{2A_T}{B^2} + C_3\frac{L}{B} + C_4\frac{S}{L} + C_5\frac{C}{L}$$

where A_i and B_i ($i = 0, \dots, 6$) and C_j ($j = 0, \dots, 5$) are tabulated in Table 10.4, together with the *residual standard errors* (S.E.). The signs of C_X have been corrected to match the definition of γ_w in Figure 10.1.

Table 10.4 Wind force parameters in surge, sway and yaw (Isherwood, 1972).

γ_w (deg)	A_0	A_1	A_2	A_3	A_4	A_5	A_6	S.E.
0	−2.152	−5.00	−0.243	−0.164	–	–	–	0.086
10	−1.714	−3.33	−0.145	−0.121	–	–	–	0.104
20	−1.818	−3.97	−0.211	−0.143	–	–	−0.033	0.096
30	−1.965	−4.81	−0.243	−0.154	–	–	−0.041	0.117
40	−2.333	−5.99	−0.247	−0.190	–	–	−0.042	0.115
50	−1.726	−6.54	−0.189	−0.173	−0.348	–	−0.048	0.109
60	−0.913	−4.68	–	−0.104	−0.482	–	−0.052	0.082
70	−0.457	−2.88	–	−0.068	−0.346	–	−0.043	0.077
80	−0.341	−0.91	–	−0.031	–	–	−0.032	0.090
90	−0.355	–	–	–	−0.247	–	−0.018	0.094
100	−0.601	–	–	–	−0.372	–	−0.020	0.096
110	−0.651	−1.29	–	–	−0.582	–	−0.031	0.090
120	−0.564	−2.54	–	–	−0.748	–	−0.024	0.100
130	−0.142	−3.58	–	−0.047	−0.700	–	−0.028	0.105
140	−0.677	−3.64	–	−0.069	−0.529	–	−0.032	0.123
150	−0.723	−3.14	–	−0.064	−0.475	–	−0.032	0.128
160	−2.148	−2.56	–	−0.081	–	1.27	−0.027	0.123
170	−2.707	−3.97	−0.175	−0.126	–	1.81	–	0.115
180	−2.529	−3.76	−0.174	−0.128	–	1.55	–	0.112
							Mean S.E.	0.103

γ_w (deg)	B_0	B_1	B_2	B_3	B_4	B_5	B_6	S.E.
10	0.096	0.22	–	–	–	–	–	0.015
20	0.176	0.71	–	–	–	–	–	0.023
30	0.225	1.38	–	0.023	–	−0.29	–	0.030
40	0.329	1.82	–	0.043	–	−0.59	–	0.054
50	1.164	1.26	0.121	–	−0.242	−0.95	–	0.055
60	1.163	0.96	0.101	–	−0.177	−0.88	–	0.049
70	0.916	0.53	0.069	–	–	−0.65	–	0.047
80	0.844	0.55	0.082	–	–	−0.54	–	0.046
90	0.889	–	0.138	–	–	−0.66	–	0.051
100	0.799	–	0.155	–	–	−0.55	–	0.050
110	0.797	–	0.151	–	–	−0.55	–	0.049
120	0.996	–	0.184	–	−0.212	−0.66	0.34	0.047
130	1.014	–	0.191	–	−0.280	−0.69	0.44	0.051
140	0.784	–	0.166	–	−0.209	−0.53	0.38	0.060
150	0.536	–	0.176	−0.029	−0.163	–	0.27	0.055
160	0.251	–	0.106	−0.022	–	–	–	0.036
170	0.125	–	0.046	−0.012	–	–	–	0.022
							Mean S.E.	0.044

Table 10.4 (*Continued*)

γ_w(deg)	C_0	C_1	C_2	C_3	C_4	C_5	S.E.
10	0.0596	0.061	–	–	–	−0.074	0.0048
20	0.1106	0.204	–	–	–	−0.170	0.0074
30	0.2258	0.245	–	–	–	−0.380	0.0105
40	0.2017	0.457	–	0.0067	–	−0.472	0.0137
50	0.1759	0.573	–	0.0118	–	−0.523	0.0149
60	0.1925	0.480	–	0.0115	–	−0.546	0.0133
70	0.2133	0.315	–	0.0081	–	−0.526	0.0125
80	0.1827	0.254	–	0.0053	–	−0.443	0.0123
90	0.2627	–	–	–	–	−0.508	0.0141
100	0.2102	–	−0.0195	–	0.0335	−0.492	0.0146
110	0.1567	–	−0.0258	–	0.0497	−0.457	0.0163
120	0.0801	–	−0.0311	–	0.0740	−0.396	0.0179
130	−0.0189	–	−0.0488	0.0101	0.1128	−0.420	0.0166
140	0.0256	–	−0.0422	0.0100	0.0889	−0.463	0.0162
150	0.0552	–	−0.0381	0.0109	0.0689	−0.476	0.0141
160	0.0881	–	−0.0306	0.0091	0.0366	−0.415	0.0105
170	0.0851	–	−0.0122	0.0025	–	−0.220	0.0057
						Mean S.E.	0.0127

Source: Isherwood, R. M. (1972). Wind Resistance of Merchant Ships. RINA Transcripts 115, 327–338. ©1972, The Royal Institution of Naval Architects

MATLAB

The wind coefficients are plotted in Figure 10.3 using the MSS toolbox example file `ExWindForce.m`. The data sets of Isherwood (1972) are programmed in the Matlab function

```
[w_wind,CX,CY,CN] = ...
    isherwood72(gamma_r,V_r,Loa,B,AFw,ALw,A_SS,S,C,M)
```

10.1.5 Wind Coefficients for Very Large Crude Carriers

Wind loads on very large crude carriers (VLCCs) in the range 150 000–500 000 dwt can be computed by applying the results of OCIMF (1977). In this work the wind coefficients are scaled using the conversion factor 1/7.6 instead of 1/2. In addition, the signs in sway and yaw must be corrected such that

$$X_{\text{wind}} = \frac{1}{7.6} C_X^{\text{OCIMF}}(\gamma_w)\rho_a V_w^2 A_{F_w} \tag{10.37}$$

$$Y_{\text{wind}} = -\frac{1}{7.6} C_Y^{\text{OCIMF}}(\gamma_w)\rho_a V_w^2 A_{L_w} \tag{10.38}$$

$$N_{\text{wind}} = -\frac{1}{7.6} C_N^{\text{OCIMF}}(\gamma_w)\rho_a V_w^2 A_{L_w} L_{oa} \tag{10.39}$$

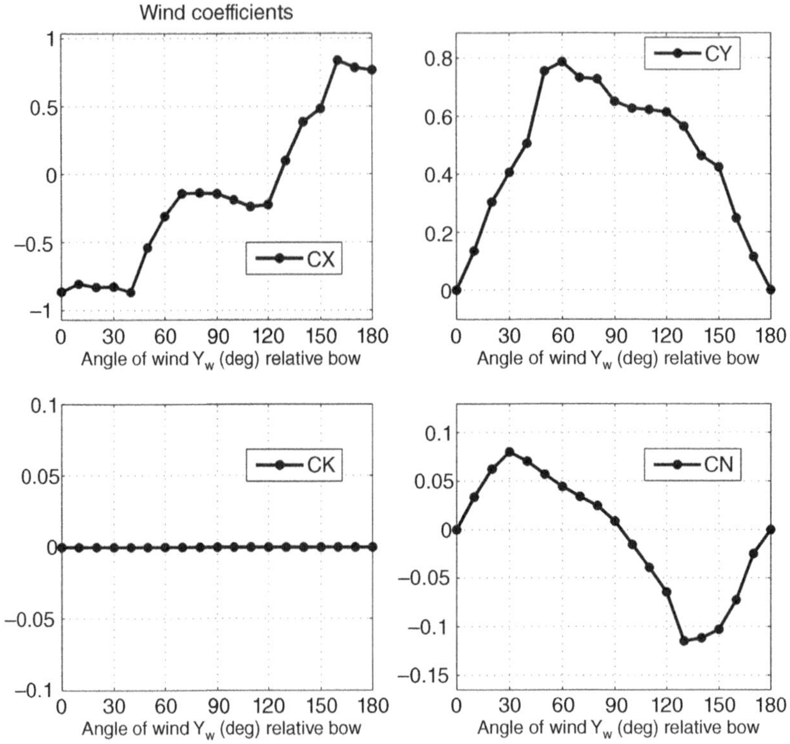

Figure 10.3 Wind coefficients for $L_{oa} = 100$, $B = 30$, $A_{L_w} = 900$, $A_{F_w} = 300$, $A_{SS} = 100$, $S = 100$, $C = 50$ and $M = 2$ using the formulae of Isherwood (1972). Source: Modified from Isherwood, R. M. (1972). Wind Resistance of Merchant Ships. RINA Transcripts 115, 327–338.

where the wind coefficients C_X^{OCIMF}, C_Y^{OCIMF} and C_N^{OCIMF} correspond to the plots given in OCIMF (1977); see Figures 10.4–10.6.

10.1.6 Wind Coefficients for Large Tankers and Medium-sized Ships

For wind resistance on large tankers in the 100 000–500 000 dwt class the reader is advised to consult Van Berlekom *et al.* (1974). Medium-sized ships of the order 600–50 000 dwt are discussed by Wagner (1967).

A detailed analysis of wind resistance using semi-empirical loading functions is given by Blendermann (1986). The data sets for seven ships are included in the report.

10.1.7 Wind Coefficients for Moored Ships and Floating Structures

Wind loads on moored ships are discussed by De Kat and Wichers (1991) while an excellent reference for huge pontoon-type floating structures is Kitamura *et al.* (1997).

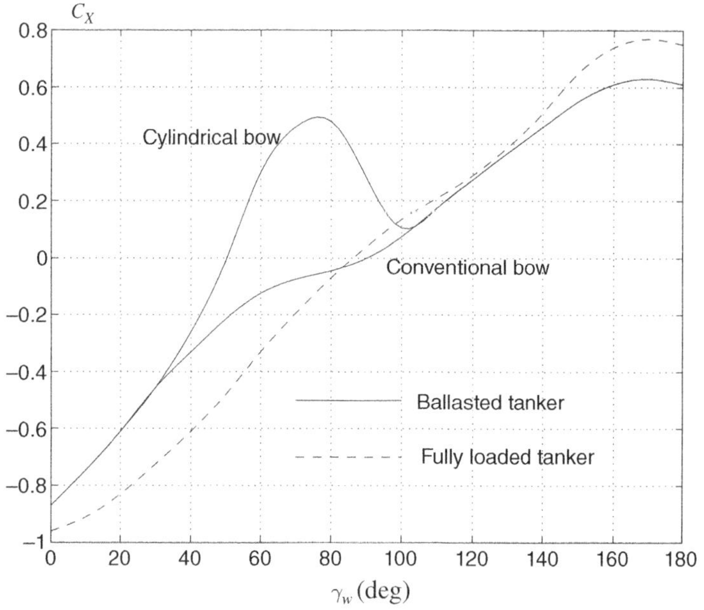

Figure 10.4 Longitudinal wind force coefficient C_X^{OCIMF} as a function of γ_w (OCIMF 1977).

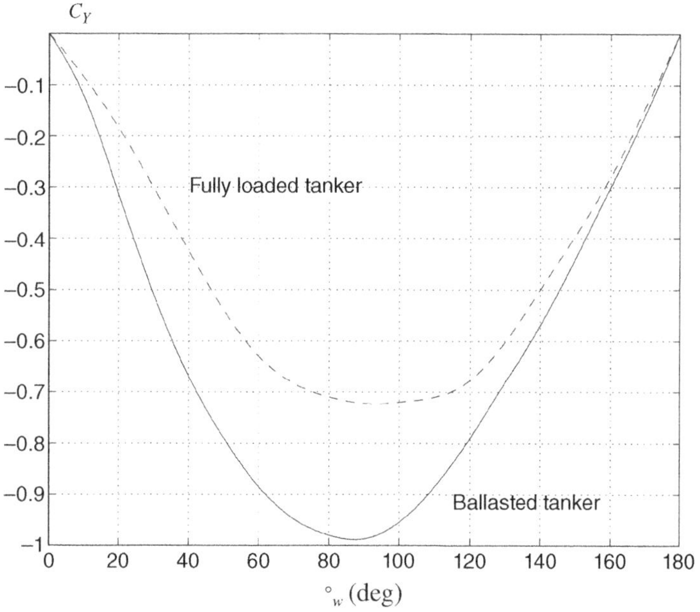

Figure 10.5 Lateral wind force coefficient C_Y^{OCIMF} as a function of γ_w (OCIMF 1977).

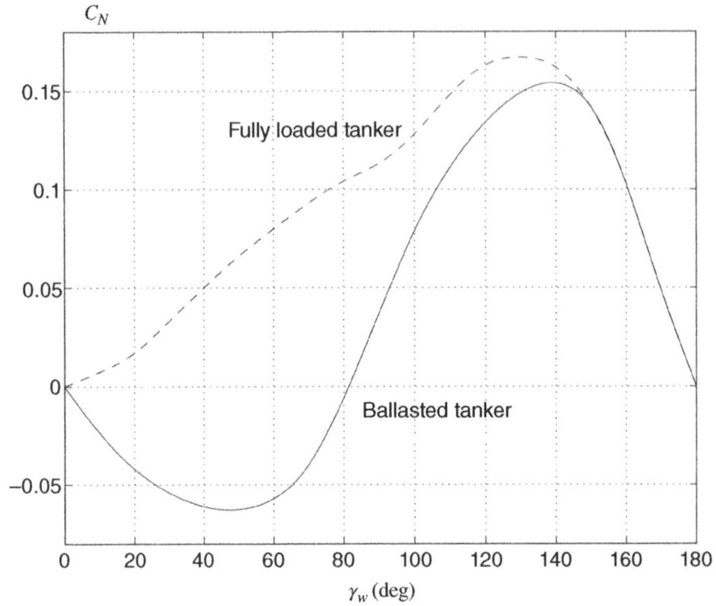

Figure 10.6 Wind moment coefficient C_N^{OCIMF} in yaw as a function of γ_w (OCIMF 1977).

10.2 Wave Forces and Moments

A motion control system can be simulated under influence of wave-induced forces by separating the *first-order* and *second-order* effects:

- **First-order wave-induced forces**: wave-frequency (WF) motion observed as zero-mean oscillatory motions.
- **Second-order wave-induced forces**: wave drift forces observed as nonzero slowly-varying components.

When designing motion control systems, it is important to evaluate robustness and performance in the presence of waves. Wave forces are observed as a mean slowly-varying component and an oscillatory component, which need to be compensated differently by a feedback control system. For instance, the mean component can be removed by using integral action while the oscillatory component usually is removed by using a cascaded notch and low-pass filter. This is usually referred to as *wave filtering*.

This section describes wave force models that can be used for prediction, observer-based wave filtering and testing of feedback control systems in the presence of waves. Both methods based on response amplitude operators (RAOs) and linear state-space models will be discussed. This includes:

- Force RAOs
- Motion RAOs
- Linear state-space models (WF models).

The first two methods require that the RAO tables are computed using a hydrodynamic program (see Section 5.1) since the wave forces depend on the geometry of the craft. The last method is preferred due to its simplicity but it is only intended for the testing of robustness and performance of control systems, that is closed-loop analysis.

The generalized wave forces can be expressed as the sum

$$\boldsymbol{\tau}_{\text{wave}} = \boldsymbol{\tau}_{\text{wave1}} + \boldsymbol{\tau}_{\text{wave2}} \tag{10.40}$$

where $\boldsymbol{\tau}_{\text{wave1}}$ and $\boldsymbol{\tau}_{\text{wave2}}$ are the first- and second-order wave-induced forces, respectively. The next sections explain how these quantities can be realized in a time-domain simulator.

10.2.1 Sea-state Descriptions

For marine craft the sea states in Table 10.5 can be characterized by the following wave spectrum parameters:

- The significant wave height H_s (the mean wave height of the one-third highest waves, also denoted as $H_{1/3}$)
- One of the following wave periods:
  - The average wave period, T_1
  - Average zero-crossing wave period, T_z
  - Peak period, T_p (this is equivalent to the modal period, T_0).

The wave periods are strongly related to each other and they are used interchangeably by naval architects. To relate the different periods to each other it is necessary to define the wave spectrum moments.

Table 10.5 Definition of sea-state (SS) codes (Price and Bishop 1974). Note that the percentage probability for SS codes 0, 1 and 2 is summarized.

			Percentage probability		
Sea state	Description	Wave height	World	North	Northern
code	of sea	observed (m)	wide	Atlantic	North Atlantic
0	Calm (glassy)	0			
1	Calm (rippled)	0–0.1	11.2486	8.3103	6.0616
2	Smooth (wavelets)	0.1–0.5			
3	Slight	0.5–1.25	31.6851	28.1996	21.5683
4	Moderate	1.25–2.5	40.1944	42.0273	40.9915
5	Rough	2.5–4.0	12.8005	15.4435	21.2383
6	Very rough	4.0–6.0	3.0253	4.2938	7.0101
7	High	6.0–9.0	0.9263	1.4968	2.6931
8	Very high	9.0–14.0	0.1190	0.2263	0.4346
9	Phenomenal	Over 14.0	0.0009	0.0016	0.0035

Wave Spectrum Moments

A wave spectrum $S(\omega)$, see Figure 10.7, can be classified by means of *wave spectrum moments*

$$m_k := \int_0^\infty \omega^k \, S(\omega)\mathrm{d}\omega \qquad (k = 0, \dots, N). \tag{10.41}$$

For $k = 0$, this yields

$$m_0 = \int_0^\infty S(\omega)\mathrm{d}\omega. \tag{10.42}$$

The instantaneous wave elevation is Gaussian distributed with zero mean and variance

$$\sigma^2 = m_0 \tag{10.43}$$

where σ is the root-mean-square (RMS) value of the spectrum.

The *modal frequency* (peak frequency) ω_0 is found by requiring that

$$\left.\frac{\mathrm{d}S(\omega)}{\mathrm{d}\omega}\right|_{\omega_0} = 0. \tag{10.44}$$

Hence, the *modal period* becomes

$$T_0 = \frac{2\pi}{\omega_0}. \tag{10.45}$$

Consequently, the maximum value of $S(\omega)$ is

$$S_{\max} = S(\omega_0). \tag{10.46}$$

Under the assumption that the wave height is Rayleigh distributed it can be shown that the significant wave height satisfies (Price and Bishop 1974)

$$H_s = 4\sigma = 4\sqrt{m_0}. \tag{10.47}$$

The *average wave period* is defined as

$$T_1 := 2\pi \frac{m_0}{m_1} \tag{10.48}$$

while the *average zero-crossings period* is defined as

$$T_z := 2\pi \sqrt{\frac{m_0}{m_2}}. \tag{10.49}$$

10.2.2 Wave Spectra

The process of wave generation due to wind starts with small wavelets appearing on the water surface. This increases the drag force, which in turn allows short waves to grow. These short waves continue to grow until they finally break and their energy is dissipated. It is observed that a *developing sea*, or storm, starts with high frequencies creating a spectrum with a peak at a relatively high frequency. A storm that has lasted for a long time is said to create a *fully developed sea*. After the wind has stopped, a low-frequency

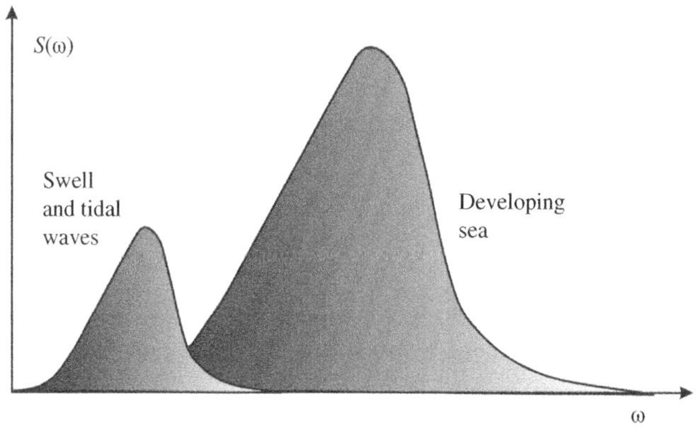

Figure 10.7 Two peaked wave spectra $S(\omega)$ where one peak is due to swell and tidal waves and the other peak is due to a developing sea.

decaying sea or swell is formed. These long waves form a wave spectrum with a low peak frequency.

If the swell from one storm interacts with the waves from another storm, a wave spectrum with two peak frequencies may be observed. In addition, tidal waves will generate a peak at a low frequency. Hence, the resulting wave spectrum might be quite complicated in cases where the weather changes rapidly (see Figure 10.7).

The state-of-the-art wave spectra will now be presented. These models are used to derive linear approximations and transfer functions for computer simulations, autopilot wave filtering and state reconstruction, which are the topics in Part II.

Neumann Spectrum

The earliest spectral formulation is due to who proposed the *one-parameter* spectrum

$$S(\omega) = C\omega^{-6} \exp(-2g^2\omega^{-2}V^{-2}) \tag{10.50}$$

where $S(\omega)$ in m s^{-2} is the wave elevation power spectral density function, C is an empirical constant, V is the wind speed and g is the acceleration of gravity. Six years later Phillips (1958) showed that the high-frequency part of the wave spectrum reached the asymptotic limit

$$\lim_{\omega \gg 1} S(\omega) = \alpha\, g^2\, \omega^{-5} \tag{10.51}$$

where α is a positive constant. This limiting function of Phillips is still used as the basis for most spectral formulations.

Bretschneider Spectrum

The spectrum of Neumann was further extended to a two-parameter spectrum by Bretschneider (1959)

$$S(\omega) = 1.25\frac{\omega_0^4 H_s^2}{4}\omega^{-5} \exp[-1.25\,(\omega_0/\omega)^4] \tag{10.52}$$

where ω_0 is the *modal* or *peak frequency* of the spectrum and H_s is the *significant wave height* (mean of the one-third highest waves). This spectrum was developed for the North Atlantic, for unidirectional seas, infinite depth, no swell and unlimited fetch. The significant wave height H_s is used to classify the type of sea in terms of sea-state codes 0, 1, ..., 9, as shown in Table 10.5.

Pierson–Moskowitz Spectrum

Pierson and Moskowitz (1963) developed a two-parameter wave spectral formulation for fully developed wind-generated seas from analyses of wave spectra in the North Atlantic Ocean

$$S(\omega) = A\omega^{-5} \exp(-B\omega^{-4}) \tag{10.53}$$

which is commonly known as the *PM spectrum* (Pierson–Moskowitz spectrum). The PM spectrum is used as the basis for several spectral formulations but with different A and B values. In its original formulation, the PM spectrum is only a one-parameter spectrum since only B changes with the sea state. The parameters are

$$A = 8.1 \times 10^{-3}\, g^2 = \text{constant} \tag{10.54}$$

$$B = 0.74\left(\frac{g}{V_{19.5}}\right)^4 = \frac{3.11}{H_s^2} \tag{10.55}$$

where $V_{19.5}$ is the wind speed at a height of 19.5 m over the sea surface.

MATLAB

The Bretschneider and PM spectra are implemented in the MSS toolbox as wave spectra 1 and 2

```
S = wavespec(1, [A,B],w,1)
S = wavespec(2,V20,w,1)
```

where `A` and `B` are the spectrum parameters, `V20` is wind speed at approximately 20 m height and `w` is the wave frequency vector.

The relationship between $V_{19.5}$ and H_s in (10.55) is based on the assumption that the waves can be represented by Gaussian random processes and that $S(\omega)$ is narrow banded. From (10.55) it is seen that

$$H_s = \frac{2.06}{g^2} V_{19.5}^2. \tag{10.56}$$

This implies that the significant wave height is proportional to the square of the wind speed. This is shown in Figure 10.8 where the *sea-state codes* and *Beaufort numbers* are plotted against each other; see Tables 10.2 and 10.5.

The *modal frequency* (peak frequency) ω_0 for the PM spectrum is found by requiring that

$$\left.\frac{\mathrm{d}S(\omega)}{\mathrm{d}\omega}\right|_{\omega_0} = 0. \tag{10.57}$$

Differentiation of (10.53) and solving for ω_0 yields

$$\omega_0 = \sqrt[4]{\frac{4B}{5}} \quad \Rightarrow \quad T_0 = 2\pi \sqrt[4]{\frac{5}{4B}} \tag{10.58}$$

where T_0 is the *modal period*. Consequently, the maximum value of $S(\omega)$ is

$$S_{max} = S(\omega_0) = \frac{5A}{4B\omega_0} \exp(-5/4) \tag{10.59}$$

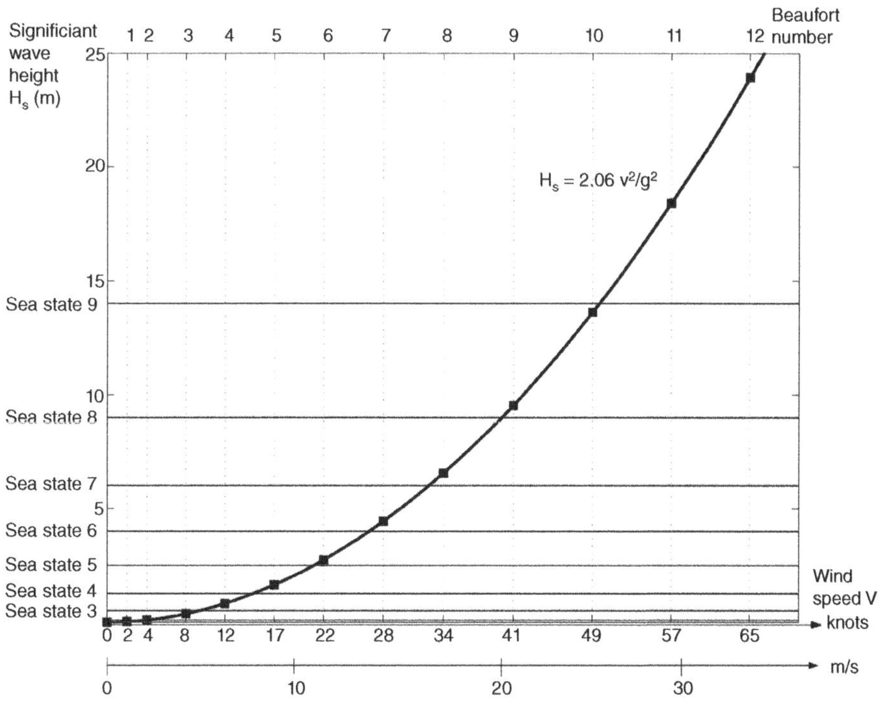

Figure 10.8 Plot showing the relationship between significant wave height, wind speed, Beaufort numbers and sea-state codes.

Modified Pierson–Moskowitz (MPM) Spectrum

In order to predict the responses of marine craft in open sea, the International Ship and Offshore Structures Congress (2nd ISSC 1964), the International Towing Tank Conferences (12th ITTC 1969 and 15th ITTC 1978) have recommended the use of a modified version of the PM spectrum (see Figure 10.9) where

$$A = \frac{4\pi^3 H_s^2}{T_z^4}, \qquad B = \frac{16\pi^3}{T_z^4}. \tag{10.60}$$

This representation of the PM spectrum has two parameters H_s and T_z, or alternatively T_0 and T_1 given by

$$T_z = 0.710\ T_0 = 0.921\ T_1. \tag{10.61}$$

MATLAB

The modified PM spectrum is implemented in the MSS toolbox as wave spectra 3 to 5

```
S = wavespec(3,[Hs,T0],w,1)
S = wavespec(4,[Hs,T1],w,1)
S = wavespec(5,[Hs,Tz],w,1)
```

where `Hs` is the significant wave height, `T0,T1` and `Tz` are the peak, average and average zero-crossing wave periods, respectively, while `w` is the wave frequency vector.

The modified PM spectrum should only be used for a fully developed sea with large (infinite) depth, no swell and unlimited fetch. For non-fully developed seas the *JONSWAP* or *Torsethaugen* spectra are recommended.

JONSWAP Spectrum

In 1968 and 1969 an extensive measurement program was carried out in the North Sea, between the island Sylt in Germany and Iceland. The measurement program is known as the *Joint North Sea Wave Project* (JONSWAP) and the results from these investigations have been adopted as an ITTC standard by the 17th ITTC (1984). Since the JONSWAP spectrum (see Figure 10.9) is used to describe *non-fully developed seas*, the spectral density function will be more peaked than those representing fully developed spectra. The proposed spectral formulation is representative for wind-generated waves under the assumption of finite water depth and limited fetch. The spectral density function is written

$$S(\omega) = 155\, \frac{H_s^2}{T_1^4}\omega^{-5} \exp\left(\frac{-944}{T_1^4}\omega^{-4}\right)\gamma^Y \tag{10.62}$$

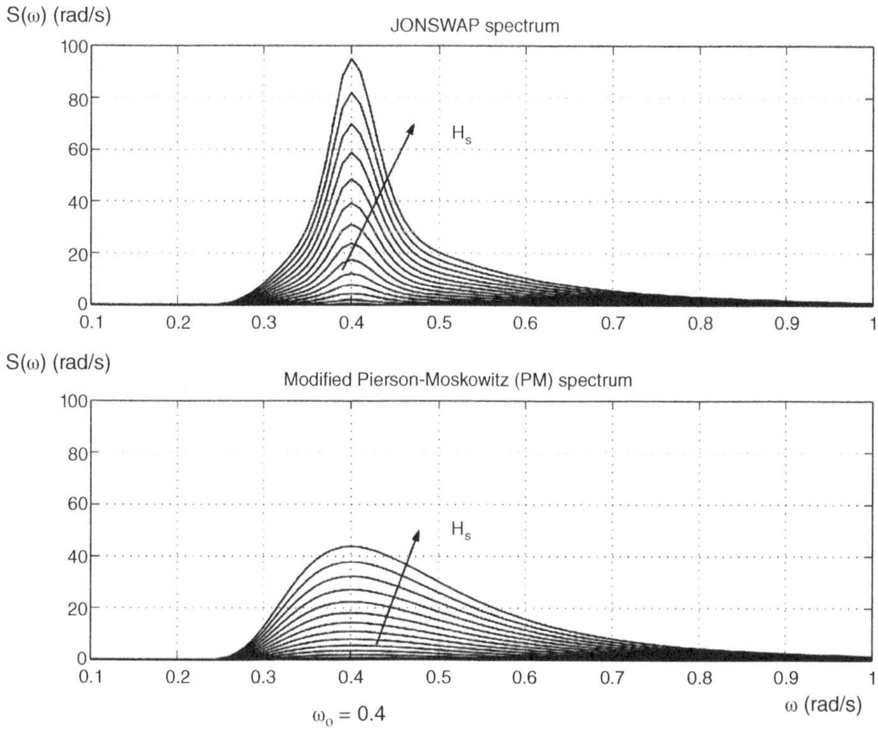

Figure 10.9 Plot showing the JONSWAP and modified Pierson–Moskowitz spectra for $\omega_0 = 0.4$ rad s^{-1} and $H_s = 3, 4, ..., 14$ m.

where Hasselmann *et al.* (1973) suggest that $\gamma = 3.3$ and

$$Y = \exp\left[-\left(\frac{0.191\omega T_1 - 1}{\sqrt{2}\sigma}\right)^2\right] \tag{10.63}$$

where

$$\sigma = \begin{cases} 0.07 & \text{for} \quad \omega \leq 5.24/T_1 \\ 0.09 & \text{for} \quad \omega > 5.24/T_1. \end{cases} \tag{10.64}$$

Alternative formulations can be derived in terms of the characteristic periods like T_0 and T_z by using

$$T_1 = 0.834\ T_0 = 1.073\ T_z. \tag{10.65}$$

MATLAB

The JONSWAP spectrum is included in the MSS toolbox as wave spectra 6 and 7

```
S = wavespec(6,[V10,fetch],w,1)
S = wavespec(7,[Hs,w0,gamma],w,1)
```

where `V10` is the wind speed at 10 m height, `Hs` is the significant wave height, `w0` is peak frequency and `w` is the wave frequency vector.

Torsethaugen Spectrum

The *Torsethaugen spectrum* is an empirical, two-peaked spectrum, which includes the effect of swell (low-frequency peak) and newly developed waves (high-frequency peak). The spectrum was developed for Norsk Hydro (Torsethaugen 1996), and standardized under the Norsok Standard (1999). The spectrum was developed using curve fitting of experimental data from the North Sea.

MATLAB

The Torsethaugen spectrum is included in the MSS toolbox as wave spectrum 7

```
S = wavespec(7,[Hs,w0],w,1)
```

where `Hs` is the significant wave height, `w0` is peak frequency and `w` is the wave frequency vector.

If the peak frequency ω_0 is chosen to be less than approximately 0.6 rad s^{-1}, the Torsethaugen spectrum reduces to a one-peak spectrum where swell dominates. For peak frequencies $\omega_0 > 0.6$ rad s^{-1} the two characteristic peaks shown in Figure 10.10 clearly appear. This is due to the fact that developing waves have energy at high frequencies compared to swell. This combined effect is very common in the North Sea, and it makes DP and autopilot design a challenging task in terms of wave filtering.

MATLAB

The different wave spectra when plotted for the same wave height and peak frequency are shown in Figure 10.11. The plots are generated by using the wave demo option in the MSS toolbox:

```
gncdemo
```

Second-order Wave Transfer Function Approximation

The nonlinear wave spectra can be approximated by a second-order transfer function

$$h(s) = \frac{\xi}{w}(s) = \frac{K_w s}{s^2 + 2\lambda\omega_0\, s + \omega_0^2} \tag{10.66}$$

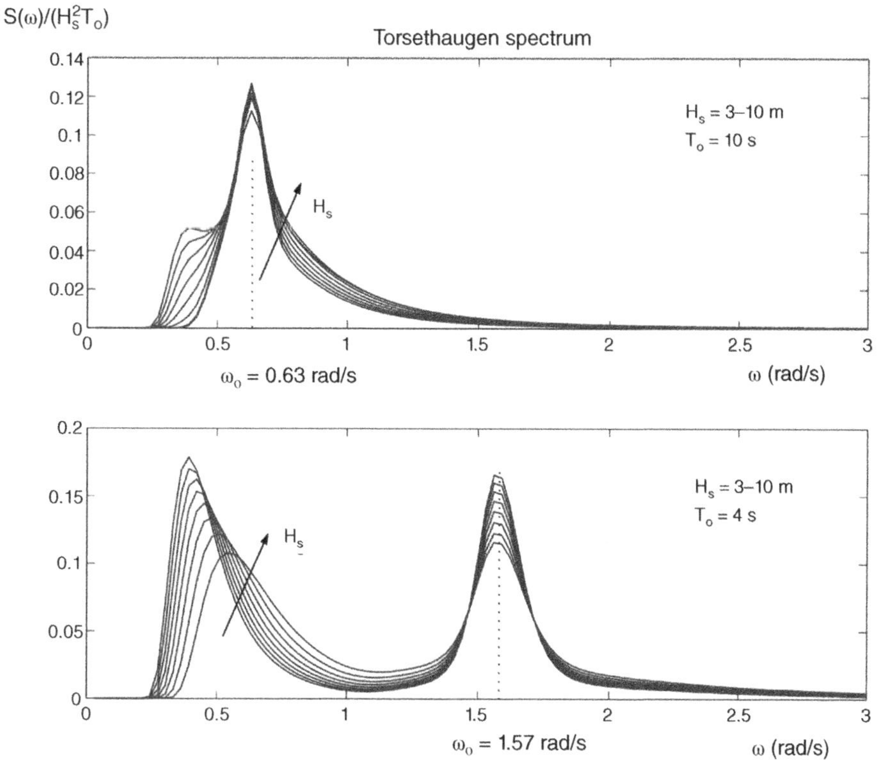

Figure 10.10 Torsethaugen spectrum: upper plot shows only one peak at $\omega_0 = 0.63$ rad s^{-1} representing swell and developing sea while the lower plot shows low-frequency swell and newly developing sea with peak frequency $\omega_0 = 1.57$ rad s^{-1}.

where w is zero-mean Gaussian white noise and ξ is the wave elevation. It is convenient to define the gain constant according to

$$K_{\rm w} = 2\lambda\,\omega_0\,\sigma \tag{10.67}$$

where σ is a constant describing the wave intensity, λ is a damping coefficient and ω_0 is the dominating wave frequency.

Consequently, substituting $s = {\rm j}\omega$ into (10.66) yields the frequency response

$$h({\rm j}\omega) = \frac{{\rm j}\,2(\lambda\,\omega_0\,\sigma)\,\omega}{(\omega_0^2 - \omega^2) + {\rm j}\,2\lambda\,\omega_0\,\omega}. \tag{10.68}$$

The magnitude of $h({\rm j}\omega)$ becomes

$$|h({\rm j}\omega)| = \frac{2(\lambda\,\omega_0\,\sigma)\,\omega}{\sqrt{(\omega_0^2 - \omega^2)^2 + 4(\lambda\,\omega_0\,\omega)^2}}. \tag{10.69}$$

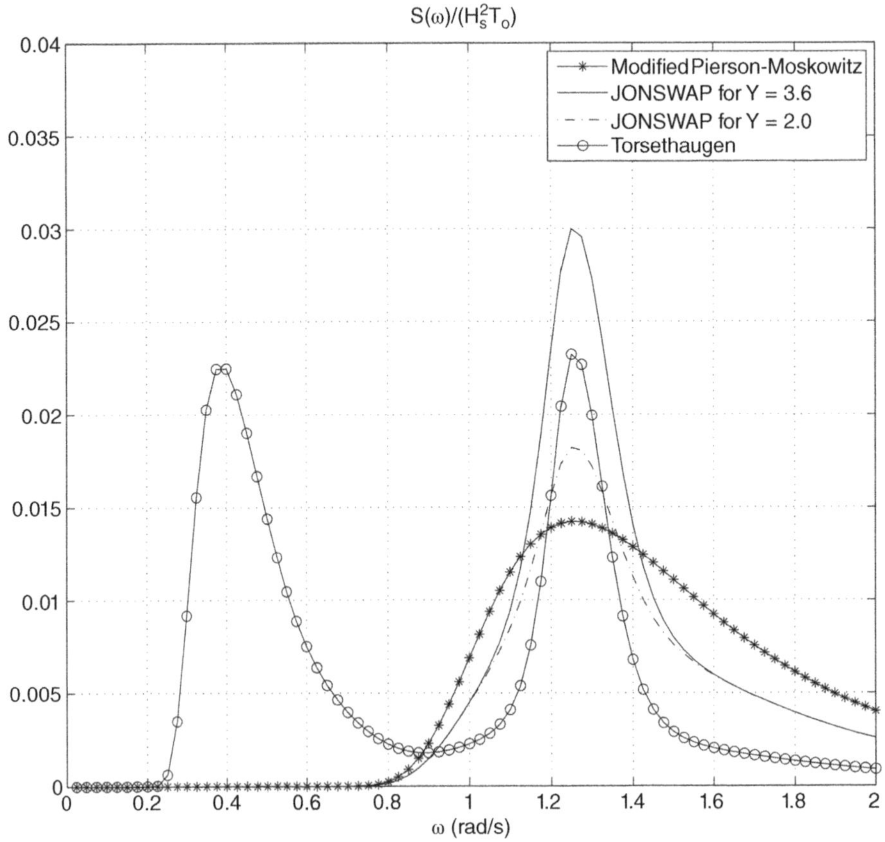

Figure 10.11 Comparison of different wave spectra.

Hence, the power spectral density (PSD) function for ξ can be computed as

$$P_{\xi\xi}(\omega) = |h(j\omega)|^2\ P_{ww}(\omega) = |h(j\omega)|^2 \tag{10.70}$$

since $P_{ww}(\omega) = 1$. The ultimate goal is to design an approximation $P_{\xi\xi}(\omega)$ to $S(\omega)$, for instance by means of nonlinear regression, such that $P_{\xi\xi}(\omega)$ reflects the energy distribution of $S(\omega)$ in the relevant frequency range. From (10.70), we obtain

$$P_{\xi\xi}(\omega) = \frac{4(\lambda\ \omega_0\ \sigma)^2\ \omega^2}{(\omega_0^2 - \omega^2)^2 + 4(\lambda\ \omega_0\ \omega)^2}. \tag{10.71}$$

Determination of σ and λ

Since the maximum value of $P_{\xi\xi}(\omega)$ and $S(\omega)$ are obtained for $\omega = \omega_0$, it follows that

$$P_{\xi\xi}(\omega_0) = S(\omega_0) \tag{10.72}$$

$$\Updownarrow$$

$$\sigma^2 = \max_{0<\omega<\infty} S(\omega). \tag{10.73}$$

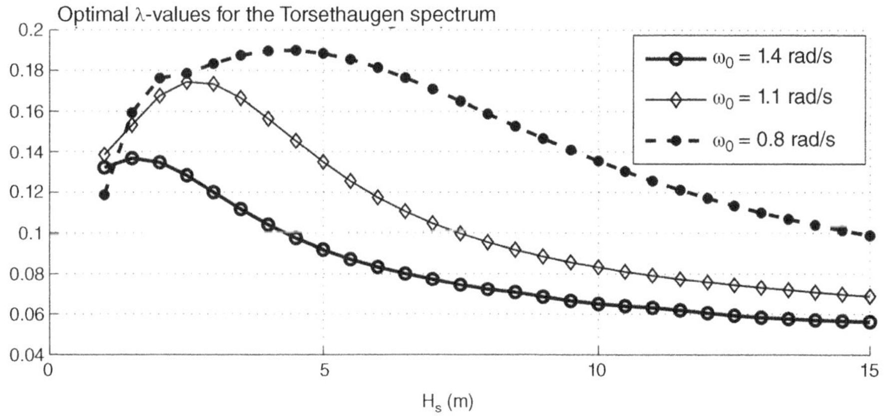

Figure 10.12 Least-squares optimal λ values for the Torsethaugen spectrum for varying H_s and ω_0 when a linear spectrum is fitted to the high-frequency peak of the spectrum.

For the PM spectrum (10.53) this implies

$$\sigma = \sqrt{\frac{A}{\omega_0^5}\exp\left(-\frac{B}{\omega_0^4}\right)} \tag{10.74}$$

while the term $\gamma^{Y(\omega_0)}$ must be included for the JONSWAP spectrum. The damping ratio λ can be computed by requiring that the energy, that is the areas under $P_{\xi\xi}(\omega)$ and $S(\omega)$ of the spectra, be equal.

An alternative approach is to use nonlinear least-squares (NLS) to compute λ such that $P_{\xi\xi}(\omega)$ fits $S(\omega)$ in a least-squares sense; see Figure 10.13. This is demonstrated in Example 10.1 using the Matlab optimization toolbox.

Example 10.1 (Nonlinear Least-Squares Optimization of Linear Spectra)
Consider the Matlab script `ExLinspec.m` *for computation of λ. The output of the nonlinear optimization process gives the following λ values for the modified PM and JONSWAP spectra:*

ω_0	*0.5*	*0.8*	*1.1*	*1.4*	*Recommended value*
λ *(MPM)*	0.2565	0.2573	0.2588	0.2606	0.26
λ *(JONSWAP)*	0.1017	0.1017	0.1017	0.1017	0.10

The λ value for both these spectra are independent of the wave height H_s. For the Torsethaugen spectrum the λ values vary with both H_s and ω_0, as shown in Figure 10.12. The results of the curve-fitting procedure for the three different spectra are shown in Figure 10.13. Since the Torsethaugen spectrum is a two-peaked spectrum a second linear spectrum should be added to fit the swell peak at low frequencies.

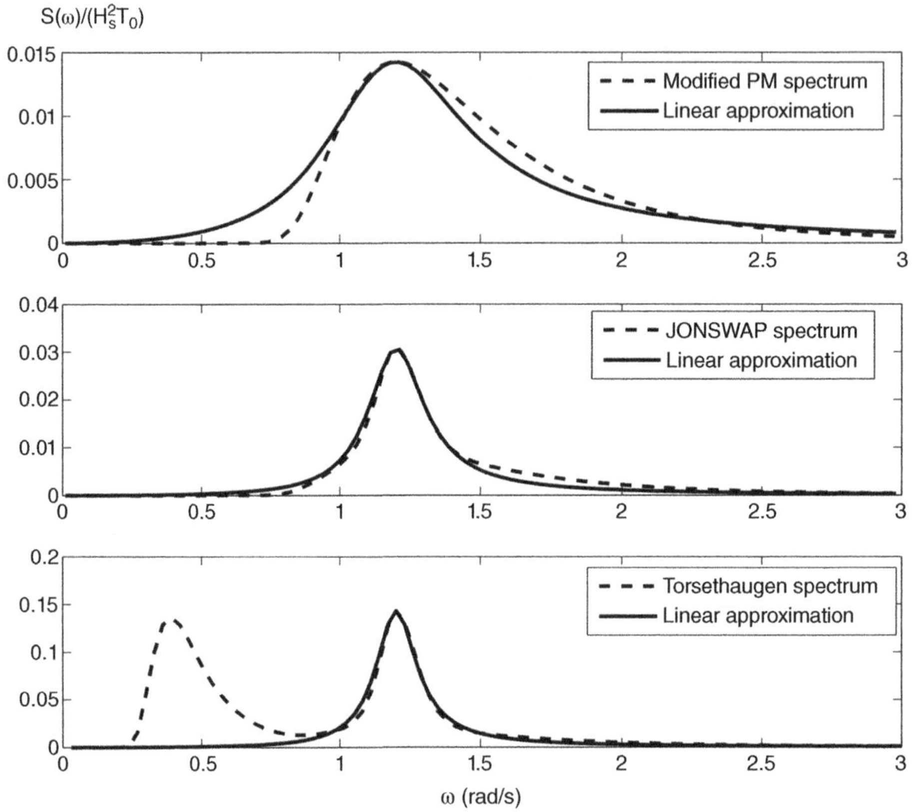

Figure 10.13 Nonlinear least-squares fit of a linear spectrum to the PM, JONSWAP and Torsethaugen spectra. Only one peak is approximated for the Torsethaugen spectrum.

MATLAB

Power spectral density function:

```
function Pyy = Slin(lambda,w)
% Pyy = Slin(lambda,w) 2nd-order linear PSD function
% w = wave spectrum frequency (rad/s)
% lambda = relative damping factor

global sigma wo
Pyy = 4*(lambda*wo*sigma)^2*w.^2 ./ ( (wo^2-w.^2).^2 + ...
    4*(lambda*wo.*w).^2 );
```

MATLAB

Nonlinear least-squares:

```
% Script for plotting the nonlinear least-squares fit
% see ExLinspec.m
global sigma wo

wo = 0.8; Hs = 10; wmax = 3;
To = 2*pi/wo;
w = (0.0001:0.01:wmax)';

% Modified PM
subplot(311)
S = wavespec(3,[Hs,To],w,1);  sigma = sqrt(max(S));
lambda = lsqcurvefit('Slin', 0.1, w, S)
hold on; plot(w,Slin(lambda,w),'linewidth',2); hold off;
legend('Modified PM spectrum','Linear approximation')

% JONSWAP
subplot(312)
S = wavespec(7,[Hs,wo,3.3],w,1);   sigma = sqrt(max(S));
lambda = lsqcurvefit('Slin', 0.1, w, S)
hold on; plot(w,Slin(lambda,w),'linewidth',2); hold off;
legend('JONSWAP spectrum','Linear approximation')

% Torsethaugen (only one peak is fitted)
subplot(313)
S = wavespec(8,[Hs,wo],w,1);   sigma = sqrt(max(S));
lambda = lsqcurvefit('Slin', 0.1, w, S)
hold on; plot(w,Slin(lambda,w),'linewidth',2); hold off;
legend('Torsethaugen spectrum','Linear approximation')
```

10.2.3 Wave Amplitude Response Model

The relationship between the wave spectrum $S(\omega_k)$ and the wave amplitude A_k for a wave component, subscript k, is (Faltinsen 1990)

$$\frac{1}{2}A_k^2 = S(\omega_k)\Delta\omega \tag{10.75}$$

where $\Delta\omega$ is a constant difference between the frequencies ω_k. Formula (10.75) can be used to compute wave-induced responses in the time domain.

Long-crested Irregular Sea

The wave elevation of a *long-crested* irregular sea in the origin o_s of of the seakeeping reference frame $\{s\}$ under the assumption of zero speed can be written as the sum of N harmonic components

$$\begin{aligned}\xi &= \sum_{k=1}^{N} A_k \cos(\omega_k t + \epsilon_k) \\ &= \sum_{k=1}^{N} \sqrt{2S(\omega_k)\Delta\omega}\ \cos(\omega_k t + \epsilon_k)\end{aligned} \tag{10.76}$$

where ϵ_k is the random phase angle of wave component number k. Since this expression repeats itself after a time $2\pi/\Delta\omega$ a large number of wave components N are needed. However, a practical way to avoid this is to choose ω_k randomly in the interval

$$\left[\omega_k - \frac{\Delta\omega}{2},\ \omega_k + \frac{\Delta\omega}{2}\right] \tag{10.77}$$

implying that good results can be obtained for N in the range 50–100.

The wave frequencies relate to the wave numbers through the *dispersion relation*

$$\omega^2 = kg\ \tanh(kh) \overset{kh\gg 1}{\approx} kg \tag{10.78}$$

where h is the water depth. It is possible to compute the wave elevation along the x_s and y_s axes for a given wave direction β (see Figure 10.14) by modifying (10.76) as

$$\xi = \sum_{k=1}^{N} \sqrt{2S(\omega_k)\Delta\omega}\ \cos(\omega_k t - k_k\ x^s\ \cos(\beta) - k_k\ y^s\ \sin(\beta) + \epsilon_k) \tag{10.79}$$

where the wave component numbers k_k are computed for each wave frequency ω_k using (10.78) and $(x^s,\ y^s)$ are the coordinates expressed in $\{s\}$.

Short-crested Irregular Sea

The most likely situation encountered at sea is *short-crested* or confused waves. This is observed as irregularities along the wave crest at right angles to the direction of the wind. The effect of short-crestedness can be modeled by a 2D wave spectrum

$$S_M(\omega, \mu) = S(\omega)M(\mu) \tag{10.80}$$

where $\mu = 0$ is the main wave propagation direction and μ will spread the energy over a certain angle contained within $[-\pi/2,\ \pi/2]$ from the wind direction. The energy is preserved by requiring that

$$\int_{-\pi/2}^{\pi/2} M(\mu) = 1. \tag{10.81}$$

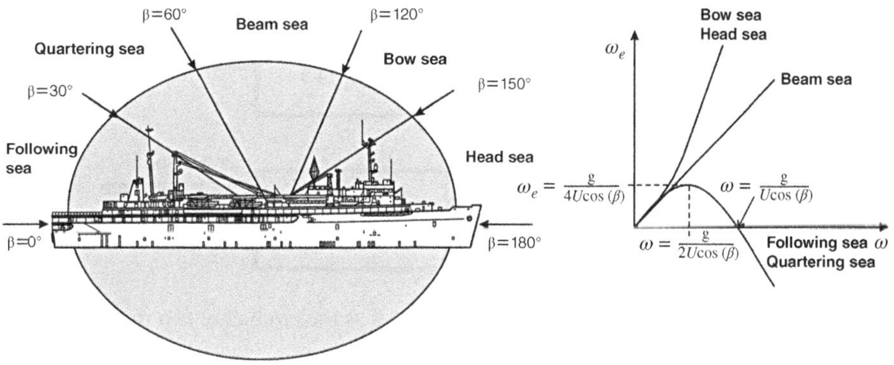

Figure 10.14 Definition of encounter angle β.

A commonly used spreading function is

$$M(\mu) = \begin{cases} \frac{2}{\pi}\cos^2(\mu), & -\pi/2 \le \mu \le \pi/2 \\ 0, & \text{elsewhere} \end{cases}. \tag{10.82}$$

For this case (10.76) becomes

$$\xi = \sum_{k=1}^{N}\sum_{i=1}^{M} \sqrt{2S_M(\omega_k, \mu_i - \beta)\Delta\omega\Delta\mu}\, \cos(\omega_k t + \epsilon_{i,k}) \tag{10.83}$$

where β is the mean wave encounter angle and μ_i is taken randomly in the interval

$$\left[\mu_i - \frac{\Delta\mu}{2},\ \mu_i + \frac{\Delta\mu}{2}\right]. \tag{10.84}$$

These equations effectively represent the first block in Figures 10.15 and 10.16.

Extension to Forward Speed Using the Frequency of Encounter

For a marine craft moving at forward speed U, the peak frequency ω_0 of the spectrum will be shifted according to

$$\omega_e(U, \omega_0, \beta) = \left|\omega_0 - \frac{\omega_0^2}{g} U\cos(\beta)\right| \tag{10.85}$$

where

ω_e encounter frequency (rad s^{-1})
ω_0 wave spectrum peak frequency (rad s^{-1})
g acceleration of gravity (m s^{-2})
U total speed of ship (m s^{-1})
β wave encounter angle (rad).

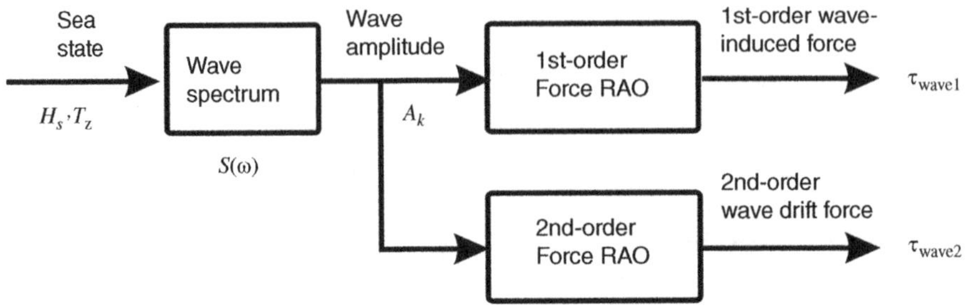

Figure 10.15 Representation of the wave-induced forces as the product of two transfer functions.

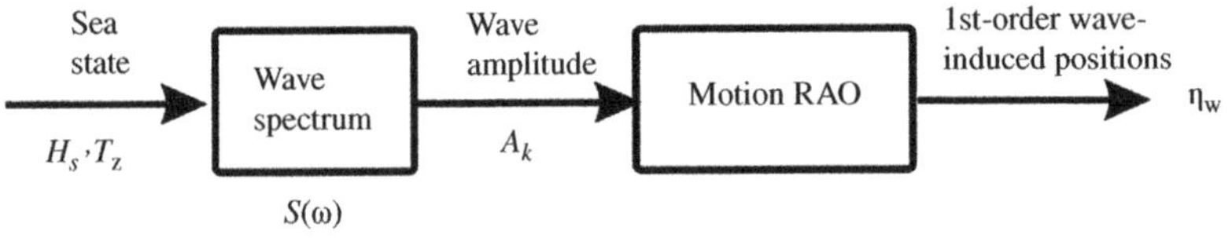

Figure 10.16 Computational setup for first-order wave-induced positions based on motion RAOs.

The definition of the encounter angle β is shown in Figure 10.14. The expression for the wave elevation (10.83) can be redefined in terms of the frequency of encounter ω_e and encounter angle β for a ship moving at forward speed $U > 0$. In other words

$$\xi = \sum_{k=1}^{N}\sum_{i=1}^{N} \sqrt{2S_M(\omega_k, \mu_i - \beta)\Delta\omega\Delta\mu} \cos\left(\omega_k t - \frac{\omega_k^2}{g} U \cos(\mu_i - \beta)\, t + \epsilon_{i,k}\right). \tag{10.86}$$

This modification is particular useful for ship maneuvering.

10.2.4 Force RAOs

Force RAOs can be computed for a particular craft using a hydrodynamic program where the hull geometry is specified in an input file. These programs are usually based on potential theory, as described in Section 5.1. Since the equations of motions of a moving craft are expressed in terms of Newton's second law

$$\boldsymbol{M}\dot{\boldsymbol{v}} = \sum_{i=1}^{K} \boldsymbol{\tau}_i \tag{10.87}$$

it is advantageous to represent the wave loads as generalized wave-induced forces

$$\boldsymbol{\tau} = \boldsymbol{\tau}_{wave1} + \boldsymbol{\tau}_{wave2}. \tag{10.88}$$

The wave force responses are computed for different sea states by using a wave spectrum $S(\omega)$ to describe the wave amplitude components A_k as discussed in Section 10.2.3. The force RAO relates the wave amplitudes to the wave-induced force, as shown in Figure 10.15. The necessary equations that are needed to represent the force RAOs and compute the wave-induced forces in the time domain are presented now.

Normalized Force RAOs

The first- and second-order wave forces for varying wave directions β_i and wave frequencies ω_k are denoted $\tau^{\{\text{dof}\}}_{\text{wave1}}(\omega_k, \beta_i)$ and $\tau^{\{\text{dof}\}}_{\text{wave2}}(\omega_k, \beta_i)$ where dof $\in \{1, 2, 3, 4, 5, 6\}$. The normalized force RAOs are complex variables (WAMIT 2010)

$$F^{\{\text{dof}\}}_{\text{wave1}}(\omega_k, \beta_i) = \left| \frac{\tau^{\{\text{dof}\}}_{\text{wave1}}(\omega_k, \beta_i)}{\rho g A_k} \right| e^{j\angle\tau^{\{\text{dof}\}}_{\text{wave1}}(\omega_k,\beta_i)} \tag{10.89}$$

$$F^{\{\text{dof}\}}_{\text{wave2}}(\omega_k, \beta_i) = \left| \frac{\tau^{\{\text{dof}\}}_{\text{wave2}}(\omega_k, \beta_i)}{\rho g A_k^2} \right| e^{j\angle\tau^{\{\text{dof}\}}_{\text{wave2}}(\omega_k,\beta_i)}. \tag{10.90}$$

The output from the hydrodynamic code is usually an ASCII file containing RAOs in table format. Let us denote the imaginary and real parts of the force RAOs by the Matlab structures: `ImRAO_wave1{dof}(k,i)` and `ReRAO_wave1{dof}(k,i)`). Similarly `ImRAO_wave2{dof}(k,i)` and `ReRAO_wave2{dof}(k,i)`. The amplitudes and phases for different frequencies ω_k and encounter angles β_i for the first-order wave-induced forces can be computed according to the formulae

$$\left|F^{\{\text{dof}\}}_{\text{wave1}}(\omega_k, \beta_i)\right| = \sqrt{(\texttt{ImRAO\_wave1\{dof\}(k,i)})^2 + (\texttt{ReRAO\_wave1\{dof\}(k,i)})^2}$$

$$\angle F^{\{\text{dof}\}}_{\text{wave1}}(\omega_k, \beta_i) = \text{atan2}(\texttt{ImRAO\_wave1\{dof\}(k,i)},\ \texttt{ReRAO\_wave1\{dof\}(k,i)}).$$

The amplitudes and phases for the second-order mean forces are

$$|F^{\{\text{dof}\}}_{\text{wave2}}(\omega_k, \beta_i)| = \texttt{ReRAO\_wave2\{dof\}(k,i)}$$

$$\angle F^{\{\text{dof}\}}_{\text{wave2}}(\omega_k, \beta_i) = 0.$$

MATLAB

The motion RAOs are processed in the MSS Hydro Matlab toolbox by using m-file commands:

```
wamit2vessel     % read and process WAMIT data
veres2vessel     % read and process ShipX (Veres) data
```

The data are represented in the workspace as Matlab structures:

```
vessel.forceRAO.w(k)                          % frequencies
vessel.forceRAO.amp{dof}(k,i,speed_no)        % amplitudes
vessel.forceRAO.phase{dof}(k,i,speed_no)      % phases
```

where `speed_no = 1` represents $U = 0$. For the mean drift forces only surge, sway and yaw are considered (`dof` $\in \{1, 2, 6\}$ where the third component corresponds to yaw)

```
vessel.driftfrc.w(k)                          % frequencies
vessel.driftfrc.amp{dof}(k,i,speed_no)        % amplitudes
```

It is possible to plot the force RAOs using

```
plotTF                                        % plot transfer function
plotWD                                        % plot wave drift
```

Generalized Wave-induced Forces

Since the first- and second-order wave forces are represented in terms of the complex variables $F_{\text{wave1}}^{\{\text{dof}\}}(\omega_k, \beta_i)$ and $F_{\text{wave2}}^{\{\text{dof}\}}(\omega_k, \beta_i)$, the responses for sinusoidal excitations can be computed using different wave spectra. When doing this, linear superposition is employed as illustrated in Figure 10.15. Let the generalized wave-induced forces in 6 DOFs be denoted by vectors

$$\boldsymbol{\tau}_{\text{wave1}} = \left[\tau_{\text{wave1}}^{\{1\}}, \tau_{\text{wave1}}^{\{2\}}, \tau_{\text{wave1}}^{\{3\}}, \tau_{\text{wave1}}^{\{4\}}, \tau_{\text{wave1}}^{\{5\}}, \tau_{\text{wave1}}^{\{6\}}\right]^{\top} \tag{10.91}$$

$$\boldsymbol{\tau}_{\text{wave2}} = \left[\tau_{\text{wave2}}^{\{1\}}, \tau_{\text{wave2}}^{\{2\}}, \tau_{\text{wave2}}^{\{3\}}, \tau_{\text{wave2}}^{\{4\}}, \tau_{\text{wave2}}^{\{5\}}, \tau_{\text{wave2}}^{\{6\}}\right]^{\top} \tag{10.92}$$

Generalized Wave-induced Forces from Force RAOs (No Spreading Function)

For the no spreading case, the encounter angle β = constant such that

$$\begin{aligned}
\tau_{\text{wave1}}^{\{\text{dof}\}} &= \sum_{k=1}^{N} \rho g |F_{\text{wave1}}^{\{\text{dof}\}}(\omega_k, \beta)| A_k \cos(\omega_e(U, \omega_k, \beta)t + \angle F_{\text{wave1}}^{\{\text{dof}\}}(\omega_k, \beta) + \epsilon_k) \\
&= \sum_{k=1}^{N} \rho g |F_{\text{wave1}}^{\{\text{dof}\}}(\omega_k, \beta)| \sqrt{2S(\omega_k)\Delta\omega} \\
&\quad \cos(\omega_e(U, \omega_k, \beta)t + \angle F_{\text{wave1}}^{\{\text{dof}\}}(\omega_k, \beta) + \epsilon_k)
\end{aligned} \tag{10.93}$$

$$\tau_{\text{wave2}}^{\{\text{dof}\}} = \sum_{k=1}^{N} \rho g |F_{\text{wave2}}^{\{\text{dof}\}}(\omega_k, \beta)| A_k^2$$
$$= \sum_{k=1}^{N} \rho g |F_{\text{wave2}}^{\{\text{dof}\}}(\omega_k, \beta)| 2S(\omega_k)\Delta\omega \tag{10.94}$$

where

$$\omega_{\text{e}}(U, \omega_k, \beta) = \omega_k - \frac{\omega_k^2}{g} U \cos(\beta) \tag{10.95}$$

Generalized Wave-induced Forces from Force RAOs (Spreading Function)

The more general case, where the spreading function (10.80) is included, can be simulated by using varying wave directions μ_i $(i = 1, .., M)$ and

$$\tau_{\text{wave1}}^{\{\text{dof}\}} = \sum_{k=1}^{N}\sum_{i=1}^{M} \rho g |F_{\text{wave1}}^{\{\text{dof}\}}(\omega_k, \mu_i - \beta)| \sqrt{2S_M(\omega_k, \mu_i - \beta)\Delta\omega\Delta\mu}$$
$$\cos\left(\omega_{\text{e}}(U, \omega_k, \mu_i - \beta)t + \angle F_{\text{wave1}}^{\{\text{dof}\}}(\omega_k, \mu_i - \beta) + \epsilon_{i,k}\right) \tag{10.96}$$

$$\tau_{\text{wave2}}^{\{\text{dof}\}} = \sum_{k=1}^{N}\sum_{i=1}^{M} \rho g |F_{\text{wave2}}^{\{\text{dof}\}}(\omega_k, \mu_i - \beta)| 2S_M(\omega_k, \mu_i - \beta)\Delta\omega\Delta\mu. \tag{10.97}$$

10.2.5 Motion RAOs

An alternative to the force RAO representation in Section 10.2.4 is to use motion RAOs for position, velocity and acceleration to compute the wave-induced motions, see Section 5.3.3. For force RAOs the response will be generalized forces as shown in Figure 10.15. However, in a linear system it is possible to move the forces through the chain of integrators to obtain generalized position. The first-order wave-induced forces, $\boldsymbol{\tau}_{\text{wave1}}$, are zero-mean oscillatory wave forces. Consider the linear system

$$(\boldsymbol{M}_{\text{RB}} + \boldsymbol{A}(\omega_{\text{e}}))\ddot{\boldsymbol{\xi}} + \boldsymbol{B}_{\text{total}}(\omega_{\text{e}})\dot{\boldsymbol{\xi}} + \boldsymbol{C}\boldsymbol{\xi} = \boldsymbol{\tau}_{\text{wave1}}. \tag{10.98}$$

By assuming harmonic motions

$$\boldsymbol{\xi} = \bar{\boldsymbol{\xi}}\, \text{e}^{\text{j}\,\omega_{\text{e}} t}, \qquad \boldsymbol{\tau}_{\text{wave1}} = \bar{\boldsymbol{\tau}}_{\text{wave1}}\, \text{e}^{\text{j}\,\omega_{\text{e}} t}. \tag{10.99}$$

Equation (10.98) can be written

$$-\omega_{\text{e}}^2[\boldsymbol{M}_{\text{RB}} + \boldsymbol{A}(\omega_{\text{e}})]\bar{\boldsymbol{\xi}} - \text{j}\omega_{\text{e}}\boldsymbol{B}_{\text{total}}(\omega_{\text{e}})\bar{\boldsymbol{\xi}} + \boldsymbol{C}\bar{\boldsymbol{\xi}} = \bar{\boldsymbol{\tau}}_{\text{wave1}}. \tag{10.100}$$

The motion responses can be evaluated as

$$\bar{\boldsymbol{\xi}} = \boldsymbol{H}(\mathrm{j}\omega_e)\bar{\boldsymbol{\tau}}_{\mathrm{wave1}} \tag{10.101}$$

where the *force-to-position* transfer function

$$\boldsymbol{H}(\mathrm{j}\omega_e) = [\boldsymbol{C} - \omega_e^2[\boldsymbol{M}_{\mathrm{RB}} + \boldsymbol{A}(\omega_e)] - \mathrm{j}\omega_e\boldsymbol{B}_{\mathrm{total}}(\omega_e)]^{-1} \tag{10.102}$$

is a low-pass filter representing the vessel dynamics. This expression confirms that the first-order wave-induced position can be computed by low-pass filtering the generalized forces $\boldsymbol{\tau}_{\mathrm{wave1}}$. Since the wave-induced forces, $\boldsymbol{\tau}_{\mathrm{wave1}}$, are computed using linear theory, the wave-induced positions, $\bar{\boldsymbol{\xi}}$, are linear responses, which can be modeled by RAOs. Notice that the motion RAOs depend on the model matrices $\boldsymbol{M}_{\mathrm{RB}}, \boldsymbol{A}(\omega_e), \boldsymbol{B}_{\mathrm{total}}(\omega_e)$ and $\boldsymbol{C}$ while force RAOs are only dependent on the wave excitations.

Hydrodynamic programs compute both the motion and force RAOs. Recall from (5.21) that $\boldsymbol{\xi} = \delta\boldsymbol{\eta}$. Hence, the perturbations due to the first-order wave-induced motions

$$\boldsymbol{\eta}_{\mathrm{w}} = \left[\eta_{\mathrm{w}}^{\{1\}},\ \eta_{\mathrm{w}}^{\{2\}},\ \eta_{\mathrm{w}}^{\{3\}},\ \eta_{\mathrm{w}}^{\{4\}},\ \eta_{\mathrm{w}}^{\{5\}},\ \eta_{\mathrm{w}}^{\{6\}}\right]^{\top} \tag{10.103}$$

implies that the total motion expressed in $\{n\}$ is

$$\boldsymbol{y} = \boldsymbol{\eta} + \boldsymbol{\eta}_{\mathrm{w}}. \tag{10.104}$$

Generalized Position from Motion RAOs (No Spreading Function)

Let $|\eta_{\mathrm{w}}^{\{\mathrm{dof}\}}(\omega_k, \beta)|$ and $\angle\eta_{\mathrm{w}}^{\{\mathrm{dof}\}}(\omega_k, \beta)$ denote the motion RAO amplitude and phase for frequency ω_k and encounter angle β. Then, the wave-induced positions (perturbations) are computed according to (see Figure 10.16)

$$\begin{aligned}
\eta_{\mathrm{w}}^{\{\mathrm{dof}\}} &= \sum_{k=1}^{N} |\eta_{\mathrm{w}}^{\{\mathrm{dof}\}}(\omega_k, \beta)| A_k \cos(\omega_e(U, \omega_k, \beta)t + \angle\eta_{\mathrm{w}}^{\{\mathrm{dof}\}}(\omega_k, \beta) + \epsilon_k) \\
&= \sum_{k=1}^{N} |\eta_{\mathrm{w}}^{\{\mathrm{dof}\}}(\omega_k, \beta)| \sqrt{2S(\omega_k)\Delta\omega} \cos\left(\omega_e(U, \omega_k, \beta)t + \angle\eta_{\mathrm{w}}^{\{\mathrm{dof}\}}(\omega_k, \beta) + \epsilon_k\right).
\end{aligned} \tag{10.105}$$

These formulae do not contain the second-order wave-induced forces. Consequently, wave drift forces must be added manually, for instance by using the wave drift force RAO to compute $\tau_{\mathrm{wave2}}^{\{\mathrm{dof}\}}$.

Generalized Position from Motion RAOs (Spreading Function)

The spreading function (10.80) is included by using varying wave directions μ_i for $i = 1, .., M$ and

$$\eta_{\rm w}^{\{\rm dof\}} = \sum_{k=1}^{N} \sum_{i=1}^{M} |\eta_{\rm w}^{\{\rm dof\}}(\omega_k, \mu_i - \beta)| \sqrt{2 S_M(\omega_k, \mu_i - \beta) \Delta\omega \Delta\mu}$$
$$\cos\left(\omega_{\rm e}(U, \omega_k, \mu_i - \beta)t + \angle\eta_{\rm w}^{\{\rm dof\}}(\omega_k, \mu_i - \beta) + \epsilon_{i,k}\right). \tag{10.106}$$

MATLAB

The motion and forces RAOs for the supply vessel in the MSS toolbox can be plotted in 6 DOFs using the following commands (see Figures 10.17 and 10.18):

```
load supply;

for DOF = 1:6
    figure(DOF);

    % Plot motion RAO
    w     = vessel.motionRAO.w;
    amp   = vessel.motionRAO.amp;
    phase = vessel.motionRAO.phase;
    subplot(411); plotRAOamp(w,amp,DOF);
    subplot(412); plotRAOphs(w,phase,DOF);

    % Plot force RAO
    w     = vessel.forceRAO.w;
    amp   = vessel.forceRAO.amp;
    phase = vessel.forceRAO.phase;
    subplot(413); plotRAOamp(w,amp,DOF);
    subplot(414); plotRAOphs(w,phase,DOF);
end

function plotRAOamp(w,amp,DOF)
    velno = 1;
    arg = amp{DOF}(:,:,velno);
    plot(w,arg(:,1),'-*k',w, arg(:,4),'-ko',w, ...
        arg(:,7),'-k<',w, arg(:,10),'-kx');
    legend('0 deg','30 deg','60 deg','90 deg');
    xlabel('wave encounter frequency (rad/s)'), grid;
end

function plotRAOphs(w,phase,DOF)
    velno = 1;
```

```
    phs = (180/pi)*unwrap(phase{DOF}(:,:,velno));
    plot(w,phs(:,1),'-*k',w, phs(:,4),'-ko',w, ...
        phs(:,7),'-k<',w, phs(:,10),'-kx');
    legend('0 deg','30 deg','60 deg','90 deg');
    xlabel('wave encounter frequency (rad/s)'), grid;
end
```

10.2.6 State-space Models for Wave Response Simulation

When simulating and testing feedback control systems it is useful to have a simple and effective way of representing the wave-induced motions. The force RAO representation discussed in Section 10.2.4 requires that the ship geometry is known a priori and that the user has access to a hydrodynamic program for numerical computation of RAO tables. This is also the case for the motion RAO approach discussed in Section 10.2.5.

An alternative approach is to represent the motion RAO formulation in Figure 10.16 by transfer functions

$$\eta_{w_i}(s) = \frac{K_{w_i}\, s}{s^2 + 2\lambda\omega_e s + \omega_e^2} w_i(s) \tag{10.107}$$

where the inputs $w_i(s)$ $(i = 1, 2, ..., 6)$ are zero-mean Gaussian white noise and K_{w_i} $(i = 1, 2, ..., 6)$ are constant tunable gains. It is assumed that the ship is sufficiently excited such that all 6 DOFs oscillate at the encounter frequency ω_e. The fixed-gain approximation produces good results in a closed-loop system where the purpose is to test robustness and performance of a feedback control system in the presence of waves. This is done by tuning of the K_{w_i} gains until realistic results are obtained. For surface craft it is common to use position test signals η_{w_1}, η_{w_2} and η_{w_3} in the magnitude of ± 1.0 m and attitude test signals η_{w_4}, η_{w_5} and η_{w_6} of magnitudes ± 5.0–10.0 degrees.

The six linear state-space models corresponding to (10.107) are $(i = 1, 2, ..., 6)$

$$\dot{\boldsymbol{x}}_{w_i} = \boldsymbol{A}_{w_i}\, \boldsymbol{x}_{w_i} + \boldsymbol{E}_{w_i}\, w_i \tag{10.108}$$

$$\eta_{w_i} = \boldsymbol{C}_{w_i}\, \boldsymbol{x}_{w_i}. \tag{10.109}$$

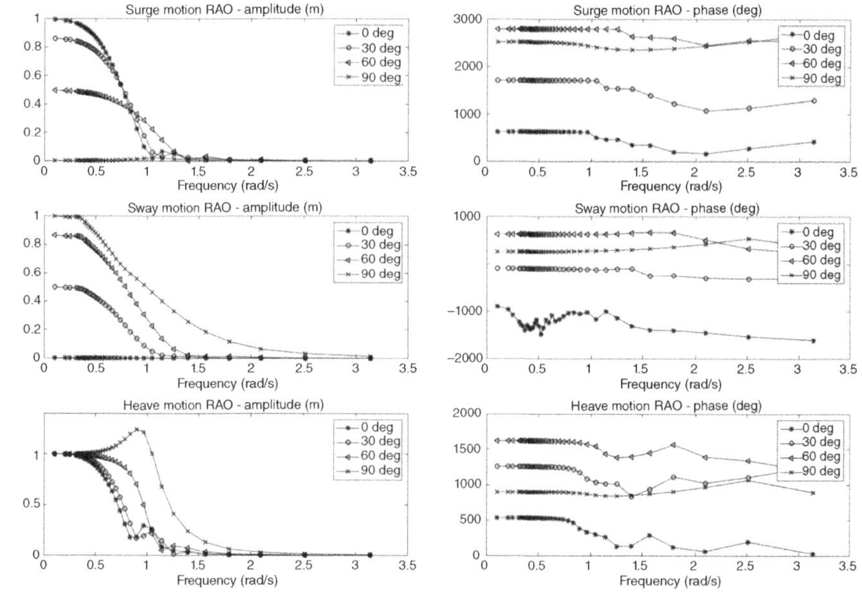

(a) Surge, sway and heave amplitudes (m) and phases (deg).

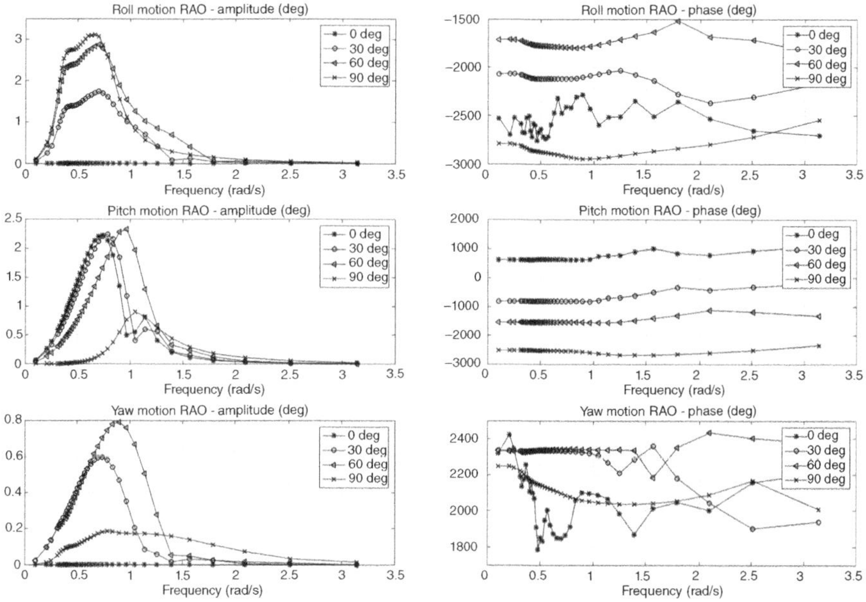

(b) Roll, pitch and yaw amplitudes (deg) and phases (deg)

Figure 10.17 Motion RAOs for the MSS supply vessel.

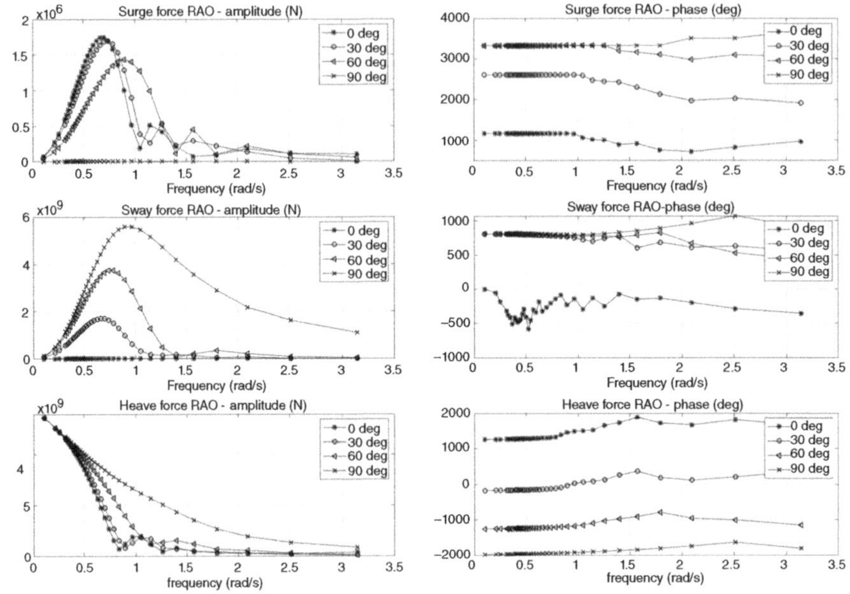

(a) Surge, sway and heave amplitudes (N) and phases (deg).

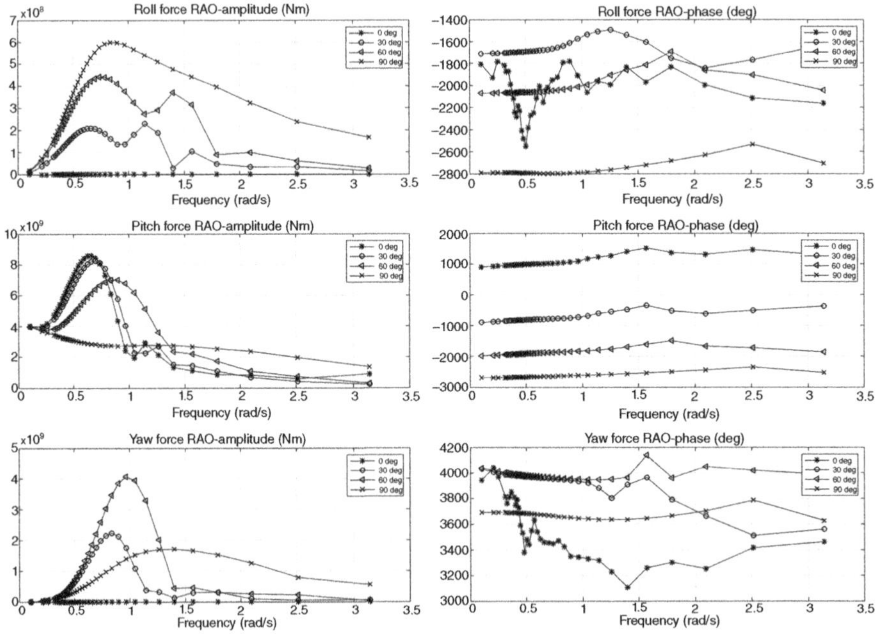

(b) Roll, pitch and yaw amplitudes (Nm) and phases (deg)

Figure 10.18 Force RAOs for the MSS supply vessel.

Expanding this expression yields

$$\dot{x}_{w_i} = \begin{bmatrix} 0 & 1 \\ -\omega_e^2 & -2\lambda\omega_e \end{bmatrix} x_{w_i} + \begin{bmatrix} 0 \\ K_{w_i} \end{bmatrix} w_i \tag{10.110}$$

$$\eta_{w_i} = \begin{bmatrix} 0 & 1 \end{bmatrix} x_{w_i}. \tag{10.111}$$

The linear wave response approximation is usually preferred by ship control systems engineers, owing to its simplicity and applicability. The first applications were reported by Balchen *et al.* (1976) who proposed modeling the *wave-frequency (WF)* motion of a dynamically positioned ship as harmonic oscillators without damping. Later Sælid *et al.* (1983) introduced a damping term λ in the wave model to fit the shape of the PM spectrum better.

Since the WF model as well as the motion RAO approach only models the first-order wave-induced motions it is necessary to include second-order wave drift forces when testing integral action in a feedback control system. The state observer must also be able to handle biased measurements.

The equations of motion, including both the WF motion $\boldsymbol{\eta}_w$ and the marine craft *low-frequency (LF)* motion $\boldsymbol{\eta}$, become (see Figure 10.19)

LF model:
$$\begin{aligned} \dot{\boldsymbol{\eta}} &= \boldsymbol{J}_\Theta(\boldsymbol{\eta})(\boldsymbol{\nu}_r + \boldsymbol{\nu}_c) \\ \boldsymbol{M}\dot{\boldsymbol{\nu}}_r + \boldsymbol{C}(\boldsymbol{\nu}_r)\boldsymbol{\nu}_r + \boldsymbol{D}(\boldsymbol{\nu}_r)\boldsymbol{\nu}_r + \boldsymbol{g}(\boldsymbol{\eta}) + \boldsymbol{g}_0 &= \boldsymbol{\tau}_{\text{wind}} + \boldsymbol{\tau}_{\text{wave2}} + \boldsymbol{\tau} \end{aligned} \tag{10.112}$$

WF model:
$$\begin{aligned} \dot{\boldsymbol{x}}_{w_i} &= \boldsymbol{A}_{w_i}\boldsymbol{x}_{w_i} + \boldsymbol{E}_{w_i} w_i, \qquad (i = 1, 2, \ldots, 6) \\ \eta_{w_i} &= \boldsymbol{C}_{w_i}\boldsymbol{x}_{w_i} \end{aligned} \tag{10.113}$$

Wave drift:
$$\begin{aligned} \dot{\boldsymbol{d}} &= \boldsymbol{w}_d \\ \boldsymbol{\tau}_{\text{wave2}} &= \boldsymbol{d} \end{aligned} \tag{10.114}$$

Measurement:
$$\boldsymbol{y} = \boldsymbol{\eta} + \boldsymbol{\eta}_w + \boldsymbol{\varepsilon} \tag{10.115}$$

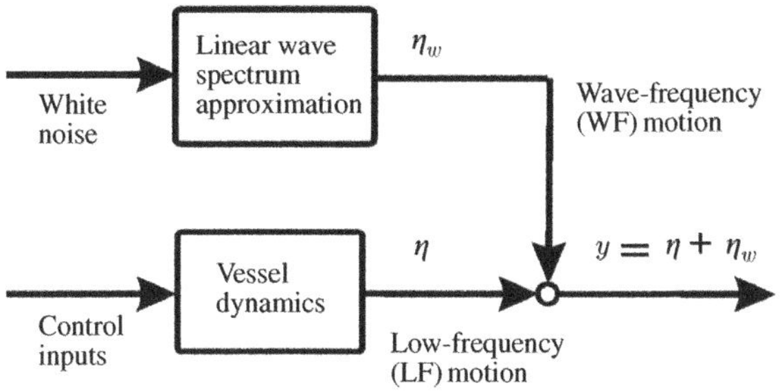

Figure 10.19 Linear approximation for computation of wave-induced positions.

where $\boldsymbol{\varepsilon} \in \mathbb{R}^6$ and $\boldsymbol{w}_d \in \mathbb{R}^6$ are zero-mean Gaussian white noise processes. The random walk process $\boldsymbol{d} \in \mathbb{R}^6$ is used to simulate slowly-varying wave drift forces. From a practical point of view, it might be necessary to saturate the d_i elements in order to avoid that the drift terms exceeds their maximum physical values. Note that the effect of $\boldsymbol{\tau}_{\text{wave1}}$ is included in $\boldsymbol{\eta}_{\text{w}}$ so this signal is not needed to represent the equations of motion.

Example 10.2 (Linear Model for First- and Second-order Wave-Induced Forces)
A marine control system can be tested under the influence of waves by separating the first- and second-order wave-induced forces. For a surface vessel in 3 DOFs ($i = 1, 2, 6$) the WF motions $\boldsymbol{\eta}_{\text{w}} = [\eta_{w_1}, \eta_{w_2}, \eta_{w_6}]^\top$ can be simulated using three transfer functions

$$\eta_{w_1}(s) = \frac{K_{w_1} s}{s^2 + 2\lambda\omega_{\text{e}} s + \omega_{\text{e}}^2} w_1(s) \tag{10.116}$$

$$\eta_{w_2}(s) = \frac{K_{w_2} s}{s^2 + 2\lambda\omega_{\text{e}} s + \omega_{\text{e}}^2} w_2(s) \tag{10.117}$$

$$\eta_{w_6}(s) = \frac{K_{w_6} s}{s^2 + 2\lambda\omega_{\text{e}} s + \omega_{\text{e}}^2} w_6(s) \tag{10.118}$$

where w_i ($i = 1, 2, 6$) are zero-mean white noise processes. The wave drift forces $\boldsymbol{\tau}_{\text{wave2}} = [d_1, d_2, d_6]^\top$ can be modeled as slowly varying bias terms (random walks)

$$\dot{d}_1 = w_{d_1} \tag{10.119}$$

$$\dot{d}_2 = w_{d_2} \tag{10.120}$$

$$\dot{d}_3 = w_{d_6} \tag{10.121}$$

where w_{d_i} ($i = 1, 2, 6$) are zero-mean white noise processes. The amplitudes K_{w_1}, K_{w_2} and K_{w_6} are adjusted such that the amplitudes η_{w_1}, η_{w_2} and η_{w_6} represent a realistic sea state. Notice that the frequency of encounter ω_{e} is used in the transfer functions since the forward speed $U > 0$. The wave spectrum parameter λ should be chosen to represent the true sea state. A good approximation is to use the λ values in Example 10.1 while a typical wave peak frequency is $\omega_0 = 0.8$ rad s^{-1}. This value together with the desired wave encounter angle β can be used to compute the encounter frequency

$$\omega_{\text{e}}(U, \omega_0, \beta) = \left| \omega_0 - \frac{\omega_0^2}{g} U \cos(\beta) \right|. \tag{10.122}$$

10.3 Ocean Current Forces and Moments

Ocean currents are horizontal and vertical circulation systems of ocean waters produced by gravity, wind friction and water density variation in different parts of the ocean. Besides *wind-generated currents*, the heat exchange at the sea surface, together with salinity changes, develop an additional sea current component, usually referred to as *thermohaline currents*. A world map showing the most major ocean surface currents is found in Defant (1961).

The oceans are conveniently divided into two water spheres, the cold and warm water spheres. Since the Earth is rotating, the Coriolis force will try to turn the major currents to the East in the northern hemisphere and West in the southern hemisphere. Finally, the major ocean circulations will also have a tidal component arising from planetary interactions like gravity. In coastal regions and fjords, tidal components can reach very high speeds, in fact speeds of 2–3 m s^{-1} or more have been measured.

Equations of Motion Including Ocean Currents

In order to simulate ocean currents and their effect on marine craft motion, the following model can be applied (Fossen 2012)

$$\underbrace{\boldsymbol{M}_{\text{RB}}\dot{\boldsymbol{\nu}} + \boldsymbol{C}_{\text{RB}}(\boldsymbol{\nu})\boldsymbol{\nu} + \boldsymbol{g}(\boldsymbol{\eta}) + \boldsymbol{g}_0}_{\text{rigid-body and hydrostatic terms}} + \underbrace{\boldsymbol{M}_{\text{A}}\dot{\boldsymbol{\nu}}_{\text{r}} + \boldsymbol{C}_{\text{A}}(\boldsymbol{\nu}_{\text{r}})\boldsymbol{\nu}_{\text{r}} + \boldsymbol{D}(\boldsymbol{\nu}_r)\boldsymbol{\nu}_{\text{r}}}_{\text{hydrodynamic terms}} = \boldsymbol{\tau}_{\text{wind}} + \boldsymbol{\tau}_{\text{wave}} + \boldsymbol{\tau} \tag{10.123}$$

where $\boldsymbol{\nu}_{\text{r}} = \boldsymbol{\nu} - \boldsymbol{\nu}_{\text{c}}$ is the relative velocity vector. The generalized ocean current velocity of an irrotational fluid is

$$\boldsymbol{\nu}_{\text{c}} = [\underbrace{u_{\text{c}},\ v_{\text{c}},\ w_{\text{c}}}_{\boldsymbol{v}_{\text{c}}^b},\ 0,\ 0,\ 0]^{\top} \tag{10.124}$$

where $u_{\text{c}}, v_{\text{c}}$ and w_{c} are expressed in $\{b\}$. Moreover, $\boldsymbol{v}_{\text{c}}^b = [u_{\text{c}},\ v_{\text{c}},\ w_{\text{c}}]^{\top}$. The ocean current velocity vectors expressed in $\{n\}$ and $\{b\}$ satisfy

$$\boldsymbol{v}_{\text{c}}^n = \boldsymbol{R}(\boldsymbol{\Theta}_{\text{nb}})\boldsymbol{v}_{\text{c}}^b. \tag{10.125}$$

Definition 10.1 (Irrotational Constant Ocean Current)
An irrotational constant ocean current expressed in $\{n\}$ is defined by

$$\dot{\boldsymbol{v}}_{\text{c}}^n = \dot{\boldsymbol{R}}(\boldsymbol{\Theta}_{\text{nb}})\boldsymbol{v}_{\text{c}}^b + \boldsymbol{R}(\boldsymbol{\Theta}_{\text{nb}})\dot{\boldsymbol{v}}_{\text{c}}^b := \boldsymbol{0} \tag{10.126}$$

where

$$\dot{\boldsymbol{R}}(\boldsymbol{\Theta}_{\text{nb}}) = \boldsymbol{R}(\boldsymbol{\Theta}_{\text{nb}})\boldsymbol{S}(\boldsymbol{\omega}_{\text{nb}}^b). \tag{10.127}$$

Consequently,

$$\dot{\boldsymbol{v}}_{\text{c}}^b = -\boldsymbol{S}(\boldsymbol{\omega}_{\text{nb}}^b)\boldsymbol{v}_{\text{c}}^b. \tag{10.128}$$

Property 10.1 (Irrotational Constant Ocean Currents)
If the Coriolis and centripetal matrix $\boldsymbol{C}_{\text{RB}}(\boldsymbol{\nu}_r)$ is parametrized independent of linear velocity $\boldsymbol{\nu}_1 = [u,\ v,\ w]^{\top}$, for instance by using (3.63), and the ocean current is irrotational and

constant (Definition 10.1), the rigid-body kinetics satisfies (Hegrenæs 2010)

$$M_{RB}\dot{\nu} + C_{RB}(\nu)\nu = M_{RB}\dot{\nu}_r + C_{RB}(\nu_r)\nu_r \tag{10.129}$$

with

$$\nu_r = \begin{bmatrix} v^b - v_c^b \\ \omega_{nb}^b \end{bmatrix}. \tag{10.130}$$

Proof. *Since the Coriolis and centripetal matrix represented by (3.63) is independent of linear velocity* $\nu_1 = [u,\ v,\ w]^\top$, *it follows that* $C_{RB}(\nu_r) = C_{RB}(\nu)$. *The property*

$$M_{RB}\dot{\nu}_c + C_{RB}(\nu_r)\nu_c = \mathbf{0} \tag{10.131}$$

is easily verified by expanding the matrices M_{RB} *and* $C_{RB}(\nu_r)$ *and corresponding acceleration and velocity vectors according to*

$$\begin{bmatrix} mI_3 & -mS(r_{bg}^b) \\ mS(r_{bg}^b) & I_b^b \end{bmatrix} \begin{bmatrix} -S(\omega_{nb}^b)v_c^b \\ 0_{3\times1} \end{bmatrix} + \begin{bmatrix} mS(\omega_{nb}^b) & -mS(\omega_{nb}^b)S(r_{bg}^b) \\ mS(r_{bg}^b)S(\omega_{nb}^b) & -S(I_b^b\omega_{nb}^b) \end{bmatrix} \begin{bmatrix} v_c^b \\ 0_{3\times1} \end{bmatrix} = \mathbf{0}.$$

Finally, it follows that

$$\begin{aligned} M_{RB}\dot{\nu} + C_{RB}(\nu)\nu &= M_{RB}[\dot{\nu}_r + \dot{\nu}_c] + C_{RB}(\nu_r)[\nu_r + \nu_c] \\ &= M_{RB}\dot{\nu}_r + C_{RB}(\nu_r)\nu_r. \end{aligned} \tag{10.132}$$

State-space Model for Relative Velocity

Property 10.1 can be used to simplify the equations of motion (10.123). In other words

$$\dot{\eta} = J_\Theta(\eta)(\nu_r + \nu_c) \tag{10.133}$$

$$\dot{\nu}_r = M^{-1}(\tau_{\text{wind}} + \tau_{\text{wave}} + \tau - C(\nu_r)\nu_r - D(\nu_r)\nu_r - g(\eta) - g_0) \tag{10.134}$$

where

$$M = M_{RB} + M_A \tag{10.135}$$

$$C(\nu_r) = C_{RB}(\nu_r) + C_A(\nu_r). \tag{10.136}$$

State-space Model for Absolute Velocity

The 6-DOF equations of motion (10.134) can also be expressed in terms of absolute velocities

$$\dot{\eta} = J_\Theta(\eta)\nu \tag{10.137}$$

$$\dot{\boldsymbol{\nu}} = \begin{bmatrix} -\boldsymbol{S}(\boldsymbol{\omega}_{nb}^b)\boldsymbol{v}_c^b \\ \boldsymbol{0}_{3\times1} \end{bmatrix} + \boldsymbol{M}^{-1}(\boldsymbol{\tau} + \boldsymbol{\tau}_{\text{wind}} + \boldsymbol{\tau}_{\text{wave}} - \boldsymbol{C}(\boldsymbol{\nu}_r)\boldsymbol{\nu}_r - \boldsymbol{D}(\boldsymbol{\nu}_r)\boldsymbol{\nu}_r - \boldsymbol{g}(\boldsymbol{\eta}) - \boldsymbol{g}_0) \tag{10.138}$$

where $\boldsymbol{\nu}_r = \boldsymbol{\nu} - [u_c, v_c, w_c, 0, 0, 0]^\top$. For 3-DOF horizontal-plane (surge, sway and yaw) applications this expression simplifies to

$$\dot{\boldsymbol{\eta}} = \boldsymbol{R}(\psi)\boldsymbol{\nu} \tag{10.139}$$

$$\dot{\boldsymbol{\nu}} = \begin{bmatrix} r v_c \\ -r u_c \\ 0 \end{bmatrix} + \boldsymbol{M}^{-1}(\boldsymbol{\tau} + \boldsymbol{\tau}_{\text{wind}} + \boldsymbol{\tau}_{\text{wave}} - \boldsymbol{C}(\boldsymbol{\nu}_r)\boldsymbol{\nu}_r - \boldsymbol{D}(\boldsymbol{\nu}_r)\boldsymbol{\nu}_r) \tag{10.140}$$

where $\boldsymbol{\nu}_r = \boldsymbol{\nu} - [u_c, v_c, 0]^\top$. A further simplification could be to assume that $r \approx 0$ (straight-line motion or stationkeeping) such that

$$\dot{\boldsymbol{\nu}} = \boldsymbol{M}^{-1}(\boldsymbol{\tau} + \boldsymbol{\tau}_{\text{wind}} + \boldsymbol{\tau}_{\text{wave}} - \boldsymbol{C}(\boldsymbol{\nu}_r)\boldsymbol{\nu}_r - \boldsymbol{D}(\boldsymbol{\nu}_r)\boldsymbol{\nu}_r.) \tag{10.141}$$

We will now turn our attention to simulation models for $\boldsymbol{\nu}_c$.

Current Speed and Direction

The ocean current speed is denoted by V_c while its direction relative to the moving craft is conveniently expressed in terms of two angles, *vertical current direction* α_{V_c} and *horizontal current direction* β_{V_c}, as shown in Figure 2.10 in Section 2.5. For computer simulations the ocean current velocity can be generated by using a first-order *Gauss–Markov process*

$$\dot{V}_c + \mu V_c = w \tag{10.142}$$

where w is Gaussian white noise and $\mu \geq 0$ is a constant. If $\mu = 0$, this model reduces to a *random walk*, corresponding to time integration of *white noise*. A saturating element is usually used in the integration process to limit the current speed to

$$V_{\min} \leq V_c(t) \leq V_{\max}. \tag{10.143}$$

The direction of the current can be fixed by specifying constant values for α_{V_c} and β_{V_c}. Time-varying ocean current directions can easily be simulated by associating dynamics to α_{V_c} and β_{V_c}.

10.3.1 3D Irrotational Ocean Current Model

A 3D ocean current model is obtained by transforming the current speed V_c from FLOW axes to NED velocities

$$\boldsymbol{v}_c^n = \boldsymbol{R}_{y,\alpha_{V_c}}^\top \boldsymbol{R}_{z,-\beta_{V_c}}^\top \begin{bmatrix} V_c \\ 0 \\ 0 \end{bmatrix} \tag{10.144}$$

where the rotation matrices $\boldsymbol{R}_{y,\alpha_{V_c}}$ and $\boldsymbol{R}_{z,-\beta_{V_c}}$ are defined in Section 2.5. Expanding (10.144) yields

$$\boldsymbol{v}_c^n = \begin{bmatrix} V_c \cos(\alpha_{V_c}) \cos(\beta_{V_c}) \\ V_c \sin(\beta_{V_c}) \\ V_c \sin(\alpha_{V_c}) \cos(\beta_{V_c}) \end{bmatrix} \tag{10.145}$$

which can be transformed to BODY using the Euler angle rotation matrix. Consequently,

$$\begin{bmatrix} u_c \\ v_c \\ w_c \end{bmatrix} = \boldsymbol{R}^{\top}(\boldsymbol{\Theta}_{nb})\boldsymbol{v}_c^n \tag{10.146}$$

and

$$\boldsymbol{\nu}_c = [u_c,\ v_c,\ w_c,\ 0,\ 0,\ 0]^{\top}. \tag{10.147}$$

10.3.2 2D Irrotational Ocean Current Model

For the 2D case (motions in the horizontal plane), the 3D equations (10.145) with $\alpha_{V_c} = 0$ reduce to

$$\boldsymbol{v}_c^n = \begin{bmatrix} V_c \cos(\beta_{V_c}) \\ V_c \sin(\beta_{V_c}) \\ 0 \end{bmatrix}. \tag{10.148}$$

Hence, from (10.146) it follows that

$$u_c = V_c \cos(\beta_{V_c} - \psi), \qquad v_c = V_c \sin(\beta_{V_c} - \psi). \tag{10.149}$$

Note that

$$V_c = \sqrt{u_c^2 + v_c^2}. \tag{10.150}$$

Example 10.3 (Maneuvering Model including Ocean Currents)
Consider the linear maneuvering model

$$\begin{bmatrix} m_{11} & m_{12} & 0 \\ m_{21} & m_{22} & 0 \\ 0 & 0 & 1 \end{bmatrix} \begin{bmatrix} \dot{v} - \dot{v}_c \\ \dot{r} \\ \dot{\psi} \end{bmatrix} + \begin{bmatrix} d_{11} & d_{12} & 0 \\ d_{21} & d_{22} & 0 \\ 0 & -1 & 0 \end{bmatrix} \begin{bmatrix} v - v_c \\ r \\ \psi \end{bmatrix}$$
$$= \begin{bmatrix} b_1 \\ b_2 \\ 0 \end{bmatrix} \delta + \begin{bmatrix} Y_{\text{wind}} \\ N_{\text{wind}} \\ 0 \end{bmatrix} + \begin{bmatrix} Y_{\text{wave}} \\ N_{\text{wave}} \\ 0 \end{bmatrix} \tag{10.151}$$

where v is the sway velocity, r is the yaw rate, ψ is the yaw angle, δ is the rudder angle and $(u_{\rm c}, v_{\rm c})$ are the current velocities given by

$$u_{\rm c} = V_{\rm c}\cos(\beta_{V_{\rm c}} - \psi) \tag{10.152}$$

$$v_{\rm c} = V_{\rm c}\sin(\beta_{V_{\rm c}} - \psi). \tag{10.153}$$

The transverse current acceleration is

$$\dot{v}_c = -ru_c. \tag{10.154}$$

Assume that the current speed $V_{\rm c}$ is a Gauss–Markov process (10.142) and the wave direction $\beta_{V_{\rm c}}$ = constant. The resulting state-space model is

$$\begin{bmatrix} \dot{v} \\ \dot{r} \\ \dot{\psi} \\ \dot{V}_{\rm c} \end{bmatrix} = \begin{bmatrix} -rV_{\rm c}\sin(\beta_{V_{\rm c}} - \psi) \\ 0 \\ 0 \\ 0 \end{bmatrix} + \begin{bmatrix} m_{11} & m_{12} & 0 & 0 \\ m_{21} & m_{22} & 0 & 0 \\ 0 & 0 & 1 & 0 \\ 0 & 0 & 0 & 1 \end{bmatrix}^{-1} \left(\begin{bmatrix} b_1 \\ b_2 \\ 0 \\ 0 \end{bmatrix} \delta + \begin{bmatrix} Y_{\rm wind} \\ N_{\rm wind} \\ 0 \\ 0 \end{bmatrix} \right.$$

$$\left. + \begin{bmatrix} Y_{\rm wave} \\ N_{\rm wave} \\ 0 \\ 0 \end{bmatrix} + \begin{bmatrix} -d_{11} & -d_{12} & 0 & 0 \\ -d_{21} & -d_{22} & 0 & 0 \\ 0 & 1 & 0 & 0 \\ 0 & 0 & 0 & -\mu \end{bmatrix} \begin{bmatrix} v \\ r \\ \psi \\ V_{\rm c} \end{bmatrix} + \begin{bmatrix} d_{11}V_{\rm c}\sin(\beta_{V_{\rm c}} - \psi) \\ d_{21}V_{\rm c}\sin(\beta_{V_{\rm c}} - \psi) \\ 0 \\ 0 \end{bmatrix} + \begin{bmatrix} 0 \\ 0 \\ 0 \\ 1 \end{bmatrix} w \right).$$

Note that the state-space model is nonlinear in ψ, $V_{\rm c}$ and $\beta_{V_{\rm c}}$ even though the ship model (10.151) was linear.

Part Two

Motion Control

De Motu Gubernando

11

Introduction to Part II

When implementing motion control systems it is important to have good software architecture to simplify software updates and maintenance. Another important aspect is cybersecurity, which puts constraints on the system architecture. As a consequence of this, the motion control system is usually constructed as three independent blocks denoted as the *guidance, navigation* and *control* (GNC) systems. These systems interact with each other through data and signal transmission as illustrated in Figure 11.1 where a control allocation block is included to transform the control forces to physical control inputs such as propeller revolutions, engine setpoints, etc.

An introduction to GNC systems including their respective definitions are presented in Section 11.1. This is accompanied with Section 11.2 discussing control allocation algorithms for motion control systems. The GNC theory and cases studies are organized as four independent chapters:

Chapter 12 (Guidance Systems): Systems for automatically guiding the path of a marine craft, usually without direct or continuous human control.

Chapter 13 (Model-Based Navigation Systems): Systems for determination of the craft's position, velocity and attitude by means of a state estimator, which is based on a mathematical model of the craft.

Chapter 14 (Inertial Navigation Systems): Systems for determination of the craft's position, velocity and attitude where sensory information from accelerometers and attitude rate sensors are fused in a state estimator, and by this avoid using a mathematical model of the craft.

Chapter 15 (Motion Control Systems): PID design methods for automatic control of position, velocity and attitude. This involves systems for stabilization, trajectory-tracking and path-following control of marine craft.

Chapter 16 (Advanced Motion Control Systems): Design of advanced motion control systems using optimal and nonlinear control theory.

Handbook of Marine Craft Hydrodynamics and Motion Control, Second Edition. Thor I. Fossen.

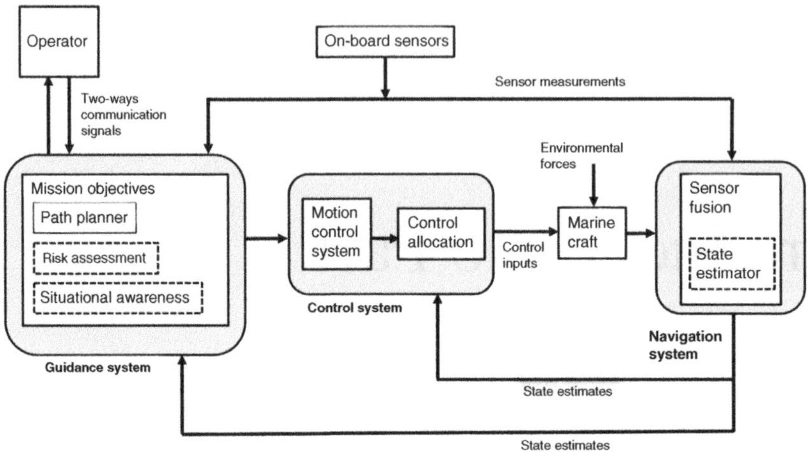

Figure 11.1 Motion control system architecture.

Each chapter contains theory and cases studies. Typical applications are:

- Ship and underwater vehicle autopilots for course-keeping and turning control
- Trajectory-tracking and target-tracking algorithms
- Waypoint tracking and path-following control systems for marine craft
- Depth autopilots for underwater vehicles
- Attitude control systems for underwater vehicles
- Dynamic positioning (DP) systems for marine craft
- Position mooring (PM) systems for floating vessels
- Fin and rudder-roll reduction systems
- Buoyancy control systems including trim and heel correction systems
- Propulsion and forward speed control systems.

In more advanced GNC systems, these blocks could be more tightly coupled and even represented by one block. Loose and tight coupling is a trade-off between modularity and high performance. From an industrial point of view it is practical to have a loosely coupled system since this allows for software updates of single blocks.

In Figure 11.1 the guidance system makes use of the estimated alternatively measured positions and velocities. This is referred to as a *closed-loop guidance system*, while a guidance system that only uses reference feedforward (no feedback) as shown in Figure 11.2 is an *open-loop guidance system*. Figure 11.2 is a state-of-the-art autopilot system consisting of a reference model (guidance system), a state estimator using a gyrocompass (navigation system) and an autopilot (control system).

11.1 Guidance, Navigation and Control Systems

In its most advanced form, the GNC blocks represent three interconnected subsystems, as shown in Figure 11.1. The tasks of the subsystems are classified according to:

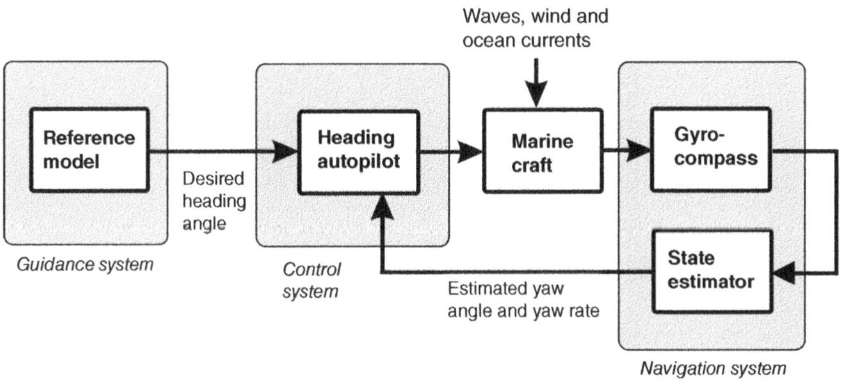

Figure 11.2 Autopilot GNC blocks where the reference model represents an open-loop guidance system.

Guidance *is the action or the system that continuously computes the reference (desired) position, velocity and attitude of a marine craft to be used by the motion control system. These data are usually provided to the human operator and the navigation system (see Figure* 11.3).

The basic components of a guidance system are motion sensors, external data such as weather data (wind speed and direction, wave height and slope, current speed and direction) and a computer. The computer collects and processes the information, and then feeds the results to the motion control system. In many cases, advanced optimization techniques are used to compute the optimal trajectory or path for the marine craft to follow. This might include sophisticated features such as fuel optimization, minimum time navigation, weather routing, collision avoidance, formation control and synchronization.

Navigation *is the science of directing a craft by determining its position, attitude, course and distance traveled. In some cases velocity and acceleration are determined as well.*

The sensor and navigation system is usually implemented as an optimal state estimator (Kalman filter) using GNSS measurements combined with motion sensors such as accelerometers and gyros. The most advanced navigation system for marine applications is the *inertial navigation system* (INS).

Navigation is derived from the Latin *navis*, "ship", and *agere*, "to drive". It originally denoted the art of ship driving, including steering and setting the sails. The skill is even more ancient than the word itself, and it has evolved over the course of many centuries into a technological science that encompasses the planning and execution of safe, timely and economical operation of ships, underwater vehicles, aircraft and spacecraft.

Control, *or more specifically motion control, is the action of determining the necessary control forces and moments to be provided by the craft in order to satisfy a certain control objective.*

Figure 11.3 Human operator monitoring the navigation data. Source: Illustration by B. Stenberg.

The desired control objective is usually seen in conjunction with the guidance system. Examples of control objectives are minimum energy, setpoint regulation, trajectory-tracking, path-following and maneuvering control. Constructing the control algorithm involves the design of feedback and feedforward control laws. The outputs from the navigation system, position, velocity and acceleration, are used for feedback control while feedforward control is implemented using signals available in the guidance system and other external sensors.

11.1.1 Historical Remarks

The history of model-based ship control starts with the invention of the *gyrocompass* in 1908, which allowed for reliable automatic yaw angle feedback. The gyrocompass was the basic instrument in the first feedback control system for heading control and today these devices are known as autopilots. The next breakthrough was the development of local positioning systems in the 1970s. Global coverage using satellite navigation systems was first made available in 1994. Position control systems opened for automatic systems for waypoint tracking, trajectory tracking and path following.

The development of local-area ship positioning systems such as hydroacoustic reference systems (*SSBL, SBL, LBL*), hyperbolic radio navigation systems (*Decca, Loran-C, Omega*), local electromagnetic distance measuring (EDM) systems (*Artemis, Autotape, Miniran, Mini-Ranger III, Syledis, Tellurometer, Trident III, Trisponder*) and taut wire in conjunction with new results in feedback control contributed to the invention and design of the first dynamic positioning (DP) systems for ships and rigs in the late 1970s; see Sections 15.3.6 and 16.1.6. The use of DP systems on a global basis in offshore applications was further strengthened by commercialization of satellite navigation systems. In 1994 *Navstar GPS* was declared fully operational (global coverage) even though the first satellite was launched in 1974 (Parkinson and Spilker 1995). GPS receivers are standard components in waypoint tracking control systems and ship positioning systems worldwide. They are used commercially and by numerous naval forces. Today, four global navigation satellite systems (GNSS) are commercially available:

- BeiDou (China)
- Galileo (European Union)
- Glonass (Russia)
- Navstar GPS (USA).

The Gyroscope and its Contributions to Ship Control

During the 1850s the French scientist *J. B. L. Foucault* conducted experiments with a wheel (rotor) mounted in gimbal rings, that is a set of rings that permit it to turn freely in any direction. The name gyroscope was adopted for this device. In the experiments Foucault noticed that the spinning wheel maintained its original orientation in space regardless of the Earth's rotation.

In *Encyclopedia Britannica* the following definition is given for a gyroscope:

> **Gyroscope**: *any device consisting of a rapidly spinning wheel set in a framework that permits it to tilt freely in any direction – that is, to rotate about any axis. The momentum of such a wheel causes it to retain its attitude when the framework is tilted; from this characteristic derive a number of valuable applications. Gyroscopes are used in such instruments as compasses and automatic pilots onboard ships and aircraft, in the steering mechanisms of torpedoes, in antiroll equipment on large ships and in inertial guidance systems.*

The first recorded construction of the gyroscope is usually credited to *C. A. Bohnenberger* in 1810, while the first electrically driven gyroscope was demonstrated in 1890 by *G. M. Hopkins* (see Allensworth 1999, Bennet 1979). The development of the electrically driven gyroscope was motivated by the need for more reliable navigation systems in steel ships and underwater warfare. A magnetic compass, as opposed to a gyro compass, is highly sensitive to magnetic disturbances, which are commonly found in steel ships and submarines equipped with electrical devices. In parallel works, *Dr H. Anschutz* of Germany and *Elmer Sperry* of the USA both worked on the practical application of the gyroscope. In 1908 Anschutz patented the first North-seeking gyrocompass, while Elmer

Sperry was granted a patent for his ballistic compass including vertical damping three years later.

The invention of the gyroscope was one of the key breakthroughs in automatic ship control since it led to the development of the *automatic pilot* (Fossen 2000a). Historic aspects in a motion control perspective are discussed by Fossen (2000b) while Fossen and Perez (2009) discuss Kalman filtering for positioning and heading control of ships and offshore rigs in conjunction with the 50th anniversary of the Kalman–Bucy filter. The pioneering work of J. G. Balchen and coauthors on ship automation and dynamic positioning is discussed by Breivik and Sand (2009).

11.1.2 Autopilots

The autopilot or *automatic pilot* is a device for controlling an aircraft, marine craft or other vehicles without constant human intervention. The earliest automatic pilots could do no more than maintain a fixed heading and they are still used to relieve the pilot on smaller boats during routine cruising. For ships, course-keeping capabilities were the first applications. Modern autopilots can, however, execute complex maneuvers, such as turning and docking operations, or enable the control of inherently unstable vessels such as submarines and some large oil tankers. Autopilots are used to steer surface ships, submarines, torpedoes, missiles, rockets and spacecraft among others.

As mentioned earlier, the work on the gyrocompass was extended to ship steering and closed-loop control by *Elmer Sperry* (1860–1930) who constructed the first automatic ship-steering mechanism (see Allensworth 1999; Bennet 1979). This device, referred to as the *Metal Mike*, was a gyroscope-guided autopilot (*gyro pilot*) or a mechanical helmsman. The first field trials of the Sperry standard gyro pilot were conducted in 1922. Metal Mike emulated much of the behavior of a skilled pilot or a helmsman, including compensating for varying sea states using feedback control and automatic gain adjustments.

Nicholas Minorsky (1885–1970) presented a detailed analysis of a position feedback control system where he formulated a three-term control law which is today known as *proportional-integral-derivative* (PID) control (Minorsky 1922). Observing the way in which a helmsman steered a ship motivated these three different behaviors. In Bennet (1979), there is an interesting analysis of the work of Sperry and Minorsky and their contributions to autopilot design.

The autopilot systems of Sperry and Minorsky were both single-input single-output (SISO) control systems, where the heading (yaw angle) of the ship was measured by a gyrocompass. Today, this signal is fed back to a computer, in which a PID control system (autopilot) is implemented in software (see Section 15.3.4). The autopilot compares the pilot setpoint (desired heading) with the measured heading and computes the rudder command, which is then transmitted to the rudder servo for corrective action.

More recently PID-type autopilots have been replaced by autopilots based on linear-quadratic Gaussian (LQG) and $\mathcal{H}_\infty$-control design techniques. One of the advantageous features of these design techniques is that they allow for frequency-dependent notch filtering of first-order wave-induced forces (see Chapter 13). Frequency components around the peak frequency of the wave spectrum in yaw must be prevented from entering the feedback loop in order to avoid wear and tear of the thruster and propeller systems. The drawback of the PID controller in cascade with a deadband, notch and/or low-pass filter is that additional phase lag and nonlinearities are introduced in the closed-loop system (see Section 13.2.1). A model-based state estimator (Kalman filter) reduces these problems. Linear-quadratic and $\mathcal{H}_\infty$ autopilot designs have been reported in the literature by a large number of authors; see Koyama (1967), Norrbin (1972), Van Amerongen and Van Nauta Lemke (1978, 1980), Donha *et al.* (1998), Tzeng (1998b) and Fossen (1994) and references therein, to mention only some.

In addition to LQG and $\mathcal{H}_\infty$ control, other design techniques have been applied to ship autopilot designs, for instance nonlinear control theory. Autopilot designs for nonlinear systems are treated in detail in Section 16.3.

11.1.3 Dynamic Positioning and Position Mooring Systems

The great successes with PID-based autopilot systems and the development of local area positioning systems suggest that three decoupled PID controllers could be used to control the horizontal motion of a ship in surge, sway and yaw exclusively by means of thrusters and propellers. The idea was tested in the 1970s, and the invention was referred to as a *dynamic positioning* (DP) system. PID designs for DP are presented in Section 15.3.6 while optimal DP is discussed in Section 16.1.6

As for the autopilot systems, a challenging problem was to prevent first-order wave-induced forces entering the feedback loop. Several techniques such as notch and low-pass filtering, and the use of dead-band techniques, were tested for this purpose, but with varying levels of success.

In 1960–1961 the *Kalman filter* was published by Kalman (1960) and Kalman and Bucy (1961). Two years later in 1963, the theory for the linear-quadratic (LQ) optimal controller was available. This motivated the application of *LQG controllers* in MIMO ship control such as DP since a state observer (Kalman filter) could be used to estimate the wave frequency (WF) and the ship low-frequency (LF) motions; see Section 13.4.6 and Figure 15.23 in Section 15.3.6. Another advantage of a MIMO control strategy was that the interactions between the surge, sway and yaw modes could be dealt with. This is not possible with three decoupled PID controllers.

The LQG design technique was first applied to DP by Balchen *et al.* (1976, 1980a, 1980b) and Grimble *et al.* (1979, 1980a). Later Grimble and coauthors suggested using $\mathcal{H}_\infty$ and μ-optimal methods for filtering and control (Katebi *et al.* 1997a). These methods were further refined by Katebi *et al.* (1997b) where the nonlinear thruster dynamics is included using describing functions.

After 1995, nonlinear PID control, passive observer design and observer backstepping designs were applied to DP by Fossen and coauthors with good results; see Grøvlen and Fossen (1996), Fossen and Grøvlen (1998), Strand (1999) and references therein. An overview of DP systems is found in Strand and Sørensen (2000) while extensions to PM systems are presented in Strand (1999). DP and PM systems are discussed in more detail in Sections 15.3.6 and 16.1.6.

11.1.4 Waypoint Tracking and Path-following Control Systems

The successful results with LQG controllers in ship autopilots and DP systems, and the availability of GNSS, resulted in a growing interest in waypoint tracking and path-following control systems; see Holzhüter and Schultze (1996), Holzhüter (1997), Fossen *et al.* (2003), Skjetne *et al.* (2004), Breivik and Fossen (2009) and references therein. The transformation of the waypoints to a feasible path or trajectory is in general a optimization problem. This is discussed in Chapter 12. Motion controllers can be designed using linear theory or by treating the control problem as nonlinear; see Sections 15.2.3–12.6.4 and 16.3.10. Guidance systems for trajectory-tracking and path-following control are discussed in Chapter 12, while maneuverability and autopilot systems are discussed in Chapters 15 and 16.

11.2 Control Allocation

Motion control systems are used to control the motion of marine craft, aircraft, unmanned vehicles, etc. by means of actuators. In this context we will distinguish between (Johansen and Fossen 2013):

- *Effectors*: mechanical devices such as control surfaces, rudders, fins, propellers, jets, engines and thrusters that can generate time-varying forces and moments to control a marine craft.
- *Actuators*: electromechanical devices that are used to control the magnitude and/or direction of forces generated by the individual effectors.

By mechanical design, there may be more effectors than strictly needed to meet the motion control objectives of a given application. Hence, for *overactuated systems*, the controllability of the chosen states and outputs would also be achieved with less control inputs. Overactuated systems are favorable due to effector redundancy allowing for fault tolerance and control reconfiguration. However, this usually involves solving an optimization problem to find the optimal actuator setpoints as illustrated by Figure 11.4.

Problem statement

For marine craft in n DOFs it is necessary to distribute the generalized control forces $\boldsymbol{\tau} \in \mathbb{R}^n$ to the actuators in terms of control inputs $\boldsymbol{u} \in \mathbb{R}^r$ as shown in Figure 11.5. For linear systems

$$\boldsymbol{\tau} = \boldsymbol{B}\boldsymbol{u} \tag{11.1}$$

where $\boldsymbol{B}$ is the input matrix. If $r > n$ this is an *overactuated control* problem while $r < n$ is referred to as *underactuated control*. For overactuated systems, the control inputs are computed using the right *Moore–Penrose pseudoinverse*

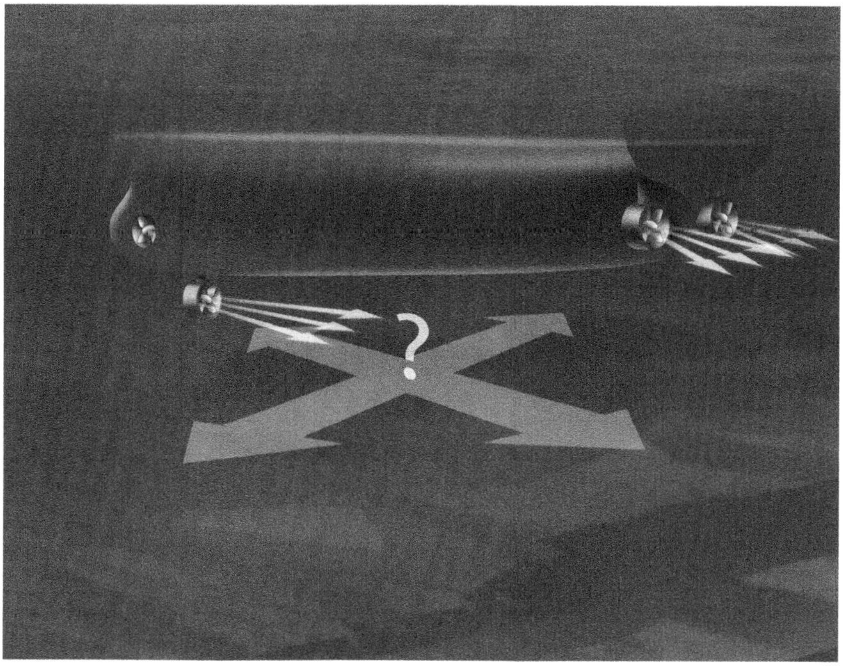

Figure 11.4 The control allocation problem.

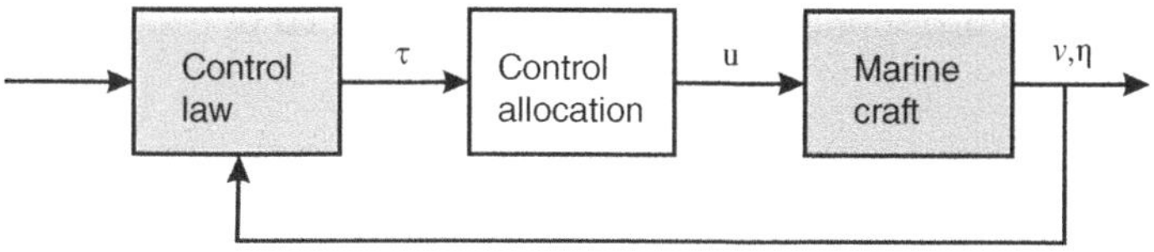

Figure 11.5 Block diagram showing the control allocation block in a feedback control system.

$$\boldsymbol{u} = \boldsymbol{B}^{\dagger}\boldsymbol{\tau} \quad \text{where} \quad \boldsymbol{B}^{\dagger} = \boldsymbol{B}^{\top}(\boldsymbol{B}\boldsymbol{B}^{\top})^{-1}. \tag{11.2}$$

For systems where the input matrix is square ($r = n$), that is the number of actuators is equal to the number of DOFs, the control inputs are computed as

$$\boldsymbol{u} = \boldsymbol{B}^{-1}\boldsymbol{\tau}. \tag{11.3}$$

In general, computation of $\boldsymbol{u}$ from $\boldsymbol{\tau}$ is a model-based optimization problem that in its simplest form is unconstrained. On the contrary, physical limitations such as input amplitude and rate saturations imply that a constrained optimization problem must be solved. Another complication is actuators that can be rotated at the same time as they produce control forces. An example is azimuth thrusters on an offshore supply vessel; see Figure 9.11. This increases the number of available controls from r to $r + p$, where p denotes the number of rotatable actuators for which additional nonlinearities are introduced. A survey of state-of-the art control allocation methods for marine craft, aircraft and automotive systems are found in Johansen and Fossen (2013).

11.2.1 Propulsion and Actuator Models

Mathematical models for propulsion systems, propellers, thrusters, control surfaces, control moment gyroscopes, moving mass systems, etc. are presented in Chapter 9. The control force F_i based on these models will in general be a nonlinear mapping

$$F_i = b(u_i) \tag{11.4}$$

where u_i is the control input depending on the propulsion system or actuator considered. However, in many cases there will be a linear relationship between the input and the force such that

$$F_i = K_i u_i \quad \Rightarrow \quad u_i = \frac{1}{K_i} F_i \tag{11.5}$$

where K_i is the force coefficient and u_i is the control input depending on the actuator considered. The linear model $F_i = K_i u_i$ can also be used to describe nonlinear monotonic control forces. For instance, if the rudder force F_i is assumed to be quadratic in rudder angle δ_i, we have that

$$F_i = K_i \; |\delta_i|\delta_i \; := K_i u_i. \tag{11.6}$$

The inverse function corresponding to $u_i = |\delta_i|\delta_i$ is computed by

$$\delta_i = \mathrm{sgn}(u_i)\sqrt{|u_i|}. \tag{11.7}$$

Generalization to Multivariable Systems

For multivariable systems we denote the control forces $\boldsymbol{F}_i^b = [F_{x_i}, F_{y_i}, F_{z_i}]^\top$ where $i = 1, 2, ..., r$. The forces and moments in 6 DOFs are (see Table 11.1)

$$\boldsymbol{\tau} = \sum_{i=1}^{r} \begin{bmatrix} \boldsymbol{F}_i^b \\ \boldsymbol{r}_{\mathrm{bp_i}}^b \times \boldsymbol{F}_i^b \end{bmatrix} = \sum_{i=1}^{r} \begin{bmatrix} F_{x_i} \\ F_{y_i} \\ F_{z_i} \\ F_{z_i} l_{y_i} - F_{y_i} l_{z_i} \\ F_{x_i} l_{z_i} - F_{z_i} l_{x_i} \\ F_{y_i} l_{x_i} - F_{x_i} l_{y_i} \end{bmatrix} \tag{11.8}$$

where $\boldsymbol{r}_{\mathrm{bp_i}}^b = [l_{x_i}, l_{y_i}, l_{z_i}]^\top$ are the lever arms, that is the perpendicular distances from the CO to the line of action of the force (subscript $\mathrm{bp_i}$). For rotatable (azimuth) thrusters the control force F_i will be a function of the rotation angle α_i and propeller revolution u_i (see Figure 9.11). Consequently, an azimuth thruster in the horizontal plane will have two force components, $F_{x_i} = F_i \cos(\alpha_i)$ and $F_{y_i} = F_i \sin(\alpha_i)$, while the main propeller aft of the ship only produces a longitudinal force $F_{x_i} = F_i$ (see Table11.1).

Thrust Configuration and Force Coefficient Matrices

For multivariable systems the r control forces $\boldsymbol{f} \in \mathbb{R}^r$ are conveniently expressed as

$$\boldsymbol{f} = \boldsymbol{K}\boldsymbol{u} \tag{11.9}$$

where $\boldsymbol{u} = [u_1, \ \dots \ , u_r]^\top$ is a vector of control inputs and $\boldsymbol{K} \in \mathbb{R}^{r \times r}$ is a diagonal force coefficient matrix given by

$$\boldsymbol{K} = \text{diag}\{K_1, \ \dots \ , K_r\} \quad \Rightarrow \quad \boldsymbol{K}^{-1} = \text{diag}\left\{\frac{1}{K_1}, \ \dots \ , \frac{1}{K_r}\right\}. \tag{11.10}$$

The control input vector $\boldsymbol{u}$ relates to the generalized force vector $\boldsymbol{\tau}$ by

$$\begin{aligned} \boldsymbol{\tau} &= \boldsymbol{T}\boldsymbol{f} \\ &= \boldsymbol{T}\boldsymbol{K}\boldsymbol{u} \quad \Rightarrow \quad \boldsymbol{B} = \boldsymbol{T}\boldsymbol{K} \end{aligned} \tag{11.11}$$

where $\boldsymbol{T} \in \mathbb{R}^{n \times r}$ is the thrust configuration matrix. For a marine craft equipped with r actuators for operation in n DOFs, the thrust configuration matrix describes the geometry or locations of the actuators specified by the vectors $\boldsymbol{r}^b_{\text{bp}_i}$ for $i = 1, 2, ..., r$.

Table 11.1 Definition of actuators and control variables.

Actuator	u_i (control input)	α_i (control input)	$\boldsymbol{F}^b_i$ (force vector)
Main propeller (longitudinal)	Pitch and rpm	–	$[F_i, \ 0, \ 0]^\top$
Tunnel thruster (transverse)	Pitch and rpm	–	$[0, \ F_i, \ 0]^\top$
Azimuth (rotatable) thruster	Pitch and rpm	Angle	$[F_i \cos(\alpha_i), \ F_i \sin(\alpha_i), \ 0]^\top$
Aft rudder	Angle	–	$[0, \ F_i, \ 0]^\top$
Stabilizing fin	Angle	–	$[0, \ 0, \ F_i]^\top$

Thrust Configuration Matrix

The trivial case refers to a marine craft equipped with non-rotatable thrusters such that

$$\boldsymbol{T} = [\boldsymbol{t}_1, \dots, \boldsymbol{t}_r]$$

is defined in terms of a set of column vectors $\boldsymbol{t}_i \in \mathbb{R}^n$. In 4 DOFs (*surge, sway, roll* and *yaw*) the column vectors for some standard actuators are

$$\underbrace{\boldsymbol{t}_i = \begin{bmatrix} 1 \\ 0 \\ 0 \\ -l_{y_i} \end{bmatrix}}_{\text{main propeller}}, \quad \boldsymbol{t}_i = \underbrace{\begin{bmatrix} 0 \\ 1 \\ -l_{z_i} \\ l_{x_i} \end{bmatrix}}_{\substack{\text{tunnel thruster} \\ \text{and aft rudder}}}, \quad \boldsymbol{t}_i = \underbrace{\begin{bmatrix} 0 \\ 0 \\ l_{y_i} \\ 0 \end{bmatrix}}_{\text{stabilizing fin}}$$

An example using this representation is found in Section 16.1.7 discussing fin and rudder control systems.

For marine craft equipped with azimuth thrusters (see Figure 9.11) the column vectors take the following form

$$\boldsymbol{t}_i = \underbrace{\begin{bmatrix} \cos(\alpha_i) \\ \sin(\alpha_i) \\ -l_{z_i}\sin(\alpha_i) \\ l_{x_i}\sin(\alpha_i) - l_{y_i}\cos(\alpha_i) \end{bmatrix}}_{\text{azimuth thruster}}$$

where α_i for $i = 1, \dots, p$ are the azimuth angles. This implies that the thrust configuration matrix

$$\boldsymbol{T}(\boldsymbol{\alpha}) = [\boldsymbol{t}_1, \dots, \boldsymbol{t}_r] \tag{11.12}$$

depends on a vector of azimuth angles $\boldsymbol{\alpha} = [\alpha_1, \dots, \alpha_p]^\top \in \mathbb{R}^p$.

Extended Thrust Configuration Matrix for Rotatable Actuators

When solving the control allocation optimization problem an alternative representation to (11.12) is the extended thrust configuration matrix. Unfortunately, (11.12) is nonlinear in the controls $\boldsymbol{\alpha}$, which makes it non-trivial to compute the inverse mapping. This usually implies that a nonlinear optimization problem must be solved. In order to avoid this, a rotatable thruster can be treated as two forces. Consider a rotatable thruster in the horizontal plane (the same methodology can be used for thrusters that can be rotated in the vertical plane)

$$\begin{aligned} F_{x_i} &= F_i \cos(\alpha_i) \\ &= K_i u_i \cos(\alpha_i) \end{aligned} \tag{11.13}$$

$$
\begin{aligned}
F_{y_i} &= F_i \sin(\alpha_i) \\
&= K_i u_i \sin(\alpha_i).
\end{aligned}
\tag{11.14}
$$

Next, the extended force vector is defined according to

$$
\boldsymbol{f}_{\rm e} := \boldsymbol{K}_{\rm e}\boldsymbol{u}_{\rm e}
\tag{11.15}
$$

such that

$$
\boldsymbol{\tau} = \boldsymbol{T}_{\rm e}\boldsymbol{K}_{\rm e}\boldsymbol{u}_{\rm e}
\tag{11.16}
$$

where $\boldsymbol{T}_{\rm e}$ and $\boldsymbol{K}_{\rm e}$ are the extended thrust configuration and coefficient matrices, respectively, and $\boldsymbol{u}_{\rm e}$ is a vector of extended control inputs or virtual controls (Sørdalen 1997b) defined by

$$
u_{i_x} := u_i \cos(\alpha_i)
\tag{11.17}
$$

$$
u_{i_y} := u_i \sin(\alpha_i).
\tag{11.18}
$$

The following example illustrates how this model can be established for a marine craft equipped with two main propellers and two azimuth thrusters in the horizontal plane (Fossen *et al.* 2009).

Example 11.1 (Thrust Configuration Matrix for a Marine Craft)
The forces and moment X, Y and N in surge, sway and yaw, respectively, for the thruster configuration shown in Figure 11.6 satisfy

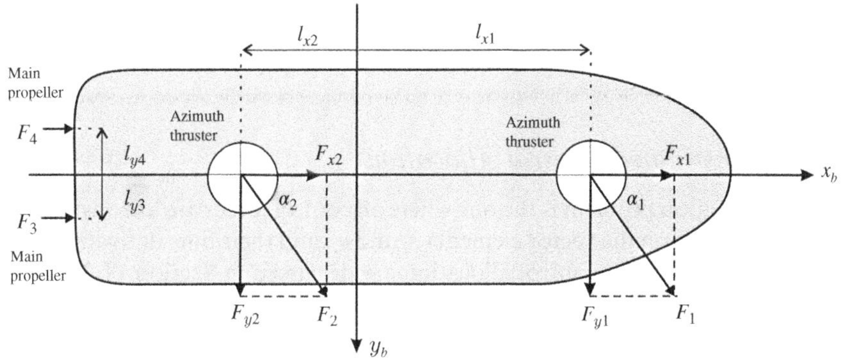

Figure 11.6 Marine craft equipped with two azimuth thrusters (forces F_1 and F_2) and two main propellers (forces F_3 and F_4). The azimuth forces are decomposed along the x_b and y_b axes.

$$\boldsymbol{\tau} = \boldsymbol{T}(\boldsymbol{\alpha})\boldsymbol{K}\boldsymbol{u} \tag{11.19}$$

$$\Updownarrow$$

$$\begin{bmatrix} X \\ Y \\ N \end{bmatrix} = \begin{bmatrix} \cos(\alpha_1) & \cos(\alpha_2) & 1 & 1 \\ \sin(\alpha_1) & \sin(\alpha_2) & 0 & 0 \\ l_{x_1}\sin(\alpha_1) & l_{x_2}\sin(\alpha_2) & -l_{y_3} & -l_{y_4} \end{bmatrix} \begin{bmatrix} K_1 & 0 & 0 & 0 \\ 0 & K_2 & 0 & 0 \\ 0 & 0 & K_3 & 0 \\ 0 & 0 & 0 & K_4 \end{bmatrix} \begin{bmatrix} u_1 \\ u_2 \\ u_3 \\ u_4 \end{bmatrix}. \tag{11.20}$$

The extended thrust vector $\boldsymbol{u}_\mathrm{e}$ *corresponding to (11.16)–(11.18) satisfies*

$$\boldsymbol{\tau} = \boldsymbol{T}_\mathrm{e}\boldsymbol{K}_\mathrm{e}\boldsymbol{u}_\mathrm{e} \tag{11.21}$$

$$\Updownarrow$$

$$\begin{bmatrix} X \\ Y \\ N \end{bmatrix} = \begin{bmatrix} 1 & 0 & 1 & 0 & 1 & 1 \\ 0 & 1 & 0 & 1 & 0 & 0 \\ 0 & l_{x_1} & 0 & l_{x_2} & -l_{y_3} & -l_{y_4} \end{bmatrix} \begin{bmatrix} K_1 & 0 & 0 & 0 & 0 & 0 \\ 0 & K_1 & 0 & 0 & 0 & 0 \\ 0 & 0 & K_2 & 0 & 0 & 0 \\ 0 & 0 & 0 & K_2 & 0 & 0 \\ 0 & 0 & 0 & 0 & K_3 & 0 \\ 0 & 0 & 0 & 0 & 0 & K_4 \end{bmatrix} \begin{bmatrix} u_{1_x} \\ u_{1_y} \\ u_{2_x} \\ u_{2_y} \\ u_3 \\ u_4 \end{bmatrix}. \tag{11.22}$$

Note that $\boldsymbol{T}_\mathrm{e}$ *is constant while* $\boldsymbol{T}(\boldsymbol{\alpha})$ *depends on* $\boldsymbol{\alpha}$*. This means that the extended control input vector* $\boldsymbol{u}_\mathrm{e}$ *can be solved directly from (11.16), for instance using the pseudoinverse (see Section 11.2.2)*

$$\boldsymbol{u}_\mathrm{e} = \boldsymbol{K}_\mathrm{e}^{-1}\boldsymbol{T}_\mathrm{w}^{\dagger}\boldsymbol{\tau} \tag{11.23}$$

This is not the case for (11.19), which represents a nonlinear optimization problem. If $\boldsymbol{u}_\mathrm{e}$ *is computed using the pseudoinverse, the azimuth controls can be derived from the extended control vector elements by mapping the pairs* (u_{1_x}, u_{1_y}) *and* (u_{2_x}, u_{2_y}) *according to*

$$u_1 = \sqrt{u_{1_x}^2 + u_{1_y}^2}, \qquad \alpha_1 = \mathrm{atan2}(u_{1_y}, u_{1_x}) \tag{11.24}$$

$$u_2 = \sqrt{u_{2_x}^2 + u_{2_y}^2}, \qquad \alpha_2 = \mathrm{atan2}(u_{2_y}, u_{2_x}). \tag{11.25}$$

The last two controls, u_3 *and* u_4*, follow directly from the pseudoinverse. Tunnel thrusters can easily be added by the following the approach shown in Example 9.1.*

11.2.2 Unconstrained Control Allocation

The simplest allocation problem is the one where all control forces are unconstrained, that is there are no bounds on the vector elements α_i and u_i, and their time derivatives. Saturating control and constrained control allocation are discussed in Section 11.2.3. Consider the generalized force vector

$$\boldsymbol{\tau} = \boldsymbol{T}_\mathrm{e}\boldsymbol{f}_\mathrm{e}. \tag{11.26}$$

For marine craft where the extended thrust configuration matrix $\boldsymbol{T}_{\text{e}}$ is square or non-square ($r \geq n$), that is there are equal or more control inputs than controllable DOFs, it is possible to find an "optimal" distribution of extended control forces $\boldsymbol{f}_{\text{e}}$ for each DOF by using an explicit method. Consider the unconstrained least-squares (LS) optimization problem (Fossen and Sagatun 1991)

$$J = \min_{\boldsymbol{f}_{\text{e}}} \{\boldsymbol{f}_{\text{e}}^{\top}\boldsymbol{W}\boldsymbol{f}_{\text{e}}\} \tag{11.27}$$

$$\text{subject to: } \boldsymbol{\tau} - \boldsymbol{T}_{\text{e}}\boldsymbol{f}_{\text{e}} = \boldsymbol{0} \tag{11.28}$$

where $\boldsymbol{W}$ is a positive definite matrix, usually diagonal, weighting the control forces. For marine craft that have both control surfaces and propellers, the elements in $\boldsymbol{W}$ should be selected so that using the control surfaces are considerably less expensive than using the propellers.

Explicit solution to the LS Optimization Problem using Lagrange Multipliers

Consider the Lagrangian

$$L(\boldsymbol{f}_{\text{e}}, \boldsymbol{\lambda}) = \boldsymbol{f}_{\text{e}}^{\top}\boldsymbol{W}\boldsymbol{f}_{\text{e}} + \boldsymbol{\lambda}^{\top}(\boldsymbol{\tau} - \boldsymbol{T}_{\text{e}}\boldsymbol{f}_{\text{e}}) \tag{11.29}$$

where $\boldsymbol{\lambda} \in \mathbb{R}^r$ is a vector of Lagrange multipliers. Consequently, differentiating the Lagrangian L with respect to $\boldsymbol{f}_{\text{e}}$ yields

$$\frac{\partial L}{\partial \boldsymbol{f}_{\text{e}}} = 2\boldsymbol{W}\boldsymbol{f}_{\text{e}} - \boldsymbol{T}_{\text{e}}^{\top}\boldsymbol{\lambda} = \boldsymbol{0} \tag{11.30}$$

$$\Downarrow$$

$$\boldsymbol{f}_{\text{e}} = \frac{1}{2}\boldsymbol{W}^{-1}\boldsymbol{T}_{\text{e}}^{\top}\boldsymbol{\lambda}. \tag{11.31}$$

Next, assume that $\boldsymbol{T}_{\text{e}}\boldsymbol{W}^{-1}\boldsymbol{T}_{\text{e}}^{\top}$ is non-singular such that

$$\boldsymbol{\tau} = \boldsymbol{T}_{\text{e}}\boldsymbol{f}_{\text{e}} = \frac{1}{2}\boldsymbol{T}_{\text{e}}\boldsymbol{W}^{-1}\boldsymbol{T}_{\text{e}}^{\top}\boldsymbol{\lambda} \tag{11.32}$$

$$\Downarrow$$

$$\boldsymbol{\lambda} = 2(\boldsymbol{T}_{\text{e}}\boldsymbol{W}^{-1}\boldsymbol{T}_{\text{e}}^{\top})^{-1}\boldsymbol{\tau}. \tag{11.33}$$

Substituting the Lagrange multipliers (11.33) into (11.31) yields

$$\boldsymbol{f}_{\text{e}} = \underbrace{\boldsymbol{W}^{-1}\boldsymbol{T}_{\text{e}}^{\top}(\boldsymbol{T}_{\text{e}}\boldsymbol{W}^{-1}\boldsymbol{T}_{\text{e}}^{\top})^{-1}}_{\boldsymbol{T}_{\text{w}}^{\dagger}}\boldsymbol{\tau} \tag{11.34}$$

where the matrix

$$\boldsymbol{T}_{\text{w}}^{\dagger} = \boldsymbol{W}^{-1}\boldsymbol{T}_{\text{e}}^{\top}(\boldsymbol{T}_{\text{e}}\boldsymbol{W}^{-1}\boldsymbol{T}_{\text{e}}^{\top})^{-1} \tag{11.35}$$

is recognized as the *generalized inverse*. For the case $\boldsymbol{W} = \boldsymbol{I}_{r+p}$, that is equally weighted control forces, (11.35) reduces to the right *Moore–Penrose pseudoinverse*

$$\boldsymbol{T}^{\dagger} = \boldsymbol{T}_{\mathrm{e}}^{\top} (\boldsymbol{T}_{\mathrm{e}}\boldsymbol{T}_{\mathrm{e}}^{\top})^{-1}. \tag{11.36}$$

Since $\boldsymbol{f}_{\mathrm{e}} = \boldsymbol{T}_{\mathrm{w}}^{\dagger} \boldsymbol{\tau}$, the extended control input vector $\boldsymbol{u}_{\mathrm{e}}$ can be computed from (11.16) as

$$\boldsymbol{u}_{\mathrm{e}} = \boldsymbol{K}_{\mathrm{e}}^{-1} \boldsymbol{T}_{\mathrm{w}}^{\dagger} \boldsymbol{\tau}. \tag{11.37}$$

MATLAB

The generalized inverse for the unconstrained control allocation case is implemented in the Matlab MSS toolbox as

```
u_e = ucalloc(K_e, T_e, W, tau)
```

11.2.3 Constrained Control Allocation

In industrial systems it is important to minimize the power consumption by taking advantage of the additional control forces in an overactuated control problem. From a critical point of view concerning safety it is also important to take into account actuator limitations such as saturation, wear and tear as well as other constraints. In general this leads to a *constrained* optimization problem.

Explicit Solution for $\tau = T_e K_e u_e$ using Piecewise Linear Functions

An explicit solution approach for parametric quadratic programming has been developed by Tøndel *et al.* (2003a) while applications to marine craft are presented by Johansen *et al.* (2005). In this work the constrained optimization problem is formulated as

$$\begin{aligned} J = \quad & \min_{\boldsymbol{f}_{\mathrm{e}}, \boldsymbol{s}, \bar{f}} \{\boldsymbol{f}_{\mathrm{e}}^{\top} \boldsymbol{W} \boldsymbol{f}_{\mathrm{e}} + \boldsymbol{s}^{\top} \boldsymbol{Q} \boldsymbol{s} + \beta \bar{f}\ \} \\ & \text{subject to:} \\ & \qquad \boldsymbol{T}_{\mathrm{e}} \boldsymbol{f}_{\mathrm{e}} = \boldsymbol{\tau} + \boldsymbol{s} \\ & \qquad \boldsymbol{f}_{\min} \leq \boldsymbol{f}_{\mathrm{e}} \leq \boldsymbol{f}_{\max} \\ & \qquad -\bar{f} \leq f_{\mathrm{e}_1}, f_{\mathrm{e}_2}, \dots, f_{\mathrm{e}_r} \leq \bar{f} \end{aligned} \tag{11.38}$$

where $\boldsymbol{s} \in \mathbb{R}^n$ is a vector of *slack variables*. The first term of the criterion corresponds to the LS criterion (11.28), while the third term is introduced to minimize the largest force $\bar{f} = \max_i |f_{e_i}|$ among the actuators. The constant $\beta \geq 0$ controls the relative weighting of the two criteria. This formulation ensures that the vector constraint $\boldsymbol{f}_{\min} \leq \boldsymbol{f}_{\mathrm{e}} \leq \boldsymbol{f}_{\max}$ is satisfied, if necessary by allowing the resulting generalized force $\boldsymbol{T}_{\mathrm{e}}\boldsymbol{f}_{\mathrm{e}}$ to deviate from its specification $\boldsymbol{\tau}$. To achieve accurate generalized force, the slack variable should be close to zero. This is obtained by choosing the weighting matrix $\boldsymbol{Q} \gg \boldsymbol{W} > 0$. Moreover, saturation is handled in an optimal manner by minimizing the combined criterion (11.38).

Letting

$$\boldsymbol{z} = \left[\boldsymbol{f}_{\mathrm{e}}^{\top}, \boldsymbol{s}^{\top}, \bar{f}\right]^{\top} \in \mathbb{R}^{r+n+1} \tag{11.39}$$

and

$$\boldsymbol{p} = \left[\boldsymbol{\tau}^{\top}, \boldsymbol{f}_{\min}^{\top}, \boldsymbol{f}_{\max}^{\top}, \beta\right]^{\top} \in \mathbb{R}^{n+2r+1} \tag{11.40}$$

it is straightforward to see that the optimization problem (11.38) can be reformulated as a QP problem

$$\begin{aligned} J = \min_{z} &\left\{\boldsymbol{z}^{\top}\boldsymbol{\Phi}\boldsymbol{z} + \boldsymbol{z}^{\top}\boldsymbol{R}\boldsymbol{p}\right\} \\ &\text{subject to:} \\ &\boldsymbol{A}_1\boldsymbol{z} = \boldsymbol{C}_1\boldsymbol{p} \\ &\boldsymbol{A}_2\boldsymbol{z} \leq \boldsymbol{C}_2\boldsymbol{p} \end{aligned} \tag{11.41}$$

where

$$\boldsymbol{\Phi} = \begin{bmatrix} \boldsymbol{W} & \boldsymbol{0}_{r\times n} & \boldsymbol{0}_{r\times 1} \\ \boldsymbol{0}_{n\times r} & \boldsymbol{Q} & \boldsymbol{0}_{n\times 1} \\ \boldsymbol{0}_{1\times r} & \boldsymbol{0}_{1\times n} & 0 \end{bmatrix}, \quad \boldsymbol{R} = \begin{bmatrix} \boldsymbol{0}_{(r+n+1)\times(n+2r)} & \begin{bmatrix} \boldsymbol{0}_{(r+n)\times 1} \\ 1 \end{bmatrix} \end{bmatrix}$$

$$\boldsymbol{A}_1 = \begin{bmatrix} \boldsymbol{T}_{\mathrm{e}} & -\boldsymbol{I}_n & \boldsymbol{0}_{n\times 1} \end{bmatrix}, \quad \boldsymbol{C}_1 = \begin{bmatrix} \boldsymbol{I}_n & \boldsymbol{0}_{n\times(2r+1)} \end{bmatrix}$$

$$\boldsymbol{A}_2 = \begin{bmatrix} -\boldsymbol{I}_r & \boldsymbol{0}_{r\times n} & \boldsymbol{0}_{r\times 1} \\ \boldsymbol{I}_r & \boldsymbol{0}_{r\times n} & \boldsymbol{0}_{r\times 1} \\ \boldsymbol{I}_r & \boldsymbol{0}_{r\times n} & \begin{bmatrix} 1 \\ 1 \\ \vdots \\ 1 \end{bmatrix} \\ \boldsymbol{I}_r & \boldsymbol{0}_{r\times n} & -\begin{bmatrix} 1 \\ 1 \\ \vdots \\ 1 \end{bmatrix} \end{bmatrix}, \quad \boldsymbol{C}_2 = \begin{bmatrix} \boldsymbol{0}_{r\times n} & -\boldsymbol{I}_r & \boldsymbol{0}_{r\times r} & \boldsymbol{0}_{r\times 1} \\ \boldsymbol{0}_{r\times n} & \boldsymbol{0}_{r\times r} & \boldsymbol{I}_r & \boldsymbol{0}_{r\times 1} \\ \boldsymbol{0}_{r\times n} & \boldsymbol{0}_{r\times r} & \boldsymbol{0}_{r\times r} & \boldsymbol{0}_{r\times 1} \\ \boldsymbol{0}_{r\times n} & \boldsymbol{0}_{r\times r} & \boldsymbol{0}_{r\times r} & \boldsymbol{0}_{r\times 1} \end{bmatrix}.$$

Since $\boldsymbol{W} > 0$ and $\boldsymbol{Q} > 0$ this is a convex quadratic program in z parametrized by $\boldsymbol{p}$. Convexity guarantees that a global solution can be found. The optimal solution $z^*(\boldsymbol{p})$ to this problem is a continuous piecewise linear function $z^*(\boldsymbol{p})$ defined on any subset

$$\boldsymbol{p}_{\min} \leq \boldsymbol{p} \leq \boldsymbol{p}_{\max} \tag{11.42}$$

of the parameter space. Moreover, an exact representation of this piecewise linear function can be computed offline using multiparametric QP (mp-QP) algorithms (Tøndel *et al.* 2003b) or the *Matlab Multi-Parametric Toolbox* (MPT) by Kvasnica *et al.* (2004). Consequently, it is not necessary to solve the QP (11.38) in real time for the current value of $\boldsymbol{\tau}$ and the parameters $\boldsymbol{f}_{\min}, \boldsymbol{f}_{\max}$ and β if they are allowed to vary. In fact, it suffices to evaluate the known piecewise linear function $z^*(\boldsymbol{p})$ as a function of the given parameter vector $\boldsymbol{p}$, which can be done efficiently with a small number of computations. For details of the implementation aspects of the mp-QP algorithm see Johansen *et al.* (2004) and references therein. An online control allocation algorithm is presented in Tøndel *et al.* (2003a).

Explicit Solution for Saturated Thrust Vectors and Forbidden Zones

An extension of the mp-QP algorithm to marine craft equipped with azimuth thrusters and rudders was given by Tøndel *et al.* (2003a). A propeller with a rudder can produce a thrust vector within a range of directions and magnitudes in the horizontal plane for low-speed maneuvering and stationkeeping. The set of attainable thrust vectors is non-convex because significant lift can be produced by the rudder only with forward thrust. The attainable thrust region can, however, be decomposed into a finite union of convex polyhedral sets. A similar decomposition can be made for azimuth thrusters including forbidden sectors. Hence, this can be formulated as a mixed-integer-like convex QP problem, and by using, arbitrarily, number, of rudders as well as thrusters, other propulsion devices can be handled. Actuator rate and position constraints are also taken into account. Using a mp-QP software, an explicit piecewise linear representation of the least-squares optimal control allocation law can be precomputed. The method has been tested on a scale model of a supply vessel by Tøndel *et al.* (2003a) and a scale model of a floating platform by Spjøtvold (2008).

Explicit Solutions Based on Minimum Norm and Null-space Methods

In flight and aerospace control systems, the problems of control allocation and saturating control have been addressed by Durham (1993, 1994a, 1994b), and Johansen and Fossen (2013). Durham also discusses an explicit solution to avoid saturation, referred to as the "direct method". By noticing that there are infinite combinations of admissible controls that generate control forces on the boundary of the closed subset of attainable controls, the "direct method" calculates admissible controls in the interior of the attainable forces as scaled-down versions of the unique solutions for force demands. Unfortunately it is not possible to minimize the norm of the control forces on the boundary or some other constraint since the solutions on the boundary are unique. The computational complexity of the algorithm is proportional to the square of the number of controls, which can be problematic in real-time applications.

In Bordignon and Durham (1995) the null-space interaction method is used to minimize the norm of the control vector, when possible, and still access the attainable forces to overcome the drawbacks of the "direct method". This method is also explicit but much more computationally intensive. For instance, 20 independent controls imply that up to 3.4 billion points have to be checked at each sample. In Durham (1999) a computationally simple and efficient method to obtain near-optimal solutions is described. The method is based on prior knowledge of the controls' effectiveness and limits such that precalculation of several generalized inverses can be done.

Iterative Constrained Control Allocation for Azimuth Thrusters

The control allocation problem for marine craft equipped with azimuth thrusters is in general a *non-convex* optimization problem that is hard to solve. The primary constraint is $\boldsymbol{\tau} = \boldsymbol{T}(\boldsymbol{\alpha})\boldsymbol{f}$ where $\boldsymbol{\alpha} \in \mathbb{R}^p$ denotes the azimuth angles. The azimuth angles must be computed at each sample together with the control inputs $\boldsymbol{u} \in \mathbb{R}^r$ which are subject to both amplitude and rate saturations. In addition, azimuth thrusters can only operate in feasible sectors $\alpha_{i,\min} \leq \alpha_i \leq \alpha_{i,\max}$ at a limiting turning rate $\dot{\alpha}_i$. Another problem is that the inverse

$$\boldsymbol{T}^{\dagger}_{\mathrm{w}}(\boldsymbol{\alpha}) = \boldsymbol{W}^{-1}\boldsymbol{T}^{\top}(\boldsymbol{\alpha})[\boldsymbol{T}(\boldsymbol{\alpha})\boldsymbol{W}^{-1}\boldsymbol{T}^{\top}(\boldsymbol{\alpha})]^{-1} \tag{11.43}$$

can be singular for certain $\boldsymbol{\alpha}$ values. The consequence of such a singularity is that no force is produced in certain directions. This may greatly reduce dynamic performance and maneuverability as the azimuth angles can be changed only slowly. This suggests that the following criterion should be minimized (Johansen *et al.* 2004)

$$\begin{aligned} J = \min_{f,\,\alpha,\,s} & \left\{ \sum_{i=1}^{r} \overline{P}_i |f_i|^{3/2} + \boldsymbol{s}^{\top}\boldsymbol{Q}\boldsymbol{s} + (\boldsymbol{\alpha}-\boldsymbol{\alpha}_0)^{\top}\boldsymbol{\Omega}(\boldsymbol{\alpha}-\boldsymbol{\alpha}_0) \right. \\ & \left. + \frac{\varrho}{\varepsilon + \det(\boldsymbol{T}(\boldsymbol{\alpha})\boldsymbol{W}^{-1}\boldsymbol{T}^{\top}(\boldsymbol{\alpha}))} \right\} \\ & \text{subject to:} \\ & \quad \boldsymbol{T}(\boldsymbol{\alpha})\boldsymbol{f} = \boldsymbol{\tau} + \boldsymbol{s} \\ & \quad \boldsymbol{f}_{\min} \leq \boldsymbol{f} \leq \boldsymbol{f}_{\max} \\ & \quad \boldsymbol{\alpha}_{\min} \leq \boldsymbol{\alpha} \leq \boldsymbol{\alpha}_{\max} \\ & \quad \Delta\boldsymbol{\alpha}_{\min} \leq \boldsymbol{\alpha} - \boldsymbol{\alpha}_0 \leq \Delta\boldsymbol{\alpha}_{\max} \end{aligned} \tag{11.44}$$

where

- $\sum_{i=1}^{r} \overline{P}_i |f_i|^{3/2}$ represents power consumption where $\overline{P}_i > 0$ $(i = 1, \ldots, r)$ are positive weights.
- $\boldsymbol{s}^{\top}\boldsymbol{Q}\boldsymbol{s}$ penalizes the error $\boldsymbol{s}$ between the commanded and achieved generalized force. This is necessary in order to guarantee that the optimization problem has a feasible solution

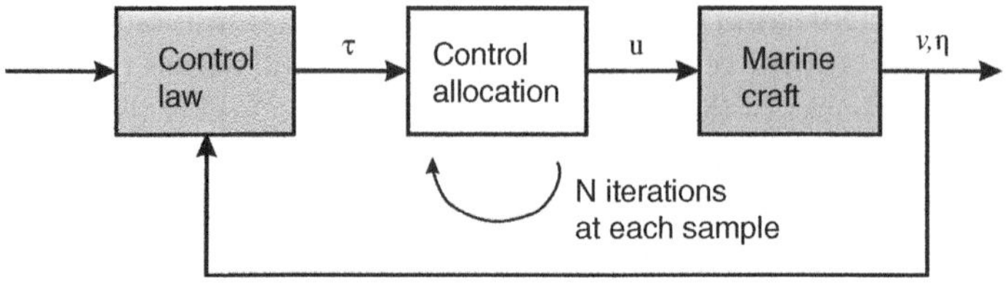

Figure 11.7 Control allocation using an iterative solution.

for any $\boldsymbol{\tau}$ and $\boldsymbol{\alpha}_0$. The weight $\boldsymbol{Q} > 0$ is chosen to be large enough so that the optimal solution is $\boldsymbol{s} \approx \boldsymbol{0}$ whenever possible.

- $\boldsymbol{f}_{\min} \leq \boldsymbol{f} \leq \boldsymbol{f}_{\max}$ is used to limit the use of force (saturation handling).
- $\boldsymbol{\alpha}_{\min} \leq \boldsymbol{\alpha} \leq \boldsymbol{\alpha}_{\max}$ denotes the feasible sectors of the azimuth angles.
- $\Delta\boldsymbol{\alpha}_{\min} \leq \boldsymbol{\alpha} - \boldsymbol{\alpha}_0 \leq \Delta\boldsymbol{\alpha}_{\max}$ ensures that the azimuth angles do not move too much within one sample, taking $\boldsymbol{\alpha}_0$ equal to the angles at the previous sample. This is equivalent to limiting $\dot{\boldsymbol{\alpha}}$, that is the turning rate of the thrusters.
- The term

$$\frac{\varrho}{\varepsilon + \det(\boldsymbol{T}(\boldsymbol{\alpha})\boldsymbol{W}^{-1}\boldsymbol{T}^{\top}(\boldsymbol{\alpha}))}$$

is introduced to avoid singular configurations caused by the problem term $\det(\boldsymbol{T}(\boldsymbol{\alpha})\boldsymbol{W}^{-1}\boldsymbol{T}^{\top}(\boldsymbol{\alpha})) = 0$. To avoid division by zero, $\varepsilon > 0$ is chosen as a small number, while $\varrho > 0$ is scalar weight. A large ϱ ensures high maneuverability at the cost of higher power consumption and vice versa.

The optimization problem (11.44) is a non-convex nonlinear program and requires a significant number of computations at each sample (Nocedal and Wright 1999). The nonlinear program is solved by using iterations as shown in Figure 11.7. The following two implementation strategies are attractive alternatives to nonlinear program efforts.

Iterative Solution using Quadratic Programming

The problem (11.44) can be locally approximated with a *convex* QP problem by assuming that:

1. The power consumption can be approximated by a quadratic term in $\boldsymbol{f}$ near the last force $\boldsymbol{f}_0$ such that $\boldsymbol{f} = \boldsymbol{f}_0 + \Delta\boldsymbol{f}$.
2. The singularity avoidance penalty can be approximated by a linear term linearized about the last azimuth angle $\boldsymbol{\alpha}_0$ such that $\boldsymbol{\alpha} = \boldsymbol{\alpha}_0 + \Delta\boldsymbol{\alpha}$.

The resulting QP criterion is (Johansen *et al.* 2004)

$$
\begin{aligned}
J = \min_{\Delta f,\, \Delta\alpha,\, s} \Bigg\{ & (f_0 + \Delta f)^{\top} P (f_0 + \Delta f) + s^{\top} Q s + \Delta\alpha^{\top} \Omega \Delta\alpha \\
& + \frac{\partial}{\partial \alpha} \left(\frac{\varrho}{\varepsilon + \det(T(\alpha) W^{-1} T^{\top}(\alpha))} \right) \Bigg|_{\alpha_0} \Delta\alpha \Bigg\} \\
\text{subject to:} & \\
& s + T(\alpha_0)\Delta f + \frac{\partial}{\partial \alpha}\left(T(\alpha_0) f \right)\big|_{\alpha_0, f_0} \Delta\alpha = \tau - T(\alpha_0) f_0 \\
& f_{\min} - f_0 \leq \Delta f \leq f_{\max} - f_0 \\
& \alpha_{\min} - \alpha_0 \leq \Delta\alpha \leq \alpha_{\max} - \alpha_0 \\
& \Delta\alpha_{\min} \leq \Delta\alpha \leq \Delta\alpha_{\max}.
\end{aligned}
\tag{11.45}
$$

The convex QP problem (11.45) can be solved by using standard software for numerical optimization, for instance the function `quadprog.m` in the Matlab optimization toolbox.

Iterative Solution using Linear Programming

Linear approximations to the thrust allocation problem have been discussed by Webster and Sousa (1999) and Lindfors (1993). In Lindfors (1993) the azimuth thrust constraints

$$
|f_i| = \sqrt{[f_i \cos(\alpha_i)]^2 + [f_i \sin(\alpha_i)]^2} \leq f_i^{\max} \tag{11.46}
$$

are represented as circles in the $(f_i \cos(\alpha_i),\, f_i \sin(\alpha_i))$ plane. The nonlinear program is transformed to a linear programming (LP) problem by approximating the azimuth thrust constraints by straight lines forming a polygon. If eight lines are used to approximate the circles (octagons), the worst case errors will be less than $\pm 4.0\%$. The criterion to be minimized is a linear combination of f, that is magnitude of force in the x and y directions, weighted against the magnitudes $|f_i|$ representing azimuth thrust. Hence, singularities and azimuth rate limitations are not weighted in the cost function. If these are important, the QP formulation should be used.

Explicit Solution using the Singular Value Decomposition and Filtering Techniques

An alternative method to solve the constrained control allocation problem is to use the singular value decomposition (SVD) and a filtering scheme to control the azimuth directions such that they are aligned with the direction where most force is required, paying attention to singularities (Sørdalen 1997b). Results from sea trials have been presented in Sørdalen (1997a). A similar technique using the damped least-squares algorithm was reported in Berge and Fossen (1997), where the results are documented by controlling a scale model of a supply vessel equipped with four azimuth thrusters.

12

Guidance Systems

This chapter describes methods for the design of *guidance systems* for marine craft. Guidance can be defined as (Shneydor 1998): "*The process for guiding the path of an object towards a given point, which in general may be moving*". Draper (1971) states: "*Guidance depends upon fundamental principles and involves devices that are similar for vehicles moving on land, on water, under water, in air, beyond the atmosphere within the gravitational field of Earth and in space outside this field*". Thus, guidance represents a basic methodology concerned with the transient motion behavior associated with the achievement of motion control objectives (Breivik and Fossen 2009).

In its simplest form, open-loop guidance systems for marine craft are used to generate a reference trajectory for time-varying *trajectory tracking* (Sections 12.1 and 12.2) or, alternatively, a path for time-invariant *path following* (Sections 12.3–12.6). A motion control system will work in close interaction with the guidance system (Siouris 2004, Yanushevsky 2008).

In the forthcoming, the following motion control scenarios are considered:

- *Setpoint regulation (point stabilization)* is a special case where the desired position and attitude are chosen to be constant.
- *Trajectory tracking*, where the objective is to force the system output $\boldsymbol{y}(t) \in \mathbb{R}^m$ to track a desired output $\boldsymbol{y}_d(t) \in \mathbb{R}^m$. The desired trajectory can be computed using reference models generated by low-pass filters, optimization methods or by simply simulating the marine craft motion using an adequate model of the craft. Feasible trajectories can be generated in the presence of both *spatial* and *temporal constraints*.
- *Target tracking* methods for motion control scenarios when there is no information about the path and there is no trajectory to track. This is particularly useful if the goal is to track a moving object such as a ship or an underwater vehicle for which no future motion information is available.
- *Path following* is following a predefined path independent of time. A popular approach is the *Dubins path* representing the shortest curve that connects two points in the 2D Euclidean plane with a constraint on the curvature of the path and with prescribed initial and terminal tangents to the path (Dubins 1957, Tsourdos *et al.* 2010). For path

Handbook of Marine Craft Hydrodynamics and Motion Control, Second Edition. Thor I. Fossen.

following, no restrictions are placed on the temporal propagation along the path. Spatial constraints can, however, be added to represent obstacles and other positional constraints if they are known in advance.

Tracking control systems can also be designed for target tracking and path tracking. For instance, a target-tracking system tracks the motion of a target that is either stationary (analogous to point stabilization) or that moves such that only its instantaneous motion is known; that is no information about the future target motion is available (Breivik and Fossen 2009).

As shown in Figure 12.1, the guidance system can use joystick or keyboard inputs, external inputs (weather data, for instance measured wind, wave and current speeds and directions), Earth topological information (digital chart, radar and sonar data), obstacle and collision avoidance data, and finally the state vector, which is available as output from the navigation and sensor systems. The required data are further processed to generate a feasible trajectory for motion control. This can be done using ad hoc techniques or sophisticated methods such as interpolation techniques, dynamic optimization or filtering techniques. A feasible trajectory means one that is consistent with the craft dynamics. For a linear system, this implies that the eigenvalues of the desired states must be chosen such that the reference model is slower than the craft dynamics.

For a ship or an underwater vehicle, the guidance and control system usually consists of the following subsystems:

- *Attitude control system*
- *Path-following control system.*

In its simplest form, the attitude control system is a heading or course autopilot, while roll and pitch are regulated to zero or left uncontrolled. The main function of the attitude feedback control system is to maintain the craft in a desired attitude on the ordered path by controlling the craft in roll, pitch and yaw. The task of the path-following controller is to keep the craft on the predescribed path with some predefined dynamics, for instance

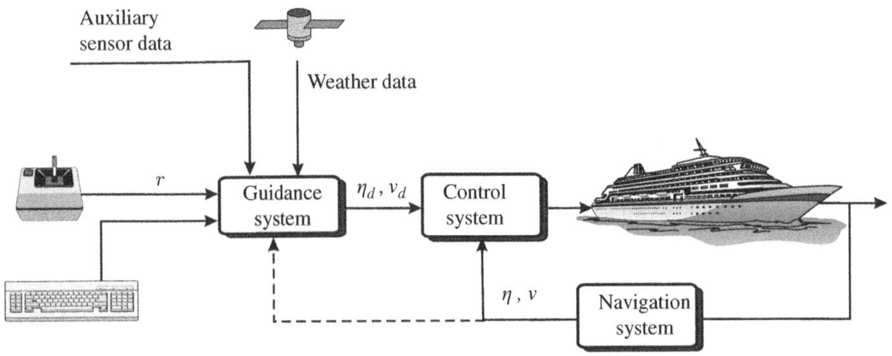

Figure 12.1 In closed-loop guidance (dotted line) the states are fed back to the guidance system while open-loop guidance only uses sensor and reference signal inputs.

a speed control system by generating orders to the attitude control system. For surface vessels it is common to use a heading controller in combination with a speed controller while aircraft and underwater vehicles also need a height/depth controller. The principles and definitions of *guidance, navigation* and *control* are further outlined in Chapter 11.

12.1 Trajectory Tracking

Guidance systems designed for tracking a smooth time-varying trajectory $\boldsymbol{y}_{\mathrm{d}} \in \mathbb{R}^m$ are useful in many applications. The desired speed and acceleration are obtained from time-differentiation of $\boldsymbol{y}_{\mathrm{d}}$ one and two times, respectively. This means that the signal $\boldsymbol{y}_{\mathrm{d}}$ defines the desired position/attitude, velocity and acceleration as a function of time t for a moving craft in 6 DOFs. We will make use of the following definition in the forthcoming.

Definition 12.1 (Trajectory Tracking)
A control system that forces the system output $\boldsymbol{y} \in \mathbb{R}^m$ to track a desired output $\boldsymbol{y}_{\mathrm{d}} \in \mathbb{R}^m$ solves a trajectory tracking problem.

This definition is consistent with Athans and Falb (1966) and later with Hauser and Hindmann (1995), Ortega *et al.* (1998), Encarnacao and Pascoal (2001) and Skjetne *et al.* (2002, 2004). In this section, methods for computation of the desired trajectory corresponding to a desired virtual target will be presented. The following methods are discussed:

- Low-pass filters for the generation of generalized position, velocity and acceleration trajectories.
- Time-domain simulation using an adequate model of the craft.
- Dynamic optimization methods.

Within this framework, it is possible to generate feasible trajectories incorporating both *spatial constraints* (obstacle avoidance and maximum velocity/acceleration) and *temporal constraints* (minimum time, on time and maximum time problems).

Trajectory-tracking Control

Trajectory-tracking control laws are classified according to the number of available effectors; see Section 11.2. This can be illustrated by considering a marine craft in *surge, sway* and *yaw*. Tracking of a *time-varying reference trajectory* $\boldsymbol{\eta}_{\mathrm{d}} = [x^n_{\mathrm{d}}, y^n_{\mathrm{d}}, \psi_{\mathrm{d}}]^\top$ is achieved by minimizing the tracking error

$$\boldsymbol{e} := \boldsymbol{\eta} - \boldsymbol{\eta}_{\mathrm{d}} = \begin{bmatrix} x^n - x^n_{\mathrm{d}} \\ y^n - y^n_{\mathrm{d}} \\ \psi - \psi_{\mathrm{d}} \end{bmatrix}. \tag{12.1}$$

Based on this interpretation, the following considerations can be made:

- **Three or more controls:** This is referred to as a *fully actuated* dynamic positioning (DP) system and typical applications are crab-wise motions (low-speed maneuvering) and stationkeeping, where the goal is to drive $\boldsymbol{e} \in \mathbb{R}^2 \times \mathbb{T} \rightarrow \mathbf{0}$. This is the standard configuration for offshore DP vessels. Feedback control laws for fully actuated vessels are discussed in Chapter 15.
- **Two controls and trajectory tracking:** Trajectory tracking in 3 DOFs, $\boldsymbol{e} \in \mathbb{R}^2 \times \mathbb{T}$, with only two controls, $\boldsymbol{u} \in \mathbb{R}^2$, is an *underactuated* control problem, which cannot be solved using linear theory. This problem has limited practical use. However, since all marine craft operate in a uniform force field due to mean wind, waves and ocean currents, it is possible to steer the craft along a path with a constant sideslip angle (given by the mean environmental force field) using only two controls, that is turning and forward speed control.
- **Two controls and weather-optimal heading:** If the ship is aligned up against the mean resulting force due to wind, waves and ocean currents, a weathervaning controller can be designed such that only two controls, $\boldsymbol{u} \in \mathbb{R}^2$, are needed to stabilize the ship positions. In this approach the heading angle is allowed to vary automatically with the mean environmental forces (Pinkster 1971, Pinkster and Nienhuis 1986, Fossen and Strand 2001) (see Section 16.3.11).
- **Two controls and path following:** It is standard procedure to define a 2D workspace (along-track and cross-track errors) and minimize the cross-track error by means of an LOS path-following controller; see Sections 12.3–12.6. Hence, it is possible to follow a path by using only two controls (surge speed and yaw moment). For a conventional ship this is achieved by using a rudder and a propeller only. Since the input and output vectors are of dimension two, the 6-DOF system model must be internally stable.
- **One control:** It is impossible to design stationkeeping systems and trajectory-tracking control systems in 3 DOFs for a marine craft using only one control.

For underwater vehicles operating in 6 DOFs it is also important to control the heave and sometimes the pitch-roll motions in addition to the surge, sway and yaw motions. However, roll and pitch can be left uncontrolled for metacentrically stable vehicles. For operation in 6 DOFs, a fully actuated vehicle must have six or more actuators producing independent forces and moments in all directions in order to track a 6-DOF time-varying reference trajectory.

12.1.1 Reference Models for Trajectory Generation

In a practical system, it is highly advantageous to keep the software as simple as possible. As a result of this, many industrial systems are designed using linear reference models for trajectory generation. This corresponds to *open-loop guidance* as described in Chapter 11 since no feedback from the states is required. The simplest form of a reference model is obtained by using a low-pass (LP) filter structure

$$\frac{x_{\rm d}}{r}(s) = h_{\rm lp}(s) \tag{12.2}$$

Figure 12.2 Joystick control system used to generate reference signals. Source: Illustration by B. Stenberg.

where x_d is the desired state and r denotes the reference signal usually specified by an operator (see Figure 12.2). The choice of filter should reflect the dynamics of the craft such that a feasible trajectory is constructed. For instance, it is important to take into account physical speed and acceleration limitations of the craft as well as input saturation. This is a non-trivial task so a compromise between performance and accurate tracking must be made by tuning the bandwidth of the reference model. It is important that the bandwidth of the reference model is chosen lower than the bandwidth of the motion control system in order to obtain satisfactory tracking performance and stability.

A frequently used method to generate a smooth reference trajectory $\boldsymbol{x}_d \in \mathbb{R}^n$ for tracking control is to use a physically motivated model. For marine craft it is convenient to use reference models motivated by the dynamics of *mass–damper–spring systems* to generate the desired state trajectories.

Velocity Reference Model

The velocity reference model should at least be of order two so as to obtain smooth reference signals for the desired velocity $\boldsymbol{v}_d$ and acceleration $\dot{\boldsymbol{v}}_d$. Let $\boldsymbol{r}^b$ denote the operator input expressed in $\{b\}$. A second-order LP filter

$$\frac{v_{d_i}}{r_i^b}(s) = \frac{\omega_{n_i}^2}{s^2 + 2\zeta_i\omega_{n_i}s + \omega_{n_i}^2} \tag{12.3}$$

where ζ_i $(i = 1, \dots, n)$ are the *relative damping ratios* and ω_{n_i} $(i = 1, \dots, n)$ are the *natural frequencies*. This implies that

$$\ddot{\boldsymbol{v}}_d + 2\boldsymbol{\Delta}\boldsymbol{\Omega}\dot{\boldsymbol{v}}_d + \boldsymbol{\Omega}^2\boldsymbol{v}_d = \boldsymbol{\Omega}^2\boldsymbol{r}^b \tag{12.4}$$

where $\boldsymbol{v}_d$ is the desired velocity, $\dot{\boldsymbol{v}}_d$ is the desired acceleration and $\ddot{\boldsymbol{v}}_d$ is interpreted as the desired "jerk". For this model, $\boldsymbol{\Delta} > 0$ and $\boldsymbol{\Omega} > 0$ are diagonal design matrices of *relative damping ratios* and *natural frequencies*

$$\boldsymbol{\Delta} = \text{diag}\{\zeta_1, \zeta_2, \dots, \zeta_n\}$$

$$\boldsymbol{\Omega} = \text{diag}\{\omega_{n_1}, \omega_{n_2}, \dots, \omega_{n_n}\}.$$

The state-space representation is

$$\dot{\boldsymbol{x}}_d = \boldsymbol{A}_d\boldsymbol{x}_d + \boldsymbol{B}_d\boldsymbol{r}^b \tag{12.5}$$

where $\boldsymbol{x}_d := [\boldsymbol{v}_d^\top, \dot{\boldsymbol{v}}_d^\top]^\top \in \mathbb{R}^{2n}$ and

$$\boldsymbol{A}_d = \begin{bmatrix} \boldsymbol{0}_{n\times n} & \boldsymbol{I}_n \\ -\boldsymbol{\Omega}^2 & -2\boldsymbol{\Delta}\boldsymbol{\Omega} \end{bmatrix}, \quad \boldsymbol{B}_d = \begin{bmatrix} \boldsymbol{0}_{n\times n} \\ \boldsymbol{\Omega}^2 \end{bmatrix}. \tag{12.6}$$

Note that a step in the command $\boldsymbol{r}^b$ will give a step in $\ddot{\boldsymbol{v}}_d$ while acceleration $\dot{\boldsymbol{v}}_d$ and velocity $\boldsymbol{v}_d$ will be low-pass filtered and therefore smooth signals in a tracking control system. We also notice that the steady-state velocity for a constant reference signal $\boldsymbol{r}^b$ satisfies

$$\lim_{t\to\infty} \boldsymbol{v}_d = \boldsymbol{r}^b. \tag{12.7}$$

Position and Attitude Reference Models

The position and attitude reference model $\boldsymbol{\eta}_d$ is typically chosen to be of third order for filtering the steps in $\boldsymbol{r}^n$. This suggests that a first-order LP filter should be cascaded with a mass–damper–spring system. Consider the transfer function

$$\frac{\eta_{d_i}}{r_i^n}(s) = \frac{\omega_{n_i}^2}{(1 + T_i s)(s^2 + 2\zeta_i\omega_{n_i}s + \omega_{n_i}^2)} \tag{12.8}$$

where a first-order LP filter with time constant $T_i = 1/\omega_{n_i} > 0$ has been added. This can also be written

$$\frac{\eta_{d_i}}{r_i^n}(s) = \frac{\omega_{n_i}^3}{s^3 + (2\zeta_i + 1)\omega_{n_i}s^2 + (2\zeta_i + 1)\omega_{n_i}^2 s + \omega_{n_i}^3} \tag{12.9}$$

which generalizes to multivariable systems as

$$\boldsymbol{\eta}_d^{(3)} + (2\boldsymbol{\Delta} + \boldsymbol{I}_n)\boldsymbol{\Omega}\ddot{\boldsymbol{\eta}}_d + (2\boldsymbol{\Delta} + \boldsymbol{I}_n)\boldsymbol{\Omega}^2\dot{\boldsymbol{\eta}}_d + \boldsymbol{\Omega}^3\boldsymbol{\eta}_d = \boldsymbol{\Omega}^3\boldsymbol{r}^n. \tag{12.10}$$

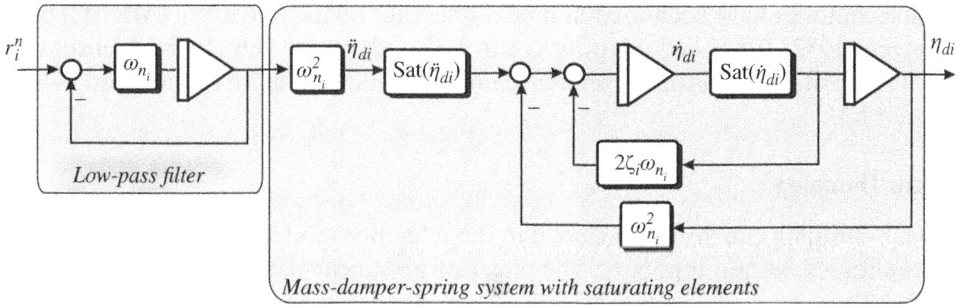

Figure 12.3 Reference model including saturating elements.

The state-space representation is

$$\dot{\boldsymbol{x}}_{\mathrm{d}} = \boldsymbol{A}_{\mathrm{d}}\boldsymbol{x}_{\mathrm{d}} + \boldsymbol{B}_{\mathrm{d}}\boldsymbol{r}^n \tag{12.11}$$

where $\boldsymbol{x}_{\mathrm{d}} := [\boldsymbol{\eta}_{\mathrm{d}}^{\top}, \dot{\boldsymbol{\eta}}_{\mathrm{d}}^{\top}, \ddot{\boldsymbol{\eta}}_{\mathrm{d}}^{\top}]^{\top} \in \mathbb{R}^{3n}$ and

$$\boldsymbol{A}_{\mathrm{d}} = \begin{bmatrix} \boldsymbol{0}_{n\times n} & \boldsymbol{I}_n & \boldsymbol{0}_{n\times n} \\ \boldsymbol{0}_{n\times n} & \boldsymbol{0}_{n\times n} & \boldsymbol{I}_n \\ -\boldsymbol{\Omega}^3 & -(2\boldsymbol{\Delta}+\boldsymbol{I}_n)\boldsymbol{\Omega}^2 & -(2\boldsymbol{\Delta}+\boldsymbol{I}_n)\boldsymbol{\Omega} \end{bmatrix}, \quad \boldsymbol{B}_{\mathrm{d}} = \begin{bmatrix} \boldsymbol{0}_{n\times n} \\ \boldsymbol{0}_{n\times n} \\ \boldsymbol{\Omega}^3 \end{bmatrix}. \tag{12.12}$$

Note that a step in the command $\boldsymbol{r}^n$ will give a step in $\boldsymbol{\eta}_d^{(3)}$ while $\ddot{\boldsymbol{\eta}}_d$, $\dot{\boldsymbol{\eta}}_d$ and $\boldsymbol{\eta}_{\mathrm{d}}$ will be low-pass filtered and therefore smooth signals in a tracking control system. We also notice that the steady-state position for a constant reference signal $\boldsymbol{r}^n$ is

$$\lim_{t\to\infty} \boldsymbol{\eta}_d = \boldsymbol{r}^n. \tag{12.13}$$

Saturating Elements

One drawback with a linear reference model is that the time constants in the model often yield a satisfactory response for one operating point of the system while the response for other amplitudes of the operator input r_i results in completely different behavior. This is due to the exponential convergence of the signals in a linear system. One way to circumvent this problem is to use amplitude gain scheduling so that the reference model design parameters $(\zeta_i, \omega_{\mathrm{n}_i})$ are scheduled with respect to the magnitude of the input signal r_i.

The performance of the linear reference model can be improved by including saturation elements for velocity and acceleration according to

$$\mathrm{sat}(x) = \begin{cases} \mathrm{sgn}(x)x^{\max} & \text{if } |x| \geq x^{\max} \\ x & \text{else} \end{cases}. \tag{12.14}$$

Hence, the saturation limits

$$|v_i| \leq v_i^{\max}, \quad |\dot{v}_i| \leq \dot{v}_i^{\max} \tag{12.15}$$

should reflect the physical limitations of the craft as illustrated in Example 12.1.

These techniques have been used in model reference adaptive control (MRAC) by Van Amerongen (1982, 1984) and adaptive control of underwater vehicles by Fjellstad *et al.* (1992). The position and attitude reference model should therefore be modified as shown in Figure 12.3.

Nonlinear Damping

Nonlinear damping can also be included in the reference model to reduce the velocity for large amplitudes or step inputs r_i. The effect of nonlinear damping is demonstrated in Example 12.1.

Example 12.1 (Nonlinear Reference Model)
Consider the mass–damper–spring reference model

$$\dot{\eta}_{\rm d} = \nu_{\rm d} \tag{12.16}$$

$$\dot{\nu}_{\rm d} + 2\zeta\omega_{\rm n}\nu_{\rm d} + \delta|\nu_{\rm d}|\nu_{\rm d} + \omega_{\rm n}^2\eta_{\rm d} = \omega_{\rm n}^2 r \tag{12.17}$$

where $\zeta = \omega_{\rm n} = 1$*. Figure 12.4 shows a comparison of responses using* $\delta = 0$, $\delta = 1$ *and a saturating element,* $\nu^{\rm max} = 1$ *for an operator step input* $r = 10$*. The Matlab example file* `ExRefMod.m` *in the MSS toolbox was used to generate the plots.*

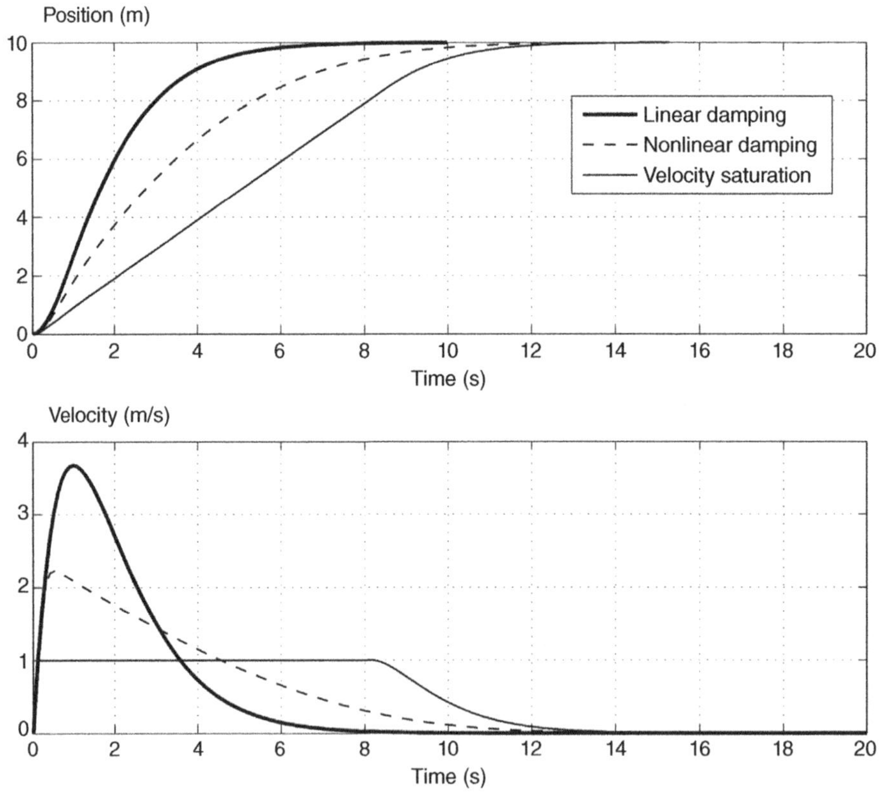

Figure 12.4 Desired position and velocity for a step input $r = 10$.

12.1.2 Trajectory Generation using a Marine Craft Simulator

The reference models in Section 12.1.1 are popular due to their simplicity. The cutoff frequency of the reference model must never exceed the closed-loop bandwidth of the system in order to guarantee that the craft is able to track the desired states. This is difficult to verify in a practical system due to factors such as nonlinearities, saturating elements and time delays. An alternative approach could be to generate a time-varying reference trajectory using a low-fidelity closed-loop model of the craft, where the time constants, relative damping ratios and natural frequencies are chosen to reflect physical limitations of the craft. For instance, the dynamic model can be chosen as

$$\dot{\boldsymbol{\eta}}_{\mathrm{d}} = \boldsymbol{J}_{\Theta}(\boldsymbol{\eta}_{\mathrm{d}})\boldsymbol{\nu}_{\mathrm{d}} \tag{12.18}$$

$$\boldsymbol{M}\dot{\boldsymbol{\nu}}_{\mathrm{d}} + \boldsymbol{D}\boldsymbol{\nu}_{\mathrm{d}} + \boldsymbol{g}(\boldsymbol{\eta}_{\mathrm{d}}) = \boldsymbol{\tau} \tag{12.19}$$

where the damping matrix is modeled as a diagonal matrix

$$\boldsymbol{D} = \mathrm{diag}\{d_1, \ldots, d_6\} > 0. \tag{12.20}$$

The system inertia matrix $\boldsymbol{M}$ is included in the model to guarantee proper scaling of the control inputs $\boldsymbol{\tau}$. Smooth reference trajectories $(\boldsymbol{\eta}_{\mathrm{d}}, \boldsymbol{\nu}_{\mathrm{d}})$ are then obtained by simulating the model under closed-loop control, for instance by using a nonlinear PD controller with feedforward (see Section 15.3)

$$\boldsymbol{\tau} = \boldsymbol{g}(\boldsymbol{\eta}_{\mathrm{d}}) - \boldsymbol{J}_{\Theta}^{\mathrm{T}}(\boldsymbol{\eta}_{\mathrm{d}})(\boldsymbol{K}_{\mathrm{p}}(\boldsymbol{\eta}_{\mathrm{d}} - \boldsymbol{\eta}_{\mathrm{ref}}) + \boldsymbol{K}_{\mathrm{d}}\dot{\boldsymbol{\eta}}_{\mathrm{d}}) \tag{12.21}$$

where $\boldsymbol{\eta}_{\mathrm{ref}}$ is the setpoint and $(\boldsymbol{\eta}_{\mathrm{d}}, \boldsymbol{\nu}_{\mathrm{d}})$ represents the desired states. The control law (12.21) is in fact a *guidance controller* since it is applied to the reference model. In addition to this, it is useful to include saturation elements for velocity and acceleration to keep these quantities within their physical limits.

Example 12.2 (Generation of Reference Trajectories using a Marine Craft Simulator)
Consider a marine craft moving at forward speed $U \gg 0$ such that $u \approx U$ and $v \approx 0$ implying that $\psi = \chi$. The desired reference trajectories are

$$\dot{x}_{\mathrm{d}}^{n} = u_{\mathrm{d}} \cos(\psi_{\mathrm{d}}) \tag{12.22}$$

$$\dot{y}_{\mathrm{d}}^{n} = u_{\mathrm{d}} \sin(\psi_{\mathrm{d}}) \tag{12.23}$$

with the surge velocity given by the model

$$(m - X_{\dot{u}})\dot{u}_{\mathrm{d}} + \frac{1}{2}\rho C_{\mathrm{d}} A |u_{\mathrm{d}}| u_{\mathrm{d}} = T_{|n|n} |n| n \tag{12.24}$$

where $u_{\mathrm{d}} \gg 0$ is the desired surge velocity, ρ is the density of water, C_{d} is the drag coefficient, A is the projected cross-sectional area of the submerged hull in the x_b direction, $(m - X_{\dot{u}})$ is the mass including the hydrodynamic added mass and $T_{|n|n}$ is the propeller force coefficient (see Section 9.1). Note that the ship is moving so fast that quadratic drag dominates and linear damping due to skin friction can be neglected. The yaw dynamics is chosen *as a first-order Nomoto model*

$$\dot{\psi}_\mathrm{d} = r_\mathrm{d} \tag{12.25}$$

$$T\dot{r}_\mathrm{d} + r_\mathrm{d} = K\delta \tag{12.26}$$

where K and T are the design parameters. The guidance system has two inputs, propeller revolution n and rudder angle δ. The guidance controllers for speed and yaw angle can be chosen of PI and PID types, respectively

$$|n|n = -K_{\mathrm{p}_u}(u_\mathrm{d} - u_\mathrm{ref}) - K_{\mathrm{i}_u}\int_0^t (u_\mathrm{d} - u_\mathrm{ref})\mathrm{d}\tau, \tag{12.27}$$

$$\delta = -K_{\mathrm{p}_\psi}(\psi_\mathrm{d} - \psi_\mathrm{ref}) - K_{\mathrm{i}_\psi}\int_0^t (\psi_\mathrm{d} - \psi_\mathrm{ref})\mathrm{d}\tau - K_{\mathrm{d}_\psi} r_\mathrm{d} \tag{12.28}$$

where ψ_{ref} is generated using an LOS algorithm (see Section 12.4)

$$\psi_\mathrm{ref} = \mathrm{atan2}\,(y_\mathrm{los} - y_\mathrm{d},\ x_\mathrm{los} - x_\mathrm{d}). \tag{12.29}$$

Numerical integration of (12.22)–(12.26) with feedback (12.27) and (12.28) yields a smooth reference trajectory $(x^n_\mathrm{d}, y^n_\mathrm{d}, \psi_\mathrm{d})$ and speed assignment $U_\mathrm{d} = u_\mathrm{d}$.

12.1.3 Optimal Trajectory Generation

Optimization methods can be used for trajectory and path generation. This gives a systematic method for inclusion of static and dynamic constraints. However, the challenge is that an optimization problem must be solved online in order to generate a feasible time-varying trajectory. Implementation and solution of optimization problems can be done using linear programming (LP), quadratic programming (QP) and nonlinear methods. All these methods require you to have a solver that can be implemented in your program; see Figure 12.5. For testing and development, the different algorithms can be implemented using the optimization toolbox in Matlab. The optimization problem can be formulated as minimum power or minimum time, for instance

$$J = \min_{\boldsymbol{\eta}_\mathrm{d},\, \boldsymbol{\nu}_\mathrm{d}} \{\text{power, time}\} \tag{12.30}$$

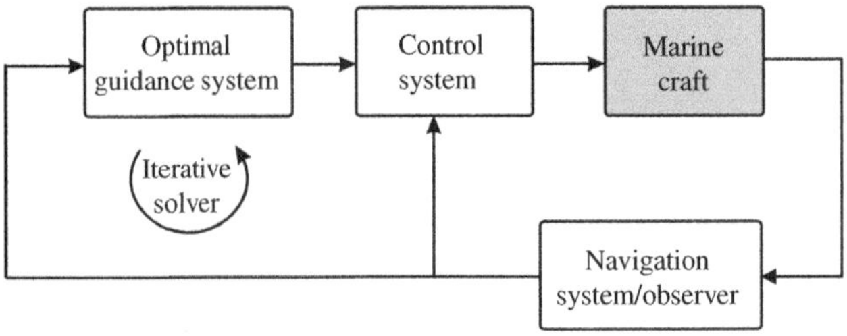

Figure 12.5 Optimal trajectory generation using an iterative solver to solve a minimum time or minimum power optimization problem.

subject to

$$U \le U^{\max} \text{ (maximum speed)}$$

$$|r| \le r^{\max} \text{ (maximum turning rate)}$$

$$|u_i| \le u_i^{\max} \text{ (saturating limit of control } u_i)$$

$$|\dot{u}_i| \le \dot{u}_i^{\max} \text{ (saturating limit of rate } \dot{u}_i)$$

which represents the constraints imposed by the vehicle dynamics. It is also possible to add constraints for obstacle avoidance and minimum fuel consumption.

12.2 Guidance Laws for Target Tracking

Sometimes no information about the path is known in advance and there is no trajectory to track. Hence, if the goal is to track a moving object, for which no future motion information is available, target-tracking methods can be applied. Guidance laws for target tracking can be used in marine operations such as underway replenishment (UNREP) operations and formation control. UNREP operations involve cargo transfer between two or more cooperating surface craft in transit. The task of the so-called *guide ship* is to maintain a steady course and speed while the *approach ship* moves up alongside the guide or target ship to receive fuel, munitions and personnel (Skejic *et al.* 2009).

For surface craft, the 2D position of the target is denoted by $\boldsymbol{p}_\mathrm{t}^n = [x_\mathrm{t}^n, y_\mathrm{t}^n]^\top$. The control objective of a target-tracking scenario can be formulated as (Breivik and Fossen 2009)

$$\lim_{t \to \infty} (\boldsymbol{p}^n - \boldsymbol{p}_\mathrm{t}^n) = \boldsymbol{0} \tag{12.31}$$

where $\boldsymbol{p}^n \in \mathbb{R}^2$ is the craft North-East positions. The target velocity is

$$\boldsymbol{v}_\mathrm{t}^n = \dot{\boldsymbol{p}}_\mathrm{t}^n = [u_\mathrm{t}^n, v_\mathrm{t}^n]^\top \tag{12.32}$$

and the speed of the target satisfies

$$U_\mathrm{t} = \|\boldsymbol{v}_\mathrm{t}^n\| = \sqrt{(u_\mathrm{t}^n)^2 + (v_\mathrm{t}^n)^2}. \tag{12.33}$$

In the missile guidance community an object that is supposed to destroy another object is referred to as a missile, an interceptor or a pursuer. Conversely, the threatened object is typically called a target or an evader. In the following, the designations *interceptor* and *target* will be used for the approach ship and target ship, respectively.

An intercepting ship can control its speed U and course angle χ using a conventional autopilot system. Assume that the control laws guarantees that $U \to U_\mathrm{d}$ and $\chi \to \chi_\mathrm{d}$. The reference signals χ_d and U_d relates to the North-East velocities by the kinematic transformations (2.125) and (2.126), which are conveniently expressed as

$$u_\mathrm{d}^n = U_\mathrm{d} \cos(\chi_\mathrm{d}) \tag{12.34}$$

$$v_\mathrm{d}^n = U_\mathrm{d} \sin(\chi_\mathrm{d}). \tag{12.35}$$

Pure pursuit and constant bearing guidance laws for target tracking compute velocity commands $(u^n_{\rm d}, v^n_{\rm d})$ expressed in $\{n\}$. However, the desired velocities (12.34) and (12.35) are easily transformed to autopilot speed and course commands by using the inverse transformations

$$\chi_{\rm d} = \text{atan2}(v^n_{\rm d}, u^n_{\rm d}) \tag{12.36}$$

$$U_{\rm d} = \sqrt{(u^n_{\rm d})^2 + (v^n_{\rm d})^2}. \tag{12.37}$$

These transformations will be used in the forthcoming sections when deriving the guidance laws for target tracking.

12.2.1 Line-of-sight Guidance Law

Line-of-sight (LOS) guidance is classified as a three-point guidance scheme forming a triangle since it involves a typically stationary reference point in addition to the interceptor and the target as shown in Figure 12.6. The LOS denotation stems from the fact that the interceptor is supposed to achieve an intercept by constraining its motion along the LOS vector between the reference point and the target. LOS guidance has typically been employed for surface-to-air missiles, often mechanized by a ground station, which illuminates the target with a beam that the guided missile is supposed to ride, also known as beam-rider guidance. The LOS guidance principle is illustrated in Figure 12.6, where the interceptor velocity vector is pointed to the LOS vector to obtain the desired velocity vector $\boldsymbol{v}^n_{\rm d} = [u^n_{\rm d}, v^n_{\rm d}]^\top$. The velocities are computed using (12.34) and (12.35) where $\chi_{\rm d}$ are taken as the sum (see Section 12.4)

$$\chi_{\rm d} = \pi_{\rm p} - \tan^{-1}\left(\frac{y^p_{\rm e}}{\Delta}\right) \tag{12.38}$$

which ensures that the velocity is directed toward a point on the path, with its location specified by the *lookahead distance* $\Delta > 0$. The angle $\pi_{\rm p}$ is the *path-tangential angle*, while $y^p_{\rm e}$ denotes the *cross-track error* expressed in the path-tangential reference frame $\{p\}$ (see Figure 12.6). Equation (12.38) is similar to the proportional LOS guidance laws used by Healey and Lienard (1993), Pettersen and Lefeber (2001), Børhaug and Pettersen (2005), Breivik and Fossen (2005b), Fredriksen and Pettersen (2006), Breivik and Fossen (2009), and Fossen and Pettersen (2014).

Computation of the Cross-track Error

For a straight line going from the reference point $(x^n_{\rm ref}, y^n_{\rm ref})$ to the target position $(x^n_{\rm t}, y^n_{\rm t})$ the path-tangential angle is computed as (see Figure 12.6)

$$\pi_p = \text{atan2}(y_t^n - y_{ref}^n, x_t^n - x_{ref}^n). \tag{12.39}$$

For a craft located at the position (x^n, y^n), the along-track and cross-track errors (x_e^p, y_e^p) are computed as the orthogonal distance to the path-tangential reference frame with origin (x_p^n, y_p^n) by setting $x_e^p \equiv 0$. In other words

$$\begin{bmatrix} 0 \\ y_e^p \end{bmatrix} = \boldsymbol{R}_p^n(\pi_p)^\top \left(\begin{bmatrix} x^n \\ y^n \end{bmatrix} - \begin{bmatrix} x_p^n \\ y_p^n \end{bmatrix} \right) \tag{12.40}$$

where

$$\boldsymbol{R}_p^n(\pi_p) = \begin{bmatrix} \cos(\pi_p) & -\sin(\pi_p) \\ \sin(\pi_p) & \cos(\pi_p) \end{bmatrix} \in SO(2) \tag{12.41}$$

Expanding (12.40) and exploiting the fact that

$$\tan(\pi_p) = \frac{y_t^n - y_p^n}{x_t^n - x_p^n} \tag{12.42}$$

gives three equations and three unknowns satisfying the linear matrix equation

$$\underbrace{\begin{bmatrix} \cos(\pi_p) & \sin(\pi_p) & 0 \\ -\sin(\pi_p) & \cos(\pi_p) & 1 \\ \tan(\pi_p) & -1 & 0 \end{bmatrix}}_{\boldsymbol{A}} \underbrace{\begin{bmatrix} x_p^n \\ y_p^n \\ y_e^p \end{bmatrix}}_{\boldsymbol{x}} = \underbrace{\begin{bmatrix} \cos(\pi_p)x^n + \sin(\pi_p)y^n \\ -\sin(\pi_p)x^n + \cos(\pi_p)y^n \\ \tan(\pi_p)x_t^n - y_t^n \end{bmatrix}}_{\boldsymbol{b}}. \tag{12.43}$$

The origin (x_p^n, y_p^n) of the path-tangential reference frame and the cross-track error y_e^n is then computed by $\boldsymbol{x} = \boldsymbol{A}^{-1}\boldsymbol{b}$.

MATLAB:

The cross-track error can be computed using the MSS toolbox function

```
[x_p, y_p, y_e] = crosstrack(x_t, y_t, x_ref, y_ref, x, y, flag)
```

where `flag = 1` is an optional input for plotting the geometry of the problem.

12.2.2 Pure-pursuit Guidance Law

Pure-pursuit (PP) guidance belongs to the two-point guidance schemes, where only the interceptor and the target are considered in the engagement geometry. The interceptor aligns its velocity along the LOS vector between the interceptor and the target by choosing

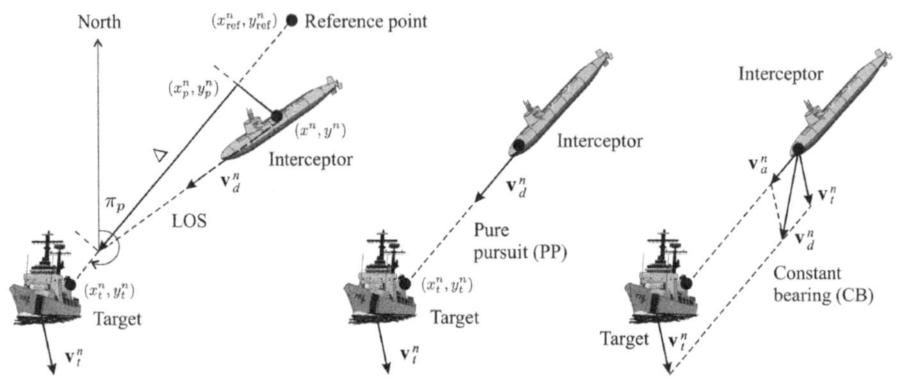

Figure 12.6 The desired interceptor approach velocity vector $\boldsymbol{v}^n_{\mathrm{d}}$ for three classical guidance principles: line-of-sight LOS, pure pursuit (PP) and constant bearing (CB). The target velocity vector is denoted $\boldsymbol{v}^n_{\mathrm{t}}$.

the desired velocity $\boldsymbol{v}^n_{\mathrm{d}} = [u^n_{\mathrm{d}}, v^n_{\mathrm{d}}]^\top$ as

$$\boldsymbol{v}^n_{\mathrm{d}} = -\kappa \frac{\tilde{\boldsymbol{p}}^n}{\|\tilde{\boldsymbol{p}}^n\|} \tag{12.44}$$

where $\kappa > 0$ and

$$\tilde{\boldsymbol{p}}^n := \boldsymbol{p}^n - \boldsymbol{p}^n_{\mathrm{t}} \tag{12.45}$$

is the LOS vector between the interceptor and the target.

This strategy is equivalent to a predator chasing a prey in the animal world, and very often results in a tail chase. PP guidance has typically been employed for air-to-surface missiles. The PP guidance principle is represented in Figure 12.6 by a vector pointing directly at the target.

Deviated pursuit guidance is a variant of PP guidance, where the velocity of the interceptor is supposed to lead the interceptor–target line of sight by a constant angle in the direction of the target movement. An equivalent term is fixed-lead navigation.

12.2.3 Constant Bearing Guidance Law

Constant bearing (CB) guidance is also a two-point guidance scheme, with the same engagement geometry as PP guidance. However, in a CB engagement, the interceptor is supposed to align the interceptor–target velocity $\boldsymbol{v}^n_{\mathrm{a}}$ along the LOS vector between the interceptor and the target. This goal is equivalent to reducing the LOS rotation rate to zero such that the interceptor perceives the target at a constant bearing, closing in on a direct collision course. CB guidance is often referred to as *parallel navigation* and has typically been employed for air-to-air missiles. Also, the CB rule has been used for centuries by mariners to avoid collisions at sea, steering away from a situation where another craft approaches at a constant bearing. Thus, guidance principles can just as well be applied to avoid collisions as to achieve them. The CB guidance principle is illustrated in Figure 12.6.

The most common method of implementing CB guidance is to make the rotation rate of the interceptor velocity directly proportional to the rotation rate of the interceptor–target LOS, which is widely known as *proportional navigation* (PN). However, CB guidance can also be implemented through the direct velocity assignment as proposed by Breivik *et al.* (2006); see Breivik (2010) for details.

The CB desired velocity is given by

$$\boldsymbol{v}_{\mathrm{d}}^{n} = \boldsymbol{v}_{\mathrm{t}}^{n} + \boldsymbol{v}_{\mathrm{a}}^{n} \tag{12.46}$$

$$\boldsymbol{v}_{\mathrm{a}}^{n} = -\kappa \frac{\tilde{\boldsymbol{p}}^{n}}{\|\tilde{\boldsymbol{p}}^{n}\|} \tag{12.47}$$

where $\boldsymbol{v}_{\mathrm{a}}^{n} = [u_{\mathrm{a}}^{n}, v_{\mathrm{a}}^{n}]^{\top}$ is the approach velocity vector specified such that the desired approach speed $U_{\mathrm{a}} = \|\boldsymbol{v}_{\mathrm{a}}^{n}\|$ is tangential to the LOS vector as shown in Figure 12.6 and

$$\kappa = U_{\mathrm{a}}^{\max} \frac{\|\tilde{\boldsymbol{p}}^{n}\|}{\sqrt{(\tilde{\boldsymbol{p}}^{n})^{\top}\tilde{\boldsymbol{p}}^{n} + \Delta^{2}}} \tag{12.48}$$

where $U_{\mathrm{a}}^{\max}$ specifies the maximum approach speed toward the target and $\Delta > 0$ affects the transient interceptor–target rendezvous behavior. The CB guidance law (12.46) computes the velocity commands needed to track the target.

Note that CB guidance becomes equal to PP guidance for a stationary target; that is the basic difference between the two guidance schemes is whether the target velocity is used as feedforward or not.

Lyapunov Stability Analysis

The convergence properties of (12.46)–(12.48) can be investigated by considering a Lyapunov function candidate (LFC)

$$V = \frac{1}{2}(\tilde{\boldsymbol{p}}^{n})^{\top}\tilde{\boldsymbol{p}}^{n} > 0, \quad \forall\, \tilde{\boldsymbol{p}}^{n} \neq \boldsymbol{0}. \tag{12.49}$$

Assume that the ship autopilot guarantees that $\boldsymbol{v}^{n} = \boldsymbol{v}_{\mathrm{d}}^{n}$. Time differentiation of V along the trajectories of $\tilde{\boldsymbol{p}}^{n} = \boldsymbol{p}^{n} - \boldsymbol{p}_{\mathrm{t}}^{n}$ gives

$$\begin{aligned}\dot{V} &= (\tilde{\boldsymbol{p}}^{n})^{\top}(\boldsymbol{v}^{n} - \boldsymbol{v}_{\mathrm{t}}^{n}) \\ &= -\kappa \frac{(\tilde{\boldsymbol{p}}^{n})^{\top}\tilde{\boldsymbol{p}}^{n}}{\|\tilde{\boldsymbol{p}}^{n}\|}\end{aligned}$$

$$= -U_a^{\max} \frac{(\tilde{p}^n)^\top \tilde{p}^n}{\sqrt{(\tilde{p}^n)^\top \tilde{p}^n + \Delta^2}}$$
$$< 0, \quad \forall\, \tilde{p}^n \neq \mathbf{0}. \tag{12.50}$$

The LFC (12.49) is positive definite and radially unbounded, while its derivative with respect to time (12.50) is negative definite when adhering to $U \geq U_a^{\max} > 0$. Hence, by standard Lyapunov arguments the origin $\tilde{p}^n = \mathbf{0}$ is USGES as shown in Appendix A.2.3.

12.3 Linear Design Methods for Path Following

This section discusses state-of-the art methods for path following. We start with the classical design methods using linear theory where the paths are generated using waypoints to connect straight lines between the waypoints. LOS guidance laws based on nonlinear control theory are presented in Sections 12.4–12.6.

12.3.1 Waypoints

Waypoints are used to indicate changes in direction, speed and altitude along a desired path. Systems for waypoint generation are used both for ships and underwater vehicles. These systems consist of a waypoint generator with a human interface. The selected waypoints are stored in a waypoint database and used for generation of a trajectory or a path for the moving craft to follow. Both trajectory and path-following control systems can be designed for this purpose. Sophisticated features such as weather routing, obstacle avoidance and mission planning can be incorporated in the design of waypoint guidance systems. Some of these features will be discussed in the forthcoming section.

Waypoint Representation

The route of a ship or an underwater vehicle is usually specified in terms of waypoints. Each waypoint is defined using Cartesian coordinates (x_i^n, y_i^n, z_i^n) for $i = 0, \ldots, n$. The waypoint *database* therefore consists of

$$\texttt{wpt.pos} = \{(x_0^n, y_0^n, z_0^n), (x_1^n, y_1^n, z_1^n), \ldots, (x_\mathrm{n}^n, y_\mathrm{n}^n, z_\mathrm{n}^n)\}.$$

For surface craft, only two coordinates (x_i^n, y_i^n) are used. Additionally, other waypoint properties such as speed and heading can be defined

$$\texttt{wpt.speed} = \{U_0, U_1, \ldots, U_\mathrm{n}\}$$
$$\texttt{wpt.heading} = \{\psi_0, \psi_1, \ldots, \psi_\mathrm{n}\}.$$

For surface craft this means that the craft should pass through the waypoint (x_i^n, y_i^n) at forward speed U_i with heading angle ψ_i. The three states (x_i^n, y_i^n, ψ_i) are also recognized as the craft's *pose* since they describe the craft's configuration. The heading angle is usually unspecified during cross-tracking, whereas it is more important during a crab-wise maneuver close to offshore installations (dynamic positioning).

The waypoint database can be generated using many criteria. These are usually based on:

- **Mission:** The craft should move from some starting point (x_0^n, y_0^n, z_0^n) to the terminal point (x_n^n, y_n^n, z_n^n) via the waypoints (x_i^n, y_i^n, z_i^n).
- **Environmental data:** Information about wind, waves and ocean currents can be used for energy optimal routing (or avoidance of bad weather for safety reasons).
- **Geographical data:** Information about shallow waters and islands should be included.
- **Obstacles:** Floating constructions and other obstacles must be avoided.
- **Collision avoidance:** Avoiding moving craft close to your own route by introducing safety margins.
- **Feasibility:** Each waypoint must be feasible, in that it must be possible to maneuver to the next waypoint without exceeding the maximum speed and turning rate.

Online replanning can be used to update the waypoint database in case of time-varying conditions such as changing weather or moving craft (collision avoidance). Optimality with regard to weather is discussed in Section 12.6.1. This is referred to as weather routing.

12.3.2 Path Generation using Straight Lines and Inscribed Circles

In practice it is common to represent the desired path using straight lines and circle arcs to connect the waypoints. This is illustrated by Figure 12.7 where the operator has specified the waypoints (x_i^n, y_i^n) and radii R_i, which relates to the radii of the inscribed circle according to

$$\bar{R}_i = R_i \tan(\alpha_i) \tag{12.51}$$

where α_i is defined in Figure 12.8. The radii R_i are used to specify the points corresponding to the turning point of the ship. These values are stored in the database as

$$\texttt{wpt.radius} = \{R_0, R_1, \ldots, R_n\}.$$

The drawback of this strategy, in comparison with a cubic interpolation strategy, for instance, is that a jump in the desired yaw rate r_d is experienced. This is due to the fact that the desired yaw rate along the straight line is $r_d = 0$ while it is $r_d =$ constant on the circle arc during steady turning. Hence, there will be a jump in the desired yaw rate during transition from the straight line to the circle arc. This produces a small offset during cross-tracking. If a smooth reference trajectory, for instance generated by interpolation, is used, these drawbacks are overcome. However, it is convenient to use straight lines and circle arcs due to their simplicity. Hence, an alternative could be to use a *Fermat spiral* to smooth the transition from zero curvature (straight line) to max curvature (circle arc) as shown by Lekkas *et al.* (2013).

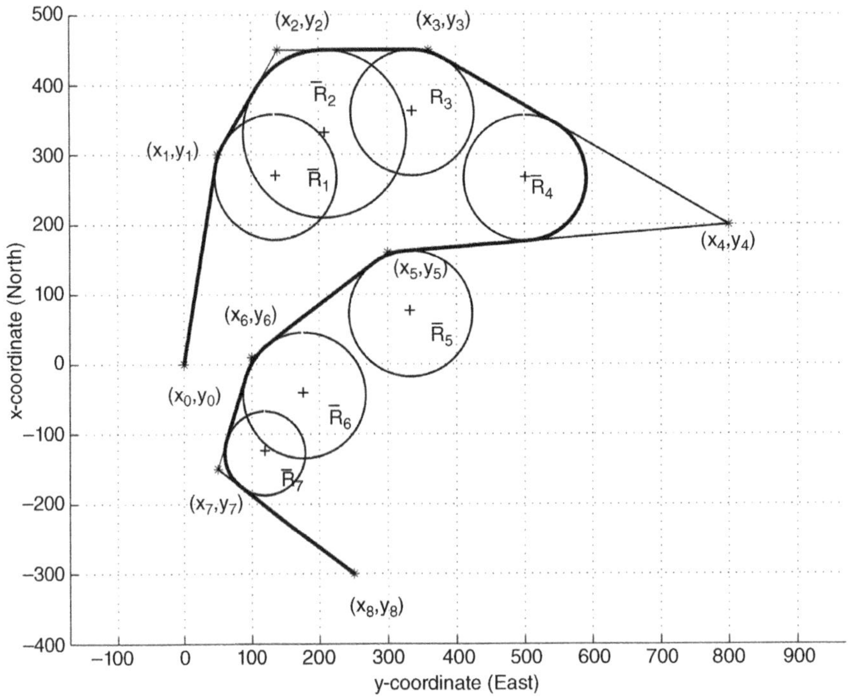

Figure 12.7 Straight lines and inscribed circles used for waypoint guidance.

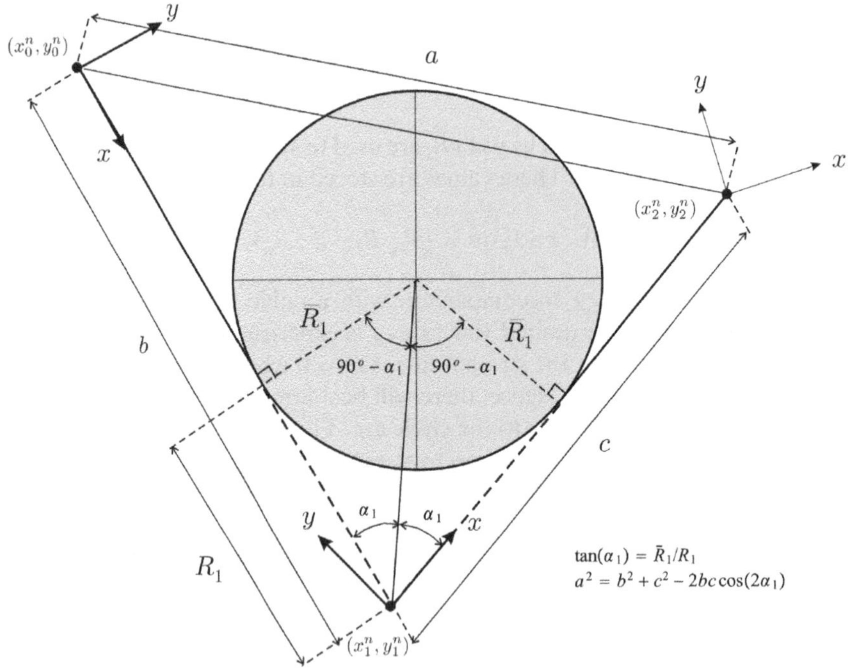

Figure 12.8 Circle with radius $\bar{R}_1$ inscribed between the points (x_0^n, y_0^n), (x_1^n, y_2^n) and (x_2^n, y_2^n).

12.3.3 Straight-line Paths Based on Circles of Acceptance

A more effective method for computer implementation is to drop the inscribed circles of Section 12.3.2 and only use straight-line segments going through the waypoints. When moving along a piecewise linear path made up of $n - 1$ line segments connected by n waypoints, a switching mechanism for selecting the next waypoint is needed. Waypoint (x^n_{i+1}, y^n_{i+1}) can be selected on the basis of whether or not the craft lies within a *circle of acceptance* with radius R_{i+1} around (x^n_{i+1}, y^n_{i+1}). In other words, if the craft positions (x^n, y^n) at time t satisfy

$$(x^n_{i+1} - x^n)^2 + (y^n_{i+1} - y^n)^2 \leq R^2_{i+1} \tag{12.52}$$

the next waypoint (x^n_{i+1}, y^n_{i+1}) should be selected. A guideline could be to choose R_{k+1} equal to two ship lengths, that is $R_{i+1} = 2L$.

A perhaps more suitable switching criterion solely involves computation of the along-track distance x^p_e (see Figure 12.9), such that if the total along-track distance $d_{i+1} = \|\boldsymbol{p}^n_{i+1} - \boldsymbol{p}^n_i\|$ between the waypoints $\boldsymbol{p}^n_i$ and $\boldsymbol{p}^n_{i+1}$, a switch is made when

$$d_{i+1} - |x^p_e| \leq R_{\text{switch}} \tag{12.53}$$

where $R_{\text{switch}} > 0$.

Computation of Along-track and Cross-track Errors using two Waypoints

The along-track and cross-track errors (x^p_e, y^p_e) are computed by using the coordinates of two waypoints $\boldsymbol{p}^n_i = [x^n_i, y^n_i]^\top$ and $\boldsymbol{p}^n_{i+1} = [x^n_{i+1}, y^n_{i+1}]^\top$. The path-tangential coordinate

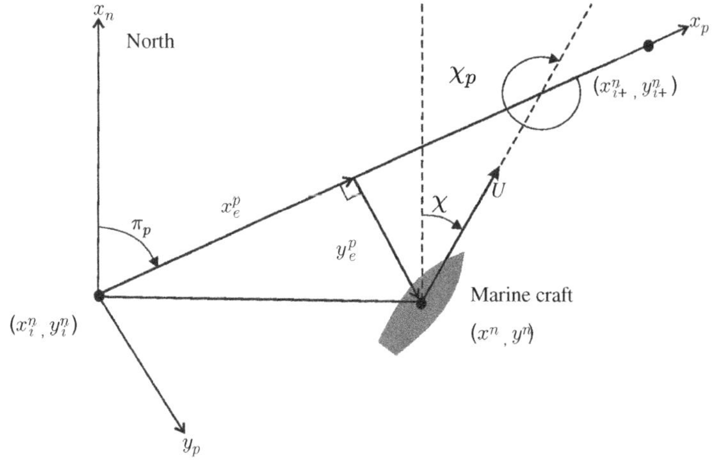

Figure 12.9 Path-tangential coordinate system $\{p\}$ going from (x^n_i, y^n_i) to (x^n_{i+1}, y^n_{i+1}). The origin is located at the waypoint (x^n_i, y^n_i) and the axes of $\{p\}$ are rotated an angle π_p with respect to $\{n\}$.

system $\{p\}$ has its origin located at $\boldsymbol{p}_i^n$ and the x_p-axis is pointing towards $\boldsymbol{p}_{i+1}^n$. The axes of $\{p\}$ are rotated an angle π_p with respect to $\{n\}$ such that (see Figure 12.9)

$$\pi_\text{p} = \text{atan2}(y_{i+1}^n - y_i^n, x_{i+1}^n - x_i^n) \tag{12.54}$$

Consequently, the along-track and cross-track errors expressed in $\{p\}$ are

$$\begin{bmatrix} x_\text{e}^p \\ y_\text{e}^p \end{bmatrix} = \boldsymbol{R}_p^n(\pi_\text{p})^\top \left(\begin{bmatrix} x^n \\ y^n \end{bmatrix} - \begin{bmatrix} x_i^n \\ y_i^n \end{bmatrix} \right) \tag{12.55}$$

where

$$\boldsymbol{R}_p^n(\pi_\text{p}) = \begin{bmatrix} \cos(\pi_\text{p}) & -\sin(\pi_\text{p}) \\ \sin(\pi_\text{p}) & \cos(\pi_\text{p}) \end{bmatrix} \in SO(2). \tag{12.56}$$

The advantage of (12.52) is that $\boldsymbol{p}^n$ does not need to enter the waypoint enclosing circle for a switch to occur; that is no restrictions are put on the cross-track error. Thus, if no intrinsic value is associated with visiting the waypoints and their only purpose is to implicitly define a piecewise linear path, there is no reason to apply the circle-of-acceptance switching criterion (12.52).

Extension to 3D Path Following for Underwater Vehicles

It is straightforward to generalize the concepts of LOS guidance to 3D maneuvering. Also for this case, the desired course angle χ_d can be chosen as (12.90) with the LOS intersection point given by (12.91) and (12.92) under the assumption that the vehicle performs slow maneuvers in the vertical plane such that a depth controller can easily achieve $z^n = z_\text{d}^n$. This works quite well for vehicles moving at low speed since it is not necessary to pitch the vehicle in order to move vertically; see Figure12.10. A typical example is a working ROV with a broad, flattened front (bluff body) moving vertically using a vertical thruster. The circle of acceptance must, however, be replaced by a *sphere of acceptance* (Healey and Lienard 1993),

$$(x_{i+1}^n - x^n)^2 + (y_{i+1}^n - y^n)^2 + (z_{i+1}^n - z^n)^2 \le R_{i+1}^2. \tag{12.57}$$

A more sophisticated approach would be to compute the LOS guidance laws for decoupled horizontal and vertical motions and use the two guidance laws to move in 3D to the next waypoint (Lekkas and Fossen 2013). The horizontal guidance law generates course commands χ_d while the vertical guidance law computes the desired pitch angle θ_d. This approach is used for "flying" vehicles equipped with fins for diving and depth control. These vehicles move at a higher speed in order to produce lifting forces (no vertical

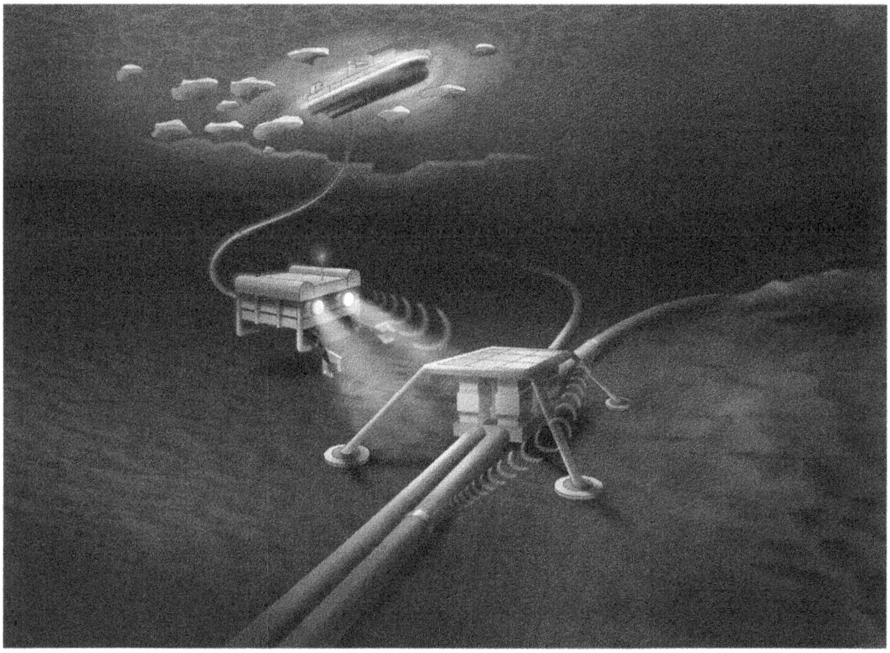

Figure 12.10 Remotely operated vehicle (ROV) performing offshore inspection and maintenance. Source: Illustration by B. Stenberg.

thrusters) and consequently they behave like an aircraft, where it is possible to control the coupled surge, heave and pitch motions (longitudinal motions).

12.3.4 Path Generation using Dubins Path

The shortest path between two poses (x_i^n, y_i^n, ψ_i) and $(x_{i+1}^n, y_{i+1}^n, \psi_{i+1})$ is based on the famous result of Dubins (1957), which can be summarized as:

> *The shortest path (minimum time) between two poses* (x_i^n, y_i^n, ψ_i) and $(x_{i+1}^n, y_{i+1}^n, \psi_{i+1})$ *of a craft moving at constant speed U is a path formed by straight lines and circular arc segments.*

The optimal path will be a combination of a right turn (R), left turn (L) or going straight (S). An optimal path will always be at least one of the six types: RSR, RSL, LSR, LSL, RLR, LRL (see Figure 12.11).

Since a craft and not a point mass is considered, the start and end configurations of the craft are specified in terms of the poses (x_i^n, y_i^n, ψ_i) while speed U is kept constant. In addition, it is assumed that there are bounds on the turning rate r or the radius. The so-called Dubins path can also be proven by using *Pontryagin's maximum principle*.

Generation of Dubins paths including obstacle avoidance are discussed by Tsourdos *et al.* (2010). Extensions to the case with turn rate and acceleration limits (convected

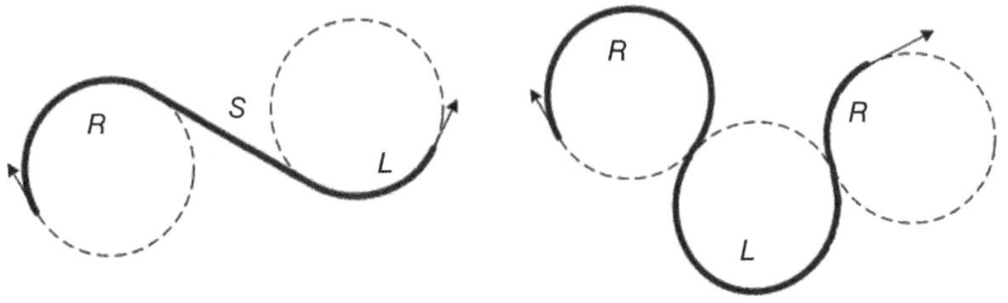

Figure 12.11 The Dubins path for the cases RSL and RLR.

Dubins path) are made by Kostov and Degtiariova-Kostova (1993) and Scheuer and Laugier (1998). Path generation for the case of uniform currents are discussed by McGee *et al.* (2006), and Techy and Woolsey (2009, 2010). In the case of time-varying speed, a dynamic optimization problem including the marine craft surge dynamics must be solved.

12.3.5 Transfer Function Models for Straight-line Path Following

Assume that the path-tangential coordinate system $\{p\}$ has its origin at the waypoint (x_i^n, y_i^n) as shown in Figure 12.9. The cross-track error can expressed in $\{p\}$ using (2.126). In other words

$$\begin{aligned} \dot{y}_e^p &= U\sin(\chi_p) \\ &\approx U\chi_p \end{aligned} \tag{12.58}$$

where χ_p is assumed to be small for straight-line path following. A GNSS receiver measures the course angle χ, which relates to χ_p by

$$\chi = \pi_p + \chi_p \tag{12.59}$$

as illustrated in Figure 12.9, while the formula for π_p is given by (12.54).

Consider the Nomoto model (see Section 7.2)

$$r = \frac{K}{Ts+1}\delta + d_r \tag{12.60}$$

where r is the yaw rate, δ is the rudder angle, and $K > 0$ and $T > 0$ are the Nomoto gain and time constants, respectively. The influence of ocean currents, wave drift and wind are modeled as an unknown drift term d_r. Since $\dot{\psi} = r$, the yaw angle transfer function becomes

$$\psi = \frac{K}{s(Ts+1)}\delta + \frac{1}{s}d_r. \tag{12.61}$$

Consequently, the course angle

$$\chi = \psi + \beta_c \tag{12.62}$$

implies that

$$\chi = \frac{K}{s(Ts+1)}\delta + d_\chi \tag{12.63}$$

where

$$d_\chi = \frac{1}{s}d_r + \beta_c. \tag{12.64}$$

Finally, the transfer function for the cross-track error is derived from (12.58) under the assumption that $U > 0$. In other words

$$y_e^p = \frac{U}{s}(\chi - \pi_p) \tag{12.65}$$

which after substitution of (12.63) and (12.64) becomes

$$y_e^p = \frac{KU}{s^2(Ts+1)}\delta + d_y \tag{12.66}$$

where

$$d_y = \frac{U}{s}(d_\chi - \pi_p). \tag{12.67}$$

The control law for δ should be chosen of PID type in order to suppress the disturbance term d_y and ensure accurate regulation of y_e^p to zero; see Section 15.2.3. If the speed U is time varying, the performance of the PID controller can be considerably improved by making the controller gains speed dependent as well.

12.4 LOS Guidance Laws for Path Following using Course Autopilots

A trajectory describes the motion of a moving object through space as a function of time. The object might be a craft, projectile or a satellite, for example. A trajectory can be described mathematically either by the geometry of the path (see Section 12.6) or as the position of the object over time. *Path following* is the task of following a predefined path independent of time; that is there are no temporal constraints. This means that no restrictions are placed on the temporal propagation along the path. Spatial constraints, however, can be added to represent obstacles and other positional constraints.

This section presents LOS guidance laws that can be used together with course autopilots; see Figure 12.12. The guidance law computes the desired course angle χ_d which is used as the autopilot reference signal. Course over ground (COG) can be measured using GNSS or computed from the position measurements (x^n, y^n). From (2.125) and (2.126) it follows that

$$\tan(\chi) = \frac{\dot{y}^n}{\dot{x}^n} = \frac{dy}{dx}. \tag{12.68}$$

Consequently, we can use position measurements (x^n, y^n) at two discrete-time data points ($x^n[k-1]$, $y^n[k-1]$) and ($x^n[k]$, $y^n[k]$) to approximate

$$\chi[k] \approx \text{atan2}(y^n[k] - y^n[k-1],\ x^n[k] - x^n[k-1]). \tag{12.69}$$

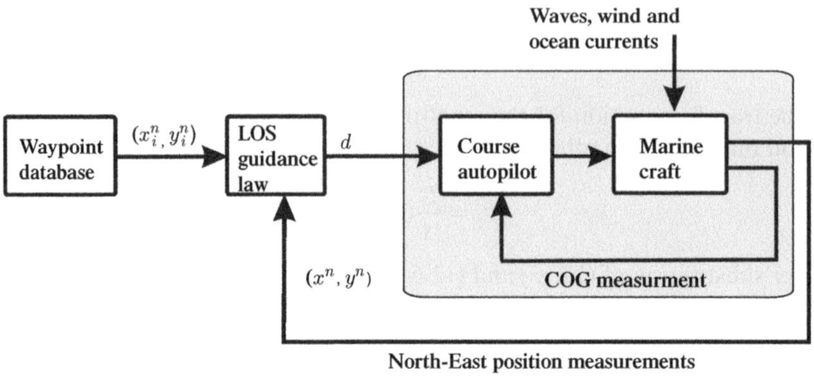

Figure 12.12 Course autopilot used for LOS path following.

12.4.1 *Vector-field Guidance Law*

A marine craft can follow a predefined path using vector-field guidance laws similar to those that are used for unmanned aerial vehicles (UAVs). The concept of path following, through the construction of vector fields surrounding the path to be followed, is based on Nelson *et al.* (2007), Lawrence *et al.* (2008), and Beard and McLain (2012). The vectors of a field provide course commands to guide the craft toward the desired path. As with other path-following methods, the objective is not to track a moving point, but to get onto the path while moving at a prescribed speed.

Consider the straight-line path shown in Figure 12.9 where

$$\chi = \pi_{\mathrm{p}} + \chi_{\mathrm{p}}. \tag{12.70}$$

To follow this path, a desired course vector field is constructed. Let y_e^p be the lateral distance of the craft from the straight line and let χ_{p} be the difference between the line and the course of the craft such that we can apply (2.126) to write the cross-track error kinematics as

$$\begin{aligned} \dot{y}_e^p &= U \sin(\chi_{\mathrm{p}}) \\ &= U \sin(\chi - \pi_{\mathrm{p}}). \end{aligned} \tag{12.71}$$

Here χ can be viewed as a virtual control input used to steer y_e^p to zero. The objective is to construct the vector field such that, when $|y_e^p|$ is large, the craft is directed to approach the path with angle $\chi_{\mathrm{p}} = \pm\chi^{\infty}$ for $\chi^{\infty} \in (0, \pi/2]$ whereas when y_e^p approaches zero, the angle $\chi_{\mathrm{p}} \to 0$ (see Figure 12.13). This can be achieved by choosing

$$\chi_{\mathrm{p}} = -\chi^{\infty} \frac{2}{\pi} \tan^{-1}(K_{\mathrm{p}}\, y_e^p) \tag{12.72}$$

where $K_{\mathrm{p}} > 0$ is the proportional gain. From (12.70), under the assumption of perfect autopilot control $\chi = \chi_{\mathrm{d}}$, it follows that the desired course angle should be chosen as

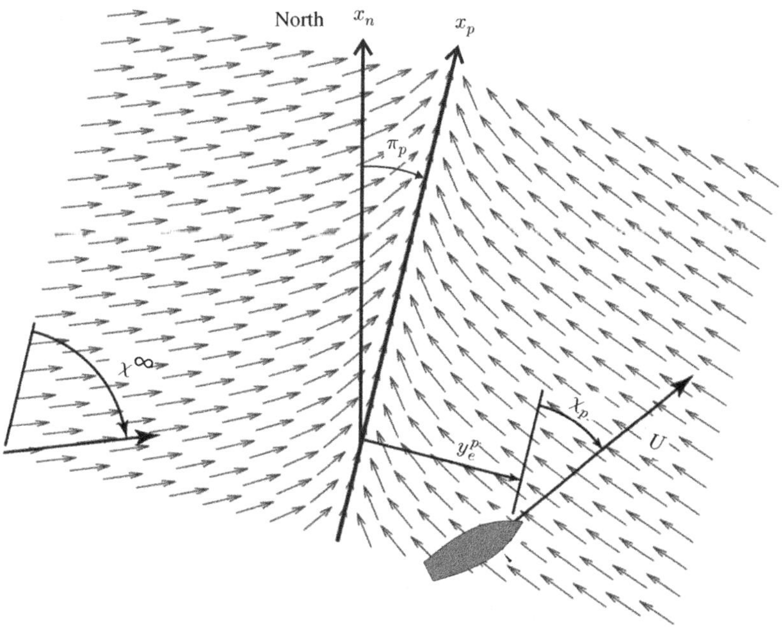

Figure 12.13 Vector field for straight-line path following. Far away from the path (large value for y_e^p), the vector field is directed with an angle χ^∞ from the path-tangential axis x_p to the path.

$$\chi_d = \pi_p - \chi^\infty \frac{2}{\pi} \tan^{-1}(K_p\, y_e^p). \tag{12.73}$$

The magnitude of K_p influences the rate of the transition from χ^∞ to zero. In view of this, the cross-track error dynamics can be written as

$$\begin{aligned} \dot{y}_e^p &= U \sin(\chi - \pi_p) \\ &= -U \sin\left(\chi^\infty \frac{2}{\pi} \tan^{-1}(K_p\, y_e^p)\right). \end{aligned} \tag{12.74}$$

The stability conditions for this system will now be established by using Lyapunov stability theory for a non-autonomous system allowing for U to be time varying.

Lyapunov Stability Analysis

It is straightforward to verify that

$$-\frac{\pi}{2} < \chi^\infty \frac{2}{\pi} \tan^{-1}(K_p\, y_e^p) < \frac{\pi}{2} \tag{12.75}$$

for $\chi^{\infty} \in (0, \pi/2]$. Consider the Lyapunov function candidate $V = 1/2(y_e^p)^2$. Time differentiation of V gives (Nelson *et al.* 2007)

$$\dot{V} = -y_e^p \; U \sin\left(\chi^{\infty}\frac{2}{\pi}\tan^{-1}(K_p \; y_e^p)\right) \tag{12.76}$$

$$< 0, \qquad \forall y_e^p \neq 0 \tag{12.77}$$

if $0 < U^{\min} \leq U \leq U^{\max}$. Consequently, the equilibrium point $y_e^p = 0$ is GAS; see Appendix A.1.3. The result is proven for a straight line but it is also valid for curved paths since the tangent to the path can replace the straight line between the two waypoints. Care must be taken when doing this since certain paths can have infinite solutions; see Section 12.6.2 for details.

12.4.2 Proportional LOS Guidance Law

The vector-field guidance law (12.73) is almost similar to a class of proportional LOS guidance laws used for marine craft path-following control; see Healey and Lienard (1993), Pettersen and Lefeber (2001), Børhaug and Pettersen (2005), Fredriksen and Pettersen (2006), Breivik and Fossen (2004b, 2005b, 2009), Breivik *et al.* (2008), and Fossen and Pettersen (2014). The difference is that we choose $\chi^{\infty} = \pi/2$ in (12.73) such that

$$\chi_d = \pi_p - \tan^{-1}(K_p \; y_e^p). \tag{12.78}$$

The proportional gain K_p is usually parametrized in terms of the lookahead distance $\Delta > 0$ according to

$$K_p = \frac{1}{\Delta}. \tag{12.79}$$

Formulas (12.78) and (12.79) are easily verified from Figure 12.14 where it is seen that $\pi_p = \chi_d + \tan^{-1}(y_e^p/\Delta)$. Assume that the path-tangential coordinate system $\{p\}$ has its origin at the waypoint $\boldsymbol{p}_i^n = [x_i^n, \; y_i^n]^{\top}$. From this it follows that cross-track error dynamics under the assumption that the course controller ensures perfect tracking, that is $\chi = \chi_d$, takes the following form

$$\begin{aligned} \dot{y}_e^p &= U \sin(\chi_p) \\ &= U \sin(\chi - \pi_p) \\ &= -U \sin(\tan^{-1}(y_e^p/\Delta)) \\ &= -\frac{U}{\sqrt{\Delta^2 + (y_e^p)^2}} \; y_e^p \end{aligned} \tag{12.80}$$

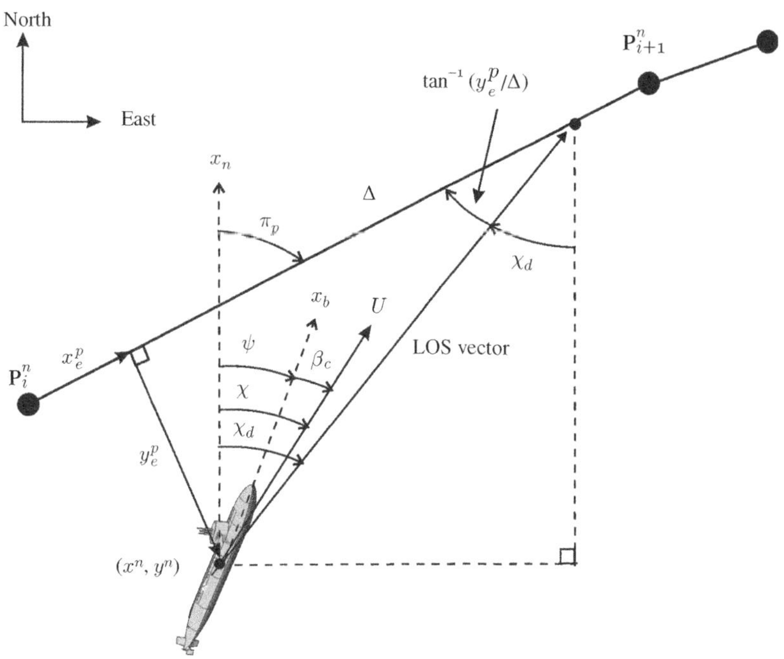

Figure 12.14 Lookahead-based LOS guidance where the desired course angle χ_d is chosen such that $\pi_p = \chi_d + \tan^{-1}(y_e^p/\Delta)$ for a user specified lookahead distance Δ.

where we have used the trigonometric identity $\sin(\tan^{-1}(x)) = x/\sqrt{1+x^2}$. Stability of the equilibrium point $y_e^p = 0$ follows from Lyapunov stability theory.

Lyapunov Stability Analysis

Consider the Lyapunov function candidate $V = 1/2(y_e^p)^2$. Time differentiation of V and substitution (12.80) gives

$$\begin{aligned} \dot{V} &= -\frac{U}{\sqrt{\Delta^2 + (y_e^p)^2}}(y_e^p)^2 \\ &< 0, \qquad \forall y_e^p \neq 0. \end{aligned} \tag{12.81}$$

Consequently, it is straightforward to show that the equilibrium point $y_e^p = 0$ is USGES by using the results in Appendix A.2.3. USGES is important from a robustness perspective. In particular, it is seen from lemmas 9.2–9.3 in Khalil (2002) that only exponential stability allows us to conclude anything about the robustness to large uniformly bounded disturbances. It should be noted that GES cannot be achieved for the proportional LOS guidance law due to the kinematic representation, which introduces saturation through the trigonometric functions.

Time-varying Lookahead Distance

To improve the convergence to the path it is advantageous to use a time-varying lookahead distance Δ, which depends on the magnitude $|y_e^p|$ of the cross-track error. Lekkas and Fossen (2012) proposed the following adaptive lookahead distance

$$\Delta = (\Delta_{max} - \Delta_{min})e^{-\gamma|y_e^p|} + \Delta_{min} \tag{12.82}$$

where Δ_{min} and Δ_{max} are the minimum and maximum allowed values for Δ respectively and, along with the convergence rate $\gamma > 0$, are the design parameters. Alternatively, $e^{-\gamma|y_e^p|}$ can be replaced by $e^{-\gamma(y_e^p)^2}$ as shown in Figure 12.15.

The idea behind (12.82) is rather simple and it can be summarized by the fact that it assigns a small Δ value when the craft is far from the desired path, (thus resulting in a more aggressive behavior that tends to decrease the cross-track error faster) and a large value for Δ when the craft is close to the path and overshooting needs to be avoided. The concepts "far" and "close" with respect to the desired path are relative and several constraints should be taken into account when determining Δ_{min}, Δ_{max} and γ, such as the speed of the craft and the maximum curvature of the path.

When the craft is close to the path (small values of $|y_e^p|$) the cross-track error dynamics (12.80) is linear

$$\dot{y}_e^p = -\frac{U}{\sqrt{\Delta^2 + (y_e^p)^2}}\, y_e^p \overset{y_e^p \text{ small}}{\approx} -\lambda\, y_e^p \tag{12.83}$$

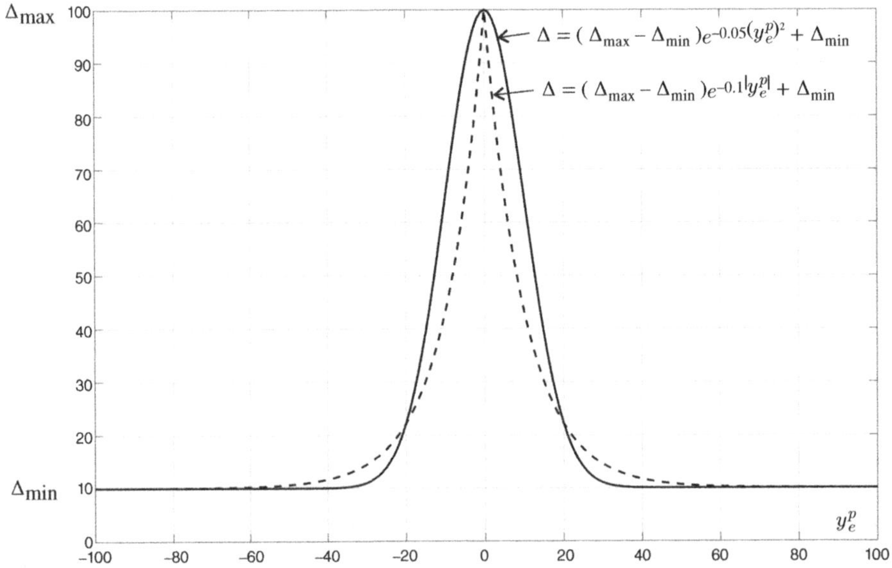

Figure 12.15 Adaptive lookahead distance for $U = 5$ m s^{-1}, $\lambda = 0.05$, $\Delta_{max} = U/\lambda$ and $\Delta_{min} = 0.1\,\Delta_{max}$.

where $\lambda = U/\Delta$ and the course angle approaches the path-tangential angle since

$$\lim_{|y_e^p|\to 0} \chi_d = \pi_p - \lim_{|y_e^p|\to 0} [\tan^{-1}(y_e^p/\Delta)]$$
$$= \pi_p. \tag{12.84}$$

This suggest that

$$\Delta_{max} = U/\lambda \tag{12.85}$$

where $\lambda > 0$ is a design parameter corresponding to the linear system $\dot{y}_e^p = -\lambda y_e^p$. When the craft is far away from the path (large values of $|y_e^p|$), the desired course angle will approach the limit

$$\lim_{|y_e^p|\to \infty} \chi_d = \pi_p - \lim_{|y_e^p|\to \infty} [\tan^{-1}(y_e^p/\Delta)]$$
$$= \pi_p - \text{sgn}(y_e^p)\frac{\pi}{2}. \tag{12.86}$$

Formula (12.86) confirms that the craft moves perpendicular to the path of travel when far away. From (12.82) we see that the lookahead distance approaches Δ_{min} for large values of $|y_e^p|$. The value for Δ_{min} can be specified according to

$$\Delta_{min} = \sigma\Delta_{max} \tag{12.87}$$

where $\sigma = \Delta_{min}/\Delta_{max}$ is the ratio between the minimum and maximum values of the lookahead distance.

12.4.3 Lookahead- and Enclosure-based LOS Steering

The proportional LOS guidance laws can be classified according to (Breivik and Fossen 2009)

- *Lookahead-based steering*
- *Enclosure-based steering.*

The two steering methods essentially operate by the same principle but, as will be made clear, the lookahead-based approach motivated by missile guidance has several advantages over the enclosure-based approach.

The guidance law (12.78) is in fact a lookahead-based steering law where the course angle assignment is separated into two parts (see Figure 12.14)

$$\chi_d = \pi_p - \tan^{-1}\left(\frac{y_e^p}{\Delta}\right) \tag{12.88}$$

where π_p is the *path-tangential angle*, while $\tan^{-1}(y_e^p/\Delta)$ is a *velocity-path relative angle*, which ensures that the velocity is directed toward a point on the path that is located a

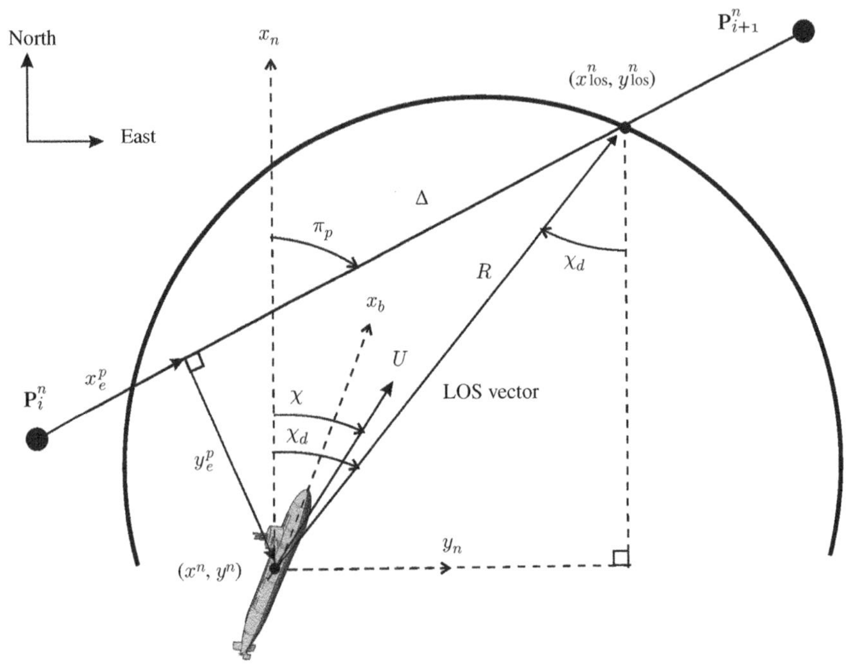

Figure 12.16 Enclosure-based LOS guidance where the desired course angle χ_d is chosen to point toward the LOS intersection point $(x_{\text{los}}^n, y_{\text{los}}^n)$ using a circle with radius $R > |y_\text{e}^p|$ enclosing the craft's position.

lookahead distance $\Delta > 0$ ahead of the direct projection of $\boldsymbol{p}^n = [x^n, y^n]^\top$ onto the path (Papoulias 1991).

Enclosure-based steering utilizes a circle with radius $R > 0$ enclosing the craft's position $\boldsymbol{p}^n$ as shown in Figure 12.16. If the circle radius is chosen sufficiently large, the circle will intersect the straight line at two points. The enclosure-based strategy for driving y_e^p to zero is then to direct the velocity toward the intersection point $\boldsymbol{p}_{\text{los}}^n = [x_{\text{los}}^n, y_{\text{los}}^n]^\top$ that corresponds to the desired direction of travel, which is implicitly defined by the sequence in which the waypoints are ordered. Such a solution involves directly assigning χ_d as shown in Figure 12.16. Since

$$\tan(\chi_\text{d}) = \frac{y_{\text{los}}^n - y^n}{x_{\text{los}}^n - x^n} \tag{12.89}$$

the desired course angle is found from

$$\chi_\text{d} = \text{atan2}(y_{\text{los}}^n - y^n, x_{\text{los}}^n - x^n). \tag{12.90}$$

In order to calculate the two unknown coordinates $(x^n_{\text{los}}, y^n_{\text{los}})$, the following two equations must be solved

$$(x^n_{\text{los}} - x^n)^2 + (y^n_{\text{los}} - y^n)^2 = R^2 \tag{12.91}$$

$$\tan(\pi_{\text{p}}) = \frac{y^n_{i+1} - y^n_i}{x^n_{i+1} - x^n_i} = \frac{y^n_{\text{los}} - y^n_i}{x^n_{\text{los}} - x^n_i} \tag{12.92}$$

where (12.91) represents the *Pythagoras theorem*, while (12.92) states that the slope of the line between the two waypoints is constant. These equations can be solved numerically or analytically for the pair $(x^n_{\text{los}}, y^n_{\text{los}})$.

As can be immediately noticed, the lookahead-based steering scheme (12.88) is less computationally intensive than the enclosure-based approach (12.90)–(12.92). It is also valid for all cross-track errors, whereas the enclosure-based strategy requires $R > |y^p_{\text{e}}|$. Furthermore, Figure 12.16 shows that

$$(y^p_{\text{e}})^2 + \Delta^2 = R^2 \tag{12.93}$$

which means that the enclosure-based approach corresponds to a lookahead-based scheme with a time variation

$$\Delta(t) = \sqrt{R^2 - y^p_{\text{e}}(t)^2} \tag{12.94}$$

where $0 < \Delta(t) \leq R$ for all t. Note that Formula (12.94) is only valid for $R > |y^p_{\text{e}}|$.

12.4.4 Integral LOS

ILOS, or integral LOS, can be used to remove steady-state offsets in χ caused by kinematic modeling errors. A typical case is neglected kinematic couplings terms due to the craft's rolling and pitching motions. This is usually done by extending the proportional LOS guidance law (12.78) to include integral action such that

$$\chi_{\text{d}} = \pi_{\text{p}} - \tan^{-1}\left(K_{\text{p}}\, y^p_{\text{e}} + K_{\text{i}} \int_0^t y^p_{\text{e}}(\tau)\text{d}\tau\right) \tag{12.95}$$

where $K_p = 1/\Delta$ and K_i are the proportional and integral gains, respectively. This is a saturated PI controller where the trigonometric function represent the saturating element. Formula (12.95) is usually implemented using trial and failure, and it can be hard to find optimal gains and satisfactory stability margins for the guidance law. An alternative representation with guaranteed stability properties is the nonlinear ILOS algorithm by Lekkas and Fossen (2014)

$$\chi_d = \pi_p - \tan^{-1}(K_p\, y_e^p + K_i\, y_{int}^p) \tag{12.96}$$

$$\dot{y}_{int}^p = \frac{U y_e^p}{\sqrt{\Delta^2 + (y_e^p + \kappa\, y_{int}^p)^2}} \tag{12.97}$$

where $K_p = 1/\Delta$, $K_i = \kappa K_p$ and $\kappa > 0$ are design parameters. For the nonlinear ILOS algorithm (12.96), the cross-track error expressed in path-tangential coordinates becomes

$$\begin{aligned} \dot{y}_e^p &= U \sin(\chi - \pi_p) \\ &= -U \frac{y_e^p + \kappa y_{int}^p}{\sqrt{\Delta^2 + (y_e^p + \kappa y_{int}^p)^2}}. \end{aligned} \tag{12.98}$$

Global asymptotic stability can be proven under the assumption that $U > 0$ by using the Lyapunov function candidate $V = 1/2(y_e^p)^2 + \kappa/2(y_{int}^p)^2$. Hence, the time-derivative of V becomes

$$\begin{aligned} \dot{V} &= y_e^p \left(-U \frac{y_e^p + \kappa y_{int}^p}{\sqrt{\Delta^2 + (y_e^p + \kappa y_{int}^p)^2}} \right) + \kappa y_{int}^p \dot{y}_{int}^p \\ &= -U \frac{(y_e^p)^2}{\sqrt{\Delta^2 + (y_e^p + \kappa y_{int}^p)^2}} + \frac{\kappa y_{int}^p \left(\dot{y}_{int}^p \sqrt{\Delta^2 + (y_e^p + \kappa y_{int}^p)^2} - U y_e\right)}{\sqrt{\Delta^2 + (y_e^p + \kappa y_{int}^p)^2}}. \end{aligned} \tag{12.99}$$

Choosing $\dot{y}_{int}^p$ as in (12.97) finally yields

$$\begin{aligned} \dot{V} &= -U \frac{(y_e^p)^2}{\sqrt{\Delta^2 + (y_e^p + \kappa y_{int}^p)^2}} \\ &< 0, \qquad \forall\, y_e^p \neq 0,\ y_{int}^p \neq 0. \end{aligned} \tag{12.100}$$

If U and Δ are constant, the equilibrium point $(y_e^p, y_{int}^p) = (0, 0)$ is GAS according to Krasovskii–LaSalle's theorem; see Appendix A.1.3.

12.5 LOS Guidance Laws for Path Following using Heading Autopilots

The LOS guidance laws in Section 12.4 are formulated as course commands χ_d, while marine craft often use compass measurements ψ to measure the heading angle. It is important to stress the differences of heading and course controlled craft. The course angle χ or COG of a marine craft is the cardinal direction in which the craft is moving. Hence, the course angle is to be distinguished from the heading angle ψ, which is the compass direction in which the craft's bow or nose is pointed. The difference between the course and heading angles is

$$\chi = \psi + \beta_c \tag{12.101}$$

where

$$\beta_c = \sin^{-1}(v/U), \qquad U > 0 \tag{12.102}$$

is the crab angle (see Section 2.5.2). Note that the crab angle, and thus the course angle, is not defined for zero speed while the yaw angle (compass direction) is well-defined for zero and forward speeds.

Since the LOS guidance law computes a course command χ_d, Equation (12.101) suggests that the heading command can be computed as

$$\begin{aligned}\psi_d &= \chi_d - \beta_c \\ &= \pi_p - \tan^{-1}(K_p y_e^p) - \beta_c\end{aligned} \tag{12.103}$$

where $K_p = 1/\Delta$. This is illustrated in Figure 12.17. Consequently, the craft can be controlled using a conventional *heading autopilot*

$$\tau = -K_{p_\psi}\tilde{\psi} - K_{d_\psi}\dot{\tilde{\psi}} - K_{i_\psi}\int_0^t \tilde{\psi}(\tau)\mathrm{d}\tau \tag{12.104}$$

where $\tilde{\psi} = \psi - \psi_d$ if the crab angle β_c in (12.103) is known.

12.5.1 *Crab Angle Compensation by Direct Measurements*

An estimate of the crab angle can be obtained by using velocity measurements. If the pair (u, v) is measured, the crab angle can be computed as

$$\beta_c = \sin^{-1}(v/\sqrt{u^2 + v^2}) \tag{12.105}$$

which gives

$$\psi_d = \pi_p - \tan^{-1}(K_p y_e^p) - \sin^{-1}(v/\sqrt{u^2 + v^2}). \tag{12.106}$$

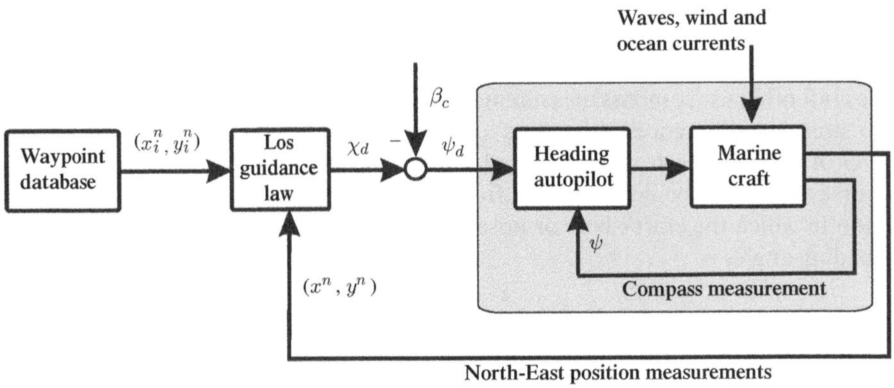

Figure 12.17 Heading autopilot used for LOS path following. The crab angle $\beta_c = \sin^{-1}(v/U)$ must be compensated for by measuring the velocities alternatively treated as an unknown drift term that is canceled by ILOS.

Even though β_c is compensated for, small offsets can still be observed in steady state since kinematic coupling terms, e.g. in roll and pitch, are still unaccounted for. Consequently, most guidance laws for marine craft are implemented using integral action instead of (12.106).

12.5.2 Integral LOS

If there are no velocity measurements, which can be used to approximate β_c, integral LOS is an alternative for compensation of the unknown drift term β_c. We observe that the lookahead-based steering law (12.78) has the same form as a saturated proportional control law. Following the same line of reasoning, it is straightforward to add integral action in order to compensate for the cross-track error caused by a constant disturbance. The conventional approach is

$$\psi_d = \pi_p - \tan^{-1}\left(K_p\, y_e^p + K_i \int_0^t y_e^p(\tau)\mathrm{d}\tau\right) \tag{12.107}$$

where $K_i > 0$ denotes the integral gain. As is usually for the case with integral control action, careful design is necessary to avoid undesired effects such as wind-up and overshooting. Unfortunately, there are no global stability results for the conventional ILOS algorithm but local stability can be proven by linearization about $y_e^p = 0$.

A more effective ILOS algorithm than the conventional ILOS algorithm has been proposed by Børhaug *et al.* (2008) using nonlinear control theory to prove UGAS/ULES. The stability analysis is based on a 3-DOF ship model (surge, sway and yaw) in cascade with a nonlinear ILOS guidance law

$$\psi_d = \pi_p - \tan^{-1}(K_p\, y_e^p + K_i\, y_{\mathrm{int}}^p) \tag{12.108}$$

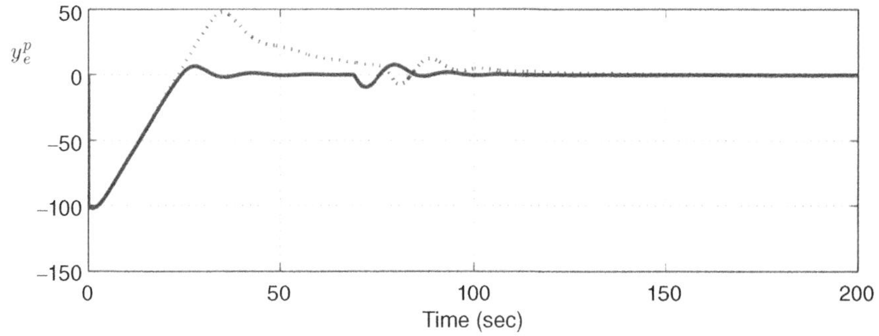

Figure 12.18 The dotted line shows the conventional ILOS algorithm while the solid line is the ILOS algorithm by Børhaug et al. (2008), which clearly has better transient behavior and performance than the conventional method. Source: Modified from Børhaug, E., A. Pavlov and K. Y. Pettersen (2008), Integral LOS control for Path Following of Underactuated Marine Surface Vessels in the presence of Constant Ocean Currents, Proc. of the 47th IEEE Conference on Decision and Control, pp. 4984–4991, Cancun, Mexico.

$$\dot{y}^p_{\text{int}} = \frac{\Delta y^p_{\text{e}}}{\Delta^2 + (y^p_{\text{e}} + \kappa\, y^p_{\text{int}})^2} \tag{12.109}$$

where $K_{\text{p}} = 1/\Delta$, $K_{\text{i}} = \kappa K_{\text{p}}$ and $\kappa > 0$ are design parameters. Equation (12.109) has been designed in a way such that the influence of the integrator diminishes when the cross-track error increases. Hence, the integral wind-up risk is reduced. The algorithms are compared in Figure 12.18. The nonlinear ILOS algorithm has been tested and verified in a large number of applications; see Caharija *et al.* (2016) and Kelasidi *et al.* (2017) for instance.

12.6 Curved-Path Path Following

This section relaxes the condition that the path consists of straight lines between the waypoints. Instead, it is assumed that the guidance systems can make use of a predefined parametrized path. The path-following controller is a *kinematic controller* that generates the desired states for the motion control system using the parametrization of the path. The drawback is that the path must be parametrized and known in advance. In many cases this is not practical and a simpler path consisting of waypoints and straight lines must be used. Section 12.6.1 discusses path generation while path-following controllers for curved parametrized paths are presented in Sections 12.6.2 and 12.6.3.

For a parametrized path, the following definitions are adopted from Skjetne *et al.* (2004):

Definition 12.2 (Parametrized Path)
A parametrized path is defined as a geometric curve $\eta_{\text{d}}(\varpi) \in \mathbb{R}^q$ with $q \geq 1$ parametrized by a continuous path variable ϖ.

For marine craft it is common to use a 3D representation

$$\boldsymbol{p}_{\mathrm{d}}^{n}(\varpi) = [x_{\mathrm{d}}(\varpi),\ y_{\mathrm{d}}(\varpi),\ z_{\mathrm{d}}(\varpi)]^{\top} \in \mathbb{R}^{3} \tag{12.110}$$

where the first two coordinates describe the position in the horizontal plane and the last coordinate is the depth. For surface craft only the coordinates $(x_{\mathrm{d}}(\varpi), y_{\mathrm{d}}(\varpi))$ are needed while underwater vehicles use all three coordinates. The first- and second-order derivatives of $\boldsymbol{p}^{n}(\varpi)$ with respect to ϖ are denoted as $\boldsymbol{p}^{n}(\varpi)'$ and $\boldsymbol{p}^{n}(\varpi)''$, respectively.

A frequently used solution of the path-following problem is to solve it as the geometric task of a *maneuvering problem*, given by the following definition:

Definition 12.3 (Maneuvering Problem)
The maneuvering problem involves solving two tasks:

1. *Geometric task: Force the position $\boldsymbol{p}^{n}$ to converge to a desired path $\boldsymbol{p}_{\mathrm{d}}^{n}(\varpi)$,*

$$\lim_{t\to\infty} [\boldsymbol{p}^{n} - \boldsymbol{p}_{\mathrm{d}}^{n}(\varpi)] = \boldsymbol{0} \tag{12.111}$$

for any continuous function ϖ.

2. *Dynamic task: Force the speed $\dot{\varpi}$ to converge to a desired speed $U_{\mathrm{d}}(\varpi)$ according to*

$$\lim_{t\to\infty} \left[\dot{\varpi} - \frac{U_{\mathrm{d}}(\varpi)}{\sqrt{x_{\mathrm{d}}'(\varpi)^2 + y_{\mathrm{d}}'(\varpi)^2}} \right] = 0. \tag{12.112}$$

The dynamic task follows from

$$U_{\mathrm{d}}(\varpi) = \sqrt{\dot{u}_{\mathrm{d}}(\varpi)^2 + \dot{v}_{\mathrm{d}}(\varpi)^2} = \sqrt{x_{\mathrm{d}}'(\varpi)^2 + y_{\mathrm{d}}'(\varpi)^2}\ \dot{\varpi}. \tag{12.113}$$

Definition 12.3 implies that the dynamics ϖ along the path can be specified independently of the error dynamics. A special case of the maneuvering problem is

$$\dot{\varpi} = 1, \quad \varpi(0) = 0 \tag{12.114}$$

which is recognized as the tracking problem since the solution of (12.114) is $\varpi = t$. A solution to the maneuvering problem for fully actuated craft is found in Skjetne *et al.* (2004).

12.6.1 Path Generation using Interpolation Methods

The path can be generated using spline or polynomial interpolation methods to generate a curve $(x_{\mathrm{d}}(\varpi),\ y_{\mathrm{d}}(\varpi))$ through a set of N predefined waypoints; see Lekkas and Fossen (2014). Notice that a trajectory $(x_{\mathrm{d}}(t),\ y_{\mathrm{d}}(t))$ is obtained by choosing $\dot{\varpi} = c$ such that $\varpi = c\ t$ where $c \in \mathbb{R}$.

Cubic Spline and Hermite Interpolation

In Matlab, several methods for interpolation are available.

> **MATLAB**
>
> The different methods for interpolation are found by typing
>
> ```
> help polyfun
> ```
>
> Two frequently used methods in Matlab for path generation are the cubic spline interpolant (`spline.m`) and the piecewise cubic Hermite interpolating polynomial (`pchip.m`).

The main difference between Hermite and cubic spline and interpolation is how the slopes at the end points are handled. For simplicity let us consider the problem of trajectory generation. The cubic Hermite interpolant ensures that the first-order derivatives $(\dot{x}_d(\varpi), \dot{y}_d(\varpi))$ are continuous. In addition, the slopes at each endpoint are chosen in such a way that $(x_d(\varpi), y_d(\varpi))$ are shape preserving and respect monotonicity.

Cubic spline interpolation is usually done by requiring that the second-order derivatives $(\ddot{x}_d(\varpi), \ddot{y}_d(\varpi))$ at the endpoints of the polynomials are equal, which gives a smooth spline. Consequently, the cubic spline will be more accurate than the Hermite interpolating polynomial if the data values are of a smooth function. The cubic Hermite interpolant, on the contrary, has less oscillations if the data are non-smooth.

The results of interpolating a set of predefined waypoints to a parametrized path $(x_d(\varpi), y_d(\varpi))$ using the cubic Hermite interpolant and cubic spline interpolation methods are shown in Figure 12.19. It is seen that different behaviors are obtained due to the conditions on the first- and second-order derivatives at the endpoints.

Polynomial Interpolation

Instead of using the Matlab functions `pchip.m` and `spline.m` a cubic spline can be interpolated through a set of waypoints by considering the *cubic polynomials*

$$x_d(\varpi) = a_3\varpi^3 + a_2\varpi^2 + a_1\varpi + a_0 \tag{12.115}$$

$$y_d(\varpi) = b_3\varpi^3 + b_2\varpi^2 + b_1\varpi + b_0 \tag{12.116}$$

where $(x_d(\varpi), y_d(\varpi))$ are the position of the craft and where ϖ is the path variable. The partial derivatives of $x_d(\varpi)$ and $y_d(\varpi)$ with respect to ϖ are

$$x'_d(\varpi) = \frac{dx_d(\varpi)}{d\varpi} = 3a_3\varpi^2 + 2a_2\varpi + a_1 \tag{12.117}$$

$$y'_{\rm d}(\varpi) = \frac{{\rm d}y_{\rm d}(\varpi)}{{\rm d}\varpi} = 3b_3\varpi^2 + 2b_2\varpi + b_1. \tag{12.118}$$

Hence, the desired speed $U_{\rm d}(\varpi)$ of the craft can be computed as

$$\dot{x}_{\rm d}(\varpi) = \frac{{\rm d}x_{\rm d}(\varpi)}{{\rm d}\varpi}\dot{\varpi} \tag{12.119}$$

$$\dot{y}_{\rm d}(\varpi) = \frac{{\rm d}y_{\rm d}(\varpi)}{{\rm d}\varpi}\dot{\varpi} \tag{12.120}$$

resulting in

$$\begin{aligned} U_{\rm d}(\varpi) &= \sqrt{\dot{x}_{\rm d}(\varpi)^2 + \dot{y}_{\rm d}(\varpi)^2} \\ &= \sqrt{x'_{\rm d}(\varpi)^2 + y'_{\rm d}(\varpi)^2}\ \dot{\varpi}. \end{aligned} \tag{12.121}$$

Similarly, an expression for the acceleration $\dot{U}_d(\varpi)$ can be found.

MATLAB

The script `ExSpline.m` generates splines thorough a set of waypoints as shown in Figure 12.19 using cubic Hermite and spline interpolation methods. The path variable is chosen as $\dot{\varpi} = 1$ such that $\varpi = t$.

```
% ExSpline - Cubic Hermite and spline interpolation of waypoints
wpt.pos.x  =  [0 100 500 700 1000];
wpt.pos.y  =  [0 100 100 200 160];
wpt.time   =  [0 40 60 80 100];

t = 0:1:max(wpt.time);                    % time
x_p = pchip(wpt.time,wpt.pos.x,t);  % cubic Hermite interpolation

y_p = pchip(wpt.time,wpt.pos.y,t);
x_s = spline(wpt.time,wpt.pos.x,t);   % spline interpolation
y_s = spline(wpt.time,wpt.pos.y,t);

subplot(311), plot(wpt.time,wpt.pos.x,'o',t,[x_p; x_s]);
subplot(312), plot(wpt.time,wpt.pos.y,'o',t,[y_p; y_s]);
subplot(313), plot(wpt.pos.y,wpt.pos.x,'o',y_p,x_{p},y_s,x_s);
```

The unknown parameters $a_0, a_1, a_2, a_3, b_0, b_1, b_2, b_3$ in (12.115) and (12.116) can also be computed using a cubic spline algorithm, as shown below.

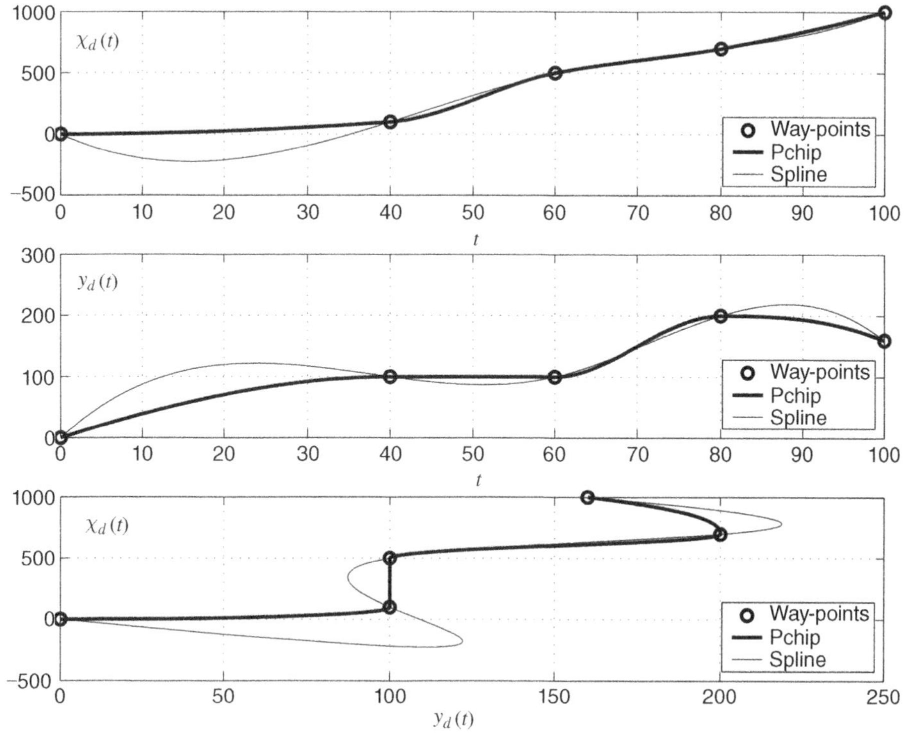

Figure 12.19 Waypoint interpolation using cubic Hermite and spline interpolation methods; see `ExSpline.m` in the MSS toolbox. The path variable is chosen as $\varpi = t$.

Cubic Spline Algorithm for Path Generation

The path through the waypoints (x_{k-1}, y_{k-1}) and (x_k, y_k) must satisfy

$$x_{\mathrm{d}}(\varpi_{k-1}) = x_{k-1}, \qquad x_{\mathrm{d}}(\varpi_k) = x_k \tag{12.122}$$

$$y_{\mathrm{d}}(\varpi_{k-1}) = y_{k-1}, \qquad y_{\mathrm{d}}(\varpi_k) = y_k \tag{12.123}$$

where $k = 1, \ldots, n$. In addition, smoothness is obtained by requiring that

$$\lim_{\varpi \to \varpi_k^-} x'_{\mathrm{d}}(\varpi_k) = \lim_{\varpi \to \varpi_k^+} x'_{\mathrm{d}}(\dot{\varpi}_k) \tag{12.124}$$

$$\lim_{\varpi \to \varpi_k^-} x''_{\mathrm{d}}(\varpi_k) = \lim_{\varpi \to \varpi_k^+} x''_{\mathrm{d}}(\varpi_k). \tag{12.125}$$

For this problem, it is possible to add only two boundary conditions (velocity or acceleration) for the x and y equations, respectively. Hence,

$$x'_{\mathrm{d}}(\varpi_0) = x'_0, \qquad x'_{\mathrm{d}}(\varpi_{\mathrm{n}}) = x'_{\mathrm{n}} \tag{12.126}$$

$$y'_{\mathrm{d}}(\varpi_0) = y'_0, \qquad y'_{\mathrm{d}}(\varpi_{\mathrm{n}}) = y'_{\mathrm{n}} \tag{12.127}$$

or

$$x''_{\rm d}(\varpi_0) = x''_0, \qquad x''_{\rm d}(\varpi_{\rm n}) = x''_{\rm n} \tag{12.128}$$

$$y''_{\rm d}(\varpi_0) = y''_0, \qquad y''_{\rm d}(\varpi_{\rm n}) = y''_{\rm n}. \tag{12.129}$$

The polynomial $x_{\rm d}(\varpi_k)$ is given by the parameters $\boldsymbol{a}_k = [a_{3k}, a_{2k}, a_{1k}, a_{0k}]^\top$, resulting in $4(n-1)$ unknown parameters. The number of constraints are also $4(n-1)$ if only velocity or acceleration constraints are chosen at the end points. The unknown parameters for n waypoints are collected into a vector

$$\boldsymbol{x} = [\boldsymbol{a}_k^\top, \dots, \boldsymbol{a}_{\rm n-1}^\top]^\top. \tag{12.130}$$

Hence, the cubic interpolation problem can be written as a linear equation

$$\boldsymbol{y} = \boldsymbol{A}(\varpi_{k-1}, \dots, \varpi_k)\boldsymbol{x}, \qquad k = 1, 2, \dots, n \tag{12.131}$$

where

$$\boldsymbol{y} = [x_{\rm start}, x_0, x_1, x_1, 0, 0, x_2, x_2, 0, 0, \dots, x_n, x_{\rm final}]^\top. \tag{12.132}$$

The start and end points can be specified in terms of velocity or acceleration constraints $x_{\rm start} \in \{x'_0, x''_0\}$ and $x_{\rm final} \in \{x'_n, x''_n\}$, respectively. This gives

$$\boldsymbol{A}(\varpi_{k-1}, \dots, \varpi_k) = \left[\begin{array}{cccc} \boldsymbol{c}_{\rm start} & \boldsymbol{0}_{1\times4} & \boldsymbol{0}_{1\times4} & \cdots\ \boldsymbol{0}_{1\times4} \\ \boldsymbol{p}(\varpi_0) & \boldsymbol{0}_{1\times4} & \boldsymbol{0}_{1\times4} & \boldsymbol{0}_{1\times4} \\ \hline \boldsymbol{p}(\varpi_1) & \boldsymbol{0}_{1\times4} & \boldsymbol{0}_{1\times4} & \boldsymbol{0}_{1\times4} \\ 0 & \boldsymbol{p}(\varpi_1) & \boldsymbol{0}_{1\times4} & \boldsymbol{0}_{1\times4} \\ -\boldsymbol{v}(\varpi_1) & \boldsymbol{v}(\varpi_1) & \boldsymbol{0}_{1\times4} & \boldsymbol{0}_{1\times4} \\ -\boldsymbol{a}(\varpi_1) & \boldsymbol{a}(\varpi_1) & \boldsymbol{0}_{1\times4} & \boldsymbol{0}_{1\times4} \\ \hline \boldsymbol{0}_{1\times4} & \boldsymbol{p}(\varpi_2) & \boldsymbol{0}_{1\times4} & \boldsymbol{0}_{1\times4} \\ \boldsymbol{0}_{1\times4} & \boldsymbol{0}_{1\times4} & \boldsymbol{p}(\varpi_2) & \boldsymbol{0}_{1\times4} \\ \boldsymbol{0}_{1\times4} & -\boldsymbol{v}(\varpi_2) & \boldsymbol{v}(\varpi_2) & \boldsymbol{0}_{1\times4} \\ \boldsymbol{0}_{1\times4} & -\boldsymbol{a}(\varpi_2) & \boldsymbol{a}(\varpi_2) & \boldsymbol{0}_{1\times4} \\ \hline \vdots & & & \ddots \\ \hline \boldsymbol{0}_{1\times4} & \boldsymbol{0}_{1\times4} & \boldsymbol{0}_{1\times4} & \boldsymbol{p}(\varpi_n) \\ \boldsymbol{0}_{1\times4} & \boldsymbol{0}_{1\times4} & \boldsymbol{0}_{1\times4} & \cdots\ \boldsymbol{c}_{\rm final} \end{array}\right] \tag{12.133}$$

where $\boldsymbol{c}_{\rm start} \in \{x'_{\rm d}(\varpi_0), x''_{\rm d}(\varpi_0)\}$, $\boldsymbol{c}_{\rm final} \in \{x'_{\rm d}(\varpi_{\rm n}), x''_{\rm d}(\varpi_{\rm n})\}$ and

$$\boldsymbol{p}(\varpi_k) = [\varpi_k^3, \varpi_k^2, \varpi_k, 1] \tag{12.134}$$

$$\boldsymbol{v}(\varpi_k) = \boldsymbol{p}'(\varpi_k) = [3\varpi_k^2, 2\varpi_k, 1, 0] \tag{12.135}$$

$$\boldsymbol{a}(\varpi_k) = \boldsymbol{p}''(\varpi_k) = [6\varpi_k, 2, 0, 0]. \tag{12.136}$$

Equation (12.131) can be solved for $\varpi_k = 0, 1, \ldots, n$ according to

$$\boldsymbol{x} = \boldsymbol{A}^{-1}\boldsymbol{y}. \tag{12.137}$$

The formulae for $\boldsymbol{b}_k = [b_{3k}, b_{2k}, b_{1k}, b_{0k}]^{\top}$ are obtained in a similar manner.

MATLAB

Formula (12.137) has been implemented in the script `ExPathGen.m` and `pva.m`. The results are for the following set of waypoints

```
wpt.pos.x = [0 200 400 700 1000]
wpt.pos.y = [0 200 500 400 1200]
```

where $\varpi = 0, \ldots, 4$ are shown in Figures 12.20 and 12.21.

Transformation of Path to Reference Trajectories using Desired Speed Profiles

In Figure 12.21 it is seen that the solution between two successive waypoints

$$x_{\mathrm{d}}(\varpi) = a_3\varpi^3 + a_2\varpi^2 + a_1\varpi + a_0 \tag{12.138}$$

$$y_{\mathrm{d}}(\varpi) = b_3\varpi^3 + b_2\varpi^2 + b_1\varpi + b_0 \tag{12.139}$$

is indeed a *time-independent* path when $x_{\mathrm{d}}(\varpi)$ is plotted against $y_{\mathrm{d}}(\varpi)$ for increasing ϖ values.

The path can be transformed to a *time-varying* trajectory by defining a *speed profile*. The speed profile assigns dynamics to ϖ such that the desired path transforms to a time-dependent reference trajectory at the same time as the desired speed and acceleration profiles are preserved. From (12.121) it is seen that

$$\dot{\varpi} = \frac{U_{\mathrm{d}}(\varpi)}{\sqrt{x'_{\mathrm{d}}(\varpi)^2 + y'_{\mathrm{d}}(\varpi)^2}} \tag{12.140}$$

where $U_{\mathrm{d}}(\varpi)$ is the desired speed profile. Let U_{ref} be the input to a first-order system

$$T\dot{U}_{\mathrm{d}} + U_{\mathrm{d}} = U_{\mathrm{ref}}, \quad T > 0. \tag{12.141}$$

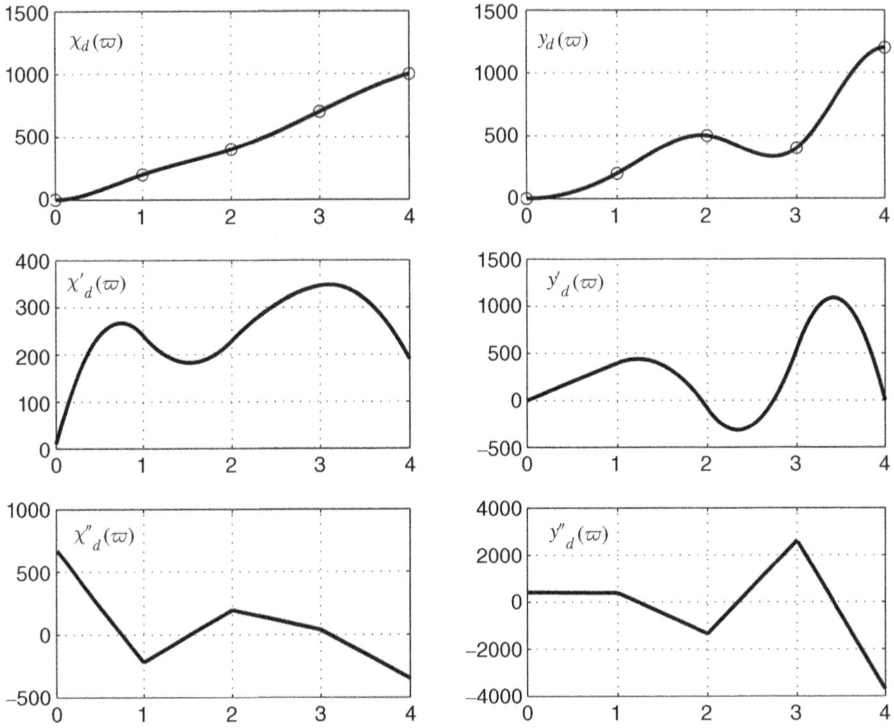

Figure 12.20 Polynomials $x_d(\varpi)$ and $y_d(\varpi)$ and their first- and second-order derivatives.

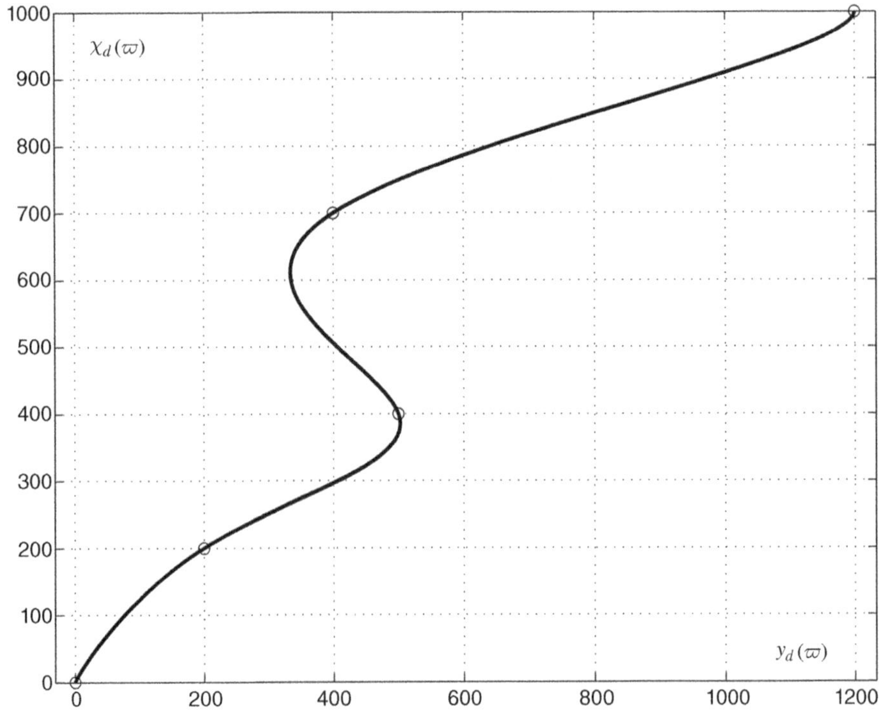

Figure 12.21 The xy plot through five waypoints using cubic spline interpolation.

A smooth transition from the desired speed U_k at waypoint k to the next waypoint $k+1$ can be made by using

$$U_{\text{ref}} = U_{k+1}. \tag{12.142}$$

This is illustrated in the following example.

Example 12.3 (Transformation of Path to Reference Trajectories)
Consider the first two waypoints in the example file `ExPathGen.m` *given by*

$$(x_0, y_0) = (0, 0)$$
$$(x_1, y_1) = (200, 200)$$

with desired speeds $U_0 = 0$ *and* $U_1 = 5$ m s^{-1}*, respectively. The cubic polynomials satisfying (12.137) are*

$$x_d(\varpi) = -29.89\,\varpi^3 + 135.63\,\varpi^2 + 94.25\,\varpi$$
$$y_d(\varpi) = 108.05\,\varpi^3 - 2.30\,\varpi^2 + 94.25\,\varpi$$

for $\varpi \in [0, 1]$*. Let the speed dynamics time constant be* $T = 10$ s*. Assume that the craft is initially at rest and that the desired speed of waypoint number 1 is* $U_{\text{ref}} = U_1$*. The numerical solutions of*

$$\dot{\varpi} = \frac{U_d}{\sqrt{x_d'(\varpi)^2 + y_d'(\varpi)^2}} \tag{12.143}$$

$$T\dot{U}_d + U_d = U_{\text{ref}} \tag{12.144}$$

for waypoints 0 and 1 corresponding to $\varpi_0 = 0$ *and* $\varpi_1 = 1$ *with* $t_0 = 0$ *and* t_1 *unknown, is shown in Figure 12.22; see* `ExPathGen.m`*. It is seen that the desired speed of* 5.0 m s^{-1} *is reached in approximately* 67 s*. Hence, the terminal time must be chosen as* $t_1 \geq 67$ s *(corresponding to* $\varpi_1 = 1$ *in order to satisfy the desired speed dynamics. If* $t_1 < 67$ s *there is not enough time to reach the desired speed of waypoint 1 unless the time constant* T *is reduced. The time constant should reflect what is physically possible for the craft. Note that the path* $(x_d(\varpi), y_d(\varpi))$ *has been transformed to a time-varying reference trajectory* $(x_d(t), y_d(t))$ *by assigning a speed profile (12.143) to be solved numerically with the path planner (12.137). This gives design flexibility since the path can be generated off-line using a waypoint database while speed is assigned to the path when the dynamics of the actual craft is considered.*

Nonlinear Constrained Optimization

Another solution to trajectory and path generation is to use nonlinear constrained optimization techniques. These methods allow an object function to be specified where minimum time and energy are design goals. In addition, the speed and acceleration constraints of the craft can be added. The drawback is that nonlinear constraint optimization problems are much harder to solve numerically than the methods described in the previous sections. The Matlab optimization toolbox will be used to demonstrate how this can be done.

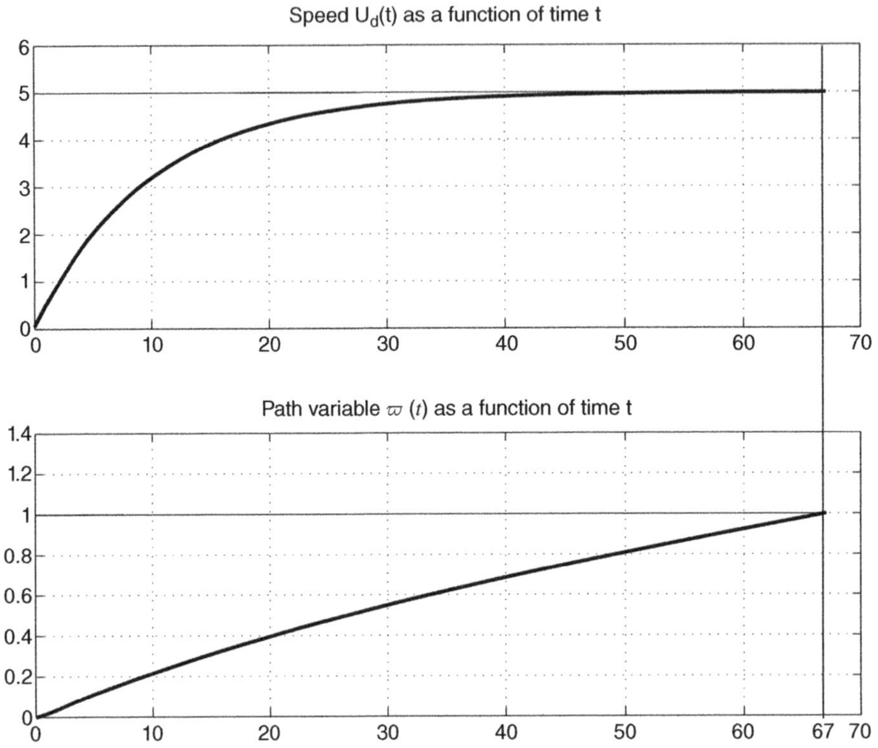

Figure 12.22 Upper plot shows that the speed U_{d} reaches the desired value of 5 m s^{-1} in approximately 67 s. The lower plot shows that the path variable ϖ is incremented from 0 to 1 during the speed transition.

In general, trajectory-tracking and path-planning problems can be formulated as

$$
\begin{aligned}
J = \min_{\boldsymbol{x}} \; \{f(\boldsymbol{x})\} \\
\text{subject to} \quad & g_k(\boldsymbol{x}) \leq 0 && (k = 1, \ldots, n_g) \\
& h_j(\boldsymbol{x}) = 0 && (j = 1, \ldots, n_h) \\
& x_i^{\min} \leq x_i \leq x_i^{\max} && (i = 1, \ldots, n_x)
\end{aligned}
\tag{12.145}
$$

where $f(\boldsymbol{x})$ should be minimized with respect to the parameter vector $\boldsymbol{x}$ with $g_i(\boldsymbol{x})$ and $h_j(\boldsymbol{x})$ as nonlinear inequality and equality constraints, respectively. A popular simplification is to use quadratic programming. Consequently,

$$
\begin{aligned}
J = \min_{x} & \left\{ \frac{1}{2} x^{\top} H x + f^{\top} x \right\} \\
\text{subject to} \quad & A x \le b \\
& x_i^{\min} \le x_i \le x_i^{\max} \quad (i = 1, \ldots, n_x).
\end{aligned}
\tag{12.146}
$$

For simplicity, consider two waypoints (x_k, y_k) and (x_{k+1}, y_{k+1}) satisfying

$$x(t_k) = x_k, \qquad y(t_k) = y_k \tag{12.147}$$

$$x(t_{k+1}) = x_{k+1}, \quad y(t_{k+1}) = y_{k+1}. \tag{12.148}$$

Choosing the speed constraints as

$$\dot{x}_{\mathrm{d}} = U_k \cos(\chi_k) \tag{12.149}$$

$$\dot{y}_{\mathrm{d}} = U_k \sin(\chi_k) \tag{12.150}$$

where $\chi_k = \operatorname{atan2}(y_{k+1} - y_k, x_{k+1} - x_k)$, that is with direction toward the next waypoint. For two waypoints this results in

$$y = A(t_k, t_{k+1}) x \tag{12.151}$$

where

$$y = [x_k, x_{k+1}, y_k, y_{k+1}, U_k \cos(\psi_k), U_k \sin(\psi_k), U_{k+1} \cos(\psi_{k+1}), U_{k+1} \sin(\psi_{k+1})]^{\top} \tag{12.152}$$

and

$$A(t_k, t_{k+1}) = \begin{bmatrix}
t_k^3 & t_k^2 & t_k & 1 & 0 & 0 & 0 & 0 \\
t_{k+1}^3 & t_{k+1}^2 & t_{k+1} & 1 & 0 & 0 & 0 & 0 \\
0 & 0 & 0 & 0 & t_k^3 & t_k^2 & t_k & 1 \\
0 & 0 & 0 & 0 & t_{k+1}^3 & t_{k+1}^2 & t_{k+1} & 1 \\
3t_k^2 & 2t_k & 1 & 0 & 0 & 0 & 0 & 0 \\
0 & 0 & 0 & 0 & 3t_k^2 & 2t_k & 1 & 0 \\
3t_{k+1}^2 & 2t_{k+1} & 1 & 0 & 0 & 0 & 0 & 0 \\
0 & 0 & 0 & 0 & 3t_{k+1}^2 & 2t_{k+1} & 0 & 0
\end{bmatrix}. \tag{12.153}$$

The criterion to minimize is

$$J = \min_{x} \{ [A(t_k, t_{k+1}) x - y]^{\top} [A(t_k, t_{k+1}) x - y] \} \tag{12.154}$$

for given pairs (t_k, t_{k+1}) of time. Expanding this expression yields

$$\bar{J} = \frac{1}{2}(J - \boldsymbol{y}^\top \boldsymbol{y}) = \min_{\boldsymbol{x}} \left\{ \frac{1}{2}\boldsymbol{x}^\top \boldsymbol{A}^\top(t_k, t_{k+1})\boldsymbol{A}(t_k, t_{k+1})\boldsymbol{x} - \boldsymbol{y}^\top \boldsymbol{A}(t_k, t_{k+1})\boldsymbol{x} \right\} \tag{12.155}$$

implying that

$$\boldsymbol{H} = \boldsymbol{A}^\top(t_k, t_{k+1})\boldsymbol{A}(t_k, t_{k+1}) \tag{12.156}$$

$$\boldsymbol{f} = -\boldsymbol{y}^\top \boldsymbol{A}(t_k, t_{k+1}). \tag{12.157}$$

In this expression, the starting time t_k is given while the arrival time t_{k+1} is unknown. The cubic polynomials (12.115) and (12.116) imply that there are eight additional unknown parameters to optimize

$$\boldsymbol{x} = [a_3, a_2, a_1, a_0, b_3, b_2, b_1, b_0]^\top \tag{12.158}$$

giving a total of nine unknown parameters. In addition, linear constraints $\boldsymbol{Ax} \le \boldsymbol{b}$ can be added. The reference trajectory can be found using quadratic programming.

MATLAB

Trajectory generation using the optimization toolbox is demonstrated by using the MSS toolbox script `ExQuadProg.m`. Consider two waypoints

$$(x_0, y_0) = (10, 10)$$

$$(x_1, y_1) = (200, 100)$$

with the speed constraint

$$U_{\rm d}(t) \le 10 \text{ m s}^{-1}.$$

The desired waypoint speeds are $U_0 = 0$ and $U_1 = 5$ m s^{-1}. The arrival time t_1 is computed in a loop by solving the quadratic optimization problem (12.146) for each time $t_1 = t_0 + {\rm d}t$ where ${\rm d}t$ is incremented by 1.0 s each time. This process is terminated when the first solution $U_{\rm d}(t) \le 10$ m s^{-1} is reached (this can be easily changed if other requirements are more important). The optimal solution

$$x_{\rm d}(t) = -0.0102\ t^3 + 0.5219\ t^2 - 4.28 \times 10^{-12}\ t + 10.0$$

$$y_{\rm d}(t) = -0.0048\ t^3 + 0.2472\ t^2 - 1.04 \times 10^{-12}\ t + 10.0$$

for $t \in [t_0, t_1]$ is obtained after 29 loops ($t_1 = 29$ s) using `quadprog.m` in the Matlab optimization toolbox. The results are shown in Figure 12.23.

Weather Routing

A weather routing or voyage planning system (VPS) computes the most efficient route using meteorological and oceanographic data, information about the craft's hull and propulsion system and shipping economics to ensure that the craft reaches port on

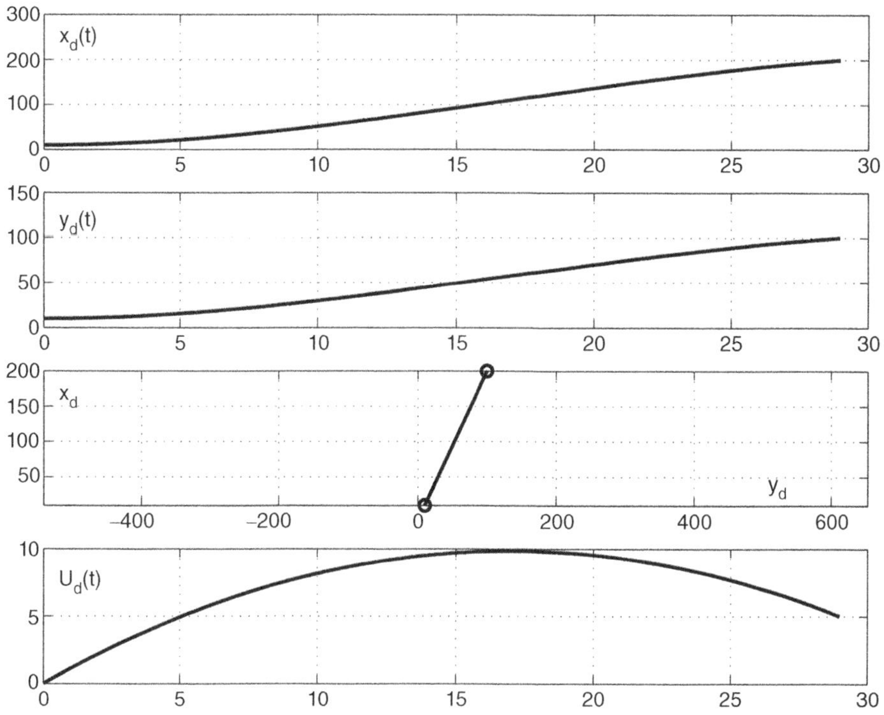

Figure 12.23 The two upper plots show the cubic polynomials $x_d(t)$ and $y_d(t)$. In the third plot $y_d(t)$ is plotted against $x_d(t)$ while the lower plot is speed $U_d(t)$.

time. The data from this analysis can be waypoints with optimal speed and heading information. The routing software of a modern weather routing system includes features such as:

- Surface analysis and forecast models
- Sea state and wind wave models
- Upper air models
- Formation description of low-pressure systems
- Hurricanes and tropical weather models
- Ocean current models
- Vessel performance models
- Cargo condition, trim, draft and deck load
- Link to internet sources for weather data
- Interface to a satellite system transmitting weather data
- Optimization of routes based on a fixed estimated time of arrival (ETA)
- Routing of vessels around hazardous weather conditions.

An optimal route is computed using a numerical optimization offline. This can be done by a computer onboard the craft or by a company onshore transmitting the results to the craft electronically on a 24 hour basis. Several companies offer continuous voyage monitoring with status reports and performance evaluations. This allows for replanning during changing weather conditions. Global weather information is available from several forecast centers.

Some useful references for weather routing of ships are Calvert (1989), Hagiwara (1989), Padadakis and Perakis (1990), and Lo and McCord (1998).

12.6.2 Proportional LOS Guidance Law for Curved Paths

This section extends the results of Section 12.4.2 to curved paths by following the approach of Fossen and Pettersen (2014). Again, we will consider the proportional LOS guidance law

$$\chi_{\mathrm{d}} = \pi_{\mathrm{p}} - \tan^{-1}(K_{\mathrm{p}}\, y_{\mathrm{e}}^{p}) \tag{12.159}$$

where $K_{\mathrm{p}} = 1/\Delta$ is the proportional gain. Let $\varpi \geq 0$ denote the path variable. Consider a two-dimensional parametrized path $(x_p^n(\varpi),\ y_p^n(\varpi))$, which is assumed to go through a set of successive waypoints $(x_i^n,\ y_i^n)$ for $i = 1, 2, ..., N$ as shown in Figure 12.24. For any point $(x_p^n(\varpi),\ y_p^n(\varpi))$ along the path, the path-tangential reference frame is rotated an angle

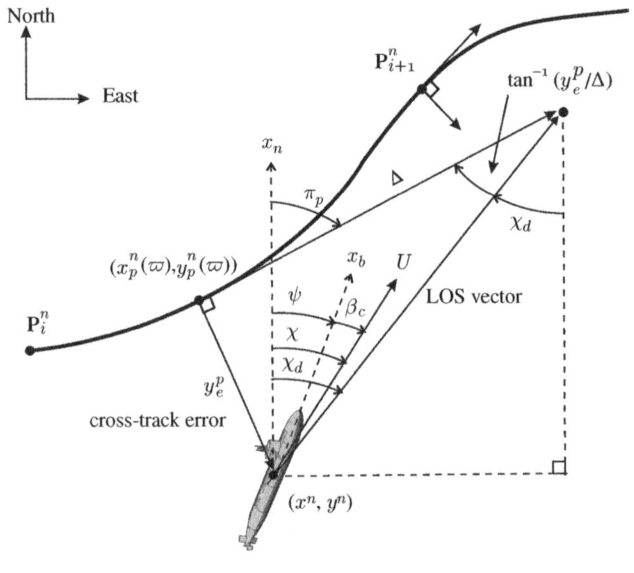

Figure 12.24 Curved paths. The desired course angle χ_{d} is determined such that the vehicle velocity is directed towards a point that is located a user specified distance Δ (lookahead distance) on the path tangent.

$$\pi_p(\varpi) = \text{atan2}\,(y_p^n(\varpi)', \; x_p^n(\varpi)') \tag{12.160}$$

with respect to the north-east reference frame. Note that for a straight line the path-tangential angle $\pi_p = \text{atan2}(y_{i+1}^n - y_i^n, \; x_{i+1}^n - x_i^n)$ is constant between the waypoints. For a craft located at the position $(x^n, \; y^n)$ the cross-track error is computed as the orthogonal distance to the path-tangential reference frame defined by the point $(x_p^n(\varpi), \; y_p^n(\varpi))$. Consequently, an expression for the cross-track error y_e^p can be found by requiring that

$$\begin{bmatrix} 0 \\ y_e^p \end{bmatrix} = \underbrace{\begin{bmatrix} \cos(\pi_p(\varpi)) & \sin(\pi_p(\varpi)) \\ -\sin(\pi_p(\varpi)) & \cos(\pi_p(\varpi)) \end{bmatrix}}_{\boldsymbol{R}_p^n(\pi_p(\varpi))^\top} \begin{bmatrix} x^n - x_p^n(\varpi) \\ y^n - y_p^n(\varpi) \end{bmatrix} \tag{12.161}$$

where $\boldsymbol{R}_p^n(\pi_p(\varpi)) \in SO(2)$. Expanding the first line of (12.161) gives the *normal line*

$$y^n - y_p^n(\varpi) = -\frac{1}{\tan(\pi_p(\varpi))}(x^n - x_p^n(\varpi)) \tag{12.162}$$

through $(x_p^n(\varpi), \; y_p^n(\varpi))$, while the second line gives a formula for the *cross-track error*

$$y_e^p = -(x^n - x_p^n(\varpi))\sin(\pi_p(\varpi)) + (y^n - y_p^n(\varpi))\cos(\pi_p(\varpi)) \tag{12.163}$$

where ϖ propagates according to (12.113). In other words

$$\dot{\varpi} = \frac{U(\varpi)}{\sqrt{(x_p^n(\varpi)')^2 + (y_p^n(\varpi)')^2}} > 0. \tag{12.164}$$

As pointed out by Samson (1992) there may be infinite solutions of (12.162) if the path is a closed curve. In the following we will assume that the path is an open curve, that is the end point is different from the start point. Definition 12.4 guarantees that there is a unique solution for the cross-track error y_e^p obtained by minimizing $\dot{\varpi}$.

Definition 12.4 (Uniqueness of solutions)
The unique solution of (12.163) is denoted $y_e^p(\varpi^*)$ and is defined by

$$\varpi^* := \arg\min_{\varpi \geq 0} \left\{ \frac{U(\varpi)^2}{(x_p^n(\varpi)')^2 + (y_p^n(\varpi)')^2} \right\} \tag{12.165}$$

subject to

$$y^n - y_p^n(\varpi) = -\frac{1}{\tan(\pi_p(\varpi))}(x^n - x_p^n(\varpi)). \tag{12.166}$$

This is a nonlinear optimization problem, which can be solved numerically. However, for many paths ϖ^* can be found by computing all possible projection candidates ϖ_i ($i = 1, \ldots, M$) given by (12.162) and choose the one closest to the previous ϖ^* value.

12.6.3 Path-following using Serret–Frenet Coordinates

Consider a two-dimensional parametrized path $(x_p^n(\varpi), y_p^n(\varpi))$, which is assumed to go through a set of successive waypoints (x_i^n, y_i^n) for $i = 1, 2, \ldots, N$. A *kinematic controller* that generates the desired states for the motion control system can be designed using a dynamic model of the craft by specifying a reference frame that moves along the path; see Figure 12.25. This reference frame is usually chosen as the *Serret–Frenet frame* (Frenet 1847, Serret 1851). During path following, the craft speed is denoted U and the kinematic controller is designed to: (i) regulate the distance e between the craft and the path to zero and (ii) regulate the angle χ_{SF} between the craft speed vector and the tangent to the path to zero (see Samson 1992, Micaelli and Samson 1993).

Virtual Target Dynamics using Serret–Frenet Coordinates

The virtual target, defined by the projection of an actual craft on to the path-tangential Serret–Frenet frame {SF}, evolves according to (Lapierre and Soetanto 2007)

$$\dot{s} = U\cos(\chi_{SF}) - (1 - \kappa e)\dot{s}_a \tag{12.167}$$

$$\dot{e} = U\sin(\chi_{SF}) - \kappa s\dot{s}_a \tag{12.168}$$

$$\dot{\chi}_{SF} = r + \dot{\beta}_c - \kappa\dot{s}_a \tag{12.169}$$

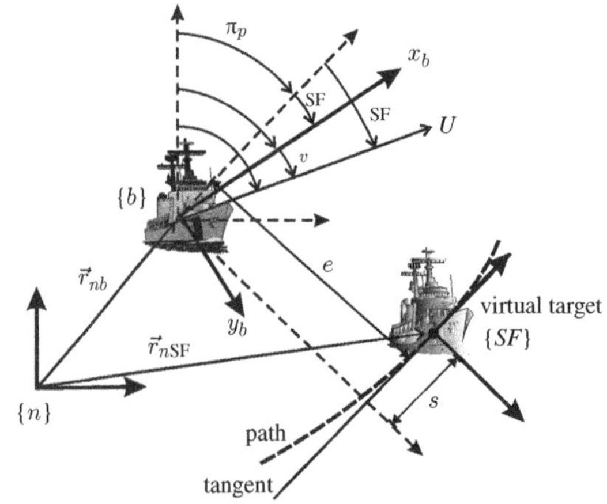

Figure 12.25 Kinematic description of the Serret–Frenet frame.

where U is the speed of the craft and (e, s) is the location on the path of the {SF} frame relative to the {b} frame. If $s = 0$, the variable e represents the closest distance between the actual craft and the origin of {SF} tangential to the path. Hence, s can be viewed as an extra controller design parameter for evolution along the path. The arc length that the target has moved along the path is denoted s_a while χ_{SF} is the angle between the x axis of {SF} and the speed vector; see Figure 12.25. Finally, κ is the path curvature.

From Figure 12.25, it is seen that the distance vectors between {n}, {b} and {SF} satisfy

$$\vec{r}_{nb} = \vec{r}_{nSF} + \vec{r}_{SFb}. \tag{12.170}$$

Hence, the time differentiation of $\vec{r}_{SFb}$ with {b} as the moving reference frame gives

$$\frac{{}^{i}\mathrm{d}}{\mathrm{d}t}\vec{r}_{SFb} = \frac{{}^{b}\mathrm{d}}{\mathrm{d}t}\vec{r}_{SFb} + \vec{\omega}_{ib} \times \vec{r}_{SFb}. \tag{12.171}$$

Assume that $\vec{\omega}_{ib} = \vec{\omega}_{nb} = \vec{\omega}_{nSF}$ such that

$$\vec{v}_{nb} = \vec{v}_{nSF} + \left(\frac{{}^{b}\mathrm{d}}{\mathrm{d}t}\vec{r}_{SFb} + \vec{\omega}_{nSF} \times \vec{r}_{SFb} \right). \tag{12.172}$$

Expressing this in {SF} gives

$$\boldsymbol{v}_{nb}^{SF} = \boldsymbol{v}_{nSF}^{SF} + \left(\frac{{}^{b}\mathrm{d}}{\mathrm{d}t}\boldsymbol{r}_{SFb}^{SF} + \boldsymbol{\omega}_{nSF}^{SF} \times \boldsymbol{r}_{SFb}^{SF} \right) \tag{12.173}$$

where $\boldsymbol{r}_{SFb}^{SF} = [s,\ e,\ 0]^\top$ and $\boldsymbol{v}_{nb}^{SF} = \boldsymbol{R}_{z,\chi_{SF}}[U,\ 0,\ 0]^\top$ is the velocity of the vehicle expressed in {SF}. From this it follows that

$$\begin{aligned} \boldsymbol{R}_{z,\chi_{SF}} \begin{bmatrix} U \\ 0 \\ 0 \end{bmatrix} &= \boldsymbol{v}_{nSF}^{SF} + \left(\frac{{}^{b}\mathrm{d}}{\mathrm{d}t}\boldsymbol{r}_{SFb}^{SF} + \boldsymbol{\omega}_{nSF}^{SF} \times \boldsymbol{r}_{SFb}^{SF} \right) \\ &= \begin{bmatrix} \dot{s}_a \\ 0 \\ 0 \end{bmatrix} + \begin{bmatrix} \dot{s} \\ \dot{e} \\ 0 \end{bmatrix} + \begin{bmatrix} 0 \\ 0 \\ \kappa \dot{s}_a \end{bmatrix} \times \begin{bmatrix} s \\ e \\ 0 \end{bmatrix}. \end{aligned} \tag{12.174}$$

Expanding this expression yields

$$U\cos(\chi_{SF}) = \dot{s} + (1 - \kappa e)\dot{s}_a \tag{12.175}$$

$$U\sin(\chi_{SF}) = \dot{e} + \kappa s \dot{s}_a \tag{12.176}$$

which proves (12.167) and (12.168). The rotation rate of the angle $\psi - \psi_{SF}$ between {n} and {SF} satisfies $\dot{\psi} - \dot{\psi}_{SF} = \kappa \dot{s}_a$ (see Figure 12.25). Since $\chi_{SF} = \psi_{SF} + \beta_c$, it follows that

$$\begin{aligned} \dot{\chi}_{SF} &= \dot{\psi} - \kappa \dot{s}_a + \dot{\beta}_c \\ &= r + \dot{\beta}_c - \kappa \dot{s}_a \end{aligned} \tag{12.177}$$

which proves (12.169).

If $\dot{s} = s = 0$, the Serret–Frenet equations (12.167)–(12.169) reduce to

$$\dot{s}_a = \frac{U\cos(\chi_{\rm SF})}{1-\kappa e} \tag{12.178}$$

$$\dot{e} = U\sin(\chi_{\rm SF}) \tag{12.179}$$

where the term $1-\kappa e$ in the denominator creates a singularity. Hence, the control law requires that the initial position of the craft must be restricted to a tube around the path with radius less than $1/\kappa^{\max}$. A discussion on the limitation of this approach is found in Breivik and Fossen (2004a). The constraint $1-\kappa e \neq 0$ is, however, removed by using (12.167) and (12.168) where an additional controller parameter s allows the origin of the {SF} frame to evolve along the path (Lapierre and Soetanto 2007).

Kinematic Controller for Path Following

The {SF} frame plays the role of the virtual target body axes and tracks the real craft. The error coordinates for control design purposes become s, e and $\tilde{\chi}_{\rm SF} = \chi_{\rm SF} - \chi_{\rm d}$ which all should be driven to zero. The desired approach angle can be chosen as a function of e according to (Micaelli and Samson 1993)

$$\chi_{\rm d} = \pi_{\rm p} - \chi_{\rm a}\frac{e^{2ke-1}}{e^{2ke+1}} \tag{12.180}$$

where $k>0$ and $0<\chi_{\rm a}<\pi/2$. An alternative approach is the proportional LOS guidance law

$$\chi_{\rm d} = \pi_{\rm p} - \tan^{-1}\left(\frac{e}{\Delta}\right) \tag{12.181}$$

where $\Delta > 0$ is the lookahead distance.

Theorem 12.1 (Kinematic Path-following Controller)
Let the yaw rate r and path tangential speed $U_{\rm d} = \dot{s}_{\rm a}$ in (12.169) be used as control variables. A feedback linearization controller (Lapierre and Soetanto 2007)

$$r = \dot{\chi}_{\rm d} - \dot{\beta}_{\rm c} + \kappa\dot{s}_{\rm a} - K_1\tilde{\chi}_{SF} \tag{12.182}$$

$$\dot{s}_{\rm a} = U\cos(\chi_{SF}) + K_2 s \tag{12.183}$$

where $\tilde{\chi}_{SF} = \chi_{SF} - \chi_{\rm d}$, renders the equilibrium point $(s, e, \tilde{\chi}_{SF}) = (0, 0, 0)$ USGES for $K_1 > 0$ and $K_2 > 0$.

Proof. Convergence and stability can be proven by noticing that the error dynamics forms a cascade of two systems. For the first system

$$\dot{\tilde{\chi}}_{\rm SF} + K_1\tilde{\chi}_{\rm SF} = 0. \tag{12.184}$$

Consequently, the angle $\chi_{\mathrm{SF}} \to \chi_{\mathrm{d}}$. For the second system in the cascade, consider the Lyapunov function candidate

$$V = \frac{1}{2}(s^2 + e^2) > 0, \qquad s \neq 0,\ e \neq 0. \tag{12.185}$$

The time derivative of V under the assumption that $\chi_{\mathrm{SF}} = \chi_{\mathrm{d}}$ is

$$\begin{aligned}\dot{V} &= s\,(U\cos(\chi_{\mathrm{d}}) - (1-\kappa e)\dot{s}_{\mathrm{a}}) + e\,(U\sin(\chi_{\mathrm{d}}) - \kappa s\dot{s}_{\mathrm{a}})\\ &= sU\cos(\chi_{\mathrm{d}}) + eU\sin(\chi_{\mathrm{d}}) - s\,(U\cos(\chi_{\mathrm{d}}) + K_2 s)\\ &= -K_2 s^2 + eU\sin(\chi_{\mathrm{d}}).\end{aligned}$$

Exploiting the fact that the desired course angle given by (12.181) satisfies

$$\sin(\chi_{\mathrm{d}}) = \frac{-e}{\sqrt{e^2 + \Delta^2}} \tag{12.186}$$

finally gives

$$\begin{aligned}\dot{V} &= -K_2 s^2 - \frac{U}{\sqrt{e^2 + \Delta^2}} e^2\\ &< 0, \qquad s \neq 0,\ e \neq 0\end{aligned} \tag{12.187}$$

for $\Delta > 0$ and $U > 0$. Since the *LFC* is positive definite and radially unbounded, while its derivative with respect to time is negative, standard Lyapunov arguments for cascaded systems proves that the equilibrium point $(s, e, \tilde{\chi}_{\mathrm{SF}}) = (0, 0, 0)$ is USGES; see Appendix A.2.3.

Remark 12.1

A differential equation for the path variable ϖ can be derived by considering the path curvature $\kappa(\varpi)$ given by

$$\kappa(\varpi) = \frac{|(x_{\mathrm{d}}^n)'(y_{\mathrm{d}}^n)'' - (y_{\mathrm{d}}^n)'(x_{\mathrm{d}}^n)''|}{(((x_{\mathrm{d}}^n)')^2 + ((y_{\mathrm{d}}^n)')^2)^{3/2}} \tag{12.188}$$

where $x_{\mathrm{d}}^n = x_{\mathrm{d}}^n(\varpi)$ and $y_{\mathrm{d}}^n = y_{\mathrm{d}}^n(\varpi)$. The arc length s_{a} satisfies

$$\mathrm{d}s_{\mathrm{a}}^2 = \mathrm{d}x^2 + \mathrm{d}y^2 \tag{12.189}$$

and by dividing by $\mathrm{d}\varpi^2$, this can be rewritten as

$$\mathrm{d}\varpi = \frac{1}{\sqrt{(x_{\mathrm{d}}^n(\varpi)')^2 + (y_{\mathrm{d}}^n(\varpi)')^2}}\mathrm{d}s_{\mathrm{a}}. \tag{12.190}$$

Hence, from (12.183) it follows that

$$\dot{\varpi} = \frac{U\cos(\chi_{SF}) + K_2 s}{\sqrt{(x_d^{n}(\varpi)')^2 + (y_d^{n}(\varpi)')^2}}. \tag{12.191}$$

12.6.4 *Case Study: Path-following Control using Serret–Frenet Coordinates*

In Encarnacao *et al.* (2000), the ocean current velocities are included in the kinematic equations of motion together with a state estimator to obtain the optimal sideslip angle during path following. This section presents a different approach where the current velocities are included in sway equation by using the crab angle β_c as state variable. Furthermore, the ocean currents are compensated for by using integral action in the kinematic controller to reduce sensitivity to time-varying currents and uncertain model parameters.

Sideslip Dynamics

For a conventional ship and underwater vehicles with no actuation in the transverse direction, the sway dynamics can be accurately described by the following maneuvering model (see Section 6.5.1)

$$(m - Y_{\dot{v}})\dot{v}_r + (m - X_{\dot{u}})u_r r - Y_v v_r - Y_r r = 0 \tag{12.192}$$

where $u_r = u - u_c$ and $v_r = v - v_c$ are the relative velocities. Assume that the relative speed U_r is slowly varying such that $\dot{U}_r \approx 0$. Hence, the relative sway acceleration can be obtained by time differentiation of (see Section 2.5.2)

$$v_r = U_r \sin(\beta) \tag{12.193}$$

which gives

$$\dot{v}_r = U_r \cos(\beta)\dot{\beta}. \tag{12.194}$$

Combining (12.192) and (12.194) gives the sideslip dynamics

$$\begin{aligned}\dot{\beta} &= \frac{1}{U_r\cos(\beta)}\dot{v}_r \\ &= \frac{1}{(m - Y_{\dot{v}})U_r\cos(\beta)}(Y_v v_r + Y_r r - (m - X_{\dot{u}})u_r r).\end{aligned} \tag{12.195}$$

Consequently,

$$\dot{\beta} = \frac{1}{(m - Y_{\dot{v}})U_r\cos(\beta)}(Y_v U_r \sin(\beta) + Y_r r - (m - X_{\dot{u}})u_r r). \tag{12.196}$$

A differential equation for the crab angle is obtained by setting the current velocities to zero. In other words

$$\dot{\beta}_c = \frac{1}{(m - Y_{\dot{v}})U\cos(\beta_c)}(Y_v U \sin(\beta_c) + Y_r r - (m - X_{\dot{u}})ur). \quad (12.197)$$

Implementation of the Kinematic Controller

When implementing the kinematic controller (12.182) and (12.183), the expression for $\dot{\beta}_c$ will depend on the model parameters. From (12.197) it follows that

$$\begin{aligned} r + \dot{\beta}_c &= r + \frac{1}{(m - Y_{\dot{v}})U\cos(\beta_c)}(Y_v U \sin(\beta_c) + Y_r r - (m - X_{\dot{u}})ur) \\ &= \left(1 - \frac{(m - X_{\dot{u}})}{(m - Y_{\dot{v}})}\frac{u}{U\cos(\beta_c)} + \frac{Y_r}{(m - Y_{\dot{v}})U\cos(\beta_c)}\right) r + \frac{Y_v}{m - Y_{\dot{v}}}\tan(\beta_c) \\ &\approx \left(1 - \frac{m - X_{\dot{u}}}{m - Y_{\dot{v}}}\right) r + \frac{Y_v}{m - Y_{\dot{v}}}\tan(\beta_c) \\ &= \dot{\chi}_d + \kappa \dot{s}_a - K_1 \chi_{SF} \end{aligned} \quad (12.198)$$

where the physical property

$$(m - Y_{\dot{v}})U\cos(\beta_c) \gg Y_r \quad (12.199)$$

has been exploited. Solving for $r = r_d$ gives the kinematic controller

$$r_d = \left(1 - \frac{m - X_{\dot{u}}}{m - Y_{\dot{v}}}\right)^{-1} \left[\dot{\chi}_d + \kappa U_d - K_1 \tilde{\chi}_{SF} - \frac{Y_v}{m - Y_{\dot{v}}}\tan(\beta_c)\right] \quad (12.200)$$

$$U_d = U\cos(\chi_{SF}) + K_2 s \quad (12.201)$$

where the desired yaw rate is denoted r_d and the desired speed $U_d = \dot{s}_a$ is the path-tangential speed. The crab angle

$$\beta_c = \sin^{-1}\left(\frac{v}{U}\right) \quad (12.202)$$

must be measured or estimated in a state observer. Alternatively, β_c can be treated as a slowly varying unknown parameter, which can be compensated for by adding an integral action. This suggests that

$$r_d = \left(1 - \frac{m - X_{\dot{u}}}{m - Y_{\dot{v}}}\right)^{-1} \left[\dot{\chi}_d + \kappa U_d - 2\lambda \tilde{\chi}_{SF} - \lambda^2 \int_0^t \tilde{\chi}_{SF}(\tau) d\tau\right] \quad (12.203)$$

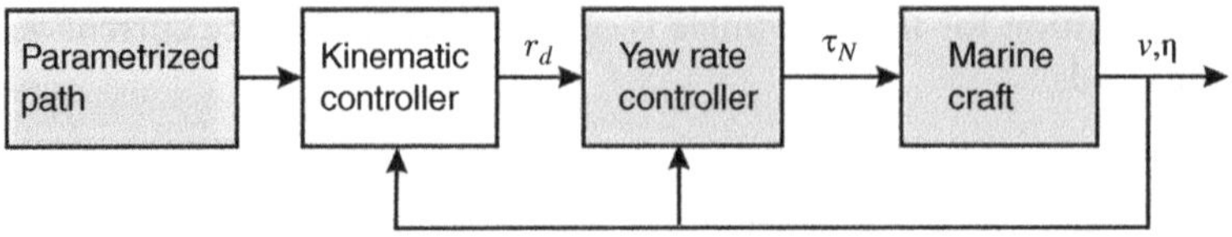

Figure 12.26 Cascaded kinematic and yaw rate controller for path-following control.

where $\lambda > 0$ is a constant parameter used to tune the bandwidth of the error system

$$\dot{\tilde{\chi}}_{\rm SF} + 2\lambda\tilde{\chi}_{\rm SF} + \lambda^2 \int_0^t \tilde{\chi}_{\rm SF}(\tau){\rm d}\tau = \frac{Y_v}{m - Y_{\dot{v}}}\tan(\beta_{\rm c}). \tag{12.204}$$

For a marine craft at constant course and constant crab angle, the integral term will balance the forcing term in the steady state such that

$$\lambda^2 \int_0^t \tilde{\chi}_{\rm SF}(\tau){\rm d}\tau = \frac{Y_v}{m - Y_{\dot{v}}}\tan(\beta_{\rm c}) \tag{12.205}$$

and $\tilde{\chi}_{\rm SF} \to 0$.

Yaw Rate Controller

Figure 12.26 shows how the kinematic controller is implemented in cascade with a yaw rate controller. Recall from Section 7.1 that the yaw dynamics of a marine craft can be modeled as

$$(I_z - N_{\dot{r}})\dot{r} - N_r r = \tau_{\rm N} \tag{12.206}$$

where $I_z - N_{\dot{r}} > 0$ and $-N_r > 0$ are constant parameters. The controller yaw moment $\tau_{\rm N}$ can easily be designed to regulate $\tilde{r} = r - r_{\rm d}$ to zero, for instance by choosing the following control law

$$\tau_{\rm N} = (I_z - N_{\dot{r}})\dot{r}_{\rm d} - N_r r_{\rm d} - K_{\rm p}(r - r_{\rm d}) \tag{12.207}$$

where $K_{\rm p} > 0$ is a design parameter and $r_{\rm d}$ is the desired yaw rate generated by the kinematic controller. This gives the closed-loop dynamics

$$(I_z - N_{\dot{r}})\dot{\tilde{r}} + (K_{\rm p} - N_{\dot{r}})\tilde{r} = 0 \tag{12.208}$$

which for

$$K_{\rm p} = (I_z - N_{\dot{r}})(N_r + \lambda) \tag{12.209}$$

reduces to an exponentially convergent system $\dot{\tilde{r}} = -\lambda\tilde{r}$ where $\lambda > 0$ is a design parameter specifying the convergence rate.

13

Model-based Navigation Systems

Conventional ship and underwater vehicle control systems are implemented with a model-based state estimator for processing of the sensor and navigation data. The quality of the raw measurements are usually monitored and handled by a signal processing unit or a program for quality check and wild-point removal. The processed measurements are transmitted to the sensor and navigation computer that uses a state estimator capable of noise filtering, prediction and reconstruction of unmeasured states. The most famous algorithm is the Kalman filter (KF) which was introduced in the 1960s (Kalman 1960, Kalman and Bucy 1961).

In a model-based KF, the craft position, velocity and attitude are states in the estimator, while linear acceleration and angular rates are generated using a mathematical model as shown in Figure 13.1. Alternatively, the model can be avoided by using accelerometers and angular rate measurements. This is referred to as an inertial navigation system (INS) and KF designs for aided INS are discussed in Chapter 14.

The drawback of the model-based approach to aided INS is model uncertainty when implemented in a KF. One obvious advantage is that additional sensors such as the inertial measurement unit (IMU) are avoided. Another benefit is that the mathematical model can be used for fault detection and isolation, as well as fault recovery. The marine craft equations of motion when implemented in a KF is in fact a predictor, which can be used to predict future motions of the craft when sensors fails or have outages for shorter periods of time.

13.1 Sensors for Marine Craft

Commercial IMUs are composed of three-axis accelerometers and attitude rate sensors. In addition some units have three-axis magnetometers, a pressure sensor and a temperature sensor. A more detailed description of the IMU sensory suit is given in Section 14.1 in conjuncture with the design of aided INS.

Handbook of Marine Craft Hydrodynamics and Motion Control, Second Edition. Thor I. Fossen.

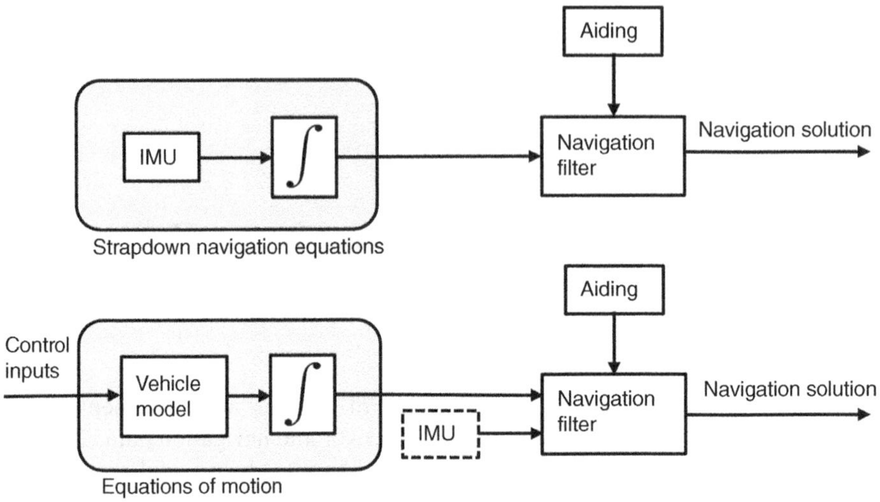

Figure 13.1 Strapdown signal-based INS (Chapter 14) versus model-based navigation filter (Chapter 13). The model-based filter uses the control inputs to compute the craft's accelerations and angular rates.

The primary measurement systems for model-based navigation filters when used onboard a surface craft are global satellite navigation systems (GNSS) and heading angle sensors. More specifically,

- GNSS position
- GNSS heading
- Magnetic compass
- Gyrocompass.

The sections below give a brief description of the sensory systems.

13.1.1 GNSS Position

GNSS position is the primary sensor for terrestrial navigation. The four commercial systems are:

- **NAVSTAR Global Positioning System (GPS):** The United States NAVSTAR GPS was started by the US Department of Defense in 1973, with the first prototype spacecraft launched in 1978 and the full constellation of 24 satellites operational in 1993 (see Hofmann-Wellenhof et al. 1994, Parkinson and Spilker 1995).
- **GLONASS:** From Russian *GLObal'naya NAvigatsionnaya Sputnikovaya Sistema*. The development of the Russian GLONASS satellite navigation system began in the Soviet Union in 1976 and the constellation was completed in 1995. After a decline in capacity in the 90s, GLONASS was restored. A full orbital constellation of 24 satellites was achieved in 2011, enabling full global coverage.

- **Galileo:** The European Union's Galileo positioning system went live in 2016. It is an independent civilian positioning system designed by European nations so they do not have to rely on GPS, GLONASS or BeiDou, which could be disabled or degraded by their operators at any time.
- **BeiDou:** Chinese for the Big Dipper or the North Star. In 2015, China launched the third generation *BeiDou* (BeiDou-3) for global navigation. BeiDou-3 consists of 35 satellites and the system has provided global services since 2020.

Integrated GNSS receivers are capable of combining signals from one or more of the GNSS systems. This also improves redundancy in marine control systems. The GNSS measurements are usually used in a motion control system that operates in the three planar degrees of freedom, namely *surge* (forward motion), *sway* (transverse motion) and *yaw* (rotation about the vertical axis, also called heading). The position of the marine craft is normally measured by differential GNSS, while the heading is measured by a gyrocompass. Additional types of sensors are usually available to ensure reliability of the positioning system, namely inertial measurement units, hydroacoustic position sensors, taut wires and laser sensors.

- **Differential and augmented GNSS:** The main idea of a *differential* GNSS system is that a fixed receiver located, for example, *on shore* with a known position is used to calculate the GNSS position errors. The position errors are then transmitted to the GNSS receiver on board the ship and used as corrections to the actual ship position. In a *differential* GNSS the horizontal positioning errors are squeezed down to less than 1 m, which is the typical accuracy of a ship positioning system today (Hofmann-Wellenhof *et al.* 1994).
- **Carrier-differential GNSS:** A GNSS receiver in lock is able to track the phase shift of the carrier and output the fractional phase measurement at each epoch. However, the overall phase measurement contains an unknown number of carrier cycles. This is called the integer ambiguity (N). This ambiguity exists because the receiver merely begins counting carrier cycles from the time a satellite signal is placed in an active track. For GPS, the precision of the phase measurement is about 0.01 cycles ($\approx 0.01 \times 19$ cm $= 1.9$ mm), and if N is determined, it allows for highly accurate position measurements. Ambiguity resolution is a very active research area, and there are several receivers known as *real-time kinematic* (RTK) receivers on the market today that utilize carrier measurements to achieve position accuracy in the order of a few centimeters. These position measurements are, however, not as robust as GNSS and differential GNSS.

13.1.2 GNSS Heading

The GNSS system can be used to determine the heading angle, even though it was not designed for this purpose. A "GNSS compass" uses a pair of antennas separated by 50 cm or more to detect the phase difference in the carrier signal from a particular GNSS satellite. Given the positions of the satellite, the position of the antenna, and the phase difference, the orientation of the two antennas can be computed. The accuracy can be further improved by using three antennas in a triangle to get three separate readings with respect to each satellite. It is also beneficial to increase the distance between the antennas. The GNSS heading solution is not subject to magnetic declination but it will be sensitive to ionospheric disturbances and multipath effects.

13.1.3 Magnetic Compass

A compass is the primary device for direction-finding on the surface of the Earth. Compasses may operate on magnetic or gyroscopic principles or by determining the direction of the Sun or a star. The discussions will be restricted to magnetic and gyroscopic compasses, since these are the primary devices onboard commercial ships and rigs.

The magnetic compass is an old Chinese invention, which probably dates back to 100 CE. Knowledge of the compass as a directional device came to western Europe sometime in the 12th century and it is today a standard unit in all commercial and navy ships.

A magnetic compass is in fact an extremely simple device (as opposed to a gyroscopic compass). It consists of a small, lightweight magnet balanced on a nearly frictionless pivot point. The magnet is generally called a needle. The magnetic field inside the Earth has its south end at the North Pole and opposite. Hence, the north end of the compass needle points towards the North Pole (opposite magnets attract). The magnetic field of the Earth is, however, not perfectly aligned along the Earth's rotational axis. It is skewed slightly off center. This skew or bias is called the *declination* and it must be compensated for (see Section 14.1.3). It is therefore common to indicate what the declination is on navigational maps. Sensitivity to magnetic variations and declination cause problems in ship navigation. These problems were overcome after the introduction of the gyroscopic compass.

13.1.4 Gyrocompass

The first recorded construction of the gyroscope is usually credited to C. A. Bohnenberger in 1810 while the first electrically driven gyroscope was demonstrated in 1890 by G. M. Hopkins (see Allensworth 1999, Bennet 1979). A gyroscope is a disk mounted on a base in such a way that the disk can spin freely on its x and y axes; that is the disk will remain in a fixed position in whatever directions the base is moved. A properly mounted gyroscope will always turn to match its plane of rotation with that of the Earth, just as a magnetic compass turns to match the Earth's magnetic field.

The large variations in the magnetic character of ships caused by electrical machinery and weapon systems made the construction of accurate declination or deviation tables for the magnetic compass very difficult. In parallel works, Dr H. Anschütz of Germany and Elmer Sperry of the USA worked on a practical application of Hopkins' gyroscope. In 1908 Anschütz patented the first north-seeking gyrocompass, while Elmer Sperry was granted a patent for his ballistic compass, which includes vertical damping, three years later.

In 1910, when the Anschütz gyro compass appeared, the problem with magnetic variations in ship navigation was eliminated. However, this compass proved to be quite unsatisfactory during rolling of the ship, since it produced an "intercardinal rolling error".

Therefore in 1912 Anschütz redesigned the compass to overcome this defect. One year later, the Sperry compass entered the market and it became a serious competitor to the Anschütz.

Today gyroscopic compasses are produced by a large number of companies for both commercial and navy ships. They are widely used for navigation because they have significant advantages over magnetic compasses. In particular they are unaffected by ferromagnetic materials, such as in a ship's steel hull, which distort the magnetic field. Another important aspect is that they are not affected by electromagnetic fields, which are generated by rotating machinery and engines moving electric charges. Unfortunately, a gyrocompass is quite expensive, which limits their use to large ships and safety-critical vehicle systems. Smaller vehicles usually navigate by using magnetic compasses, course over ground or GNSS heading.

13.2 Wave Filtering

Wave filtering is one of the most important issues to take into account when designing ship control systems (Fossen 1994, Fossen and Perez 2009). Environmental forces due to waves, wind and ocean currents are considered disturbances to the motion control system. These forces, which can be described in stochastic terms, are conceptually separated into low-frequency (LF) and wave-frequency (WF) components; see Figure 13.2.

Waves produce a pressure change on the hull surface, which in turn induces forces. These pressure-induced forces have an oscillatory component that depends linearly on the wave elevation. Hence, these forces have the same frequency as that of the waves and are therefore referred to as *wave-frequency forces*. Wave forces also have a component that depends nonlinearly on the wave elevation (Newman 1977, Faltinsen 1990). Nonlinear wave forces are due to the quadratic dependence of the pressure on the fluid-particle velocity induced by the passing of the waves. If, for example, two sinusoidal incident waves have different frequencies, then their quadratic relationship gives pressure forces with frequencies at both the sum and difference of the incident wave frequencies. They also contribute with zero-frequency or mean forces. Hence, the nonlinear wave forces have a wider frequency range than the incident waves. The mean wave forces cause the craft to drift. The forces with a frequency content at the difference of the wave frequencies can have LF content, which can lead to resonance in the horizontal motion of marine craft with mooring lines or under positioning control. The high-frequency wave-pressure-induced forces, which have a frequency content at the sum of the wave frequencies, are normally too high to be considered in ship-motion control, but these forces can contribute to structural vibration in the hull. For further details about wave loads and their effects on ship motion see Newman (1977) and Faltinsen (1990).

In low-to-medium sea states, the frequency of oscillations of the linear wave forces do not normally affect the operational performance of the craft. Hence, controlling only LF motion avoids correcting the motion for every single wave, which can result in unacceptable operational conditions for the propulsion system due to power consumption and potential wear of the actuators. Operations that require the control of only the LF motion include dynamic positioning, heading autopilots and thruster-assisted position mooring. Dynamic positioning refers to the use of the propulsion system to regulate the horizontal

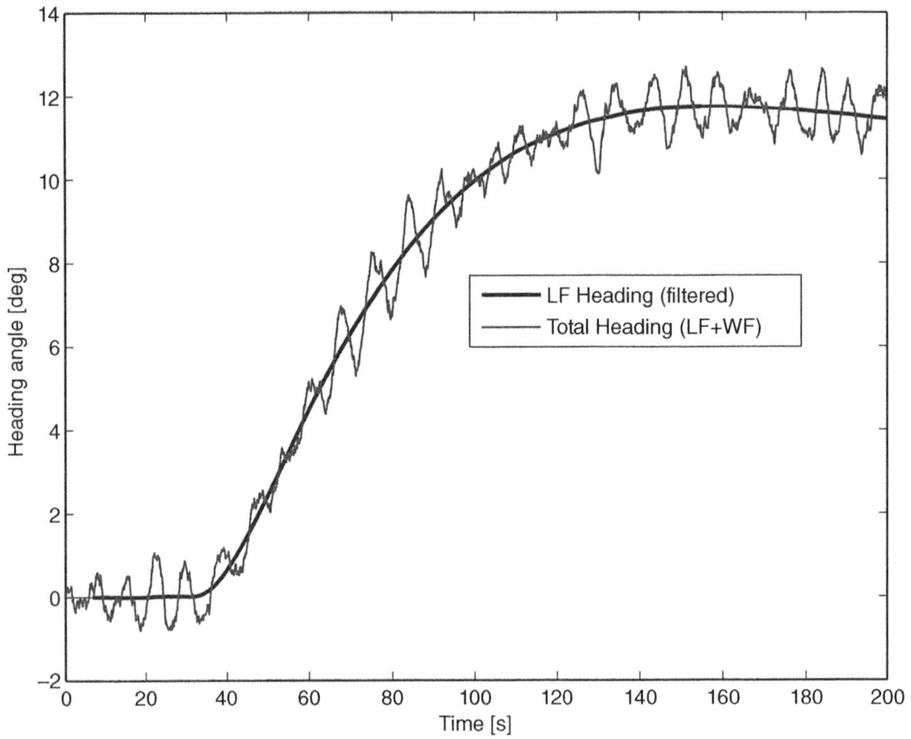

Figure 13.2 Separation of the total motion of a marine craft into LF and WF motion components.

position and heading of the craft. In thruster-assisted position mooring, the propulsion system is used to reduce the mean loading on the mooring lines. Additional operations that require the control of only the LF motion include slow maneuvers that arise, for example, from following underwater remotely operated vehicles. Operations that require the control of only the WF motions include heave compensation for deploying loads on the sea floor (Perez and Steinmann 2007) as well as ride control of passenger vessels, where reducing roll and pitch motion helps avoid motion sickness (Perez 2005).

It is important that only the slowly-varying forces are counteracted by the steering and propulsion systems while the oscillatory motion due to the waves (first-order wave-induced forces) should be prevented from entering the feedback loop. This is done by using *wave filtering* techniques (Balchen *et al.* 1976). A wave filter is usually a model-based state estimator that separates the position and heading measurements into LF and WF position and heading signals; see Figure 13.2. In the forthcoming we will use the following definition for wave filtering.

Definition 13.1 (Wave Filtering)
Wave filtering can be defined as the reconstruction of the LF motion components from wave-induced noisy measurements of position, heading and in some cases velocity and acceleration by means of a state estimator or a filter.

Remark 13.1
If a state estimator such as the Kalman filter is applied, estimates of the WF motion components (first-order wave-induced forces) can also be computed.

Wave filtering is crucial in ship motion control systems since the WF part of the motion should not be compensated for by the control system unless wave-induced vibration damping is an issue. This is the case for high-speed craft. If the WF part of the motion enters the feedback loop, this will cause unnecessary use of the actuators (thrust modulation) and reduce the tracking performance, which, again, results in increased fuel consumption.

In this chapter, model-based wave filtering and observer design using linear wave response models are discussed. This is one of the most important features of a high-precision ship control system. The best commercial autopilot and DP systems all have some kind of wave filtering in order to reduce wear and tear on the steering machine, as well as thrust modulation.

13.2.1 Low-pass Filtering

Low-pass and notch filters can be used to reduce motions induced by ocean waves in the feedback loop. This assumes that the filters can be implemented in series, as shown in Figure 13.3. For wave periods in the interval $5 \text{ s} < T_0 < 20$ s, the dominating wave frequency (modal frequency) f_0 of a wave spectrum will be in the range (see Section 10.2)

$$0.05 \text{ Hz} < f_0 < 0.2 \text{ Hz}. \tag{13.1}$$

The circular frequency $\omega_0 = 2\pi f_0$ corresponding to periods $T_0 > 5$ s is

$$\omega_0 < 1.3 \text{ rad s}^{-1}. \tag{13.2}$$

Waves within the frequency band (13.1) can be accurately described by first- and second-order wave theory. The first-order wave forces produce large *oscillations* about

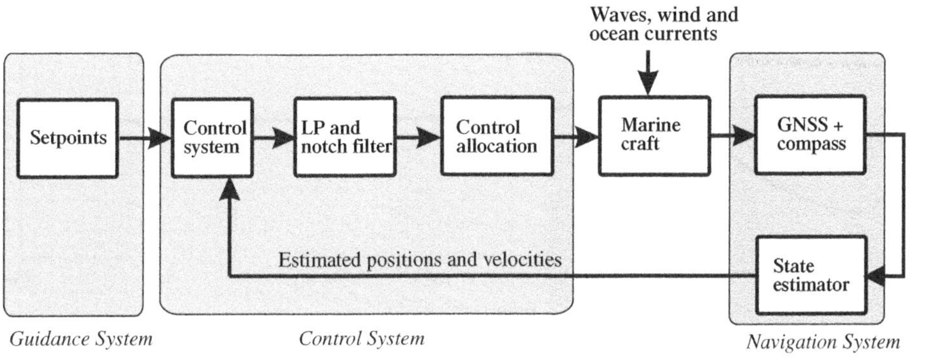

Figure 13.3 LP and notch filters in series with the control system.

a *mean* wave force, which can be computed from second-order wave theory (see Figure 13.2). The mean wave (drift) force is slowly varying and is usually compensated for by using *integral action* in the control law, while *wave filtering* must be performed to remove first-order components from the feedback loop.

For instance, first-order wave forces around $f_0 = 0.1$ Hz can be close to or outside the control bandwidth of the marine craft depending of the craft considered. For a large oil tanker, the crossover frequency can be as low as 0.01 rad s^{-1}, as shown in Figure 13.4, while smaller vessels such as cargo ships and the Mariner class vessel are close to 0.05 rad s^{-1}.

A feedback control system will typically move the bandwidth of these vessels up to 0.1 rad s^{-1}, which still is below the wave spectrum shown in Figure 13.4. However, the wave forces will be inside the bandwidth of the servos and actuators of the craft. Hence, the wave forces must be filtered out before feedback is applied in order to avoid unnecessary control action. In other words, we do not want the rudder and thruster actuators of the ship to compensate for the first-order WF motion. This is usually referred to as *wave filtering*.

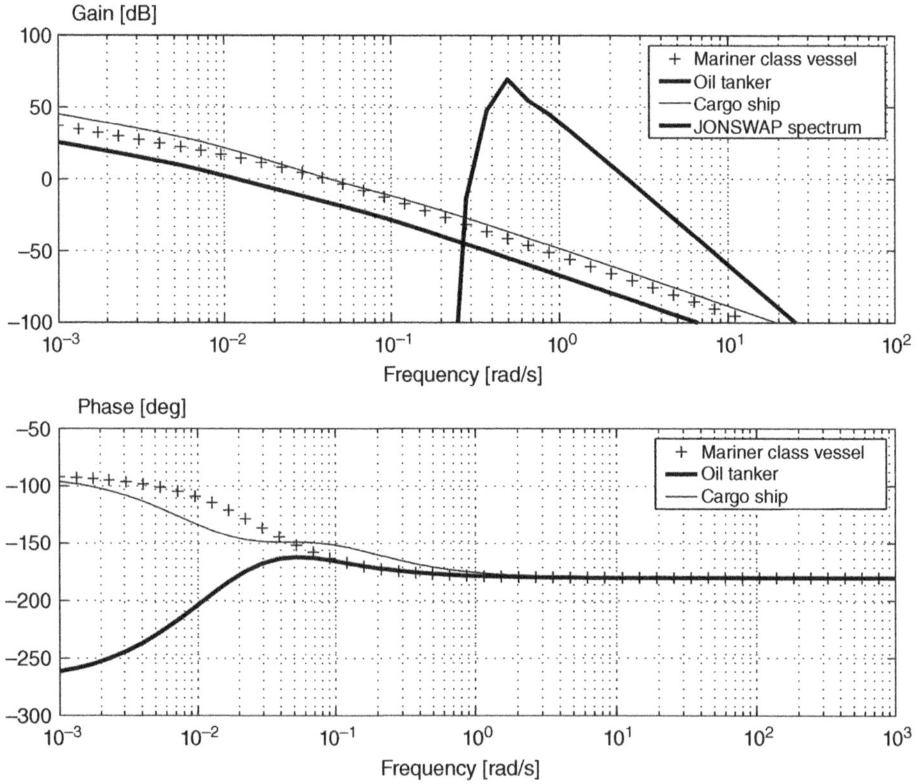

Figure 13.4 Bode plots showing $\psi/\delta(s)$ for three different vessels and the JONSWAP wave spectrum for $\omega_0 = 0.5$ rad s^{-1} and $H_s = 5$ m.

For sea states where the WF motion is much higher than the bandwidth ω_b of the controller, a low-pass filter can be used to filter out the WF motions if ω_b satisfies

$$\omega_b \ll \omega_e \tag{13.3}$$

where

$$\omega_e = \left| \omega_0 - \omega_0^2 \frac{U}{g} \cos(\beta) \right| \tag{13.4}$$

is the frequency of encounter (see Figure 10.14). This is typically the case for large vessels such as oil tankers. In the autopilot case, the design objective can be understood by considering the measurement equation

$$y(s) = \underbrace{h_{\text{ship}}(s)\delta(s)}_{\psi(s)} + \underbrace{h_{\text{wave}}(s)w(s)}_{\psi_w(s)} \tag{13.5}$$

where $y(s)$ is the compass measurement, $w(s)$ is zero-mean Gaussian white noise and $\delta(s)$ is the rudder input. The signal $\psi(s)$ represents the LF motion, while $\psi_w(s)$ is the WF motion. Linear theory suggests that, see (10.107) and (7.20),

$$h_{\text{wave}}(s) = \frac{K_w s}{s^2 + 2\lambda\omega_0 s + \omega_0^2} \tag{13.6}$$

$$h_{\text{ship}}(s) = \frac{K(T_3 s + 1)}{s(T_1 s + 1)(T_2 s + 1)}. \tag{13.7}$$

Feedback directly from y will therefore include the WF motion. For a large tanker, proper *wave filtering* can be obtained by using a low-pass filter to produce an estimate of $\psi(s)$ such that

$$\hat{\psi}(s) = h_{\text{lp}}(s)y(s). \tag{13.8}$$

Consequently, the feedback control law δ should be a function of $\hat{\psi}$ and not y in order to avoid first-order wave-induced rudder motions. For instance, a first-order low-pass filter with time constant T_f can be designed according to

$$h_{\text{lp}}(s) = \frac{1}{T_f s + 1}, \qquad \omega_b < \frac{1}{T_f} < \omega_e \tag{13.9}$$

This filter will suppress forces over the frequency $1/T_f$. This criterion is obviously hard to satisfy for smaller craft since ω_b can be close to or even larger than ω_e.

Higher-order low-pass filters can be designed by using a *Butterworth filter*, for instance. The nth-order Butterworth filter

$$h_{\text{lp}}(s) = \frac{1}{p(s)} \tag{13.10}$$

is found by solving the Butterworth polynomial

$$p(s)p(-s) = 1 + (s/j\omega_f)^{2n} \tag{13.11}$$

for $p(s)$. Here n denotes the order of the filter while ω_f is the cutoff frequency. For $n = 1, \dots, 4$ the solutions are

$$(n = 1) \quad h_{lp}(s) = \frac{1}{\frac{s}{\omega_f} + 1}$$

$$(n = 2) \quad h_{lp}(s) = \frac{\omega_f^2}{s^2 + 2\zeta\omega_f s + \omega_f^2}; \qquad \zeta = \sin(45^\circ)$$

$$(n = 3) \quad h_{lp}(s) = \frac{\omega_f^2}{s^2 + 2\zeta\omega_f s + \omega_f^2} \cdot \frac{1}{\frac{s}{\omega_f} + 1}; \qquad \zeta = \sin(30^\circ)$$

$$(n = 4) \quad h_{lp}(s) = \prod_{i=1}^{2} \frac{\omega_f^2}{s^2 + 2\zeta_i\omega_f s + \omega_f^2}; \qquad \zeta_1 = \sin(22.5^\circ), \quad \zeta_2 = \sin(67.5^\circ).$$

A higher-order low-pass filter implies better disturbance suppression of the price of additional phase lags (see Figure 13.5).

13.2.2 Cascaded Low-pass and Notch Filtering

For smaller craft the bandwidth of the controller ω_b can be close to or within the range $\omega^{min} < \omega_e < \omega^{max}$ of the wave spectrum. This problem can be handled by using a low-pass filter in cascade with a notch filter

$$\hat{\psi}(s) = h_{lp}(s)h_n(s)y(s) \tag{13.12}$$

where

$$h_n(s) = \frac{s^2 + 2\zeta\omega_n s + \omega_n^2}{(s + \omega_n)^2}. \tag{13.13}$$

Here $0 < \zeta < 1$ is a design parameter used to control the magnitude of the notch while the notch frequency ω_n should be chosen equal to the peak frequency ω_0 of the spectrum for a marine craft at zero speed (dynamic positioning). The low-pass and notch filters are shown in Figure 13.6 for different values of ζ.

For a marine craft moving at forward speed U the optimal notch frequency will be

$$\omega_n = \omega_e. \tag{13.14}$$

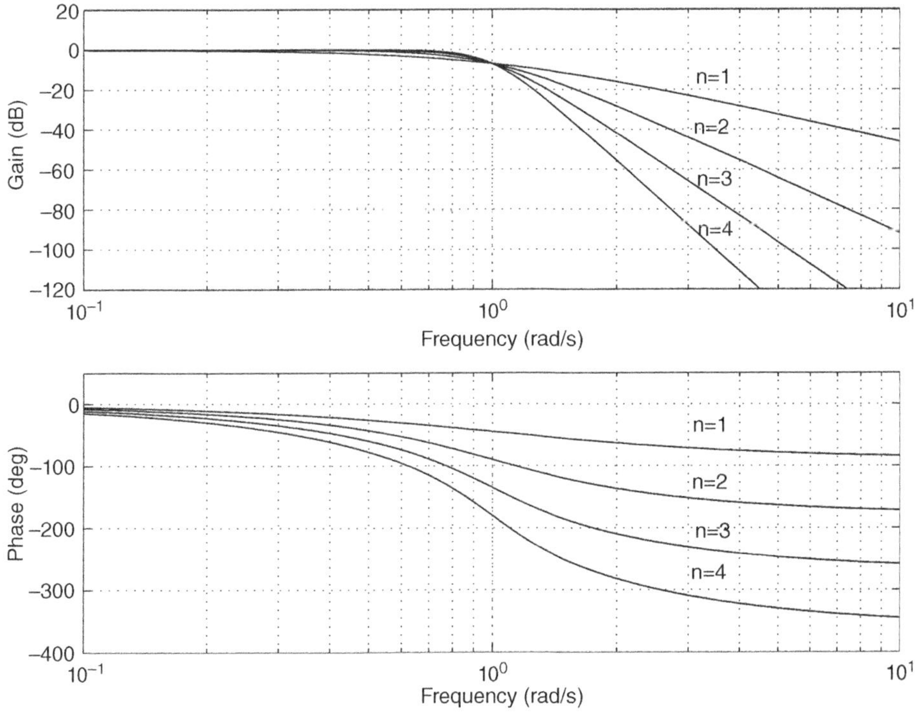

Figure 13.5 Bode plot showing the Butterworth filter for $n = 1, \ldots, 4$ with cutoff frequency $\omega_f = 1.0$ rad s^{-1}.

This frequency can be computed online by using a frequency tracker, adaptive filtering techniques or the fast Fourier transform (FFT, see Section 13.2.3). Since the estimate of ω_n can be poor and one single-notch filter only covers a small part of the actual frequency range of the wave spectrum, an alternative filter structure consisting of three cascaded notch filters with fixed center frequencies has been suggested; see Grimble and Johnson (1989)

$$h_n(s) = \prod_{i=1}^{3} \frac{s^2 + 2\zeta\omega_i s + \omega_i^2}{(s + \omega_i)^2}. \tag{13.15}$$

The center frequencies of the notch filters are typically chosen as $\omega_1 = 0.4$ rad s^{-1}, $\omega_2 = 0.63$ rad s^{-1} and $\omega_3 = 1.0$ rad s^{-1}. This is shown in Figure 13.6. Note that additional phase lag is introduced when using a cascaded notch filter.

13.2.3 Wave-frequency Estimation

Several techniques can be used for estimation of the wave encounter frequency (see Figure 10.14) for a marine craft moving at speed $U > 0$ with wave encounter angle β.

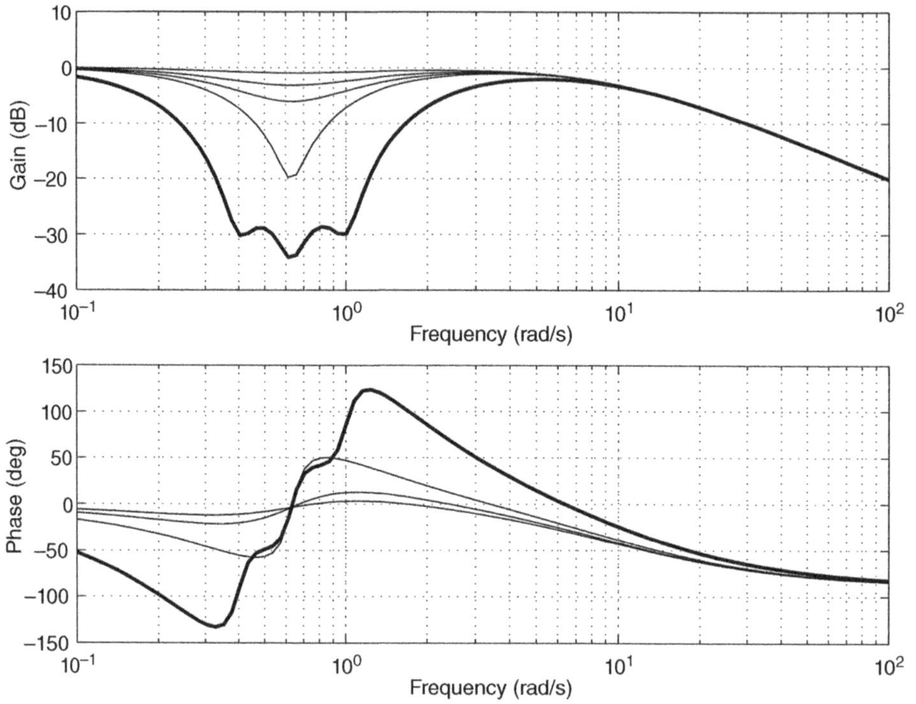

Figure 13.6 Bode plot showing the notch filter for relative damping ratios $\zeta \in \{0.1, 0.5, 0.9\}$ and $\omega_0 = 0.63$ rad s^{-1} in cascade with a low-pass filter with time constant $T_f = 0.1$ s. The thick line represents three cascaded notch filters at $\omega_1 = 0.4$ rad s^{-1}, $\omega_2 = 0.63$ rad s^{-1} and $\omega_3 = 1.0$ rad s^{-1}.

If U and β are constant, the FFT can be applied to a moving window of data. For the time-varying case, it is necessary to construct a state estimator for ω_e such that changes can be tracked during turning.

FFT for Computation of the Heave-response Spectrum

The classical method computes the heave-response spectrum from FFT spectral analysis (Enshaei and Brimingham 2012). Unfortunately, creating a FFT frequency spectrum takes time and consequently it results in back-dated information when estimating the time-varying wave encounter frequency. This is due to the moving window necessary for applying the FFT algorithm. However, it is possible to estimate ω_e if U and β in (10.85) are constant for a period of time (typically 30 min). The best results are obtained by using the heave response, which can be logged by using an accelerometer mounted onboard the craft. Good results are also obtained for pitch response data. The following example illustrates how response data can be processed in Matlab to obtain en estimate of ω_e.

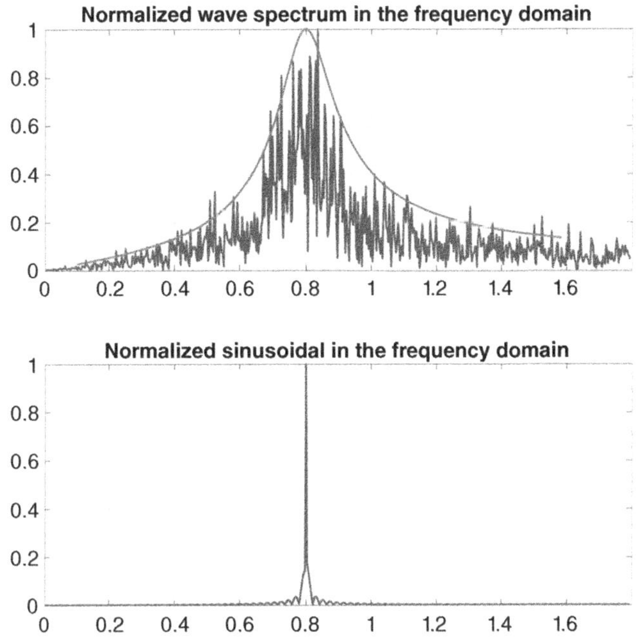

Figure 13.7 The FFT applied to a moving window of data generated by using a wave spectrum and a sinusoidal wave, respectively. The wave encounter frequency $\omega_e = 0.8$ rad s^{-1} is observed as the peak frequency of both data sets.

MATLAB

The Matlab example file `ExFFT.m` in the MSS toolbox shows how the wave encounter frequency can be estimated from response data using FFT. The heave acceleration time series are generated using the wave spectrum

$$a_z = \frac{10\,s}{s^2 + 2\lambda\omega_e s + \omega_e^2}\,w \tag{13.16}$$

where w is Gaussian white noise. This is compared to the data of a regular wave $a_z = A\cos(\omega_e t)$. The unknown peak frequency is chosen as $\omega_e = 0.8$ rad s^{-1} and the peaks are easily observed in the FFT plots of Figure 13.7.

```
we = 0.8;                          % peak frequency [rad/s]

fs = 500;                          % IMU sampling frequency [Hz]
h = 1/fs;                          % sampling time [s]
N = 30 * 60 * fs;                  % 30 minutes data
t = (0:N-1) * h;                   % time vector [s]
```

```
% Wave spectrum data and sinusoidal (regular) waves
Kw = 10; lambda = 0.1;                                % wave spectrum
sys = tf([Kw 0],[1 2*lambda*we we*we]);
[mag,phase,wout] = bode(sys,logspace(-1,0.2,1000));
mag = reshape(mag(1,:),1,1000);
x1 = lsim(sys,randn(1,length(t)),t,0,'zoh')';   % time responses
x2 = cos(we * t);
X = [x1; x2];

% Fast Fourier transform (FFT)
n = 2^nextpow2(N);        % pad the input with trailing zeros
Y = fft(X,n,2);           % compute the FFT
P2 = abs(Y/N);            % double-sided spectrum of each signal
P1 = P2(:,1:n/2+1);       % signle-sided spectrum of each signal
P1(:,2:end-1) = 2*P1(:,2:end-1);

% Plots
f = 0:(fs/n):(fs/2-fs/n); w = 2*pi*f;   % frequency vectors
M = 600;                                % no of samples to plot
subplot(2,1,1);
plot(w(1:M),P1(1,1:M)/max(P1(1,1:M)),wout,mag/max(mag));
title(['Normalized wave spectrum in the frequency domain']);
subplot(2,1,2);
plot(w(1:M),P1(2,1:M));
title(['Normalized sinusoidal in the frequency domain']);
```

Nonlinear Observer for Online Estimation of the Wave Encounter Frequency

The signal-based approach is an observer capable of estimating the frequency of a sinusoid with unknown frequency, amplitude and phase (Belleter *et al.* 2015). Hence, the measured heave acceleration or pitch angle can be used to estimate the dominating frequency of the waves. This method is motivated by the algorithm of Aranovskiy *et al.* (2007), which is modified to include an adaptive gain-switching mechanism.

The key assumption when designing the observer is that the peak frequency of the wave spectrum can be estimated by considering a regular wave

$$y = A\sin(\omega_e t + \epsilon) \tag{13.17}$$

with A the unknown amplitude, ω_e the unknown frequency and ϵ the unknown phase, and reconstruct the frequency ω_e from y. This is a good assumption for higher sea states where the wave spectrum clearly has a dominating peak.

Assume that y is the measured response of the craft. The amplitude can be estimated by low-pass filtering the squared signal

$$y^2 = \frac{A^2}{2}(1 - \cos(2\omega_e t + 2\epsilon)) \tag{13.18}$$

to obtain the amplitude $A^2/2$ of the squared signal y^2. For instance,

$$\chi = \frac{1}{Ts+1} y^2 \tag{13.19}$$

$$\hat{A} = \sqrt{2\chi} \tag{13.20}$$

where $T > 0$ is the filter time constant. The wave encounter frequency estimator is chosen as

$$\dot{x}_1 = x_2 \tag{13.21}$$

$$\dot{x}_2 = \omega_f^2 (y - x_1) - 2\omega_f x_2 \tag{13.22}$$

$$\dot{\hat{\theta}}_w = k_f x_1 (\dot{x}_2 - \hat{\theta}_w x_1) \tag{13.23}$$

where $\hat{\theta}_w := -\hat{\omega}_e^2$ is the estimated parameter. The adaption gain k_f is made time varying to improve convergence. The algorithms is

$$T_f \dot{k}_f + k_f = k(\hat{A}) \tag{13.24}$$

where $T_f > 0$ is the filter time constant and

$$k(\hat{A}) = \begin{cases} k_{init} & \text{if } t \leq t_{init} \\ k_{min} & \text{if } t > t_{init} \text{ and } \hat{A} > A_0 \\ k_{max} & \text{if } t > t_{init} \text{ and } \hat{A} \leq A_0 \end{cases} . \tag{13.25}$$

Here $k_{init} \geq k_{min} > 0$ is the initial gain used to increase the convergence rate at start up. During normal operation the gain is switched between the positive gains k_{min} and k_{max}. Moreover, the gain $k(\hat{A})$ will switch to the high value if the amplitude $\hat{A} \leq A_0$ and to the low gain when $\hat{A} > A_0$. This is illustrated in Figure 13.8.

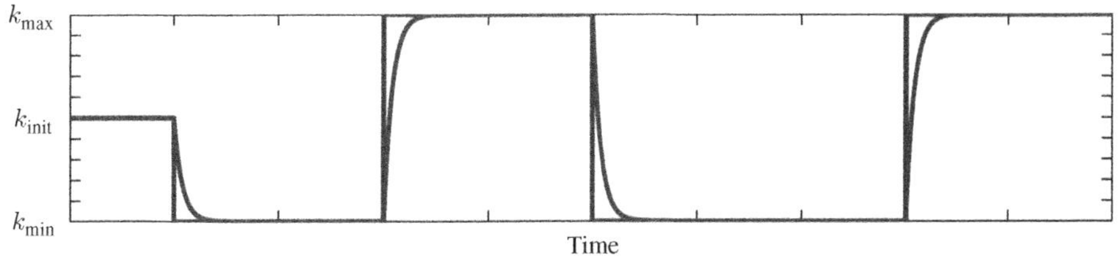

Figure 13.8 Illustration of low-pass filtered gain changes for step inputs $k(\hat{A})$.

The choice of the cut-off frequency ω_f should be made based on desired performance, that is convergence rate and steady-state error as well as noise filtering capabilities. The performance of the estimator (13.21)–(13.23) has been tested on experimental and full-scale data using heave and pitch data from a 281 m long container ship. The towing tank tests with a 1:45 scale model is shown in Figure 13.9, which is adopted from Belleter *et al.* (2015).

Global Exponential Stability Result

The regular wave (13.17) has a positive amplitude $0 < A_{\min} \leq A$ for all $t \geq 0$ and frequency $\omega_e > 0$. The solution of

$$\frac{x_1}{y}(s) = \frac{\omega_f^2}{(s+\omega_f)^2} \tag{13.26}$$

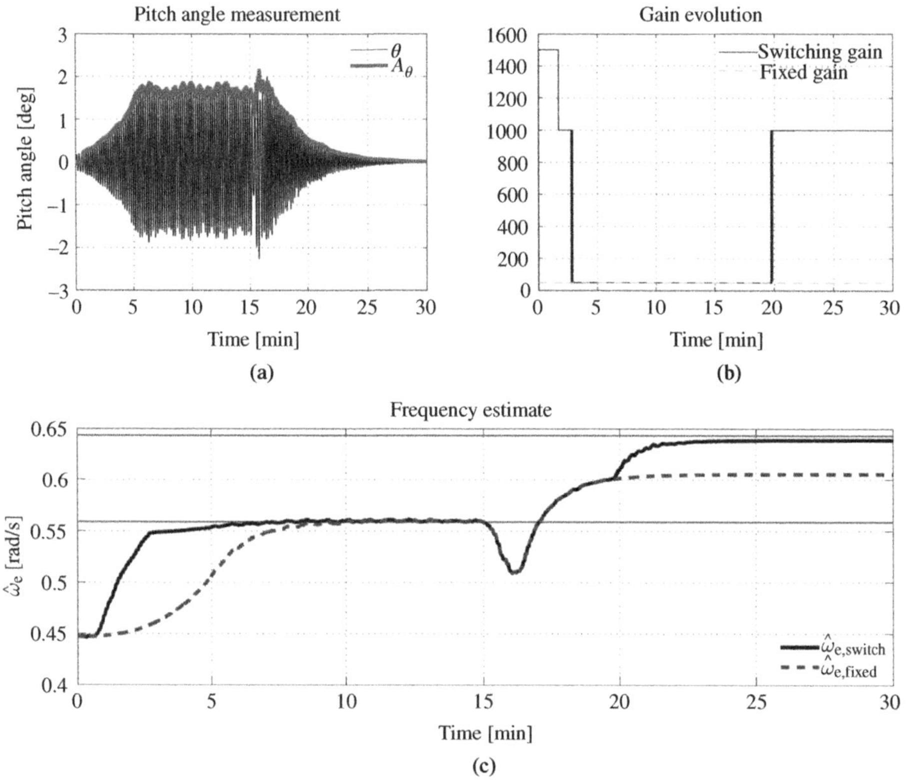

Figure 13.9 Comparison of the frequency estimator with and without the gain switching mechanism. The convergence rate is significantly improved when using $k_f(\hat{A})$. Source: Adopted from Belleter, D. J., R. Galeazzi and T. I. Fossen (2015). Experimental Verification of a Globally Exponentially Stable Nonlinear Wave Encounter Frequency Estimator. Ocean Engineering 97(15), 48–56.

for a sinusoidal input (13.17) and frequencies $\omega_e < \omega_f$, is $x_1 = A\sin(\omega_e t + \epsilon_1)$ where ϵ_1 is the phase. The signal x_1 is persistently exciting (PE) since there exist a positive μ and T such that

$$\mu \leq \int_t^{t+T} x_1^2(\tau)\mathrm{d}\tau, \quad \forall t \geq 0. \tag{13.27}$$

The PE condition guarantees that the equilibrium point $\tilde{\theta}_w = 0$ of the estimation error dynamics

$$\dot{\tilde{\theta}}_w = -k_f x_1^2 \tilde{\theta}_w \tag{13.28}$$

is GES for $\tilde{\theta}_w = \theta_w - \hat{\theta}_w$ under the assumption that $\theta_w = -\omega_e^2$ is constant (Belleter *et al.* 2015, Theorem 1).

13.3 Fixed-gain Observer Design

The simplest state estimator is designed as a fixed-gain observer where the ultimate goal of the observer is to reconstruct the unmeasured state vector $\hat{\boldsymbol{x}}$ from the measurements $\boldsymbol{u}$ and $\boldsymbol{y}$ of a dynamical system (see Figure 13.10). In order for this to succeed, the system must be *observable*.

13.3.1 Observability

Observability can be understood as a measure for how well internal states $\boldsymbol{x}$ of a system can be inferred by knowledge of its external outputs $\boldsymbol{u}$ and $\boldsymbol{y}$ (Gelb *et al.* 1988, Chen 2012). The *observability* and *controllability* of a system are mathematical duals. More specifically, a system is said to be observable if, for any possible sequence of state and control vectors, the current state can be determined in finite time using only the outputs. In other words, this means that from the outputs of the system it is possible to determine the behavior of the entire system. If a system is not observable, this means that the current values of some of its states cannot be determined through output sensors. This implies that their value is unknown to the controller and, consequently, that it will be unable to fulfill the control specifications referred to these outputs.

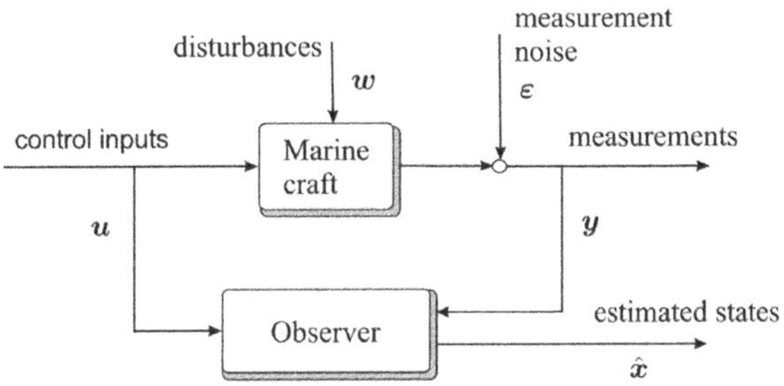

Figure 13.10 Block diagram showing the system model and the observer signal flow.

For linear time-invariant (LTI) systems, a convenient observability test that guarantees that the state $\boldsymbol{x}$ can be reconstructed from the output $\boldsymbol{y}$ and the input $\boldsymbol{u}$ is given by the following definition.

Definition 13.2 (Observability in LTI Systems)
Consider the linear time-invariant system

$$\dot{\boldsymbol{x}} = \boldsymbol{A}\boldsymbol{x} + \boldsymbol{B}\boldsymbol{u} \tag{13.29}$$

$$\boldsymbol{y} = \boldsymbol{C}\boldsymbol{x} + \boldsymbol{D}\boldsymbol{u}. \tag{13.30}$$

The pair $(\boldsymbol{A}, \boldsymbol{C})$ is observable if and only if the observability matrix

$$\mathcal{O} = [\boldsymbol{C}^{\top} \mid \boldsymbol{A}^{\top}\boldsymbol{C}^{\top} \mid \cdots \mid (\boldsymbol{A}^{\top})^{n-1}\boldsymbol{C}^{\top}] \tag{13.31}$$

has full column rank or (at least) a left inverse exists.

Observability in linear time-varying (LTV) systems is guaranteed by the following definition.

Definition 13.3 (Observability in LTV Systems)
Consider the linear time-varying system

$$\dot{\boldsymbol{x}} = \boldsymbol{A}(t)\boldsymbol{x} + \boldsymbol{B}(t)\boldsymbol{u} \tag{13.32}$$

$$\boldsymbol{y} = \boldsymbol{C}(t)\boldsymbol{x} + \boldsymbol{D}(t)\boldsymbol{u}. \tag{13.33}$$

The pair $(\boldsymbol{A}(t), \boldsymbol{C}(t))$ is observable at time t_0 if and only if there exists a finite $t_1 > t_0$ such that the $n \times n$ the observability Gramian

$$\boldsymbol{W}(t_0, t_1) = \int_0^{\infty} \boldsymbol{\Phi}^{\top}(t_1, \tau)\boldsymbol{C}^{\top}(\tau)\boldsymbol{C}(\tau)\boldsymbol{\Phi}(t_1, \tau)\mathrm{d}\tau \tag{13.34}$$

is non-singular when $\boldsymbol{\Phi}(t, \tau)$ is the state transition matrix of the system $\dot{\boldsymbol{x}} = \boldsymbol{A}(t)\boldsymbol{x}$.

For discrete-time systems, we can use the following definition.

Definition 13.4 (Observability in Discrete-time Systems)
Consider the linear discrete-time system

$$\boldsymbol{x}[k+1] = \boldsymbol{A}_{\mathrm{d}}\boldsymbol{x}[k] + \boldsymbol{B}_{\mathrm{d}}\boldsymbol{u}[k] \tag{13.35}$$

$$\boldsymbol{y}[k] = \boldsymbol{C}_{\mathrm{d}}\boldsymbol{x}[k] + \boldsymbol{D}_{\mathrm{d}}\boldsymbol{u}[k]. \tag{13.36}$$

The pair $(\boldsymbol{A}_\mathrm{d}, \boldsymbol{C}_\mathrm{d})$ *is observable if and only if the unique solution of*

$$\boldsymbol{A}_\mathrm{d}^\top \boldsymbol{W}_\mathrm{d} \boldsymbol{A}_\mathrm{d} - \boldsymbol{W}_\mathrm{d} = -\boldsymbol{C}_\mathrm{d}^\top \boldsymbol{C}_\mathrm{d} \tag{13.37}$$

is positive definite and given by the discrete observability Gramian

$$\boldsymbol{W}_\mathrm{d} = \sum_{m=0}^{\infty} (\boldsymbol{A}_\mathrm{d}^\top)^m \boldsymbol{C}_\mathrm{d}^\top \boldsymbol{C}_\mathrm{d} \boldsymbol{A}_\mathrm{d}^m. \tag{13.38}$$

13.3.2 Luenberger Observer

A state observer can be used to provide an estimate of the internal state $\boldsymbol{x}$ of a system from the measurements of the input $\boldsymbol{u}$ and the output $\boldsymbol{y}$. Observers, such as the *Luenberger observer*, are derived from deterministic models, which neglects process and measurements noise. However, an observer will still work when adding Gaussian white noise to the system if the gains are tuned properly. Consider the LTI system

$$\dot{\boldsymbol{x}} = \boldsymbol{A}\boldsymbol{x} + \boldsymbol{B}\boldsymbol{u} \tag{13.39}$$

$$\boldsymbol{y} = \boldsymbol{C}\boldsymbol{x} + \boldsymbol{D}\boldsymbol{u}. \tag{13.40}$$

An observer copying the dynamics (13.39) and (13.40) is

$$\dot{\hat{\boldsymbol{x}}} = \boldsymbol{A}\hat{\boldsymbol{x}} + \boldsymbol{B}\boldsymbol{u} + \boldsymbol{K}(\boldsymbol{y} - \hat{\boldsymbol{y}}) \tag{13.41}$$

$$\hat{\boldsymbol{y}} = \boldsymbol{C}\hat{\boldsymbol{x}} + \boldsymbol{D}\boldsymbol{u} \tag{13.42}$$

where $\boldsymbol{K}$ is an *observer gain matrix* to be constructed such that $\hat{\boldsymbol{x}} \to \boldsymbol{x}$ as $t \to \infty$. Equations (13.41) and (13.42) are also known as the *Luenberger observer*. Note that the states of an observer are commonly denoted by a "hat" to distinguish them from the variables of the equations satisfied by the physical system.

Defining the estimation error as $\tilde{\boldsymbol{x}} = \boldsymbol{x} - \hat{\boldsymbol{x}}$ implies that the error dynamics takes the form

$$\dot{\tilde{\boldsymbol{x}}} = (\boldsymbol{A} - \boldsymbol{K}\boldsymbol{C})\tilde{\boldsymbol{x}}. \tag{13.43}$$

Asymptotic convergence of $\tilde{\boldsymbol{x}}$ to zero can be obtained for a constant $\boldsymbol{K}$ if the pair $(\boldsymbol{A}, \boldsymbol{C})$ is observable (see Section 13.3.1).

MATLAB

If the observability matrix $\mathcal{O}$ is non-singular, the poles of the error dynamics can be placed in the left half-plane. The rank of $\mathcal{O}$ is checked by `rank(obsv(A,C))` while the observer gain matrix $\boldsymbol{K}$ is computed using

```
p = [p1,...,pn]'        % vector of distinct observer error poles
K = place(A',C,p)'      % observer gain matrix
```

Note that both `K` and `A` are transposed, since the dual problem of the regulator problem is solved.

Examples 13.1 and 13.2 in Section 13.3.3 demonstrate how the Luenberger observer can be used in ship control when only compass measurements are available. Emphasis is placed on wave filtering and the estimation of the yaw rate.

13.3.3 Case Study: Luenberger Observer for Heading Autopilot

An alternative to LP and notch filtering of wave-induced forces is to apply an observer. An observer can be designed to separate the LF components of the wave-induced measurements by using a mathematical model of the ship and the waves. In fact, a model-based wave filter is well suited to separate the LF and WF motions from each other, even for marine craft, where the control bandwidth is close to or higher than the encounter frequency. It will now be shown how this can be done by considering a ship autopilot for heading control. It is assumed that the heading angle ψ is measured using a gyrocompass while angular rate is left unmeasured, even though it is possible to use an attitude rate sensor to measure the yaw rate $\dot{\psi}$.

Example 13.1 (Nomoto Ship Model Exposed to Wind, Waves and Ocean Currents)
Assume that a first-order Nomoto model

$$\dot{\psi} = r \tag{13.44}$$

$$\dot{r} = -\frac{1}{T}r + \frac{K}{T}(\delta - b) + w_1 \tag{13.45}$$

$$\dot{b} = w_2 \tag{13.46}$$

describes the LF motion of the ship where the rudder offset b is modeled as Gaussian random walk. The rudder bias model is needed to counteract slowly-varying moments on the ship due to wave-drift forces, LF wind and ocean currents. Consequently, the bias term b ensures that $\delta = b$ gives $r = 0$ and $\psi =$ constant in the steady state for $w_1 = w_2 = 0$. The linear wave model (10.110)–(10.111) can be used to model the wave response

$$\dot{\xi}_{\rm w} = \psi_{\rm w} \tag{13.47}$$

$$\dot{\psi}_{\rm w} = -\omega_0^2 \xi_{\rm w} - 2\lambda\omega_0\psi_{\rm w} + K_{\rm w} w_3. \tag{13.48}$$

The process noise terms w_1, w_2 and w_3 are modeled as zero-mean Gaussian white noise processes. By combining the ship and wave models, the compass measurement equation can be expressed by the sum

$$y = \psi + \psi_{\rm w} + \varepsilon \tag{13.49}$$

where ε represents zero-mean Gaussian measurement noise. Note that neither the yaw rate r nor the wave states $\xi_{\rm w}$ and $\psi_{\rm w}$ are measured. The resulting SISO state-space model for $u = \delta$, $\boldsymbol{x} = [\xi_{\rm w},\ \psi_{\rm w},\ \psi,\ r,\ b]^{\rm T}$ and $\boldsymbol{w} = [w_1,\ w_2,\ w_3]^{\rm T}$ becomes

$$\dot{\boldsymbol{x}} = \boldsymbol{A}\boldsymbol{x} + \boldsymbol{b}u + \boldsymbol{E}\boldsymbol{w} \tag{13.50}$$

$$y = \boldsymbol{c}^{\rm T}\boldsymbol{x} + \varepsilon \tag{13.51}$$

where

$$\boldsymbol{A} = \left[\begin{array}{cc|ccc} 0 & 1 & 0 & 0 & 0 \\ -\omega_0^2 & -2\lambda\omega_0 & 0 & 0 & 0 \\ \hline 0 & 0 & 0 & 1 & 0 \\ 0 & 0 & 0 & -\frac{1}{T} & -\frac{K}{T} \\ 0 & 0 & 0 & 0 & 0 \end{array}\right], \quad \boldsymbol{b} = \left[\begin{array}{c} 0 \\ 0 \\ \hline 0 \\ \frac{K}{T} \\ 0 \end{array}\right] \tag{13.52}$$

$$\boldsymbol{E} = \left[\begin{array}{c|cc} 0 & 0 & 0 \\ K_{\rm w} & 0 & 0 \\ \hline 0 & 0 & 0 \\ 0 & 1 & 0 \\ 0 & 0 & 1 \end{array}\right], \quad \boldsymbol{c}^{\rm T} = [0,\ 1,\ 1,\ 0,\ 0] \tag{13.53}$$

and $K_{\rm w} = 2\lambda\omega_0\sigma$.

MATLAB

The following example shows how the Luenberger observer gains of a ship autopilot system can be computed in Matlab.

Example 13.2 (Luenberger Observer Gains)
It is straightforward to see that the autopilot model with WF, wind and ocean current model (13.52)–(13.53) is observable from the input δ to the compass measurement y. Let $K = 1$, $T = 50$, $\lambda = 0.1$ and $\omega_0 = 1$, then

```
K = 1; T = 50; lambda = 0.1; w0 = 1;

A = [0 1 0 0 0
     -w0^2 -2*lambda*w0 0 0 0
     0 0 0 1 0
     0 0 0 -1/T -K/T
     0 0 0 0 0];

c = [0 1 1 0 0]';

n = rank(obsv(A,c'))
```

results in $n = 5$ *corresponding to* $\text{rank}(\mathcal{O}) = 5$. *Hence, the system is observable according to Definition 13.2, implying that the states* r, b, ψ_w *and* ξ_w *can be reconstructed from a single measurement* $y = \psi + \psi_w + \varepsilon$ *using a Luenberger observer*

$$\dot{\hat{x}} = A\hat{x} + bu + k(y - \hat{y}) \tag{13.54}$$

$$\hat{y} = c^{\top}\hat{x}. \tag{13.55}$$

The filter gains can be computed by pole placement, for instance

```
k = place(A',c,[p1,p2,p3,p4,p5])'
```

where `p1,p2,p3,p4` *and* `p5` *are the desired closed-loop poles of the error dynamics (13.43).*

13.4 Kalman Filter Design

The Kalman filter (KF) is an efficient recursive filter that estimates the state of a linear or nonlinear dynamic system from a series of noisy measurements. It is widely used in sensor and navigation systems since it can reconstruct unmeasured states as well as remove white and colored noise from the state estimates. It is also possible to include wild-point removal capabilities. In cases of temporarily loss of measurements, the filter equations behave such as a predictor. As soon as new measurements are available, the predictor is corrected and updated online to give the minimum variance estimate. This feature is particularly useful when satellite signals are lost since the KF can predict the motion using only gyros and accelerometers. Inertial navigation systems and state estimators for inertial measurement units are discussed in Chapter 14.

Together with the linear-quadratic regulator (LQR), the KF solves the linear-quadratic Gaussian (LQG) control problem; see Section 16.1. This section summarizes the most useful results for the design of discrete-time KFs for marine craft.

The key assumption when designing a KF is that the system model is observable. This is necessary in order to obtain convergence of the estimated states $\hat{x}$ to x. Moreover, if the system model is *observable* (see Definition 13.2), the state vector $x \in \mathbb{R}^n$ can be reconstructed recursively through the measurement vector $y \in \mathbb{R}^m$ and the control input vector $u \in \mathbb{R}^p$ as shown in Figure 13.10.

13.4.1 Discrete-time Kalman Filter

Consider the LTV state-space model

$$\dot{x} = A(t)x + B(t)u + E(t)w \tag{13.56}$$

$$y = C(t)x + D(t)u + \varepsilon \tag{13.57}$$

where w is zero-mean Gaussian process noise and ε is zero-mean Gaussian measurement noise. The discrete-time KF (Kalman 1960) is defined in terms of the discretized system model

$$x[k+1] = A_{\mathrm{d}}[k]x[k] + B_{\mathrm{d}}[k]u[k] + E_{\mathrm{d}}[k]w[k] \tag{13.58}$$

$$y[k] = C_{\mathrm{d}}[k]x[k] + D_{\mathrm{d}}[k]u[k] + \varepsilon[k]. \tag{13.59}$$

For LTI systems the discretized system matrices are constant (see Appendix B.1.1)

$$A_{\mathrm{d}}[k] = \Phi \tag{13.60}$$

$$B_{\mathrm{d}}[k] = A^{-1}(\Phi - I_n)B \tag{13.61}$$

$$C_{\mathrm{d}}[k] = C \tag{13.62}$$

$$D_{\mathrm{d}}[k] = D \tag{13.63}$$

$$E_{\mathrm{d}}[k] = A^{-1}(\Phi - I_n)E. \tag{13.64}$$

The transition matrix is computed as

$$\Phi \approx I_n + Ah + \frac{1}{2}(Ah)^2 + \cdots + \frac{1}{N!}(Ah)^N. \tag{13.65}$$

Note that Euler integration implies choosing $N = 1$ such that $A_{\mathrm{d}} = I_n + Ah$ where h is the sampling time. If A^{-1} is singular, the matrices B_{d} and E_{d} can be computed using the methods in Appendix B.1.1.

MATLAB

The discretized system matrices for an LTI systems are computed in Matlab by

```
[Ad,Bd] = c2d(A,B,h)
[Ad,Ed] = c2d(A,E,h)
```

The linear discrete-time KF algorithm for the system (13.58)–(13.59) is given in Table 13.1. The KF is implemented as an iterative loop, see Figure 13.11. A Matlab template script is included in the MSS toolbox as `ExKF.m` where a second-order ship model is used as case study for estimation of the yaw rate from a compass measurement.

Table 13.1 Discrete-time Kalman filter.

Initial values	$\hat{\boldsymbol{x}}^-[0] = \boldsymbol{x}_0$ $\hat{\boldsymbol{P}}^-[0] = \mathrm{E}[(\boldsymbol{x}[0] - \hat{\boldsymbol{x}}^-[0])(\boldsymbol{x}[0] - \hat{\boldsymbol{x}}^-[0])^\mathrm{T}] = \boldsymbol{P}_0$
KF gain	$\boldsymbol{K}[k] = \hat{\boldsymbol{P}}^-[k]\boldsymbol{C}_\mathrm{d}^\mathrm{T}[k](\boldsymbol{C}_\mathrm{d}[k]\hat{\boldsymbol{P}}^-[k]\boldsymbol{C}_\mathrm{d}^\mathrm{T}[k] + \boldsymbol{R}_\mathrm{d}[k])^{-1}$
State corrector	$\hat{\boldsymbol{x}}[k] = \hat{\boldsymbol{x}}^-[k] + \boldsymbol{K}[k](\boldsymbol{y}[k] - \boldsymbol{C}_\mathrm{d}[k]\hat{\boldsymbol{x}}^-[k] - \boldsymbol{D}_\mathrm{d}[k]\boldsymbol{u}[k])$
Covariance corrector	$\hat{\boldsymbol{P}}[k] = (\boldsymbol{I}_n - \boldsymbol{K}[k]\boldsymbol{C}_\mathrm{d}[k])\hat{\boldsymbol{P}}^-[k](\boldsymbol{I}_n - \boldsymbol{K}[k]\boldsymbol{C}_\mathrm{d}[k])^\mathrm{T}$ $+ \boldsymbol{K}[k]\boldsymbol{R}_\mathrm{d}[k]\boldsymbol{K}^\mathrm{T}[k]$
State predictor	$\hat{\boldsymbol{x}}^-[k+1] = \boldsymbol{A}_\mathrm{d}[k]\hat{\boldsymbol{x}}[k] + \boldsymbol{B}_\mathrm{d}[k]\boldsymbol{u}[k]$
Covariance predictor	$\hat{\boldsymbol{P}}^-[k+1] = \boldsymbol{A}_\mathrm{d}[k]\hat{\boldsymbol{P}}[k]\boldsymbol{A}_\mathrm{d}^\mathrm{T}[k] + \boldsymbol{E}_\mathrm{d}[k]\boldsymbol{Q}_\mathrm{d}[k]\boldsymbol{E}_\mathrm{d}^\mathrm{T}[k]$

where

$\boldsymbol{Q}_\mathrm{d}$, $\boldsymbol{R}_\mathrm{d}$ Covariance matrices for the process and measurement noises
$\hat{\boldsymbol{x}}^-$, $\hat{\boldsymbol{P}}^-$ *A priori* state and covariance matrix estimates (before update)
$\hat{\boldsymbol{x}}$, $\hat{\boldsymbol{P}}$ *A posteriori* state and covariance matrix estimates (after update)

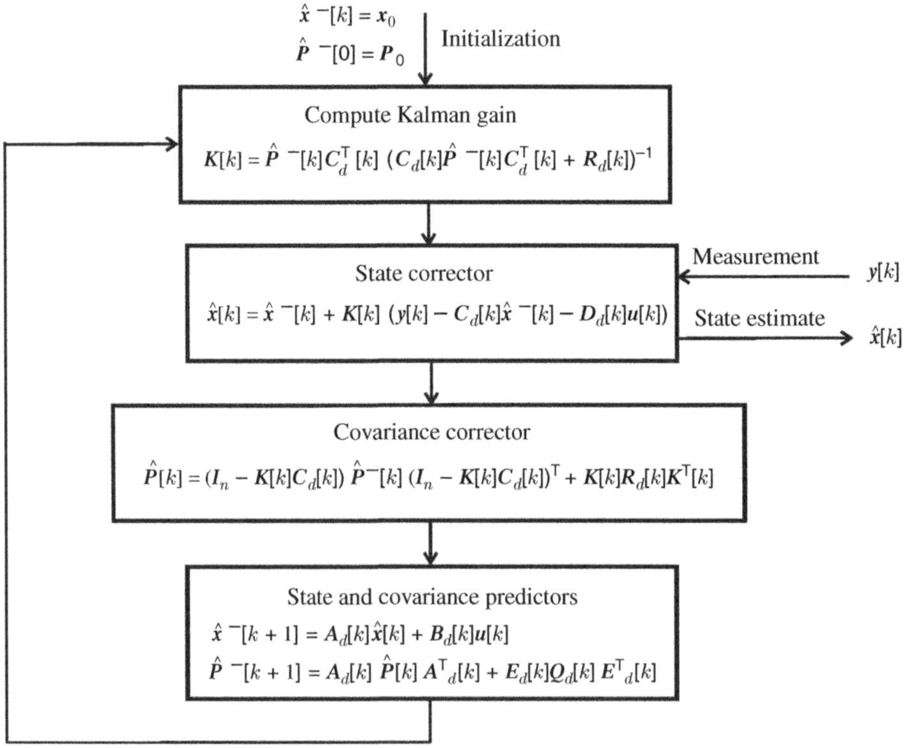

Figure 13.11 Kalman filter loop.

MATLAB

The pseduocode for the KF loop in Figure 13.11 is:

```
% initialization
x_prd = x0;    Qd = constant;
P_prd = P0;    Rd = constant;

% MAIN LOOP
for i=1:N

    % KF gain: K[k]
    K = P_prd * Cd' * inv( Cd * P_prd * Cd' + Rd );
    IKC = eye(n) - K * Cd;

    % Measurement: y[k]
    y = ...

    % Corrector: x_hat[k] and P_hat[k]
    x_hat = x_prd + K * ( y - Cd * x_prd - Dd * u );
    P_hat = IKC * P_prd * IKC' + K * Rd * K';

    % Predictor: x_prd[k+1] and P_prd[k+1]
    x_prd = Ad * x_hat + Bd * u;
    P_prd = Ad * P_hat * Ad' + Ed * Qd * Ed';

end
```

13.4.2 Discrete-time Extended Kalman Filter

The Kalman filter can also be applied to nonlinear systems with non-additive input and process noise

$$\dot{\boldsymbol{x}} = \boldsymbol{f}(\boldsymbol{x}, \boldsymbol{u}, \boldsymbol{w}) \tag{13.66}$$

$$\boldsymbol{y} = \boldsymbol{h}(\boldsymbol{x}, \boldsymbol{u}) + \boldsymbol{\varepsilon} \tag{13.67}$$

where $\boldsymbol{f}(\boldsymbol{x}, \boldsymbol{u}, \boldsymbol{w})$ and $\boldsymbol{h}(\boldsymbol{x}, \boldsymbol{u})$ are nonlinear vector fields, $\boldsymbol{w}$ is white Gaussian process noise and $\boldsymbol{\varepsilon}$ is white Gaussian measurement noise. If the system (13.66)–(13.67) is observable, the state vector can be estimated using the discrete-time extended Kalman filter (EKF) algorithm of Table 13.2.

A *discrete-time predictor* based on the state-space model (13.66)–(13.67) is

$$\boldsymbol{x}[k+1] = \boldsymbol{x}[k] + h\boldsymbol{f}(\boldsymbol{x}[k], \boldsymbol{u}[k], \boldsymbol{0}) \tag{13.68}$$

$$\boldsymbol{y}[k] = \boldsymbol{h}(\boldsymbol{x}[k], \boldsymbol{u}[k]) \tag{13.69}$$

Table 13.2 Discrete-time extended Kalman filter.

Initial values	$\hat{\boldsymbol{x}}^{-}[0] = \boldsymbol{x}_0$ $\hat{\boldsymbol{P}}^{-}[0] = \mathrm{E}[(\boldsymbol{x}[0] - \hat{\boldsymbol{x}}^{-}[0])(\boldsymbol{x}[0] - \hat{\boldsymbol{x}}^{-}[0])^{\mathsf{T}}] = \boldsymbol{P}_0$
KF gain	$\boldsymbol{K}[k] = \hat{\boldsymbol{P}}^{-}[k]\boldsymbol{C}_{\mathrm{d}}^{\mathsf{T}}[k](\boldsymbol{C}_{\mathrm{d}}[k]\hat{\boldsymbol{P}}^{-}[k]\boldsymbol{C}_{\mathrm{d}}^{\mathsf{T}}[k] + \boldsymbol{R}_{\mathrm{d}}[k])^{-1}$
State corrector	$\hat{\boldsymbol{x}}[k] = \hat{\boldsymbol{x}}^{-}[k] + \boldsymbol{K}[k](\boldsymbol{y}[k] - \boldsymbol{h}[k](\hat{\boldsymbol{x}}^{-}[k], \boldsymbol{u}[k]))$
Covariance corrector	$\hat{\boldsymbol{P}}[k] = (\boldsymbol{I}_n - \boldsymbol{K}[k]\boldsymbol{C}_{\mathrm{d}}[k])\hat{\boldsymbol{P}}^{-}[k](\boldsymbol{I}_n - \boldsymbol{K}[k]\boldsymbol{C}_{\mathrm{d}}[k])^{\mathsf{T}}$ $+ \boldsymbol{K}[k]\boldsymbol{R}_{\mathrm{d}}[k]\boldsymbol{K}^{\mathsf{T}}[k]$
State predictor	$\hat{\boldsymbol{x}}^{-}[k+1] = \hat{\boldsymbol{x}}[k] + h\boldsymbol{f}(\hat{\boldsymbol{x}}[k], \boldsymbol{u}[k], \boldsymbol{0})$
Covariance predictor	$\hat{\boldsymbol{P}}^{-}[k+1] = \boldsymbol{A}_{\mathrm{d}}[k]\hat{\boldsymbol{P}}[k]\boldsymbol{A}_{\mathrm{d}}^{\mathsf{T}}[k] + \boldsymbol{E}_{\mathrm{d}}[k]\boldsymbol{Q}_{\mathrm{d}}[k]\boldsymbol{E}_{\mathrm{d}}^{\mathsf{T}}[k]$

The discrete-time system matrices are defined by the *Jacobians*

$$\boldsymbol{A}_{\mathrm{d}}[k] = \boldsymbol{I}_n + h \left.\frac{\partial \boldsymbol{f}(\boldsymbol{x}[k], \boldsymbol{u}[k], \boldsymbol{w}[k])}{\partial \boldsymbol{x}[k]}\right|_{\boldsymbol{x}[k]=\hat{\boldsymbol{x}}[k],\ \boldsymbol{w}[k]=\boldsymbol{0}}$$

$$\boldsymbol{E}_{\mathrm{d}}[k] = h \left.\frac{\partial \boldsymbol{f}(\boldsymbol{x}[k], \boldsymbol{u}[k], \boldsymbol{w}[k])}{\partial \boldsymbol{w}[k]}\right|_{\boldsymbol{x}[k]=\hat{\boldsymbol{x}}[k],\ \boldsymbol{w}[k]=\boldsymbol{0}}$$

$$\boldsymbol{C}_{\mathrm{d}}[k] = \left.\frac{\partial \boldsymbol{h}(\boldsymbol{x}[k], \boldsymbol{u}[k])}{\partial \boldsymbol{x}[k]}\right|_{\boldsymbol{x}[k]=\hat{\boldsymbol{x}}^{-}[k]}$$

where the Gaussian white noise terms are set to zero and *forward Euler* integration has been used to discretize the vector field $\boldsymbol{f}(\boldsymbol{x}, \boldsymbol{u}, \boldsymbol{w})$; see Appendix B.1. Equation (13.68) should be replaced by an more accurate formula if numerical stability is an issue.

Disadvantages

The linear KF is an optimal estimator and it is easy to establish stability and convergence properties thanks to linear system theory. Unfortunately, optimality is lost when applying the EKF since it relays on linearization. In addition, if the initial estimate of the state is wrong, or if the process is modeled incorrectly, the filter may quickly diverge, owing to its linearization. Another problem with the EKF is that the estimated covariance matrix tends to underestimate the true covariance matrix and therefore risks becoming inconsistent in the statistical sense. Care should also be taken with respect to covariance blow-up and instability. However, the EKF gives excellent performance in most navigation systems at it is the *de facto standard* in aided inertial navigation.

13.4.3 Modification for Euler Angles to Avoid Discontinuous Jumps

Care must be taken when implementing attitude controllers and state estimators using Euler angles since the *roll, pitch* and *yaw* angles are confined to the interval $[0, 2\pi)$ or $[-\pi, \pi)$, also known as the 1-sphere or the topological space $\mathbb{S}^1$ corresponding to a circle in the plane.

For instance, when implementing a heading control system for a marine craft, it is crucial that the angle difference is mapped to the *smallest signed angle* (SSA) between the current heading and the reference (Coates *et al.* 2021). To illustrate why, consider a craft with an actual heading of 0° and a heading setpoint of 355°. A naive controller implementation would calculate a heading error of -355°s, thus commanding a near full rotation, going clockwise, although the setpoint is only 5° away in the opposite direction. This fact is often disregarded in the literature, and when implemented in practical motion control systems, it is usually ignored in the analysis, where the heading error is rather treated as a real number.

Definition 13.5 (Smallest Signed Angle (SSA))
The operator ssa: $\mathbb{R} \rightarrow [-\pi, \pi)$ *maps the unconstrained angle* $\tilde{x} = x - x_0 \in \mathbb{R}$ *representing the difference between the two angles* x *and* x_0 *to the smallest difference between the angles*

$$\tilde{x}_s = \text{ssa}(\tilde{x}) \tag{13.70}$$

where $\tilde{x}_s \in \mathbb{S}^1$.

MATLAB

MSS Matlab function `ssa.m` for the smallest signed angle.

```
function angle = ssa(angle,unit)
% SSA is the "smallest signed angle" or the smallest difference
% between two angles. Examples:
% >> angle = ssa(angle) maps an angle in rad to [-pi pi)
% >> angle = ssa(angle,'deg') maps an angle in deg to
% [-180 180)
%
% Note that in many languages (C, C++, C#, JavaScript), the
% operator mod(x,y) returns a value with the same sign as x.
% For these use a custom mod function: mod(x,y)= x-floor(x/y)*y
% For the Unity game engine use: Mathf.DeltaAngle

if (nargin == 1)
    angle = mod( angle + pi, 2 * pi ) - pi;
elseif strcmp(unit,'deg')
    angle = mod( angle + 180, 360 ) - 180;
end
```

PD Control using the Smallest Signed Angle to Avoid Discontinuous Jumps

Consider the yaw angle dynamics (see Section 7.2.2)

$$\dot{\psi} = r \tag{13.71}$$

$$T\dot{r} + r = K\delta \tag{13.72}$$

where $K > 0$ and $T > 0$ are the Nomoto gain and time constants, respectively. Assume that the craft is controlled by a PD controller with reference feedforward

$$\delta = T\dot{r}_{\rm d} + r_{\rm d} - K_{\rm p}\tilde{\psi} - K_{\rm d}\tilde{r} \tag{13.73}$$

for $K_{\rm p} > 0$ and $K_{\rm d} > 0$ where $\tilde{\psi} = \psi - \psi_{\rm d}$ and $\tilde{r} = r - r_{\rm d}$. Then the error dynamics becomes

$$\ddot{\tilde{\psi}} + \frac{(1 + KK_{\rm d})}{T}\dot{\tilde{\psi}} + \frac{KK_{\rm p}}{T}\tilde{\psi} = 0. \tag{13.74}$$

It is tempting to claim that the equilibrium point $(\tilde{\psi}, \dot{\tilde{\psi}}) = (0, 0)$ of the linear second-order system (13.74) is GES since both poles are in the left-half plane. However, as shown by Bhat and Bernstein (2000), systems with rotational degrees of motion cannot be globally stabilized by continuous feedback due to the topological obstruction imposed by SO(3), that is $\tilde{\psi}$ is defined on the 1-sphere $\mathbb{S}^1$ and not on $\mathbb{R}$.

When implementing the heading controller (13.73) it is necessary to map the error angle $\tilde{\psi}$ to $[-\pi, \pi)$ to avoid steps and numerical instability of the closed-loop system. Therefore we modify the PD controller using the ssa mapping

$$\delta = T\dot{r}_{\rm d} + r_{\rm d} - K_{\rm p}\,{\rm ssa}(\tilde{\psi}) - K_{\rm d}\tilde{r}. \tag{13.75}$$

The error dynamics (13.74) is modified accordingly

$$\ddot{\tilde{\psi}} + \frac{(1 + KK_{\rm d})}{T}\dot{\tilde{\psi}} + \frac{KK_{\rm p}}{T}\tilde{\psi}_{\rm s} = 0 \tag{13.76}$$

where $\tilde{\psi}_{\rm s} = {\rm ssa}(\tilde{\psi})$. Stability of the system (13.76) is guaranteed by Coates *et al.* (2021) where it is shown that the equilibrium point $(\tilde{\psi}_{\rm s},\ \dot{\tilde{\psi}}) = (0,\ 0)$ is exponentially stable and globally attractive for the the entire domain $\mathbb{S}^1 \times \mathbb{R}$.

State Estimation using the Smallest Signed Angle to Avoid Discontinuous Jumps

The controller and the state estimator represent dual problems. Hence, the state estimator must also be modified to use the SSA between a measured angle and an estimated angle. Consider the discrete-time Nomoto model for a marine craft

$$\psi[k+1] = \psi[k] + hr[k] \tag{13.77}$$

$$r[k+1] = r[k] + \frac{h}{T}(-r[k] + K\delta[k] + w[k]) \tag{13.78}$$

$$y[k] = \psi[k] + \varepsilon[k] \tag{13.79}$$

where h is the sampling time, $y[k]$ is the measured yaw angle, and $w[k]$ and $\varepsilon[k]$ are Gaussian white noise processes. The corrector of the discrete-time KF (see Table 13.1) must be modified to include the ssa mapping in order to avoid discontinuous jumps. For the system (13.77)–(13.79) this implies that

$$\hat{\psi}[k] = \hat{\psi}^-[k] + K_1\ \mathrm{ssa}(y[k] - \hat{y}^-[k]) \tag{13.80}$$

$$\hat{r}[k] = \hat{r}^-[k] + K_2\ \mathrm{ssa}(y[k] - \hat{y}^-[k]) \tag{13.81}$$

$$\hat{y}^-[k] = \hat{\psi}^-[k] \tag{13.82}$$

where the injection term $\tilde{y}[k] = y[k] - \hat{y}^-[k]$ is mapped to $[-\pi,\ \pi)$.

For the continuous system where the estimation error is denoted $\tilde{y} = y - \hat{y}$ it follows from Coates *et al.* (2021) that the equilibrium point of the estimation error $(\tilde{y}_s,\ \dot{\tilde{y}}) = (0,\ 0)$ is exponentially stable and globally attractive for the the entire domain $\mathbb{S}^1 \times \mathbb{R}$.

13.4.4 Modification for Asynchronous Measurement Data

When implementing a discrete-time KF it is practical to choose the sampling frequency $f_s = 1/h$ of the system model equal to the measurement frequency such that the states can be propagated from time t_k to time $t_{k+1} = ht_k$ where h is the sampling time. This obviously only works when the measurements are synchronous.

In embedded computer systems the measurements can be received at different frequencies than the sampling frequency of the system. For instance, GNSS measurements are typically sampled at $f_{\mathrm{gnss}} = 1$ Hz while inertial sensors such as accelerometers and gyros can be sampled at a user specified frequency $f_{\mathrm{imu}} \gg f_{\mathrm{gnss}}$. In fact, MEMS-based IMUs can operate at frequencies much larger than 1000 Hz. Consequently, it is necessary to modify the KF algorithm to deal with asynchronous measurement data.

Assume that the sampling frequency is chosen equal to the frequency of the fastest measurement frequency

$$f_s = f_{\mathrm{imu}} \tag{13.83}$$

and let the integer Z denote the ratio between the sampling frequency and the slower GNSS measurement frequency. In other words

$$Z = \frac{f_s}{f_{\mathrm{gnss}}} > 1. \tag{13.84}$$

It is practical to choose Z as an integer such that the slow measurement appear each Z time in the KF logical loop. In addition, it is necessary to set the corrector equations equal to the predicted values when there are no measurements

$$\hat{\boldsymbol{x}}[k] = \hat{\boldsymbol{x}}^-[k] \tag{13.85}$$

$$\hat{\boldsymbol{P}}[k] = \hat{\boldsymbol{P}}^-[k]. \tag{13.86}$$

This ensures that the state vector and covariance matrix are not corrected before a new measurement arrives. The Matlab example file `ExKF.m` in the MSS toolbox illustrates how this modification can be used in a discrete-time KF when applied to a ship model.

MATLAB

Pseudocode for implementation of a discrete-time KF with asynchronous measurement data:

```
f_s    = 100;          % sampling frequency [Hz]
f_imu  = f_s;          % IMU (fast) measurement frequency [Hz]
f_gnss = 1;            % GNSS (slow) measurement frequency [Hz]

h = 1/f_s;             % sampling time
h_gnss = 1/f_gnss;     % GNSS sampling time

% MAIN LOOP
for i=1:N

   % GNSS measurements are Z = f_s/f_gnss times slower than
   % the sampling frequency f_s
   if mod( t, h_gnss ) == 0
      y = ...;                        % new measurement: y[k]

      % KF gain: K[k]
      K = P_prd * Cd' * inv( Cd * P_prd * Cd' + Rd );
      IKC = eye(n) - K * Cd;

      % Corrector: x_hat[k] and P_hat[k]
      x_hat = x_prd + K * ( y - Cd * x_prd - Dd * u );
      P_hat = IKC * P_prd * IKC' + K * Rd * K';
   else
        x_hat = x_prd;                % no measurement
        P_hat = P_prd;
   end

   % Predictor: x_prd[k+1] and P_prd[k+1]
   x_prd = Ad * x_hat + Bd * u;
   P_prd = Ad * P_hat * Ad' + Ed * Qd * Ed';

end
```

13.4.5 Case Study: Kalman Filter Design for Heading Autopilots

In this section we will revisit Case Study 13.3.3 and design a KF for a ship heading autopilot with wave filtering capabilities. The main sensor components for a heading controlled ship are (see Section 13.1),

- Magnetic and/or gyroscopic compasses measuring the yaw angle ψ.
- Attitude rate sensor (ARS) measuring the yaw rate $\dot{\psi} = r$.

In many commercial systems only the compass is used for feedback control since the yaw rate can be estimated quite well by a state estimator. Hence, we will assume that ψ is the only measurement. Again, consider the ship-wave model in Example 13.3.3 with compass measurement

$$\dot{\boldsymbol{x}} = \boldsymbol{A}\boldsymbol{x} + \boldsymbol{b}u + \boldsymbol{E}\boldsymbol{w} \tag{13.87}$$

$$y = \boldsymbol{c}^{\top}\boldsymbol{x} + \varepsilon \tag{13.88}$$

where $u = \delta$, $\boldsymbol{x} = [\xi_{\mathrm{w}},\ \psi_{\mathrm{w}},\ \psi,\ r,\ b]^{\top}$, $\boldsymbol{w} = [w_1,\ w_2,\ w_3]^{\top}$ and

$$\boldsymbol{A} = \left[\begin{array}{cc|ccc} 0 & 1 & 0 & 0 & 0 \\ -\omega_0^2 & -2\lambda\omega_0 & 0 & 0 & 0 \\ \hline 0 & 0 & 0 & 1 & 0 \\ 0 & 0 & 0 & -\frac{1}{T} & -\frac{K}{T} \\ 0 & 0 & 0 & 0 & 0 \end{array}\right], \quad \boldsymbol{b} = \left[\begin{array}{c} 0 \\ 0 \\ \hline 0 \\ \frac{K}{T} \\ 0 \end{array}\right] \tag{13.89}$$

$$\boldsymbol{E} = \left[\begin{array}{c|cc} 0 & 0 & 0 \\ K_{\mathrm{w}} & 0 & 0 \\ \hline 0 & 0 & 0 \\ 0 & 1 & 0 \\ 0 & 0 & 1 \end{array}\right], \quad \boldsymbol{c}^{\top} = [0,\ 1,\ 1,\ 0,\ 0] \tag{13.90}$$

with $K_{\mathrm{w}} = 2\lambda\omega_0\sigma$.

In order to implement a KF for a heading autopilot, the discrete-time system model (13.58)–(13.59), as presented in Section 13.4.1, is applied. This suggests that

$$\begin{array}{ll} \boldsymbol{A}_{\mathrm{d}} = \boldsymbol{I}_5 + h\boldsymbol{A}, & \boldsymbol{c}_{\mathrm{d}} = \boldsymbol{c} \\ \boldsymbol{b}_{\mathrm{d}} = h\boldsymbol{b}, & \boldsymbol{E}_{\mathrm{d}} = h\boldsymbol{E} \end{array}.$$

The main problem in the realization of the KF is that the parameters T, K, ω_0 and λ are uncertain. The parameters T and K can be estimated from ship maneuvering tests performed in calm water while the parameters ω_0 and λ of the first-order WF model and the covariance of the driving noise w_1 can be estimated from maneuvering trials or parameter estimation. Holzhüter (1992) claims that the damping coefficient in the wave model can be chosen rather arbitrarily as long as it is low (typically $\lambda = 0.01$–0.1), whereas the wave frequency ω_0 can be treated as a tunable or gain-scheduled parameter. In some cases it can be advantageous to estimate ω_0 online by applying a wave-frequency estimator (Belleter *et al.* 2015) or simply use the FFT to compute the peak frequency of the wave spectrum from the heave position or pitch angle (see Section 13.2.3). The FFT algorithm can run offline by using data collected for a given period (typically 30 min) to adjust to varying sea states.

Example 13.3 (Discrete-time KF for Ship Autopilots)

The example illustrates how the KF gains can be computed in Matlab for a ship exposed to waves. The KF must be modified to handle yaw angle measurements $y = \psi + \psi_{\mathrm{w}} + \varepsilon$ *on the intervals* $[0°, 360°)$ *or* $[-180°, 180°)$. *The yaw angle injection term is the difference between the measured and the estimated yaw angles. It is necessary to map this signal to* $[-\pi, \pi)$ *to avoid discontinuous jumps in the estimates. The tool for this the MSS function* `ssa.m` *(see Section 13.4.3), which is used to modify the corrector.*

MATLAB

Pseudocode for simulating a discrete-time KF for a ship exposed to first-order WF motions. The only measurement is a gyrocompass but the 5-state ship-wave model is observable so all states can be reconstructed from a single measurement (see Example 13.3.3).

```
x = x0; x_prd = x0;                      % initialization
P_prd = P0;

Qd = diag([q11 q22 q33]);                % covariance matrices
Rd = r11;

A = [ 0 1 0 0 0                          % continious-time ship model
      -w0^2 -2*lambda*w0 0 0 0
      0 0 0 1 0
      0 0 0 -1/T -K/T
      0 0 0 0 0 ];
B = [0 0 0 K/T 0]';
C = [0 1 1 0 0];
E = [0 0 0; Kw 0 0; 0 0 0; 0 1 0; 0 0 1];

Ad = eye(5) + h * A; Bd = h * B;          % discrete-time KF model
Cd = C; Ed = h * E

% MAIN LOOP
for i=1:N

    % KF gain: K[k]
    K = P_prd * Cd' * inv( Cd * P_prd * Cd' + Rd );
    IKC = eye(5) - K * Cd;

    % Control input and measurement: u[k] and y[k]
    u = delta;            % control system
    y = C * x + v;        % compass measurement with white noise v

    % Corrector: x_hat[k] and P_hat[k]
    x_hat = x_prd + K * ssa( y - Cd * x_prd );% ssa modification
    P_hat = IKC * P_prd * IKC' + K * Rd * K';

    % Predictor: x_prd[k+1] and P_prd[k+1]
    x_prd = Ad * x_hat + Bd * u;
```

```
        P_prd = Ad * P_hat * Ad' + Ed * Qd * Ed';

        % Ship-wave simulator: x[k+1]
        x = x + h * ( A * x + B * u + E * w );

    end
```

The tuning of the KF is done by choosing the four design parameters `r11`, `q11`, `q22` *and* `q33`. *The first of these,* `r11`, *represents the compass covariance, which can be computed by logging a time series* `psi` $= \psi(t)$ *of the compass at a constant heading. Hence, the Matlab command,* `r11 = cov(psi)`, *gives a good estimate of the measurement noise. The disadvantage with the KF approach is that information about the process noise* w_1, w_2 *and* w_3 *represented by the weights* `q11`, `q22` *and* `q33` *are necessary. These three quantities are usually found by trial and error. The variance of the process and measurement noise will vary with each sea state, implying that several sets of KF gains must be computed.*

Example 13.4 (Kalman-filter-based Wave Filter for the Mariner Class Vessel)
To illustrate the performance of KF-based wave filtering, we consider the case study in Fossen and Perez (2009) of an autopilot application taken from the MSS toolbox.

The marine craft considered is a 160 m *Mariner class vessel with a nominal service speed of* 15 knots, *or* 7.7 m s^{-1}. *The parameters of a complete and validated nonlinear model for the Mariner class vessel are given in Fossen (1994). From the step tests performed on the nonlinear model, a first-order Nomoto model is identified with the parameters* $K = 0.185$ s^{-1} *and* $T = 107.3$ s. *Based on the time constant, a sampling period of* 0.5 s *is chosen for the implementation of the KF. The standard deviation of the noise of the compass is* 0.5°. *From a record of heading motion while the rudder is kept constant, the parameters of the first-order wave-induced motion model are estimated, namely* $\lambda = 0.1$, $\omega_0 = 1.2$ rad s^{-1}, *and the standard deviation of the noise driving the filter is* $\sigma_{w_1} = \sqrt{300}$ rad s^{-1}.

Figure 13.12 demonstrates the performance of the KF. The two upper plots show the true LF heading angle and rate together with the KF estimates while the lower plot shows the first-order wave-induced heading angle and its estimate.

13.4.6 Case Study: Kalman Filter for Dynamic Positioning Systems

Dynamic positioning (DP) systems have been commercially available for marine craft since the 1960s. The first DP systems were designed using conventional PID controllers in cascade with low-pass and/or notch filters to suppress the wave-induced motion components. From the middle of the 1970s more advanced filtering techniques were available thanks to the KF. This motivated Balchen and coauthors to develop optimal wave filtering and state estimation; see Balchen et al. (1976, 1980a, 1980b). KF-based wave filtering has also been discussed by Grimble et al. (1980a, 1980b), Fung and Grimble (1981, 1983), Fotakis *et al.* (1982), Sælid and Jenssen (1983), Sælid *et al.* (1983),

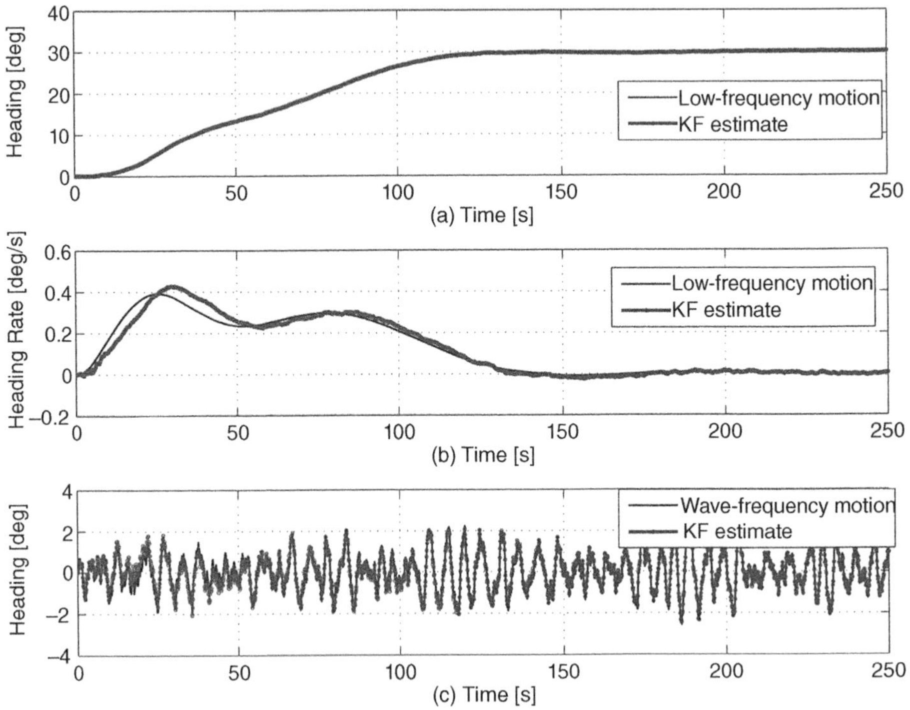

Figure 13.12 Kalman filter performance for a heading autopilot designed for the Mariner class cargo ship: (a) shows the true LF heading ψ and estimate $\hat{\psi}$, (b) shows the true LF heading rate r and estimate $\hat{r}$, and (c) shows the WF component of the heading $\psi_{\rm w}$ and its estimate $\hat{\psi}_{\rm w}$.

Reid *et al.* (1984), Holzhüter and Strauch (1987), Holzhüter (1992), Sørensen et al. (1995, 1996, 2000), Fossen and Strand (2000) and Fossen and Perez (2009).

In this section, the *Kalman filter* is presented for DP applications. Positioning feedback systems are described more closely in Sections 15.3.6 and 16.1.6.

System Model for Dynamic Positioning KF Design

The WF motions are modeled as three colored-noise processes approximating the zero-speed wave responses in surge, sway and yaw. The linear spectra are chosen as

$$x^n_{\rm w}(s) = \frac{K_{{\rm w}_1} s}{s^2 + 2\lambda\omega_0 s + \omega_0^2}\, w_1(s) \tag{13.91}$$

$$y^n_{\rm w}(s) = \frac{K_{{\rm w}_2} s}{s^2 + 2\lambda\omega_0 s + \omega_0^2}\, w_2(s) \tag{13.92}$$

$$\psi_{\rm w}(s) = \frac{K_{{\rm w}_3} s}{s^2 + 2\lambda\omega_0 s + \omega_0^2}\, w_3(s) \tag{13.93}$$

where the sea state is specified in terms of λ and the wave spectrum peak frequency ω_0; see Section 10.2.6. It is assumed that the craft is excited sufficiently by the waves such that all DOFs oscillate with the same frequency ω_0 in steady state. The corresponding state-space model is

$$\dot{\xi} = A_w \xi + E_w w_1 \tag{13.94}$$

$$\eta_w = C_w \xi \tag{13.95}$$

where the output $\eta_w = [x_w^n, \ y_w^n, \ \psi_w]^\top$ represents the three linear wave response models in surge, sway and yaw with state vector $\xi \in \mathbb{R}^6$. The quantities $A_w \in \mathbb{R}^{6\times6}$, $E_w \in \mathbb{R}^{6\times3}$ and $C_w \in \mathbb{R}^{3\times6}$ are constant matrices of appropriate dimensions for the system (13.91)–(13.93). The Gaussian white noise terms are collected into the vector $w_1 = [w_1, \ w_2, \ w_3]^\top$.

Combing the WF model (13.94)–(13.95) with the 3-DOF linear DP model in Section 6.7.3 gives the resulting model for dynamic positioning

$$\dot{\xi} = A_w \xi + E_w w_1 \tag{13.96}$$

$$\dot{\eta} = R(t) v \tag{13.97}$$

$$\dot{b} = w_2 \tag{13.98}$$

$$M \dot{v} = -D v + R^\top(t) b + \tau + \tau_{\text{wind}} + w_3 \tag{13.99}$$

$$y = \eta + C_w \xi + \varepsilon \tag{13.100}$$

where $\eta = [x^n, \ y^n, \ \psi]^\top$ and $v = [u, \ v, \ r]^\top$. The measurement y is the sum of the LF and WF components representing the GNSS position and compass measurements.

The process noise terms w_1, w_2 and w_3 are assumed to be zero-mean Gaussian white noise while unmodeled nonlinear dynamics and disturbances caused by

- Second-order wave drift forces
- Ocean currents
- Mean wind forces

are lumped into the bias term b which is driven by w_2. Note that the estimate of b will be non-physical since it contains several components. Many DP operators call this *"DP current"* since it is experienced as a drift force.

The kinematic nonlinearity due to the rotation matrix in yaw is conveniently removed by assuming that

$$R(\psi(t)) := R(t) \tag{13.101}$$

is known for all $t \geq 0$, e.g. by using a gyrocompass, a magnetic compass or GNSS heading measurements; see Section 13.1. This implies that the dynamics associated with the motion in heave, roll and pitch are neglected.

Discrete-time Kalman Filter for Dynamic Positioning

The control forces usually have two components

$$\boldsymbol{\tau} = -\hat{\boldsymbol{\tau}}_{\text{wind}} + \boldsymbol{B}_u \boldsymbol{u} \tag{13.102}$$

where $\hat{\boldsymbol{\tau}}_{\text{wind}}$ is an estimate of the wind forces implemented by using feedforward compensation and $\boldsymbol{B}_u\boldsymbol{u}$ represents actuator forces (see Chapter 9). The wind feedforward term, which is proportional to the square of the measured wind velocity, depends on the craft's projected area in the direction of the wind (see Section 10.1). The vector $\boldsymbol{u}$ is the command to the actuators, which are assumed to have a much faster dynamic response than the craft; thus the coefficient $\boldsymbol{B}_u$ represents the mapping from the actuator command to the force generated by the actuator (see Section 11.2). For example, if the command to a propeller is the rotation speed, then the corresponding coefficient in $\boldsymbol{B}_u$ maps the speed to the generated thrust.

The resulting state-space model for a DP Kalman filter design is the 15th-order LTV model

$$\dot{\boldsymbol{x}} = \boldsymbol{A}(t)\boldsymbol{x} + \boldsymbol{B}\boldsymbol{u} + \boldsymbol{E}\boldsymbol{w} \tag{13.103}$$

$$\boldsymbol{y} = \boldsymbol{C}\boldsymbol{x} + \boldsymbol{\varepsilon} \tag{13.104}$$

where $\boldsymbol{x} = [\boldsymbol{\xi}^{\top}, \boldsymbol{\eta}^{\top}, \boldsymbol{b}^{\top}, \boldsymbol{\nu}^{\top}]^{\top} \in \mathbb{R}^{15}$ is the state vector, $\boldsymbol{u} \in \mathbb{R}^{p}$ ($p \geq 3$) is the control vector, $\boldsymbol{w} = [\boldsymbol{w}_1^{\top}, \boldsymbol{w}_2^{\top}, \boldsymbol{w}_3^{\top}]^{\top} \in \mathbb{R}^{9}$ represents the process noise vector and $\boldsymbol{\varepsilon} \in \mathbb{R}^{3}$ is a vector of Gaussian white measurement noise. The system matrices are

$$\boldsymbol{A}(t) = \left[\begin{array}{c|ccc} \boldsymbol{A}_{\text{w}} & \boldsymbol{0}_{6\times3} & \boldsymbol{0}_{6\times3} & \boldsymbol{0}_{6\times3} \\ \hline \boldsymbol{0}_{3\times6} & \boldsymbol{0}_{3\times3} & \boldsymbol{0}_{3\times3} & \boldsymbol{R}(t) \\ \boldsymbol{0}_{3\times6} & \boldsymbol{0}_{3\times3} & \boldsymbol{0}_{3\times3} & \boldsymbol{0}_{3\times3} \\ \boldsymbol{0}_{3\times6} & \boldsymbol{0}_{3\times3} & \boldsymbol{M}^{-1}\boldsymbol{R}^{\top}(t) & -\boldsymbol{M}^{-1}\boldsymbol{D} \end{array}\right], \quad \boldsymbol{B} = \left[\begin{array}{c} \boldsymbol{0}_{6\times p} \\ \hline \boldsymbol{0}_{3\times p} \\ \boldsymbol{0}_{3\times p} \\ \boldsymbol{M}^{-1}\boldsymbol{B}_u \end{array}\right] \tag{13.105}$$

$$\boldsymbol{E} = \left[\begin{array}{c|cc} \boldsymbol{E}_{\text{w}} & \boldsymbol{0}_{6\times3} & \boldsymbol{0}_{6\times3} \\ \hline \boldsymbol{0}_{3\times3} & \boldsymbol{0}_{3\times3} & \boldsymbol{0}_{3\times3} \\ \boldsymbol{0}_{3\times3} & \boldsymbol{I}_3 & \boldsymbol{0}_{3\times3} \\ \boldsymbol{0}_{3\times3} & \boldsymbol{0}_{3\times3} & \boldsymbol{M}^{-1} \end{array}\right], \qquad \boldsymbol{C} = \left[\boldsymbol{C}_{\text{w}} \mid \boldsymbol{I}_3 \ \boldsymbol{0}_{3\times3} \ \boldsymbol{0}_{3\times3}\right]. \tag{13.106}$$

The system is observable since `rank(obsv(A,C)) = 15`. In order to implement the discrete-time KF of Table 13.1, the state-space model (13.103)–(13.104) is discretized using Euler's method (see Appendix B.1.1)

$$\boldsymbol{x}[k+1] = \boldsymbol{A}_{\text{d}}[k]\boldsymbol{x}[k] + \boldsymbol{B}_{\text{d}}\boldsymbol{u}[k] + \boldsymbol{E}_{\text{d}}\boldsymbol{w}[k] \tag{13.107}$$

$$\boldsymbol{y}[k] = \boldsymbol{C}_{\text{d}}\boldsymbol{x}[k] + \boldsymbol{\varepsilon}[k] \tag{13.108}$$

where the system matrices are

$$\boldsymbol{A}_{\rm d}[k] \approx \boldsymbol{I}_{15} + h\boldsymbol{A}(t_k), \qquad \boldsymbol{C}_{\rm d} = \boldsymbol{C}$$
$$\boldsymbol{B}_{\rm d} \approx h\boldsymbol{B}, \qquad \boldsymbol{E}_{\rm d} \approx h\boldsymbol{E}.$$

MATLAB

Pseudocode showing the computation of the Kalman gain and the corrector-predictor with ssa modification.

```
% KF gain: K[k]
K = P_prd * Cd' * inv( Cd * P_prd * Cd' + Rd );
IKC = eye(15) - K * Cd;

% Corrector: x_hat[k] and P_hat[k]
e = y - Cd * x_prd;                       % estimation error
e(3) = ssa(e(3));                         % ssa modification
x_hat = x_prd + K * e;
P_hat = IKC * P_prd * IKC' + K * Rd * K';

% Predictor: x_prd[k+1] and P_prd[k+1]
x_prd = Ad * x_hat + Bd * u;
P_prd = Ad * P_hat * Ad' + Ed * Qd * Ed';
```

Dead Reckoning

Dead reckoning refers to the case where there are no updates, for instance GNSS position and/or compass signal losses for a period of time. During sensor failures, the best thing to do is to trust the model without any updates. Consequently, the corrector in the KF is bypassed by setting $\boldsymbol{K}[k] = \boldsymbol{0}$ and prediction is based on the system model only. During dead reckoning (signal loss) the KF must be modified according to

Corrector:

$$\hat{\boldsymbol{x}}[k] = \hat{\boldsymbol{x}}^-[k] \tag{13.109}$$

$$\hat{\boldsymbol{P}}[k] = \hat{\boldsymbol{P}}^-[k]. \tag{13.110}$$

Predictor:

$$\hat{\boldsymbol{x}}^-[k+1] = \boldsymbol{A}_{\rm d}[k]\hat{\boldsymbol{x}}[k] + \boldsymbol{B}_{\rm d}[k]\boldsymbol{u}[k] \tag{13.111}$$

$$\hat{\boldsymbol{P}}^-[k+1] = \boldsymbol{A}_{\rm d}[k]\hat{\boldsymbol{P}}[k]\boldsymbol{A}_{\rm d}^{\top}[k] + \boldsymbol{E}_{\rm d}[k]\boldsymbol{Q}_{\rm d}[k]\boldsymbol{E}_{\rm d}^{\top}[k]. \tag{13.112}$$

13.5 Passive Observer Design

The drawback of the Kalman filter is that it is difficult and time-consuming to tune the state estimator, which is a stochastic system with 15 states and 120 covariance equations. The main reason for this is that the numerous covariance tuning parameters may be difficult to relate to physical quantities. This results in an ad hoc tuning procedure for the process covariance matrix $\boldsymbol{Q}_{\mathrm{d}}[k]$ while the measurement covariance matrix $\boldsymbol{R}_{\mathrm{d}}[k]$ usually is well defined in terms of sensor noise specifications.

Another drawback with KF-based design techniques is that a relatively large number of parameters must be determined through experimental testing of the craft. This motivated the research of a nonlinear passivity-based observer, since passivity arguments simplify the tuning procedure significantly (Fossen and Strand 1999b). Hence, the time needed for sea trials and tuning can be drastically reduced. The nonlinear passive observer, guarantees convergence of all estimation errors (including the bias terms) to zero. Hence, only one set of observer gains is needed to cover the whole state space. In addition, the number of observer tuning parameters is significantly reduced and the wave filter parameters are directly coupled to the dominating wave frequency. Passivity implies that the phase of the error dynamics is limited by 90°, which gives excellent stability properties. Passivity theory also proved to be a new tool with respect to accurate tuning of the observer. The proposed nonlinear observer opens the way for new controller designs that are more in line with the actual structure of the physical system, for instance by using a nonlinear *separation principle* (Loria *et al.* 2000).

For extensions to adaptive wave filtering, see Strand and Fossen (1999), while extensions to position mooring systems are found in Strand (1999).

13.5.1 Case Study: Passive Observer for Dynamic Positioning using GNSS and Compass Measurements

Passivity is a property of engineering systems, most commonly used in electronic engineering and control systems. A passive component, may be either a component that consumes (but does not produce) energy, or a component that is incapable of power gain. A component that is not passive is called an active component. An electronic circuit consisting entirely of passive components is called a passive circuit (and has the same properties as a passive component). A transfer function $h(s)$ must have phase greater than $-90°$ in order to be passive. For definitions on passivity see, for instance, Sepulchre *et al.* (1997), Ortega *et al.* (1998) or Lozano *et al.* (2000).

The passive observer in this section is based on Fossen and Strand (1999b). The following assumptions are necessary to prove passivity:

Assumption P1: $\boldsymbol{w} = \boldsymbol{0}$ and $\boldsymbol{\varepsilon} = \boldsymbol{0}$. The zero-mean Gaussian white noise terms are omitted in the deterministic stability analysis of the observer. If they are included in the Lyapunov function analysis the error dynamics will be uniformly ultimated bounded (UUB) instead of asymptotical/exponential stable.

Assumption P2: $\boldsymbol{R}(\psi(t)) := \boldsymbol{R}(t)$ is known for all $t \geq 0$. This is a good assumption since the yaw angle is usually measured by a gyrocompass.

The following model properties of the inertia and damping matrices will be exploited in the passivation design

$$M = M^{\top} > 0, \quad \dot{M} = \mathbf{0}, \quad D > 0.$$

System Model for the Nonlinear Passive Observer

The application of assumptions P1 and P2 to (13.96)–(13.100) gives the following DP observer model

$$\dot{\xi} = A_{w}\xi \tag{13.113}$$

$$\dot{\eta} = R(t)v \tag{13.114}$$

$$\dot{b} = -T^{-1}b \tag{13.115}$$

$$M\dot{v} = -Dv + R^{\top}(t)b + \tau + \tau_{\text{wind}} \tag{13.116}$$

$$y = \eta + C_{w}\xi. \tag{13.117}$$

The matrix $T = \text{diag}\{T_1, T_2, T_3\} > 0$ of bias time constants is included as additional tuning parameters to obtain passivity. For notational simplicity (13.113), (13.114) and (13.117) are written in state-space form

$$\dot{\eta}_0 = A_0\eta_0 + B_0R(t)v \tag{13.118}$$

$$y = C_0\eta_0 \tag{13.119}$$

where $\eta_0 = [\xi^{\top}, \eta^{\top}]^{\top}$ and

$$A_0 = \begin{bmatrix} A_{w} & \mathbf{0}_{6\times3} \\ \mathbf{0}_{3\times6} & \mathbf{0}_{3\times3} \end{bmatrix}, \quad B_0 = \begin{bmatrix} \mathbf{0}_{6\times3} \\ I_3 \end{bmatrix}, \quad C_0 = \begin{bmatrix} C_{w} & I_3 \end{bmatrix}. \tag{13.120}$$

Observer Equations

The observer equations are chosen to copy the dynamics (13.113)–(13.117) resulting in 15 ODEs, as shown in Figure 13.13. In other words,

$$\dot{\hat{\xi}} = A_{w}\hat{\xi} + K_1(\omega_0)\tilde{y} \tag{13.121}$$

$$\dot{\hat{\eta}} = R(t)\hat{v} + K_2\tilde{y} \tag{13.122}$$

$$\dot{\hat{b}} = -T^{-1}\hat{b} + K_3\tilde{y} \tag{13.123}$$

$$M\dot{\hat{v}} = -D\hat{v} + R^{\top}(t)\hat{b} + \tau + \tau_{\text{wind}} + R^{\top}(t)K_4\tilde{y} \tag{13.124}$$

$$\hat{y} = \hat{\eta} + C_{w}\hat{\xi} \tag{13.125}$$

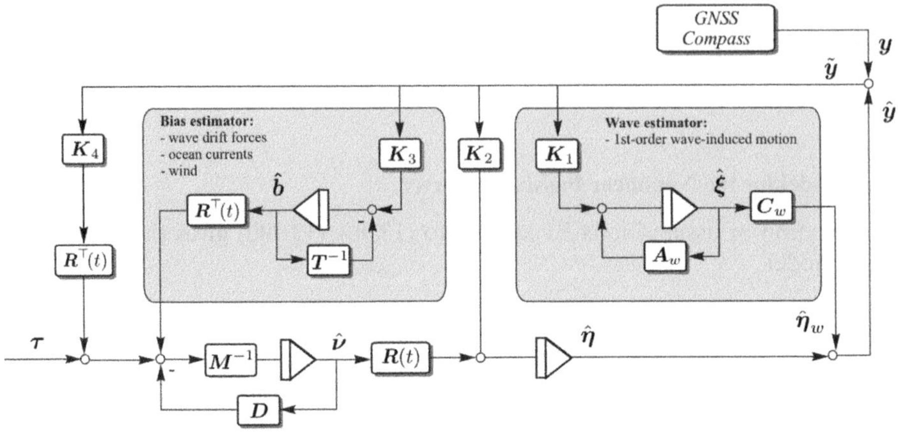

Figure 13.13 Block diagram showing the passive DP observer.

where $\tilde{y} = y - \hat{y}$ is modified to include the ssa function (see Section 13.4.3),

$$\tilde{y}_s = \begin{bmatrix} y_1 - \hat{y}_1 \\ y_2 - \hat{y}_2 \\ \text{ssa}(y_3 - \hat{y}_3) \end{bmatrix} \tag{13.126}$$

when implementing the observer. The matrices $\boldsymbol{K}_1(\omega_0) \in \mathbb{R}^{6\times3}$ and $\boldsymbol{K}_{2,3,4} \in \mathbb{R}^{3\times3}$ are observer gain matrices to be interpreted later. Note that $\boldsymbol{K}_1(\omega_0)$ is a function of the wave spectrum peak frequency ω_0, allowing for gain scheduling.

Observer Estimation Errors

As for (13.118)–(13.119), the system (13.121), (13.122) and (13.125) is written in state-space form

$$\dot{\hat{\boldsymbol{\eta}}}_0 = \boldsymbol{A}_0\hat{\boldsymbol{\eta}}_0 + \boldsymbol{B}_0\boldsymbol{R}(t)\hat{\boldsymbol{\nu}} + \boldsymbol{K}_0(\omega_0)\tilde{\boldsymbol{y}} \tag{13.127}$$

$$\hat{\boldsymbol{y}} = \boldsymbol{C}_0\hat{\boldsymbol{\eta}}_0 \tag{13.128}$$

where $\hat{\boldsymbol{\eta}}_0 = [\hat{\boldsymbol{\xi}}^\top, \hat{\boldsymbol{\eta}}^\top]^\top$ and

$$\boldsymbol{K}_0(\omega_0) = \begin{bmatrix} \boldsymbol{K}_1(\omega_0) \\ \boldsymbol{K}_2 \end{bmatrix}. \tag{13.129}$$

The estimation errors are defined as $\tilde{\boldsymbol{\nu}} = \boldsymbol{\nu} - \hat{\boldsymbol{\nu}}$, $\tilde{\boldsymbol{b}} = \boldsymbol{b} - \hat{\boldsymbol{b}}$ and $\tilde{\boldsymbol{\eta}}_0 = \boldsymbol{\eta}_0 - \hat{\boldsymbol{\eta}}_0$. Hence, the error dynamics can be written

$$\dot{\tilde{\boldsymbol{\eta}}}_0 = (\boldsymbol{A}_0 - \boldsymbol{K}_0(\omega_0)\boldsymbol{C}_0)\tilde{\boldsymbol{\eta}}_0 + \boldsymbol{B}_0\boldsymbol{R}(t)\tilde{\boldsymbol{\nu}} \tag{13.130}$$

$$\dot{\tilde{\boldsymbol{b}}} = -\boldsymbol{T}^{-1}\tilde{\boldsymbol{b}} - \boldsymbol{K}_3\tilde{\boldsymbol{y}} \tag{13.131}$$

$$\boldsymbol{M}\dot{\tilde{\boldsymbol{\nu}}} = -\boldsymbol{D}\tilde{\boldsymbol{\nu}} + \boldsymbol{R}^\top(t)\tilde{\boldsymbol{b}} - \boldsymbol{R}^\top(t)\boldsymbol{K}_4\tilde{\boldsymbol{y}}. \tag{13.132}$$

The dynamics of the velocity estimation error (13.132) is

$$M\dot{\tilde{v}} = -D\tilde{v} - R^{\top}(t)\tilde{z} \tag{13.133}$$

where

$$\tilde{z} = K_4\tilde{y} - \tilde{b}. \tag{13.134}$$

By defining a new state vector

$$\tilde{x} = \begin{bmatrix} \tilde{\eta}_0 \\ \tilde{b} \end{bmatrix} \tag{13.135}$$

Equations (13.130), (13.131) and (13.134) can be written in compact form as

$$\dot{\tilde{x}} = A\tilde{x} + BR(t)\tilde{v} \tag{13.136}$$

$$\tilde{z} = C\tilde{x} \tag{13.137}$$

where the system matrices are

$$A = \begin{bmatrix} A_0 - K_0(\omega_0)C_0 & 0_{9\times 9} \\ -K_3C_0 & -T^{-1} \end{bmatrix}, \quad B = \begin{bmatrix} B_0 \\ 0_{3\times 3} \end{bmatrix}, \quad C = \begin{bmatrix} K_4C_0 & -I_3 \end{bmatrix}. \tag{13.138}$$

In Figure 13.14 the error signals ε_z and ε_v are defined according to

$$\varepsilon_z := -R^{\top}(t)\,\tilde{z} \tag{13.139}$$

$$\varepsilon_v := R(t)\tilde{v}. \tag{13.140}$$

Thus, the observer error system can be viewed as two linear blocks $\mathcal{H}_1$ and $\mathcal{H}_2$, interconnected through the bounded transformation matrix $R(t)$; that is

$$\mathcal{H}_1 : \{M\dot{\tilde{v}} = -D\tilde{v} + \varepsilon_z \tag{13.141}$$

$$\mathcal{H}_2 : \begin{cases} \dot{\tilde{x}} = A\tilde{x} + B\varepsilon_v \\ \tilde{z} = C\tilde{x} \end{cases}. \tag{13.142}$$

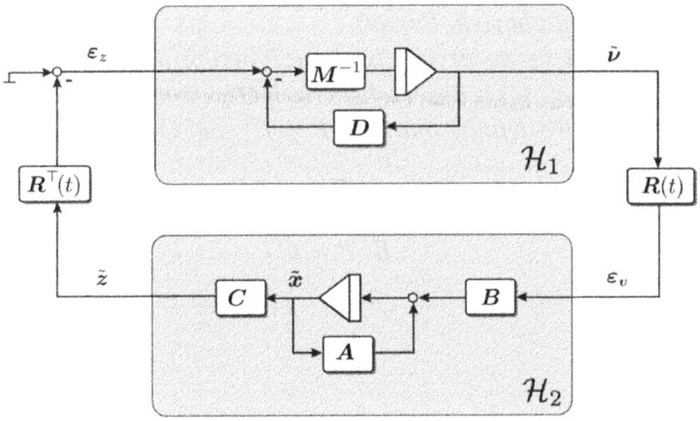

Figure 13.14 Block diagram showing the dynamics of the position/bias and velocity estimation errors.

Stability Analysis

A nonlinear system $\boldsymbol{u} \mapsto \boldsymbol{y}$ is said to be *passive* if there exists a continuously differentiable positive definite function $V(\boldsymbol{x})$ (called storage function) such that (Khalil 2002, Definition 6.3),

$$\boldsymbol{u}^{\top}\boldsymbol{y} \geq \dot{V} \tag{13.143}$$

and *strictly passive* if

$$\boldsymbol{u}^{\top}\boldsymbol{y} \geq \dot{V} + \boldsymbol{x}^{\top}\boldsymbol{Q}\boldsymbol{x} \tag{13.144}$$

for some $\boldsymbol{Q} = \boldsymbol{Q}^{\top} > 0$. Based on the physical properties of the marine craft dynamics, the following statement can be made.

Proposition 13.1 (Strictly Passive Velocity Error Dynamics)
The mapping $\mathcal{H}_1$ is strictly passive.

Proof. *Let,*

$$S_1 = \frac{1}{2}\tilde{\boldsymbol{\nu}}^{\top}\boldsymbol{M}\tilde{\boldsymbol{\nu}} \tag{13.145}$$

be a positive definite storage function. Time differentiation of S_1 along the trajectories of $\tilde{\boldsymbol{\nu}}$ yields

$$\dot{S}_1 = -\frac{1}{2}\tilde{\boldsymbol{\nu}}^{\top}(\boldsymbol{D} + \boldsymbol{D}^{\top})\tilde{\boldsymbol{\nu}} - \tilde{\boldsymbol{z}}^{\top}\boldsymbol{R}(t)\tilde{\boldsymbol{\nu}}. \tag{13.146}$$

Using the fact that $\boldsymbol{\varepsilon}_z = -\boldsymbol{R}^{\top}(t)\tilde{\boldsymbol{z}}$, yields

$$\boldsymbol{\varepsilon}_z^{\top}\tilde{\boldsymbol{\nu}} = \dot{S}_1 + \frac{1}{2}\tilde{\boldsymbol{\nu}}^{\top}(\boldsymbol{D} + \boldsymbol{D}^{\top})\tilde{\boldsymbol{\nu}}. \tag{13.147}$$

Hence, according to (13.144), the mapping $\boldsymbol{\varepsilon}_z \mapsto \tilde{\boldsymbol{\nu}}$ corresponding to $\mathcal{H}_1$ is strictly passive.

The next step is choose the observer gain matrices such that the block $\mathcal{H}_2$ is strictly passive. The tool for this is the *Kalman–Yakubovich–Popov lemma.*

Lemma 13.1 (Kalman–Yakubovich–Popov)
Let $\boldsymbol{H}(s) = \boldsymbol{C}(s\boldsymbol{I}_m - \boldsymbol{A})^{-1}\boldsymbol{B}$ be an $m \times m$ transfer function matrix, where $\boldsymbol{A}$ is Hurwitz, $(\boldsymbol{A}, \boldsymbol{B})$ is controllable and $(\boldsymbol{A}, \boldsymbol{C})$ is observable. Then $\boldsymbol{H}(s)$ is strictly positive real (SPR) if and only if there exist positive definite matrices $\boldsymbol{P} = \boldsymbol{P}^{\top}$ and $\boldsymbol{Q} = \boldsymbol{Q}^{\top}$ such that

$$\boldsymbol{P}\boldsymbol{A} + \boldsymbol{A}^{\top}\boldsymbol{P} = -\boldsymbol{Q} \tag{13.148}$$

$$\boldsymbol{B}^{\top}\boldsymbol{P} = \boldsymbol{C}. \tag{13.149}$$

Proof. *See Khalil (2002).*

Theorem 13.1 (Strictly Passive Observer Error Dynamics)
Assume that the observer gains $\boldsymbol{K}_i$ $(i = 1, \ldots, 4)$ are chosen such that (13.142) satisfies the KYP lemma. Then the interconnected system (13.141) and (13.142) is strictly passive.

Proof. *Since it is established that $\mathcal{H}_1$ is strictly passive (Proposition 13.1) and $\mathcal{H}_2$, which is given by the matrices $(\boldsymbol{A}, \boldsymbol{B}, \boldsymbol{C})$, can be made SPR by choosing the gain matrices $\boldsymbol{K}_i$ $(i = 1, \dots, 4)$ according to the KYP lemma, the interconnected system (13.141) and (13.142) is strictly passive. Note that post-multiplication with the bounded transformation matrix $\boldsymbol{R}(t)$ and pre-multiplication by its transpose will not affect the passivity properties of the blocks $\mathcal{H}_1$ and $\mathcal{H}_2$.*

The equilibrium point of the passive observer is also exponentially stable. This follows from the following theorem.

Theorem 13.2 (Exponential Stability)
Under assumptions P1 and P2 the nonlinear observer given by (13.121)–(13.125) is exponentially stable. The heading angle error is computed as the smallest signed angle belonging to the set $[-\pi, \pi)$. Hence, the result will be global on $\mathbb{S}^1$ (circle in the plane) but not on the set $\mathbb{R}$.

Proof. *Consider the following Lyapunov function candidate*

$$V = \tilde{\boldsymbol{v}}^\top \boldsymbol{M} \tilde{\boldsymbol{v}} + \tilde{\boldsymbol{x}}^\top \boldsymbol{P} \tilde{\boldsymbol{x}}. \tag{13.150}$$

Differentiation of V along the trajectories of $\tilde{\boldsymbol{v}}$ and $\tilde{\boldsymbol{x}}$ yields

$$\dot{V} = -\tilde{\boldsymbol{v}}^\top (\boldsymbol{D} + \boldsymbol{D}^\top) \tilde{\boldsymbol{v}} + \tilde{\boldsymbol{x}}^\top (\boldsymbol{P}\boldsymbol{A} + \boldsymbol{A}^\top \boldsymbol{P}) \tilde{\boldsymbol{x}} + 2\tilde{\boldsymbol{v}}^\top \boldsymbol{R}^\top(t) \boldsymbol{B}^\top \boldsymbol{P} \tilde{\boldsymbol{x}} - 2\tilde{\boldsymbol{v}}^\top \boldsymbol{R}^\top(t) \tilde{\boldsymbol{z}}. \tag{13.151}$$

Application of the KYP lemma with $\boldsymbol{B}^\top \boldsymbol{P} \tilde{\boldsymbol{x}} = \boldsymbol{C} \tilde{\boldsymbol{x}} = \tilde{\boldsymbol{z}}$ yields

$$\dot{V} = -\tilde{\boldsymbol{v}}^\top (\boldsymbol{D} + \boldsymbol{D}^\top) \tilde{\boldsymbol{v}} - \tilde{\boldsymbol{x}}^\top \boldsymbol{Q} \tilde{\boldsymbol{x}} < 0, \quad \forall \tilde{\boldsymbol{x}} \neq 0, \ \tilde{\boldsymbol{v}} \neq \boldsymbol{0}. \tag{13.152}$$

Consequently, asymptotic stability follows from LaSalle–Yoshizawa's theorem (see Appendix A.2.2). Since the error dynamics of the two blocks $\mathcal{H}_1$ and $\mathcal{H}_2$ are linear, asymptotic stability implies exponential stability.

Determination of the Observer Gains

In practice it is easy to find a set of gain matrices $\boldsymbol{K}_i$ $(i = 1, \dots, 4)$ satisfying the KYP lemma. Note that the mapping $\boldsymbol{\varepsilon}_v \mapsto \tilde{\boldsymbol{z}}$ (block $\mathcal{H}_2$) describes three decoupled systems in surge, sway and yaw. This suggests that the observer gain matrices should have a diagonal structure

$$\boldsymbol{K}_1(\omega_0) = \begin{bmatrix} \text{diag}\{K_{11}(\omega_0), \ K_{12}(\omega_0), \ K_{13}(\omega_0)\} \\ \text{diag}\{K_{14}(\omega_0), \ K_{15}(\omega_0), \ K_{16}(\omega_0)\} \end{bmatrix} \tag{13.153}$$

$$\boldsymbol{K}_2 = \text{diag}\{K_{21}, \ K_{22}, \ K_{23}\} \tag{13.154}$$

$$\boldsymbol{K}_3 = \text{diag}\{K_{31}, \ K_{32}, \ K_{33}\} \tag{13.155}$$

$$\boldsymbol{K}_4 = \text{diag}\{K_{41}, \ K_{42}, \ K_{43}\}. \tag{13.156}$$

Consequently, three decoupled transfer functions can be found

$$H(s) = \mathrm{diag}\{h_1(s),\ h_2(s),\ h_3(s)\} \tag{13.157}$$

such that

$$\begin{aligned} \tilde{z}(s) &= H(s)\varepsilon_{\nu}(s) \\ &:= H_0(s)H_{\mathrm{B}}(s)\varepsilon_{\nu}(s) \end{aligned} \tag{13.158}$$

where

$$H_0(s) = C_0(sI_3 + A_0 - K_0(\omega_0)C_0)^{-1}B_0$$
$$H_{\mathrm{B}}(s) = K_4 + (sI_3 + T^{-1})^{-1}K_3.$$

The diagonal structure of $H(s)$ is illustrated in Figure 13.15. The transfer functions $h_{0i}(s)$ and $h_{\mathrm{B}i}(s)$ $(i = 1, \ldots, 3)$ corresponding to $H_0(s)$ and $H_{\mathrm{B}}(s)$, respectively, become

$$h_{0i}(s) = \frac{s^2 + 2\lambda\omega_0 s + \omega_0^2}{s^3 + (K_{1(i+3)} + K_{2i} + 2\lambda\omega_0)s^2 + (\omega_0^2 + 2\lambda\omega_0 K_{2i} - K_{1i}\omega_0^2)s + K_{2i}\omega_0^2} \tag{13.159}$$

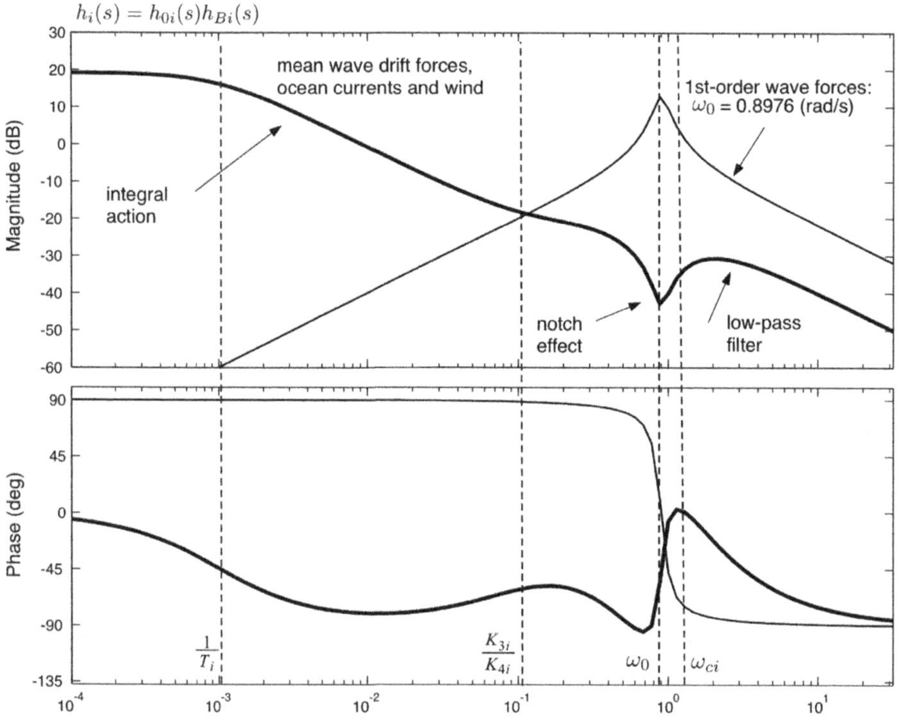

Figure 13.15 Bode plots showing the transfer function $h_i(s)$ in surge $(i = 1)$ when $1/T_i \ll K_{3i}/K_{4i} < \omega_0 < \omega_{ci}$; see ExPassiveObs.m in the MSS toolbox.

$$h_{Bi}(s) = K_{4i}\frac{s + \left(\frac{1}{T_i} + \frac{K_{3i}}{K_{4i}}\right)}{s + \frac{1}{T_i}}$$

$$\overset{T_i \gg 1}{\approx} K_{4i}\frac{s + \frac{K_{3i}}{K_{4i}}}{s + \frac{1}{T_i}}. \tag{13.160}$$

In order to obtain the desired notch effect (wave filtering) of the observer, the desired shape of $h_{0i}(s)$ is specified as

$$h_{di}(s) = \frac{s^2 + 2\lambda\omega_0 s + \omega_0^2}{(s^2 + 2\zeta_{ni}\omega_0 s + \omega_0^2)(s + \omega_{ci})} \tag{13.161}$$

where $\zeta_{ni} > \lambda$ determines the notch and $\omega_{ci} > \omega_0$ is the filter cutoff frequency. Typically $\zeta_{ni} = 1.0$ and $\lambda = 0.1$. Equating (13.159) and (13.161) yields the following formulae for the filter gains in $\boldsymbol{K}_1(\omega_0)$ and $\boldsymbol{K}_2$,

$$K_{1i}(\omega_0) = -2(\zeta_{ni} - \lambda)\frac{\omega_{ci}}{\omega_0} \tag{13.162}$$

$$K_{1(i+3)}(\omega_0) = 2\omega_0(\zeta_{ni} - \lambda) \tag{13.163}$$

$$K_{2i} = \omega_{ci} \tag{13.164}$$

Note that the filter gains can be gain-scheduled with respect to the dominating wave frequencies ω_0 if desired. Algorithms for wave frequency estimation are given in Section 13.2.3. In Figure 13.15 the transfer function $h_i(s) = h_{Bi}(s)h_{0i}(s)$ is illustrated when all filter gains are properly selected. It is important that the three decoupled transfer functions $h_i(s)$ all have a phase greater than $-90°$ in order to meet the SPR requirement. It turns out that the KYP lemma and therefore the SPR requirement can easily be satisfied if the following tuning rules for T_i, K_{3i} and K_{4i} are applied

$$1/T_i \ll K_{3i}/K_{4i} < \omega_0 < \omega_{ci} \qquad (i = 1, \ldots, 3). \tag{13.165}$$

Here ω_0 is the wave peak frequency and $T_i \gg 1$ $(i = 1, \ldots, 3)$ are the bias time constants used to specify the limited integral effect in the bias estimator.

Computer Simulations and Experimental Results

A combination of computer simulations and full-scale experiments have been used to evaluate the performance and robustness of the nonlinear passive observer.

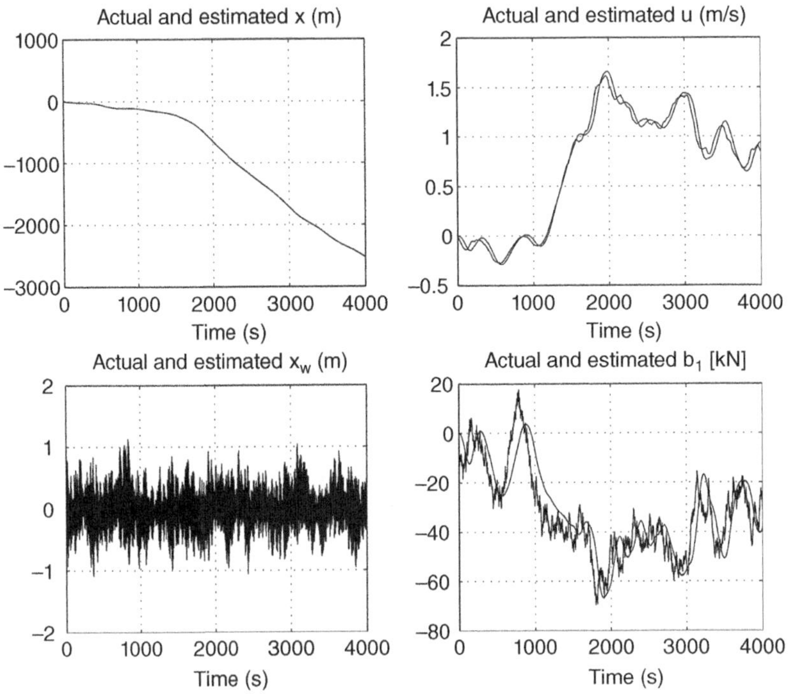

Figure 13.16 Simulation study: LF and WF position, velocity, bias and their estimates in surge.

Example 13.5 (Passive DP Observer)

The case studies are based on the following models of the ship–bias–wave system (Fossen and Strand 1999b)

$$\boldsymbol{M} = \begin{bmatrix} 5.3122 \times 10^6 & 0 & 0 \\ 0 & 8.2831 \times 10^6 & 0 \\ 0 & 0 & 3.7454 \times 10^9 \end{bmatrix} \tag{13.166}$$

$$\boldsymbol{D} = \begin{bmatrix} 5.0242 \times 10^4 & 0 & 0 \\ 0 & 2.7229 \times 10^5 & -4.3933 \times 10^6 \\ 0 & -4.3933 \times 10^6 & 4.1894 \times 10^8 \end{bmatrix} \tag{13.167}$$

with the origin CO of the $\{b\}$ frame is located in the CG. In the experiments the bias time constants were chosen as

$$\boldsymbol{T} = \text{diag}\{1000,\ 1000,\ 1000\}. \tag{13.168}$$

The wave model parameters were chosen as $\lambda = 0.1$ and $\omega_0 = 0.8976$ rad s^{-1}, corresponding to a wave period of 7.0 s. *The notch filter parameters were chosen as $\zeta_{ni} = 1.0$ and $\omega_{ci} = 1.2255\,\omega_0 = 1.1$ rad s^{-1}. From (13.162)–(13.164) we get (see the MSS toolbox* *script* `ExPassiveObs.m`)

$$\boldsymbol{K}_1 = \begin{bmatrix} -\text{diag}\{2.2059,\ 2.2059,\ 2.2059\} \\ \text{diag}\{1.6157,\ 1.6157,\ 1.6157\} \end{bmatrix} \tag{13.169}$$

$$\boldsymbol{K}_2 = \text{diag}\{1.1,\ 1.1,\ 1.1\}. \tag{13.170}$$

The loop transfer function $h_i(s) - h_{Bi}(s)h_{0i}(s)$ *for*

$$\boldsymbol{K}_3 = 0.1\ \boldsymbol{K}_4 \tag{13.171}$$

$$\boldsymbol{K}_4 = \text{diag}\{0.1,\ 0.1,\ 0.01\} \tag{13.172}$$

is plotted in Figure 13.15

The simulation study was performed with non-zero noise terms $\boldsymbol{\varepsilon}$ and $\boldsymbol{w}$ even though these terms were assumed to be zero in the Lyapunov analysis. This was done to demonstrate the excellent performance of the deterministic observer in the presence of stochastic noise.

The results of the computer simulations are shown in Figures 13.16 and 13.17. The plots illustrate that all state estimates converge to their true values. In Figures 13.18 amd 13.19 full-scale experimental results with the same observer are reported. Again, excellent convergence and performance in surge, sway and yaw are observed. In the full-scale experiment it was not possible to verify that the velocity estimates converged to their true values; see the lower plots in Figure 13.19. The main reason for this was that only GNSS position measurements were available. However, simulation studies indicate that the velocity estimates converge to their true values as well.

13.5.2 Case Study: Passive Observer for Heading Autopilots using only Compass Measurements

The DP observer in Section 13.5.1 can be reduced to one DOF and used in autopilot designs. For this purpose, the autopilot model in Section 13.4.5 is revisited. In the 1-DOF case, the compass measurement is taken as the sum of the LF and WF signals

$$y = \psi + \psi_{\text{w}}. \tag{13.173}$$

The corresponding system model in yaw is

$$\dot{\xi}_{\text{w}} = \psi_{\text{w}} \tag{13.174}$$

$$\dot{\psi}_{\text{w}} = -\omega_0^2\ \xi_{\text{w}} - 2\lambda\omega_0\ \psi_{\text{w}} \tag{13.175}$$

$$\dot{\psi} = r \tag{13.176}$$

$$\dot{r} = -\frac{1}{T}r + \frac{1}{m}(\tau_{\text{wind}} + \tau_{\text{N}}) + b \tag{13.177}$$

$$\dot{b} = -\frac{1}{T_{\text{b}}}b \tag{13.178}$$

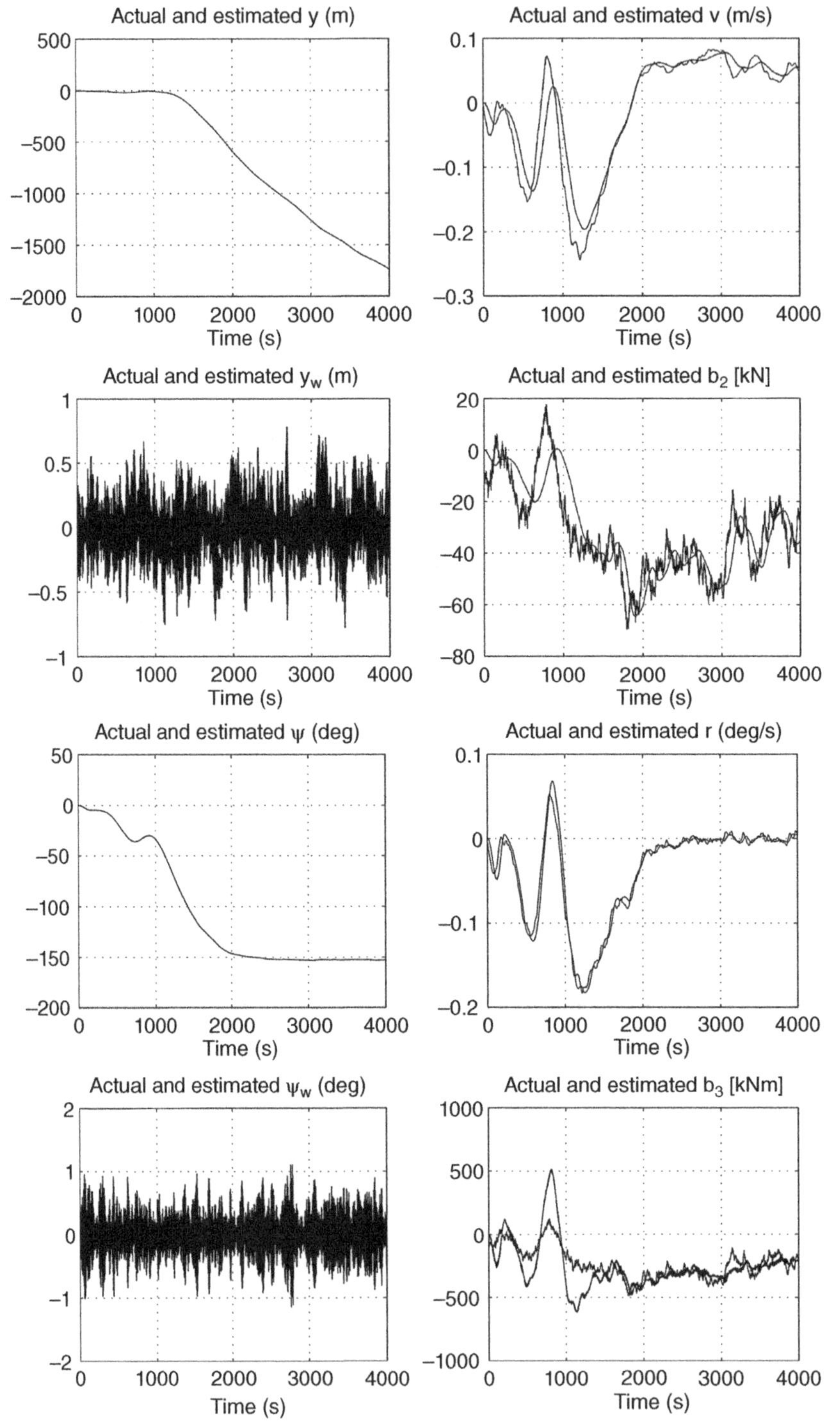

Figure 13.17 Simulation study: LF and WF position, velocity, bias and their estimates in sway and yaw.

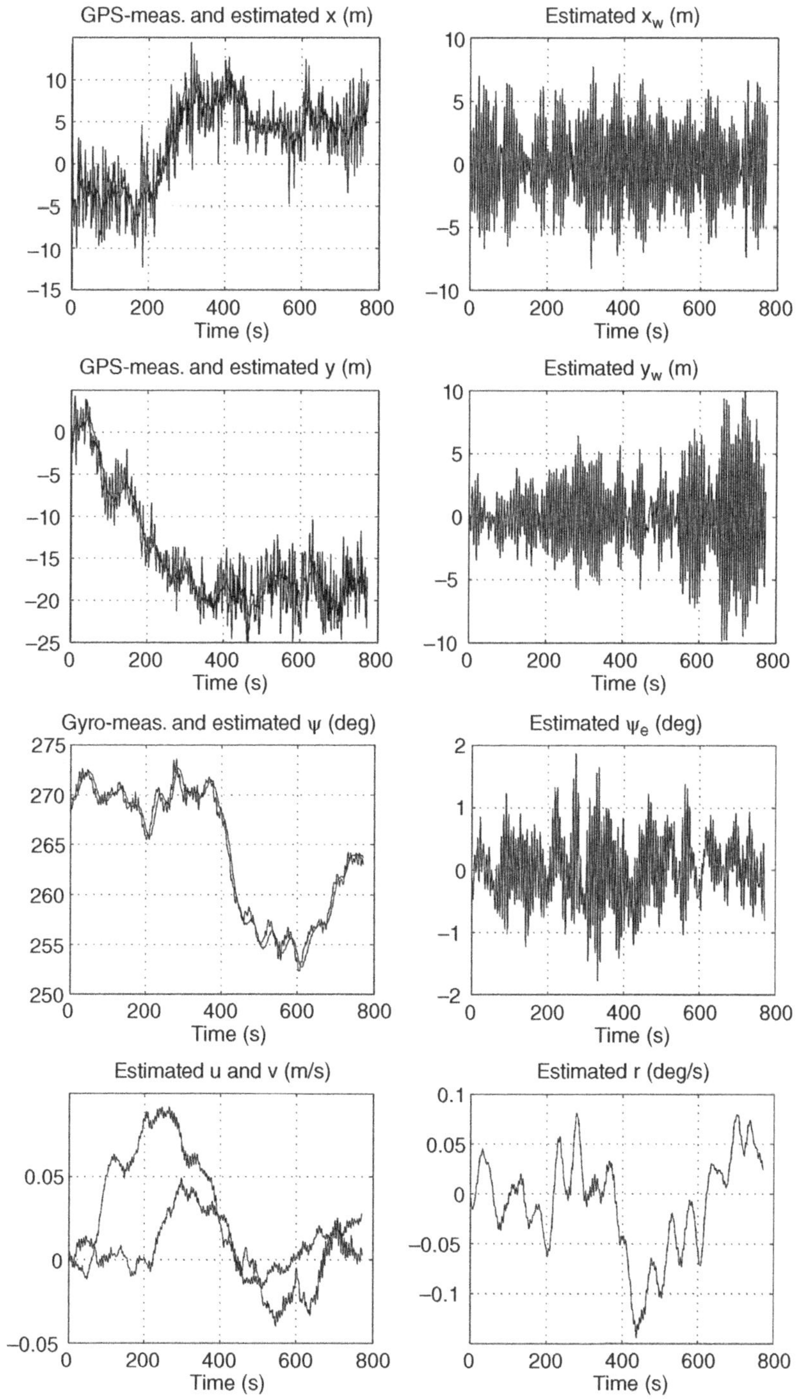

Figure 13.18 Experimental data. Three upper plots: actual position (LF+WF) with estimates of the LF and WF positions in surge, sway and yaw. Lower plots: estimates of the LF velocities.

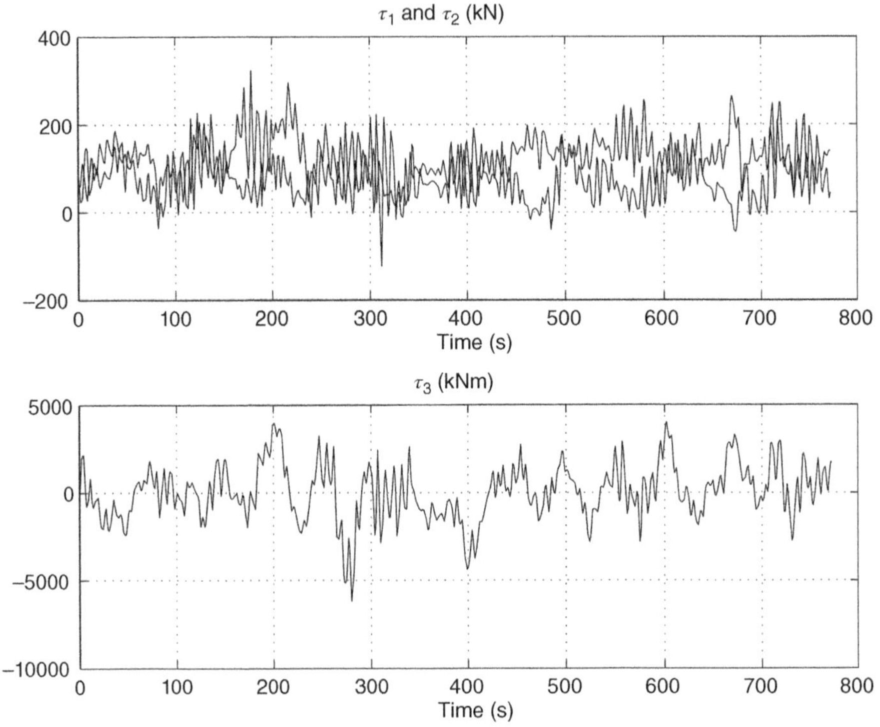

Figure 13.19 Experimental data: control inputs in surge, sway and yaw.

where λ and ω_0 are the relative damping ratio and peak frequency of the wave spectrum, respectively. The constant $m = I_z - N_{\dot{r}}$ is introduced for convenience such that the rudder angle δ generates a yaw moment $\tau_{\rm N}$ given by

$$\begin{aligned}\tau_{\rm N} &= m\frac{K}{T}\delta \\ &= N_{\delta}\delta \end{aligned} \tag{13.179}$$

while $\tau_{\rm wind}$ represents an optional term for wind feedforward. Note that neither the yaw rate r nor the wave states $\xi_{\rm w}$ and $\psi_{\rm w}$ are measured. The resulting state-space model is

$$\dot{\boldsymbol{x}} = \boldsymbol{A}\boldsymbol{x} + \boldsymbol{b}u \tag{13.180}$$

$$y = \boldsymbol{c}^{\top}\boldsymbol{x} \tag{13.181}$$

where $\boldsymbol{x} = [\xi_{\rm w},\ \psi_{\rm w},\ \psi,\ r,\ b]^{\top}, u = \tau_{\rm wind} + \tau_{\rm N}$ and

$$\boldsymbol{A} = \left[\begin{array}{cc|ccc} 0 & 1 & 0 & 0 & 0 \\ -\omega_0^2 & -2\lambda\omega_0 & 0 & 0 & 0 \\ \hline 0 & 0 & 0 & 1 & 0 \\ 0 & 0 & 0 & -1/T & 1 \\ 0 & 0 & 0 & 0 & -1/T_{\rm b} \end{array}\right], \quad \boldsymbol{b} = \left[\begin{array}{c} 0 \\ 0 \\ \hline 0 \\ 1/m \\ 0 \end{array}\right] \tag{13.182}$$

$$\boldsymbol{c}^{\top} = [0,\ 1,\ 1,\ 0,\ 0]. \tag{13.183}$$

The passive observer copying the dynamics (13.180) is chosen as

$$\dot{\hat{\boldsymbol{x}}} = \boldsymbol{A}\hat{\boldsymbol{x}} + \boldsymbol{b}u + \boldsymbol{k}\ \text{ssa}(y - \boldsymbol{c}^{\top}\hat{\boldsymbol{x}}) \tag{13.184}$$

where the ssa modification has been applied to the yaw angle estimation errros (see Section 13.4.3). Expanding this expression gives

$$\dot{\hat{\xi}}_{\text{w}} = \hat{\psi}_{\text{w}} + K_1\ \text{ssa}(y - \hat{\psi} - \hat{\psi}_{\text{w}}) \tag{13.185}$$

$$\dot{\hat{\psi}}_{\text{w}} = -\omega_0^2\,\hat{\xi}_{\text{w}} - 2\lambda\omega_0\hat{\psi}_{\text{w}} + K_2\ \text{ssa}(y - \hat{\psi} - \hat{\psi}_{\text{w}}) \tag{13.186}$$

$$\dot{\hat{\psi}} = \hat{r} + K_3\ \text{ssa}(y - \hat{\psi} - \hat{\psi}_{\text{w}}) \tag{13.187}$$

$$\dot{\hat{r}} = -\frac{1}{T}\hat{r} + \frac{1}{m}(\tau_{\text{wind}} + \tau_{\text{N}}) + \hat{b} + K_4\ \text{ssa}(y - \hat{\psi} - \hat{\psi}_{\text{w}}) \tag{13.188}$$

$$\dot{\hat{b}} = -\frac{1}{T_{\text{b}}}\hat{b} + K_5\ \text{ssa}(y - \hat{\psi} - \hat{\psi}_{\text{w}}). \tag{13.189}$$

The observer gains K_1, K_2, K_3, K_4 and K_5 can be computed by noticing that the observer error dynamics can be reformulated as two subsystems for yaw angle/rudder bias and yaw rate. These systems form a *passive interconnection* if the observer gains are chosen according to

$$\boldsymbol{k} = \begin{bmatrix} -2(1-\lambda)\dfrac{\omega_{\text{c}}}{\omega_0} \\ 2\omega_0(1-\lambda) \\ \omega_{\text{c}} \\ K_4 \\ K_5 \end{bmatrix} \tag{13.190}$$

where $\omega_{\text{c}} > \omega_0$ is the filter cutoff frequency and the remaining gains must satisfy

$$0 < 1/T_{\text{b}} < K_5/K_4 < \omega_0 < \omega_{\text{c}}. \tag{13.191}$$

The design problem is now reduced to choosing K_4 and K_5 such that the ratio K_5/K_4 satisfies the passive gain constraint (13.191). A more detailed analysis of the passive observer is done in Section 13.5.1, which discusses applications to ship positioning in 3 DOFs.

Example 13.6 (Passive Wave Filtering)

Consider the Mariner class cargo ship with $K = 0.185\ \text{s}^{-1}$*, equivalent time constant* $T = T_1 + T_2 - T_3 = 107.3\ \text{s}$ *and input* $\tau_{\text{N}}/m = (K/T)\delta$*, where* δ *is the rudder angle (Chislett* *and Strøm-Tejsen 1965a). The bias time constant is chosen to be rather large, for instance* $T_b = 100$ s. *The wave response model is modeled by a linear approximation to the JONSWAP spectrum with* $\lambda = 0.1$ *and* $\omega_0 = 1.2 \text{ rad s}^{-1}$ *(see Section 10.2.6). Hence, (13.89) and (13.90) become*

$$A = \left[\begin{array}{cc|ccc} 0 & 1 & 0 & 0 & 0 \\ -1.44 & -0.24 & 0 & 0 & 0 \\ \hline 0 & 0 & 0 & 1 & 0 \\ 0 & 0 & 0 & -0.0093 & 1 \\ 0 & 0 & 0 & 0 & -0.01 \end{array}\right], \quad b = \left[\begin{array}{c} 0 \\ 0 \\ \hline 0 \\ 0.0017 \\ 0 \end{array}\right] \tag{13.192}$$

$$E = \left[\begin{array}{c|cc} 0 & 0 & 0 \\ 0.24\,\sigma & 0 & 0 \\ \hline 0 & 0 & 0 \\ 0 & 1 & 0 \\ 0 & 0 & 1 \end{array}\right], \quad c^{\top} = [0, 1, 1, 0, 0] \tag{13.193}$$

where $\sigma > 0$ *defines the sea state. Using passivity as a tool for filter design with cutoff frequency* $\omega_c = 1.1\ \omega_0$ *yields*

$$k = \left[\begin{array}{c} K_1 \\ K_2 \\ \hline K_3 \\ K_4 \\ K_5 \end{array}\right] = \left[\begin{array}{c} -2(1-\lambda)\dfrac{\omega_c}{\omega_0} \\ 2\omega_0(1-\lambda) \\ \hline \omega_c \\ K_4 \\ K_5 \end{array}\right] = \left[\begin{array}{c} -1.98 \\ 1.80\ \omega_0 \\ \hline 1.10\ \omega_0 \\ K_4 \\ K_5 \end{array}\right]. \tag{13.194}$$

This clearly shows that the gains should be adjusted with varying frequency ω_0. *Choosing* $K_4 = 0.1$ *and* $K_5 = 0.01$ *such that* $K_5/K_4 = 0.1$ *yields the transfer functions shown later in Figure 13.21. Note that the notch effect at* ω_0 *for* $h_3(s)$ *and* $h_4(s)$ *represents the state estimates* $\hat{\psi}$ *and* $\hat{r}$. *We also see that high-frequency motion components above* ω_c *are low-pass filtered. Finally, the transfer function* $h_2(s)$ *representing reconstruction of the WF motion* $\hat{\psi}_w$ *filters out signals on the outside of the wave response spectrum, while signals close to* ω_0 *pass through the filter with unity gain, that is* 0 dB. *The time series for* $\sigma = 6.25$ *are shown in Figure 13.20.*

Wave Filter Frequency Analysis

Consider the state estimator

$$\dot{\hat{x}} = A\hat{x} + bu + k(y - c^{\top}\hat{x}). \tag{13.195}$$

It is then straightforward to show that

$$\hat{x}(s) = (sI_5 - A + kc^{\top})^{-1}(ky(s) + bu(s)). \tag{13.196}$$

Assume that $u(s) = 0$ (no feedback) such that

$$h(s) = [h_1,\ h_2,\ h_3,\ h_4,\ h_5]^{\top} = (sI_5 - A + kc^{\top})^{-1}k. \tag{13.197}$$

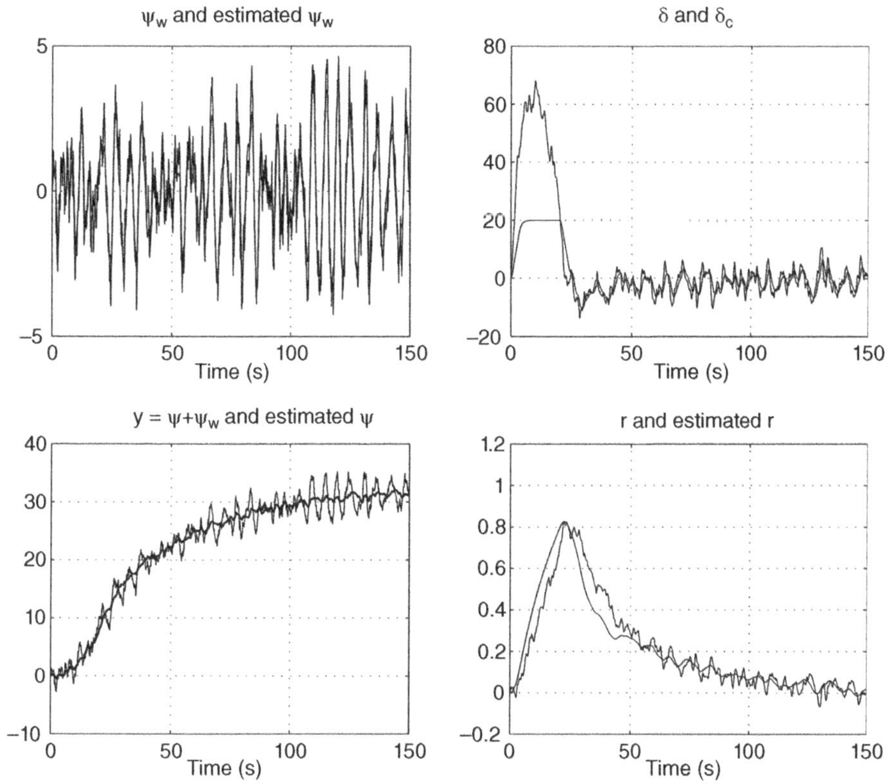

Figure 13.20 Time series showing the performance of the passive wave filter.

The states of interest are

$$\hat{\psi}_w(s) = h_2(s)y(s) \tag{13.198}$$

$$\hat{\psi}(s) = h_3(s)y(s) \tag{13.199}$$

$$\hat{r}(s) = h_4(s)y(s) \tag{13.200}$$

where $h_3(s)$ represents a notch filter with a low-pass filter in cascade

$$h_3(s) = h_{\text{notch}}(s)\, h_{\text{lp}}(s). \tag{13.201}$$

The filter $h_4(s)$ also possesses notch filtering in cascade with a second filter representing a limited differentiator for generation of $\hat{r}(s)$ from $y(s)$. Note that $h_2(s)$ is close to 1 (0 dB) in a band around the wave spectrum, while lower and higher frequencies are suppressed in order to reconstruct $\psi_w(s)$ from $y(s)$. This can be seen from the Bode plots in Figure 13.21. These results have also been theoretically verified by Grimble (1978). In this work Grimble showed that the *stationary Kalman filter* for the ship positioning problem will be approximately equivalent to a notch filter in cascade with a second filter, typically a low-pass filter.

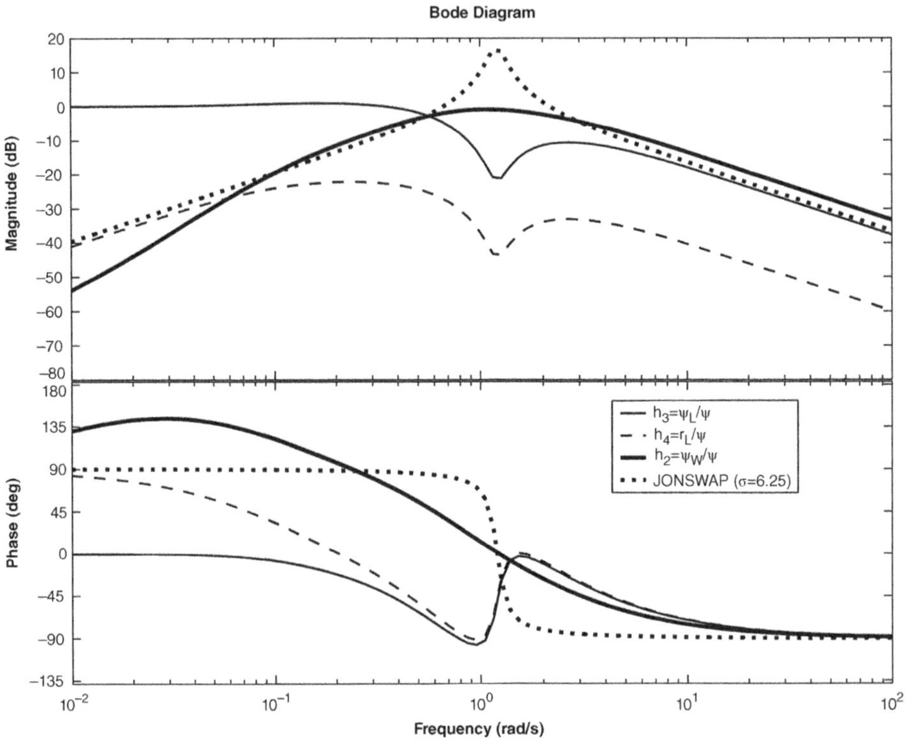

Figure 13.21 Bode plots showing the wave filter transfer functions and the JONSWAP spectrum.

When including the feedback term $u(s)$ in the analysis, it is well known that application of an observer is superior to notch and low-pass filtering in cascade, since the observer uses the input $u(s)$ for prediction in addition to filtering the measured output $y(s)$. In fact, this input signal reduces the problems associated with additional phase lag in the filtered signal, which is the main problem with most standard filters (low-pass, high-pass and notch). Simulation results verifying these observations have been documented by Grimble (1978).

13.5.3 Case Study: Passive Observer for Heading Autopilots using both Compass and Angular Rate Sensor Measurements

In this section the design of the previous section is modified to include an angular rate sensor (ARS) in addition to the compass (see Section 14.1). This is advantageous since the ARS can be integrated with the compass in an optimal manner, resulting in less variance and better accuracy of the state estimates. One simple way to do this is to treat the ARS measurement as an input u_{ars} to the system model by writing the yaw dynamics according to

$$\dot{\psi} = u_{\text{ars}} + b_{\text{ars}} \tag{13.202}$$

where $b_{\rm ars}$ denotes the ARS bias. The WF model is similar to (13.174) and (13.175). This model will give proper wave filtering of the state ψ. However, the estimate of $\dot{\psi} = r$ is not wave filtered, since this signal is taken directly from the ARS measurement. This can be improved by filtering $u_{\rm ars}$ with a notch filter $h_{\rm notch}(s)$ and a low-pass filter $h_{\rm lp}(s)$ to the cost of some phase lag

$$u_{\rm f} = h_{\rm notch}(s)\, h_{\rm lp}(s)\, u_{\rm ars}. \tag{13.203}$$

The resulting observer equations are

$$\dot{\hat{\xi}}_{\rm w} = \hat{\psi}_{\rm w} + K_1\ {\rm ssa}(y - \hat{\psi} - \hat{\psi}_{\rm w}) \tag{13.204}$$

$$\dot{\hat{\psi}}_{\rm w} = -\omega_0^2\hat{\xi}_{\rm w} - 2\lambda\omega_0\hat{\psi}_{\rm w} + K_2\ {\rm ssa}(y - \hat{\psi} - \hat{\psi}_{\rm w}) \tag{13.205}$$

$$\dot{\hat{\psi}} = u_f - \hat{b}_{\rm ars} + K_3\ {\rm ssa}(y - \hat{\psi} - \hat{\psi}_{\rm w}) \tag{13.206}$$

$$\dot{\hat{b}}_{\rm ars} = -\frac{1}{T_b}\hat{b}_{\rm ars} + K_4\ {\rm ssa}(y - \hat{\psi} - \hat{\psi}_{\rm w}) \tag{13.207}$$

where $T_{\rm b} \gg 0$. Note that the ARS bias must be estimated online since it will vary with temperature and possible scale factor/misalignment errors when mounted onboard the ship. This is a slowly-varying process so the gain K_4 can be chosen quite small, reflecting a large bias time constant. If passivity-based pole placement (13.190) is used, the formulae for K_1, K_2 and K_3 become

$$K_1 = -2\frac{\omega_{\rm c}}{\omega_0}(1 - \lambda) \tag{13.208}$$

$$K_2 = 2\omega_0(1 - \lambda) \tag{13.209}$$

$$K_3 = \omega_{\rm c}. \tag{13.210}$$

The observer gains can be adjusted for varying sea states by estimating ω_0 online as described in Section 13.2.3. Gain-scheduling techniques are used in nonlinear control theory to obtain satisfactory performance for different operating points of the system.

14

Inertial Navigation Systems

Today inertial measurement technology is available for commercial users thanks to a significant reduction in price over the last few decades. As a consequence of this, low-cost *microelectromechanical system* (MEMS) sensors can be integrated with a satellite navigation system using a conventional Kalman filter.

The inertial navigation algorithms depend on sensors for acceleration and angular rates to compute an estimate of the position, velocity and attitude of the craft. An inertial navigation system (INS) consists of the following:

- **Inertial measurement unit (IMU)** containing a cluster of sensors. The sensors are rigidly mounted to a common base to maintain the same relative orientations.
- **Navigation computer** to calculate the gravitational acceleration (not measured by accelerometers) and integrate the acceleration and angular rate measurements to maintain an estimate of the position, velocity and attitude of the craft.

There are two primary types of INS:

- **Gimbal-mounted systems** using accelerometers or an IMU in a platform, that is isolated from the craft's rotations through a set of gimbals, thus maintaining a fixed orientation in space. Three gimbals would technically suffice, but most systems use four in order to avoid what is called gimbal-lock, which happens when the axis of two of the gimbals are driven in the same direction, disabling their isolation capabilities. Mechanized systems use rotation sensing devices (angular rate sensors), feedback control and an actuated platform to keep the IMU in a fixed orientation in space.
- **Strapdown systems** use an IMU, which is strapped to the craft. Consequently, the IMU will pick up the motions of the craft. This implies that the *strapdown navigation equations* must be integrated online in a state estimator to accurately describe the motions of the IMU and the marine craft in order to separate these.

Gimbal-mounted systems are expensive compared to strapdown INS, while strapdown systems require more computational power than gimbal-mounted systems. The development of solid-state sensors based on MEMS technology make strapdown INSs very

Handbook of Marine Craft Hydrodynamics and Motion Control, Second Edition. Thor I. Fossen.

affordable but strapdown navigation systems can have significant error, especially in low-cost systems. This is usually handled by using feedback from accurate positioning systems to remove drift. Techniques for aided INS will be discussed later in this chapter, when designing the state estimators. Useful references are Farrell and Barth (1998), Titterton and Weston (1997), and Grewal *et al.* (2001).

14.1 Inertial Measurement Unit

An INS uses a computer, accelerometers and attitude rate sensors (ARS) to continuously calculate by dead reckoning the position, velocity and attitude of a moving craft without the need for external reference signals. The key sensory component is the IMU, which is composed of the following sensors:

- Three-axis ARS or gyroscopes
- Three-axis accelerometer.

This is referred to as an 6-DOF IMU. Some MEMS-based IMUs have additional measurements such as:

- Three-axis magnetometer
- One-axis barometric pressure (altitude) sensor.

The complete sensor suite will then be 10 DOFs. In addition to this, it is common to include a temperature sensor that can be used for temperature compensation. This is usually implemented as a temperature-indexed lookup table with offset values.

IMU Measurement Frame

The IMU is mounted onboard the craft in a measurement frame $\{m_{\mathrm{I}}\}$ with coordinate origin CM_{I} located at $\boldsymbol{r}^b_{\mathrm{b}m_{\mathrm{I}}} = [x_{m_{\mathrm{I}}}, y_{\mathrm{m_I}}, z_{\mathrm{m_I}}]^{\top}$ with respect to the $\{b\}$ frame coordinate origin CO (see Figure 14.1). It is assumed that the axes of $\{m_{\mathrm{I}}\}$ and $\{b\}$ point in the same directions. Since the IMU is rigidly attached to the body and a lightweight digital computer is used to propagate the navigation equations this is called a *strapdown system*. Thus the need for a mechanical-gimbal system is eliminated.

Instead of transforming the IMU measurements to the coordinate origin CO, the state estimator is formulated in the CM_{I} and the estimated states $\hat{\boldsymbol{v}}^b_{\mathrm{nm_I}}$ and $\hat{\boldsymbol{\omega}}^b_{\mathrm{nm_I}}$ are transformed to the CO using the lever-arm vector $\boldsymbol{r}^b_{\mathrm{b}m_I}$. The tool for this is the transformation matrix in Appendix C, which can be applied in the following manner

$$\begin{bmatrix} \hat{\boldsymbol{v}}^b_{\mathrm{nb}} \\ \hat{\boldsymbol{\omega}}^b_{\mathrm{nb}} \end{bmatrix} = \boldsymbol{H}^{-1}(\boldsymbol{r}^b_{\mathrm{b}m_{\mathrm{I}}}) \begin{bmatrix} \hat{\boldsymbol{v}}^b_{\mathrm{nm_I}} \\ \hat{\boldsymbol{\omega}}^b_{\mathrm{nm_I}} \end{bmatrix} \tag{14.1}$$

$$\Updownarrow$$

$$\hat{\boldsymbol{v}} = \boldsymbol{H}^{-1}(\boldsymbol{r}^b_{\mathrm{b}m_{\mathrm{I}}})\hat{\boldsymbol{v}}_{\mathrm{m_I}}. \tag{14.2}$$

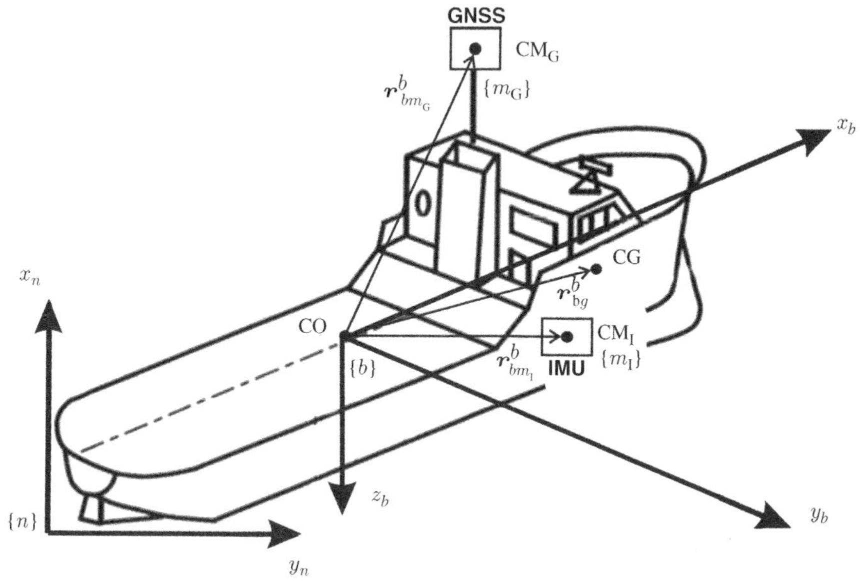

Figure 14.1 The GNSS and IMU measurement frames frames $\{m_\mathrm{G}\}$ and $\{m_\mathrm{I}\}$, and their coordinate origins $\mathrm{CM_G}$ and $\mathrm{CM_I}$, respectively.

It is also possible to formulate the state estimator in the CG by transforming the specific force measurement to the CG. However, this approach involves numerical differentiation of the angular rate measurement to obtain $\dot{\boldsymbol{\omega}}^b_{\mathrm{nb}}$. Assume that the CG is located at $\boldsymbol{r}^b_{\mathrm{bg}} = [x_g, y_g, z_g]^\top$ with respect to the CO. Since, $\boldsymbol{r}^b_{\mathrm{bg}} = \boldsymbol{r}^b_{bm_\mathrm{I}} + \boldsymbol{r}^b_{\mathrm{m_I}g}$, the vector from the $\mathrm{CM_I}$ to the CG is (see Figure 14.1),

$$\boldsymbol{r}^b_{\mathrm{m_I}g} = \boldsymbol{r}^b_{\mathrm{b}g} - \boldsymbol{r}^b_{bm_\mathrm{I}} = \begin{bmatrix} x_g - x_{\mathrm{m_I}} \\ y_g - y_{\mathrm{m_I}} \\ z_g - z_{\mathrm{m_I}} \end{bmatrix}. \tag{14.3}$$

The linear velocity in the CG is

$$\begin{aligned} \boldsymbol{v}^b_{\mathrm{ng}} &= \boldsymbol{v}^b_{\mathrm{nm_I}} + \boldsymbol{\omega}^b_{\mathrm{nm_I}} \times \boldsymbol{r}^b_{\mathrm{m_I}g} \\ &= \boldsymbol{v}^b_{\mathrm{nm_I}} + \boldsymbol{\omega}^b_{\mathrm{nb}} \times \boldsymbol{r}^b_{\mathrm{m_I}g} \end{aligned} \tag{14.4}$$

where we have exploited that $\boldsymbol{\omega}^b_{\mathrm{nm_I}} = \boldsymbol{\omega}^b_{\mathrm{nb}}$ for a rigid body rotating about the approximate inertial frame $\{n\}$. Time differentiation of $\boldsymbol{v}^b_{\mathrm{ng}}$ gives the specific force (linear acceleration) in the CG

$$\boldsymbol{f}^b_{\mathrm{ng}} = \boldsymbol{f}^b_{\mathrm{nm_I}} + \dot{\boldsymbol{\omega}}^b_{\mathrm{nb}} \times \boldsymbol{r}^b_{\mathrm{m_I}g} + \boldsymbol{\omega}^b_{\mathrm{nb}} \times (\boldsymbol{\omega}^b_{\mathrm{nb}} \times \boldsymbol{r}^b_{\mathrm{m_I}g}) \tag{14.5}$$

confirming that the angular acceleration $\dot{\boldsymbol{\omega}}^b_{\mathrm{nb}}$ is needed for lever-arm compensation of linear acceleration. Since this signal is not available as a measurement it is more practical to use $\{m_\mathrm{I}\}$ as reference frame when integrating the strapdown navigation equations.

14.1.1 Attitude Rate Sensors

The classic attitude rate sensor (ARS) is a gyro, that is a spinning wheel that utilizes conservation of momentum to detect rotation. For strapdown applications, optical gyros such as ring-laser gyros (RLG) and fiber-optic gyros (FOG) have been used for some time, and are also expected to be the standard for high-accuracy strapdown INS for the foreseeable future. For low- and medium-cost applications, ARS based on MEMS are dominating (Barbour and Schmidt 1998).

The IMU measurement equation for a three-axis ARS is

$$\boldsymbol{\omega}^b_{\text{imu}} = \boldsymbol{\omega}^b_{\text{nb}} + \boldsymbol{b}^b_{\text{ars}} + \boldsymbol{w}^b_{\text{ars}} \tag{14.6}$$

$$\dot{\boldsymbol{b}}^b_{\text{ars}} = \boldsymbol{w}^b_{\text{b, ars}} \tag{14.7}$$

where the ARS bias vector is denoted as $\boldsymbol{b}^b_{\text{ars}}$. Additive zero-mean Gaussian white noise terms $\boldsymbol{w}^{\text{b}}_{\text{ars}}$ and $\boldsymbol{w}^b_{\text{b, ars}}$ are used to model the measurement and bias noise, respectively. It is necessary to estimate $\boldsymbol{b}^b_{\text{ars}}$ online since the ARS bias will grow over time.

The ARS measurement (14.6) is only valid for low-speed applications such as a marine craft moving on the surface of the Earth since it assumes that $\{n\}$ is non-rotating, that is $\boldsymbol{\omega}^b_{\text{ib}} \approx \boldsymbol{\omega}^b_{\text{nb}}$. For terrestrial navigation the Earth rotation will affect the results and it is necessary to use the inertial frame $\{i\}$ instead of the approximate frame $\{n\}$.

14.1.2 Accelerometers

There are several different types of accelerometer. Two of these are mechanical and vibratory accelerometers. The mechanical accelerometer can be a pendulum, which in its simplest form is based on Newton's second law of motion. The vibratory accelerometers are usually based on measurement of frequency shifts due to increased or decreased tension in a string. The operation is similar to that of a violin. When a violin string is tightened, the frequency goes up. Similarly, when the accelerometer proof mass attached to a quartz beam is loaded, the frequency of the quartz beam increases. The difference in frequency is measured, and is proportional to the applied acceleration. In addition to quartz technology, vibrating beam accelerometers using silicon are also being developed.

The IMU accelerometer is a device that measure three-axis specific force

$$\boldsymbol{f}^b_{\text{imu}} := \boldsymbol{f}^b_{\text{nm}_\text{I}}. \tag{14.8}$$

The specific force is defined as the non-gravitational force per unit mass m,

$$\begin{aligned} \boldsymbol{f}^b_{\text{imu}} &:= \frac{1}{m} \sum \boldsymbol{f}^b_{\text{non-gravitational}} \\ &= \frac{1}{m}\left(\boldsymbol{f}^b_{\text{total}} - \sum \boldsymbol{f}^b_{\text{gravitational}}\right). \end{aligned} \tag{14.9}$$

Specific force, also called g-force, is not actually a force but a type of acceleration.

If we consider a marine craft experiencing hydrodynamic, buoyancy, gravitational and thrust forces (subscripts h, b, g and t, respectively), the IMU measures

$$\begin{aligned} \boldsymbol{f}^b_{\text{imu}} &= \frac{1}{m}(\boldsymbol{f}^b_{\text{total}} - \boldsymbol{f}^b_{\text{g}}) \\ &= \frac{1}{m}((\boldsymbol{f}^b_{\text{h}} + \boldsymbol{f}^b_{\text{b}} + \boldsymbol{f}^b_{\text{g}} + \boldsymbol{f}^b_{\text{t}}) - \boldsymbol{f}^b_{\text{g}}) \\ &= \frac{1}{m}(\boldsymbol{f}^b_{\text{h}} + \boldsymbol{f}^b_{\text{t}} + \boldsymbol{f}^b_{\text{b}}). \end{aligned} \tag{14.10}$$

Newton's second law expressed in the rotating reference frame $\{m_{\text{I}}\}$ is (see Section 3.2),

$$m\boldsymbol{a}^b_{\text{nm}_\text{I}} = \boldsymbol{f}^b_{\text{total}} \tag{14.11}$$

where linear velocity and acceleration depend on the IMU distance vector $\boldsymbol{r}^b_{bm_\text{I}}$ (see Figure 14.1) according to

$$\boldsymbol{v}^b_{\text{nm}_\text{I}} = \boldsymbol{v}^b_{\text{nb}} + \boldsymbol{\omega}^b_{\text{nb}} \times \boldsymbol{r}^b_{bm_\text{I}} \tag{14.12}$$

$$\boldsymbol{a}^b_{\text{nm}_\text{I}} = \dot{\boldsymbol{v}}^b_{\text{nb}} + \boldsymbol{\omega}^b_{\text{nb}} \times \boldsymbol{v}^b_{\text{nb}} + \dot{\boldsymbol{\omega}}^b_{\text{nb}} \times \boldsymbol{r}^b_{bm_\text{I}} + \boldsymbol{\omega}^b_{\text{nb}} \times (\boldsymbol{\omega}^b_{\text{nb}} \times \boldsymbol{r}^b_{bm_\text{I}}). \tag{14.13}$$

Combining (14.9) and (14.11) gives the formula for specific force

$$\boldsymbol{f}^b_{\text{imu}} = \boldsymbol{a}^b_{\text{nm}_\text{I}} - \boldsymbol{g}^b \tag{14.14}$$

where $\boldsymbol{g}^b = (1/m)\boldsymbol{f}^b_{\text{g}}$. An expression for the specific force measurements in terms of the body-fixed velocities (u, v, w) and accelerations $(\dot{u}, \dot{v}, \dot{w})$ is obtained by combining (14.13) and (14.14) together with $\boldsymbol{g}^b = \boldsymbol{R}^\top(\boldsymbol{\Theta}_{\text{nb}})\boldsymbol{g}^n$,

$$\begin{aligned} f_x &= \dot{u} - vr + wq - x_{\text{m}_\text{I}}(q^2 + r^2) + y_{\text{m}_\text{I}}(pq - \dot{r}) + z_{\text{m}_\text{I}}(pr + \dot{q}) + g\sin(\theta) \\ f_y &= \dot{v} - wp + ur - y_{\text{m}_\text{I}}(r^2 + p^2) + z_{\text{m}_\text{I}}(qr - \dot{p}) + x_{\text{m}_\text{I}}(qp + \dot{r}) - g\cos(\theta)\sin(\phi) \\ f_z &= \dot{w} - uq + vp - z_{\text{m}_\text{I}}(p^2 + q^2) + x_{\text{m}_\text{I}}(rp - \dot{q}) + y_{\text{m}_\text{I}}(rq + \dot{p}) - g\cos(\theta)\cos(\phi). \end{aligned} \tag{14.15}$$

When designing Kalman filters for inertial navigation we will avoid using (14.15) since this formula requires information about the linear accelerations $(\dot{u}, \dot{v}, \dot{w})$ and angular velocities $(\dot{p}, \dot{q}, \dot{r})$. Instead, we integrate linear acceleration $\ddot{\boldsymbol{p}}^{\text{n}}_{\text{nm}_\text{I}} = \boldsymbol{a}^n_{\text{nm}_\text{I}}$ twice to obtain an estimate of the position $\boldsymbol{p}^n_{\text{nm}_\text{I}}$. From the formula for specific force (14.14), it follows that the measurement equation for a three-axis accelerometer is

$$\boldsymbol{f}^b_{\text{imu}} = \boldsymbol{R}^\top(\boldsymbol{\Theta}_{\text{nb}})(\boldsymbol{a}^n_{\text{nm}_\text{I}} - \boldsymbol{g}^n) + \boldsymbol{b}^b_{\text{acc}} + \boldsymbol{w}^b_{\text{acc}} \tag{14.16}$$

$$\dot{\boldsymbol{b}}^b_{\text{acc}} = \boldsymbol{w}^b_{\text{b, acc}} \tag{14.17}$$

where $\boldsymbol{g}^n = [0, 0, g]^\top$ is the gravity vector and $\boldsymbol{R}(\boldsymbol{\Theta}_{nb})$ is the Euler angle rotation matrix. Alternatively, the quaternion rotation matrix $\boldsymbol{R}(\boldsymbol{q}_b^n)$ can be used. The accelerometer bias is denoted as $\boldsymbol{b}_{\mathrm{acc}}^b$, while $\boldsymbol{w}_{\mathrm{acc}}^b$ and $\boldsymbol{w}_{\mathrm{b,\,acc}}^b$ are additive Gaussian white measurement and bias noise, respectively.

Note that the accelerometers, if leveled ($\phi = \theta = 0$) and at rest $\boldsymbol{a}_{\mathrm{nm_I}}^n = \boldsymbol{0}$ under the assumptions that the bias is zero and there are no measurement noise, will measure

$$\boldsymbol{f}_{\mathrm{imu}}^b = -\boldsymbol{g}^n. \tag{14.18}$$

Thus the acceleration biases can be estimated by requiring that the averaged acceleration measurement satisfies $E(\boldsymbol{f}_{\mathrm{imu}}^b) = -\boldsymbol{g}^n$ before moving. However, the biases will grow over time so it is necessary to estimate them during operation as well unless the mission is quite limited in time. Online bias compensation will introduce additional states in the state estimator and it is in general a complicated problem, which depends on the quality of the IMU.

The specific force measurement (14.16) is only valid for low-speed applications such as a marine craft moving on the surface of the Earth since it assumes that $\{n\}$ is non-rotating (approximative inertial frame).

Earth-gravity Model

The gravity of Earth in $\{n\}$ is modeled as a constant vector

$$\boldsymbol{g}^n = [0, 0, g(\mu)]^\top \tag{14.19}$$

where $g(\mu)$ is the acceleration of gravity as a function of latitude μ. Gravity increases from $\mathbb{G}_e = 9.780\,325\,335\,9$ m s^{-2} at the equator to $\mathbb{G}_p = 9.8321849378$ m s^{-2} at the poles. The nominal "average" value at the surface of the Earth, known as standard gravity, is, by definition of the ISO/IEC 8000, $\mathbb{G}_s = 9.80665$ m s^{-2}.

We can compute gravity by using the World Geodetic System (1984) ellipsoidal gravity formula

$$g(\mu) = \mathbb{G}_e \frac{1 + k\sin^2(\mu)}{\sqrt{1 - e^2\sin^2(\mu)}} \tag{14.20}$$

where

$$k = \frac{r_p\mathbb{G}_p - r_e\mathbb{G}_e}{r_e\mathbb{G}_e}, \qquad e^2 = 1 - (r_p/r_e)^2 \tag{14.21}$$

are two constants. Eccentricity e, equatorial radius r_e and polar radius r_p of the ellipsoid are given by Table 2.2. The numerical representation of (14.20) is

$$g(\mu) = 9.7803253359 \frac{1 + 0.001931850400\sin^2(\mu)}{\sqrt{1 - 0.006694384442\sin^2(\mu)}}. \tag{14.22}$$

The WGS-84 gravity model is implemented in the MSS toolbox as: `gravity(mu)`.

14.1.3 *Magnetometer*

The magnetic field of the Earth is similar to a simple bar magnet. The magnetic field is a magnetic dipole that has its field lines originating at a point near the South Pole and terminating at a point near the North Pole. The field lines vary in both strength and direction about the face of the Earth. At each location on the Earth, the field lines intersect the Earth's surface at a specific angle of inclination. Near the equator, the field lines are approximately parallel to the Earth's surface and thus the inclination angle in this region is 0° (see Figure 14.2). As one travels north from the equator the field lines become progressively steeper. At the magnetic pole, the field lines are directed almost straight down into the Earth and the inclination is 90°.

The heading angle is the sum of the the magnetic heading measurement ψ_{m} and the declination angle δ. In other words

$$\psi = \psi_{\mathrm{m}} + \delta. \tag{14.23}$$

The declination angle δ for a given longitude l and latitude μ can be calculated using the World Magnetic Model (WMM), which is a joint project by the United States' National Geospatial-Intelligence Agency (NGA) and the United Kingdom's Defence Geographic Centre (DGC). The magnetic model comes with C software and executables.

The magnetic field strength can be measured by a three-axis magnetometer, usually included in the sensor suite of commercial available IMUs. Mathematically this can be represented as

$$\boldsymbol{m}^b_{\mathrm{mag}} = \boldsymbol{R}^{\top}(\boldsymbol{\Theta}_{\mathrm{nb}})\boldsymbol{m}^n + \boldsymbol{b}^b_{\mathrm{mag}} + \boldsymbol{w}^b_{\mathrm{mag}} \tag{14.24}$$

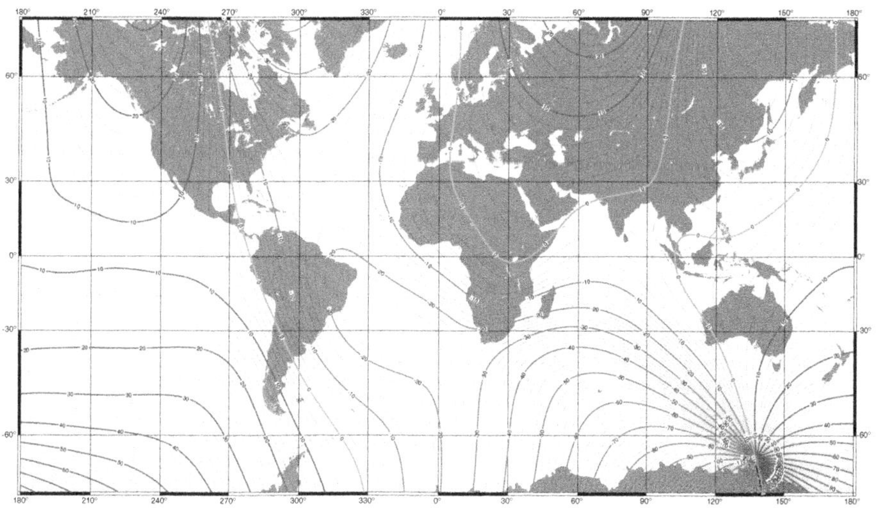

Figure 14.2 Magnetic field declination δ according to the US/UK World Magnetic Model (2020).

where $\boldsymbol{m}^n \in \mathbb{R}^3$ is the strength and direction of Earth's magnetic field expressed in $\{n\}$ and $\boldsymbol{w}^b_{\mathrm{mag}} \in \mathbb{R}^3$ is additive zero-mean white noise expressed in $\{b\}$.

The measurement $\boldsymbol{m}^b_{\mathrm{mag}}$ is affected by a largely time-invariant bias $\boldsymbol{b}^b_{\mathrm{mag}}$, induced by local magnetic disturbances. It is necessary to perform a hard-iron magnetometer calibration to derive an estimate $\hat{\boldsymbol{b}}^b_{\mathrm{mag}}$ for $\boldsymbol{b}^b_{\mathrm{mag}}$, which is then subtracted from all future measurements. However, care must be taken since magnetometers also are prone to time-varying electric currents from motors and transmission wires. As a result of this, ships use gyrocompasses for safe navigation. Gyrocompasses have the advantage that they find true north as determined by the axis of the Earth's rotation. This is different from, and navigationally more useful than, magnetic north. In addition, gyrocompasses are unaffected by ferromagnetic materials, such as in a ship's steel hull, and electromagnetic-field disturbances which distort the magnetic field.

Magnetic Heading

The magnetic heading ψ_{m} can be determined from measurements of the magnetic field strength along the body-frame axes. Assume that the magnetic measurement vector $\boldsymbol{m}^b = [m_x, m_y, m_z]^{\top}$ has been calibrated for hard-iron disturbances. If the roll and pitch angles are known the magnetometer measurements can be transformed to the horizontal plane according to

$$\begin{bmatrix} h_x \\ h_y \\ h_z \end{bmatrix} = \boldsymbol{R}_{y,\theta}\boldsymbol{R}_{x,\phi} \begin{bmatrix} m_x \\ m_y \\ m_z \end{bmatrix} \tag{14.25}$$

or

$$\begin{bmatrix} h_x \\ h_y \\ h_z \end{bmatrix} = \begin{bmatrix} \cos(\theta) & 0 & \sin(\theta) \\ 0 & 1 & 0 \\ -\sin(\theta) & 0 & \cos(\theta) \end{bmatrix} \begin{bmatrix} 1 & 0 & 0 \\ 0 & \cos(\phi) & -\sin(\phi) \\ 0 & \sin(\phi) & \cos(\phi) \end{bmatrix} \begin{bmatrix} m_x \\ m_y \\ m_z \end{bmatrix}. \tag{14.26}$$

The magnetic compass heading is deduced from the horizontal components h_x and h_y according to

$$\psi_{\mathrm{m}} = -\mathrm{atan2}(h_y, h_x). \tag{14.27}$$

One method to determine the roll and pitch angles is to use inclinometers (see Section 14.2.1) or the INS attitude estimates.

If the magnetometer is sitting in a local horizontal plane leveled to the surface of the Earth such that $\phi = \theta = 0$, the *magnetic* heading angle is equivalent to the direction planar with the surface of the Earth, for which

$$\psi_{\mathrm{m}} = -\mathrm{atan2}(m_y, m_x). \tag{14.28}$$

14.2 Attitude Estimation

In many practical applications it is useful to determine the attitude of the craft or an object located onboard a moving craft. A complete navigation solution will estimate positions, velocities and attitude. This section, however, presents static and dynamic solutions to the attitude estimation problem without using information about the position. Commercial systems implementing these algorithms are usually called *attitude heading reference systems* (AHRS).

14.2.1 Static Mapping from Specific Force to Roll and Pitch Angles

This section presents the core algorithm for mapping the three-axis specific force from an IMU to roll and pitch angles. The principle for this is that the angle between the acceleration and gravity vectors can be computed using trigonometry. This is a static mapping that suffers from inaccuracies when performing high-acceleration maneuvers. For ships this works quite well but aircraft and other highly maneuverable vehicles should use other methods.

Recall from (14.16) that the IMU specific force measurement expressed in $\{b\}$ is

$$\boldsymbol{f}^b_{\text{imu}} = \boldsymbol{R}^\top(\boldsymbol{\Theta}_{\text{nb}})(\boldsymbol{a}^n_{\text{nm}_\text{I}} - \boldsymbol{g}^n) + \boldsymbol{b}^b_{\text{acc}} + \boldsymbol{w}^b_{\text{acc}}. \tag{14.29}$$

From this formula we can compute the static roll and pitch angles by noticing that for three orthogonal accelerometers onboard a craft at rest, $\boldsymbol{a}^n_{\text{nm}_\text{I}} = \boldsymbol{0}$. Then the measurement equation (14.29) reduces to

$$\boldsymbol{f}^b_{\text{imu}} = -\boldsymbol{R}^\top(\boldsymbol{\Theta}_{\text{nb}})\boldsymbol{g}^n + \boldsymbol{b}^b_{\text{acc}} + \boldsymbol{w}^b_{\text{acc}}. \tag{14.30}$$

The initial accelerometer biases $\boldsymbol{b}^b_{\text{acc}}$ are usually removed by calibrating the accelerometer in a laboratory for varying temperatures. This can be implemented as a look-up table in combination with a temperature sensor. It is also necessary to remove the dynamic drift, for instance by recalibrating the sensor when the craft is at rest. In addition, the measurement noise $\boldsymbol{w}^b_{\text{acc}}$ should be removed by low-pass filtering before the roll and pitch angles are computed. The key assumption is to assume that the average acceleration with respect to the environment during some period of time is zero, for instance 10–20 s. For aircraft this assumption does not hold since they can generate significant accelerations lasting longer than the maximum time.

Proper calibration and filtering implies that

$$\boldsymbol{f}^b := \boldsymbol{f}^b_{\text{imu}} - \boldsymbol{b}^b_{\text{acc}} \approx -\boldsymbol{R}^\top(\boldsymbol{\Theta}_{\text{nb}})\boldsymbol{g}^n \tag{14.31}$$

$$\Updownarrow$$

$$\begin{bmatrix} f_x \\ f_y \\ f_z \end{bmatrix} \approx -\boldsymbol{R}^\top(\boldsymbol{\Theta}_{\text{nb}}) \begin{bmatrix} 0 \\ 0 \\ g \end{bmatrix} = \begin{bmatrix} g\sin(\theta) \\ -g\cos(\theta)\sin(\phi) \\ -g\cos(\theta)\cos(\phi) \end{bmatrix}. \tag{14.32}$$

Taking the ratios

$$\frac{f_y}{f_z} \approx \tan(\phi), \quad \frac{f_x}{g} \approx \sin(\theta), \quad \frac{f_y^2 + f_z^2}{g^2} \approx \cos^2(\theta) \tag{14.33}$$

it follows that

$$\phi \approx \tan^{-1}\left(\frac{f_y}{f_z}\right) \tag{14.34}$$

$$\theta \approx \tan^{-1}\left(\frac{f_x}{\sqrt{f_y^2 + f_z^2}}\right). \tag{14.35}$$

Note that the transformation is singular for $\theta = \pm 90^\circ$. When combined with a compass measuring the yaw angle ψ the attitude vector $\boldsymbol{\Theta}_{\text{nb}} = [\phi, \theta, \psi]^\top$ is completely determined. The vector $\boldsymbol{\Theta}_{\text{nb}}$ of Euler angles can easily be transformed to unit quaternions $\boldsymbol{q}_b^n = [\eta, \varepsilon_1, \varepsilon_2, \varepsilon_3]^\top$ by using Formula 2.91 in Section 2.2.3. The unit quaternion representation is advantageous when implementing the attitude observer in a computer.

14.2.2 *Vertical Reference Unit (VRU) Transformations*

The hardware and software needed to compute ϕ and θ and sometimes the heave position z^{n} from inertial sensors are referred to as a *vertical reference unit* (VRU). The roll and pitch angles can be computed using the static solution (14.34) and (14.35) or dynamically in a state estimator. The performance of state-of-the-art VRUs has been evaluated by Ingram *et al.* (1996).

A VRU is particularly useful if you want to transform the GNSS position and velocity measurements

$$\boldsymbol{p}_{\text{gnss}}^n := \boldsymbol{p}_{\text{nm}_\text{G}}^n = [x^{\text{n}}, y^{\text{n}}, z^{\text{n}}]^\top \tag{14.36}$$

$$\boldsymbol{v}_{\text{gnss}}^n := \boldsymbol{v}_{\text{nm}_\text{G}}^n = [\dot{x}^{\text{n}}, \dot{y}^{\text{n}}, \dot{z}^{\text{n}}]^\top \tag{14.37}$$

for a GNSS receiver located in a measurement frame $\{m_\text{G}\}$ with coordinate origin CM_G (see Figure 14.1). The coordinate vector from the CO to the CM_G is denoted $\boldsymbol{r}_{\text{b}m_\text{G}}^b = [x_{\text{m}_\text{G}}, y_{\text{m}_\text{G}}, z_{\text{m}_\text{G}}]^\top$. Then it follows that

$$\boldsymbol{p}_{\text{nb}}^n + \boldsymbol{R}(\boldsymbol{\Theta}_{\text{nb}})\boldsymbol{r}_{\text{b}m_\text{G}}^b = \boldsymbol{p}_{\text{nm}_\text{G}}^n. \tag{14.38}$$

Since the IMU is located at $\boldsymbol{r}^b_{bm_I} = [x_{m_I}, y_{m_I}, z_{m_I}]^\top$ with respect to the CO, the GNSS position measurement with respect to the IMU measurement frame is

$$\begin{aligned}\boldsymbol{p}^n_{nm_I} &= \boldsymbol{p}^n_{nb} + \boldsymbol{R}(\boldsymbol{\Theta}_{nb})\boldsymbol{r}^b_{bm_I} \\ &= \boldsymbol{p}^n_{nm_G} - \boldsymbol{R}(\boldsymbol{\Theta}_{nb})\boldsymbol{r}^b_{bm_G} + \boldsymbol{R}(\boldsymbol{\Theta}_{nb})\boldsymbol{r}^b_{bm_I} \\ &= \boldsymbol{p}^n_{gnss} + \boldsymbol{R}(\boldsymbol{\Theta}_{nb})(\boldsymbol{r}^b_{bm_I} - \boldsymbol{r}^b_{bm_G}). \end{aligned} \tag{14.39}$$

In order to compute the NED velocity $\boldsymbol{v}^n_{nm_I}$ expressed in $\{n\}$, we will exploit that $\dot{\boldsymbol{r}}^b_{bm_I} = \boldsymbol{0}$ and $\dot{\boldsymbol{r}}^b_{bm_G} = \boldsymbol{0}$; that is the positions of the IMU and the GNSS receiver are constant in $\{b\}$. Hence, time differentiation of (14.39) gives

$$\boldsymbol{v}^n_{nm_I} = \dot{\boldsymbol{p}}^n_{nm_I} = \dot{\boldsymbol{p}}^n_{gnss} + \dot{\boldsymbol{R}}(\boldsymbol{\Theta}_{nb})(\boldsymbol{r}^b_{bm_I} - \boldsymbol{r}^b_{bm_G}). \tag{14.40}$$

Application of $\dot{\boldsymbol{p}}^n_{gnss} = \boldsymbol{v}^n_{gnss}$ and $\dot{\boldsymbol{R}}(\boldsymbol{\Theta}_{nb}) = \boldsymbol{R}(\boldsymbol{\Theta}_{nb})\boldsymbol{S}(\boldsymbol{\omega}^b_{nb})$; see Theorem 2.2 in Section 2.2.1, finally yields

$$\boldsymbol{v}^n_{nm_I} = \boldsymbol{v}^n_{gnss} + \boldsymbol{R}(\boldsymbol{\Theta}_{nb})\boldsymbol{S}(\boldsymbol{\omega}^b_{nb})(\boldsymbol{r}^b_{bm_I} - \boldsymbol{r}^b_{bm_G}). \tag{14.41}$$

14.2.3 Nonlinear Attitude Observer using Reference Vectors

Nonlinear attitude estimation has received significant attention as a stand-alone problem; see Shuster and Oh (1981), Salcudean (1991), Vik and Fossen (2001), Mahony *et al.* (2008), Batista et al. (2011a, 2011b) and Grip *et al.* (2012). An extensive survey of attitude estimation methods is given by Crassidis *et al.* (2007).

The first nonlinear quaternion-based attitude observer was derived by Salcudean (1991). The proposed observer was based on the availability of an independent attitude measurement. For instance, a quaternion measurement could be constructed by measuring the Euler angles or by mapping specific force to roll and pitch angles. Vik and Fossen (2001) expanded the attitude observer of Salcudean (1991) to include ARS bias estimation, and used this observer as part of a complete GNSS/INS integration observer. In most cases, however, an attitude measurement is not available, and the attitude observer must be combined with position and velocity estimation in a more complicated way.

Hua (2010) used the vector-based attitude observer of Hamel and Mahony (2006), and Mahony *et al.* (2008) as the basis for an attitude and velocity observer that depends only on inertial, GNSS velocity and magnetometer measurements. This observer exploits the fact that the vehicle's acceleration vector in the navigation frame is implicitly available in the derivative of the GNSS velocity, and it can therefore be compared to the accelerometer measurement to help determine the attitude.

This section presents the nonlinear attitude observer of Mahony *et al.* (2008), which is less computationally demanding than commercial algorithms such as the MEKF (see Section 14.4.3). In addition, the attitude observer renders the equilibrium point of the error dynamics semi-globally exponentially stable. The nonlinear attitude observer uses accelerometer and magnetometer measurements directly to update the states. This is achieved by constructing a nonlinear injection term that compares the directions of the measurement vectors in the body-fixed frame $\{b\}$ to reference vectors in $\{n\}$, which is assumed to be an approximative inertial frame.

Attitude Observer with ARS Bias Correction

The quaternion-based attitude observer is shown in Figure 14.3 where $\boldsymbol{\omega}^b_{\text{mes}}$ is a nonlinear injection term based on reference vectors. The corresponding observer equations are (Mahony *et al.* 2008)

$$\boldsymbol{\omega}^b_{\text{mes}} = -\text{vex}\left(\sum_{i=1}^{n} \frac{k_i}{2}(\boldsymbol{v}_i^b(\hat{\boldsymbol{v}}_i^b)^\top - \hat{\boldsymbol{v}}_i^b(\boldsymbol{v}_i^b)^\top)\right) \tag{14.42}$$

$$\dot{\hat{\boldsymbol{q}}}^n_b = \boldsymbol{T}(\hat{\boldsymbol{q}}^n_b)(\boldsymbol{\omega}^b_{\text{imu}} - \hat{\boldsymbol{b}}^b_{\text{ars}} + \boldsymbol{K}_{\text{p}}\,\boldsymbol{\omega}^b_{\text{mes}}) \tag{14.43}$$

$$\dot{\hat{\boldsymbol{b}}}^b_{\text{ars}} = -\boldsymbol{K}_i\,\boldsymbol{\omega}^b_{\text{mes}} \tag{14.44}$$

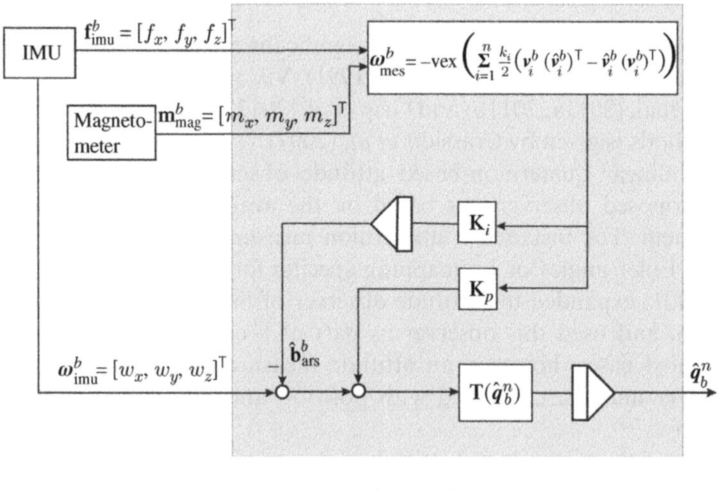

Figure 14.3 The nonlinear quaternion-based attitude observer Mahony et al. (2008). Source: Modified from Mahony, R., T. Hamel and J.M. Pflimlin (2008). Nonlinear Complementary Filters on the Special Orthogonal Group. IEEE Transactions on Automatic Control 53(5), 1203–1218.

where $\boldsymbol{v}_i^b$ and $\hat{\boldsymbol{v}}_i^b$ $(i = 1, \ldots, n)$ are sets of n true and estimated reference vectors to be defined later, and $\hat{\boldsymbol{b}}_{\text{ars}}^b$ is the estimate of the ARS bias expressed in $\{b\}$. The reference vectors are weighted by tunable gains $k_i > 0$ for $i = 1, .., n$. The operator vex: SO(3) $\rightarrow \mathbb{R}^3$ denotes the inverse of the cross-product operator $\boldsymbol{S}(\boldsymbol{a})$. In other words

$$\boldsymbol{a} \times \boldsymbol{b} = \boldsymbol{S}(\boldsymbol{a})\boldsymbol{b} \tag{14.45}$$

$$\text{vex}(\boldsymbol{S}(\boldsymbol{a})) = \boldsymbol{a}. \tag{14.46}$$

Hence, the following expression corresponding to (14.42) can be derived

$$\text{vex}(\boldsymbol{a}\boldsymbol{b}^\top - \boldsymbol{b}\boldsymbol{a}^\top) = \begin{bmatrix} a_3 b_2 - a_2 b_3 \\ a_1 b_3 - a_3 b_1 \\ a_2 b_1 - a_1 b_2 \end{bmatrix}. \tag{14.47}$$

Following the notation of Grip *et al.* (2016), the injection term $\boldsymbol{\omega}_{\text{mes}}^b$ can also be expressed as a sum of cross products

$$\boldsymbol{\sigma} = \sum_{i=1}^{\text{n}} k_i \boldsymbol{v}_i^b \times \boldsymbol{R}^\top(\hat{\boldsymbol{q}}_b^n)\boldsymbol{v}_{0i}^n \tag{14.48}$$

$$\dot{\hat{\boldsymbol{q}}}_b^n = \boldsymbol{T}(\hat{\boldsymbol{q}}_b^n)(\boldsymbol{\omega}_{\text{imu}}^b - \hat{\boldsymbol{b}}_{\text{ars}}^b + \boldsymbol{\sigma}) \tag{14.49}$$

$$\dot{\hat{\boldsymbol{b}}}_{\text{ars}}^b = -\boldsymbol{K}_i\, \boldsymbol{\sigma} \tag{14.50}$$

where $\boldsymbol{v}_{0i}^n$ $(i = 1, \ldots, n)$ denote a set of n known inertial directions. The reference vectors are expressed in $\{b\}$ by applying the following transformation

$$\boldsymbol{v}_i^b = \boldsymbol{R}^\top(\boldsymbol{q}_b^n)\boldsymbol{v}_{0i}^n. \tag{14.51}$$

Since only the direction of the measurement is relevant to the observer, it is assumed that all measurements are normalized such that $\|\boldsymbol{v}_{0i}^n\| = 1$. The associated estimate of $\boldsymbol{v}_i^b$ is computed as

$$\hat{\boldsymbol{v}}_i^b = \boldsymbol{R}^\top(\hat{\boldsymbol{q}}_b^n)\boldsymbol{v}_{0i}^n \tag{14.52}$$

where $\hat{\boldsymbol{q}}_b^n$ is the estimate of the unit quaternion vector. Hence, the difference between (14.51) and (14.52) is zero if $\hat{\boldsymbol{q}}_b^n = \boldsymbol{q}_b^n$.

Grip *et al.* (2016) has extend the observer (14.48)–(14.50) to terrestrial navigation by using reference vectors in the Earth-fixed Earth-centered reference frame $\{e\}$. Hence, the rotation and curvature of the Earth as well as the gravity vector are taken into account.

Specific Force Measurements

The IMU specific force $\boldsymbol{f}_{\text{imu}}^b$ measurement is given by (14.16). Hence, by assuming that $\boldsymbol{a}_{\text{nm}_\text{I}}^n$ is small, the unbiased specific force vector expressed in $\{b\}$ can be computed as

$$\boldsymbol{f}^b := \boldsymbol{f}_{\text{imu}}^b - \hat{\boldsymbol{b}}_{\text{acc}}^b = -\boldsymbol{R}^\top(\boldsymbol{\Theta}_{\text{nb}})\boldsymbol{g}^n \tag{14.53}$$

where the estimated acceleration bias $\hat{\boldsymbol{b}}_{\text{acc}}^b$ is an optional correction term.

The normalized specific force vectors are chosen as

$$\boldsymbol{v}_1^b = -\frac{\boldsymbol{f}^b}{g(\mu)}, \qquad \hat{\boldsymbol{v}}_1^b = \boldsymbol{R}^\top(\hat{\boldsymbol{q}}_b^n)\begin{bmatrix}0\\0\\1\end{bmatrix} \tag{14.54}$$

where the World Geodetic System (1984) ellipsoidal gravity formula (14.20) is used to compute $g(\mu)$. The reference vector $\boldsymbol{v}_{01}^n = [0,\ 0,\ 1]^\top$ is chosen as the normalized gravity vector, pointing downwards.

Magnetic Field Measurements

The magnetometer $\boldsymbol{m}_{\text{mag}}^b$ measurement is given by (14.24). Hence, the unbiased magnetic field vector expressed in $\{b\}$ is

$$\boldsymbol{m}^b := \boldsymbol{m}_{\text{mag}}^b - \hat{\boldsymbol{b}}_{\text{mag}}^b \tag{14.55}$$

where the estimated magnetometer bias $\hat{\boldsymbol{b}}_{\text{mag}}^b$ is an optional correction term.

The normalized magnetic field vectors are chosen as (Hua *et al.* 2014)

$$\boldsymbol{v}_2^b = \frac{\Pi_{\boldsymbol{v}_1^b}\ \boldsymbol{m}^b}{\|\Pi_{\boldsymbol{v}_{01}^n}\ \boldsymbol{m}^n\|}, \qquad \hat{\boldsymbol{v}}_2^b = \boldsymbol{R}^\top(\hat{\boldsymbol{q}}_b^n)\frac{\Pi_{\boldsymbol{v}_{01}^n}\ \boldsymbol{m}^n}{\|\Pi_{\boldsymbol{v}_{01}^n}\ \boldsymbol{m}^n\|} \tag{14.56}$$

where $\Pi_{\boldsymbol{x}} = ||\boldsymbol{x}||^2\boldsymbol{I}_3 - \boldsymbol{x}\boldsymbol{x}^\top$, $\forall \boldsymbol{x} \in \mathbb{R}^3$ denotes the orthogonal projection on the plan orthogonal to $\boldsymbol{x}$.

The strength and direction of the magnetic field $\boldsymbol{m}^n$ expressed in $\{n\}$ can be assumed to be constant for a given location and period of time. The numerical values can be found by averaging the magnetometer measurements for a given period while keeping the roll and pitch angles close to zero. If this procedure is impractical, an alternative is to use the compass measurement to compute a reference vector. This method is outlined below.

Compass Measurements

For marine craft, an alternative to the three-axis magnetometer is using a gyrocompass or a magnetic compass as reference vector (see Section 13.1). This is based on the assumption that the roll and pitch angles are small such that a unit vector $\boldsymbol{e}_3^n = [1,\ 0,\ 0]^\top$ pointing northwards satisfies

$$\boldsymbol{e}_3^b \approx \boldsymbol{R}_{z,\psi}^\top\ \boldsymbol{e}_3^n = \begin{bmatrix}\cos(\psi)\\-\sin(\psi)\\0\end{bmatrix} \tag{14.57}$$

where $\psi = \psi_{\mathrm{m}} + \delta$ denotes the true magnetic north, which is obtained by correcting the measurement ψ_{m} using the declination angle δ (see Section 14.1.3). Thus we can approximate

$$\boldsymbol{v}_3^b = \begin{bmatrix} \cos(\psi) \\ -\sin(\psi) \\ 0 \end{bmatrix}, \qquad \hat{\boldsymbol{v}}_3^b = \boldsymbol{R}^{\mathrm{T}}(\hat{\boldsymbol{q}}_b^n) \begin{bmatrix} 1 \\ 0 \\ 0 \end{bmatrix}. \tag{14.58}$$

Stability Analysis

The stability proof of Mahony *et al.* (2008) assumes that the reference vectors $\boldsymbol{v}_{0i}^n$ are constant. This guarantees almost-global asymptotic stability. The constant reference vector assumption has been removed by Grip *et al.* (2012) who introduced a parameter projection in the bias estimate

$$\dot{\hat{\boldsymbol{b}}}_{\mathrm{ars}}^b = \mathrm{Proj}(\hat{\boldsymbol{b}}_{\mathrm{ars}}^b, -\boldsymbol{K}_i\, \boldsymbol{\omega}_{\mathrm{mes}}^b) \tag{14.59}$$

which restricts the parameter estimate $\hat{\boldsymbol{b}}_{\mathrm{ars}}^b$ to the compact set defined by $||\hat{\boldsymbol{b}}_{\mathrm{ars}}^b|| \leq M_{\hat{\theta}}$ where $M_{\hat{\theta}}$ is a positive constant chosen slighter larger than M_θ. From this work it is concluded that the equilibrium point of the quaternion error dynamics is semi-globally exponentially stable for $n \geq 2$ independent inertial directions $\boldsymbol{v}_{0i}^{\mathrm{n}}$ if

$$\mathrm{Proj}(\hat{\boldsymbol{\theta}}, \boldsymbol{\tau}) := \begin{cases} (1 - c(\hat{\boldsymbol{\theta}}))\boldsymbol{\tau} & \text{if } \| \hat{\boldsymbol{\theta}} \| > M_\theta \text{ and } \hat{\boldsymbol{\theta}}^{\mathrm{T}}\boldsymbol{\tau} > 0 \\ \boldsymbol{\tau} & \text{otherwise} \end{cases} \tag{14.60}$$

where $c(\hat{\boldsymbol{\theta}}) = \min\{1, (\| \hat{\boldsymbol{\theta}} \|^2 - M_\theta^2)/(M_{\hat{\theta}}^2 - M_\theta^2)\}$. This is a special case of the parameter projection algorithm of Krstic *et al.* (1995, Appendix E).

14.3 Direct Filters for Aided INS

In a direct filter design, the marine craft position and velocity are states in the estimator, while accelerometer and angular rate measurements are used as inputs to the *strapdown navigation equations*. Alternatively, the accelerometer and angular rate measurements can be replaced by a marine craft model as shown in Figure 13.1. The model-based approach (see Chapter 13) will of course suffer from model uncertainty when implemented in a state estimator. Hence, it is advantageous to use a signal-based KF based on the strapdown navigation equations. Kinematic equations are not prone to parameter uncertainty since they are geometrical descriptions.

The measurements available from a typical IMU are three-axis specific force, angular rate and magnetic field intensity (see Section 14.1). An open-loop IMU solution, where the acceleration measurements are integrated twice and the ARS measurements are integrated once to obtain positions and attitude, respectively, will drift due to sensor biases, misalignments and temperature variations (see Figure 14.4). Hence, an estimator

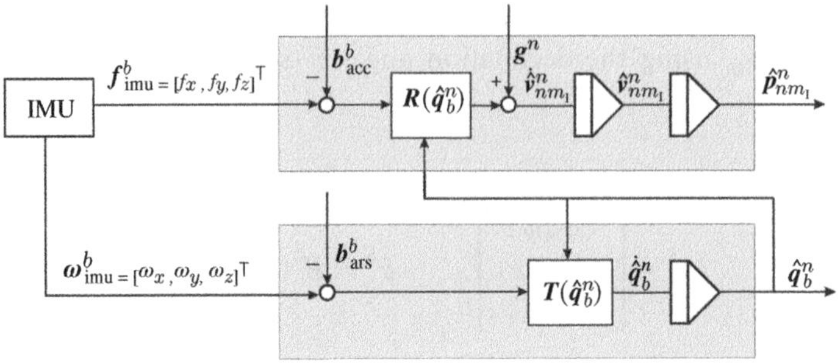

Figure 14.4 The principle for integration of IMU sensor data. The position and quaternion outputs will drift due to the bias terms.

providing feedback and compensation of bias drift terms is needed. The technique for this is an GNSS-aided INS (see Section 14.4) where the kinematic equations (strapdown navigation equations) are integrated numerically using high-rate IMU measurements and the drift is removed by using low-rate GNSS measurements in a feedback interconnection.

The position and velocity accuracies will mainly depend on the position quality while acceleration and attitude depend on the quality of the accelerometers, ARS and magnetometers. If a low-cost IMU is used, the position and attitude estimates will drift rapidly during position shortages while a more expensive unit will have better standalone capabilities. Construction of integrated GNSS/INS navigation systems, their performance and standalone capabilities are described more closely by Farrell and Barth (1998), Titterton and Weston (1997) and Grewal *et al.* (2001), to mention only some.

The goal of this section is to present low-cost IMU/GNSS integration techniques for marine craft navigation by neglecting the Earth's rotation and assuming that the GNSS signals are available all the time. Consequently, the *north-east-down* reference frame $\{n\}$ is assumed to be the inertial reference frame even though the Earth is moving relatively to a star-fixed reference frame. This is, indeed, a good approximation for a marine craft navigating on the surface of the Earth.

14.3.1 Fixed-gain Observer using Attitude Measurements

In this section we will assume that the attitude $\boldsymbol{\Theta}_{nb} = [\phi, \theta, \psi]^{\top}$ is known and exploit this when designing an observer for integration of IMU and GNSS measurements. In other words, it is assumed that $\boldsymbol{\Theta}_{nb}$ are measured for all $t \geq 0$ using an AHRS such that

$$\boldsymbol{R}(\boldsymbol{\Theta}_{nb}) := \boldsymbol{R}^n_b(t), \qquad \boldsymbol{T}(\boldsymbol{\Theta}_{nb}) := \boldsymbol{T}^n_b(t). \tag{14.61}$$

The expression for the linear acceleration expressed in $\{n\}$ is given by (14.16). Consequently,

$$\dot{\boldsymbol{v}}^n_{\text{nm}_\text{I}} = \boldsymbol{R}^n_b(t)(\boldsymbol{f}^b_{\text{imu}} - \boldsymbol{b}^b_{\text{acc}} - \boldsymbol{w}^b_{\text{acc}}) + \boldsymbol{g}^n. \tag{14.62}$$

Since the measurement noise $E(\boldsymbol{w}^b_{\text{acc}}) = \boldsymbol{0}$, the velocity observer will be designed using a deterministic approach by assuming that the zero-mean Gaussian white noise term $\boldsymbol{w}^b_{\text{acc}}$ can be neglected when analyzing the stability properties of the error dynamics.

Further, it is assumed that the IMU is mounted onboard the craft in a measurement frame $\{m_\text{I}\}$ with coordinate origin CM_I located at $\boldsymbol{r}^b_{\text{bm}_\text{I}} = [x_{\text{m}_\text{I}}, y_{\text{m}_\text{I}}, z_{\text{m}_\text{I}}]^\top$ with respect to the CO (see Figure 14.1) such that $\boldsymbol{f}^b_{\text{imu}} := \boldsymbol{f}^b_{\text{nm}_\text{I}}$ as explained in Section 14.1.

Integration of IMU and GNSS Position Measurements

Assume that the GNSS receiver is located in a measurement frame $\{m_\text{G}\}$ with coordinate origin CM_G located at $\boldsymbol{r}^b_{\text{bm}_\text{G}} = [x_{\text{m}_\text{G}}, y_{\text{m}_\text{G}}, z_{\text{m}_\text{G}}]^\top$ with respect to the CO (see Figure 14.1). The GNSS position measurement

$$\boldsymbol{p}^n_{\text{gnss}} := \boldsymbol{p}^n_{\text{nm}_\text{G}} \tag{14.63}$$

is transformed to $\{m_\text{I}\}$ by applying the lever-arm correction (14.39) such that

$$\boldsymbol{p}^n_{\text{nm}_\text{I}} = \boldsymbol{p}^\text{n}_{\text{gnss}} + \boldsymbol{R}^n_b(t)(\boldsymbol{r}^b_{\text{bm}_\text{I}} - \boldsymbol{r}^b_{\text{bm}_\text{G}}). \tag{14.64}$$

Consequently, the translational dynamics including acceleration bias states is a 9-state system with GNSS position measurement $\boldsymbol{y}_1 = \boldsymbol{p}^\text{n}_{\text{nm}_\text{I}}$ given by (see Figure 14.4)

$$\dot{\boldsymbol{p}}^n_{\text{nm}_\text{I}} = \boldsymbol{v}^n_{\text{nm}_\text{I}} \tag{14.65}$$

$$\dot{\boldsymbol{v}}^n_{\text{nm}_\text{I}} = \boldsymbol{R}^n_b(t)(\boldsymbol{f}^b_{\text{imu}} - \boldsymbol{b}^b_{\text{acc}}) + \boldsymbol{g}^n \tag{14.66}$$

$$\dot{\boldsymbol{b}}^b_{\text{acc}} = \boldsymbol{0} \tag{14.67}$$

$$\boldsymbol{y}_1 = \boldsymbol{p}^n_{\text{nm}_\text{I}}. \tag{14.68}$$

A nonlinear design method for simultaneously linear and angular velocity estimation has been proposed by Hua (2010) using IMU/GNSS measurements. This method discusses the stability of accelerated vehicles where the linear and angular dynamics are coupled. However, since the kinematic nonlinearities due to the transformations matrices (14.61) are known, the KF method discussed in Section 13.4.1 can be used for this purpose. Since the state-space model is a LTV system the time-varying Riccati equations must be numerically integrated to compute the Kalman gain. This increases the number of states from 9 to 45 in a practical implementation.

An alternative approach is to use a decoupled fixed-gain observer exploiting that the attitude vector $\boldsymbol{\Theta}_{\text{nb}}$ is measured using an AHRS when estimating the linear velocity $\hat{\boldsymbol{v}}^n_{\text{nm}_\text{I}}$. Algorithms for computation of $\boldsymbol{\Theta}_{\text{nb}}$ are presented in Section 14.2. The deterministic observer is chosen as (see Figure 14.4)

$$\dot{\hat{\boldsymbol{p}}}^n_{\mathrm{nm_I}} = \hat{\boldsymbol{v}}^n_{\mathrm{nm_I}} + \boldsymbol{K}_1\tilde{\boldsymbol{y}}_1 \tag{14.69}$$

$$\dot{\hat{\boldsymbol{v}}}^n_{\mathrm{nm_I}} = \boldsymbol{R}^n_b(t)(\boldsymbol{f}^b_{\mathrm{imu}} - \hat{\boldsymbol{b}}^b_{\mathrm{acc}}) + \boldsymbol{g}^n + \boldsymbol{K}_2\tilde{\boldsymbol{y}}_1 \tag{14.70}$$

$$\dot{\hat{\boldsymbol{b}}}^b_{\mathrm{acc}} = \boldsymbol{K}_3\boldsymbol{R}^n_b(t)^\top\tilde{\boldsymbol{y}}_1 \tag{14.71}$$

$$\hat{\boldsymbol{y}}_1 = \hat{\boldsymbol{p}}^n_{\mathrm{nm_I}} \tag{14.72}$$

where $\tilde{\boldsymbol{y}}_1 = \boldsymbol{y}_1 - \hat{\boldsymbol{y}}_1 = \boldsymbol{p}^n_{\mathrm{nm_I}} - \hat{\boldsymbol{p}}^n_{\mathrm{nm_I}}$ is the injection term. The estimated position $\hat{\boldsymbol{p}}^n_{\mathrm{nm_I}}$ and velocity $\hat{\boldsymbol{v}}^n_{\mathrm{nm_I}}$ are transformed to the CO (coordinate origin used by the motion control system) to obtain $\hat{\boldsymbol{p}}^n_{\mathrm{nb}} = [\hat{x}^n, \hat{y}^n, \hat{z}^n]^\top$ and $\hat{\boldsymbol{v}}^b_{\mathrm{nb}} = [\hat{u}, \hat{v}, \hat{w}]^\top$. The transformations are

$$\begin{aligned}\hat{\boldsymbol{p}}^n_{\mathrm{nb}} &= \hat{\boldsymbol{p}}^n_{\mathrm{nm_I}} + \boldsymbol{r}^n_{\mathrm{m_Ib}} \\ &= \hat{\boldsymbol{p}}^n_{\mathrm{nm_I}} - \boldsymbol{R}^n_b(t)\boldsymbol{r}^b_{\mathrm{bm_I}}\end{aligned} \tag{14.73}$$

$$\begin{aligned}\hat{\boldsymbol{v}}^b_{\mathrm{nb}} &= \hat{\boldsymbol{v}}^b_{\mathrm{nm_I}} + \boldsymbol{\omega}^b_{\mathrm{nm_I}} \times \boldsymbol{r}^b_{\mathrm{m_Ib}} \\ &= \boldsymbol{R}^n_b(t)^\top\hat{\boldsymbol{v}}^n_{\mathrm{nm_I}} - \boldsymbol{\omega}^b_{\mathrm{nb}} \times \boldsymbol{r}^b_{\mathrm{bm_I}}\end{aligned} \tag{14.74}$$

where we have exploited that $\boldsymbol{r}^b_{\mathrm{bm_I}} = -\boldsymbol{r}^b_{\mathrm{m_Ib}}$ and $\boldsymbol{\omega}^b_{\mathrm{nb}} = \boldsymbol{\omega}^b_{\mathrm{nm_I}}$. The observer error dynamics becomes

$$\begin{bmatrix}\dot{\tilde{\boldsymbol{p}}}^n_{\mathrm{nm_I}} \\ \dot{\tilde{\boldsymbol{v}}}^n_{\mathrm{nm_I}} \\ \dot{\tilde{\boldsymbol{b}}}^b_{\mathrm{acc}}\end{bmatrix} = \begin{bmatrix}-\boldsymbol{K}_1 & \boldsymbol{I}_3 & \boldsymbol{0}_{3\times3} \\ -\boldsymbol{K}_2 & \boldsymbol{0}_{3\times3} & -\boldsymbol{R}^n_b(t) \\ -\boldsymbol{K}_3\boldsymbol{R}^n_b(t)^\top & \boldsymbol{0}_{3\times3} & \boldsymbol{0}_{3\times3}\end{bmatrix}\begin{bmatrix}\tilde{\boldsymbol{p}}^n_{\mathrm{nm_I}} \\ \tilde{\boldsymbol{v}}^n_{\mathrm{nm_I}} \\ \tilde{\boldsymbol{b}}^b_{\mathrm{acc}}\end{bmatrix} \tag{14.75}$$

$$\Updownarrow$$

$$\dot{\boldsymbol{x}} = \boldsymbol{A}(t)\boldsymbol{x}. \tag{14.76}$$

The gains $\boldsymbol{K}_1$, $\boldsymbol{K}_2$ and $\boldsymbol{K}_3$ can be chosen such that $\boldsymbol{x}$ converges exponentially to zero. This is not straightforward since the matrix $\boldsymbol{A}(t)$ is time dependent.

For marine craft a practical solution to this problem can be found by noticing that the angular rate vector $\boldsymbol{\omega}^b_{\mathrm{nb}} = [p, q, r]^\top$ is quite small. This is the key assumption in order to apply the result of Lindegaard and Fossen (2001a). Consider the transformation

$$\boldsymbol{x} = \boldsymbol{M}(t)\boldsymbol{z} \tag{14.77}$$

where $\boldsymbol{M}(t)$ is a transformation matrix

$$\boldsymbol{M}(t) = \mathrm{diag}\{\boldsymbol{R}^n_b(t), \boldsymbol{R}^n_b(t), \boldsymbol{I}_3\} \tag{14.78}$$

satisfying $\boldsymbol{M}^{-1}(t) = \boldsymbol{M}^{\top}(t)$. Assume that the angular velocity $\boldsymbol{\omega}_{nb}^{b}$ is small such that $\dot{\boldsymbol{M}}(t) = \boldsymbol{0}$. Hence, from a practical point of view it is sufficient to check stability of the system

$$\dot{\boldsymbol{z}} = \boldsymbol{M}^{\top}(t)\boldsymbol{A}(t)\boldsymbol{M}(t)\boldsymbol{z}. \tag{14.79}$$

A pole-placement algorithm can be derived by using the following property.

Property 14.1 (Commuting Matrices)
A matrix $\boldsymbol{K}(t) \in \mathbb{R}^{3\times3}$ is said to commute *with the rotation matrix $\boldsymbol{R}_b^n(t)$ if*

$$\boldsymbol{K}(t)\boldsymbol{R}_b^n(t) = \boldsymbol{R}_b^n(t)\boldsymbol{K}(t). \tag{14.80}$$

Examples of $\boldsymbol{K}(t)$ matrices satisfying Property 14.1 are linear combinations

$$\boldsymbol{K}(t) = a_1\boldsymbol{R}_b^n(t) + a_2\boldsymbol{I}_3 + a_3\boldsymbol{k}\boldsymbol{k}^{\top} \tag{14.81}$$

where $\boldsymbol{k} = [0,\ 0,\ 1]^{\top}$ is the axis of rotation and a_i for $i = 1, \ldots, 3$ are scalars.

If the observer gain matrices $\boldsymbol{K}_i$ for $i = 1, \ldots, 3$ are chosen to commute with the rotation matrix $\boldsymbol{R}_b^n(t)$, Property 14.1 implies that the error dynamics (14.79) can be written as a linear system

$$\dot{\boldsymbol{z}} = \boldsymbol{F}\boldsymbol{z} \tag{14.82}$$

where $\boldsymbol{F} = \boldsymbol{M}^{\top}(t)\boldsymbol{A}(t)\boldsymbol{M}(t)$ is a constant system matrix

$$\boldsymbol{F} = \begin{bmatrix} -\boldsymbol{K}_1 & \boldsymbol{I}_3 & \boldsymbol{0}_{3\times3} \\ -\boldsymbol{K}_2 & \boldsymbol{0}_{3\times3} & -\boldsymbol{I}_3 \\ -\boldsymbol{K}_3 & \boldsymbol{0}_{3\times3} & \boldsymbol{0}_{3\times3} \end{bmatrix}. \tag{14.83}$$

One way to satisfy the commuting matrix property is to choose the matrices $\boldsymbol{K}_i$ with the following diagonal structure

$$\boldsymbol{K}_i = \text{diag}\{k_i,\ k_i,\ l_i\}, \quad i = 1, 2, 3 \tag{14.84}$$

where surge and sway have the same gains $k_i > 0$ and heave can be tuned independently by $l_i > 0$. This clearly satisfies (14.81) since $a_1 = 0$, $a_2 > 0$ and $a_3 > 0$. Hence, stability can be checked by computing the eigenvalues of $\boldsymbol{F}$. A necessary condition for exponential stability is that the eigenvalues of $\boldsymbol{F}$ lie in the left half-plane, that is $\boldsymbol{F}$ must be *Hurwitz*.

Integration of IMU and GNSS Position and Velocity Measurements

It is straightforward to modify the observer (14.69)–(14.72) to include the GNSS velocity measurement, $\boldsymbol{y}_2 = \boldsymbol{v}_{\text{nm}_\text{I}}^n$. In other words

$$\dot{\hat{\boldsymbol{p}}}_{\text{nm}_\text{I}}^n = \hat{\boldsymbol{v}}_{\text{nm}_\text{I}}^n + \boldsymbol{K}_{11}\tilde{\boldsymbol{y}}_1 + \boldsymbol{K}_{21}\tilde{\boldsymbol{y}}_2 \tag{14.85}$$

$$\dot{\hat{\boldsymbol{v}}}^n_{\mathrm{nm_I}} = \boldsymbol{R}^n_b(t)(\boldsymbol{f}^b_{\mathrm{imu}} - \hat{\boldsymbol{b}}^b_{\mathrm{acc}}) + \boldsymbol{g}^n + \boldsymbol{K}_{12}\tilde{\boldsymbol{y}}_1 + \boldsymbol{K}_{22}\tilde{\boldsymbol{y}}_2 \tag{14.86}$$

$$\dot{\hat{\boldsymbol{b}}}^b_{\mathrm{acc}} = \boldsymbol{K}_{13}\boldsymbol{R}^n_b(t)^\top\tilde{\boldsymbol{y}}_1 + \boldsymbol{K}_{23}\boldsymbol{R}^n_b(t)^\top\tilde{\boldsymbol{y}}_2 \tag{14.87}$$

$$\hat{\boldsymbol{y}}_1 = \hat{\boldsymbol{p}}^n_{\mathrm{nm_I}} \tag{14.88}$$

$$\hat{\boldsymbol{y}}_2 = \hat{\boldsymbol{v}}^n_{\mathrm{nm_I}} \tag{14.89}$$

where $\tilde{\boldsymbol{y}}_i = \boldsymbol{y}_i - \hat{\boldsymbol{y}}_i$ $(i = 1, 2)$ results in the error dynamics

$$\begin{bmatrix} \dot{\tilde{\boldsymbol{p}}}^n_{\mathrm{nm_I}} \\ \dot{\tilde{\boldsymbol{v}}}^n_{\mathrm{nm_I}} \\ \dot{\tilde{\boldsymbol{b}}}^b_{\mathrm{acc}} \end{bmatrix} = \begin{bmatrix} -\boldsymbol{K}_{11} & \boldsymbol{I}_3 - \boldsymbol{K}_{21} & \boldsymbol{0}_{3\times3} \\ -\boldsymbol{K}_{12} & -\boldsymbol{K}_{22} & -\boldsymbol{R}^n_b(t) \\ -\boldsymbol{K}_{13}\boldsymbol{R}^n_b(t)^\top & -\boldsymbol{K}_{23}\boldsymbol{R}^n_b(t)^\top & \boldsymbol{0}_{3\times3} \end{bmatrix} \begin{bmatrix} \tilde{\boldsymbol{p}}^n_{\mathrm{nm_I}} \\ \tilde{\boldsymbol{v}}^n_{\mathrm{nm_I}} \\ \tilde{\boldsymbol{b}}^b_{\mathrm{acc}} \end{bmatrix}$$

$$\Updownarrow$$

$$\dot{\boldsymbol{x}} = \boldsymbol{A}(t)\boldsymbol{x}. \tag{14.90}$$

Choosing the gains $\boldsymbol{K}_{ij}$ $(i = 1, 2, 3,\ j = 1, 2)$ according to Property 14.1 such that they commute with $\boldsymbol{R}^n_b(t)$ and assuming that the angular rate vector $\boldsymbol{\omega}^b_{\mathrm{nb}}$ is small, gives the following error dynamics $\dot{\boldsymbol{z}} = \boldsymbol{F}\boldsymbol{z}$ where

$$\boldsymbol{F} = \begin{bmatrix} -\boldsymbol{K}_{11} & \boldsymbol{I}_3 - \boldsymbol{K}_{21} & \boldsymbol{0}_{3\times3} \\ -\boldsymbol{K}_{12} & -\boldsymbol{K}_{22} & -\boldsymbol{I}_3 \\ -\boldsymbol{K}_{13} & -\boldsymbol{K}_{23} & \boldsymbol{0}_{3\times3} \end{bmatrix}. \tag{14.91}$$

Hence, exponential convergence of $\boldsymbol{z}$ to zero is guaranteed if the gains $\boldsymbol{K}_{ij}$ are chosen such that $\boldsymbol{F}$ is *Hurwitz*.

14.3.2 Direct Kalman Filter using Attitude Measurements

Assume that the IMU is mounted onboard the craft in a measurement frame $\{m_\mathrm{I}\}$ with coordinate origin $\mathrm{CM_I}$ located at $\boldsymbol{r}^b_{bm_\mathrm{I}} = [x_{\mathrm{m_I}}, y_{\mathrm{m_I}}, z_{\mathrm{m_I}}]^\top$ with respect to the CO (see Figure 14.1). The measurements are $\boldsymbol{f}^b_{\mathrm{imu}} := \boldsymbol{f}^b_{\mathrm{nm_I}}$ and $\boldsymbol{\omega}^b_{\mathrm{imu}} := \boldsymbol{\omega}^b_{\mathrm{nb}}$; see Section 14.1. Further assume that

$$\boldsymbol{R}(\boldsymbol{\Theta}_{\mathrm{nb}}) := \boldsymbol{R}^n_b(t), \qquad \boldsymbol{T}(\boldsymbol{\Theta}_{\mathrm{nb}}) := \boldsymbol{T}^n_b(t) \tag{14.92}$$

are known for all $t \geq 0$ by using an AHRS to measure the Euler angles $\boldsymbol{\Theta}_{\mathrm{nb}}$. This assumption is removed in Section 14.3.3 where position, velocity and attitude are estimated all together. The *strapdown navigation equations* are

$$\dot{\boldsymbol{p}}^n_{\text{nm}_\text{I}} = \boldsymbol{v}^n_{\text{nm}_\text{I}} \tag{14.93}$$

$$\dot{\boldsymbol{v}}^n_{\text{nm}_\text{I}} = \boldsymbol{R}^n_b(t)(\boldsymbol{f}^b_{\text{imu}} - \boldsymbol{b}^b_{\text{acc}} - \boldsymbol{w}^b_{\text{acc}}) + \boldsymbol{g}^n \tag{14.94}$$

$$\dot{\boldsymbol{b}}^b_{\text{acc}} = \boldsymbol{w}^b_{\text{b, acc}} \tag{14.95}$$

$$\dot{\boldsymbol{\Theta}}_{\text{nb}} = \boldsymbol{T}^n_b(t)(\boldsymbol{\omega}^b_{\text{imu}} - \boldsymbol{b}^b_{\text{ars}} - \boldsymbol{w}^b_{\text{ars}}) \tag{14.96}$$

$$\dot{\boldsymbol{b}}^b_{\text{ars}} = \boldsymbol{w}^b_{\text{b, ars}} \tag{14.97}$$

where the biases are modeled as random walk processes.

The KF has 15 states

$$\boldsymbol{x} = [(\boldsymbol{p}^n_{\text{nm}_\text{I}})^\top, (\boldsymbol{v}^n_{\text{nm}_\text{I}})^\top, (\boldsymbol{b}^b_{\text{acc}})^\top, \boldsymbol{\Theta}^\top_{\text{nb}}, (\boldsymbol{b}^b_{\text{ars}})^\top]^\top \tag{14.98}$$

and three inputs (IMU measurements and gravity vector)

$$\boldsymbol{u} = [(\boldsymbol{f}^b_{\text{imu}})^\top, (\boldsymbol{\omega}^b_{\text{imu}})^\top, (\boldsymbol{g}^n)^\top]^\top \tag{14.99}$$

while the process noise vector is

$$\boldsymbol{w} = [(\boldsymbol{w}^b_{\text{acc}})^\top, (\boldsymbol{w}^b_{\text{b, acc}})^\top, (\boldsymbol{w}^b_{\text{ars}})^\top, (\boldsymbol{w}^b_{\text{b, ars}})^\top]^\top. \tag{14.100}$$

The measurement vector is chosen as

$$\boldsymbol{y} = [(\boldsymbol{p}^n_{\text{nm}_\text{I}})^\top, \boldsymbol{\Theta}^\top_{\text{nb}}]^\top. \tag{14.101}$$

The LTV state-space model corresponding to (14.93)–(14.97) with (14.92) takes the following form

$$\dot{\boldsymbol{x}} = \boldsymbol{A}(t)\boldsymbol{x} + \boldsymbol{B}(t)\boldsymbol{u} + \boldsymbol{E}(t)\boldsymbol{w} \tag{14.102}$$

$$\boldsymbol{y} = \boldsymbol{C}\boldsymbol{x} + \boldsymbol{\varepsilon} \tag{14.103}$$

where the IMU measurements are treated as input signals and

$$\boldsymbol{A}(t) = \begin{bmatrix} \boldsymbol{0}_{3\times3} & \boldsymbol{I}_3 & \boldsymbol{0}_{3\times3} & \boldsymbol{0}_{3\times3} & \boldsymbol{0}_{3\times3} \\ \boldsymbol{0}_{3\times3} & \boldsymbol{0}_{3\times3} & -\boldsymbol{R}^n_b(t) & \boldsymbol{0}_{3\times3} & \boldsymbol{0}_{3\times3} \\ \boldsymbol{0}_{3\times3} & \boldsymbol{0}_{3\times3} & \boldsymbol{0}_{3\times3} & \boldsymbol{0}_{3\times3} & \boldsymbol{0}_{3\times3} \\ \boldsymbol{0}_{3\times3} & \boldsymbol{0}_{3\times3} & \boldsymbol{0}_{3\times3} & \boldsymbol{0}_{3\times3} & -\boldsymbol{T}^n_b(t) \\ \boldsymbol{0}_{3\times3} & \boldsymbol{0}_{3\times3} & \boldsymbol{0}_{3\times3} & \boldsymbol{0}_{3\times3} & \boldsymbol{0}_{3\times3} \end{bmatrix}, \quad \boldsymbol{B}(t) = \begin{bmatrix} \boldsymbol{0}_{3\times3} & \boldsymbol{0}_{3\times3} & \boldsymbol{0}_{3\times3} \\ \boldsymbol{R}^n_b(t) & \boldsymbol{0}_{3\times3} & \boldsymbol{I}_3 \\ \boldsymbol{0}_{3\times3} & \boldsymbol{0}_{3\times3} & \boldsymbol{0}_{3\times3} \\ \boldsymbol{0}_{3\times3} & \boldsymbol{T}^n_b(t) & \boldsymbol{0}_{3\times3} \\ \boldsymbol{0}_{3\times3} & \boldsymbol{0}_{3\times3} & \boldsymbol{0}_{3\times3} \end{bmatrix} \tag{14.104}$$

$$\boldsymbol{C} = \begin{bmatrix} \boldsymbol{I}_3 & \boldsymbol{0}_{3\times3} & \boldsymbol{0}_{3\times3} & \boldsymbol{0}_{3\times3} & \boldsymbol{0}_{3\times3} \\ \boldsymbol{0}_{3\times3} & \boldsymbol{0}_{3\times3} & \boldsymbol{0}_{3\times3} & \boldsymbol{I}_3 & \boldsymbol{0}_{3\times3} \end{bmatrix}, \quad \boldsymbol{E}(t) = \begin{bmatrix} \boldsymbol{0}_{3\times3} & \boldsymbol{0}_{3\times3} & \boldsymbol{0}_{3\times3} & \boldsymbol{0}_{3\times3} \\ -\boldsymbol{R}^n_b(t) & \boldsymbol{0}_{3\times3} & \boldsymbol{0}_{3\times3} & \boldsymbol{0}_{3\times3} \\ \boldsymbol{0}_{3\times3} & \boldsymbol{I}_3 & \boldsymbol{0}_{3\times3} & \boldsymbol{0}_{3\times3} \\ \boldsymbol{0}_{3\times3} & \boldsymbol{0}_{3\times3} & -\boldsymbol{T}^n_b(t) & \boldsymbol{0}_{3\times3} \\ \boldsymbol{0}_{3\times3} & \boldsymbol{0}_{3\times3} & \boldsymbol{0}_{3\times3} & \boldsymbol{I}_3 \end{bmatrix}. \tag{14.105}$$

The linear discrete-time KF, Table 13.1 in Section 13.4.1, will estimate all states thanks to the observability result below.

Observability Analysis

It is possible to determine if the acceleration and attitude rate sensor biases are observable for the system (14.93)–(14.97) by using the result of Chen (2013, theorem 6.012). The pair $(\boldsymbol{A}(t), \boldsymbol{C}(t))$ of a LTV system of dimension n is observable if

$$\text{rank}\begin{bmatrix} \boldsymbol{N}_0(t) \\ \boldsymbol{N}_1(t) \\ \vdots \\ \boldsymbol{N}_{n-1}(t) \end{bmatrix} = n \tag{14.106}$$

where

$$\boldsymbol{N}_0(t) = \boldsymbol{C}(t) \tag{14.107}$$

$$\boldsymbol{N}_{\mathrm{m}+1}(t) = \boldsymbol{N}_{\mathrm{m}}(t)\boldsymbol{A}(t) + \frac{\mathrm{d}}{\mathrm{d}t}\boldsymbol{N}_{\mathrm{m}}(t), \quad m = 0, 1, \ldots, n-1. \tag{14.108}$$

The system (14.93)–(14.97) can be seen as two decoupled systems for attitude and translational motions (see Figure 14.5) thanks to the AHRS measurement. The attitude system

$$\boldsymbol{A}_1 = \begin{bmatrix} \boldsymbol{0}_{3\times3} & -\boldsymbol{T}_b^n(t) \\ \boldsymbol{0}_{3\times3} & \boldsymbol{0}_{3\times3} \end{bmatrix}, \quad \boldsymbol{C}_1 = \begin{bmatrix} \boldsymbol{I}_3 & \boldsymbol{0}_{3\times3} \end{bmatrix} \tag{14.109}$$

is observable since

$$\text{rank}\begin{bmatrix} \boldsymbol{N}_0(t) \\ \boldsymbol{N}_1(t) \end{bmatrix} = \text{rank}\begin{bmatrix} \boldsymbol{I}_3 & \boldsymbol{0}_{3\times3} \\ \boldsymbol{0}_{3\times3} & -\boldsymbol{T}_b^n(t) \end{bmatrix} = 6 \tag{14.110}$$

except for the singular point $\theta = \pm 90°$. The dynamics of the translational motion is

$$\boldsymbol{A}_2 = \begin{bmatrix} \boldsymbol{0}_{3\times3} & \boldsymbol{I}_3 & \boldsymbol{0}_{3\times3} \\ \boldsymbol{0}_{3\times3} & \boldsymbol{0}_{3\times3} & -\boldsymbol{R}_b^n(t) \\ \boldsymbol{0}_{3\times3} & \boldsymbol{0}_{3\times3} & \boldsymbol{0}_{3\times3} \end{bmatrix}, \quad \boldsymbol{C}_2 = \begin{bmatrix} \boldsymbol{I}_3 & \boldsymbol{0}_{3\times3} & \boldsymbol{0}_{3\times3} \end{bmatrix} \tag{14.111}$$

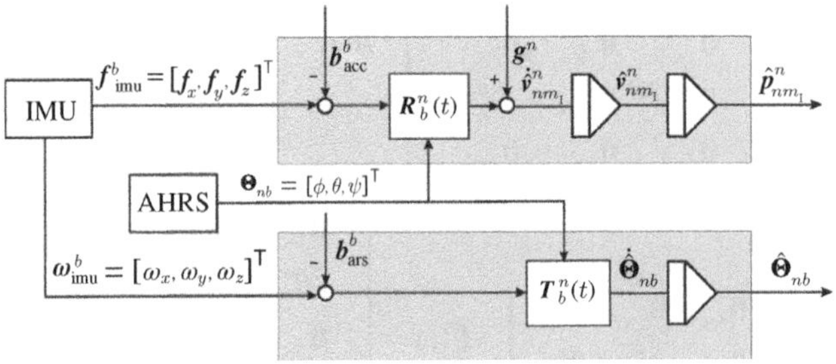

Figure 14.5 The principle for decoupled estimation of translational motion and attitude using IMU and AHRS measurements.

for which

$$\text{rank}\begin{bmatrix} N_0(t) \\ N_1(t) \\ N_2(t) \end{bmatrix} = \text{rank}\begin{bmatrix} I_3 & 0_{3\times3} & 0_{3\times3} \\ 0_{3\times3} & I_3 & 0_{3\times3} \\ 0_{3\times3} & 0_{3\times3} & -R_b^n(t) \end{bmatrix} = 9. \tag{14.112}$$

Again observability is guaranteed since the rotation matrix has orthogonal columns and therefore $\text{rank}(R_b^n(t)) = 3$.

14.3.3 Direct Kalman Filter with Attitude Estimation

In Section 14.3.2 it was shown that the direct KF can estimate both the ARS and acceleration biases if the attitude is known. This follows from the observability proof. Unfortunately, this result does not generalize to the case when attitude is estimated along with position and linear velocity. For this system to be observable the craft must undergo certain maneuvers to make the acceleration bias observable. In other words, the rolling and pitching motion of the craft must persistently excite the system in order to accurately estimate the acceleration bias terms.

An EKF for position, velocity and attitude estimation can be designed using unit quaternions q_b^n or Euler angles θ_{nb}. The unit quaternion solution is known as the multiplicative extended Kalman filer (MEKF), for which the quaternion estimation error is represented by the Schur product (Crassidis *et al.* 2007). The MEKF has the advantage that representation singularities and blow-up of the estimate error covariance matrix is avoided. This can be a major problem when using the Euler angle representation. The state-of-the art MEKF solution is the error-state Kalman filter presented in Section 14.4.3.

This section presents the direct EKF using Euler angles θ_{nb} to represent attitude. Since the estimation of the acceleration bias is in general a hard problem, when estimated along with the attitude vector, this state should be treated as optional. For short-time applications it is common to calibrate the acceleration measurements and remove the bias before the mission. The ARS bias, however, is much easier to estimate so this state is normally included in the EKF.

The *strapdown navigation equations* are

$$\dot{p}^n_{\text{nm}_\text{I}} = v^n_{\text{nm}_\text{I}} \tag{14.113}$$

$$\dot{v}^n_{\text{nm}_\text{I}} = R(\Theta_{\text{nb}})(f^b_{\text{imu}} - b^b_{\text{acc}} - w^b_{\text{acc}}) + g^n \tag{14.114}$$

$$\dot{b}^b_{\text{acc}} = w^b_{\text{b, acc}} \tag{14.115}$$

$$\dot{\Theta}_{\text{nb}} = T(\Theta_{\text{nb}})(\omega^b_{\text{imu}} - b^b_{\text{ars}} - w^b_{\text{ars}}) \tag{14.116}$$

$$\dot{b}^b_{\text{ars}} = w^b_{\text{b, ars}}. \tag{14.117}$$

The EKF state vector contains 15 states

$$x = [(p^n_{\text{nm}_\text{I}})^\top, (v^n_{\text{nm}_\text{I}})^\top, (b^b_{\text{acc}})^\top, \Theta^\top_{\text{nb}}, (b^b_{\text{ars}})^\top]^\top. \tag{14.118}$$

The state-space model (14.113)–(14.117) has non-additive input and process noise terms due to the state-dependent rotation matrix. Consequently, the system (14.113)–(14.117) is in the form

$$\dot{\boldsymbol{x}} = \boldsymbol{f}(\boldsymbol{x}, \boldsymbol{u}, \boldsymbol{w}) \tag{14.119}$$

$$\boldsymbol{y} = \boldsymbol{h}(\boldsymbol{x}, \boldsymbol{u}) + \boldsymbol{\varepsilon} \tag{14.120}$$

where $\boldsymbol{u} = [(\boldsymbol{f}^b_{\text{imu}})^\top, (\boldsymbol{\omega}^b_{\text{imu}})^\top, (\boldsymbol{g}^n)^\top]^\top$, $\boldsymbol{w} = [(\boldsymbol{w}^b_{\text{acc}})^\top, (\boldsymbol{w}^b_{\text{b, acc}})^\top, (\boldsymbol{w}^b_{\text{ars}})^\top, (\boldsymbol{w}^b_{\text{b, ars}})^\top]^\top$ and

$$\boldsymbol{f}(\boldsymbol{x}, \boldsymbol{u}, \boldsymbol{w}) = \begin{bmatrix} \boldsymbol{v}^n_{\text{nm}_\text{I}} \\ \boldsymbol{R}(\boldsymbol{\Theta}_{\text{nb}})(\boldsymbol{f}^b_{\text{imu}} - \boldsymbol{b}^b_{\text{acc}} - \boldsymbol{w}^b_{\text{acc}}) + \boldsymbol{g}^n \\ \boldsymbol{w}^b_{\text{b, acc}} \\ \boldsymbol{T}(\boldsymbol{\Theta}_{\text{nb}})(\boldsymbol{\omega}^b_{\text{imu}} - \boldsymbol{b}^b_{\text{ars}} - \boldsymbol{w}^b_{\text{ars}}) \\ \boldsymbol{w}^b_{\text{b, ars}} \end{bmatrix}. \tag{14.121}$$

The measurement vector must contain the position vector $\boldsymbol{p}^n_{\text{nm}_\text{I}}$ for proper aiding. In addition at least two reference vectors are needed to determine the attitude. The reference vectors can be chosen as specific force and magnetic field, alternatively compass direction; see Section 14.2.3. Consequently,

$$\boldsymbol{y} = \begin{bmatrix} \boldsymbol{p}^n_{\text{nm}_\text{I}} \\ \boldsymbol{v}^n_{\text{nm}_\text{I}}(\text{optionally}) \\ \boldsymbol{f}^b_{\text{imu}}(\text{optionally}) \\ \boldsymbol{m}^b_{\text{mag}}(\text{or compass}) \end{bmatrix}. \tag{14.122}$$

The corresponding vector field is

$$\boldsymbol{h}(\boldsymbol{x}, \boldsymbol{u}) = \begin{bmatrix} \boldsymbol{p}^n_{\text{nm}_\text{I}} \\ \boldsymbol{v}^n_{\text{nm}_\text{I}}(\text{optionally}) \\ \boldsymbol{S}(\boldsymbol{\omega}^b_{\text{imu}} - \boldsymbol{b}^b_{\text{ars}})\boldsymbol{R}^\top(\boldsymbol{\Theta}_{\text{nb}})\boldsymbol{v}^n_{\text{nm}_\text{I}} - \boldsymbol{R}^\top(\boldsymbol{\Theta}_{\text{nb}})\boldsymbol{g}^n + \boldsymbol{b}^b_{\text{acc}}(\text{optionally}) \\ \boldsymbol{R}^\top(\boldsymbol{\Theta}_{\text{nb}})\boldsymbol{m}^n \end{bmatrix}. \tag{14.123}$$

The reason for all the terms in the optional specific force measurement equation, row three, when compared to the static solution $\boldsymbol{f}^b_{\text{imu}} = -\boldsymbol{R}^\top(\boldsymbol{\Theta}_{\text{nb}})\boldsymbol{g}^n$ is formula (14.16) which is more accurate in a maneuvering situation. Consider (14.16) under the assumption that the accelerometer white noise term is negligible

$$\begin{aligned} \boldsymbol{f}^b_{\text{imu}} &= \boldsymbol{R}^\top(\boldsymbol{\Theta}_{\text{nb}})(\boldsymbol{a}^n_{\text{nm}_\text{I}} - \boldsymbol{g}^n) + \boldsymbol{b}^b_{\text{acc}} \\ &= \dot{\boldsymbol{v}}^b_{\text{nm}_\text{I}} + \boldsymbol{S}(\boldsymbol{\omega}^b_{\text{nb}})\boldsymbol{v}^b_{\text{nm}_\text{I}} - \boldsymbol{R}^\top(\boldsymbol{\Theta}_{\text{nb}})\boldsymbol{g}^n + \boldsymbol{b}^b_{\text{acc}}. \end{aligned} \tag{14.124}$$

Hence, by assuming that $\dot{\boldsymbol{v}}^b_{\text{nm}_\text{I}} \approx \boldsymbol{0}$ we arrive at

$$\boldsymbol{f}^b_{\text{imu}} \approx \boldsymbol{S}(\boldsymbol{\omega}^b_{\text{imu}} - \boldsymbol{b}^b_{\text{ars}})\boldsymbol{R}^\top(\boldsymbol{\Theta}_{\text{nb}})\boldsymbol{v}^n_{\text{nm}_\text{I}} - \boldsymbol{R}^\top(\boldsymbol{\Theta}_{\text{nb}})\boldsymbol{g}^n + \boldsymbol{b}^b_{\text{acc}}. \tag{14.125}$$

Note that the accelerometer measurement vector is used as input to the EKF but it is also possible to add the accelerometers (14.125) in the measurement equation as shown in (14.123). This is, however, based on the assumption that $\dot{\boldsymbol{v}}^b_{\text{nm}_\text{I}}$ is small such that (14.125) holds. A craft moving at constant or slowly-varying speed clearly satisfies this assumption. The advantage of including (14.125) as a measurement is that it is possible to estimate the roll, pitch and yaw angles without a magnetometer. Hence, the INS can be used in compass-denied environments. For this to work, the system must be sufficient excited. For zero-speed applications (stationkeeping) such as DP vessels, ROVs, etc. some sort of aiding sensor related to the attitude is always necessary.

Since the model (14.113)–(14.117) is nonlinear it is necessary to compute the discrete-time *Jacobians*

$$\boldsymbol{A}_{\mathrm{d}}[k] = \boldsymbol{I}_{15} + h \left. \frac{\partial \boldsymbol{f}(\boldsymbol{x}[k], \boldsymbol{u}[k], \boldsymbol{w}[k])}{\partial \boldsymbol{x}[k]} \right|_{\boldsymbol{x}[k]=\hat{\boldsymbol{x}}[k],\ \boldsymbol{w}[k]=\boldsymbol{0}} \tag{14.126}$$

$$\boldsymbol{E}_{\mathrm{d}}[k] = h \left. \frac{\partial \boldsymbol{f}(\boldsymbol{x}[k], \boldsymbol{u}[k], \boldsymbol{w}[k])}{\partial \boldsymbol{w}[k]} \right|_{\boldsymbol{x}[k]=\hat{\boldsymbol{x}}[k],\ \boldsymbol{w}[k]=\boldsymbol{0}} \tag{14.127}$$

$$\boldsymbol{C}_{\mathrm{d}}[k] = \left. \frac{\partial \boldsymbol{h}(\boldsymbol{x}[k], \boldsymbol{u}[k])}{\partial \boldsymbol{x}[k]} \right|_{\boldsymbol{x}[k]=\hat{\boldsymbol{x}}^{-}[k]} \tag{14.128}$$

in order to implement the EKF algorithm in Table 13.2. This gives the algorithm

Kalman gain matrix:

$$\boldsymbol{K}[k] = \hat{\boldsymbol{P}}^{-}[k]\boldsymbol{C}_{\mathrm{d}}^{\top}[k](\boldsymbol{C}_{\mathrm{d}}[k]\hat{\boldsymbol{P}}^{-}[k]\boldsymbol{C}_{\mathrm{d}}^{\top}[k] + \boldsymbol{R}_{\mathrm{d}}[k])^{-1} \tag{14.129}$$

Corrector:

$$\hat{\boldsymbol{x}}[k] = \hat{\boldsymbol{x}}^{-}[k] + \boldsymbol{K}[k](\boldsymbol{y}[k] - \boldsymbol{h}(\hat{\boldsymbol{x}}^{-}[k], \boldsymbol{u}[k])) \tag{14.130}$$

$$\hat{\boldsymbol{P}}[k] = (\boldsymbol{I}_{15} - \boldsymbol{K}[k]\boldsymbol{C}_{\mathrm{d}}[k])\hat{\boldsymbol{P}}^{-}[k](\boldsymbol{I}_{15} - \boldsymbol{K}[k]\boldsymbol{C}_{\mathrm{d}}[k])^{\top} + \boldsymbol{K}[k]\boldsymbol{R}_{\mathrm{d}}[k]\boldsymbol{K}^{\top}[k] \tag{14.131}$$

Predictor:

$$\hat{\boldsymbol{x}}^{-}[k+1] = \hat{\boldsymbol{x}}[k] + h\boldsymbol{f}(\hat{\boldsymbol{x}}[k], \boldsymbol{u}[k], \boldsymbol{0}) \tag{14.132}$$

$$\hat{\boldsymbol{P}}^{-}[k+1] = \boldsymbol{A}_{\mathrm{d}}[k]\hat{\boldsymbol{P}}[k]\boldsymbol{A}_{\mathrm{d}}^{\top}[k] + \boldsymbol{E}_{\mathrm{d}}[k]\boldsymbol{Q}_{\mathrm{d}}[k]\boldsymbol{E}_{\mathrm{d}}^{\top}[k]. \tag{14.133}$$

14.4 Indirect Filters for Aided INS

Indirect filtering implies that the KF filter is formulated as an error-state filter. In this context, the INS state $\boldsymbol{x}_{\mathrm{ins}}[k]$ and error state $\delta\boldsymbol{x}[k]$ are related to the true state $\boldsymbol{x}[k]$ by

$$\boldsymbol{x}[k] = \boldsymbol{x}_{\mathrm{ins}}[k] + \delta\boldsymbol{x}[k]. \tag{14.134}$$

It should be noted that $\boldsymbol{x}[k] = \boldsymbol{x}_{\mathrm{ins}}[k] - \delta\boldsymbol{x}[k]$ is also a possible choice. The position, velocity and attitude errors together with sensor bias and misalignment errors, etc. are used as states $\delta\boldsymbol{x}[k]$ in the KF, while the INS state, $\boldsymbol{x}_{\mathrm{ins}}[k]$, is based on integration of the *strapdown*

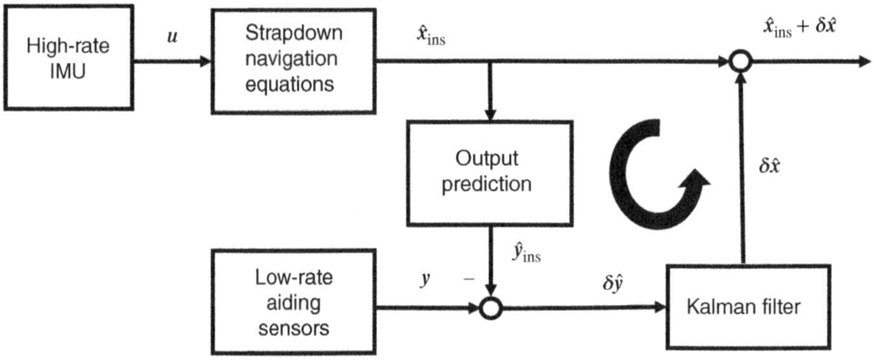

Figure 14.6 Indirect feedforward filter for INS. The strapdown navigation equations are propagated to obtain the estimate $\hat{\boldsymbol{x}}_{\text{ins}}$, which is corrected by the KF error-state estimate $\delta\hat{\boldsymbol{x}}$. The correction is added as a feedforward term to $\hat{\boldsymbol{x}}_{\text{ins}}$.

navigation equations (14.113)–(14.117) with accelerometer and angular rate measurement as driving terms. The measurement equations will be the difference of the numerically integrated navigation equations (open-loop integration) and external measurements used to corrected the INS. The technical term for this is *aided INS* and typical aiding sensors are GNSS, compass, DVL, magnetometer, etc.

The indirect filter can be represented as a *feedforward* or a *feedback* filter. The reason for the term feedforward filter can be seen in Figure 14.6 where the error corrections are added to the INS state as a feedforward compensation. However, for the feedback filter the error estimates are used to update the INS estimates directly in order to prevent the INS errors to grow (see Figure 14.8). This is achieved by regulating the error $\delta\boldsymbol{x}$ to zero such that $\boldsymbol{x}_{\text{ins}}[k] \to \boldsymbol{x}[k]$, analogues to a feedback control system.

The error-state KF filter is based on a discrete-time state-space model

$$\delta\boldsymbol{x}[k+1] = \boldsymbol{A}_{\text{d}}[k]\delta\boldsymbol{x}[k] + \boldsymbol{B}_{\text{d}}[k]\boldsymbol{u}[k] + \boldsymbol{E}_{\text{d}}[k]\boldsymbol{w}[k] \tag{14.135}$$

$$\delta\boldsymbol{y}[k] = \boldsymbol{C}_{\text{d}}[k]\delta\boldsymbol{x}[k] + \boldsymbol{D}_{\text{d}}[k]\boldsymbol{u}[k] + \boldsymbol{\varepsilon}[k] \tag{14.136}$$

where $\boldsymbol{A}_{\text{d}}$, $\boldsymbol{B}_{\text{d}}$, $\boldsymbol{C}_{\text{d}}$, $\boldsymbol{D}_{\text{d}}$ and $\boldsymbol{E}_{\text{d}}$ are the discrete-time system matrices while $\boldsymbol{w}[k]$ and $\boldsymbol{\varepsilon}[k]$ are Gaussian white noise vectors. The error-state KF equations for the system (14.135) and (14.136) is given by Table 14.1.

For the feedforward representation of the filter the estimated error states $\delta\hat{\boldsymbol{x}}[k]$ are added to the INS states $\hat{\boldsymbol{x}}_{\text{ins}}[k]$ to produce the state estimate

$$\hat{\boldsymbol{x}}[k] = \hat{\boldsymbol{x}}_{\text{ins}}[k] + \delta\hat{\boldsymbol{x}}[k]. \tag{14.137}$$

Note that for the true state vector, $\boldsymbol{x}[k] = \hat{\boldsymbol{x}}_{\text{ins}}[k] + \delta\boldsymbol{x}[k]$. This implies that we want to design the KF such that $\delta\hat{\boldsymbol{x}}[k] \to \delta\boldsymbol{x}[k]$.

The advantage of the indirect filter to the direct filter is that the estimator will continue to work if the KF update fails due to faulty measurements (no aiding). During

Table 14.1 Discrete-time error-state Kalman filter.

Initial values	$\delta\hat{x}^-[0] = \delta x_0$ $\hat{P}^-[0] = \mathrm{E}[(\delta x[0] - \delta\hat{x}^-[0])(\delta x[0] - \delta\hat{x}^-[0])^\top] = P_0$
KF gain	$K[k] = \hat{P}^-[k]C_d^\top[k](C_d[k]\hat{P}^-[k]C_d^\top[k] + R_d[k])^{-1}$
State corrector Covariance corrector	$\delta\hat{x}[k] = \delta\hat{x}^-[k] + K[k]\,(\delta y[k] - C_d[k]\delta\hat{x}^-[k] - D_d[k]u[k])$ $\hat{P}[k] = (I_n - K[k]C_d[k])\hat{P}^-[k](I_n - K[k]C_d[k])^\top$ $+K[k]R_d[k]K^\top[k]$
State predictor Covariance predictor	$\delta\hat{x}^-[k+1] = A_d[k]\,\delta\hat{x}[k] + B_d[k]\,u[k]$ $\delta\hat{P}^-[k+1] = A_d[k]\hat{P}[k]A_d^\top[k] + E_d[k]Q_d[k]E_d^\top[k]$

where

Q_d, R_d	Covariance matrices for the process and measurement noises
$\delta\hat{x}^-$, $\hat{P}^-$	A priori error state and covariance matrix estimates (before update)
$\delta\hat{x}$, $\hat{P}$	A posteriori error state and covariance matrix estimates (after update)

this situation the $\delta\hat{x}[k]$ state vector is not updated while the open-loop estimate $\hat{x}_{\text{ins}}[k]$ propagates thanks to the IMU measurements. The failure situation with no position measurements is called *dead reckoning*. However, during dead reckoning the error states will continue to grow since $\delta\hat{x}[k]$ will be constant.

14.4.1 *Introductory Example*

A step-by-step procedure for derivation of the error-state KF for a 2-DOF example system with decoupled linear acceleration and orientation will now be presented. The generalization to 6 DOFs with and without attitude measurements are presented in Sections 14.4.2 and 14.4.3, respectively where the latter is the MEKF algorithm.

Strapdown navigation equations: Consider the strapdown navigation equations

$$\dot{p} = v \tag{14.138}$$

$$\dot{v} = a \tag{14.139}$$

$$\dot{\theta} = \omega \tag{14.140}$$

where p is the position, v is the velocity, a is the acceleration, θ is the orientation and ω is the angular rate.

IMU measurements: The inputs to the strapdown navigation equations are the IMU specific force and ARS measurements

$$f_{\text{imu}} = a - g + b_{\text{acc}} + w_{\text{acc}} \tag{14.141}$$

$$\dot{b}_{\text{acc}} = w_{b,\,\text{acc}} \tag{14.142}$$

$$\omega_{\text{imu}} = \omega + b_{\text{ars}} + w_{\text{ars}} \tag{14.143}$$

$$\dot{b}_{\text{ars}} = w_{b,\,\text{ars}}. \tag{14.144}$$

The process noise terms w_{acc}, w_{ars}, $w_{b,\,acc}$ and $w_{b,\,ars}$ are all assumed to be Gaussian white noise.

Aiding: The measurements used for aiding are assumed to be

$$y_1 = p + \varepsilon_1 \tag{14.145}$$

$$y_2 = \theta + \varepsilon_2 \tag{14.146}$$

where the measurement noise terms ε_1 and ε_2 are Gaussian white noise.

The INS estimates are obtained by integrating the strapdown navigation equations (14.138)–(14.140) with high-rate IMU measurements (14.141)–(14.144) for which all white-noise terms are set to zero. This gives the following open-loop estimator

Inertial navigation system (INS):

$$\dot{\hat{p}}_{ins} = \hat{v}_{ins} \tag{14.147}$$

$$\dot{\hat{v}}_{ins} = f_{imu} - \hat{b}_{acc,\,ins} + g \tag{14.148}$$

$$\dot{\hat{b}}_{acc,\,ins} = 0 \tag{14.149}$$

$$\dot{\hat{\theta}}_{ins} = \omega_{imu} - \hat{b}_{ars,\,ins} \tag{14.150}$$

$$\dot{\hat{b}}_{ars,\,ins} = 0. \tag{14.151}$$

The next step is derive the error states $\delta \boldsymbol{x} = \boldsymbol{x} - \hat{\boldsymbol{x}}_{ins}$ to be used in the KF. This is mathematically equivalent to

$$\delta p = p - \hat{p}_{ins} \tag{14.152}$$

$$\delta v = v - \hat{v}_{ins} \tag{14.153}$$

$$\delta b_{acc} = b_{acc} - \hat{b}_{acc,\,ins} \tag{14.154}$$

$$\delta \theta = \theta - \hat{\theta}_{ins} \tag{14.155}$$

$$\delta b_{ars} = b_{ars} - \hat{b}_{ars,\,ins}. \tag{14.156}$$

The differential equations for the translational motion is obtained by time differentiation of (14.152) and (14.153)

$$\begin{aligned}
\delta \dot{p} &= \dot{p} - \dot{\hat{p}}_{ins} \\
&= v - \hat{v}_{ins} = \delta v
\end{aligned} \tag{14.157}$$

$$\begin{aligned}
\delta \dot{v} &= (f + g) - \dot{\hat{v}}_{ins} \\
&= (f_{imu} - b_{acc} - w_{acc} + g) - (f_{imu} - \hat{b}_{acc,\,ins} + g)
\end{aligned}$$

$$
\begin{aligned}
&= -b_{\text{acc}} - w_{\text{acc}} + \hat{b}_{\text{acc, ins}} \\
&= -\delta b_{\text{acc}} - w_{\text{acc}}. \qquad (14.158)
\end{aligned}
$$

Similarly, the model (14.155) for the attitude angle is differentiated

$$
\begin{aligned}
\delta\dot{\theta} &= \omega - \dot{\hat{\theta}}_{\text{ins}} \\
&= (\omega_{\text{imu}} - b_{\text{ars}} - w_{\text{ars}}) - (\omega_{\text{imu}} - \hat{b}_{\text{ars, ins}}) \\
&= -b_{\text{ars}} - w_{\text{ars}} + \hat{b}_{\text{ars, ins}} \\
&= -\delta b_{\text{ars}} - w_{\text{ars}}. \qquad (14.159)
\end{aligned}
$$

The errors in the INS due to drift and noise on the accelerometer and ARS measurements can be accurately described by first-order bias models

$$
\delta\dot{b}_{\text{acc}} := -\frac{1}{T_{\text{acc}}}\delta b_{\text{acc}} + w_{\text{b, acc}} \qquad (14.160)
$$

$$
\delta\dot{b}_{\text{ars}} := -\frac{1}{T_{\text{ars}}}\delta b_{\text{ars}} + w_{\text{b, ars}} \qquad (14.161)
$$

where T_{acc} and T_{ars} are user specified time constants, and $w_{\text{b, acc}}$ and $w_{\text{b, ars}}$ are Gaussian white noise. The time constants ensure that the bias errors go exponentially to zero during dead reckoning, that is situations where there are no aiding.

The aiding position and attitude measurements p and θ relate to the error states as follow

$$
\begin{aligned}
\delta y_1 &= (p + \varepsilon_1) - \hat{p}_{\text{ins}} \\
&= \delta p + \varepsilon_1 \qquad (14.162)
\end{aligned}
$$

$$
\begin{aligned}
\delta y_2 &= (\theta + \varepsilon_2) - \hat{\theta}_{\text{ins}} \\
&= \delta\theta + \varepsilon_2 \qquad (14.163)
\end{aligned}
$$

where ε_1 and ε_2 are Gaussian white measurement noise. Thus, the resulting error-state model becomes

Error-state model:

$$
\delta\dot{\boldsymbol{x}} = \boldsymbol{A}\delta\boldsymbol{x} + \boldsymbol{E}\boldsymbol{w} \qquad (14.164)
$$

$$
\delta\boldsymbol{y} = \boldsymbol{C}\delta\boldsymbol{x} + \boldsymbol{\varepsilon} \qquad (14.165)
$$

where $\delta\boldsymbol{x} = [\delta p,\ \delta v,\ \delta b_{\text{acc}},\ \delta\theta,\ \delta b_{\text{ars}}]^{\top}$, $\boldsymbol{w} = [w_{\text{acc}},\ w_{\text{b,acc}},\ w_{\text{ars}},\ w_{\text{b, ars}}]^{\top}$, $\boldsymbol{\varepsilon} = [\varepsilon_1,\ \varepsilon_2]^{\top}$ and

$$
\boldsymbol{A} = \begin{bmatrix} 0 & 1 & 0 & 0 & 0 \\ 0 & 0 & -1 & 0 & 0 \\ 0 & 0 & -\frac{1}{T_{\text{acc}}} & 0 & 0 \\ 0 & 0 & 0 & 0 & -1 \\ 0 & 0 & 0 & 0 & -\frac{1}{T_{\text{ars}}} \end{bmatrix}, \quad \boldsymbol{E} = \begin{bmatrix} 0 & 0 & 0 & 0 \\ -1 & 0 & 0 & 0 \\ 0 & 1 & 0 & 0 \\ 0 & 0 & -1 & 0 \\ 0 & 0 & 0 & 1 \end{bmatrix}, \quad \boldsymbol{C} = \begin{bmatrix} 1 & 0 & 0 & 0 & 0 \\ 0 & 0 & 0 & 1 & 0 \end{bmatrix}. \qquad (14.166)
$$

The discretized system matrices are (see Appendix B.1.1)

$$A_\mathrm{d} \approx I_5 + hA, \qquad C_\mathrm{d} = C, \qquad E_\mathrm{d} \approx hE \tag{14.167}$$

with h being the sample time. The predictor then becomes (see Table 14.1)

$$\delta\hat{x}^-[k+1] = A_\mathrm{d}[k]\, \delta\hat{x}[k] \tag{14.168}$$

and the error states are updated using the corrector

$$\delta\hat{x}[k] = \delta\hat{x}^-[k] + K[k](\delta y[k] - C_\mathrm{d}[k]\delta\hat{x}^-[k]). \tag{14.169}$$

Note that the INS model integrates acceleration twice to generate a position estimate $\hat{p}_\mathrm{ins}[k]$ and angular rate once to obtain $\hat{\theta}_\mathrm{ins}[k]$. The INS position and attitude will both drift because of numerical errors and measurement noise. Consequently, the position and angle correction are added to the computed INS position and angle as feedforward terms. The error signals are used as KF measurements of the error state-space model. The KF algorithm will estimate the position and angle errors, which again are used to correct the INS estimates as illustrated in Figure 14.6.

Also note that the IMU measurements f_imu and ω_imu are not used by the indirect KF in the predictor (14.168). Instead, the IMU measurements are used as inputs to drive the INS model.

14.4.2 Error-state Kalman Filter using Attitude Measurements

This section generalizes the 2-DOF introductory example in Section 14.4.1 to 6 DOFs. The error-state filter will first be formulated as a feedforward filter. Later a reset is introduced to prevent the filter error states to grow to large values for long-time applications. The reset technique is equivalent to a feedback filter which regulates the error $\delta \boldsymbol{x}$ to zero such that $\boldsymbol{x}_\mathrm{ins} \rightarrow \boldsymbol{x}$.

Assume that the Euler angles are measured using an AHRS such that

$$\boldsymbol{R}(\boldsymbol{\Theta}_\mathrm{nb}) := \boldsymbol{R}_b^n(t), \qquad \boldsymbol{T}(\boldsymbol{\Theta}_\mathrm{nb}) := \boldsymbol{T}_b^n(t) \tag{14.170}$$

are known for all $t \geq 0$. This assumption is removed in Section 14.4.3 where the MEKF is used to estimate the attitude.

Strapdown navigation equations: The strapdown navigation equations are

$$\dot{\boldsymbol{p}}^n_{\mathrm{nm_I}} = \boldsymbol{v}^n_{\mathrm{nm_I}} \tag{14.171}$$

$$\dot{\boldsymbol{v}}^n_{\mathrm{nm_I}} = \boldsymbol{a}^n_{\mathrm{nm_I}} \tag{14.172}$$

$$\dot{\boldsymbol{\Theta}}_\mathrm{nb} = \boldsymbol{T}_b^n(t)\boldsymbol{\omega}^b_\mathrm{nb}. \tag{14.173}$$

IMU measurements: The inputs to the strapdown navigation equations are IMU specific force and ARS measurements

$$f^b_{\text{imu}} = R^n_b(t)^\top(a^n_{\text{nm}_\text{I}} - g^n) + b^b_{\text{acc}} + w^b_{\text{acc}} \tag{14.174}$$

$$\dot{b}^b_{\text{acc}} = w^b_{\text{b, acc}} \tag{14.175}$$

$$\omega^b_{\text{imu}} = \omega^b_{\text{nb}} + b^b_{\text{ars}} + w^b_{\text{ars}} \tag{14.176}$$

$$\dot{b}^b_{\text{ars}} = w^b_{\text{b, ars}}. \tag{14.177}$$

The process noise terms w^b_{acc}, w^b_{ars}, $w^b_{b,\text{acc}}$ and $w^b_{b,\text{ars}}$ are all assumed to be Gaussian white noise.

Aiding: The measurements used for aiding are assumed to be

$$y_1 = p^n_{\text{nm}_\text{I}} + \varepsilon_1 \tag{14.178}$$

$$y_2 = v^n_{\text{nm}_\text{I}} + \varepsilon_2 \quad \text{(optionally)} \tag{14.179}$$

$$y_3 = \Theta_{\text{nb}} + \varepsilon_3 \tag{14.180}$$

where the measurement noise terms ε_i for $i = 1, \ldots, 3$ are Gaussian white noise.

Inertial navigation system (INS):

$$\dot{\hat{p}}^n_{\text{ins}} = \hat{v}^n_{\text{ins}} \tag{14.181}$$

$$\dot{\hat{v}}^n_{\text{ins}} = R^n_b(t) f^b_{\text{ins}} + g^n \tag{14.182}$$

$$\dot{\hat{b}}^b_{\text{acc, ins}} = \mathbf{0} \tag{14.183}$$

$$\dot{\hat{\theta}}_{\text{ins}} = T^n_b(t)\omega^b_{\text{ins}} \tag{14.184}$$

$$\dot{\hat{b}}^b_{\text{ars, ins}} = \mathbf{0} \tag{14.185}$$

where the INS inputs f^b_{ins} and ω^b_{ins} are bias compensated IMU measurements

$$f^b_{\text{ins}} := f^b_{\text{imu}} - \Delta f^b \tag{14.186}$$

$$\omega^b_{\text{ins}} := \omega^b_{\text{imu}} - \Delta\omega^b. \tag{14.187}$$

$$\begin{aligned} &\text{Feedforward filter:} && \Delta f^b = \mathbf{0} \\ &&& \Delta\omega^b = \mathbf{0} \\ &\text{Feedback filter:} && \Delta f^b = \hat{b}^b_{\text{acc, ins}} \\ &&& \Delta\omega^b = \hat{b}^b_{\text{ars, ins}}. \end{aligned}$$

Error-state model: The error-state model is a 15-states LTV system

$$\delta\dot{x} = A(t)\delta x + E(t)w \tag{14.188}$$

$$\delta y = C\delta x + \varepsilon \tag{14.189}$$

with state and process noise vectors

$$\delta \boldsymbol{x} = [(\delta \boldsymbol{p}^n)^\top, (\delta \boldsymbol{v}^n)^\top, (\delta \boldsymbol{b}^b_{\text{acc}})^\top, \delta \boldsymbol{\Theta}^\top_{\text{nb}}, (\delta \boldsymbol{b}^b_{\text{ars}})^\top]^\top \tag{14.190}$$

$$\boldsymbol{w} = [\boldsymbol{w}^\top_{\text{acc}}, \boldsymbol{w}^\top_{b,\text{acc}}, \boldsymbol{w}^\top_{\text{ars}}, \boldsymbol{w}^\top_{b,\text{ars}}]^\top. \tag{14.191}$$

The measurement vector is $\delta \boldsymbol{y} = \boldsymbol{y} - \hat{\boldsymbol{y}}_{\text{ins}}$ and the system matrices are

$$\boldsymbol{A}(t) = \begin{bmatrix} \boldsymbol{0}_{3\times3} & \boldsymbol{I}_3 & \boldsymbol{0}_{3\times3} & \boldsymbol{0}_{3\times3} & \boldsymbol{0}_{3\times3} \\ \boldsymbol{0}_{3\times3} & \boldsymbol{0}_{3\times3} & -\boldsymbol{R}^n_b(t) & \boldsymbol{0}_{3\times3} & \boldsymbol{0}_{3\times3} \\ \boldsymbol{0}_{3\times3} & \boldsymbol{0}_{3\times3} & -\frac{1}{T_{\text{acc}}}\boldsymbol{I}_3 & \boldsymbol{0}_{3\times3} & \boldsymbol{0}_{3\times3} \\ \boldsymbol{0}_{3\times3} & \boldsymbol{0}_{3\times3} & \boldsymbol{0}_{3\times3} & \boldsymbol{0}_{3\times3} & -\boldsymbol{T}^n_b(t) \\ \boldsymbol{0}_{3\times3} & \boldsymbol{0}_{3\times3} & \boldsymbol{0}_{3\times3} & \boldsymbol{0}_{3\times3} & -\frac{1}{T_{\text{ars}}}\boldsymbol{I}_3 \end{bmatrix} \tag{14.192}$$

$$\boldsymbol{E}(t) = \begin{bmatrix} \boldsymbol{0}_{3\times3} & \boldsymbol{0}_{3\times3} & \boldsymbol{0}_{3\times3} & \boldsymbol{0}_{3\times3} \\ -\boldsymbol{R}^n_b(t) & \boldsymbol{0}_{3\times3} & \boldsymbol{0}_{3\times3} & \boldsymbol{0}_{3\times3} \\ \boldsymbol{0}_{3\times3} & \boldsymbol{I}_3 & \boldsymbol{0}_{3\times3} & \boldsymbol{0}_{3\times3} \\ \boldsymbol{0}_{3\times3} & \boldsymbol{0}_{3\times3} & -\boldsymbol{T}^n_b(t) & \boldsymbol{0}_{3\times3} \\ \boldsymbol{0}_{3\times3} & \boldsymbol{0}_{3\times3} & \boldsymbol{0}_{3\times3} & \boldsymbol{I}_3 \end{bmatrix} \tag{14.193}$$

$$\boldsymbol{C} = \begin{bmatrix} \boldsymbol{I}_3 & \boldsymbol{0}_{3\times3} & \boldsymbol{0}_{3\times3} & \boldsymbol{0}_{3\times3} & \boldsymbol{0}_{3\times3} \\ \boldsymbol{0}_{3\times3} & \boldsymbol{I}_3 & \boldsymbol{0}_{3\times3} & \boldsymbol{0}_{3\times3} & \boldsymbol{0}_{3\times3} \\ \boldsymbol{0}_{3\times3} & \boldsymbol{0}_{3\times3} & \boldsymbol{0}_{3\times3} & \boldsymbol{I}_3 & \boldsymbol{0}_{3\times3} \end{bmatrix} \tag{14.194}$$

where the bias models are chosen as

$$\delta \dot{\boldsymbol{b}}^b_{\text{acc}} = -\frac{1}{T_{\text{acc}}}\delta \boldsymbol{b}^b_{\text{acc}} + \boldsymbol{w}^b_{\text{b, acc}} \tag{14.195}$$

$$\delta \dot{\boldsymbol{b}}^b_{\text{ars}} = -\frac{1}{T_{\text{ars}}}\delta \boldsymbol{b}^b_{\text{ars}} + \boldsymbol{w}^b_{\text{b, ars}}. \tag{14.196}$$

Here T_{acc} and T_{ars} are user specified time constants, and $\boldsymbol{w}^b_{\text{b, acc}}$ and $\boldsymbol{w}^b_{\text{b, ars}}$ are Gaussian white noise.

For the indirect feedforward filter the estimated state is computed as

$$\hat{\boldsymbol{x}} = \hat{\boldsymbol{x}}_{\text{ins}} + \delta \hat{\boldsymbol{x}} \tag{14.197}$$

where $\delta \hat{\boldsymbol{x}}$ is estimated using an LTV Kalman filter and

$$\hat{\boldsymbol{x}}_{\text{ins}} = \begin{bmatrix} \hat{\boldsymbol{p}}^n_{\text{ins}} \\ \hat{\boldsymbol{v}}^n_{\text{ins}} \\ \boldsymbol{0} \\ \hat{\boldsymbol{\Theta}}_{\text{ins}} \\ \boldsymbol{0} \end{bmatrix}. \tag{14.198}$$

Since the error dynamics is a LTV system it is possible to show that the pair $(\boldsymbol{A}(t), \boldsymbol{C})$ is observable (see Section 14.3.2). This implies that the estimated biases

$$\hat{\boldsymbol{b}}^b_{\text{acc}} = \boldsymbol{0} + \delta\hat{\boldsymbol{b}}^b_{\text{acc}} \tag{14.199}$$

$$\hat{\boldsymbol{b}}^b_{\text{ars}} = \boldsymbol{0} + \delta\hat{\boldsymbol{b}}^b_{\text{ars}} \tag{14.200}$$

converge to their true values $\boldsymbol{b}^b_{\text{acc}}$ and $\boldsymbol{b}^b_{\text{ars}}$, respectively. If the attitude $\boldsymbol{\Theta}_{\text{nb}}$ is estimated using the expressions $\boldsymbol{R}(\boldsymbol{\Theta}_{\text{nb}})$ and $\boldsymbol{T}(\boldsymbol{\Theta}_{\text{nb}})$ in an MEKF, observability of angular rate bias is still guaranteed but the acceleration bias estimates might diverge unless $\boldsymbol{\Theta}_{\text{nb}}$ is PE. This is a well-known problem when designing inertial navigation systems.

Summary: Indirect feedforward Kalman filter using attitude measurements

Discrete-time system matrices:

$$\boldsymbol{A}_{\text{d}} \approx \boldsymbol{I}_{15} + h\boldsymbol{A} \qquad \boldsymbol{C}_{\text{d}} = \boldsymbol{C}, \qquad \boldsymbol{E}_{\text{d}} \approx h\boldsymbol{E}. \tag{14.201}$$

State vector:

$$\delta\boldsymbol{x}[k] = [\delta\boldsymbol{p}^n_{\text{nm}_\text{I}}[k]^\top,\ \delta\boldsymbol{v}^n_{\text{nm}_\text{I}}[k]^\top,\ \delta\boldsymbol{b}^b_{\text{acc}}[k]^\top,\ \delta\boldsymbol{\Theta}^\top_{\text{nb}}[k],\ \delta\boldsymbol{b}^b_{\text{ars}}[k]^\top]^\top. \tag{14.202}$$

Kalman gain:

$$\boldsymbol{K}[k] = \hat{\boldsymbol{P}}^-[k]\boldsymbol{C}^\top_{\text{d}}[k](\boldsymbol{C}_{\text{d}}[k]\hat{\boldsymbol{P}}^-[k]\boldsymbol{C}^\top_{\text{d}}[k] + \boldsymbol{R}_{\text{d}}[k])^{-1}. \tag{14.203}$$

State estimate:

$$\hat{\boldsymbol{x}}_{\text{ins}}[k] = [\hat{\boldsymbol{p}}^n_{\text{ins}}[k]^\top,\ \hat{\boldsymbol{v}}^n_{\text{ins}}[k]^\top,\ \boldsymbol{0}^\top, \hat{\boldsymbol{\Theta}}^\top_{\text{ins}}[k],\ \boldsymbol{0}^\top]^\top \tag{14.204}$$

$$\hat{\boldsymbol{x}}[k] = \hat{\boldsymbol{x}}_{\text{ins}}[k] + \delta\hat{\boldsymbol{x}}[k]. \tag{14.205}$$

Predictor:

$$\delta\hat{\boldsymbol{x}}^-[k+1] = \boldsymbol{A}_{\text{d}}[k]\,\delta\hat{\boldsymbol{x}}[k] \tag{14.206}$$

$$\hat{\boldsymbol{P}}^-[k+1] = \boldsymbol{A}_{\text{d}}[k]\hat{\boldsymbol{P}}[k]\boldsymbol{A}^\top_{\text{d}}[k] + \boldsymbol{E}_{\text{d}}[k]\boldsymbol{Q}_{\text{d}}[k]\boldsymbol{E}^\top_{\text{d}}[k]. \tag{14.207}$$

Corrector:

$$\delta\hat{\boldsymbol{x}}[k] = \delta\hat{\boldsymbol{x}}^-[k] + \boldsymbol{K}[k](\delta\boldsymbol{y}[k] - \boldsymbol{C}_{\text{d}}[k]\,\delta\hat{\boldsymbol{x}}^-[k]) \tag{14.208}$$

$$\hat{\boldsymbol{P}}[k] = (\boldsymbol{I}_{15} - \boldsymbol{K}[k]\boldsymbol{C}_{\text{d}}[k])\hat{\boldsymbol{P}}^-[k](\boldsymbol{I}_{15} - \boldsymbol{K}[k]\boldsymbol{C}_{\text{d}}[k])^\top + \boldsymbol{K}[k]\boldsymbol{R}_{\text{d}}[k]\boldsymbol{K}^\top[k]. \tag{14.209}$$

INS propagation:

$$\hat{\boldsymbol{p}}^n_{\text{ins}}[k+1] = \hat{\boldsymbol{p}}^n_{\text{ins}}[k] + h\hat{\boldsymbol{v}}^n_{\text{ins}}[k] \tag{14.210}$$

$$\hat{\boldsymbol{v}}^n_{\text{ins}}[k+1] = \hat{\boldsymbol{v}}^n_{\text{ins}}[k] + h(\boldsymbol{R}^n_b[k](\boldsymbol{f}^b_{\text{imu}}[k] - \boldsymbol{0}) + \boldsymbol{g}^n) \tag{14.211}$$

$$\hat{\boldsymbol{\Theta}}_{\text{ins}}[k+1] = \hat{\boldsymbol{\Theta}}_{\text{ins}}[k] + h\boldsymbol{T}^n_b[k](\boldsymbol{\omega}^b_{\text{imu}}[k] - \boldsymbol{0}). \tag{14.212}$$

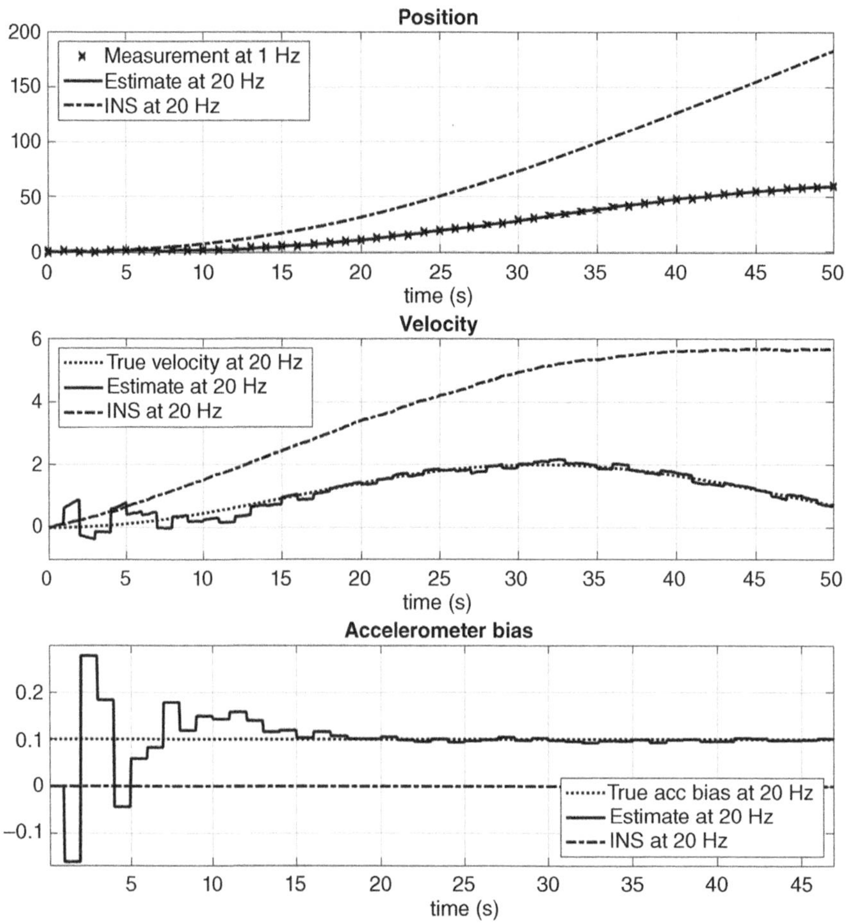

Figure 14.7 Position, velocity and accelerometer bias for an indirect feedforward Kalman filter. The INS is aided by GNSS position measurements.

Figure 14.7 shows the position, velocity and accelerometer bias estimates as a function of time for the indirect feedforward KF when considering 1 DOF. As expected the INS position and velocities estimates grow fast when specific force is integrated in open loop. The main reason for this is the uncompensated accelerometer bias. However, the KF error-state filter succeeds to estimate the drift and the resulting estimate $\hat{\boldsymbol{x}} = \hat{\boldsymbol{x}}_{\text{ins}} + \delta\hat{\boldsymbol{x}}$ converges to the true state $\boldsymbol{x}$.

An open-loop IMU solution where the IMU measurements are integrated to obtain positions and attitude (see Figure 14.6), will always drift due to sensor biases, misalignments and temperature variations and for long-endurance applications this is undesirable behavior. Hence, the feedforward filter has limited practical use. To overcome this limitation the feedforward filer should be modified to compensate for bias drift terms by feedback. The technique for this is a reset strategy, which is the topic for the next section.

Indirect Feedback Kalman Filter

For the feedback version of the indirect KF, the error estimates are used to update the INS estimates directly in order to avoid drift. This prevents the INS errors from growing. This is done by regulating the error $\delta\boldsymbol{x}$ to zero such that $\hat{\boldsymbol{x}}_{\text{ins}} \rightarrow \boldsymbol{x}$, analogues to a feedback control system. Hence, the term feedback filter. The KF feedback structure is shown in Figure 14.8.

A feedback filter for INS can be designed by using the estimates $\delta\hat{\boldsymbol{b}}^b_{\text{acc}}$ and $\delta\hat{\boldsymbol{b}}^b_{\text{ars}}$ to calibrate the IMU accelerometer and ARS. This is referred to as sensor pretension and historically it has been implemented by using an analog sensor input.

Algorithm 14.1 (Feedback Filter Reset Algorithm)
The feedback loop (Figure 14.8) will remove the INS error by regulating $\delta\boldsymbol{x}$ to zero. This can be obtained by modifying the feedforward filter according to:

1. *Slow measurements: after every slow measurement (position and velocity), the INS states*

$$\hat{\boldsymbol{x}}_{\text{ins}} = [(\hat{\boldsymbol{p}}^n_{\text{ins}})^\top, (\hat{\boldsymbol{v}}^n_{\text{ins}})^\top, (\hat{\boldsymbol{b}}^b_{\text{acc, ins}})^\top, \hat{\boldsymbol{\Theta}}^\top_{\text{ins}}, (\hat{\boldsymbol{b}}^b_{\text{ars, ins}})^\top]^\top \tag{14.213}$$

are corrected by setting the error-state vector to zero. This is mathematically equivalent to

$$\hat{\boldsymbol{x}}_{\text{ins}}[k] \leftarrow \hat{\boldsymbol{x}}_{\text{ins}}[k] + \delta\hat{\boldsymbol{x}}[k]. \tag{14.214}$$

2. *Fast measurements: Step 1 ensures that error-state vector will be zero before the fast measurements (accelerometer and ARS) are applied. Therefore, the state predictor of the feedforward filter (14.168) becomes redundant since*

$$\delta\hat{\boldsymbol{x}}^-[k+1] = \boldsymbol{A}_\text{d}[k]\boldsymbol{0} \equiv \boldsymbol{0}. \tag{14.215}$$

It should, however, be noted that $\boldsymbol{A}_\text{d}[k]$ is still a necessary to compute since $\boldsymbol{A}_\text{d}[k]$ is used in the KF covariance predication step.

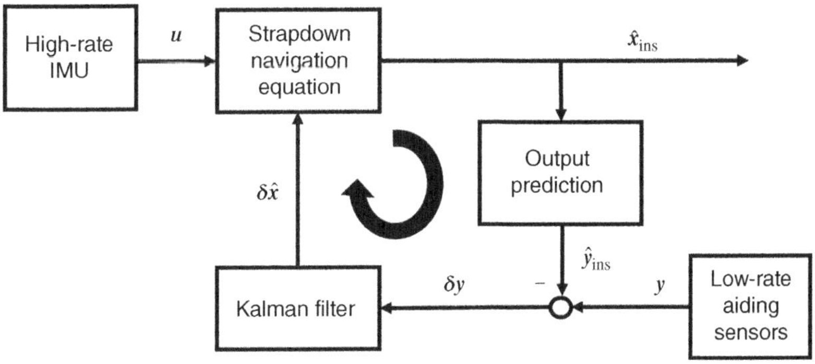

Figure 14.8 Indirect feedback filter for INS. The strapdown navigation equations are propagated to obtain the estimate $\hat{\boldsymbol{x}}_{\text{ins}}$. The KF error-state estimate $\delta\hat{\boldsymbol{x}}$ is fed back (reset mechanism) to ensure that $\hat{\boldsymbol{x}}_{\text{ins}} \rightarrow \boldsymbol{x}$.

As observed from Figure 14.8 and the error state model (14.206), the inertial measurements $\boldsymbol{f}^b_{\text{imu}}$ and $\boldsymbol{\omega}^b_{\text{imu}}$ do not enter the state-space model as an input signal. On the contrary, this is the case for the direct filter representation in Section 14.3. The resulting effect is that the position and attitude measurements are low-pass filtered and that the accelerometer and ARS measurement are high-pass filtered. The latter effect reduces drift due to the slowly-varying bias terms, $\boldsymbol{b}^b_{\text{acc}}$ and $\boldsymbol{b}^b_{\text{ars}}$; see Farrell (2008, section 5.10.5.3) for more details.

Summary: Indirect feedback Kalman filter using attitude measurements

Discrete-time system matrices:

$$\boldsymbol{A}_{\text{d}} \approx \boldsymbol{I}_{15} + h\boldsymbol{A}, \qquad \boldsymbol{C}_{\text{d}} = \boldsymbol{C}, \qquad \boldsymbol{E}_{\text{d}} \approx h\boldsymbol{E} \tag{14.216}$$

State vectors:

$$\delta\hat{\boldsymbol{x}}[k] = [\delta\hat{\boldsymbol{p}}^n_{\text{nm}_\text{I}}[k]^\top, \delta\hat{\boldsymbol{v}}^n_{\text{nm}_\text{I}}[k]^\top, \delta\hat{\boldsymbol{b}}^b_{\text{acc}}[k]^\top, \delta\hat{\boldsymbol{\Theta}}_{\text{nb}}[k]^\top, \delta\hat{\boldsymbol{b}}^b_{\text{ars}}[k]^\top]^\top \tag{14.217}$$

$$\hat{\boldsymbol{x}}_{\text{ins}}[k] = [\hat{\boldsymbol{p}}^n_{\text{ins}}[k]^\top, \hat{\boldsymbol{v}}^n_{\text{ins}}[k]^\top, \hat{\boldsymbol{b}}^b_{\text{acc, ins}}[k]^\top, \hat{\boldsymbol{\Theta}}_{\text{ins}}[k]^\top, \hat{\boldsymbol{b}}^b_{\text{ars, ins}}[k]^\top]^\top \tag{14.218}$$

Kalman gain:

$$\boldsymbol{K}[k] = \hat{\boldsymbol{P}}^-[k]\boldsymbol{C}^\top_{\text{d}}[k](\boldsymbol{C}_{\text{d}}[k]\hat{\boldsymbol{P}}^-[k]\boldsymbol{C}^\top_{\text{d}}[k] + \boldsymbol{R}_{\text{d}}[k])^{-1} \tag{14.219}$$

INS reset and state estimate:

$$\hat{\boldsymbol{x}}_{\text{ins}}[k] = \hat{\boldsymbol{x}}_{\text{ins}}[k] + \delta\hat{\boldsymbol{x}}[k] \tag{14.220}$$

Predictor:

$$\hat{\boldsymbol{P}}^-[k+1] = \boldsymbol{A}_{\text{d}}[k]\hat{\boldsymbol{P}}[k]\boldsymbol{A}^\top_{\text{d}}[k] + \boldsymbol{E}_{\text{d}}[k]\boldsymbol{Q}_{\text{d}}[k]\boldsymbol{E}^\top_{\text{d}}[k] \tag{14.221}$$

Corrector:

$$\delta\hat{\boldsymbol{x}}[k] = \boldsymbol{K}[k](\boldsymbol{y}[k] - \boldsymbol{C}_{\text{d}}[k]\,\hat{\boldsymbol{x}}_{\text{ins}}[k]) \tag{14.222}$$

$$\hat{\boldsymbol{P}}[k] = (\boldsymbol{I}_{15} - \boldsymbol{K}[k]\boldsymbol{C}_{\text{d}}[k])\hat{\boldsymbol{P}}^-[k](\boldsymbol{I}_{15} - \boldsymbol{K}[k]\boldsymbol{C}_{\text{d}}[k])^\top + \boldsymbol{K}[k]\boldsymbol{R}_{\text{d}}[k]\boldsymbol{K}^\top[k] \tag{14.223}$$

INS propagation:

$$\hat{\boldsymbol{p}}^n_{\text{ins}}[k+1] = \hat{\boldsymbol{p}}^n_{\text{ins}}[k] + h\hat{\boldsymbol{v}}^n_{\text{ins}}[k] \tag{14.224}$$

$$\hat{\boldsymbol{v}}^n_{\text{ins}}[k+1] = \hat{\boldsymbol{v}}^n_{\text{ins}}[k] + h(\boldsymbol{R}^n_b[k](\boldsymbol{f}^b_{\text{imu}}[k] - \hat{\boldsymbol{b}}^b_{\text{acc, ins}}[k]) + \boldsymbol{g}^n) \tag{14.225}$$

$$\hat{\boldsymbol{\Theta}}_{\text{ins}}[k+1] = \hat{\boldsymbol{\Theta}}_{\text{ins}}[k] + h\boldsymbol{T}^n_b[k](\boldsymbol{\omega}^b_{\text{imu}}[k] - \hat{\boldsymbol{b}}^b_{\text{ars, ins}}[k]). \tag{14.226}$$

MATLAB

The pseudocode for the 15-states indirect feedback KF is given below. The stand-alone Matlab function for the INS is included in the MSS toolbox as:

```
[x_ins,P_prd] = ins_ahrs(x_ins, P_prd, mu, h, Qd, Rd, f_imu, ...
    w_imu, y_ahrs, y_pos, y_vel)
```

The toolbox also includes a user editable script `ExINS_AHRS.m`, which demonstrates how the KF loop can be implemented using real-time measurements.

```
% Initialization
Qd = constant; Rd = constant;        % KF covariance matrices
g_n = [ 0 0 gravity(mu) ]';          % g as a function of latitude
P_prd = P0;

p_ins = zeros(3,1); v_ins = zeros(3,1); b_acc_ins = zeros(3,1);
theta_ins = zeros(3,1); b_ars_ins = zeros(3,1);

% MAIN LOOP
for i=1:N

  % IMU measurements
  f_imu = ...                       % specific force: f_imu[k]
  w_imu = ...                       % attitude rate:  w_imu[k]

  % AHRS measurements: phi[k], theta[k], psi[k]
  % If only compass measurement psi[k], use static solution
  % [phi, theta] = acc2rollpitch(f_imu)
  R = Rzyx( phi, theta, psi );
  T = Tzyx( phi, theta );

  % Kalman filter matrices
  Ad = ...
  Cd = ...
  Ed = ...

  if (new_measurement)

     y = ...       % GNSS measurement: y[k] = [p[k]; v[k]]'

     % KF gain: K[k]
     K = P_prd * Cd' * inv( Cd * P_prd * Cd' + Rd );
     IKC = eye(15) - K * Cd;

     % Corrector: delta_x_hat[k] and P_hat[k]
     delta_x_hat = K * ( y - Cd * x_ins );
     P_hat = IKC * P_prd * IKC' + K * Rd * K';

     % INS reset: x_ins[k]
```

```
            p_ins = p_ins + delta_x_hat(1:3);             % position
            v_ins = v_ins + delta_x_hat(4:6);             % velocity
            b_acc_ins = b_acc_ins + delta_x_hat(7:9);     % acc bias
            theta_ins = theta_ins + delta_x_hat(10:12);   % attitude
            b_ars_ins = b_ars_ins + delta_x_hat(13:15);   % ars bias

     else

         P_hat = P_prd;                         % no measurement

     end

     % Predictor: P_prd[k+1]
     P_prd = Ad * P_hat * Ad' + Ed * Qd * Ed';

     % INS propagation: x_ins[k+1]
     p_ins = p_ins + h * v_ins;
     v_ins = v_ins + h * ( R * ( f_imu - b_acc_ins) + g_n );
     theta_ins = theta_ins + h * T * ( w_imu - b_ars_ins );

end
```

14.4.3 Error-state Extended Kalman Filter with Attitude Estimation

The multiplicative extended Kalman filter (MEKF) is an error-state extended Kalman filter (EKF) where attitude is parametrized by a four-dimensional *unit quaternion*. The *attitude error*, however, is uniquely defined by three parameters, which is the minimal representation for the 3-DOF rotational motion of a rigid body (Crassidis *et al.* 2007, Markley and Crassidis 2014).

It is well known that is impossible to represent attitude by three parameters globally. Three-parameter representations such as the Euler angles do all have singular points. It is tempting to use the four-parameter unit quaternion in the EKF in order to avoid singular points. However, the covariance matrix will then be rank deficient because of the unit constraint. Another problem is the EKF injection term. The standard error-state EKF with additive-error injection cannot be used to create an unbiased estimator with the unit quaternion in the estimated state, as the additive-error injection would violate the quaternion unit constraint as well.

To overcome these difficulties the MEKF is used. The main idea of the MEKF is that the unit quaternion error

$$\delta \boldsymbol{q}_b^n = \begin{bmatrix} \delta\eta \\ \delta\boldsymbol{\varepsilon} \end{bmatrix} \tag{14.227}$$

is parametrized using a three-parameter attitude representation where $\delta\eta$ and $\delta\boldsymbol{\varepsilon}$ are the real and imaginary components of the unit quaternion $\delta \boldsymbol{q}_b^n$ (see Section 2.2.2). There are many candidates for representation of the unit quaternion error and three frequently used representations are presented below.

1. **Rodrigues parameters (Gibbs vector)**

$$\delta \boldsymbol{a}_{\mathrm{g}} = \frac{\delta \boldsymbol{\varepsilon}}{\delta \eta}, \qquad \delta \boldsymbol{q}_b^n = \frac{1}{\sqrt{1 + \delta \boldsymbol{a}_{\mathrm{g}}^\top \delta \boldsymbol{a}_{\mathrm{g}}}} \begin{bmatrix} 1 \\ \delta \boldsymbol{a}_{\mathrm{g}} \end{bmatrix}. \tag{14.228}$$

It is convenient to scale the Gibbs vector by a factor 2 such that the KF covariance estimates will be given in radians squared, which is equivalent to angle errors using a first-order approximation. In other words

$$\delta \boldsymbol{a} := 2\delta \boldsymbol{a}_{\mathrm{g}}, \qquad \delta \boldsymbol{q}_b^n = \frac{1}{\sqrt{4 + \delta \boldsymbol{a}^\top \delta \boldsymbol{a}}} \begin{bmatrix} 2 \\ \delta \boldsymbol{a} \end{bmatrix}. \tag{14.229}$$

The Gibbs vector has the property that the elements approach infinity as the rotation approaches 180°. This means that insensible attitude errors of more than 180° cannot exist.

2. **Modified Rodrigues parameters**

$$\delta \boldsymbol{a}_{\mathrm{m}} = \frac{\delta \boldsymbol{\varepsilon}}{1 + \delta \eta}, \qquad \delta \boldsymbol{q}_b^n = \frac{1}{1 + \delta \boldsymbol{a}_{\mathrm{m}}^\top \delta \boldsymbol{a}_{\mathrm{m}}} \begin{bmatrix} 1 - \delta \boldsymbol{a}_{\mathrm{m}}^\top \delta \boldsymbol{a}_{\mathrm{m}} \\ 2\delta \boldsymbol{a}_{\mathrm{m}} \end{bmatrix}. \tag{14.230}$$

By scaling the modified Rodrigues parameters vector by the factor 4, the covariance estimates will be given in radians squared. This gives

$$\delta \boldsymbol{a} := 4\delta \boldsymbol{a}_{\mathrm{m}}, \qquad \delta \boldsymbol{q}_b^n = \frac{1}{16 + \delta \boldsymbol{a}^\top \delta \boldsymbol{a}} \begin{bmatrix} 16 - \delta \boldsymbol{a}^\top \delta \boldsymbol{a} \\ 8\delta \boldsymbol{a} \end{bmatrix}. \tag{14.231}$$

3. **Truncated Taylor-series expansion.** The Rodrigues parametrization (14.229) with $\delta \boldsymbol{a}^\top \delta \boldsymbol{a} = \boldsymbol{0}$ is equivalent to the truncated Taylor-series expansion of the error quaternion (Solà 2016)

$$\delta \boldsymbol{q}_b^n = \begin{bmatrix} 1 \\ \frac{1}{2}\delta \boldsymbol{a} \end{bmatrix} + O(\|\delta \boldsymbol{a}\|^2). \tag{14.232}$$

Three-parameter Attitude Dynamics

The linearized differential equations for the three-parameter attitude representations will all be equal. However, the mapping $\delta \boldsymbol{a} \rightarrow \delta \boldsymbol{q}_b^n$ will have different numerical properties when implemented in the EKF. Consequently, and without loss of generality, we will use the formula for Gibbs vector to derive the differential equation for the error state.

The unit quaternion error $\delta \boldsymbol{q}_b^n$ satisfies (Solà 2016)

$$\boldsymbol{q}_b^n = \hat{\boldsymbol{q}}_{\mathrm{ins}} \otimes \delta \boldsymbol{q}_b^n \tag{14.233}$$

where $\hat{\boldsymbol{q}}_{\mathrm{ins}} \equiv \hat{\boldsymbol{q}}_b^n$. The product of two quaternions (*Hamiltonian product*) is defined as

$$\boldsymbol{q}_1 \otimes \boldsymbol{q}_2 = \begin{bmatrix} \eta_{q_1}\eta_{q_2} - \boldsymbol{\varepsilon}_{q_1}^\top \boldsymbol{\varepsilon}_{q_2} \\ \eta_{q_1}\boldsymbol{\varepsilon}_{q_2} + \eta_{q_2}\boldsymbol{\varepsilon}_{q_1} + \boldsymbol{S}(\boldsymbol{\varepsilon}_{q_1})\boldsymbol{\varepsilon}_{q_2} \end{bmatrix}. \tag{14.234}$$

This implies that

$$\begin{aligned} \boldsymbol{R}(\boldsymbol{q}_b^n) &= \boldsymbol{R}(\hat{\boldsymbol{q}}_{\text{ins}} \otimes \delta\boldsymbol{q}_b^n) \\ &= \boldsymbol{R}(\hat{\boldsymbol{q}}_{\text{ins}})\boldsymbol{R}(\delta\boldsymbol{q}_b^n) \\ &\approx \boldsymbol{R}(\hat{\boldsymbol{q}}_{\text{ins}})(\boldsymbol{I}_3 + \boldsymbol{S}(\delta\boldsymbol{a})) \end{aligned} \tag{14.235}$$

It is also possible to define the quaternion product with the sign flipped in front of $\boldsymbol{S}(\boldsymbol{\varepsilon}_{q_1})$ and modify (14.235) accordingly; see Markley and Crassidis (2014).

Note that for quaternion operations $\boldsymbol{q}_1 \otimes \boldsymbol{q}_2 \neq \boldsymbol{q}_2 \otimes \boldsymbol{q}_1$ but $\boldsymbol{q}_1 + \boldsymbol{q}_2 = \boldsymbol{q}_2 + \boldsymbol{q}_1$. The quaternion product is implemented in the MSS toolbox as `quatprod(q1,q2)`. Since a unit quaternions satisfies (2.76), we can write

$$\dot{\boldsymbol{q}}_b^n = \frac{1}{2}\boldsymbol{q}_b^n \otimes \begin{bmatrix} 0 \\ \boldsymbol{\omega}_{\text{nb}}^b \end{bmatrix}, \quad \dot{\hat{\boldsymbol{q}}}_{\text{ins}} = \frac{1}{2}\hat{\boldsymbol{q}}_{\text{ins}} \otimes \begin{bmatrix} 0 \\ \boldsymbol{\omega}_{\text{ins}}^b \end{bmatrix}. \tag{14.236}$$

The time derivative of $\delta\boldsymbol{a}$ is obtained by combining (14.229) with (14.236)

$$\begin{aligned} \delta\dot{\boldsymbol{a}} &= \left(\boldsymbol{I}_3 + \frac{1}{4}\delta\boldsymbol{a}\delta\boldsymbol{a}^\top\right)(\boldsymbol{\omega}_{\text{nb}}^b - \boldsymbol{\omega}_{\text{ins}}^b) - \frac{1}{2}\boldsymbol{S}(\boldsymbol{\omega}_{\text{nb}}^b + \boldsymbol{\omega}_{\text{ins}}^b)\delta\boldsymbol{a} \\ &\approx \boldsymbol{\omega}_{\text{nb}}^b - \boldsymbol{\omega}_{\text{ins}}^b - \frac{1}{2}\boldsymbol{S}(\boldsymbol{\omega}_{\text{nb}}^b + \boldsymbol{\omega}_{\text{ins}}^b)\delta\boldsymbol{a} \end{aligned} \tag{14.237}$$

where we have neglected the quadratic term in $\delta\boldsymbol{a}$. The INS uses the estimated bias vector $\hat{\boldsymbol{b}}_{\text{ars}}^b$ to correct the MU measurement. This is mathematically equivalent to

$$\begin{aligned} \boldsymbol{\omega}_{\text{ins}}^b &:= \boldsymbol{\omega}_{\text{imu}}^b - \hat{\boldsymbol{b}}_{\text{ars}}^b \\ &= (\boldsymbol{\omega}_{\text{nb}}^b + \boldsymbol{b}_{\text{ars}}^b + \boldsymbol{w}_{\text{ars}}^b) - \hat{\boldsymbol{b}}_{\text{ars}}^b \\ &= \boldsymbol{\omega}_{\text{nb}}^b + \delta\boldsymbol{b}_{\text{ars}}^b + \boldsymbol{w}_{\text{ars}}^b. \end{aligned} \tag{14.238}$$

Combining (14.237) and (14.238) gives

$$\begin{aligned} \delta\dot{\boldsymbol{a}} &\approx -\delta\boldsymbol{b}_{\text{ars}}^b - \boldsymbol{w}_{\text{ars}}^b - \frac{1}{2}\boldsymbol{S}(2\boldsymbol{\omega}_{\text{ins}}^b - \delta\boldsymbol{b}_{\text{ars}}^b - \boldsymbol{w}_{\text{ars}}^b)\delta\boldsymbol{a} \\ &= -\delta\boldsymbol{b}_{\text{ars}}^b - \boldsymbol{w}_{\text{ars}}^b - \boldsymbol{S}(\boldsymbol{\omega}_{\text{ins}}^b)\delta\boldsymbol{a} + \frac{1}{2}\boldsymbol{S}(\delta\boldsymbol{b}_{\text{ars}}^b + \boldsymbol{w}_{\text{ars}}^b)\delta\boldsymbol{a}. \end{aligned} \tag{14.239}$$

Finally, neglecting the quadratic cross-product terms $\boldsymbol{S}(\delta\boldsymbol{b}_{\text{ars}}^b)\delta\boldsymbol{a}$ and $\boldsymbol{S}(\boldsymbol{w}_{\text{ars}}^b)\delta\boldsymbol{a}$ gives

$$\delta\dot{\boldsymbol{a}} \approx -\boldsymbol{S}(\boldsymbol{\omega}_{\text{ins}}^b)\delta\boldsymbol{a} - \delta\boldsymbol{b}_{\text{ars}}^b - \boldsymbol{w}_{\text{ars}}^b \tag{14.240}$$

$$\delta\dot{\boldsymbol{b}}_{\text{ars}}^b = -\frac{1}{T_{\text{ars}}}\delta\boldsymbol{b}_{\text{ars}}^b + \boldsymbol{w}_{\text{b, ars}}^b \tag{14.241}$$

where $\boldsymbol{\omega}_{\text{ins}}^b := \boldsymbol{\omega}_{\text{imu}}^b - \hat{\boldsymbol{b}}_{\text{ars}}^b$ is the bias compensated IMU measurement. The ARS bias is modeled as a first-order system with user specified time constant T_{ars} and Gaussian white noise $\boldsymbol{w}_{\text{b, ars}}^b$ as driving input. The exponential convergence of $\delta\boldsymbol{b}_{\text{ars}}^b$ to zero is important during sensor failure (dead reckoning).

Translational Motion Error Dynamics

The expressions for the translational motion errors are

$$\delta\dot{\boldsymbol{p}}^n_{\mathrm{nm_I}} = \delta\boldsymbol{v}^n_{\mathrm{nm_I}} \tag{14.242}$$

$$\begin{aligned}
\delta\dot{\boldsymbol{v}}^n_{\mathrm{nm_I}} &= \boldsymbol{R}(\boldsymbol{q}^n_b)(\boldsymbol{f}^b_{\mathrm{imu}} - \hat{\boldsymbol{b}}^b_{\mathrm{acc}} - \delta\boldsymbol{b}^b_{\mathrm{acc}} - \boldsymbol{w}^b_{\mathrm{acc}}) + \boldsymbol{g}^n - \boldsymbol{R}(\hat{\boldsymbol{q}}^n_b)(\boldsymbol{f}^b_{\mathrm{imu}} - \hat{\boldsymbol{b}}^b_{\mathrm{acc}}) - \boldsymbol{g}^n \\
&= (\boldsymbol{R}(\boldsymbol{q}^n_b) - \boldsymbol{R}(\hat{\boldsymbol{q}}^n_b))(\boldsymbol{f}^b_{\mathrm{imu}} - \hat{\boldsymbol{b}}^b_{\mathrm{acc}}) - \boldsymbol{R}(\boldsymbol{q}^n_b)(\delta\boldsymbol{b}^b_{\mathrm{acc}} + \boldsymbol{w}^b_{\mathrm{acc}}) \\
&\approx \boldsymbol{R}(\hat{\boldsymbol{q}}_{\mathrm{ins}})\boldsymbol{S}(\delta\boldsymbol{a})(\boldsymbol{f}^b_{\mathrm{imu}} - \hat{\boldsymbol{b}}^b_{\mathrm{acc}}) - \boldsymbol{R}(\hat{\boldsymbol{q}}_{\mathrm{ins}})(\boldsymbol{I}_3 + \boldsymbol{S}(\delta\boldsymbol{a}))(\delta\boldsymbol{b}^b_{\mathrm{acc}} + \boldsymbol{w}^b_{\mathrm{acc}})
\end{aligned} \tag{14.243}$$

where we have used (14.235). Finally, we assume that the second-order error term $\boldsymbol{S}(\delta\boldsymbol{a})\delta\boldsymbol{b}^b_{acc} \approx \boldsymbol{0}$ such that

$$\delta\dot{\boldsymbol{p}}^n_{\mathrm{nm_I}} = \delta\boldsymbol{v}^n_{\mathrm{nm_I}} \tag{14.244}$$

$$\delta\dot{\boldsymbol{v}}^n_{\mathrm{nm_I}} \approx -\boldsymbol{R}(\hat{\boldsymbol{q}}_{\mathrm{ins}})\boldsymbol{S}(\boldsymbol{f}^b_{\mathrm{ins}})\delta\boldsymbol{a} - \boldsymbol{R}(\hat{\boldsymbol{q}}_{\mathrm{ins}})(\delta\boldsymbol{b}^b_{\mathrm{acc}} + \boldsymbol{w}^b_{\mathrm{acc}}) \tag{14.245}$$

$$\delta\dot{\boldsymbol{b}}^b_{\mathrm{acc}} = -\frac{1}{T_{\mathrm{acc}}}\delta\boldsymbol{b}^b_{\mathrm{acc}} + \boldsymbol{w}^b_{\mathrm{b,\,acc}} \tag{14.246}$$

where $\boldsymbol{f}^b_{\mathrm{ins}} := \boldsymbol{f}^b_{\mathrm{imu}} - \hat{\boldsymbol{b}}^b_{\mathrm{acc}}$ is the bias compensated IMU measurement. The specific force bias is modeled as a first-order system with user specified time constant T_{acc} and Gaussian white noise $\boldsymbol{w}^b_{\mathrm{b,\,acc}}$ as driving input. The exponential convergence of $\delta\boldsymbol{b}^b_{\mathrm{acc}}$ to zero is important during sensor failure (dead reckoning).

Aiding

GNSS is the primary sensor used for aiding of surface vessels. For underwater vehicles a hydroacoustic positioning system will have the same functionality. The error-measurement equations for position and velocity are

$$\begin{aligned}
\delta\boldsymbol{y}_{\mathrm{p}} &= (\boldsymbol{p}^n_{\mathrm{nm_I}} + \boldsymbol{\varepsilon}_{\mathrm{p}}) - \hat{\boldsymbol{p}}^n_{\mathrm{nm_I}} \\
&= \delta\boldsymbol{p}^n_{\mathrm{nm_I}} + \boldsymbol{\varepsilon}_{\mathrm{p}}
\end{aligned} \tag{14.247}$$

$$\begin{aligned}
\delta\boldsymbol{y}_{\mathrm{v}} &= (\boldsymbol{v}^n_{\mathrm{nm_I}} + \boldsymbol{\varepsilon}_{\mathrm{v}}) - \hat{\boldsymbol{v}}^n_{\mathrm{nm_I}} \\
&= \delta\boldsymbol{v}^n_{\mathrm{nm_I}} + \boldsymbol{\varepsilon}_{\mathrm{v}}
\end{aligned} \tag{14.248}$$

where $\boldsymbol{\varepsilon}_{\mathrm{p}}$ and $\boldsymbol{\varepsilon}_{\mathrm{v}}$ are assumed to be Gaussian white measurement noise. In order to successfully estimate the unit quaternion for attitude determination, a heading reference is needed to guarantee observability (see Section 14.2). For low-cost applications, the obvious reference vectors are the gravity vector (roll and pitch) in combination with a magnetic field measurement (yaw). Ships, however, usually replace the magnetometer with a high-quality gyrocompass to guarantee safe operation.

Assume that $\boldsymbol{v}_i^b$ is a measured vector expressed in $\{b\}$. Consequently, the estimated vector is $\hat{\boldsymbol{v}}_i^b = \boldsymbol{R}^\top(\hat{\boldsymbol{q}}_b^n)\boldsymbol{v}_{0i}^n$ where $\boldsymbol{v}_{0i}^{\mathrm{n}}$ is the *reference vector*. Then,

$$
\begin{aligned}
\delta \boldsymbol{y}_i &= (\boldsymbol{v}_i^{\mathrm{b}} + \boldsymbol{\varepsilon}_i) - \boldsymbol{R}^\top(\hat{\boldsymbol{q}}_b^n)\boldsymbol{v}_{0i}^n \\
&= \boldsymbol{R}^\top(\boldsymbol{q}_b^n)\boldsymbol{v}_{0i}^n + \boldsymbol{\varepsilon}_i - \boldsymbol{R}^\top(\hat{\boldsymbol{q}}_b^n)\boldsymbol{v}_{0i}^n \\
&\approx (\boldsymbol{I}_3 - \boldsymbol{S}(\delta\boldsymbol{a}))\boldsymbol{R}^\top(\hat{\boldsymbol{q}}_{\mathrm{ins}})\boldsymbol{v}_{0i}^n - \boldsymbol{R}^\top(\hat{\boldsymbol{q}}_{\mathrm{ins}})\boldsymbol{v}_{0i}^n + \boldsymbol{\varepsilon}_i \\
&= -\boldsymbol{S}(\delta\boldsymbol{a})\boldsymbol{R}^\top(\hat{\boldsymbol{q}}_{\mathrm{ins}})\boldsymbol{v}_{0i}^n + \boldsymbol{\varepsilon}_i \\
&= \boldsymbol{S}(\boldsymbol{R}^\top(\hat{\boldsymbol{q}}_{\mathrm{ins}})\boldsymbol{v}_{0i}^n)\delta\boldsymbol{a} + \boldsymbol{\varepsilon}_i
\end{aligned}
\tag{14.249}
$$

where we have used (14.235) and $\hat{\boldsymbol{q}}_b^n = \hat{\boldsymbol{q}}_{\mathrm{ins}}$ on line three.

The compass measurement can be included in the error-state filter by parameterizing the rotation matrix in terms of Gibbs parameters $\boldsymbol{a}_{\mathrm{g}} = [g_1, g_2, g_3]^\top$. From (2.96) and Wie (1998) it follows that

$$
\psi = \tan^{-1}\left(\frac{R_{21}}{R_{11}}\right) = \tan^{-1}\left(\frac{2(g_1 g_2 + g_3)}{1 + g_1^2 - g_2^2 - g_3^2}\right). \tag{14.250}
$$

Since (14.229) implies that $\boldsymbol{a} = 2\boldsymbol{a}_{\mathrm{g}}$, we get

$$
\psi = h(\boldsymbol{a}) = \tan^{-1}\left(\frac{2(a_1 a_2 + 2a_3)}{4 + a_1^2 - a_2^2 - a_3^2}\right) \tag{14.251}
$$

where $\boldsymbol{a} = [a_1, a_2, a_3]^\top$. Linearization about $\boldsymbol{a} = \hat{\boldsymbol{a}}$ gives

$$
\delta y_\psi = \psi - h(\hat{\boldsymbol{a}}) \approx \left.\frac{\mathrm{d}h(\boldsymbol{a})}{\mathrm{d}\boldsymbol{a}}\right|_{\boldsymbol{a}=\hat{\boldsymbol{a}}}^\top \delta\hat{\boldsymbol{a}}. \tag{14.252}
$$

The gradient can be computed by using the chain rule

$$
\frac{\mathrm{d}h(\boldsymbol{a})}{\mathrm{d}\boldsymbol{a}} = \frac{\partial \tan^{-1}(u)}{\partial u}\frac{\partial u}{\partial \boldsymbol{a}} \tag{14.253}
$$

$$
= \frac{1}{1+u^2}\frac{\partial u}{\partial \boldsymbol{a}} \tag{14.254}
$$

where

$$
u := \frac{2(a_1 a_2 + 2a_3)}{4 + a_1^2 - a_2^2 - a_3^2} \tag{14.255}
$$

and

$$\frac{\partial u}{\partial \boldsymbol{a}} = \left[\frac{-2((a_1^2 + a_3^2 - 4)a_2 + a_2^3 + 4a_1a_3)}{(4 + a_1^2 - a_2^2 - a_3^2)^2}, \right.$$

$$\left. \frac{2((a_2^2 - a_3^2 + 4)a_1 + a_1^3 + 4a_2a_3)}{(4 + a_1^2 - a_2^2 - a_3^2)^2}, \frac{4(a_3^2 + a_1a_2a_3 + a_1^2 - a_2^2 + 4)}{(4 + a_1^2 - a_2^2 - a_3^2)^2} \right]^\top. \tag{14.256}$$

The estimate â is computed from the unit quaternion $\hat{\boldsymbol{q}}_{\text{ins}} = \hat{\boldsymbol{q}}_b^n \equiv [\hat{q}_1, \hat{q}_2, \hat{q}_3, \hat{q}_4]^\top$ by multiplying Gibbs vector by the factor two. This is mathematically equivalent to

$$\hat{\boldsymbol{a}} = 2\hat{\boldsymbol{a}}_g = \frac{2}{\hat{\eta}} \begin{bmatrix} \hat{\varepsilon}_1 \\ \hat{\varepsilon}_2 \\ \hat{\varepsilon}_3 \end{bmatrix} = \frac{2}{\hat{q}_1} \begin{bmatrix} \hat{q}_2 \\ \hat{q}_3 \\ \hat{q}_4 \end{bmatrix}. \tag{14.257}$$

Finally, the gradient

$$\boldsymbol{c}_\psi(\hat{\boldsymbol{q}}_{\text{ins}}) := \left. \frac{\mathrm{d}h(\boldsymbol{a})}{\mathrm{d}\boldsymbol{a}} \right|_{\delta \boldsymbol{a} = \delta \hat{\boldsymbol{a}}} \tag{14.258}$$

is included in the KF measurement matrix as

$$\boldsymbol{C}_d[k] = \begin{bmatrix} \cdots & \cdots & \cdots \\ \boldsymbol{0}_{1\times 9} & \boldsymbol{c}_\psi^\top(\hat{\boldsymbol{q}}_{\text{ins}}[k]) & \boldsymbol{0}_{1\times 3} \end{bmatrix}. \tag{14.259}$$

We are now ready to present the mathematical models for the error-state filter. This includes the strapdown navigation equations, IMU measurement equations and the differential equations used for INS state propagation.

Strapdown navigation equations: The strapdown navigation equations are

$$\dot{\boldsymbol{p}}^n_{\text{nm}_\text{I}} = \boldsymbol{v}^n_{\text{nm}_\text{I}} \tag{14.260}$$

$$\dot{\boldsymbol{v}}^n_{\text{nm}_\text{I}} = \boldsymbol{a}^n_{\text{nm}_\text{I}} \tag{14.261}$$

$$\dot{\boldsymbol{q}}^n_{\text{b}} = \boldsymbol{T}(\boldsymbol{q}^n_b)\boldsymbol{\omega}^b_{\text{nb}}. \tag{14.262}$$

IMU measurements: The strapdown navigation equations are driven by the IMU specific force and ARS measurements

$$\boldsymbol{f}^b_{\text{imu}} = \boldsymbol{R}^\top(\boldsymbol{q}^n_b)(\boldsymbol{a}^n_{\text{nm}_\text{I}} - \boldsymbol{g}^n) + \boldsymbol{b}^b_{\text{acc}} + \boldsymbol{w}^b_{\text{acc}} \tag{14.263}$$

$$\boldsymbol{\omega}^b_{\text{imu}} = \boldsymbol{\omega}^b_{\text{nb}} + \boldsymbol{b}^b_{\text{ars}} + \boldsymbol{w}^b_{\text{ars}} \tag{14.264}$$

with random walk bias models for the sensors

$$\dot{\boldsymbol{b}}^b_{\text{acc}} = \boldsymbol{w}^b_{\text{b, acc}} \qquad \dot{\boldsymbol{b}}^b_{\text{ars}} = \boldsymbol{w}^b_{\text{b, ars}}. \tag{14.265}$$

The process noise terms $\boldsymbol{w}^b_{\text{acc}}$, $\boldsymbol{w}^b_{\text{ars}}$, $\boldsymbol{w}^b_{b,\text{acc}}$ and $\boldsymbol{w}^b_{b,\text{ars}}$ are all assumed to be Gaussian white noise.

Inertial navigation system (INS):

$$\dot{\hat{\boldsymbol{p}}}^n_{\text{ins}} = \hat{\boldsymbol{v}}^n_{\text{ins}} \tag{14.266}$$

$$\dot{\hat{\boldsymbol{v}}}^n_{\text{ins}} = \boldsymbol{R}(\hat{\boldsymbol{q}}_{\text{ins}})\boldsymbol{f}^b_{\text{ins}} + \boldsymbol{g}^n \tag{14.267}$$

$$\dot{\hat{\boldsymbol{b}}}^b_{\text{ins, acc}} = \boldsymbol{0} \tag{14.268}$$

$$\dot{\hat{\boldsymbol{q}}}_{\text{ins}} = \boldsymbol{T}(\hat{\boldsymbol{q}}_{\text{ins}})\boldsymbol{\omega}^b_{\text{ins}} \tag{14.269}$$

$$\dot{\hat{\boldsymbol{b}}}^b_{\text{ins, ars}} = \boldsymbol{0} \tag{14.270}$$

where the INS inputs $\boldsymbol{f}^b_{\text{ins}}$ and $\boldsymbol{\omega}^b_{\text{ins}}$ are bias-compensated IMU measurements

$$\boldsymbol{f}^b_{\text{ins}} := \boldsymbol{f}^b_{\text{imu}} - \hat{\boldsymbol{b}}^b_{\text{acc,ins}} \tag{14.271}$$

$$\boldsymbol{\omega}^b_{\text{ins}} := \boldsymbol{\omega}^b_{\text{imu}} - \hat{\boldsymbol{b}}^b_{\text{ars, ins}}. \tag{14.272}$$

Error-state model: The error-state model is a 15-states nonlinear system

$$\delta\dot{\boldsymbol{x}} = \boldsymbol{f}(\delta\boldsymbol{x}, \boldsymbol{u}, \boldsymbol{w}) \tag{14.273}$$

$$\delta\boldsymbol{y} = \boldsymbol{h}(\delta\boldsymbol{x}, \boldsymbol{u}) + \boldsymbol{\varepsilon} \tag{14.274}$$

with state and process noise vectors

$$\delta\boldsymbol{x} = [(\delta\boldsymbol{p}^n_{\text{nm}_\text{I}})^\top, (\delta\boldsymbol{v}^n_{\text{nm}_\text{I}})^\top, (\delta\boldsymbol{b}^b_{\text{acc}})^\top, \delta\boldsymbol{a}^\top, (\delta\boldsymbol{b}^b_{\text{ars}})^\top]^\top \tag{14.275}$$

$$\boldsymbol{w} = [(\boldsymbol{w}^b_{\text{acc}})^\top, (\boldsymbol{w}^b_{\text{b, acc}})^\top, (\boldsymbol{w}^b_{\text{ars}})^\top, (\boldsymbol{w}^b_{\text{b, ars}})^\top]^\top. \tag{14.276}$$

Note that $\delta\boldsymbol{a}$ replaces the unit quaternion as error state in (14.275). The measurement vector is given by

$$\delta\boldsymbol{y} = \begin{bmatrix} \delta\boldsymbol{y}_\text{p} & \text{position} \\ \delta\boldsymbol{y}_\text{v} & \text{velocity (optionally)} \\ \delta\boldsymbol{v}_1 & \text{gravity reference vector} \\ \delta\boldsymbol{v}_2 & \text{magnetometer reference vector} \\ \delta y_\psi & \text{compass (alternative to } \delta\boldsymbol{v}_2\text{)} \end{bmatrix}. \tag{14.277}$$

At every time step the nonlinear model (14.273) and (14.274) is linearized about $\delta\boldsymbol{x}[k] = \boldsymbol{0}$ and $\boldsymbol{w}[k] = \boldsymbol{0}$. The expressions for the discrete-time system matrices then become (see Table 13.2)

$$\boldsymbol{A}_\text{d}[k] \approx \boldsymbol{I}_{15} + h\left.\frac{\partial \boldsymbol{f}(\delta\boldsymbol{x}[k], \boldsymbol{u}[k], \boldsymbol{w}[k])}{\partial \delta\boldsymbol{x}[k]}\right|_{\delta\boldsymbol{x}[k]=\boldsymbol{0},\, \boldsymbol{w}[k]=\boldsymbol{0}}$$

$$\approx \boldsymbol{I}_{15} + h\begin{bmatrix} \boldsymbol{0}_{3\times3} & \boldsymbol{I}_3 & \boldsymbol{0}_{3\times3} & \boldsymbol{0}_{3\times3} & \boldsymbol{0}_{3\times3} \\ \boldsymbol{0}_{3\times3} & \boldsymbol{0}_{3\times3} & -\boldsymbol{R}(\hat{\boldsymbol{q}}_{\text{ins}}[k]) & -\boldsymbol{R}(\hat{\boldsymbol{q}}_{\text{ins}}[k])\boldsymbol{S}(\boldsymbol{f}^b_{\text{ins}}[k]) & \boldsymbol{0}_{3\times3} \\ \boldsymbol{0}_{3\times3} & \boldsymbol{0}_{3\times3} & -\frac{1}{T_{\text{acc}}}\boldsymbol{I}_3 & \boldsymbol{0}_{3\times3} & \boldsymbol{0}_{3\times3} \\ \boldsymbol{0}_{3\times3} & \boldsymbol{0}_{3\times3} & \boldsymbol{0}_{3\times3} & -\boldsymbol{S}(\boldsymbol{\omega}^b_{\text{ins}}[k]) & -\boldsymbol{I}_3 \\ \boldsymbol{0}_{3\times3} & \boldsymbol{0}_{3\times3} & \boldsymbol{0}_{3\times3} & \boldsymbol{0}_{3\times3} & -\frac{1}{T_{\text{ars}}}\boldsymbol{I}_3 \end{bmatrix} \tag{14.278}$$

$$C_d[k] \approx \left.\frac{\partial \boldsymbol{h}(\delta \boldsymbol{x}[k], \boldsymbol{u}[k])}{\partial \delta \boldsymbol{x}[k]}\right|_{\delta \boldsymbol{x}[k]=\boldsymbol{0}}$$

$$\approx \begin{bmatrix} \boldsymbol{I}_3 & \boldsymbol{0}_{3\times3} & \boldsymbol{0}_{3\times3} & \boldsymbol{0}_{3\times3} & \boldsymbol{0}_{3\times3} \\ \boldsymbol{0}_{3\times3} & \boldsymbol{I}_3 & \boldsymbol{0}_{3\times3} & \boldsymbol{0}_{3\times3} & \boldsymbol{0}_{3\times3} \\ \boldsymbol{0}_{3\times3} & \boldsymbol{0}_{3\times3} & \boldsymbol{0}_{3\times3} & \boldsymbol{S}(\boldsymbol{R}^\top(\hat{\boldsymbol{q}}_{\text{ins}}[k])\boldsymbol{v}^n_{01}) & \boldsymbol{0}_{3\times3} \\ \boldsymbol{0}_{3\times3} & \boldsymbol{0}_{3\times3} & \boldsymbol{0}_{3\times3} & \boldsymbol{S}(\boldsymbol{R}^\top(\hat{\boldsymbol{q}}_{\text{ins}}[k])\boldsymbol{v}^n_{02}) & \boldsymbol{0}_{3\times3} \\ \boldsymbol{0}_{1\times3} & \boldsymbol{0}_{1\times3} & \boldsymbol{0}_{1\times3} & \boldsymbol{c}^\top_{\psi}(\hat{\boldsymbol{q}}_{\text{ins}}[k]) & \boldsymbol{0}_{1\times3} \end{bmatrix} \tag{14.279}$$

$$\boldsymbol{E}_d[k] \approx h \left.\frac{\partial \boldsymbol{f}(\delta \boldsymbol{x}[k], \boldsymbol{u}[k], \boldsymbol{w}[k])}{\partial \boldsymbol{w}[k]}\right|_{\delta \boldsymbol{x}[k]=\boldsymbol{0},\ \boldsymbol{w}[k]=\boldsymbol{0}}$$

$$\approx h \begin{bmatrix} \boldsymbol{0}_{3\times3} & \boldsymbol{0}_{3\times3} & \boldsymbol{0}_{3\times3} & \boldsymbol{0}_{3\times3} \\ -\boldsymbol{R}(\hat{\boldsymbol{q}}_{\text{ins}}[k]) & \boldsymbol{0}_{3\times3} & \boldsymbol{0}_{3\times3} & \boldsymbol{0}_{3\times3} \\ \boldsymbol{0}_{3\times3} & \boldsymbol{I}_3 & \boldsymbol{0}_{3\times3} & \boldsymbol{0}_{3\times3} \\ \boldsymbol{0}_{3\times3} & \boldsymbol{0}_{3\times3} & -\boldsymbol{I}_3 & \boldsymbol{0}_{3\times3} \\ \boldsymbol{0}_{3\times3} & \boldsymbol{0}_{3\times3} & \boldsymbol{0}_{3\times3} & \boldsymbol{I}_3 \end{bmatrix}. \tag{14.280}$$

Summary: Indirect feedback Kalman filter with MEKF attitude estimation

State vectors:

$$\delta\hat{\boldsymbol{x}}[k] = [\delta\hat{\boldsymbol{p}}^n_{\text{nm}_\text{I}}[k]^\top, \delta\hat{\boldsymbol{v}}^n_{\text{nm}_\text{I}}[k]^\top, \delta\hat{\boldsymbol{b}}^b_{\text{acc}}[k]^\top, \delta\hat{\boldsymbol{a}}[k]^\top, \delta\hat{\boldsymbol{b}}^b_{\text{ars}}[k]^\top]^\top \tag{14.281}$$

$$\hat{\boldsymbol{x}}_{\text{ins}}[k] = [\hat{\boldsymbol{p}}^n_{\text{ins}}[k]^\top, \hat{\boldsymbol{v}}^n_{\text{ins}}[k]^\top, \hat{\boldsymbol{b}}^b_{\text{acc, ins}}[k]^\top, \hat{\boldsymbol{q}}_{\text{ins}}[k]^\top, \hat{\boldsymbol{b}}^b_{\text{ars, ins}}[k]^\top]^\top \tag{14.282}$$

Kalman gain:

$$\boldsymbol{K}[k] = \hat{\boldsymbol{P}}^-[k]\boldsymbol{C}^\top_d[k](\boldsymbol{C}_d[k]\hat{\boldsymbol{P}}^-[k]\boldsymbol{C}^\top_d[k] + \boldsymbol{R}_d[k])^{-1} \tag{14.283}$$

INS reset and state estimates:

$$\hat{\boldsymbol{p}}^n_{\text{ins}}[k] = \hat{\boldsymbol{p}}^n_{\text{ins}}[k] + \delta\hat{\boldsymbol{p}}^n_{\text{nm}_\text{I}}[k] \tag{14.284}$$

$$\hat{\boldsymbol{v}}^n_{\text{ins}}[k] = \hat{\boldsymbol{v}}^n_{\text{ins}}[k] + \delta\hat{\boldsymbol{v}}^n_{\text{nm}_\text{I}}[k] \tag{14.285}$$

$$\hat{\boldsymbol{b}}^b_{\text{acc, ins}}[k] = \hat{\boldsymbol{b}}^b_{\text{acc, ins}}[k] + \delta\hat{\boldsymbol{b}}^b_{\text{acc}}[k] \tag{14.286}$$

$$\hat{\boldsymbol{b}}^b_{\text{ars, ins}}[k] = \hat{\boldsymbol{b}}^b_{\text{ars, ins}}[k] + \delta\hat{\boldsymbol{b}}^b_{\text{ars}}[k] \tag{14.287}$$

$$\hat{\boldsymbol{q}}_{\text{ins}}[k] = \hat{\boldsymbol{q}}_{\text{ins}}[k] \otimes \delta\hat{\boldsymbol{q}}^n_b[k] \quad \text{(Schur product)} \tag{14.288}$$

$$\hat{\boldsymbol{q}}_{\text{ins}}[k] = \hat{\boldsymbol{q}}_{\text{ins}}[k]\ /\|\hat{\boldsymbol{q}}_{\text{ins}}[k]\| \quad \text{(Normalization)} \tag{14.289}$$

Predictor:

$$\hat{\boldsymbol{P}}^-[k+1] = \boldsymbol{A}_d[k]\hat{\boldsymbol{P}}[k]\boldsymbol{A}^\top_d[k] + \boldsymbol{E}_d[k]\boldsymbol{Q}_d[k]\boldsymbol{E}^\top_d[k] \tag{14.290}$$

Corrector:

$$\delta\hat{\boldsymbol{x}}[k] = \boldsymbol{K}[k](\boldsymbol{y}[k] - \boldsymbol{C}_{\mathrm{d}}[k]\,\hat{\boldsymbol{x}}_{\mathrm{ins}}[k]) \tag{14.291}$$

$$\Longrightarrow \delta\hat{\boldsymbol{q}}_b^n[k] = \frac{1}{\sqrt{4+\delta\hat{\boldsymbol{a}}[k]^{\top}\,\delta\hat{\boldsymbol{a}}[k]}}\begin{bmatrix} 2 \\ \delta\hat{\boldsymbol{a}}[k] \end{bmatrix} \quad \text{(Gibbs vector)} \tag{14.292}$$

$$\hat{\boldsymbol{P}}[k] = (\boldsymbol{I}_{15} - \boldsymbol{K}[k]\boldsymbol{C}_{\mathrm{d}}[k])\hat{\boldsymbol{P}}^{-}[k](\boldsymbol{I}_{15} - \boldsymbol{K}[k]\boldsymbol{C}_{\mathrm{d}}[k])^{\top} + \boldsymbol{K}[k]\boldsymbol{R}_{\mathrm{d}}[k]\boldsymbol{K}^{\top}[k] \tag{14.293}$$

INS propagation:

$$\hat{\boldsymbol{p}}^n_{\mathrm{ins}}[k+1] = \hat{\boldsymbol{p}}^n_{\mathrm{ins}}[k] + h\hat{\boldsymbol{v}}^n_{\mathrm{ins}}[k] \tag{14.294}$$

$$\hat{\boldsymbol{v}}^n_{\mathrm{ins}}[k+1] = \hat{\boldsymbol{v}}^n_{\mathrm{ins}}[k] + h(\boldsymbol{R}^n_b(\hat{\boldsymbol{q}}_{\mathrm{ins}}[k])(\boldsymbol{f}^b_{\mathrm{imu}}[k] - \hat{\boldsymbol{b}}^b_{\mathrm{acc,\,ins}}[k]) + \boldsymbol{g}^n) \tag{14.295}$$

$$\hat{\boldsymbol{q}}_{\mathrm{ins}}[k+1] = \hat{\boldsymbol{q}}_{\mathrm{ins}}[k] + h\boldsymbol{T}^n_b(\hat{\boldsymbol{q}}_{\mathrm{ins}}[k])(\boldsymbol{\omega}^b_{\mathrm{imu}}[k] - \hat{\boldsymbol{b}}^b_{\mathrm{ars,\,ins}}[k]) \tag{14.296}$$

$$\hat{\boldsymbol{q}}_{\mathrm{ins}}[k+1] = \hat{\boldsymbol{q}}_{\mathrm{ins}}[k+1]\,/\,\|\hat{\boldsymbol{q}}_{\mathrm{ins}}[k+1]\| \quad \text{(Normalization).} \tag{14.297}$$

MATLAB

The pseudocode for the 15-state indirect feedback KF where attitude is parametrized using unit quaternions/Gibbs vector is given below. The unit quaternion error-state representation is similar to the MEKF for attitude estimation. The stand-alone Matlab functions for the INS are included in the MSS toolbox as:

```
[x_ins,P_prd] = ins_MEKF(x_ins, P_prd, mu, h, Qd, Rd, f_imu, ...
    w_imu, m_imu, m_ref, y_pos, y_vel)

[x_ins,P_prd] = ins_MEKF_psi(x_ins, P_prd, mu, h, Qd, Rd, ...
    f_imu, w_imu, y_psi, y_pos, y_vel)
```

where the first function is for magnetometer measurements (see pseudocode below). The second function is a modification where the three-axis magnetometer measurements are replaced by a single compass measurement (yaw angle). The toolbox also includes a user editable script `ExINS_MEKF.m`, which demonstrates how the KF loop can be implemented using real-time measurements.

```
% Initialization
Qd = constant; Rd = constant;     % KF covariance matrices
g_n = [ 0 0 gravity(mu) ]';       % g as a function of latitude
P_prd = P0;

p_ins = zeros(3,1); v_ins = zeros(3,1); b_acc_ins = zeros(3,1);
b_ars_ins = zeros(3,1); q_ins = euler2q(0, 0, 0);

% MAIN LOOP
for i=1:N
```

```
% IMU measurements
f_imu = ...                    % specific force: f_imu[k]
w_imu = ...                    % attitude rate:  w_imu[k]
m_imu = ...                    % magentic field: m_imu[k]

% Attitude matrices
R = Rquat(q_ins);
T = Tquat(q_ins);

% Bias compensated IMU measurements
f_ins = f_imu - b_acc_ins;
w_ins = w_imu - b_ars_ins;

% Normalized gravity vectors
v10 = [0 0 1]';               % NED
v1 = -f_ins/g;                % BODY
v1 = v1 / sqrt( v1' * v1 );

% Normalized magnetic field vectors
v20 = m_ref / sqrt( m_ref' * m_ref );     % NED
v2  = m_imu / sqrt( m_ref' * m_ref );     % BODY

% Kalman filter matrices
Ad = ...
Cd = ...
Ed = ...

if (new_measurement)

   y_pos = ...        % GNSS position measurement: y_pos[k]
   y_vel = ...        % GNSS velocity measurement: y_vel[k]

   % KF gain: K[k]
   K = P_prd * Cd' * inv( Cd * P_prd * Cd' + Rd );
   IKC = eye(15) - K * Cd;

   % Estimation error: eps[k]
   eps_pos = y_pos - p_ins;
   eps_vel = y_vel - v_ins;
   eps_g   = v1 - R'* v10';
   eps_psi = v2 - R'* v20';
   eps = [eps_pos; eps_vel; eps_g; eps_psi];

   % Corrector: delta_x_hat[k] and P_hat[k]
   delta_x_hat = K * eps;
   P_hat = IKC * P * IKC' + K * Rd * K';

   % Error quaternion (2 x Gibbs vector): delta_q_hat[k]
   delta_a = delta_x_hat(10:12);
   delta_q_hat = 1/sqrt(4 + delta_a'*delta_a)*[2 delta_a']';

   % INS reset: x_ins[k]
   p_ins = p_ins + delta_x_hat(1:3);          % position
   v_ins = v_ins + delta_x_hat(4:6);          % velocity
```

```
        b_acc_ins = b_acc_ins + delta_x_hat(7:9);    % acc bias
        b_ars_ins = b_ars_ins + delta_x_hat(13:15); % ars bias

        q_ins = quatprod(q_ins, delta_q_hat); % Schur product
        q_ins = q_ins / sqrt(q_ins' * q_ins); % normalization

    else

        P_hat = P_prd;                        % no measurement

    end

    % Predictor: P_prd[k+1]
    P_prd = Ad * P_hat * Ad' + Ed * Qd * Ed';

    % INS propagation: x_ins[k+1]
    p_ins = p_ins + h * v_ins;
    v_ins = v_ins + h * ( R * f_ins + g_n );
    q_ins = q_ins + h * T * w_ins;
    q_ins = q_ins / sqrt( q_ins' * q_ins )   % normalization

end
```

The performance of the indirect feedback Kalman filter is shown in Figures 14.9 and 14.10 where the MSS toolbox file `ExINS_MEKF.m` is used to call the Matlab function `ins_mekf.m` in a loop. The latter is the MEKF implementation of the unit quaternion representation, which is parametrized using Gibbs vector. The state estimator runs at 100 Hz, which is the chosen IMU specific force and ARS measurement frequency. It is straightforward to use larger values such as 1000–2000 Hz if desirable. The INS is aided by GNSS position measurements, which arrives at 1 Hz.

The MSS toolbox function `ins_mekf.m` is the implementation of an 9-DOF IMU with a three-axis magnetometer. This is the standard sensor suite for small vehicles. For ships and boats, however, it is recommended to use an 6-DOF IMU together with a gyrocompass since the gyrocompass is resistant to magnificent field disturbances.

Another possibility is to use GNSS heading measurements. The MSS toolbox implementation for compass aiding is `ins_mekf_psi.m`, which is based on (14.252) and (14.259). Both the magnetometer and the compass implementations can be tested by specifying the configuration parameters in the example script `ExINS_MEKF.m`. It is also possible to test GNSS velocity aiding by changing the parameters.

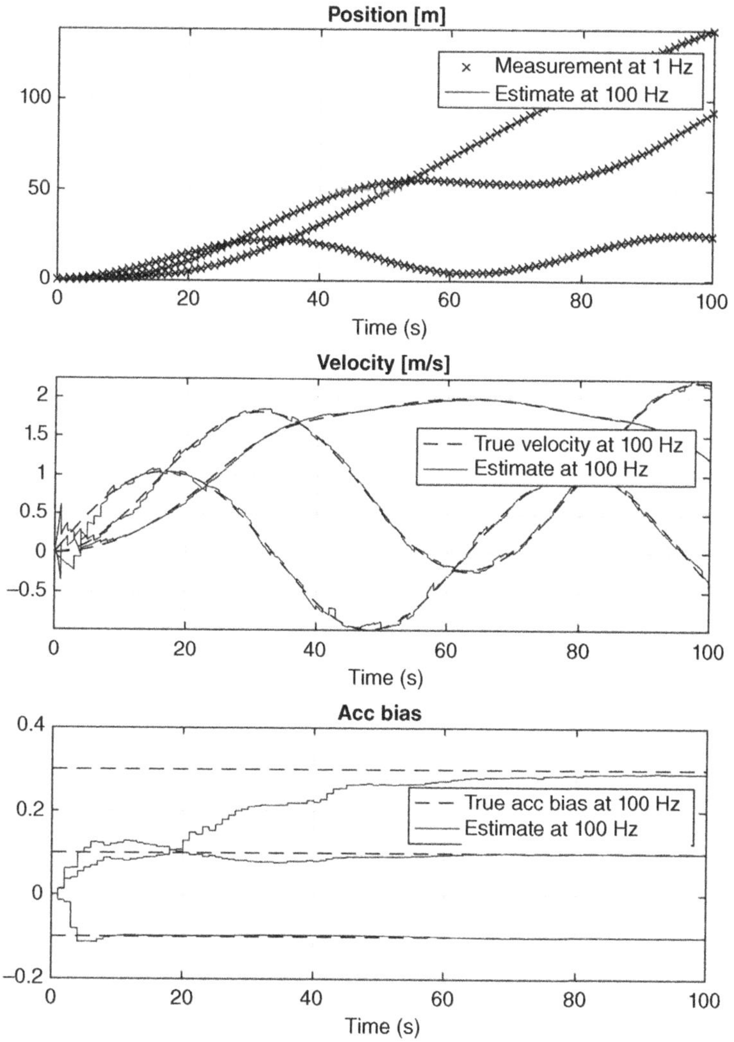

Figure 14.9 Upper plot: INS position estimates $\hat{\boldsymbol{p}}^{\rm n}_{\rm ins}$ (solid lines) together with true GNSS positions (*). Middle plot: INS velocity estimates $\hat{\boldsymbol{v}}^{\rm n}_{\rm ins}$ (solid lines) together with true velocities (dashed lines). Lower plot: INS acceleration bias estimates $\hat{\rm a}^{\rm b}_{\rm acc,\,ins}$ (solid lines) together with the true values (dashed lines).

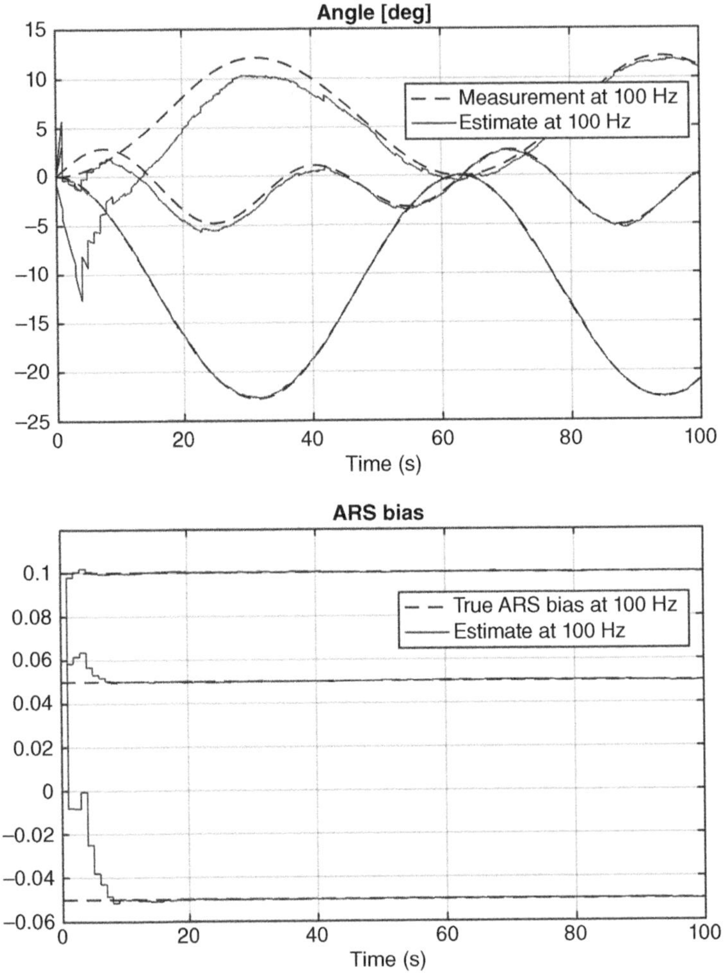

Figure 14.10 Upper plot: INS attitude estimates $\hat{\boldsymbol{q}}_{\text{ins}}$ transformed to Euler angles (solid lines) together with true Euler angles (dashed lines). Lower plot: INS angular rate sensor bias estimates $\hat{a}^{b}_{\text{ars, ins}}$ (solid line) together with the true values (dashed lines).

Figures 14.9 and 14.10 show that the acceleration biases converge to their true values in 1–2 min, while the angular rate sensor biases needs less than 10 s to converge. The plots also show that the position, velocity and attitude states converge to their true values. The attitude plot is obtained by transforming the estimated unit quaternions to Euler angles using `q2euler.m`.

15

Motion Control Systems

Motion control systems for marine craft have been an active field of research since the first mechanical autopilot was constructed by *Elmer Sperry* in 1911. Modern control systems are based on a variety of design techniques such as PID control, linear-quadratic optimal and stochastic control, $\mathcal{H}_\infty$ control methods, fuzzy systems, neural networks and nonlinear control theory, to mention only some. In the first part of the book, models for simulation of marine craft were presented. In this chapter, dynamic models are used to design model-based control systems. The dynamic properties and limitation of the craft are incorporated into the design process to obtain robust performance. Many of the presented design methods have been successfully implemented and tested on board ships, underwater vehicles and offshore structures.

Chapter 15 covers state-of-the-art PID control and successive loop closure methods for setpoint regulation, trajectory tracking and path-following control of marine craft. This includes autopilot design, stationkeeping, position mooring systems, cross-tracking control systems and LOS path-following control systems. Advanced methods such as linear-quadratic optimal control, sliding mode control, state-feedback linearization and integrator backstepping are discussed in Chapter 16.

Preview of the Chapter

This chapter starts with open-loop analysis and maneuverability (Section 15.1) followed by state-of-the-art successive loop closure methods (Section 15.2) and linear PID design methods (Section 15.3). Conventional PID control systems have their origin in SISO linear systems theory. However, it is possible to generalize this to nonlinear MIMO systems by using results from robotics (Fossen 1991). This requires that the marine craft equations of motion are expressed in matrix-vector form

$$\dot{\boldsymbol{\eta}} = \boldsymbol{J}_\Theta(\boldsymbol{\eta})\boldsymbol{\nu} \tag{15.1}$$

$$\boldsymbol{M}\dot{\boldsymbol{\nu}} + \boldsymbol{C}(\boldsymbol{\nu})\boldsymbol{\nu} + \boldsymbol{D}(\boldsymbol{\nu})\boldsymbol{\nu} + \boldsymbol{g}(\boldsymbol{\eta}) = \boldsymbol{\tau} + \boldsymbol{w}. \tag{15.2}$$

Handbook of Marine Craft Hydrodynamics and Motion Control, Second Edition. Thor I. Fossen.

For this model class, MIMO nonlinear PID control systems can be designed by exploiting the fact that the mass system inertia matrix is positive definite and constant ($\boldsymbol{M} = \boldsymbol{M}^{\top} > 0$, $\dot{\boldsymbol{M}} = \boldsymbol{0}$), the Coriolis and centripetal matrix $\boldsymbol{C}(\boldsymbol{\nu}) = -\boldsymbol{C}^{\top}(\boldsymbol{\nu})$ is skew-symmetrical and the damping matrix $\boldsymbol{D}(\boldsymbol{\nu}) > 0$ is strictly positive.

15.1 Open-Loop Stability and Maneuverability

When designing a motion control system a compromise between stability and maneuverability must be made. More specifically,

- *Stability* of an uncontrolled marine craft can be defined as the ability to return to an equilibrium point after a disturbance, without any corrective action of the control system.
- *Maneuverability*, on the other hand, is defined as the capability of the craft to carry out specific maneuvers.

It is well known that a craft that is easy to maneuver, for instance a fighter aircraft or a high-speed watercraft, can be marginally stable or even unstable in open loop. On the other hand, excessive stability implies that the control effort will be excessive in a maneuvering situation whereas a marginally stable ship is easy to maneuver. Consequently, a compromise between stability and maneuverability must be made (see Figure 15.1).

Figure 15.1 Maneuverability versus stability. Source: Illustration by B. Stenberg.

15.1.1 Straight-line, Directional and Positional Motion Stability

For marine craft it is common to distinguish between three types of stability, namely

- Straight-line stability
- Directional or course stability
- Positional motion stability.

This can be explained using open- and closed-loop stability analyzes. In order to understand the different types of stability one can consider the following test system

$$\dot{x}^n = U\cos(\psi) \tag{15.3}$$

$$\dot{y}^n = U\sin(\psi) \tag{15.4}$$

$$\dot{\psi} = r \tag{15.5}$$

$$T\dot{r} + r = K\delta + w \tag{15.6}$$

where w is the external disturbances and U = constant is the cruise speed. The first two equations represent the $(x^n, \ y^n)$ positions of the ship while the last two equations describe the yaw dynamics modeled by Nomoto's first-order model (see section 7.2.2). For simplicity, it is assumed that the yaw motion of the craft is stabilized by a PD controller

$$\delta = -K_p(\psi - \psi_d) - K_d r \tag{15.7}$$

where ψ_d = constant denotes the desired heading angle and K_p and K_d are two positive regulator gains. Substituting the control law (15.7) into Nomoto's first-order model (15.6) yields the closed-loop system

$$\underbrace{T}_{m}\ \ddot{\psi} + \underbrace{(1 + KK_d)}_{d}\ \dot{\psi} + \underbrace{KK_p}_{k}\ \psi = \underbrace{KK_p\psi_d + w}_{f(t)}. \tag{15.8}$$

The closed-loop system (15.8) represents a second-order mass–damper–spring system

$$m\ddot{\psi} + d\dot{\psi} + k\psi = f(t) \tag{15.9}$$

with driving input $f(t) = k\psi_d + w$. The eigenvalues $\lambda_{1,2}$, the natural frequency ω_n and the relative damping ratio ζ for the mass–damper–spring system are

$$\lambda_{1,2} = \frac{-d \pm \sqrt{d^2 - 4km}}{2m}, \quad \omega_n = \sqrt{\frac{k}{m}}, \quad \zeta = \frac{d}{2}\frac{1}{\sqrt{km}}. \tag{15.10}$$

MATLAB

The test system (15.8) is simulated in Matlab for varying model parameters using the MSS toolbox script

```
StabDemo
```

The simulation results and the stability analysis are presented on the next pages. This includes the following cases:

- Instability
- Straight-line stability
- Directional stability
- Positional motion stability.

Instability: For uncontrolled marine craft ($K_p = K_d = 0$) instability occurs when

$$\lambda_1 = -\frac{d}{m} = -\frac{1}{T} > 0, \qquad \lambda_2 = 0$$

which simply states that $T < 0$. This is common for large tankers.

Straight-line stability: Consider an uncontrolled marine craft ($K_p = K_d = 0$) moving in a straight path. If the new path is straight after a disturbance w in yaw the craft is said to have straight-line stability. The direction of the new path will usually differ from the initial path because no restoring forces are present ($k = 0$). This corresponds to

$$\lambda_1 = -\frac{d}{m} = -\frac{1}{T} < 0, \qquad \lambda_2 = 0.$$

Consequently, the requirement $T > 0$ implies straight-line stability for the uncontrolled craft ($\delta = 0$).

Directional stability (stability on course): Directional stability is a much stronger requirement than straight-line stability (see Figure 15.2). Directional stability requires the final path to be parallel to the initial path that is obtained for $K_p > 0 \Rightarrow k > 0$. Additional damping is added through $K_d > 0$. This corresponds to PD control. A marine craft is said to be directionally stable if both eigenvalues have negative real parts, that is $\text{Re}\{\lambda_{1,2}\} < 0$. The following two types of directional stability are observed:

No oscillations ($d^2 - 4km \geq 0$): This implies that both eigenvalues are negative and real, that is $\zeta \geq 1$ such that

$$\lambda_{1,2} = \frac{-d \pm \sqrt{d^2 - 4km}}{2m} = \left(-\zeta \pm \sqrt{\zeta^2 - 1}\right) \omega_n < 0$$

For a critically damped system $\zeta = 1.0$, such that $\lambda_{1,2} = -1/2(d/m) = -\omega_n$.

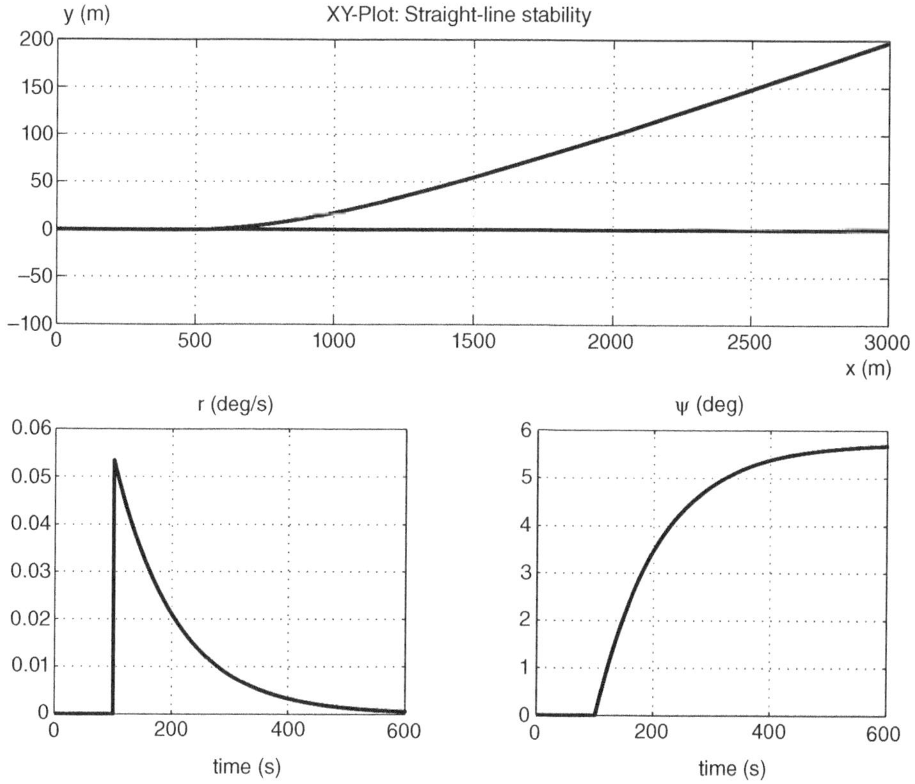

Figure 15.2 Straight-line stability for a ship when an impulse $w(t)$ is injected at $t = 100$ s.

Damped oscillator ($d^2 - 4km < 0$): This corresponds to two imaginary eigenvalues $\lambda_{1,2}$ with negative real parts ($\zeta < 1$), that is

$$\lambda_{1,2} = \frac{-d \pm j\sqrt{4km - d^2}}{2m} = \left(-\zeta \pm j\sqrt{1-\zeta^2}\right)\omega_n.$$

Directional stability for a critically damped ($\zeta = 1.0$) and underdamped craft ($\zeta = 0.1$) is shown in Figures 15.3 and 15.4. Note the oscillations in both positions and yaw angle in Figure 15.4. Directional stability requires feedback control since there are no restoring forces in yaw. However, in heave, roll and pitch where metacentric restoring forces are present ($k > 0$) no feedback is required to damp out the oscillations.

Positional motion stability: Positional motion stability implies that the ship should return to its original path after a disturbance (see Figure 15.5). This can be achieved by including integral action in the controller. Hence, a PID controller can be designed to compensate for the unknown disturbance term w while a PD controller will generally result in a steady-state offset.

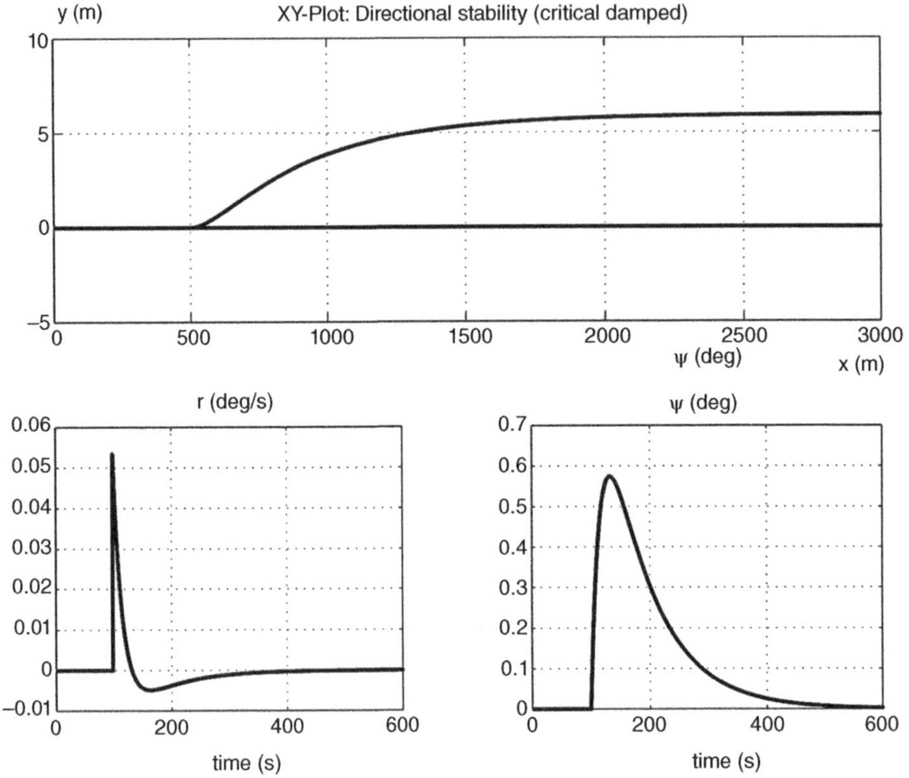

Figure 15.3 Directional stability for a critically damped ship ($\zeta = 1.0$) when an impulse $w(t)$ is injected at $t = 100$ s.

Example 15.1 (Straight-Line Stability)
Consider the cargo ship and oil tanker of Example 7.1. Recall that the equivalent time constant in Nomoto's first-order model was defined as (see Section 7.2.2)

$$T := T_1 + T_2 - T_3.$$

Hence, the uncontrolled cargo ship has an equivalent time constant

$$T_{cargo\,ship} = 118.0 + 7.8 - 18.5 = 107.3 \text{ s} > 0$$

while the oil tanker has an equivalent time constant

$$T_{oil\,tanker} = -124.1 + 16.4 - 46.0 = -153.7 \text{ s} < 0.$$

This implies that the cargo ship is straight-line stable while the oil tanker is unstable.

Criteria for Straight-line Stability

Recall that a ship is said to be dynamically straight-line stable if it returns to a straight-line motion after a disturbance in yaw without any corrective action from the

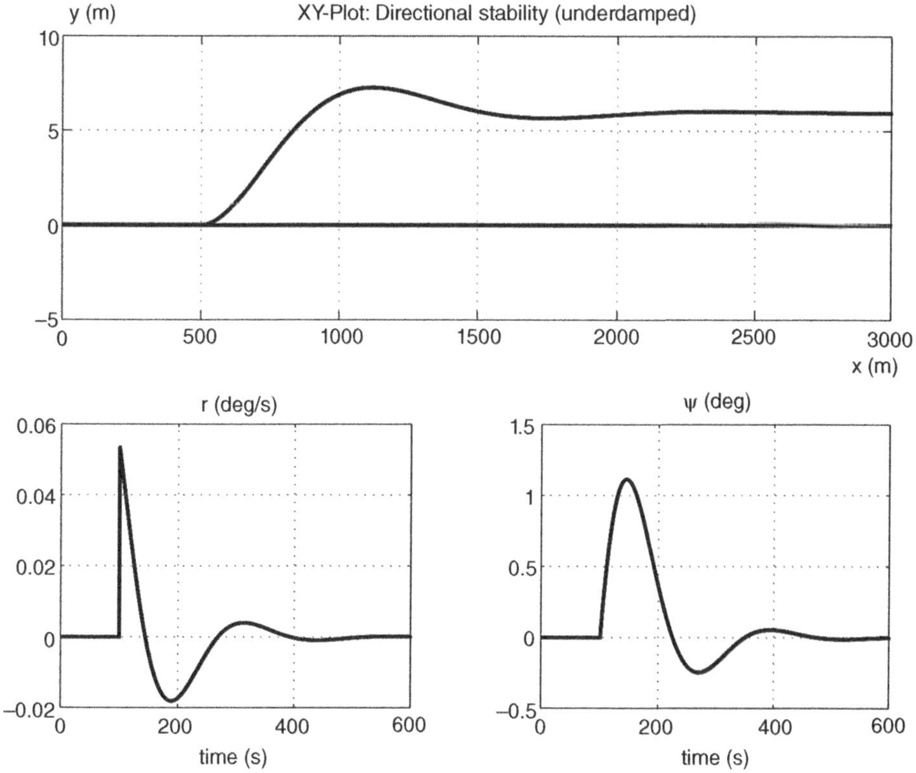

Figure 15.4 Directional stability for an underdamped ship ($\zeta = 0.1$) when an impulse $w(t)$ is injected at $t = 100$ s.

rudder. Consequently, instability refers to the case when the ship goes into a starboard or port turn without any rudder deflections. For Nomoto's first-order model straight-line motion was guaranteed for a positive time constant T. Similarly, it is possible to derive a criterion for straight-line stability for the state-space model (6.138)

$$\boldsymbol{M}\dot{\boldsymbol{v}} + \boldsymbol{N}\boldsymbol{v} = \boldsymbol{b}\,\delta \tag{15.11}$$

where $\boldsymbol{v} = [v,\ r]^{\top}$. Applications of *Laplace's transformation* to the linear model (15.11) with $\boldsymbol{v}(0) = \boldsymbol{0}$ yields

$$(\boldsymbol{M}s + \boldsymbol{N})\,\boldsymbol{v}(s) = \boldsymbol{b}\,\delta(s). \tag{15.12}$$

Consequently,

$$\boldsymbol{v}(s) = (\boldsymbol{M}s + \boldsymbol{N})^{-1}\,\boldsymbol{b}\,\delta(s) = \frac{\mathrm{adj}(\boldsymbol{M}s + \boldsymbol{N})}{\det(\boldsymbol{M}s + \boldsymbol{N})}\boldsymbol{b}\,\delta(s). \tag{15.13}$$

The characteristic equation is

$$\det(\boldsymbol{M}\sigma + \boldsymbol{N}) = A\sigma^2 + B\sigma + C = 0 \tag{15.14}$$

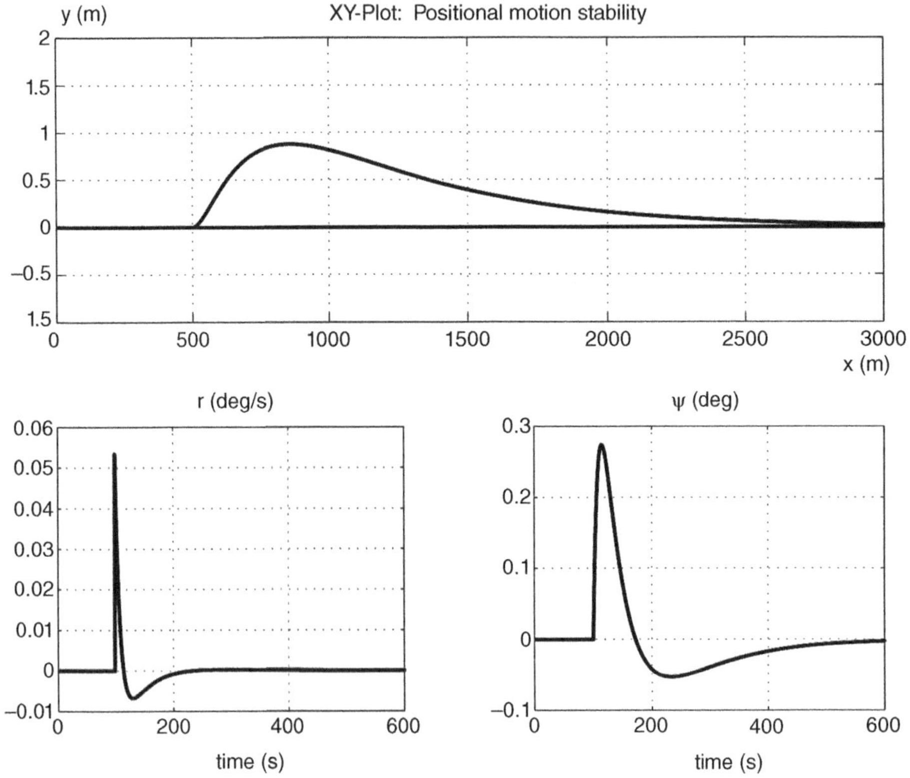

Figure 15.5 Positional motion stability for a PID-controlled ship when an impulse $w(t)$ is injected at $t = 100$ s.

where

$$\begin{aligned} A &= \det(\boldsymbol{M}) \\ B &= n_{11}m_{22} + n_{22}m_{11} - n_{12}m_{21} - n_{21}m_{12} \\ C &= \det(\boldsymbol{N}). \end{aligned} \tag{15.15}$$

The two roots $\sigma_{1,2}$ of (15.14), both of which must have negative real parts for open-loop stability, are

$$\text{Re}\{\sigma_{1,2}\} = \text{Re}\left\{ \frac{-B \pm \sqrt{B^2 - 4AC}}{2A} \right\} < 0. \tag{15.16}$$

The quantities $\sigma_{1,2}$ are often referred to as the control-fixed stability indices for straight-line stability. Alternatively, the Routh stability criterion can be applied.

Table 15.1 Routh array.

λ^n	a_n	a_{n-2}	a_{n-4}	...
λ^{n-1}	a_{n-1}	a_{n-3}	a_{n-5}	...
λ^{n-2}	b_1	b_2	b_3	...
λ^{n-3}	c_1	c_2	c_3	...
λ^{n-4}	d_1	d_2	d_3	...
$\vdots$	...			

Theorem 15.1 (The Routh Stability Criterion)
Consider the characteristic equation

$$a_n\lambda^n + a_{n-1}\lambda^{n-1} + a_{n-2}\lambda^{n-2} + \cdots + a_0 = 0. \tag{15.17}$$

To apply the Routh criterion, the Routh array shown in Table 15.1 must be constructed. The coefficients a_i are the coefficients of the characteristic equation (15.17) and b_i, c_i, d_i, ... are defined as

$$\begin{array}{lll} b_1 = (a_{n-1}a_{n-2} - a_na_{n-3})/a_{n-1} & b_2 = (a_{n-1}a_{n-4} - a_na_{n-5})/a_{n-1} & \cdots \\ c_1 = (b_1a_{n-3} - a_{n-1}b_2)/b_1 & c_2 = (b_1a_{n-5} - a_{n-1}b_3)/b_1 & \cdots \\ d_1 = (c_1b_2 - c_2b_1)/c_1 & \ldots . & \end{array}$$

Necessary and sufficient conditions for the system to be stable are:

1. All the coefficients of the characteristic equation (15.17) must be non-zero and have the same sign.
2. All the coefficients of the first column of the Routh array must have the same sign.

If Condition 2 is violated, the number of sign changes will indicate how many roots of the characteristic equation will have positive real parts. Hence, the system will be unstable.

Proof. *See Routh (1877).*

According to the Routh stability criterion, necessary and sufficient conditions for a ship given by (15.11) with characteristic equation (15.14) to be stable are

$$A,\ B,\ C > 0. \tag{15.18}$$

The first condition $A = \det(\boldsymbol{M}) > 0$ is automatically satisfied since the inertia matrix $\boldsymbol{M}$ is always positive definite for a marine craft. Condition $B > 0$ implies that

$$n_{11}m_{22} + n_{22}m_{11} > n_{12}m_{21} + n_{21}m_{12}. \tag{15.19}$$

Consequently, the products of the diagonal elements of $\boldsymbol{M}$ and $\boldsymbol{N}$ must be larger than the products of the off-diagonal elements. This is is satisfied for most ships. Consequently, condition (15.18) reduces to

$$C = \det(\boldsymbol{N}) > 0. \tag{15.20}$$

This condition has also been verified by Abkowitz (1964), who stated the following theorem.

Theorem 15.2 (Straight-Line Stability (Abkowitz 1964))
A ship is dynamically stable in straight-line motion if the hydrodynamic derivatives satisfy

$$\begin{aligned}\det(\boldsymbol{N}) &= \det\begin{bmatrix}-Y_{\mathrm{v}} & mU - Y_{\mathrm{r}} \\ -N_{\mathrm{v}} & mx_{\mathrm{g}}U - N_{\mathrm{r}}\end{bmatrix} \\ &= Y_{\mathrm{v}}(N_{\mathrm{r}} - mx_{\mathrm{g}}U) - N_{\mathrm{v}}(Y_{\mathrm{r}} - mU) > 0.\end{aligned} \tag{15.21}$$

Proof. *This is seen as a consequence of (15.20) and (15.21).*

It is interesting to notice that making C more positive will improve stability and thus reduce the ship's maneuverability, and the other way around. Straight-line stability implies that the new path of the ship will be a straight line after a disturbance in yaw. The direction of the new path will usually differ from the initial path. Contrary to this, unstable ships will go into a starboard or port turn without any rudder deflection. It should be noted that most modern large tankers are slightly unstable. For such ships, the criterion (15.21) corresponds to one of the poles being in the right half-plane.

Straight-line Stability in Terms of Time Constants

The criterion (15.18) can be related to Nomoto's second-order model (7.20) by noticing that

$$T_1 T_2 = \frac{A}{C} > 0, \quad T_1 + T_2 = \frac{B}{C} > 0. \tag{15.22}$$

Consequently, straight-line stability is guaranteed if $T_1 > 0$ and $T_2 > 0$. This can also be seen from

$$\sigma_{1,2} = -\frac{1}{T_{1,2}} = \mathrm{Re}\left\{\frac{-B \pm \sqrt{B^2 - 4AC}}{2A}\right\} < 0. \tag{15.23}$$

Criteria for Directional Stability

Dynamic stability on course, or directional stability, cannot be obtained without activating the rudder. Usually a PID control system is used to generate the necessary rudder action to stabilize the ship. For simplicity, consider a PD controller

$$\delta = -K_p(\psi - \psi_d) - K_d r \tag{15.24}$$

which after substitution into Nomoto's second-order model (7.22) yields the closed-loop dynamics

$$T_1 T_2 \psi^{(3)} + (T_1 + T_2 + T_3 K K_d)\ddot{\psi} + (1 + K K_d + T_3 K K_p)\dot{\psi} + K K_p \psi = K K_p \psi_d. \tag{15.25}$$

From this expression, the cubic characteristic equation

$$A\sigma^3 + B\sigma^2 + C\sigma + D = 0 \tag{15.26}$$

is recognized, where

$$A = T_1 T_2 \tag{15.27}$$

$$B = T_1 + T_2 + T_3 K K_d \tag{15.28}$$

$$C = 1 + K K_d + T_3 K K_p \tag{15.29}$$

$$D = K K_p. \tag{15.30}$$

The requirement for directional stability is

$$\mathrm{Re}\{\sigma_{1,2,3}\} < 0. \tag{15.31}$$

This can be checked by forming the Routh array

$$\begin{matrix} A & C \\ B & D \\ \frac{BC-AD}{B} & 0 \\ D & \end{matrix}. \tag{15.32}$$

Consequently, sufficient and necessary conditions for the ship to be dynamically stable on course are

$$A, B, C, D > 0 \tag{15.33}$$

$$BC - AD > 0. \tag{15.34}$$

This again implies that the controller gains K_p and K_d must be chosen such that the conditions (15.33) and (15.34) are satisfied.

15.1.2 *Maneuverability*

Several ship maneuvers can be used to evaluate the robustness, performance and limitations of a ship. This is usually done by defining a criterion in terms of a *maneuvering index* or by simply using a *maneuvering characteristic* to compare the maneuverability of the test ship with previously obtained results from other ships. A frequently used measure of maneuverability is the turning index of Norrbin (1965).

The Norrbin Measure of Maneuverability

Norrbin (1965) defines the *turning index* as

$$P := \frac{\psi'(t'=1)}{\delta'(t'=1)} \tag{15.35}$$

where $t' = t\,(U/L)$ is the non-dimensional time. Similar ψ' and δ' are the non-dimensional yaw and rudder angles, respectively (see Appendix D). P is a measure of turning ability or maneuverability since it can be interpreted as the heading change per unit rudder angle in one ship length traveled at $U = 1$ m s^{-1}. An expression for P can be found by solving the ODE

$$T'\ddot{\psi}' + \dot{\psi}' = K'\delta' \tag{15.36}$$

with $\delta' =$ constant. This results in

$$\psi'(t') = K'\left[t' - T' + T'\mathrm{e}^{-\frac{t'}{T'}}\right]\delta'(t'). \tag{15.37}$$

A second-order Taylor expansion of $\mathrm{e}^{-t'/T'}$ is

$$\mathrm{e}^{-\frac{t'}{T'}} = 1 - \frac{t'}{T'} + \frac{(t')^2}{2(T')^2} + O(3) \tag{15.38}$$

such that

$$\frac{\psi'(t')}{\delta'(t')} \approx K'\left[t' - T' + T'\left(1 - \frac{t'}{T'} + \frac{(t')^2}{2(T')^2}\right)\right] = K\frac{(t')^2}{2T'} \tag{15.39}$$

$$\frac{\psi'(t'=1)}{\delta'(t'=1)} \approx K'\left[\frac{(t')^2}{2T'}\right]_{t'=1} = \frac{K'}{2T'}. \tag{15.40}$$

Consequently,

$$P \approx \frac{1}{2}\frac{K'}{T'}. \tag{15.41}$$

The P number is a good measure of maneuverability for course-stable ships. Norrbin concludes that $P > 0.3$ guarantees a reasonable standard of course-change quality for most ships while $P > 0.2$ seems to be sufficient for large oil tankers. For poorly stable ships it is recommended to use P together with another maneuverability index, for instance the slope $\mathrm{d}r'/\mathrm{d}\delta'$ or the width of the r'–δ' loop (see Figure 15.12 later).

Maneuvering Characteristics

A maneuvering characteristic can be obtained by changing or keeping a predefined course and speed of the ship in a systematic manner by means of active controls. For most surface vessels these controls are rudders, fins, propellers and thrusters. However, since ship maneuverability depends on the water depth, environmental forces, ship speed and hydrodynamic derivatives care must be taken when performing a full-scale maneuvering test. A guide for sea trials describing how these maneuvers should be performed is found in SNAME (1989). The following standard ship maneuvers have been proposed by the International Towing Tank Conference (ITTC).

- **Turning circle:** This trial is mainly used to calculate the ship's steady turning radius and to check how well the steering machine performs under course-changing maneuvers.
- **Kempf's zigzag maneuver:** The zigzag test is a standard maneuver used to compare the maneuvering properties and control characteristic of a ship with those of other ships. Another feature is that the experimental results of the test can be used to calculate the K and T values of Nomoto's first-order model.
- **Pull-out maneuver:** The pull-out maneuver can be used to check whether the ship is straight-line stable or not. The maneuver can also be used to indicate the degree of stability.
- **Dieudonné's spiral maneuver:** The spiral maneuver is also used to check straight-line stability. The maneuver gives an indication of the range of validity of the linear theory.
- **Bech's reverse spiral maneuver:** The reverse spiral maneuver can be used for unstable ships to produce a nonlinear maneuvering characteristic. The results from the test indicate which rudder corrections are required to stabilize an unstable ship.
- **Stopping trials:** Crash stops and low-speed stopping trials can be used to determine the ship's head reach and maneuverability during emergency situations.

Turning Circle

This is probably the oldest maneuvering test. The test can be used as an indication of how well the steering machine and rudder control performs during course-changing maneuvers. It is also used to calculate standard measures of maneuverability such as *tactical diameter, advance* and *transfer* shown in Figure 15.6; see Gertler and Hagen (1960) for a detailed description.

MATLAB

The turning circle for the Mariner class vessel is computed using the MSS toolbox script `ExTurnCircle.m`, where

```
t_final = 700;                % final simulation time (s)
t_rudderexecute = 100;        % time rudder is executed (s)
h = 0.1;                      % sampling time (s)

% Mariner class cargo ship, cruise speed U0 = 7.7 m/s
x  = zeros(7,1);      % x=[u v r x y psi delta]' (initial values)
ui = -15*pi/180;      % delta_c=-delta_R at time t=t_rudderexecute

[t,u,v,r,x,y,psi,U] = ...
    turncircle('mariner', x, ui, t_final, t_rudderexecute, h)
```

The results are plotted in Figure 15.6. Similar results are obtained by replacing `mariner.m` with the container ship `container.m`; see `ExTurnCircle.m`. The maneuvering characteristics for the Mariner class vessel were computed to be:

Rudder execute (x_b coordinate)	769 m
Steady turning radius	711 m
Maximum transfer	1315 m
Maximum advance	947 m
Transfer at 90° heading	534 m
Advance at 90° heading	943 m
Tactical diameter at 180° heading	1311 m

The *steady turning radius* R is perhaps the most interesting quantity obtained from the turning trials. In the maneuvering trial code of the 14th ITTC (1975) it is proposed to turn the ship over at maximum speed and with a rudder angle of minimum 15° to obtain the turning circle. The rudder angle δ should be held constant such that a constant rate of turn is reached (in practice a turning circle of 540° may be necessary).

The output from a positioning system is used to calculate the tactical diameter, steady turning radius, maximum advance and maximum transfer. A typical turning circle corresponding to a negative rudder angle is shown in Figure 15.6.

For a constant rudder angle δ the ship will move in a circle with constant turning radius R and speed U in the steady state, that is $\dot{\boldsymbol{v}} = \boldsymbol{0}$. Solving (15.11) for the steady-state solution of $\boldsymbol{v} = [v,\ r]^\top$ yields

$$\boldsymbol{N}\boldsymbol{v} = \boldsymbol{b}\,\delta \quad \Longrightarrow \quad \boldsymbol{v} = \boldsymbol{N}^{-1}\boldsymbol{b}\,\delta. \tag{15.42}$$

The equation for r in this expression becomes

$$r = \frac{(Y_v N_\delta - N_v Y_\delta)}{Y_v(N_r - m x_g U) - N_v(Y_r - mU)}\,\delta. \tag{15.43}$$

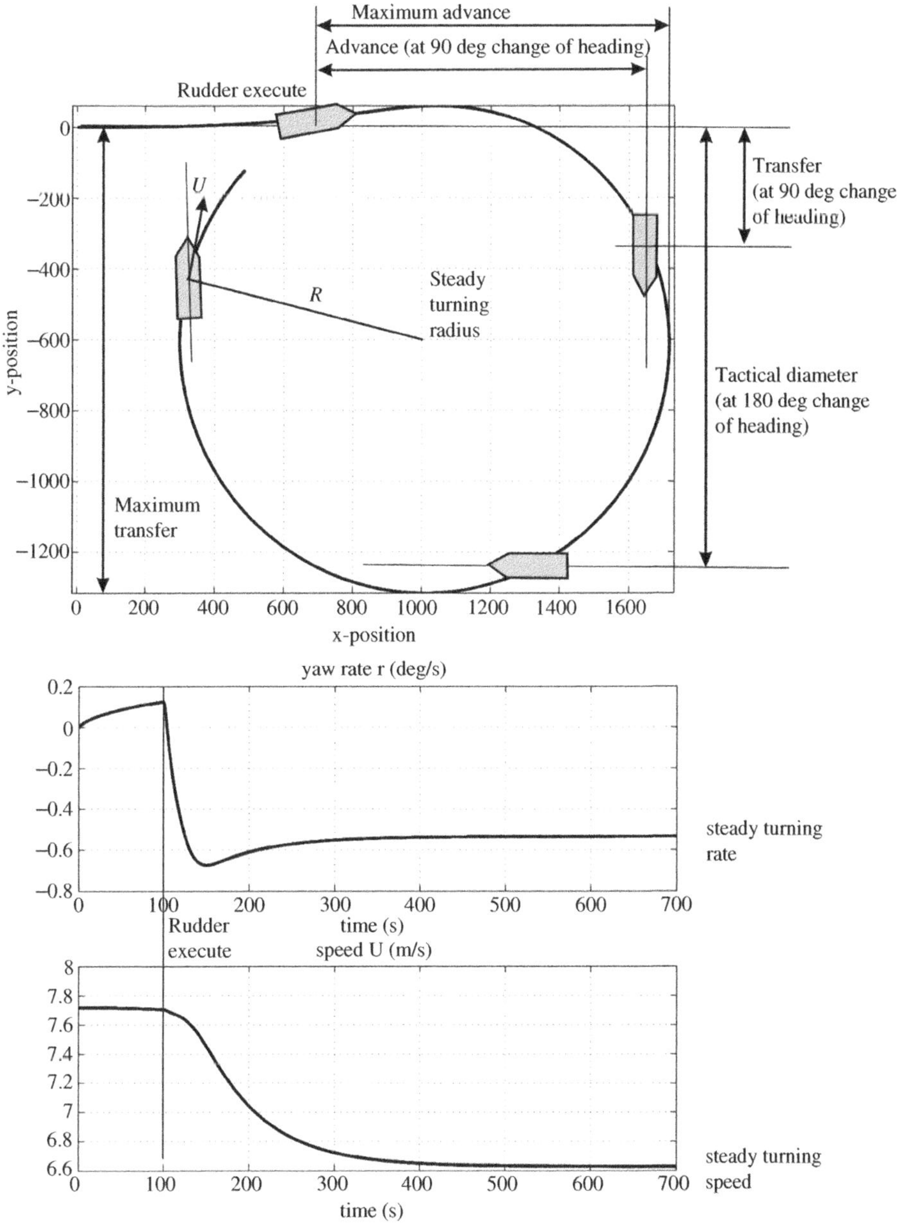

Figure 15.6 Turning circle, yaw rate and speed for the Mariner class vessel for a constant rudder angle $\delta_R = -15°$ applied at $t = 100$ s.

The ship's turning radius R is defined as

$$R := \frac{U}{r} \quad \text{where} \quad U = \sqrt{u^2 + v^2}. \tag{15.44}$$

Introducing the length $L = L_{pp}$ of the ship, the following expression for the ratio (R/L) is obtained

$$\left(\frac{R}{L}\right) = \left(\frac{U}{L}\right) \frac{C}{(Y_v N_\delta - N_v Y_\delta)} \frac{1}{\delta}, \qquad \delta \neq 0 \tag{15.45}$$

where

$$C = \det(N) = Y_v(N_r - m x_g U) - N_v(Y_r - mU) > 0 \qquad \text{(stable ship)}$$

is recognized as one of the stability derivatives in the straight-line stability criterion discussed in Section 15.1.1. From (15.45) it is seen that increased stability (large C) implies that the turning radius will increase. Consequently, a highly stable ship requires more maneuvering effort than a marginally stable one. The ratio (R/L) can also be written in terms of non-dimensional quantities by (see Appendix D)

$$\left(\frac{R}{L}\right) = \frac{Y'_v(N'_r - m' x'_g) - N'_v(Y'_r - m')}{Y'_v N'_\delta - N'_v Y'_\delta} \frac{1}{\delta}, \qquad \delta \neq 0. \tag{15.46}$$

This formula is independent of the ship speed. It should be noted that the formula for the turning radius are based on linear theory, which assumes that δ is small and accordingly that R is large.

Another feature of the turning test is that the Nomoto gain and time constants can be determined. This is illustrated in the following example where a cargo ship is considered.

Example 15.2 (Determination of the Nomoto Gain and Time Constants)
The Nomoto gain and time constants can be computed from a turning test by using nonlinear least-squares curve fitting, for instance. Solving the ODE

$$T\dot{r} + r = K\delta \tag{15.47}$$

for a step input $\delta = \delta_0 =$ constant yields

$$r(t) = e^{-\frac{t}{T}} r(0) + (1 - e^{-\frac{t}{T}}) K \delta_0 \tag{15.48}$$

where K and T are unknowns. The Matlab MSS toolbox script `ExKT.m` *fits this model to a simulated step response of the model* `mariner.m`*, which is a nonlinear model of the Mariner class vessel.*

The results for a step $\delta_0 = 5^\circ$ and $U = 7.7 \text{ m s}^{-1} = 15$ knots, are (see Figure 15.7)

$$K = 0.09 \text{ s}^{-1}, \qquad T = 22.6 \text{ s}. \tag{15.49}$$

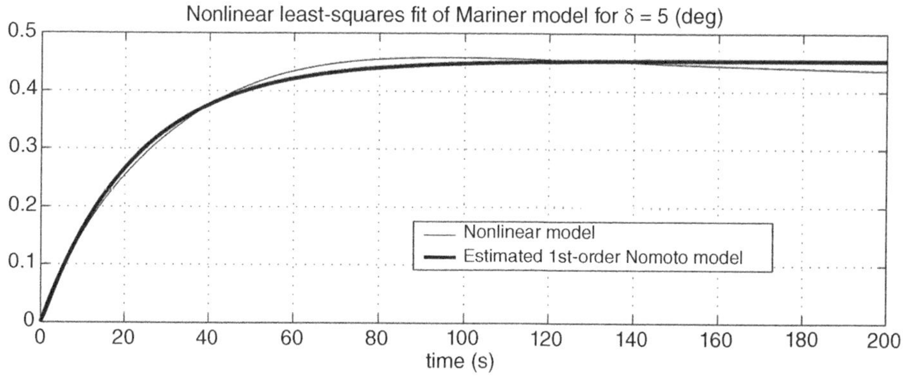

Figure 15.7 Plot showing the estimated linear model and the nonlinear Mariner model for a step $\delta = \delta_0 = 5°$.

The Norrbin measure of maneuverability becomes

$$P = \frac{1}{2}\frac{K'}{T'} = \frac{1}{2}\frac{K}{T}\left(\frac{L}{U}\right)^2 = \frac{1}{2}\left(\frac{0.09}{22.6}\right)\left(\frac{160.9}{7.7}\right)^2 = 0.87 \tag{15.50}$$

which guarantees good maneuverability since $P > 0.3$. *The turning circle is shown in Figure 15.6, indicating that the steady-state turning radius is* 711 m.

MATLAB

```
% ExKT Script for computation of Nomoto gain and time constants
% using nonlinear least squares.
N = 2000;                  % number of samples
h = 0.1;                   % sample time

xout = zeros(N,2);
x = zeros(7,1);
delta_R = 5 * (pi/180);                  % rudder angle step input

for i=1:N
    xout(i,:) = [(i-1)*h, x(3)];
    xdot = mariner(x,delta_R);           % nonlinear Mariner model
    x = euler2(xdot,x,h);                % Euler integration
end

% time-series
tdata = xout(:,1);
rdata = xout(:,2)*180/pi;

% nonlinear least-squares parametrization: x(1)=1/T and x(2)=K
```

```
x0 = [0.01 0.1]'
F = inline('exp(-tdata*x(1))*0 + ...
            x(2)*(1-exp(-tdata*x(1)))*5','x','tdata')
x = lsqcurvefit(F,x0, tdata, rdata);

% plots
plot(tdata,rdata,'g',tdata,exp(-tdata*x(1))*0 + ...
     x(2)*(1-exp(-tdata*x(1)))*5,'r'),grid
title('NLS fit of Mariner model for \delta = 5 (deg)')
xlabel('time (s)')
legend('Nonlinear model','Estimated 1st-order Nomoto model')
```

Kempf's Zigzag Maneuver

The zigzag test was first proposed by Kempf (1932). Comprehensive test results of 75 freighters are published in Kempf (1944). The zigzag time response (see Figures 15.8

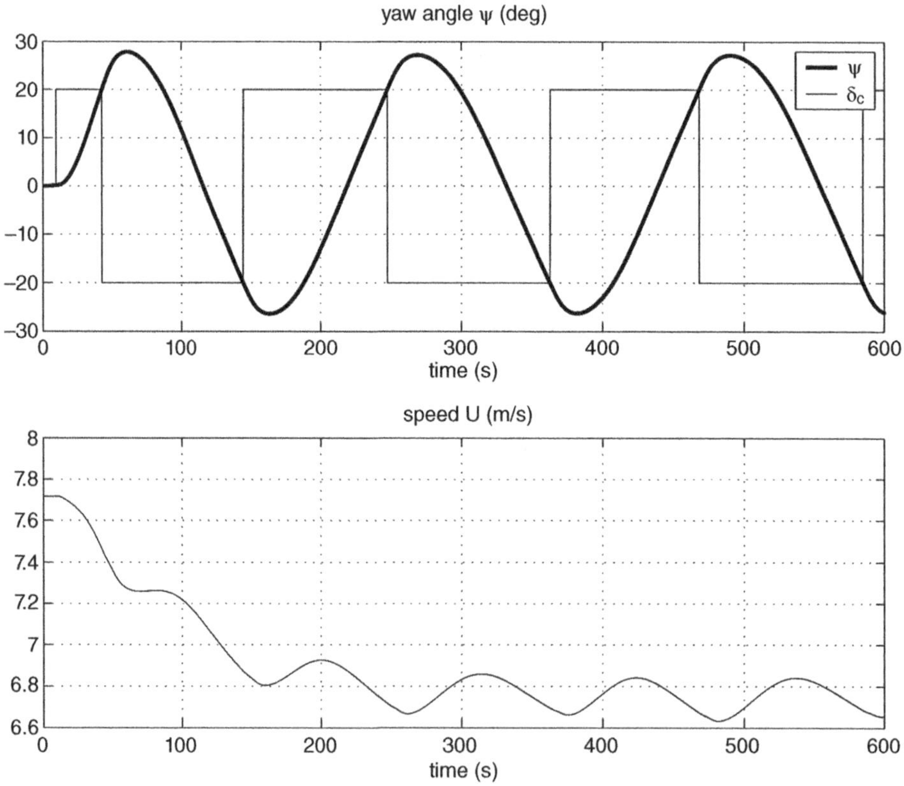

Figure 15.8 A 20°–20° maneuver for the Mariner class vessel.

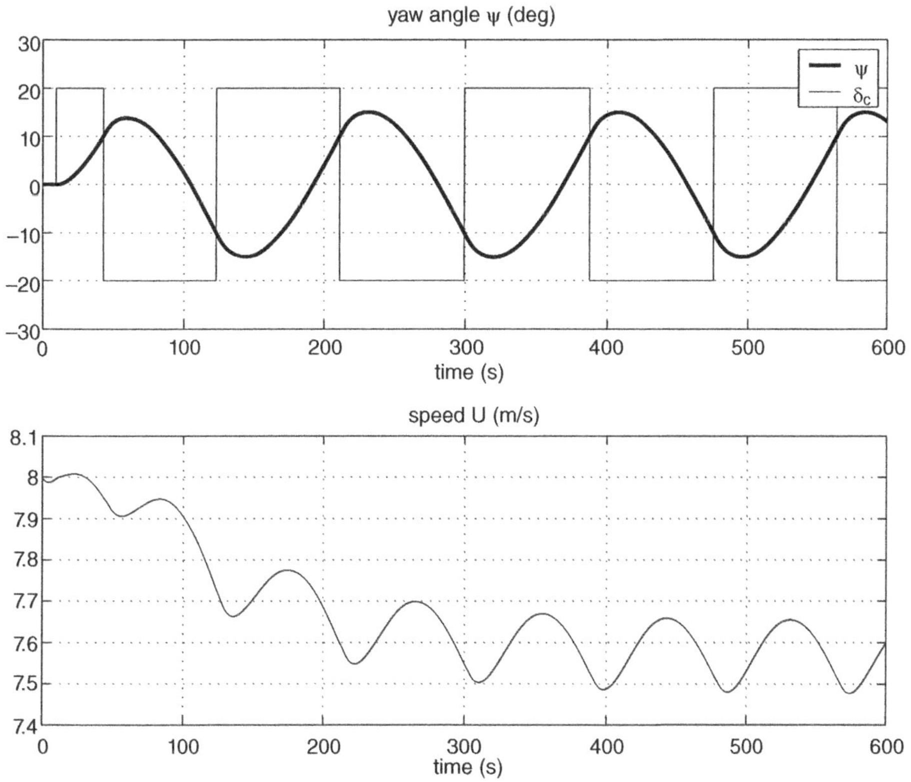

Figure 15.9 A 20°–10° maneuver for the container ship.

and 15.9) is obtained by moving the rudder 20° to starboard from an initially straight course. The rudder setting is kept constant until the heading is changed 20°, and then the rudder is reversed 20° to port. Again, this rudder setting is maintained until the ship's heading has reached 20° in the opposite direction. This process continues until a total of five rudder step responses have been completed. This test is usually referred to as a 20°–20° maneuver; the first angle refers to the actual rudder settings while the second angle denotes how much the heading angle should change before the rudder is reversed.

The zigzag maneuver was standardized by the International Towing Tank Conference (ITTC) in 1963. For larger ships, ITTC has recommended the use of a 10°–10° or a 20°–10° maneuver to reduce the time and waterspace required.

The only apparatus required to perform the test is a compass and a stopwatch. Alternatively, a computer interfaced for real-time logging of compass data can be used. The results from the zigzag maneuver can be used to compare the maneuvering properties of different ships. Maneuvering trials are also used in the design process since it is possible to test scale models in towing tanks to see how well they perform. In addition, maneuvering characteristics can be computed using hull parameters and by performing computer simulations based on seakeeping and maneuvering models.

Example 15.3 (Zigzag Maneuvering Trials)
Both the Mariner class vessel (`mariner.m`) and the container ship (`container.m`) are simulated for a 20°–20° and a 20°–10° zigzag maneuver, respectively, by using the Matlab script `ExZigZag.m`. The simulation results for the two vessels are shown in Figures 15.8 and 15.9.

MATLAB

```
t_final = 600;              % final simulation time (s)
t_rudderexecute = 10;       % time rudder is executed (s)
h = 0.1;                    % sampling time (s)

% 20-20 zigzag maneuver for the Mariner class cargo ship
% cruise speed U0 = 7.7 m/s (see mariner.m)
x = zeros(7,1); % x = [u v r x y psi delta]' (initial values)
ui = 0;           % delta_c = 0 for time t < t_rudderexecute
[t,u,v,r,x,y,psi,U] = ...
    zigzag('mariner',x,ui,t_final,t_rudderexecute,h,[20,20]);

% 20-10 zigzag maneuver for a container ship
% cruise speed 8.0 m/s (see container.m)
x = [8.0 0 0 0 0 0 0 0 0 70]';  % x = [u v r x y psi delta n]'
delta_c = 0;      % delta_c = 0 for time t < t_rudderexecute
n_c = 80;         % n_c = propeller revolution in rpm
ui = [delta_c, n_c];
[t,u,v,r,x,y,psi,U] = ...
    zigzag('container',x,ui,t_final,t_rudderexecute,h,[20,10]);
```

Pull-out Maneuver

In 1969 Roy Burcher proposed a simple test procedure to determine whether a ship is straight-line stable or not. This test is referred to as the pull-out maneuver (12th ITTC 1969). The pull-out maneuver involves a pair of maneuvers in which a rudder angle of approximately 20° is applied and returned to zero after steady turning has been attained. Both a port and a starboard turn should be performed.

During the test the ship's rate of turn must be measured or at least calculated by numerical derivation of the measured compass heading. If the ship is straight-line stable the rate of turn will decay to the same value for both the starboard and port turns (see Figure 15.10). The ship is unstable if the steady rate of turn from the port and starboard turns differ (see Figure15.11). The difference between these two steady rates of turn corresponds exactly to the height of Dieudonné's spiral loop.

Example 15.4 (Pull-out Maneuver for a Stable and an Unstable Ship)
Both the Mariner class vessel (`mariner.m`) and the Esso Osaka tanker (`tanker.m`) are simulated under a pullout maneuver by using the Matlab script `ExPullout.m`.

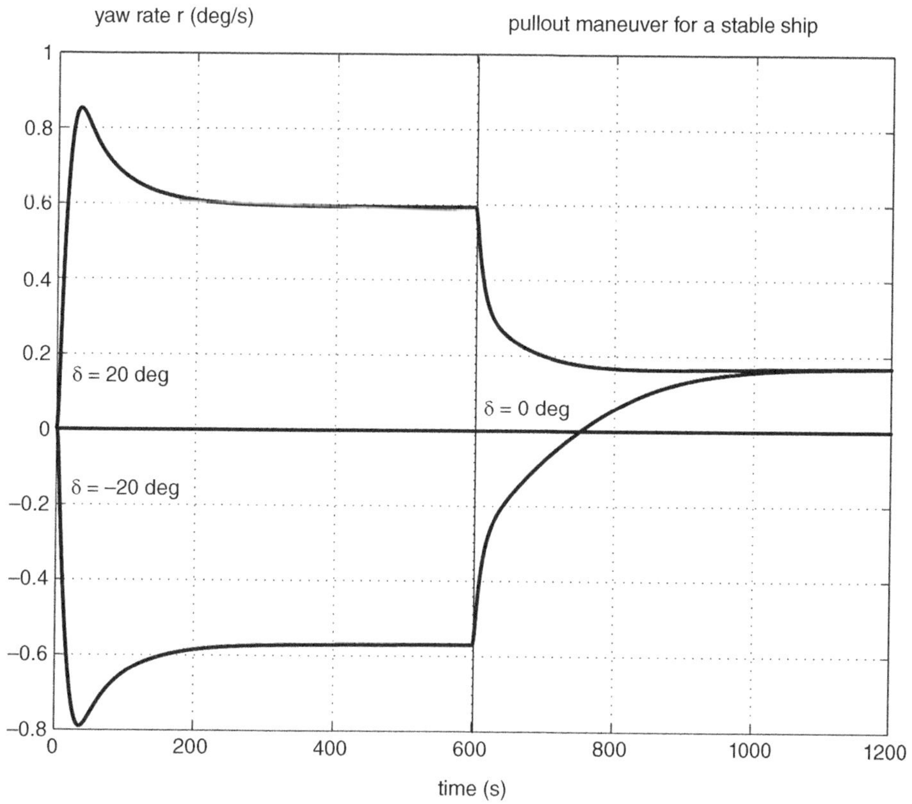

Figure 15.10 Pull-out maneuver for the Mariner class vessels. Notice that the positive and negative curves meet for the stable ship.

MATLAB

```
delta_c = 20*pi/180;       % rudder angle for maneuver (rad)
h = 0.1;                   % sampling time (s)

% Mariner class cargo ship, speed U0 = 7.7 m/s (see mariner.m)
x = zeros(7,1); % x = [ u v r x y psi delta ]' (initial values)
ui = delta_c;    % ui = delta_c
[t,r1,r2] = pullout('mariner',x,ui,h);

% The Esso Osaka tanker (see tanker.m)
n = 80;
U = 8.23;
x = [ U 0 0 0 0 0 0 n ]';    % x = [ u v r x y psi delta n ]'
n_c = 80;                    % n_c = propeller revolution in rpm
```

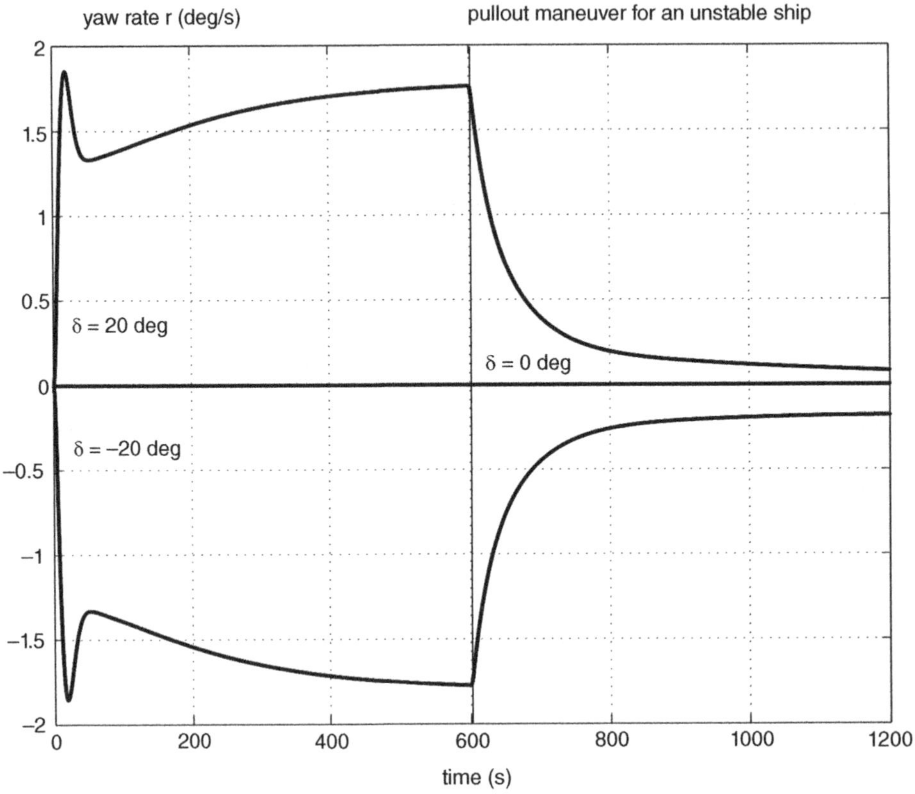

Figure 15.11 Pullout maneuver for the Esso Osaka tanker. Note that the positive and negative curves do not meet.

```
depth = 200;                    % water depth
ui = [delta_c, n_c, depth];
[t,r1,r2] = pullout('tanker',x,ui,h);
```

The results are shown in Figures 15.10 and 15.11 where the curves meet for the stable ship (Mariner class vessel) while there is an offset between the curves for the unstable model of the Esso Osaka tanker.

Dieudonné's Spiral Maneuver

The direct spiral test was published first in 1949–1950 by the French scientist Jean Dieudonné. An English translation is found in Dieudonné (1953). The direct spiral maneuver is used to check straight-line stability. As seen from Figure 15.12, the

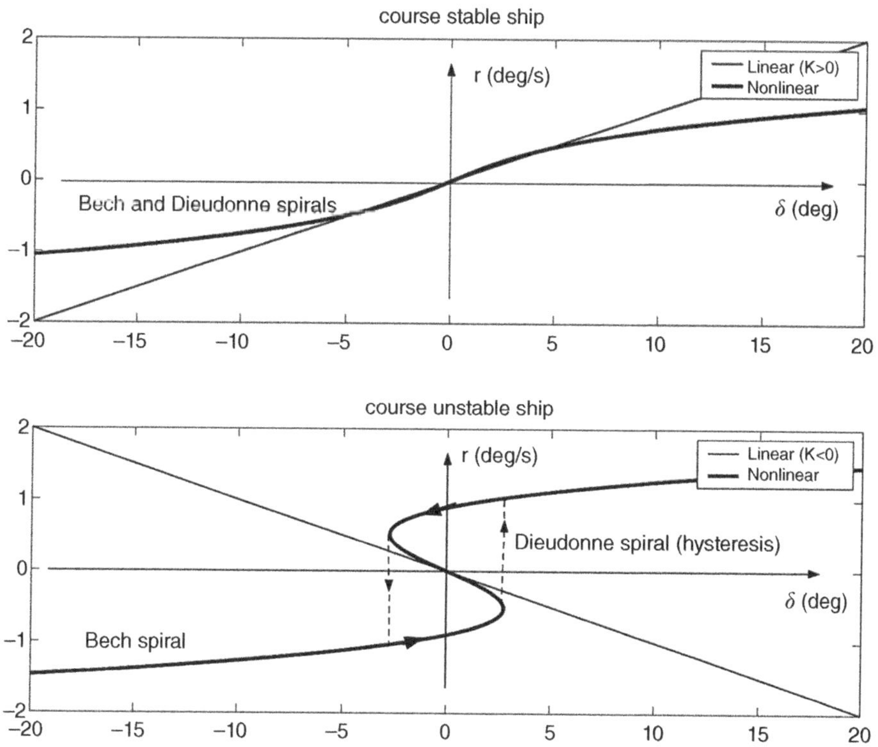

Figure 15.12 An r–δ diagram showing the Dieudonne and Bech spirals for both a stable and course-unstable ship. Notice the hysteresis loop in the Dieudonne spiral for the unstable ship.

maneuver also gives an indication of the degree of stability and the range of validity of the linear theory.

To perform the test the ship should initially be held on a straight course. The rudder angle is then put to 25° starboard and held until a steady yawing rate is obtained. After this the rudder angle is decreased in steps of 5° and again held until constant yawing rates are obtained for all the rudder angles. The procedure is performed for all rudder angles between 25° starboard and 25° port. In the range around zero rudder angle the step of 5° rudder should be reduced to obtain more precise values.

The results are plotted in an r–δ diagram, as shown in Figure 15.12. It should be noted that the spiral maneuver should be performed in still air and calm water to obtain the best results.

For straight-line unstable ships it is recommended to use Bech's reverse spiral maneuver.

Bech's Reverse Spiral Maneuver

For stable ships both Dieudonné's direct and Bech's reverse spiral tests can be used. For unstable ships within the limits indicated by the pull-out maneuver Bech's reverse spiral should be applied. The reverse spiral test was first published by Mogens Bech in 1966 and later in 1968 (Bech 1968). Since then the reverse spiral test has been quite popular, because of the simplicity and reliability of the method. The reverse spiral test is also less time-consuming than Dieudonné's spiral test.

By observing that the ship steering characteristic is nonlinear outside a limited area, Bech (1968) suggested that one describes the *mean* value of the required rudder deflection δ_{ss} to steer the ship at a constant rate of turn r_{ss} as a nonlinear function

$$\delta_{ss} = H_B(r_{ss}) \tag{15.51}$$

where $H_B(r_{ss})$ is a nonlinear function (7.31) describing the maneuvering characteristic. This can be understood by considering Nomoto's second-order model (see Section (7.2.3))

$$T_1 T_2 \ddot{r} + (T_1 + T_2)\dot{r} + K H_B(r) = K(\delta + T_3 \dot{\delta}) \tag{15.52}$$

where the linear term r has been replaced with a function $H_B(r)$. Assuming that $r = r_{ss}$ is constant in the steady state, that is $\ddot{r} = \dot{r} = \dot{\delta} = 0$, directly gives (15.51).

This implies that the r–δ curve will be a single-valued (one-to-one) function of r for both the stable and unstable ship (see Figure 15.12). If the conventional spiral test is applied to an unstable ship a hysteresis loop will be observed.

The full-scale test is performed by measuring the necessary rudder action required to bring the ship into a desired rate of turn. For an unstable ship this implies that the rudder angle will oscillate about a mean rudder angle. The amplitude of the rudder oscillations should be kept to a minimum. After some time a *balance condition* is reached and both the mean rudder angle and rate of turn can be calculated. Care should be taken for large ships since they will require some more time to converge to their "balance condition".

15.2 Autopilot Design Using Successive Loop Closure

The marine craft equations of motion will in general be nonlinear and coupled. In many cases the couplings between the modes can be neglected. The key assumption is that certain modes of a marine craft can be controlled by using a linear model (transfer function) and treat couplings and unmodeled dynamics as disturbances, which can be compensated for by using integral action. Section 15.2 covers successive loop closure designs for marine craft with emphasis placed on

- Heading autopilot design
- Path-following control
- Depth autopilot design.

The feedback control loops will be designed using pole-placement algorithms.

15.2.1 Successive Loop Closure

The basic idea behind successive loop closure is to close several simple feedback loops in succession around the open-loop plant dynamics rather than designing a single (presumably more complicated) control system. To illustrate how this approach can be applied,

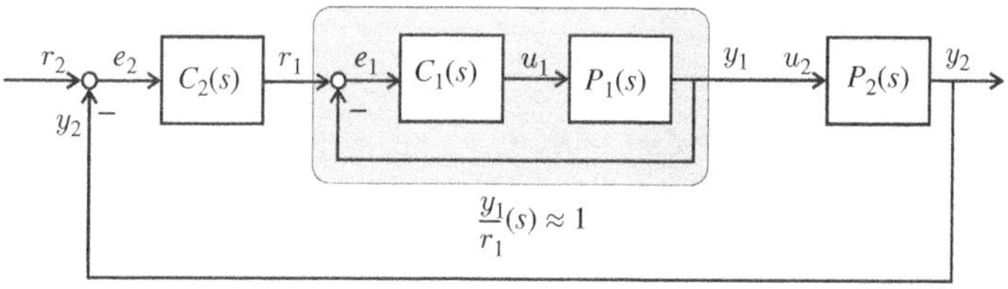

Figure 15.13 Successive loop closure design with inner loop modeled as a unity gain.

consider the cascade of two transfer functions $P(s) = P_1(s)P_2(s)$ as shown in Figure 15.13 for which

$$y_2 = P_2(s)u_2, \qquad u_2 = y_1 \tag{15.53}$$

$$y_1 = P_1(s)u_1. \tag{15.54}$$

If the inner-loop control law is chosen as

$$u_1 = C_1(s)e_1, \qquad e_1 = r_1 - y_1 \tag{15.55}$$

where $C_1(s)$ is the transfer function of the control system, the closed-loop system becomes

$$\frac{y_1}{r_1}(s) = \frac{P_1(s)C_1(s)}{1 + P_1(s)C_1(s)} \overset{C_1(s) \gg 1}{\approx} 1. \tag{15.56}$$

Since $C_1(s)$ guarantees that $y_1 \approx r_1$ up to the bandwidth ω_1, the outer-loop control law $u_2 = C_2(s)e_2$ can be designed independently of the inner loop to achieve

$$\frac{y_2}{r_2}(s) \approx \frac{P_2(s)C_2(s)}{1 + P_2(s)C_2(s)}. \tag{15.57}$$

To ensure proper stability of the successive-loop-closure design, it is essential that there is sufficient bandwidth separation between the inner and outer feedback loops. Adequate separation can be achieved by choosing the bandwidth ω_2 of the outer loop smaller than the bandwidth ω_1 of the inner loop, for instance

$$\omega_2 = \frac{1}{W}\,\omega_1 \tag{15.58}$$

where adequate values for the successive loop bandwidth factor W are 5–10.

Pole-placement Algorithm

The pole-placement algorithm is based on Beard and McLain (2012). Consider the second-order open-loop system

$$\frac{y}{u}(s) = \frac{b_0}{s^2 + a_1 s + a_0} \tag{15.59}$$

and the PD control law

$$u = -K_\mathrm{p} e_y - K_\mathrm{d}\,\dot{y}, \qquad e_y = y - y_\mathrm{d} \tag{15.60}$$

where K_p and K_d are feedback gains to be decided, and y_d is the reference signal. The closed-loop transfer function is

$$\frac{y}{y_\mathrm{d}}(s) = \frac{K_\mathrm{p} b_0}{s^2 + (a_1 + b_0 K_\mathrm{d})s + (a_0 + b_0 K_\mathrm{p})}. \tag{15.61}$$

Let the maximum amplitude of the control $u = -K_\mathrm{p} e_y - K_\mathrm{d}\,\dot{y}$ be saturated for $\dot{y} = 0$ such that

$$K_\mathrm{p} = \frac{u^{\max}}{e_y^{\max}}. \tag{15.62}$$

The maximum control effort $u^{\max}$ and step error $e_y^{\max}$ from a step input of nominal size are user inputs. The desired closed-loop poles can be chosen as the roots of

$$s^2 + 2\zeta\omega_\mathrm{n} s + \omega_\mathrm{n}^2 = 0. \tag{15.63}$$

Equating the coefficients corresponding to the $2\zeta\omega_\mathrm{n}$ and ω_n^2 gives

$$\omega_\mathrm{n} = \sqrt{a_0 + b_0 \frac{u^{\max}}{e_y^{\max}}} \tag{15.64}$$

$$K_\mathrm{d} = \frac{2\zeta\omega_\mathrm{n} - a_1}{b_0}. \tag{15.65}$$

15.2.2 Case Study: Heading Autopilot for Marine Craft

Consider the Nomoto model (see Section 7.2)

$$r = \frac{K}{Ts + 1}\delta_\mathrm{R} + d_\mathrm{r} \tag{15.66}$$

where r is the yaw rate, δ_R is the stern rudder angle, and $K > 0$ and $T > 0$ are the Nomoto gain and time constants, respectively. The influence of ocean currents, wave drift and wind are modeled as an unknown drift term d_r. Since $\dot{\psi} = r$, the yaw angle transfer function becomes

$$\psi = \frac{K}{s(Ts + 1)}\delta_\mathrm{R} + \frac{1}{s} d_\mathrm{r}. \tag{15.67}$$

Following the approach of Section 15.2.1, we see that $a_0 = 0$, $a_1 = 1/T$ and $b_0 = K/T$. The PD control law is chosen as

$$\delta_{\mathrm{R}} = -K_{\mathrm{p}_\psi} e_\psi - K_{\mathrm{d}_\psi} \dot{\psi}, \qquad e_\psi = \psi - \psi_{\mathrm{d}} \tag{15.68}$$

Next, we specify the maximum rudder angle $\delta_R^{\max}$ (typically 30°) and leave the step error $e_\psi^{\max}$ as a tunable parameter. This gives the formulae

$$K_{\mathrm{p}_\psi} = \frac{\delta_R^{\max}}{e_\psi^{\max}}, \qquad K_{\mathrm{d}_\psi} = \frac{2\zeta_\psi \omega_\psi T - 1}{K}, \qquad \omega_\psi = \sqrt{\frac{K}{T}\frac{\delta_R^{\max}}{e_\psi^{\max}}} \tag{15.69}$$

Note that the value for ω_ψ is the upper limit of the bandwidth of the closed-loop system. Typical values for marine craft are 0.01–0.1. Hence, it is important to treat $e_\psi^{\max}$ as a tunable parameter such that ω_ψ does not exceed its maximum value limited by the rudder servo.

A marine craft will be exposed to wind, waves and ocean currents. Consequently, it is necessary to modify the PD control law (15.68) to include integral action. From a practical point of view, it is also important to map the error signal e_ψ to $[-\pi, \pi)$ using the SSA to avoid large steps in the control input (see Section 13.4.3). The resulting control law is the PID controller

$$\delta_{\mathrm{R}} = -K_{\mathrm{p}_\psi}\,\mathrm{ssa}(e_\psi) - K_{\mathrm{d}_\psi}\dot{\psi} - K_{\mathrm{i}_\psi}\int_0^t \mathrm{ssa}(e_\psi(\tau))\mathrm{d}\tau \tag{15.70}$$

where ssa : $\mathbb{R} \to [-\pi, \pi)$ maps the unconstrained angle $e_\psi \in \mathbb{R}$ to the smallest difference between the angles (see Section 13.4.3). A *rule-of-thumb* is to choose

$$K_{\mathrm{i}_\psi} = \frac{\omega_\psi}{10} K_{\mathrm{p}_\psi}. \tag{15.71}$$

This guarantees that the integrator is 10 times slower than the natural frequency ω_ψ.

15.2.3 Case Study: Path-following Control System for Marine Craft

Often it is of primary importance to steer a ship, a submersible or a rig along a desired *path* with a prescribed *speed*. The path is usually defined in terms of *waypoints* and straight-line segments going through the waypoints (see Section 12.3).

Assume that the path-tangential coordinate system $\{p\}$ has its origin at the waypoint (x_i^n, y_i^n) as shown in Figure 12.9. The cross-track error can expressed in $\{p\}$ using (2.126). This gives

$$\dot{y}_{\mathrm{e}}^p = U \sin(\chi_{\mathrm{p}}) \approx U\chi_{\mathrm{p}} \tag{15.72}$$

where χ_{p} is assumed to be small for straight-line path following. Since GNSS measures the course angle χ it is necessary to use the formula for π_{p}, given by (12.54), to compute the state $\chi_{\mathrm{p}} = \chi - \pi_{\mathrm{p}}$. The Nomoto model (15.67) can be expressed in terms of the course angle $\chi = \psi + \beta_{\mathrm{c}}$ according to

$$\chi = \frac{K}{s(Ts+1)}\delta_{\mathrm{R}} + d_{\chi} \tag{15.73}$$

where

$$d_{\chi} = \frac{1}{s}d_{\mathrm{r}} + \beta_{\mathrm{c}}. \tag{15.74}$$

This suggests that the inner-loop course controller should be chosen as

$$\delta_{\mathrm{R}} = -K_{\mathrm{p}_{\chi}}\mathrm{ssa}(e_{\chi}) - K_{\mathrm{d}_{\chi}}\dot{\chi}, \qquad e_{\chi} = \chi - \chi_{\mathrm{d}} \tag{15.75}$$

where the gains are given by the pole-placement formulae (15.62), (15.64) and (15.65)

$$K_{\mathrm{p}_{\chi}} = \frac{\delta_{R}^{\max}}{e_{\chi}^{\max}}, \qquad K_{\mathrm{d}_{\chi}} = \frac{2\zeta_{\chi}\omega_{\chi}T - 1}{K}, \qquad \omega_{\chi} = \sqrt{\frac{K}{T}\frac{\delta_{R}^{\max}}{e_{\chi}^{\max}}}. \tag{15.76}$$

The closed-loop system is

$$\chi = \underbrace{\frac{\frac{KK_{\mathrm{p}_{\chi}}}{T}}{s^2 + \left(\frac{1}{T} + \frac{KK_{\mathrm{d}_{\chi}}}{T}\right)s + \frac{KK_{\mathrm{p}_{\chi}}}{T}}}_{\approx 1}\chi_{\mathrm{d}} + \underbrace{\frac{1}{s^2 + \left(\frac{1}{T} + \frac{KK_{\mathrm{d}_{\chi}}}{T}\right)s + \frac{KK_{\mathrm{p}_{\chi}}}{T}}}_{\approx \frac{T}{KK_{\mathrm{p}_{\chi}}}}d_{\chi}. \tag{15.77}$$

The PD controller will not cancel the drift term d_{χ} in the inner loop but the steady-state gain, $T/(KK_{\mathrm{p}_{\chi}})$, is reduced by increasing $K_{\mathrm{p}_{\chi}} > 1$. Integral action is instead included in the outer loop in order to avoid two integral controllers in cascade. Additional integral action will reduce the phase margin of the system, which again can lead to stability problems. In our case, the control objective is $y_{\mathrm{e}}^{p} = 0$, while small errors in e_{χ} is acceptable as long y_{e}^{p} converges to zero. Since $\chi \approx \chi_{\mathrm{d}}$, the cross-track error becomes (see Figure 15.14)

$$y_{\mathrm{e}}^{p} = \frac{U}{s}\chi_{\mathrm{p}} \approx \frac{U}{s}(\chi_{\mathrm{d}} - \pi_{\mathrm{p}}). \tag{15.78}$$

Hence, the outer loop path-following controller can be chosen of PI type

$$\chi_{\mathrm{d}} = \pi_{\mathrm{p}} - K_{\mathrm{p}_y}e_y - K_{\mathrm{i}_y}\int_0^t e_y(\tau)\mathrm{d}\tau, \qquad e_y = y_{\mathrm{e}}^{p} - 0. \tag{15.79}$$

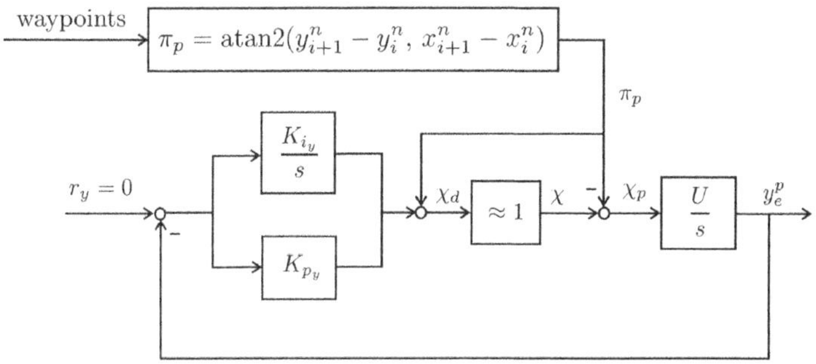

Figure 15.14 Path-following control using successive loop closure.

The closed-loop transfer function is

$$\frac{y^p_e}{r_y}(s) = \frac{K_{p_y}Us + K_{i_y}U}{s^2 + K_{p_y}Us + K_{i_y}U}. \tag{15.80}$$

The control objective is to regulate e_y to zero. Pole-placement gives

$$K_{p_y} = \frac{2\zeta_y\omega_y}{U}, \qquad K_{i_y} = \frac{\omega_y^2}{U}. \tag{15.81}$$

Bandwidth separation is achieved by choosing

$$\omega_y = \frac{1}{W_y}\omega_\chi \tag{15.82}$$

with W_y between 5 and 10.

15.2.4 Case Study: Diving Autopilot for Underwater Vehicles

Recall from Section 8.2.1 that the decoupled pitch equation for an underwater vehicle is

$$(I_y - M_{\dot{q}})\ddot{\theta} - M_q\dot{\theta} + \mathrm{BG}_z W\theta = M_{\delta_S}\delta_S \tag{15.83}$$

where δ_S is the dive plane deflection. The dive planes are located in the stern of the vehicle (see Section 9.7). The associated depth equation is

$$\dot{z}^n = -U\theta + d_z \tag{15.84}$$

where d_z is an unknown disturbance term due to ocean currents and unmodeled dynamics. The pitch angle transfer function is

$$\frac{\theta}{\delta_S}(s) = \frac{\frac{M_{\delta_S}}{I_y - M_{\dot{q}}}}{s^2 + \frac{-M_q}{I_y - M_{\dot{q}}}s + \frac{\mathrm{BG}_z W}{I_y - M_{\dot{q}}}}. \tag{15.85}$$

Following the approach of Section 15.2.1, we note that $a_0 = \mathrm{BG}_z W/(I_y - M_{\dot{q}})$, $a_1 = -M_q/(I_y - M_{\dot{q}})$ and $b_0 = M_{\delta_S}/(I_y - M_{\dot{q}})$. This suggests that the inner-loop pitch controller should be chosen as

$$\delta_S = -K_{p_\theta}\, \mathrm{ssa}(e_\theta) - K_{d_\theta}\, \dot{\theta}, \qquad e_\theta = \theta - \theta_d \tag{15.86}$$

where ssa: $\mathbb{R} \rightarrow [-\pi, \pi)$ maps the unconstrained angle $e_\theta \in \mathbb{R}$ to the smallest difference between the angles (see Section 13.4.3). The gains are given by the pole-placement formulae (15.62), (15.64) and (15.65)

$$K_{p_\theta} = \frac{\delta_S^{\max}}{e_\theta^{\max}}, \qquad K_{d_\theta} = \frac{2\zeta_\theta\omega_\theta(I_y - M_{\dot{q}}) + M_q}{M_{\delta_S}} \tag{15.87}$$

where

$$\omega_\theta = \sqrt{\frac{\mathrm{BG}_z W}{I_y - M_{\dot{q}}} + \frac{M_{\delta_S}}{I_y - M_{\dot{q}}}\frac{\delta_S^{\max}}{e_\theta^{\max}}}. \tag{15.88}$$

The closed-loop system is

$$\frac{\theta}{\theta_d}(s) = \frac{\frac{M_{\delta_S} K_{p_\theta}}{I_y - M_{\dot{q}}}}{s^2 + \left(\frac{-M_q}{I_y - M_{\dot{q}}} + \frac{M_{\delta_S} K_{d_\theta}}{I_y - M_{\dot{q}}}\right)s + \left(\frac{\mathrm{BG}_z W}{I_y - M_{\dot{q}}} + \frac{M_{\delta_S} K_{p_\theta}}{I_y - M_{\dot{q}}}\right)} \overset{s=0}{=} \underbrace{\frac{M_{\delta_S} K_{p_\theta}}{\mathrm{BG}_z W + M_{\delta_S} K_{p_\theta}}}_{K_{DC}} \tag{15.89}$$

where K_{DC} is the non-unity DC gain in pitch. Since $\theta = K_{DC}\, \theta_d$ in steady state, the outer loop is governed by the dynamics

$$z = -\frac{U}{s}\theta \approx -\frac{K_{DC}\, U}{s}\theta_d. \tag{15.90}$$

The steady-state offset introduced by the DC gain can be compensated for by adding integral action in the outer loop using the PI control law (see Figure 15.15)

$$\theta_d = -K_{p_z} e_z - K_{i_z}\int_0^t e_z(\tau)\mathrm{d}\tau, \qquad e_z = z^n - z_d^n. \tag{15.91}$$

This gives the closed-loop transfer function

$$\frac{z^n}{z_d^n}(s) = \frac{K_{DC}K_{p_z}Us + K_{DC}K_{i_z}U}{s^2 + K_{DC}K_{p_z}Us + K_{DC}K_{i_z}U}. \tag{15.92}$$

Pole-placement results in the following formulae

$$K_{p_z} = \frac{2\zeta_z\omega_z}{K_{DC}\, U}, \qquad K_{i_z} = \frac{\omega_z^2}{K_{DC}\, U}. \tag{15.93}$$

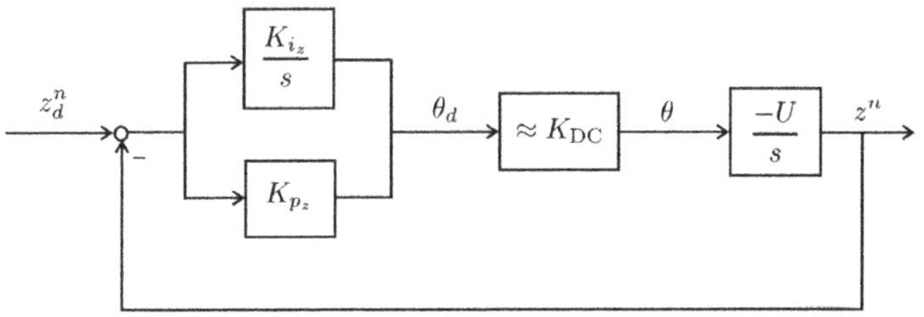

Figure 15.15 Depth control using successive loop closure.

Finally, bandwidth separation is achieved by choosing

$$\omega_z = \frac{1}{W_z}\omega_\theta \tag{15.94}$$

with W_z between 5 and 10.

15.3 PID Pole-Placement Algorithms

This section discusses nonlinear PID control for SISO and MIMO systems using pole-placement algorithms. Lyapunov stability analyses for optimal shaping of the kinetic and potential energies of a marine craft are the main tool to derive the pole-placement algorithms. The PID controllers can be applied to a large number of industrial motion control systems including dynamic positioning systems, autopilots for steering and diving as well as path-following control systems.

15.3.1 Linear Mass–Damper–Spring Systems

Consider the following two equivalent systems

$$m\ddot{x} + d\dot{x} + kx = 0 \tag{15.95}$$

$$\ddot{x} + 2\zeta\omega_n\dot{x} + \omega_n^2 x = 0. \tag{15.96}$$

From (15.95) and (15.96) it follows that

$$2\zeta\omega_n = \frac{d}{m}, \qquad \omega_n^2 = \frac{k}{m} \tag{15.97}$$

where ω_n is the natural frequency and ζ is the relative damping factor.

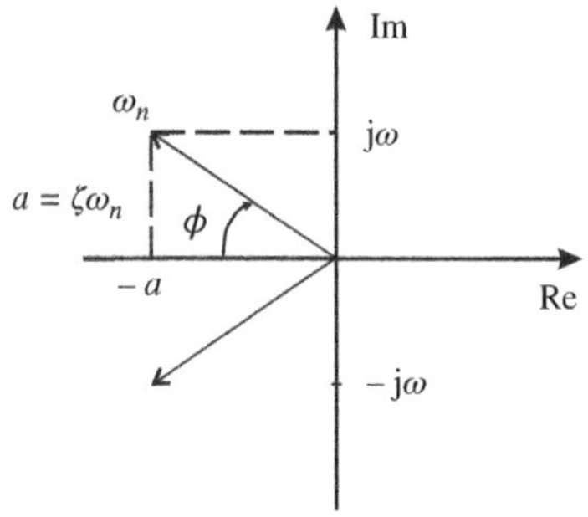

Figure 15.16 Graphical illustration of natural frequency ω_n, frequency of the damped system ω and absolute damping factor a.

Damped Oscillator

For the damped system $d > 0$, the frequency of the oscillation will be smaller than the natural frequency. This can be explained by considering the eigenvalues of the mass–damper–spring system (15.96),

$$\lambda_{1,2} = -\underbrace{\zeta\omega_n}_{a} \pm j\omega. \tag{15.98}$$

From Figure 15.16 it is seen that

$$a^2 + \omega^2 = \omega_n^2, \qquad \zeta = \frac{a}{\omega_n} = \cos(\phi). \tag{15.99}$$

MATLAB

The step responses in Figure 15.17 is computed using (see `ExMDS.m`):

```
wn = 1;            % natural frequency

subplot(211)
t = 0:0.01:20;
z = 0.5; sys = tf([wn*wn],[1 2*z*wn wn*wn]); step(sys,t)
hold on
z = 1.0; sys = tf([wn*wn],[1 2*z*wn wn*wn]); step(sys,t)
z = 2.0; sys = tf([wn*wn],[1 2*z*wn wn*wn]); step(sys,t)
hold off

subplot(212)
t = 0:0.01:50;
z = 0.1; sys = tf([wn*wn],[1 2*z*wn wn*wn]); step(sys,t)
hold on
sys = tf([wn*wn],[1 0 wn*wn]); step(sys,t)
hold off
```

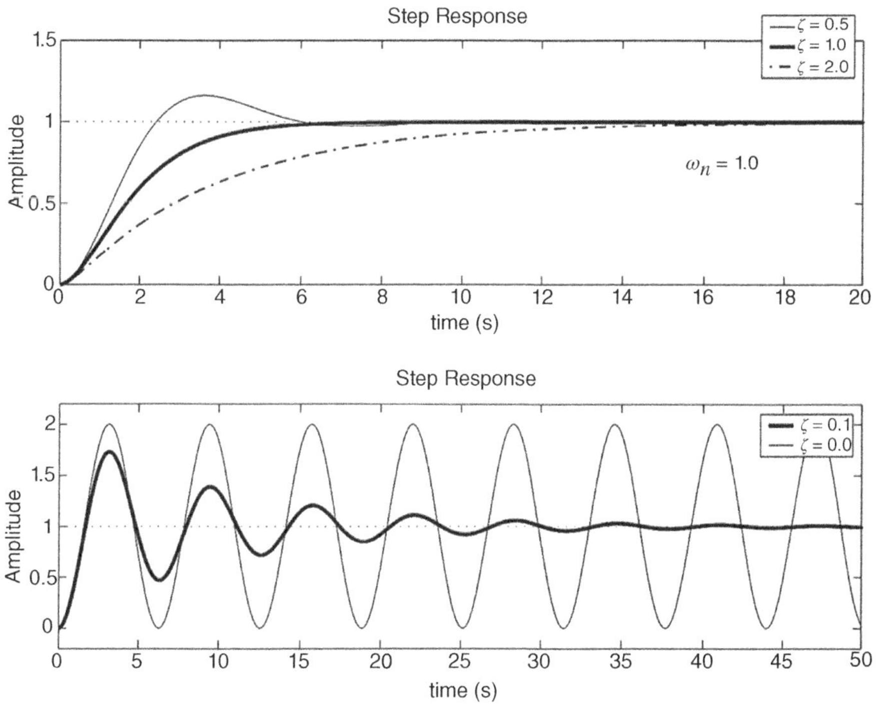

Figure 15.17 The upper plot shows a mass–damper–spring system for different relative damping ratios. The lower plot shows the undamped oscillator together with a damped oscillator. The plots are generated by `ExMDS.m`.

where a is the absolute damping factor and ω is the frequency of oscillation (damped system). The undamped oscillator is obtained by choosing $a = 0$. It is convenient to set

$$\omega = r\omega_{\rm n} \tag{15.100}$$

where r is a reduction factor denoting the ratio between the natural frequency $\omega_{\rm n}$ and the frequency ω of the linearly damped system. For marine craft a reduction of 0.5 % in the natural frequency is common (Faltinsen 1990). Hence,

$$r = 1 - \frac{0.5}{100} = 0.995. \tag{15.101}$$

From (15.99) and (15.100) it is seen that

$$a^2 + (r\omega_{\rm n})^2 = \omega_{\rm n}^2 \tag{15.102}$$

$$\Downarrow$$

$$a = \underbrace{\sqrt{1 - r^2}}_{\zeta}\ \omega_{\rm n}. \tag{15.103}$$

For $r = 0.995$ we obtain $\zeta = 0.1$, which is quite typical for a ship with bilge keels while the heave and pitch motions usually are more damped, for instance $\zeta = 0.2$.

Heave, roll and pitch damping: For the *mass–damper–spring* system (15.97) we obtain the following formula for linear damping

$$d = 2\zeta\sqrt{km}, \qquad \zeta = \sqrt{1-r^2} \tag{15.104}$$

where r is the design parameter. This formula is quite useful to determine the linear damping in *heave, roll* and *pitch*. The mass parameter m and spring coefficient k are easily obtained by other methods; see Chapters 3–5.

Surge, sway and yaw damping: Damping in *surge, sway* and *yaw*, however, cannot be determined by formula (15.104) since $k = 0$ in a pure *mass–damper* system. Linear damping for such a system can be found by specifying the time constant $T > 0$ corresponding to

$$m\ddot{x} + d\dot{x} = \tau \tag{15.105}$$

which for the design parameter $T = m/d$ is equivalent to

$$T\ddot{x} + \dot{x} = \frac{1}{d}\tau. \tag{15.106}$$

This yields the following design formula for linear damping

$$d = \frac{m}{T}. \tag{15.107}$$

Example 15.5 (Linear Damping in Roll and Pitch for Submarines)
Consider the linear pitch equation (8.35)

$$(I_y - M_{\dot{q}})\ddot{\theta} - M_q\dot{\theta} + \mathrm{BG}_z W\theta = \tau_5.$$

Hence, the linear damping coefficient can be computed from (15.104), which gives

$$-M_q = 2\sqrt{1-r_\theta^2}\sqrt{\mathrm{BG}_z W(I_y - M_{\dot{q}})} > 0.$$

Here $M_{\dot{q}}$, W and BG_z are assumed to be known and $r_\theta > 0$ is a design parameter. For roll (see (8.43)) a similar expression is obtained for $r_\phi > 0$

$$-K_p = 2\sqrt{1-r_\phi^2}\sqrt{\mathrm{BG}_z W(I_x - K_{\dot{p}})} > 0.$$

Example 15.6 (Linear Damping in Yaw for Ships and Underwater Vehicles)
Consider the Nomoto model (see Section 6.5.5)

$$(I_z - N_{\dot{r}})\dot{r} - N_r r = N_\delta \delta. \tag{15.108}$$

Assume that the moment of inertia $I_z - N_{\dot{r}}$ is known. The linear damping coefficient N_r can be estimated by specifying the time constant $T_\psi > 0$ in yaw. From (15.107) the unknown *hydrodynamic derivative is obtained as*

$$-N_r = \frac{I_z - N_{\dot{r}}}{T_\psi} \tag{15.109}$$

15.3.2 SISO Linear PID Control

Consider the control law

$$\tau = \underbrace{kx_d}_{\substack{\text{reference}\\ \text{feedforward}}} - \underbrace{\left(K_p\tilde{x} + K_d\dot{x} + K_i\int_0^t \tilde{x}(\tau)\mathrm{d}\tau\right)}_{\text{PID controller}} \tag{15.110}$$

with gains $K_p > 0$, $K_d > 0$ and $K_i > 0$ and tracking error $\tilde{x} = x - x_d$ where x_d = constant. The control law can be applied to the mass–damper–spring system (15.95). For simplicity, assume that $K_i = 0$ such that the closed-loop system becomes

$$m\ddot{x} + (d + K_d)\dot{x} + (k + K_p)\tilde{x} = 0 \tag{15.111}$$

such that

$$\zeta = \frac{d + K_d}{2m\omega_n}, \qquad \omega_n = \sqrt{\frac{k + K_p}{m}}. \tag{15.112}$$

Pole placement of the mass–damper–spring system suggests that K_p and K_d can be computed by specifying ω_n and ζ in (15.112). Solving for K_p and K_d, yields

$$K_p = m\omega_n^2 - k \tag{15.113}$$

$$K_d = 2\zeta\omega_n m - d. \tag{15.114}$$

Consequently, the closed-loop transfer function becomes

$$\frac{x}{x_d}(s) = \frac{\omega_n^2}{s^2 + 2\zeta\omega_n s + \omega_n^2}. \tag{15.115}$$

Integral action $K_i > 0$ should be added to compensate for constant and slowly varying disturbances. Let the PID controller (15.110) be written as

$$\tau = \underbrace{kx_d}_{\substack{\text{reference}\\ \text{feedforward}}} \underbrace{- K_p\left(1 + T_d s + \frac{1}{T_i s}\right)\tilde{x}}_{\text{PID}} \tag{15.116}$$

where $T_d = K_d/K_p$ and $T_i = K_p/K_i$ are the derivative and integral time constants, respectively. A *rule-of-thumb* is to choose

$$\frac{1}{T_i} \approx \frac{\omega_n}{10} \tag{15.117}$$

which guarantees that the integrator is 10 times slower than the natural frequency ω_n. This gives the following formula for the integral gain

$$K_i = \frac{\omega_n}{10} K_p = \frac{\omega_n}{10}(m\omega_n^2 - k). \tag{15.118}$$

The natural frequency ω_n can be related to the system bandwidth ω_b by using the following definition.

Definition 15.1 (Control Bandwidth)
The control bandwidth of the system $y = h(s)u$ in Figure 15.18 with negative unity feedback is defined as the frequency ω_b at which the loop transfer function $l(s) = h(s) \cdot 1$ satisfies

$$|l(j\omega)|_{\omega=\omega_b} = \frac{\sqrt{2}}{2}$$

or equivalently

$$20 \log |l(j\omega)|_{\omega=\omega_b} = -3 \text{ dB}.$$

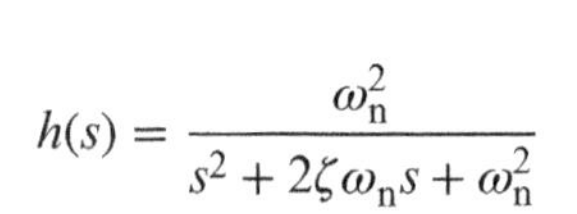

Figure 15.18 Closed-loop feedback system.

From this definition it can be shown that the control bandwidth of a second-order system

$$h(s) = \frac{\omega_n^2}{s^2 + 2\zeta\omega_n s + \omega_n^2} \tag{15.119}$$

with negative unity feedback is (see Figure 15.18)

$$\omega_b = \omega_n \sqrt{1 - 2\zeta^2 + \sqrt{4\zeta^4 - 4\zeta^2 + 2}}. \tag{15.120}$$

For a critically damped system, $\zeta = 1.0$, this expression reduces to

$$\omega_b = \omega_n \sqrt{\sqrt{2} - 1} \approx 0.64\ \omega_n. \tag{15.121}$$

Algorithm 15.1 summarizes the pole-placement algorithm for SISO motion control systems.

Algorithm 15.1 (SISO PID Pole-Placement Algorithm)

1. Specify the bandwidth:	$\omega_b > 0$
2. Specify the relative damping ratio:	$\zeta > 0$
3. Compute the natural frequency:	$\omega_n = \frac{1}{\sqrt{1-2\zeta^2+\sqrt{4\zeta^4-4\zeta^2+2}}}\,\omega_b$
4. Compute the P gain:	$K_p = m\omega_n^2 - k$
5. Compute the D gain:	$K_d = 2\zeta\omega_n m - d$
6. Compute the I gain:	$K_i = \frac{\omega_n}{10} K_p$

Example 15.7 (Ship Autopilot Design)

Consider the Nomoto model (Nomoto et al. *1957)*

$$T\ddot{\psi} + \dot{\psi} = K\delta \tag{15.122}$$

where ψ is the yaw angle and δ is the rudder angle. From (15.95) it is seen that

$$m = \frac{T}{K}, \quad d = \frac{1}{K}, \quad k = 0. \tag{15.123}$$

Application of Algorithm 15.1 gives

$$w_n = \frac{1}{\sqrt{1 - 2\zeta^2 + \sqrt{4\zeta^4 - 4\zeta^2 + 2}}}\,\omega_b \tag{15.124}$$

and the PID controller gains

$$K_p = \omega_n^2 \frac{T}{K}, \qquad K_d = \frac{2\zeta\omega_n T - 1}{K}, \qquad K_i = \omega_n^3 \frac{T}{10K} \tag{15.125}$$

where ω_b and ζ are the design parameters.

15.3.3 MIMO Nonlinear PID Control

Algorithm 15.1 can be generalized to nonlinear MIMO systems by exploiting the kinematic equations of motion in the design. Consider the nonlinear model

$$\dot{\boldsymbol{\eta}} = \boldsymbol{J}_\Theta(\boldsymbol{\eta})\boldsymbol{\nu} \tag{15.126}$$

$$\boldsymbol{M}\dot{\boldsymbol{\nu}} + \boldsymbol{C}(\boldsymbol{\nu})\boldsymbol{\nu} + \boldsymbol{D}(\boldsymbol{\nu})\boldsymbol{\nu} + \boldsymbol{g}(\boldsymbol{\eta}) = \boldsymbol{\tau} \tag{15.127}$$

where $\boldsymbol{\eta}$ and $\boldsymbol{\nu}$ are assumed to be measured. The nonlinear controller is chosen as

$$\boldsymbol{\tau} = \boldsymbol{g}(\boldsymbol{\eta}) + \boldsymbol{J}_\Theta^\top(\boldsymbol{\eta})\boldsymbol{\tau}_{\text{PID}} \tag{15.128}$$

where $\boldsymbol{g}(\boldsymbol{\eta})$ compensates for gravity and

$$\boldsymbol{\tau}_{\text{PID}} = -\boldsymbol{K}_p\tilde{\boldsymbol{\eta}} - \boldsymbol{K}_{\text{d}}\dot{\boldsymbol{\eta}} - \boldsymbol{K}_{\text{i}}\int_0^t \tilde{\boldsymbol{\eta}}(\tau)\text{d}\tau. \tag{15.129}$$

For simplicity, assume that $\boldsymbol{K}_{\text{i}} = \boldsymbol{0}$. This yields the closed-loop system

$$\boldsymbol{M}\dot{\boldsymbol{\nu}} + [\boldsymbol{C}(\boldsymbol{\nu}) + \boldsymbol{D}(\boldsymbol{\nu}) + \boldsymbol{K}_{\text{d}}^*(\boldsymbol{\eta})]\boldsymbol{\nu} + \boldsymbol{J}_\Theta^\top(\boldsymbol{\eta})\boldsymbol{K}_p\tilde{\boldsymbol{\eta}} = \boldsymbol{0} \tag{15.130}$$

where $\tilde{\boldsymbol{\eta}} = \boldsymbol{\eta} - \boldsymbol{\eta}_{\text{d}}$,

$$\boldsymbol{K}_{\text{d}}^*(\boldsymbol{\eta}) = \boldsymbol{J}_\Theta^\top(\boldsymbol{\eta})\boldsymbol{K}_{\text{d}}\boldsymbol{J}_\Theta(\boldsymbol{\eta}). \tag{15.131}$$

Lyapunov Stability Analysis

In the Lyapunov stability analysis it is assumed that $\dot{\boldsymbol{\eta}}_{\text{d}} = \boldsymbol{0}$, that is regulation of $\boldsymbol{\eta}$ to $\boldsymbol{\eta}_{\text{d}} =$ constant. A Lyapunov function candidate for the system (15.130) is

$$V = \underbrace{\frac{1}{2}\boldsymbol{\nu}^\top\boldsymbol{M}\boldsymbol{\nu}}_{\substack{\text{kinetic}\\ \text{energy}}} + \underbrace{\frac{1}{2}\tilde{\boldsymbol{\eta}}^\top\boldsymbol{K}_p\tilde{\boldsymbol{\eta}}}_{\substack{\text{potential}\\ \text{energy}}} \tag{15.132}$$

where $\boldsymbol{K}_p = \boldsymbol{K}_p^\top > 0$. Time differentiation of (15.132) along the trajectories of $\boldsymbol{\nu}$ and $\tilde{\boldsymbol{\eta}}$ yields

$$\begin{aligned}\dot{V} &= \boldsymbol{\nu}^\top\boldsymbol{M}\dot{\boldsymbol{\nu}} + \dot{\boldsymbol{\eta}}^\top\boldsymbol{K}_p\tilde{\boldsymbol{\eta}} \\ &= \boldsymbol{\nu}^\top[\boldsymbol{M}\dot{\boldsymbol{\nu}} + \boldsymbol{J}_\Theta^\top(\boldsymbol{\eta})\boldsymbol{K}_p\tilde{\boldsymbol{\eta}}]\end{aligned} \tag{15.133}$$

where $\dot{\tilde{\boldsymbol{\eta}}} = \dot{\boldsymbol{\eta}} - \dot{\boldsymbol{\eta}}_{\text{d}} = \dot{\boldsymbol{\eta}}$ and $\dot{\boldsymbol{\eta}}^\top = \boldsymbol{\nu}^\top\boldsymbol{J}_\Theta^\top(\boldsymbol{\eta})$. Substituting (15.130) into (15.133) yields

$$\begin{aligned}\dot{V} &= -\boldsymbol{\nu}^\top[\boldsymbol{C}(\boldsymbol{\nu}) + \boldsymbol{D}(\boldsymbol{\nu}) + \boldsymbol{K}_{\text{d}}^*(\boldsymbol{\eta})]\boldsymbol{\nu} \\ &= -\boldsymbol{\nu}^\top[\boldsymbol{D}(\boldsymbol{\nu}) + \boldsymbol{K}_{\text{d}}^*(\boldsymbol{\eta})]\boldsymbol{\nu}\end{aligned} \tag{15.134}$$

since $\boldsymbol{\nu}^\top\boldsymbol{C}(\boldsymbol{\nu})\boldsymbol{\nu} = 0$ for all $\boldsymbol{\nu}$; see Property 8.2 in Section 8.1.

Krasovskii–LaSalle's Theorem A.2 in Appendix A.1 can be used to prove that the system (15.126) – (15.127) with nonlinear PD control ($\boldsymbol{K}_{\mathrm{i}} = \boldsymbol{0}$) is *globally asymptotically stable* (GAS) if $\boldsymbol{J}_{\Theta}(\boldsymbol{\eta})$ is defined for all $\boldsymbol{\eta}$ (no representation singularity). Moreover, the trajectories will converge to the set Ω found from

$$\dot{V}(\boldsymbol{x}) = -\boldsymbol{\nu}^{\top}[\boldsymbol{D}(\boldsymbol{\nu}) + \boldsymbol{K}_{\mathrm{d}}^{*}(\boldsymbol{\eta})]\boldsymbol{\nu} \equiv 0 \tag{15.135}$$

which is true for $\boldsymbol{\nu} = \boldsymbol{0}$. Therefore,

$$\Omega = \{(\tilde{\boldsymbol{\eta}}, \boldsymbol{\nu}) : \boldsymbol{\nu} = \boldsymbol{0})\}. \tag{15.136}$$

Now, $\boldsymbol{\nu} \equiv \boldsymbol{0}$ implies that $\boldsymbol{M}\dot{\boldsymbol{\nu}} = -\boldsymbol{J}_{\Theta}^{\top}(\boldsymbol{\eta})\boldsymbol{K}_{p}\tilde{\boldsymbol{\eta}}$, which is non-zero as long as $\tilde{\boldsymbol{\eta}} \neq \boldsymbol{0}$. Hence, the system cannot get "stuck" at an equilibrium point value other than $\tilde{\boldsymbol{\eta}} = \boldsymbol{0}$. Since the equilibrium point $(\tilde{\boldsymbol{\eta}}, \boldsymbol{\nu}) = (\boldsymbol{0}, \boldsymbol{0})$ is the largest invariant set M in Ω, the equilibrium point is GAS according to Theorem A.2.

If integral action is included with $\boldsymbol{K}_{\mathrm{i}} > 0$ (PID control), it is possible to prove local asymptotic stability. This result is well known from robotics (Arimoto and Miyazaki 1984). A drift term can also be removed by using parameter adaptation (Fossen *et al.* 2001).

MIMO Pole-placement Algorithm

The MIMO pole-placement algorithm is based on the closed-loop system

$$\boldsymbol{M}\dot{\boldsymbol{\nu}} + [\boldsymbol{C}(\boldsymbol{\nu}) + \boldsymbol{D}(\boldsymbol{\nu}) + \boldsymbol{K}_{\mathrm{d}}^{*}(\boldsymbol{\eta})]\boldsymbol{\nu} + \boldsymbol{J}_{\Theta}^{\top}(\boldsymbol{\eta})\boldsymbol{K}_{p}\tilde{\boldsymbol{\eta}} = \boldsymbol{0}. \tag{15.137}$$

Assume that $\dot{\boldsymbol{J}}_{\Theta}(\boldsymbol{\eta}) \approx \boldsymbol{0}$ and premultiply (15.137) with $\boldsymbol{J}_{\Theta}^{-\top}$ such that

$$\boldsymbol{J}_{\Theta}^{-\top}(\boldsymbol{\eta})\boldsymbol{M}\boldsymbol{J}_{\Theta}^{-1}(\boldsymbol{\eta})\ddot{\boldsymbol{\eta}} + \boldsymbol{J}_{\Theta}^{-\top}(\boldsymbol{\eta})[\boldsymbol{C}(\boldsymbol{\nu}) + \boldsymbol{D}(\boldsymbol{\nu})]\boldsymbol{J}_{\Theta}^{-1}(\boldsymbol{\eta})\dot{\boldsymbol{\eta}} + \boldsymbol{K}_{\mathrm{d}}\dot{\boldsymbol{\eta}} + \boldsymbol{K}_{p}\tilde{\boldsymbol{\eta}} = \boldsymbol{0}. \tag{15.138}$$

The next step is to compare this expression to

$$\boldsymbol{M}^{*}(\boldsymbol{\eta})[\ddot{\boldsymbol{\eta}} + 2\boldsymbol{Z}\boldsymbol{\Omega}_{\mathrm{n}}\dot{\boldsymbol{\eta}} + \boldsymbol{\Omega}_{\mathrm{n}}^{2}\tilde{\boldsymbol{\eta}}] = \boldsymbol{0} \tag{15.139}$$

where $\boldsymbol{Z}$ and $\boldsymbol{\Omega}_{\mathrm{n}}$ are two design matrices to be specified later, and

$$\boldsymbol{M}^{*}(\boldsymbol{\eta}) := \boldsymbol{J}_{\Theta}^{-\top}(\boldsymbol{\eta})\boldsymbol{M}\boldsymbol{J}_{\Theta}^{-1}(\boldsymbol{\eta}). \tag{15.140}$$

Equating the above equations gives the gain requirements

$$2\boldsymbol{M}^{*}(\boldsymbol{\eta})\boldsymbol{Z}\boldsymbol{\Omega}_{\mathrm{n}} := \boldsymbol{J}_{\Theta}^{-\top}(\boldsymbol{\eta})[\boldsymbol{C}(\boldsymbol{\nu}) + \boldsymbol{D}(\boldsymbol{\nu})]\boldsymbol{J}_{\Theta}^{-1}(\boldsymbol{\eta}) + \boldsymbol{K}_{\mathrm{d}} \tag{15.141}$$

$$\boldsymbol{M}^{*}(\boldsymbol{\eta})\boldsymbol{\Omega}_{\mathrm{n}}^{2} := \boldsymbol{K}_{p} \tag{15.142}$$

which can be solved for $\boldsymbol{K}_{p}$ and $\boldsymbol{K}_{\mathrm{d}}$. Algorithm 15.2 summarizes the pole-placement algorithm for MIMO nonlinear systems.

Algorithm 15.2 (MIMO Nonlinear PID Pole-placement Algorithm)

1. *Specify a matrix of bandwidths:*	$\boldsymbol{\Omega}_b = \text{diag}\{\omega_{b_1}, \omega_{b_2}, ..., \omega_{b_6}\} > 0$
2. *Specify a matrix of relative damping ratios:*	$\boldsymbol{Z} = \text{diag}\{\zeta_1, \zeta_2, ..., \zeta_6\} > 0$
3. *Compute the natural frequencies:*	$\omega_{n_i} = \frac{1}{\sqrt{1-2\zeta_i^2+\sqrt{4\zeta_i^4-4\zeta_i^2+2}}}\omega_{b_i}$
	$\boldsymbol{\Omega}_n = \text{diag}\{\omega_{n_1}, \omega_{n_2}, ..., \omega_{n_6}\} > 0$
4. *Compute the P gain matrix:*	$\boldsymbol{K}_p = \boldsymbol{M}^*(\boldsymbol{\eta})\boldsymbol{\Omega}_{\rm n}^2$
5. *Compute the D gain matrix:* *(compensation of* $\boldsymbol{C}(\boldsymbol{\nu}) + \boldsymbol{D}(\boldsymbol{\nu})$ *is optional)*	$\boldsymbol{K}_{\rm d} = 2\boldsymbol{M}^*(\boldsymbol{\eta})\boldsymbol{Z}\boldsymbol{\Omega}_{\rm n} - \boldsymbol{J}_\Theta^{-\top}(\boldsymbol{\eta})[\boldsymbol{C}(\boldsymbol{\nu}) + \boldsymbol{D}(\boldsymbol{\nu})]\boldsymbol{J}_\Theta^{-1}(\boldsymbol{\eta})$
6. *Compute the I gain matrix:*	$\boldsymbol{K}_{\rm i} = \frac{1}{10}\boldsymbol{K}_p\boldsymbol{\Omega}_n$

Note that $\boldsymbol{K}_p$ is time varying in Algorithm 15.2, which violates (15.132). However, the Lyapunov proof is based on the assumption that $\dot{\boldsymbol{J}}_\Theta(\boldsymbol{\eta}) \approx \boldsymbol{0}$ and thus $\dot{\boldsymbol{K}}_p \approx \boldsymbol{0}$.

15.3.4 Case Study: Heading Autopilot for Marine Craft

The principal blocks of a heading angle autopilot system, shown in Figure 15.19, are

Control system: The feedback control system provides the necessary commands to track the desired yaw angle $\psi_{\rm d}$. The output is the yaw moment $\tau_{\rm N}$.

Control allocation: This module distributes the output from the feedback control system, usually the yaw moment $\tau_{\rm N}$, to the actuators (rudders and in some cases propellers and thrusters) in an optimal manner (see Section 11.2). For single-screw ships the controller yaw moment $\tau_{\rm N}$ will simply be a function of the rudder command $\delta_{\rm c}$.

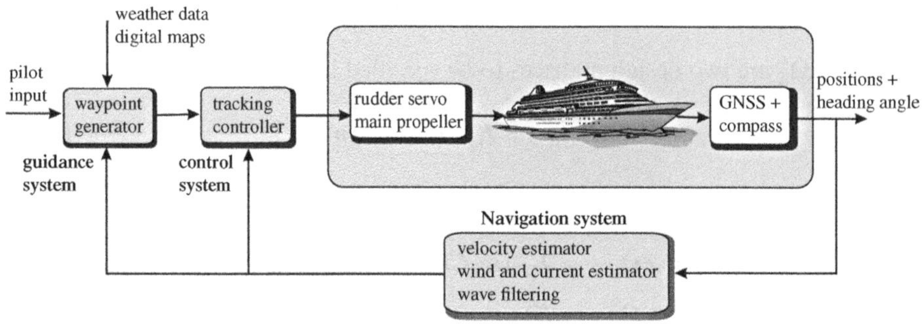

Figure 15.19 Block diagram of a heading autopilot system.

Reference model: The autopilot reference model computes smooth trajectories ψ_d, r_d and $\dot{r}_d$ needed for *course-changing* maneuvers. *Course-keeping* is the special case then $\psi_d =$ constant and $r_d = \dot{r}_d = 0$ (see Section 12.1.1).

Compass and yaw rate sensor: The compass measures the yaw angle ψ which is needed for feedback (see Section 13.1). In some cases a yaw rate sensor is available for yaw rate feedback, that is feedback from $r = \dot{\psi}$ (see Section 14.1).

State estimator with wave filter: In its simplest form the first-order wave-induced motion components ψ_w and r_w are filtered out from the raw measurements $y_1 = \psi + \psi_w$ and $y_2 = r + r_w$, and consequently prevented from entering the feedback loop. This is known as *wave filtering*, where the output of the filter is the LF motion components ψ and r. This is necessary to avoid excessive rudder action. In cases where y_2 is not measured the wave filter must be constructed as a state estimator so that r can be estimated from the yaw angle measurement y_1; see Sections 13.4.5, 13.5.2 and 13.5.3.

Wind feedforward: In cases where a wind sensor is available for *wind speed* and *direction*, a wind model (see Section 10.1) can be used for wind feedforward. This is often advantageous since the integral action term in the PID controller does not have to integrate up the wind disturbance term. However, an accurate model of the wind force and moment as a function of ship speed and wind direction is needed to implement wind feedforward.

The different autopilot blocks of Figure 15.19 needed to implement a PID control law based on the Nomoto model will now be discussed.

Autopilot Reference Model

A modern autopilot must have both course-keeping and turning capabilities. This can be obtained in one design by using a reference model to compute the desired states ψ_d, r_d and $\dot{r}_d$ needed for turning. A third-order filter for this purpose was derived in Section 12.1.1. Consider

$$\frac{\psi_d}{\psi_r}(s) = \frac{\omega_n^3}{(s+\omega_n)(s^2 + 2\zeta\omega_n s + \omega_n^2)} \tag{15.143}$$

where the reference ψ_r is the operator input, ζ is the relative damping ratio and ω_n is the natural frequency. Note that

$$\lim_{t\to\infty} \psi_d(t) = \psi_r \tag{15.144}$$

and that $\dot{\psi}_d$ and $\ddot{\psi}_d$ are smooth and bounded for steps in ψ_r. This is the main motivation for choosing a third-order model since a second-order model will result in steps in $\ddot{\psi}_d$ for steps in the commands ψ_r.

In many cases it is advantageous to limit the desired yaw rate $|r_d| \leq r^{max}$ during turning. This can be done by including a saturating element in the reference model (Van Amerongen 1982, 1984). The yaw acceleration $a_d = \ddot{\psi}_d$ can also be limited such that $|a_d| \leq a^{max}$

by using a second saturating element. The resulting state-space model including velocity and acceleration saturating elements becomes

$$\dot{\psi}_{\rm d} = {\rm sat}(r_{\rm d}) \tag{15.145}$$

$$\dot{r}_{\rm d} = {\rm sat}(a_{\rm d}) \tag{15.146}$$

$$\dot{a}_{\rm d} = -(2\zeta + 1)\omega_{\rm n}\, {\rm sat}(a_{\rm d}) - (2\zeta + 1)\omega_{\rm n}^2\, {\rm sat}(r_{\rm d}) + \omega_{\rm n}^3(\psi_{\rm r} - \psi_{\rm d}). \tag{15.147}$$

The saturating element is defined as

$$ {\rm sat}(x) := \begin{cases} {\rm sgn}(x)x^{\max} & \text{if } |x| \geq x^{\max} \\ x & \text{else} \end{cases}. \tag{15.148}$$

The autopilot reference model has been simulated in Matlab with yaw rate limitation $r^{\max} = 1.0^\circ\ s^{-1}$, acceleration limit $a^{\max} = 0.5^\circ\ s^{-1}$ and command $\psi_{\rm r} = 30^\circ$. The results are shown in Figure 15.20. Note that the unlimited (linear) case yields unsatisfactorily high values for $r_{\rm d}$.

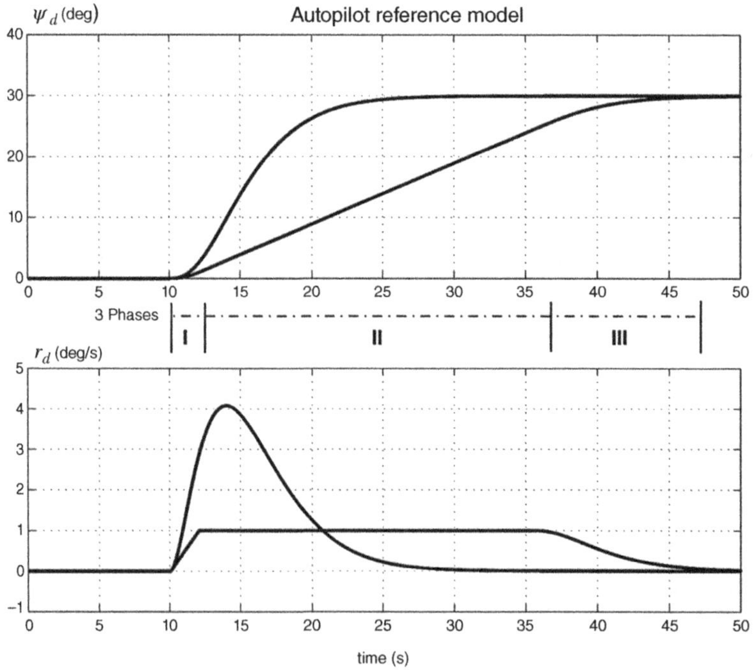

Figure 15.20 The plots show the effect of including a rate limiter of $r_{\max} = 1^\circ\ s^{-1}$ in a third-order reference model for heading. Notice that $r_{\rm d}$ becomes very high in the linear case while $\psi_{\rm d}$ looks satisfactory in both cases.

The main motivation for using a rate-limiting element in the reference model is that the course-changing maneuver will be described by three phases (positive turn):

I: Start of turn, acceleration ($r_d > 0$ and $0 < \dot{r}_d \leq a^{max}$)
II: Steady turning ($r_d = r^{max}$ and $\dot{r}_d = 0$)
III: End of turn, deceleration ($r_d > 0$ and $-a^{max} \leq \dot{r}_d < 0$).

For a negative turn the signs of the turning rate and acceleration must be changed. The three phases are advantageous when performing a large change in course. The effect of a saturating element and nonlinear damping in a reference model are also demonstrated in Example 12.1 in Section 12.1.1.

A more sophisticated method for generating heading reference signals could be the expense of a more complicated software algorithm to be implemented in real time.

PID Control with Wind Feedforward

Consider the Nomoto model of Section 7.2 in the following form

$$(I_z - N_{\dot{r}})\dot{r} - N_r r = \tau_{wind} + \tau_N \tag{15.149}$$

where τ_{wind} is an optional input for wind feedforward and τ_N is the yaw moment generated by the controller. The constants $m = I_z - N_{\dot{r}}$, $d = -N_r$ and

$$T = \frac{m}{d} = \frac{I_z - N_{\dot{r}}}{-N_r} \tag{15.150}$$

are introduced such that

$$\dot{r} + \frac{1}{T}r = \frac{1}{m}(\tau_{wind} + \tau_N). \tag{15.151}$$

The yaw moment can be generated by a single rudder

$$\tau_N = N_\delta \delta \tag{15.152}$$

or several actuators $u_i (i = 1, \ldots, r)$ satisfying

$$\tau_N = \boldsymbol{b}^\top \boldsymbol{u}, \qquad \boldsymbol{u} = [u_1, \ldots, u_r]^\top. \tag{15.153}$$

Assume that both ψ and r are measured by using a compass and an ARS. A PID controller for heading control is (see Section 15.3.2)

$$\tau_{\rm N}(s) = -\hat{\tau}_{\rm wind} + \tau_{\rm FF}(s) \underbrace{-K_{\rm p}\left(1 + T_{\rm d}s + \frac{1}{T_{\rm i}s}\right)\tilde{\psi}(s)}_{\tau_{\rm PID}} \tag{15.154}$$

where $\tau_{\rm FF}$ is a feedforward term to be decided and $\tilde{\psi} = \psi - \psi_{\rm d}$ is the heading error. The wind feedforward term $\hat{\tau}_{\rm wind}$ is an estimate of the wind moment $\tau_{\rm wind}$ using wind coefficients and an anemometer measuring wind speed $V_{\rm w}$ and direction $\beta_{V_{\rm w}}$. An estimate of the wind yaw moment can be computed according to (see Section 10.1)

$$\hat{\tau}_{\rm wind} = \frac{1}{2}\rho_a V_{\rm rw}^2 C_N(\gamma_{\rm rw}) A_{\rm Lw} L_{\rm oa} \tag{15.155}$$

where the relative wind speed and angle of attack are

$$V_{\rm rw} = \sqrt{u_{\rm rw}^2 + v_{\rm rw}^2} \tag{15.156}$$

$$\gamma_{\rm rw} = -{\rm atan2}(v_{\rm rw}, u_{\rm rw}). \tag{15.157}$$

The relative velocities depend on the heading angle ψ, wind direction $\beta_{V_{\rm w}}$ and wind speed $V_{\rm w}$ according to

$$u_{\rm rw} = u - u_{\rm w} = u - V_{\rm w}\cos(\beta_{V_{\rm w}} - \psi) \tag{15.158}$$

$$v_{\rm rw} = v - v_{\rm w} = v - V_{\rm w}\sin(\beta_{V_{\rm w}} - \psi). \tag{15.159}$$

When wind feedforward is implemented it is important that the wind measurements are low-pass filtered to avoid rapid changes in heading command. Wind feedforward is an optional term since the integrator in the PID control law can compensate for a slowly varying wind moment as well. The main difference will be the response time. In general, wind feedforward will be much faster than integral action since the integrator needs several minutes to remove a large wind component during the start-up of an autopilot system. Integral action works fairly well during fixed heading (stationkeeping and transit) while in a maneuvering situation large course deviations might be expected. Consequently, it is advantageous to implement wind feedforward to reduce the loads on the integrator and obtain maximum performance during start-up and in maneuvering situations. However, if the wind coefficients are poorly known, the closed-loop system can be destabilized by the wind feedforward term so care must be taken.

A continuous-time representation of the controller (15.154) is

$$\tau_{\rm N} = -\hat{\tau}_{\rm wind} + \tau_{\rm FF} - K_{\rm p}\,{\rm ssa}(\tilde{\psi}) - \underbrace{K_{\rm p}T_{\rm d}}_{K_{\rm d}}\,\tilde{r} - \underbrace{\frac{K_{\rm p}}{T_{\rm i}}}_{K_{\rm i}}\int_0^t {\rm ssa}(\tilde{\psi}(\tau))\,{\rm d}\tau \tag{15.160}$$

where $\tilde{r} := r - r_{\mathrm{d}}$ and $\tilde{\psi} := \psi - \psi_{\mathrm{d}}$. The operator ssa: $\mathbb{R} \rightarrow [-\pi, \pi)$ maps the unconstrained angle $\tilde{\psi} \in \mathbb{R}$ to the smallest difference between the angles (see Section 13.4.3). The controller gains can be found by pole placement; see Algorithm 15.1. By specifying the control bandwidth ω_{b}, we get

$$\omega_{\mathrm{n}} = \frac{1}{\sqrt{1 - 2\zeta^2 + \sqrt{4\zeta^4 - 4\zeta^2 + 2}}}\omega_{\mathrm{b}} \tag{15.161}$$

and

$$K_{\mathrm{p}} = m\omega_{\mathrm{n}}^2$$
$$K_{\mathrm{d}} = m\left(2\zeta\omega_{\mathrm{n}} - \frac{1}{T}\right) \overset{T\gg 0}{\approx} 2\zeta\omega_{\mathrm{n}}m$$
$$K_{\mathrm{i}} = \frac{\omega_{\mathrm{n}}}{10}K_{\mathrm{p}}.$$

The relative damping ratio ζ is usually chosen in the range 0.8–1.0, which means that the only tunable parameter is the control bandwidth ω_{b} (typically 0.01 rad s^{-1} for large tankers and 0.1 rad s^{-1} for smaller ships and underwater vehicles). This makes the system very easy to tune. However, it is important to have a good estimate of $m = I_z - N_{\dot{r}}$ to obtain good performance.

Control Allocation

For a rudder-controlled craft, the input command is computed from (15.152), implying that

$$\delta = \frac{1}{N_\delta}\tau_{\mathrm{N}}. \tag{15.162}$$

In the case of several actuators, the generalized inverse can be used to compute $\boldsymbol{u}$ from (15.153) if the scalar $\boldsymbol{b}^{\top}\boldsymbol{b} \neq 0$ (see Section 11.2). This gives

$$\boldsymbol{u} = \boldsymbol{b}(\boldsymbol{b}^{\top}\boldsymbol{b})^{-1}\tau_{\mathrm{N}}. \tag{15.163}$$

Reference Feedforward

The *feedforward* term τ_{FF} in (15.154) is determined such that perfect tracking during course-changing maneuvers is obtained. Using Nomoto's first-order model (15.151) as a basis for feedforward control, suggests that *reference feedforward* should be implemented according to

$$\tau_{\mathrm{FF}} = m\left(\dot{r}_{\mathrm{d}} + \frac{1}{T}r_{\mathrm{d}}\right). \tag{15.164}$$

Substituting (15.164) and (15.154) into (15.151), the error dynamics becomes

$$\ddot{e} + \frac{1}{T}\dot{e} = \frac{1}{m}\tau_{\text{PID}} \tag{15.165}$$

where $e = \psi - \psi_{\text{d}}$. Since this system is linear, the closed-loop system can be analyzed in the frequency plane by using *Bode* plots. Consider the transfer function

$$h(s) = \frac{e}{\tau_{\text{PID}}}(s) = \frac{\frac{T}{m}}{s(Ts+1)} \tag{15.166}$$

and let

$$\begin{aligned} h_{\text{PID}}(s) &= K_{\text{p}}\left(1 + T_{\text{d}}s + \frac{1}{T_{\text{i}}s}\right) \\ &= K_{\text{p}}\frac{T_{\text{i}}T_{\text{i}}s^2 + T_{\text{i}}s + 1}{T_{\text{i}}s}. \end{aligned} \tag{15.167}$$

Hence, the loop transfer function becomes

$$\begin{aligned} l(s) &= h(s)h_{\text{PID}}(s) \\ &= \frac{T}{m}\frac{K_{\text{p}}}{T_{\text{i}}}\frac{(T_{\text{i}}T_{\text{d}}s^2 + T_{\text{i}}s + 1)}{s^2(Ts+1)}. \end{aligned} \tag{15.168}$$

15.3.5 Case Study: LOS Path-following Control for Marine Craft

A line-of-sight (LOS) path-following controller can be designed for conventional craft by representing the desired path by waypoints, as described in Section 12.4. This is particularly useful for underwater vehicles and surface vessels in transit operations where the user can specify the path by straight lines using a digital chart. For curved paths, the approaches of Section 12.6.2 can be used.

If the craft is equipped with a conventional heading autopilot, an outer feedback loop representing the guidance system can be designed as shown in Figure 15.21. This is practical since a commercial autopilot system can be treated as a black box where the outer-loop LOS algorithm computes heading commands to the autopilot. For this purpose, the guidance laws of Section 12.6.2 can be used to steer along the LOS vector which again forces the craft to follow the path.

LOS Path-following Guidance Law

In Section 12.5 it was shown that the desired heading angle needed to follow a straight line is

$$\psi_{\text{d}} = \pi_{\text{p}} - \tan^{-1}(K_{\text{p}}y_{\text{e}}^{p}) - \beta_{\text{c}} \tag{15.169}$$

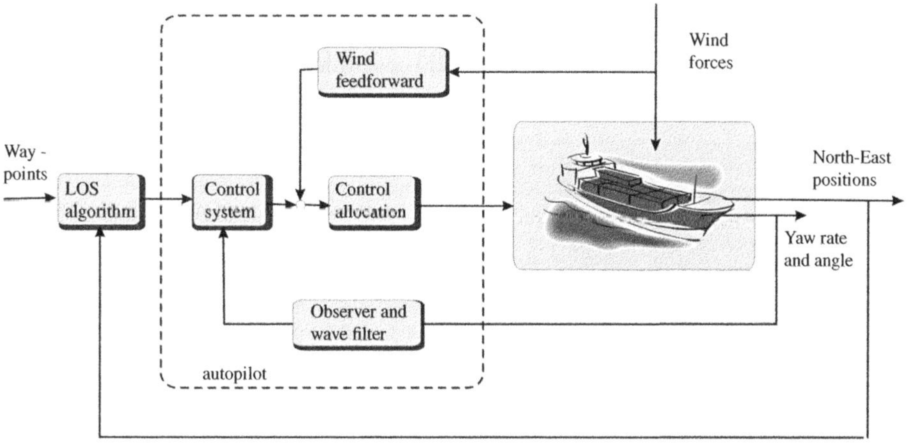

Figure 15.21 Conventional autopilot (inner loop) used in conjuncture with an LOS guidance algorithm in the outer loop.

where

$$\pi_p = \text{atan2}(y_{i+1}^n - y_i^n, \; x_{i+1}^n - x_i^n) \tag{15.170}$$

is the angle from the north axes to the path-tangential coordinate system $\{p\}$ defined by the straight line through the waypoints $\boldsymbol{p}_i^n = [x_i^n, \; y_i^n]^\top$ and $\boldsymbol{p}_{i+1}^n = [x_{i+1}^n, \; y_{i+1}^n]^\top$; see Figure 12.9 in Section 12.3.3.

The cross-track error expressed in $\{p\}$ for a marine craft located at the position $(x^n, \; y^n)$ is computed by

$$y_e^p = -\sin(\pi_p)(x^n - x_i^n) + \cos(\pi_p)(y^n - y_i^n). \tag{15.171}$$

This expression is obtained from the second line of Formula (12.55). The crab angle β_c and unmodeled dynamics need to be compensated for in order to follow the path without steady-state errors. This is achieved by replacing (15.169) with the ILOS guidance law (see Section 12.5.2)

$$\psi_d = \pi_p - \tan^{-1}(K_p \, y_e^p + K_i \, y_{int}^p) \tag{15.172}$$

$$\dot{y}_{int}^p = \frac{\Delta y_e^p}{\Delta^2 + (y_e^p + \kappa \, y_{int}^p)^2} \tag{15.173}$$

where $K_p = 1/\Delta$, $K_i = K_p\kappa$ and $\kappa > 0$ are design parameters.

Heading Autopilot

The heading autopilot is usually chosen as a PID controller with wind and reference feed-forward (see Section 15.3.4). In other words

$$\tau_{\mathrm{N}} = -\hat{\tau}_{\mathrm{wind}} + \tau_{\mathrm{FF}} - K_{\mathrm{p}}\ \mathrm{ssa}(\tilde{\psi}) - K_{\mathrm{d}}\dot{\tilde{\psi}} - K_i \int_0^t \mathrm{ssa}(\tilde{\psi}(\tau))\mathrm{d}\tau \tag{15.174}$$

$$\tau_{\mathrm{FF}} = m\left(\dot{r}_{\mathrm{d}} + \frac{1}{T}r_{\mathrm{d}}\right) \tag{15.175}$$

where $\tilde{\psi} = \psi - \psi_{\mathrm{d}}$, and $K_{\mathrm{p}} > 0$, $K_{\mathrm{d}} > 0$ and $K_{\mathrm{i}} > 0$ are the controller gains. The error signal $\tilde{\psi}$ is mapped to $[-\pi, \pi)$ using the SSA to avoid large steps in the control input (see Section 13.4.3).

Waypoint Switching Mechanism

When moving along a piecewise linear path made up of N straight-line segments connected by $N + 1$ waypoints, a switching mechanism for selecting the next waypoint is needed. Waypoint (x^n_{i+1}, y^n_{i+1}) can be selected on a basis of whether the craft lies within a *circle of acceptance* with radius R_{i+1} around (x^n_{i+1}, y^n_{i+1}). Hence, if the craft position (x^n, y^n) at time t satisfies

$$(x^n_{i+1} - x^n)^2 + (y^n_{i+1} - y^n)^2 \leq R^2_{i+1} \tag{15.176}$$

the next waypoint (x^n_{i+1}, y^n_{i+1}) should be selected. This is described more closely in Section 12.3.3, which also discusses extensions from 2D to 3D path-following control.

15.3.6 Case Study: Dynamic Positioning System for Surface Vessels

Control systems for stationkeeping and low-speed maneuvering are commonly known as dynamic positioning (DP) systems. The classification society DNV defines a DP vessel according to:

> *Dynamically positioned vessel*: a free-floating vessel which maintains its position (fixed location or predetermined track) exclusively by means of thrusters.

It is, however, possible to exploit rudder forces in DP also by using the propeller to generate rudder lift forces (Lindegaard and Fossen 2003).

For ships that are anchored, additional spring forces are introduced into the control model. These systems are referred to as position mooring (PM) systems (see Section 15.3.7). Optimality with respect to changing weather conditions will be discussed in Section 16.3.11 using the concept of weather optimal positioning control (WOPC).

DP and PM Systems

In the 1960s, systems for automatic control of the horizontal position, in addition to the heading, were developed. Systems for the simultaneous control of the three horizontal motions (surge, sway and yaw) are today commonly known as *DP systems* and are used in a wide range of marine operations such as stationkeeping, drilling and offloading, as illustrated in Figure 15.22. More recently anchored positioning systems or *PM systems* have been designed; see Section 15.3.7. For a free-floating vessel the thrusters are the prime actuators for stationkeeping, while for a PM system the assistance of thrusters are only complementary since most of the position-keeping is provided by a deployed anchor system. Different DP applications are described more closely in Strand and Sørensen (2000).

DP systems have traditionally been a *low-speed* application, where the basic DP functionality is either to keep a fixed position and heading or to move slowly from one location to another (*marked positioning*). In addition, specialized tracking functions for cable and pipe-layers, and operations of ROVs have been included. The traditional *autopilot* and *waypoint-tracking* functionalities have also been included in modern DP systems. The trend today is that *high-speed* operation functionality merges with classical DP functionality, resulting in a *unified system* for all speed ranges and types of operations.

The first DP systems were designed using conventional PID controllers in cascade with low-pass and/or notch filters to suppress the wave-induced motion components. This was based on the assumption that the interactions were negligible (Sargent and Cowgill 1976, and Morgan 1978). From the middle of the 1970s a new model-based control concept utilizing stochastic optimal control theory and Kalman filtering techniques was employed with the DP problem by Balchen *et al.* (1976). The Kalman filter is used to separate the LF and WF motion components such that only feedback from the LF motion components is used (see Chapter 13). Later extensions and modifications of this work have been proposed by numerous authors; see Balchen et al. (1980a, 1980b), Grimble et al. (1980a, 1980b), Fung and Grimble (1983), Sælid *et al.* (1983) and more lately Fossen *et al.* (1996), Sørensen et al. (1996, 2000), Fossen and Grøvlen (1998) and Fossen and Strand (1999a).

Roll and Pitch Damping in DP

Traditionally DP systems have been designed for 3-DOF low-speed trajectory-tracking control by means of thrusters and propellers. However, extensions to 5-DOF control for the purpose of roll and pitch damping of semi-submersibles has been proposed by Sørensen and Strand (1998). It is well known that for marine structures with a small waterplane area and low metacentric height, which results in relatively low hydrostatic restoration compared to the inertia forces, an unintentional coupling phenomenon between the vertical and the horizontal planes through the thruster action can be invoked. Examples are found in semi-submersibles and SWATHs, which typically have natural periods in roll and pitch in the range of 35–65 s. If the inherent vertical damping properties are small, the amplitudes of roll and pitch may be emphasized by the thruster's induction by up to 2°–5° in the resonance range. These oscillations have caused discomfort to the vessel's

Figure 15.22 Dynamically positioned supply vessel used in offshore offloading. Source: Illustration by B. Stenberg.

crew and have in some cases limited the operation. Hence, the motions in both the horizontal and vertical planes should be considered in the controller design, as proposed in Sørensen and Strand (2000).

Optimal Setpoint Chasing in DP for Drilling and Intervention Vessels

Further extension in the development of DP systems includes extended functionality adapted for the particular marine operation considered. In Sørensen *et al.* (2001) a function for optimal setpoint chasing in DP of drilling and intervention vessels is proposed in order to minimize riser angle offsets at the sea bed and on the vessel.

DP Control Design

For DP systems a low-frequency (LF) design model will be employed for feedback control. This is based on the assumption that the wave-frequency (WF) motions are filtered out of the feedback loop (see Section 13.2). Recall from Section 6.7.3 that

$$\dot{\boldsymbol{\eta}} = \boldsymbol{R}(t)\boldsymbol{\nu} \tag{15.177}$$

$$\boldsymbol{M}\dot{\boldsymbol{\nu}} + \boldsymbol{D}\boldsymbol{\nu} = \boldsymbol{R}^{\top}(t)\boldsymbol{b} + \boldsymbol{\tau} + \boldsymbol{\tau}_{\text{wind}} + \boldsymbol{\tau}_{\text{wave}} \tag{15.178}$$

$$\dot{\boldsymbol{b}} = \boldsymbol{0} \tag{15.179}$$

where $\boldsymbol{R}(t) := \boldsymbol{R}_{z,\psi}(\psi(t))$ is assumed known thanks to gyrocompass measurements. Note that the ocean currents, modeled by the bias term $\boldsymbol{b}$, are assumed constant in $\{n\}$ and therefore transformed to $\{b\}$ by $\boldsymbol{R}^{\top}(t)\boldsymbol{b}$.

DP Control System

A surface craft is exposed to *waves, ocean currents* and *wind.* The state estimator and controller must be robust and compensate for environmental forces and unmodeled dynamics. These are the most important design requirements in an industrial vessel control system since a full-state feedback controller will not work in bad weather unless the environmental forces are included in the design specifications. In commercial DP systems it is therefore necessary to include the following features:

- *Integral action* to compensate for slowly-varying forces (bias term $\boldsymbol{b}$) due to ocean currents, second-order wave drift forces and unmodeled dynamics.
- *Wind feedforward control* to compensate for *mean* wind forces. Wind gust cannot be compensated for since the actuators do not have the capacity for moving a large vessel in the frequency range of the wind gust.

- *Wave filtering* to avoid where first-order wave-induced oscillations are fed back to the control system as explained in Section 13.2. This is an important feature since the actuators cannot move a large vessel fast enough to suppress the disturbances.
- *State estimator* for noise filtering and estimation of unmeasured states, for instance linear and angular velocities. The main tool for this is the Kalman filter (see Section 13.4.6), alternatively nonlinear and passive observers as described in Section 13.5.
- *Optimal allocation of thrust* where the main goal is to compute optimal setpoints for thrusters, rudders and other actuators based on the force and moment commands generated by the DP control system. This is treated in detail in Section 11.2.

The different blocks in a closed-loop DP system are shown in Figure 15.23. The control system can be designed as a MIMO nonlinear PID controller using the results in Section 15.3.3. This is mathematically equivalent to

$$\boldsymbol{\tau} = -\hat{\boldsymbol{\tau}}_{\text{wind}} + \boldsymbol{R}^{\top}(t)\boldsymbol{\tau}_{\text{PID}} \tag{15.180}$$

where $\hat{\boldsymbol{\tau}}_{\text{wind}}$ is an estimate of the generalized wind forces and the PID controller is expressed in $\{n\}$ according to

$$\boldsymbol{\tau}_{\text{PID}} = -\boldsymbol{K}_p\tilde{\boldsymbol{\eta}} - \boldsymbol{K}_{\text{d}}\dot{\boldsymbol{\eta}} - \boldsymbol{K}_{\text{i}}\int_0^t \tilde{\boldsymbol{\eta}}(\tau)\text{d}\tau. \tag{15.181}$$

By combining (15.180) and (15.181), the DP control law becomes

$$\boldsymbol{\tau} = -\hat{\boldsymbol{\tau}}_{\text{wind}} - \boldsymbol{R}^{\top}(t)\boldsymbol{K}_p\tilde{\boldsymbol{\eta}} - \underbrace{\boldsymbol{R}^{\top}(t)\boldsymbol{K}_{\text{d}}\boldsymbol{R}(t)}_{\boldsymbol{K}_{\text{d}}^{*}}\boldsymbol{\nu} - \boldsymbol{R}^{\top}(t)\boldsymbol{K}_i\int_0^t \tilde{\boldsymbol{\eta}}(\tau)\text{d}\tau \tag{15.182}$$

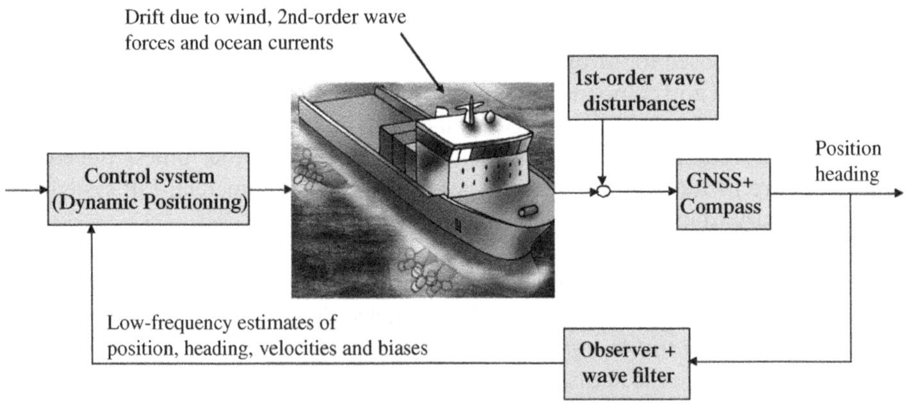

Figure 15.23 Dynamic positioning system. The state estimator can be implemented as a Kalman filter or an observer. Source: Illustration by B. Stenberg.

where $\boldsymbol{K}_{\rm d}^* := \boldsymbol{R}^\top(t)\boldsymbol{K}_{\rm d}\boldsymbol{R}(t)$. It is common to choose $\boldsymbol{K}_{\rm d}$ as a diagonal matrix and thus $\boldsymbol{K}_{\rm d}^* \equiv \boldsymbol{K}_{\rm d}$. For the full-state feedback case, asymptotic stability follows using Lyapunov arguments (see Section 15.3.3). However, in order to implement the nonlinear PID controller a state estimator and wave filter must be designed. This is straightforward for the LTV model (15.177)–(15.179) where additional states for the WF motions can be augmented and used directly in a Kalman filter (see Section 13.4.6). An alternative approach could be to use the nonlinear passive observer (see Section 13.5)

$$\dot{\hat{\boldsymbol{\xi}}} = \boldsymbol{A}_w\hat{\boldsymbol{\xi}} + \boldsymbol{K}_1(\omega_0)\tilde{\boldsymbol{y}} \tag{15.183}$$

$$\dot{\hat{\boldsymbol{\eta}}} = \boldsymbol{R}(t)\hat{\boldsymbol{\nu}} + \boldsymbol{K}_2\tilde{\boldsymbol{y}} \tag{15.184}$$

$$\dot{\hat{\boldsymbol{b}}} = -\boldsymbol{T}^{-1}\hat{\boldsymbol{b}} + \boldsymbol{K}_3\tilde{\boldsymbol{y}} \tag{15.185}$$

$$\boldsymbol{M}\dot{\hat{\boldsymbol{\nu}}} = -\boldsymbol{D}\hat{\boldsymbol{\nu}} + \boldsymbol{R}^\top(t)\hat{\boldsymbol{b}} + \boldsymbol{\tau} + \hat{\boldsymbol{\tau}}_{\rm wind} + \boldsymbol{R}^\top(t)\boldsymbol{K}_4\tilde{\boldsymbol{y}} \tag{15.186}$$

$$\hat{\boldsymbol{y}} = \hat{\boldsymbol{\eta}} + \boldsymbol{C}_w\hat{\boldsymbol{\xi}} \tag{15.187}$$

where drift is estimated using the bias term $\hat{\boldsymbol{b}}$. For the DP controller (15.182), the drift forces have been compensated for by adding integral action in the controller.

Wind Feedforward

Wind feedforward is implemented using measurements of wind speed and direction. Different wind models are presented in Section 10.1, suggesting that

$$\hat{\boldsymbol{\tau}}_{\rm wind} = \frac{1}{2}\rho_a V_{\rm rw}^2 \begin{bmatrix} C_X(\gamma_{\rm rw})A_{\rm F_w} \\ C_Y(\gamma_{\rm rw})A_{\rm L_w} \\ C_N(\gamma_{\rm rw})A_{\rm L_w}L_{\rm oa} \end{bmatrix} \tag{15.188}$$

where the relative wind speed and angle of attack are

$$V_{\rm rw} = \sqrt{u_{\rm rw}^2 + v_{\rm rw}^2} \tag{15.189}$$

$$\gamma_{\rm rw} = -{\rm atan2}(v_{\rm rw}, u_{\rm rw}). \tag{15.190}$$

The relative velocity components depend on the heading angle ψ, wind direction $\beta_{V_{\rm w}}$ and wind speed $V_{\rm w}$ according to

$$u_{\rm rw} = u - u_{\rm w} = u - V_{\rm w}\cos(\beta_{V_{\rm w}} - \psi) \tag{15.191}$$

$$v_{\rm rw} = v - v_{\rm w} = v - V_{\rm w}\sin(\beta_{V_{\rm w}} - \psi). \tag{15.192}$$

When wind feedforward is implemented, it is important that the wind measurements are low-pass filtered to avoid rapid changes in the actuator commands. Wind feedforward is an optional term since the integrator in the DP system can compensate for slowly varying wind forces as well. The main difference will be the response time. In general, wind feedforward will be much faster than integral action since the integrator needs several minutes to remove a large wind component during the start-up of the DP system.

15.3.7 Case Study: Position Mooring System for Surface Vessels

Figure 15.24 illustrates different mooring strategies for ships and floating structures. The results of Section 15.3.6 can be generalized to PM systems by adding a spring to the model. For an anchored vessel the heading angle is slowly varying such that $\dot{\boldsymbol{R}}(t) \approx \boldsymbol{0}$. Hence, $\boldsymbol{\nu} \approx \boldsymbol{R}^{\top}(t)\dot{\boldsymbol{\eta}}$ and $\dot{\boldsymbol{\nu}} \approx \boldsymbol{R}^{\top}(t)\ddot{\boldsymbol{\eta}}$. This implies that the model (15.178) can be written as

$$\underbrace{\boldsymbol{R}(t)\boldsymbol{M}\boldsymbol{R}^{\top}(t)}_{\boldsymbol{M}^{*}(t)}\,\ddot{\boldsymbol{\eta}} + \underbrace{\boldsymbol{R}(t)\boldsymbol{D}\boldsymbol{R}^{\top}(t)}_{\boldsymbol{D}^{*}(t)}\,\dot{\boldsymbol{\eta}} + \boldsymbol{K}\boldsymbol{\eta} = \boldsymbol{b} + \boldsymbol{R}(t)\boldsymbol{\tau} + \boldsymbol{R}(t)\boldsymbol{\tau}_{\text{wind}} + \boldsymbol{R}(t)\boldsymbol{\tau}_{\text{wave}} \tag{15.193}$$

where $\boldsymbol{\nu} = [u,\ v,\ r]^{\top}$ and $\boldsymbol{\eta} = [x^n, y^n, \psi]^{\top}$. For this system

$$\boldsymbol{M} = \boldsymbol{M}^{\top} = \begin{bmatrix} m_{11} & 0 & 0 \\ 0 & m_{22} & m_{23} \\ 0 & m_{32} & m_{33} \end{bmatrix} \tag{15.194}$$

$$\boldsymbol{D} = \boldsymbol{D}^{\top} = \begin{bmatrix} d_{11} & 0 & 0 \\ 0 & d_{22} & d_{23} \\ 0 & d_{32} & d_{33} \end{bmatrix} \tag{15.195}$$

$$\boldsymbol{K} = \text{diag}\{k_{11}, k_{22}, k_{33}\}. \tag{15.196}$$

The additional spring $\boldsymbol{K}\boldsymbol{\eta}$ due to the mooring system adds spring stiffness in surge, sway and yaw described by the parameters $k_{11} > 0, k_{22} > 0$ and $k_{33} \geq 0$. With this in mind, two different design philosophies for mooring systems are quite common.

- *Turret mooring systems* have cables that are connected to the turret via bearings. This allows the vessel to rotate around the anchor legs. In this case, the rotational spring can be neglected such that $k_{33} = 0$. The turret can be mounted either internally or externally. An external turret is fixed, with appropriate reinforcements, to the bow or stern of the ship. In the internal case the turret is placed within the hull in a moon pool. A moon pool is a wet porch, that is an opening in the floor or base of the hull giving access to the water below, allowing technicians or researchers to lower tools and instruments into the sea. Turret mooring systems allow the vessel to rotate in the horizontal plane (yaw)

Figure 15.24 Mooring systems for a submersible, FPSO and platform. Source: Illustration by B. Stenberg.

into the direction where environmental loading due to wind, waves and ocean currents is minimal. This is referred to as weathervaning.

- *Spread mooring systems* are used to moor floating production, storage and offloading (FPSO) units, tankers and floating platforms (see Figure 15.24). The system consists of mooring lines attached somewhere to the vessel. The drawback with a spread mooring system is that it restrains the vessel from rotating ($k_{33} > 0$) and hence weathervaning is impossible. On the other hand, it is relatively inexpensive to equip an existing vessel with mooring lines that can be attached directly to the hull.

For thruster-assisted PM systems the thrusters are complementary to the mooring system and the main idea is to provide the system with additional damping, for instance by using a D controller

$$\begin{aligned}\boldsymbol{\tau} &= -\boldsymbol{R}^{\top}(t)\boldsymbol{K}_{\mathrm{d}}\dot{\boldsymbol{\eta}} \\ &= -\boldsymbol{R}^{\top}(t)\boldsymbol{K}_{\mathrm{d}}\boldsymbol{R}(t)\boldsymbol{\nu}. \end{aligned} \tag{15.197}$$

This gives the closed-loop system

$$M^*(t)\ddot{\eta} + (D^*(t) + K_\text{d})\dot{\eta} + K\eta = b + R(t)\tau_\text{wind} + R(t)\tau_\text{wave}. \tag{15.198}$$

The mooring term $K\eta$ is a passive P controller but additional spring forces can be included by position feedback if necessary. Integral action is not used in PM systems, since the vessel is only allowed to move within a limited radius from the equilibrium point or *field-zero point* (FZP). If the vessel moves outside the specified radius of the mooring system, a stabilizing control system of PD type can be used to drive the vessel inside the circle again. This is usually done in an energy perspective since it is important to reduce the fuel consumption of PM systems. Consequently, in bad weather it will be more optimal to use additional thrust to stay on the circle rather than move the vessel to the FZP. In good weather, no control action is needed since the vessel is free to move within the circle.

PM systems have been commercially available since the 1980s, and provide a flexible solution for floating structures for drilling and oil and gas exploitation on the smaller and marginal fields (Sørensen *et al.* 2000). Modeling and control of turret-moored ships are complicated problems since the mooring forces and moments are inherently nonlinear (Strand *et al.* 1998). The control design of PM systems using nonlinear theory is addressed by Strand (1999).

16

Advanced Motion Control Systems

State-of-the-art motion control systems are usually designed using PID control methods, as described in Chapter 15. This chapter presents more advanced methods for optimal and nonlinear control of marine craft. The main motivation for this is design simplicity and performance. Nonlinear control theory can often yield a more intuitive design than linear theory. Linearization destroys model properties and the results can be a more complicated design process with limited physical insight. Chapter 16 is written for the advanced user who wants to exploit a more advanced model and use this model to improve the performance of the control system. Readers of this chapter need a background in optimal and nonlinear control theory.

Preview of the Chapter

Chapter 16 starts with linear-quadratic optimal control theory (Section 16.1) with the focus on regulation, trajectory-tracking control and disturbance feedforward. Optimal motion control systems are designed by considering the linear state-space model

$$\dot{x} = Ax + Bu + Ew. \tag{16.1}$$

For a marine craft, the linear model (16.1) is based on several assumptions such as zero or constant cruise speed U together with the assumptions that the velocities v, w, p, q and r are small. In addition, the kinematic equation $\dot{\boldsymbol{\eta}} = \boldsymbol{J}_{\Theta}(\boldsymbol{\eta})\boldsymbol{\nu}$ must be linearized under a set of assumptions on the Euler angles ϕ, θ and ψ.

When linearizing the equations of motion, several model properties such as symmetry of the inertia matrix $\boldsymbol{M}$, skew-symmetry of the Coriolis and centripetal matrix $\boldsymbol{C}(\boldsymbol{\nu})$ and positiveness of the damping matrix $\boldsymbol{D}(\boldsymbol{\nu})$ are destroyed, and this often complicates the control design. Also physical properties that are important tools for good engineering

Handbook of Marine Craft Hydrodynamics and Motion Control, Second Edition. Thor I. Fossen.

judgment are lost. This is seen by comparing the LQ design procedure with the nonlinear techniques in Sections 16.2–16.4. It is also demonstrated how the nonlinear controllers can be related to the PID control design methods in Chapter 15 in particular, under the assumption of setpoint regulation. Often it is useful to think about the nonlinear controller as a PID control system where additional terms are added to obtain global stability results. Keeping this in mind, it is also possible to derive a nonlinear controller using advanced methods and then use engineering insight to simplify the representation of the controller. The resulting controller should be as simple as possible but still contain the most important terms when implementing the algorithm into a computer. In fact, a so-called simplified nonlinear controller will be recognized as a PID controller with additional terms. Many nonlinear methods are popular due to their simplicity and design flexibility. The assumptions on u, v, w, p, q, r and ϕ, θ, ψ which are needed when linearizing the models are also avoided.

The nonlinear design methods in this chapter are based on the robot-inspired model of Fossen (1991), which can be expressed as

$$\dot{\boldsymbol{\eta}} = \boldsymbol{J}_{\Theta}(\boldsymbol{\eta})\boldsymbol{\nu} \tag{16.2}$$

$$\boldsymbol{M}\dot{\boldsymbol{\nu}} + \boldsymbol{C}(\boldsymbol{\nu})\boldsymbol{\nu} + \boldsymbol{D}(\boldsymbol{\nu})\boldsymbol{\nu} + \boldsymbol{g}(\boldsymbol{\eta}) = \boldsymbol{\tau} + \boldsymbol{w}. \tag{16.3}$$

It is important to understand the physical properties of the model in order to know which terms in the model can be omitted when deriving a model-based nonlinear controller. This is an important question since model inaccuracies can destabilize a feedback control system. Often better results are obtained when uncertain terms are chosen to be zero in the controller.

16.1 Linear-quadratic Optimal Control

Optimal control deals with the problem of finding a control law for a given system such that a certain optimality criterion is achieved. This is usually a cost function that depends on the state and control variables. The optimal control law is a set of differential equations that minimize the cost functional and it can be derived using Pontryagin's maximum principle (a necessary condition) or by solving the Hamilton–Jacobi–Bellman equation (a sufficient condition). We will limit our discussion to linear systems and quadratic cost functions. This is referred to as linear-quadratic (LQ) optimal control theory (Athans and Falb 1966).

16.1.1 Linear-quadratic Regulator

A fundamental design problem is the regulator problem, where it is necessary to regulate the outputs $\boldsymbol{y} \in \mathbb{R}^m$ of the system to zero or a constant value while ensuring that they satisfy time-response specifications. A linear-quadratic regulator (LQR) can be designed for this purpose by considering the state-space model

$$\dot{x} = Ax + Bu \tag{16.4}$$

$$y = Cx \tag{16.5}$$

where $x \in \mathbb{R}^n, u \in \mathbb{R}^r$ and $y \in \mathbb{R}^m$. In order to design a linear optimal control law the system (A, B, C) must be controllable while observability (see Definition 13.2 in Section 13.3.3) is necessary if some of the states must be estimated. Controllability for linear time-invariant systems is given by the following definition:

Definition 16.1 (Controllability)
The state and input matrix $(A,\ B)$ *must satisfy the* controllability condition *to ensure that there exists a control* $u(t)$ *that can drive any arbitrary state* $x(t_0)$ *to another arbitrary state* $x(t_1)$ *for* $t_1 > t_0$. *The* controllability condition *requires that the matrix (Gelb* et al. *1988)*

$$C = [B \mid AB \mid \cdots \mid (A)^{n-1}B] \tag{16.6}$$

must be of full row rank such that a right inverse exists.

The feedback control law for the system (16.4) and (16.5) is found by minimizing the quadratic cost function

$$\begin{aligned} J &= \min_u \left\{ \frac{1}{2}\int_0^T (y^\top Q y + u^\top R u)\mathrm{d}t \right. \\ &= \left. \frac{1}{2}\int_0^T (x^\top C^\top Q C x + u^\top R u)\,\mathrm{d}t \right\} \end{aligned} \tag{16.7}$$

where $R = R^\top > 0$ and $Q = Q^\top \geq 0$ are the weighting matrices. The steady-state solution to this problem is (Athans and Falb 1966)

$$u = \underbrace{-R^{-1}B^\top P_\infty}_{G} x \tag{16.8}$$

$$P_\infty A + A^\top P_\infty - P_\infty B R^{-1} B^\top P_\infty + C^\top Q C = 0 \tag{16.9}$$

where $P_\infty = \lim_{t\to\infty} P(t)$. The optimal feedback control system is illustrated in Figure 16.1.

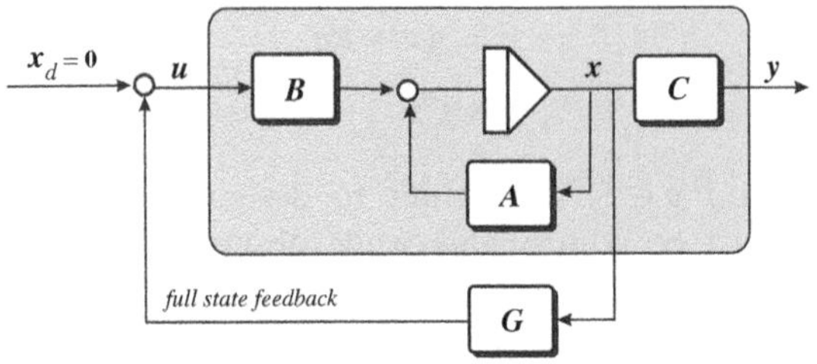

Figure 16.1 Block diagram showing the linear-quadratic regulator (LQR).

MATLAB

The steady-state LQR feedback control law is computed as (see the script `ExLQR.m`)

```
% Weights
Q = diag([1]);          % tracking error weights (dim m x m)
R = diag([1]);          % input weights (dim r x r)

% System matrices
A = [0 1; -1 -2];       % state matrix (dim n x n)
B = [0; 1];             % input matrix (dim n x r)
C = [1 0];              % output matrix (dim m x n)

% Compute the optimal feedback gain matrix G
[K,P,E] = lqr(A, B, C'*Q*C, R);
G = -K
```

The Matlab function `lqr.m` also returns the eigenvalues of the closed-loop system

$$\dot{\boldsymbol{x}} = (\boldsymbol{A} + \boldsymbol{B}\boldsymbol{G})\boldsymbol{x} \tag{16.10}$$

denoted by the symbol `E`.

16.1.2 LQR Design for Trajectory Tracking and Integral Action

The linear-quadratic regulator can be redesigned to track a time-varying reference trajectory $\boldsymbol{x}_{\mathrm{d}} \in \mathbb{R}^{\mathrm{n}}$ for a large class of mechanical systems possessing certain structural properties. This section presents a simple solution to this problem while a more general solution is presented in Section 16.1.3.

Transformation of the LQ Tracker to a Setpoint Regulation Problem

In order to transform a trajectory-tracking problem to a setpoint regulation problem *reference feedforward* can be used. Unmeasured slowly-varying or constant disturbances are compensated for by including integral action. This is usually done by augmenting an integral state to the system model. A mass–damper–spring system will be used to demonstrate the design methodology.

Example 16.1 (Mass–Damper–Spring Trajectory-Tracking Problem)
Consider the mass–damper–spring system

$$\dot{x} = v$$
$$m\dot{v} + dv + kx = \tau.$$

Let

$$\tau = \tau_{FF} + \tau_{LQ} \tag{16.11}$$

where the feedforward term is chosen as

$$\tau_{FF} = m\dot{v}_{\mathrm{d}} + dv_{\mathrm{d}} + kx_{\mathrm{d}} \tag{16.12}$$

such that

$$m\ddot{e} + d\dot{e} + ke = \tau_{LQ} \tag{16.13}$$

where $e = x - x_{\mathrm{d}}$ and $\dot{e} = v - v_{\mathrm{d}}$. The desired states x_{d} and v_{d} can be computed using a reference model. The trajectory-tracking control problem has now been transformed to an LQ setpoint regulation problem given by (16.13), which can be written in state-space form as

$$\dot{\boldsymbol{x}} = \underbrace{\begin{bmatrix} 0 & 1 \\ -\frac{k}{m} & -\frac{d}{m} \end{bmatrix}}_{\boldsymbol{A}} \boldsymbol{x} + \underbrace{\begin{bmatrix} 0 \\ \frac{1}{m} \end{bmatrix}}_{\boldsymbol{B}} u$$
$$e = \underbrace{\begin{bmatrix} 1 & 0 \end{bmatrix}}_{\boldsymbol{C}} \boldsymbol{x}$$

where $\boldsymbol{x} = [e,\ \dot{e}]^{\top}$ and $u = \tau_{LQ}$ is computed by (16.8) and (16.9).

Integral Action

In Example 16.1 it was shown that a feedforward term τ_{FF} could transform the LQ trajectory-tracking problem to an LQR problem. For the system model

$$\dot{\boldsymbol{x}} = \boldsymbol{A}\boldsymbol{x} + \boldsymbol{B}\boldsymbol{u} \tag{16.14}$$

integral action is obtained by augmenting the integral state

$$\dot{\boldsymbol{z}} = \boldsymbol{y} = \boldsymbol{C}\boldsymbol{x} \tag{16.15}$$

to the state vector. The matrix $\boldsymbol{C}$ is used to extract potential integral states from the $\boldsymbol{x}$ vector. This system is a standard LQR problem

$$\dot{\boldsymbol{x}}_{\mathrm{a}} = \boldsymbol{A}_{\mathrm{a}}\boldsymbol{x}_{\mathrm{a}} + \boldsymbol{B}_{\mathrm{a}}\boldsymbol{u} \tag{16.16}$$

where $\boldsymbol{x}_a = [\boldsymbol{z}^\top, \boldsymbol{x}^\top]^\top$ is the augmented state vector and

$$\boldsymbol{A}_a = \begin{bmatrix} \mathbf{0} & \boldsymbol{C} \\ \mathbf{0} & \boldsymbol{A} \end{bmatrix}, \quad \boldsymbol{B}_a = \begin{bmatrix} \mathbf{0} \\ \boldsymbol{B} \end{bmatrix}. \tag{16.17}$$

The control objective is regulation of $\boldsymbol{x}_a$ to zero using $\boldsymbol{u}$. This is obtained by choosing the performance index

$$J = \min_{\boldsymbol{u}} \left\{ \frac{1}{2} \int_0^t (\boldsymbol{x}_a^\top \boldsymbol{Q}_a \boldsymbol{x}_a + \boldsymbol{u}^\top \boldsymbol{R} \boldsymbol{u}) \mathrm{d}\tau \right\} \tag{16.18}$$

where $\boldsymbol{R} = \boldsymbol{R}^\top > 0$ and $\boldsymbol{Q}_a = \boldsymbol{Q}_a^\top \geq 0$ are the weighting matrices. Hence, the solution of the integral LQR setpoint regulation problem is (see Section16.1.1)

$$\begin{aligned} \boldsymbol{u} &= -\boldsymbol{R}^{-1} \boldsymbol{B}_a^\top \boldsymbol{P}_\infty \boldsymbol{x}_a \\ &= -\boldsymbol{R}^{-1} [\mathbf{0}\ \boldsymbol{B}^\top] \begin{bmatrix} \boldsymbol{P}_{11} & \boldsymbol{P}_{12} \\ \boldsymbol{P}_{21} & \boldsymbol{P}_{22} \end{bmatrix} \begin{bmatrix} \boldsymbol{z} \\ \boldsymbol{x} \end{bmatrix} \\ &= \underbrace{-\boldsymbol{R}^{-1} \boldsymbol{B}^\top \boldsymbol{P}_{12}}_{\boldsymbol{K}_i} \boldsymbol{z} \underbrace{- \boldsymbol{R}^{-1} \boldsymbol{B}^\top \boldsymbol{P}_{22}}_{\boldsymbol{K}_p} \boldsymbol{x} \end{aligned} \tag{16.19}$$

where $\boldsymbol{P}_{12}$ and $\boldsymbol{P}_{22}$ are found by solving the *algebraic Riccati equation* (ARE)

$$\boldsymbol{P}_\infty \boldsymbol{A}_a + \boldsymbol{A}_a^\top \boldsymbol{P}_\infty - \boldsymbol{P}_\infty \boldsymbol{B}_a \boldsymbol{R}^{-1} \boldsymbol{B}_a^\top \boldsymbol{P}_\infty + \boldsymbol{Q}_a = \mathbf{0}. \tag{16.20}$$

Note that the feedback term $\boldsymbol{u}$ now includes feedback from the state vector $\boldsymbol{x}$ and the integral state

$$\boldsymbol{z} = \int_0^t \boldsymbol{y}(\tau) \mathrm{d}\tau. \tag{16.21}$$

16.1.3 General Solution of the LQ Trajectory-tracking Problem

Consider the state-space model

$$\dot{\boldsymbol{x}} = \boldsymbol{A}\boldsymbol{x} + \boldsymbol{B}\boldsymbol{u} \tag{16.22}$$

$$\boldsymbol{y} = \boldsymbol{C}\boldsymbol{x}. \tag{16.23}$$

The LQ trajectory-tracking control problem is addressed under the assumption that both the state vector $\boldsymbol{x}$ is measured or at least obtained by state estimation. If the estimated values are used for $\boldsymbol{x}$, stability can be proven by applying a *separation principle*. This is known as LQG control in the literature and involves the design of a *Kalman filter* for reconstruction of the unmeasured states, which again requires that the system is *observable*. For simplicity, full-state feedback is assumed in this chapter. The interested reader is recommended to consult the extensive literature on LQG control for output feedback control; see Athans and Falb (1966) and Brian *et al.* (1989), for instance.

Control Objective

The control objective is to design a linear-quadratic optimal trajectory-tracking controller using a time-varying smooth reference trajectory $\boldsymbol{x}_{\mathrm{d}}$. Assume that $\boldsymbol{x}_{\mathrm{d}}$ and thus the desired output $\boldsymbol{y}_{\mathrm{d}} = \boldsymbol{C}\boldsymbol{x}_{\mathrm{d}}$ is known for all time $t \in [0,\ T]$, where T is the final time. Define the error signal

$$\begin{aligned} \boldsymbol{e} &:= \boldsymbol{y} - \boldsymbol{y}_{\mathrm{d}} \\ &= \boldsymbol{C}(\boldsymbol{x} - \boldsymbol{x}_{\mathrm{d}}). \end{aligned} \tag{16.24}$$

The goal is to design an optimal trajectory-tracking controller that tracks the desired output, that is regulates the error $\boldsymbol{e}$ to zero while minimizing

$$\begin{gathered} J = \min_{\boldsymbol{u}} \left\{ \frac{1}{2}\boldsymbol{e}^{\top}(T)\boldsymbol{Q}_{\mathrm{f}}\boldsymbol{e}(T) + \frac{1}{2}\int_0^T (\boldsymbol{e}^{\top}\boldsymbol{Q}\boldsymbol{e} + \boldsymbol{u}^{\top}\boldsymbol{R}\boldsymbol{u})\,\mathrm{d}t \right\} \\ \text{subject to } \dot{\boldsymbol{x}} = \boldsymbol{A}\boldsymbol{x} + \boldsymbol{B}\boldsymbol{u}, \qquad \boldsymbol{x}(0) = \boldsymbol{x}_0 \end{gathered} \tag{16.25}$$

where $\boldsymbol{R} = \boldsymbol{R}^{\top} > 0$ and $\boldsymbol{Q} = \boldsymbol{Q}^{\top} \geq 0$ are the tracking error and control weighting matrices, respectively. The weight matrix $\boldsymbol{Q}_{\mathrm{f}} = \boldsymbol{Q}_{\mathrm{f}}^{\top} \geq 0$ can be included to add penalty to the final state. Note that this is a *finite time-horizon* optimal control problem and it has to be solved by using the *differential Riccati equation* (DRE); see Athans and Falb (1966, pp. 793–801).

Solution of the LQ Trajectory-tracking Problem

It can be shown that the optimal control law solving (16.25) is (Brian *et al.* 1989)

$$\boldsymbol{u} = -\boldsymbol{R}^{-1}\boldsymbol{B}^{\top}\,(\boldsymbol{P}\boldsymbol{x} + \boldsymbol{h}) \tag{16.26}$$

where $\boldsymbol{P}$ and $\boldsymbol{h}$ originate from the *Hamiltonian* system. The matrix $\boldsymbol{P}$ accounts for the feedback part and $\boldsymbol{h}$ accounts for the feedforward part due to the time-varying nature of the reference signal $\boldsymbol{y}_{\mathrm{d}}$. The equations that need to be numerically integrated are

$$\dot{\boldsymbol{P}} = -\boldsymbol{P}\boldsymbol{A} - \boldsymbol{A}^{\top}\boldsymbol{P} + \boldsymbol{P}\boldsymbol{B}\boldsymbol{R}^{-1}\boldsymbol{B}^{\top}\boldsymbol{P} - \boldsymbol{C}^{\top}\boldsymbol{Q}\boldsymbol{C} \tag{16.27}$$

$$\dot{\boldsymbol{h}} = -(\boldsymbol{A} - \boldsymbol{B}\boldsymbol{R}^{-1}\boldsymbol{B}^{\top}\boldsymbol{P})^{\top}\boldsymbol{h} + \boldsymbol{C}^{\top}\boldsymbol{Q}\boldsymbol{C}\boldsymbol{x}_{\mathrm{d}} \tag{16.28}$$

with final conditions

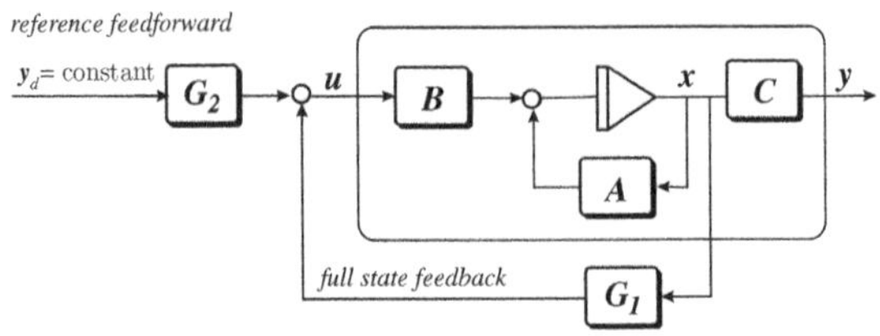

Figure 16.2 Block diagram showing the full state feedback LQ tracker solution with reference feedforward.

$$P(T) = C^{\mathrm{T}} Q_{\mathrm{f}} C \tag{16.29}$$

$$h(T) = -C^{\mathrm{T}} Q_{\mathrm{f}} C x_{\mathrm{d}}(T). \tag{16.30}$$

Equations (16.27) and (16.28) represent two differential equations: a matrix DRE and a vector differential equation (adjoint operators), respectively. Note that the initial conditions for these equations are not known, but rather the final conditions are known. Consequently, they have to be integrated *backward* in time a priori to find the initial conditions, and then be executed forward in time again with the closed-loop plant from $[0,\ T]$.

There are different ways of doing this. A frequently used method is to discretize the system and run the resulting difference equation backward. A simple Euler integration routine for (16.27) is given below, where δ is set as a small *negative* sampling time. The solution is the first-order Taylor expansion

$$P(t+\delta) \approx P(t) + \delta\{-PA - A^{\mathrm{T}}P + PBR^{-1}B^{\mathrm{T}}P - C^{\mathrm{T}}QC\} \tag{16.31}$$

with $P(T) = C^{\mathrm{T}} Q_{\mathrm{f}} C$, which is iterated to produce $P(0)$. Another procedure is to simulate backwards in time. The system

$$\dot{x} = f(x, t) + G(x, t)u, \qquad t \in [T,\ 0] \tag{16.32}$$

can be simulated backwards in time by the following change of integration variable $t = T - \tau$ with $\mathrm{d}t = -\mathrm{d}\tau$. Thus

$$-\frac{\mathrm{d}x(T-\tau)}{\mathrm{d}\tau} = f(x(T-\tau), T-\tau) + G(x(T-\tau), T-\tau)u(T-\tau). \tag{16.33}$$

Let $z(\tau) = x(T-\tau)$; then

$$\frac{\mathrm{d}z(\tau)}{\mathrm{d}\tau} = -f(z(\tau), T-\tau) - G(z(\tau), T-\tau)u(T-\tau). \tag{16.34}$$

This system (16.34) can now be simulated forward in time with the initial condition $z(0) = x(T)$. The method is demonstrated in Example 16.2, where it is assumed that x_{d} is time varying but known for all future t. A special case dealing with constant values for x_{d} will be studied later.

Example 16.2 (Optimal Time-varying LQ Trajectory-tracking Problem)
Consider a mass–damper–spring system

$$m\ddot{x} + d\dot{x} + kx = u \tag{16.35}$$

where m is the mass, d is the damping coefficient, k is the spring stiffness coefficient and u is the input. Choosing the states as $x_1 = x$ *and* $x_2 = \dot{x}$*, the following state-space realization is obtained*

$$\begin{bmatrix} \dot{x}_1 \\ \dot{x}_2 \end{bmatrix} = \begin{bmatrix} 0 & 1 \\ -\frac{k}{m} & -\frac{d}{m} \end{bmatrix} \begin{bmatrix} x_1 \\ x_2 \end{bmatrix} + \begin{bmatrix} 0 \\ \frac{1}{m} \end{bmatrix} u. \tag{16.36}$$

For simplicity, assume that $m = k = 1$ *and* $d = 2$ *such that*

$$\dot{\boldsymbol{x}} = \begin{bmatrix} 0 & 1 \\ -1 & -2 \end{bmatrix} \boldsymbol{x} + \begin{bmatrix} 0 \\ 1 \end{bmatrix} u \tag{16.37}$$

$$y = \begin{bmatrix} 1 & 0 \end{bmatrix} \boldsymbol{x} \tag{16.38}$$

where $\boldsymbol{x} = [x_1,\ x_2]^{\top}$*. The reference signal is assumed to be known for all future time and is given by*

$$\dot{\boldsymbol{x}}_{\mathrm{d}} = \begin{bmatrix} 0 & 1 \\ -1 & -1 \end{bmatrix} \boldsymbol{x}_{\mathrm{d}} + \begin{bmatrix} 0 \\ 1 \end{bmatrix} r \tag{16.39}$$

$$y_{\mathrm{d}} = \begin{bmatrix} 1 & 0 \end{bmatrix} \boldsymbol{x}_{\mathrm{d}} \tag{16.40}$$

where $r = \sin(t)$ *is the command. The Matlab MSS toolbox script* `ExLQFinHor.m` *demonstrates how forward and backward integration can be implemented for the mass–damper–spring system. The simulation results are shown in Figure 16.3.*

Approximate Solution for x_{d} = constant

A special case of (16.25) is the one with no weight on the final state; that is $\boldsymbol{Q}_{\mathrm{f}} = \boldsymbol{0}$, resulting in the quadratic performance index

$$J = \min_{\boldsymbol{u}} \left\{ \frac{1}{2} \int_0^T (\boldsymbol{e}^{\top}\boldsymbol{Q}\boldsymbol{e} + \boldsymbol{u}^{\top}\boldsymbol{R}\boldsymbol{u})\, \mathrm{d}t \right\}. \tag{16.41}$$

Unfortunately, the theory dealing with the limiting case $T \to \infty$ is not available. This solution is very useful since it represents a steady-state solution of the LQ trajectory-tracking problem. Fortunately, this problem can be circumvented by assuming that T is large but still limited; that is

$$0 \ll T_1 \leq T < \infty \tag{16.42}$$

where T_1 is a large constant. For $T \to \infty$ the solution of (16.27) will tend to the constant matrix $\boldsymbol{P}_{\infty}$ satisfying the ARE

$$\boldsymbol{P}_{\infty}\boldsymbol{A} + \boldsymbol{A}^{\top}\boldsymbol{P}_{\infty} - \boldsymbol{P}_{\infty}\boldsymbol{B}\boldsymbol{R}^{-1}\boldsymbol{B}^{\top}\boldsymbol{P}_{\infty} + \boldsymbol{C}^{\top}\boldsymbol{Q}\boldsymbol{C} = \boldsymbol{0}. \tag{16.43}$$

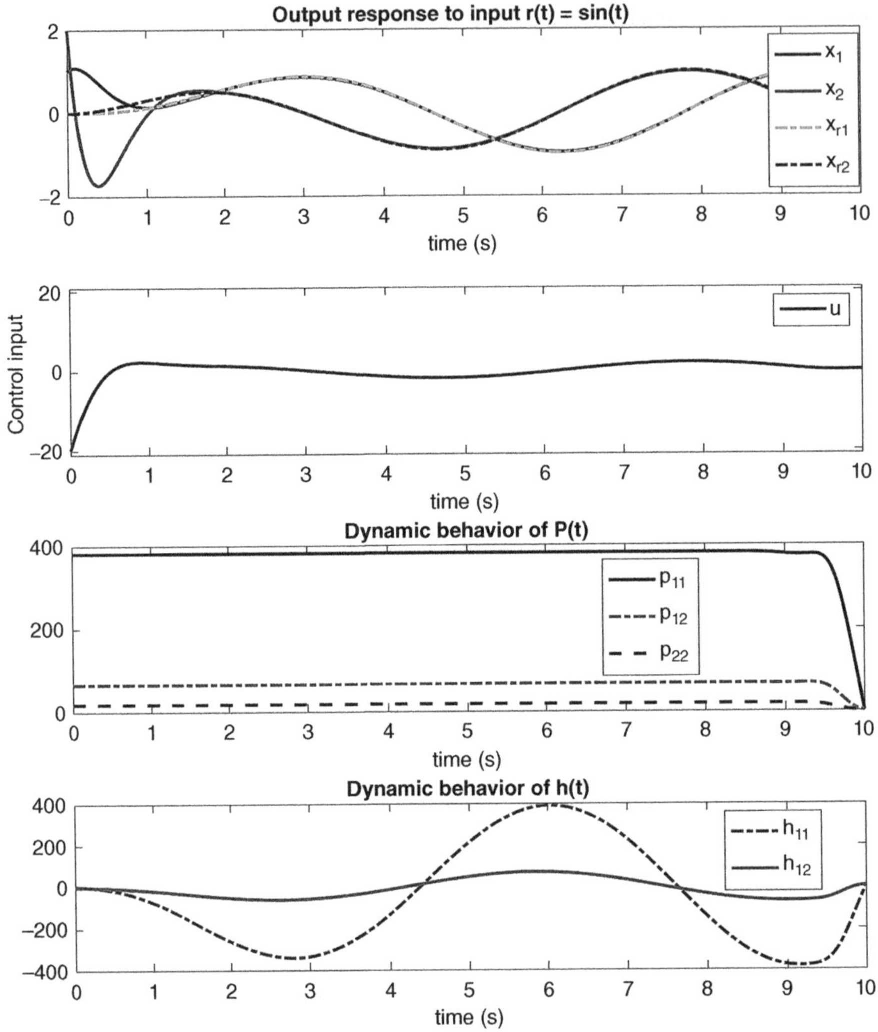

Figure 16.3 First plot: states x_1 and x_2 and the reference trajectories as a function of time. Second plot: optimal control u as a function of time. Third and fourth plots: optimal solutions of the elements in $\boldsymbol{P}$ and $\boldsymbol{h}$ as a function of time.

This solution is interpreted as the steady-state solution of (16.27) where $\boldsymbol{P}(t) \approx \boldsymbol{P}_\infty$ for all $t \in [0,\ T_1]$. Furthermore, it is assumed that $\boldsymbol{x}_\mathrm{d}$ = constant for all $t \in [0,\ T_1]$. In practice the assumption that $\boldsymbol{x}_\mathrm{d}$ is constant can be relaxed with $\boldsymbol{x}_\mathrm{d}$ being slowly varying compared to the state dynamics. Next, if the eigenvalues of the matrix

$$\boldsymbol{A}_\mathrm{c} = \boldsymbol{A} + \boldsymbol{B}\boldsymbol{G}_1 \quad \text{where} \quad \boldsymbol{G}_1 = -\boldsymbol{R}^{-1}\boldsymbol{B}^\mathrm{T}\boldsymbol{P}_\infty \tag{16.44}$$

have negative real parts $\lambda_i(A_c) < 0$ ($i = 1, \ldots, n$), the steady-state solution for $\boldsymbol{h}$ in (16.28) becomes

$$\boldsymbol{h}_\infty = (\boldsymbol{A} + \boldsymbol{B}\boldsymbol{G}_1)^{-\top}\boldsymbol{C}^\top\boldsymbol{Q}\boldsymbol{C}\boldsymbol{x}_d. \tag{16.45}$$

Substitution of (16.45) into (16.26) yields the steady-state optimal control law (see Figure 16.2)

$$\boldsymbol{u} = \boldsymbol{G}_1\,\boldsymbol{x} + \boldsymbol{G}_2\,\boldsymbol{y}_d \tag{16.46}$$

where $\boldsymbol{y}_d$ = constant and

$$\boldsymbol{G}_1 = -\boldsymbol{R}^{-1}\boldsymbol{B}^\top\boldsymbol{P}_\infty \tag{16.47}$$

$$\boldsymbol{G}_2 = -\boldsymbol{R}^{-1}\boldsymbol{B}^\top(\boldsymbol{A} + \boldsymbol{B}\boldsymbol{G}_1)^{-\top}\boldsymbol{C}^\top\boldsymbol{Q}. \tag{16.48}$$

MATLAB

The function `lqtracker.m` is implemented in the MSS toolbox for computation of the matrices $\boldsymbol{G}_1$ and $\boldsymbol{G}_2$.

```
function [G1,G2] = lqtracker(A,B,C,Q,R)
[K,P,E] = lqr(A,B,C'*Q*C,R);
G1 = -inv(R) * B' * P;
G2 = -inv(R) * B' * inv((A+B*G1)') * C' * Q;
```

For a mass–damper–spring system the optimal trajectory tracking controller can be designed is using the script `ExLQtrack.m`:

```
% Weights
Q = diag([1]);          % tracking error weights
R = diag([1]);          % input weights

% System matrices
A = [0 1; -1 -2];       % state matrix
B = [0; 1];             % input matrix
C = [1 0];              % output matrix

% Optimal gain matrices
[G1,G2] = lqtracker(A,B,C,Q,R)
```

16.1.4 Operability and Motion Sickness Incidence Criteria

Operability criteria for manual and intellectual work as well as motion sickness are important design criteria for the evaluation of autopilot and roll damping systems. Sea-sickness is especially important in high-speed craft and ships with high vertical accelerations.

Human operability limiting criteria in roll: Operability limiting criteria with regard to vertical and lateral accelerations, and roll angle for the effectiveness of the crew and the passengers are given in Table 16.1. This gives an indication on what type of work that can be expected to be carried out for different roll angles/sea states.

Table 16.1 Criteria for effectiveness of the crew (Faltinsen 1990).

Standard deviation (root mean square) criteria			
Vertical acceleration ($\dot{w}$)	Lateral acceleration ($\dot{v}$)	Roll angle (ϕ)	Description of work
0.20 g	0.10 g	6.0°	Light manual work
0.15 g	0.07 g	4.0°	Heavy manual work
0.10 g	0.05 g	3.0°	Intellectual work
0.05 g	0.04 g	2.5°	Transit passengers
0.02 g	0.03 g	2.0°	Cruise liner

ISO 2631-1 (1997) criterion for motion sickness incidence: In addition to operability, limiting criteria passenger comfort can be evaluated with respect to motion sickness. The most important factors for seasickness are vertical (heave) accelerations a_z (m s^{-2}), exposure time t (h) and encounter frequency ω_e (rad s^{-1}). The ISO standard criterion for MSI proposes an MSI of 10%, which means that 10% of the passengers become seasick during t h. The MSI curves as a function of exposure time are shown in Figure 16.4, where

$$a_z(t,\omega_e) = \begin{cases} 0.5\sqrt{2/t} & \text{for } \ 0.1 \text{ Hz} < \frac{\omega_e}{2\pi} \le 0.315 \text{ Hz} \\ 0.5\sqrt{2/t} \cdot 6.8837\left(\frac{\omega_e}{2\pi}\right)^{1.67} & \text{for } \ 0.315 \text{ Hz} \le \frac{\omega_e}{2\pi} \le 0.63 \text{ Hz}. \end{cases} \tag{16.49}$$

The MSI curves (16.49) as functions of the exposure time are implemented in the Matlab MSS toolbox as `[a_z,w_e] = ISOmsi(t)`. Figure 16.4 is generated by using the example script `ExMSI.m`. The main limitation of the ISO criterion is that it only predicts the exceedence of the 10% MSI point. It is also assumed that the accelerations in the CG are representative for the entire ship and that a representative wave period can be used instead of the actual wave. In many cases it is advantageous to use the extended sickness method for more accurate predictions. This method is presented below.

Probability integral method for MSI: The O'Hanlon and McCauley (1974) probability integral method is convenient to use since it produces an MSI criterion in percentage for combinations of heave acceleration a_z (m s^{-2}) and frequency of encounter ω_e (rad s^{-1}).

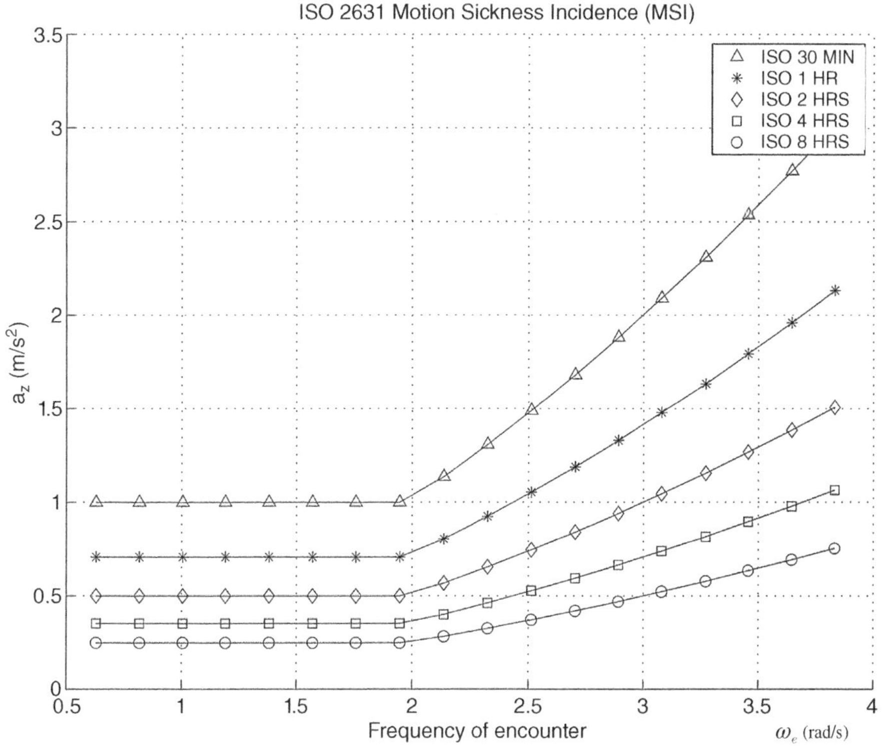

Figure 16.4 Heave acceleration a_z (m s^{-2}) as a function of frequency of encounter ω_e (rad s^{-1}) for different exposure times. The ISO curves represent an MSI of 10%.

The MSI index is defined as the number of sea sick people in percentage for an exposure time of 2 h (Lloyd 1989, Lewis 1989). The criterion is as follows

$$\mathrm{MSI} = 100\left[0.5 \pm \mathrm{erf}\left(\frac{\pm \log_{10}(a_z/g) \mp \mu_{\mathrm{MSI}}}{0.4}\right)\right] (\%) \tag{16.50}$$

where

$$\mu_{\mathrm{MSI}} = -0.819 + 2.32\ (\log_{10}\omega_e)^2 \tag{16.51}$$

$$\mathrm{erf}(x) = \mathrm{erf}(-x) = \frac{1}{\sqrt{2\pi}}\int_0^x \exp\left(-\frac{z^2}{2}\right)\mathrm{d}z. \tag{16.52}$$

The Matlab MSS toolbox function `msi = HMmsi(a_z,w_e)` can be used for computation of the MSI. Note that the erf function in `HMmsi.m` is scaled differently from the Matlab function `erf.m`. The MSI curves in Figure 16.5 are plotted for different a_z and ω_e using the example script `ExMSI.m`.

The major drawback of the O'Hanlon and McCauley method is that it only applies to a two hours exposure time. Another effect to take into account is that the O'Hanlon and

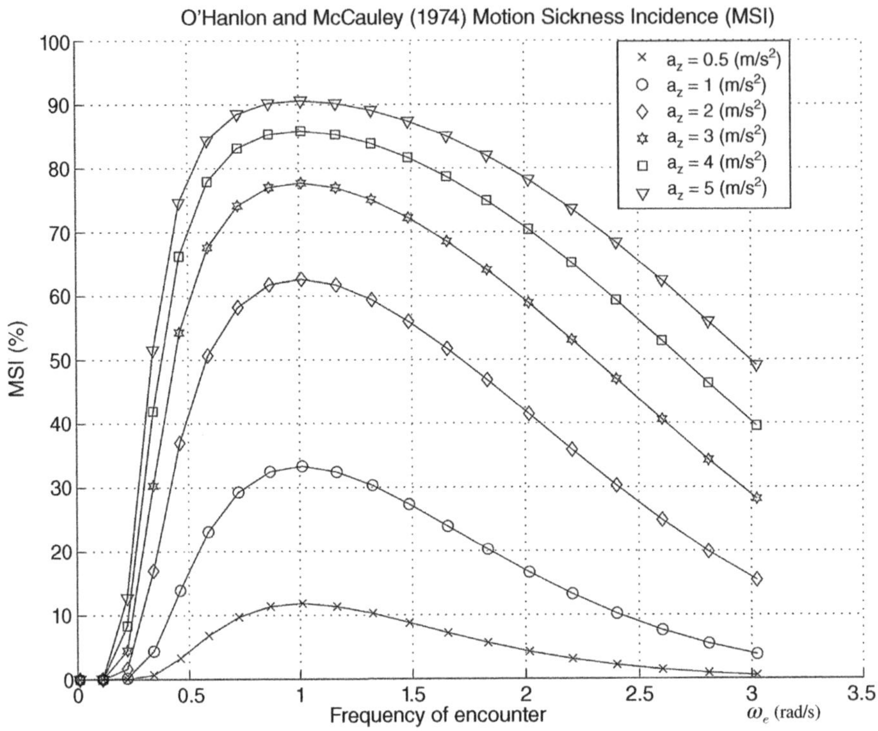

Figure 16.5 MSI is the number of motion sick persons in percentage during a two hour exposure time as a function of encounter frequency ω_e (rad s^{-1}) and heave acceleration a_z (ms^{-2}).

McCauley MSI criterion is derived from tests with young men seated separately in insulated cabins. According to ISO 2631-1, the MSI number is about 1.5 higher among women and children, suggesting that the actual MSI number for passengers of average age and sex distribution should be at least 1.25 times higher.

16.1.5 Case Study: Optimal Heading Autopilot for Marine Craft

Autopilots for rudder-controlled ships and underwater vehicles can be designed by considering a linear-quadratic optimization problem

$$J = \min_{\delta} \left\{ \frac{\alpha}{T} \int_0^T (e^2 + \lambda_1 r^2 + \lambda_2 \delta^2) \mathrm{d}\tau \right\} \tag{16.53}$$

where α is a constant to be interpreted later, $e = \psi_{\mathrm{d}} - \psi$ is the heading error, δ is the actual rudder angle and λ_1 and λ_2 are two factors weighting the cost of heading errors e and heading rate r against the control effort δ. This criterion can also be reformulated to describe marine craft that not are turned using a single rudder by replacing the quadratic term δ^2 with other control inputs. In the forthcoming, we will restrict our analysis to a single input.

For marine craft, operation in restricted waters usually requires accurate control, while the minimization of fuel consumption is more important in open seas. This can be obtained by changing the weights λ_1 and λ_2. We will discuss several criteria for control weighting that have all been derived by considering a ship. However, the same principles apply for underwater vehicles.

The Steering Criterion of Koyama

The first criterion was derived by Koyama (1967) who observed that the ship's swaying motion y could be approximated by a sinusoid

$$y = \sin(et) \quad \Rightarrow \quad \dot{y} = e\cos(et) \tag{16.54}$$

during autopilot control. The length of one arch L_a of the sinusoid is

$$L_a = \int_0^{\pi} \sqrt{(1+\dot{y}^2)}\,\mathrm{d}\tau = \int_0^{\pi} \sqrt{1+e^2\cos^2(e\tau)}\,\mathrm{d}\tau \approx \pi\left(1+\frac{e^2}{4}\right). \tag{16.55}$$

Hence, the relative elongation due to a sinusoidal course error is

$$\frac{\Delta L}{L} = \frac{L_a - L}{L} = \frac{\pi(1+e^2/4)-\pi}{\pi} = \frac{e^2}{4}. \tag{16.56}$$

This suggests that the percentage loss of speed during course control can be calculated by using the elongation in distance due to a sinusoidal course error. Consequently, Koyama (1967) proposed minimizing the speed loss term $e^2/4$ against the increased resistance due to steering given by the quadratic term δ^2. This motivates the following criterion

$$J = \min_{\delta}\left\{100\left(\frac{\pi}{180}\right)^2 \frac{1}{T}\int_0^T \left[\frac{e^2}{4} + \lambda_2\delta^2\right]\mathrm{d}\tau \approx \frac{0.0076}{T}\int_0^T [e^2 + \lambda_2\delta^2]\mathrm{d}\tau\right\}. \tag{16.57}$$

In this context (16.53) can be interpreted as

$$J = \text{loss of speed (\%)} \tag{16.58}$$

$$\alpha = 0.0076. \tag{16.59}$$

Note that $\lambda_1 = 0$ for this method. In practice it might be desirable to penalize r^2 by choosing $\lambda_1 > 0$. For ships, Koyama suggested a λ_2 factor of approximately 8–10. Experiments show that such high values for λ_2 avoid large rudder angles, and thus high turning rates. Therefore, $\lambda_2 = 10$ will be a good choice in bad weather, where it is important to suppress high-frequency rudder motions.

Norrbin's Steering Criterion

Another approach for computation of λ_2 was proposed by Norrbin (1972). Consider the surge dynamics of a rudder-controlled marine craft in the form

$$(m - X_{\dot{u}})\dot{u} = X_{|u|u}|u|u + (1-t)T + T_{\text{loss}} \tag{16.60}$$

where

$$T_{\text{loss}} = (m + X_{vr})vr + X_{cc\delta\delta}c^2\,\delta^2 + (X_{rr} + mx_g)r^2 + X_{\text{ext}}. \tag{16.61}$$

Norrbin (1972) suggested minimizing the loss term T_{loss} to obtain maximum forward speed u. Consequently, the controller should minimize the centripetal term vr, the square rudder angle δ^2 and the square heading rate r^2, while the unknown disturbance term X_{ext} is neglected in the analysis. The assumptions in doing this are as follows.

1. The sway velocity v is approximately proportional to r. From Section 7.2 it follows that

$$v(s) = \frac{K_{\text{v}}(T_v s + 1)}{K(Ts + 1)} r(s) \approx \frac{K_{\text{v}}}{K} r(s) \tag{16.62}$$

   if $T_v \approx T$. Hence, the centripetal term vr will be approximately proportional to the square of the heading rate; that is $vr \approx (K_{\text{v}}/K)r^2$.
2. The ship's yawing motion is periodic under autopilot control and the maximum yawing rate is $r^{\max} = \omega_{\text{r}}\, e^{\max}$ where ω_{r} is the frequency of the sinusoidal yawing.

These two assumptions suggest that the loss term T_{loss} can be minimized by minimizing e^2 and δ^2 which is the same result obtained in Koyama's analysis. The only difference between the criteria of Norrbin and Koyama is that the λ_2 values arising from Norrbin's approach will be different when computed for the same ship. The performance of the controller also depends on the sea state. This suggests that a trade-off between the λ_2 values proposed by Koyama and Norrbin could be made according to

$$\underbrace{\text{(calm sea)}}_{\text{Norrbin}} \qquad 0.1 \leq \lambda_2 \leq 10 \qquad \underbrace{\text{(rough sea)}}_{\text{Koyama}}. \tag{16.63}$$

Van Amerongen and Van Nauta Lemke's Steering Criterion

Experiments with the steering criteria of Koyama and Norrbin soon showed that the performance could be further improved by considering the squared yaw rate r^2, in addition to e^2 and δ^2 (Van Amerongen and Van Nauta Lemke 1978). Consequently, the following criterion was proposed

$$J = \min_{\delta}\left\{\frac{0.0076}{T}\int_0^T (e^2 + \lambda_1 r^2 + \lambda_2\delta^2)\,\mathrm{d}\tau\right\}. \tag{16.64}$$

For a tanker and a cargo ship, Van Amerongen and Van Nauta Lemke (1978, 1980) gave the following values for the weighting factors λ_1 and λ_2 corresponding to the data set of Norrbin (1972)

$$\begin{aligned} \text{Tanker:} \quad & L_{pp} = 300\ \text{m},\ \lambda_1 = 15\,000,\ \lambda_2 = 8.0 \\ \text{Cargo ship:} \quad & L_{pp} = 200\ \text{m},\ \lambda_1 = 1600,\ \lambda_2 = 6.0 \end{aligned}$$

The solution of the optimal steering criteria (16.64) is found by considering Nomoto's first-order model in non-dimensional form (see Appendix D.1.2)

$$\dot{\psi} = r \tag{16.65}$$

$$T'\dot{r} + (U/L)r = (U/L)^2 K'\ \delta. \tag{16.66}$$

Straightforward application of optimal control theory to the criterion of Van Amerongen and Van Nauta Lempke (1978) yields (see Section 16.1.3)

$$\delta = -K_p(\psi - \psi_d) - K_d r \tag{16.67}$$

where the controller gains are computed using the steady-state solution (16.8) and (16.9). This gives

$$K_p = \sqrt{\frac{1}{\lambda_2}} \tag{16.68}$$

$$K_d = \frac{L}{U} \frac{\sqrt{1 + 2K_p K'T' + K'^2 (U/L)^2\ (\lambda_1/\lambda_2)} - 1}{K'}. \tag{16.69}$$

An extension to Nomoto's second-order model is found by considering the state-space model (see Section 7.2)

$$\dot{\boldsymbol{x}} = \boldsymbol{A}\boldsymbol{x} + \boldsymbol{B}\boldsymbol{u} \tag{16.70}$$

$$\boldsymbol{y} = \boldsymbol{C}\boldsymbol{x} \tag{16.71}$$

where $\boldsymbol{x} = [v,\ r,\ \psi]^\top$, $\boldsymbol{u} = \delta$, $\boldsymbol{y} = [r,\ \psi]^\top$ and

$$\boldsymbol{A} = \begin{bmatrix} a_{11} & a_{12} & 0 \\ a_{21} & a_{22} & 0 \\ 0 & 1 & 0 \end{bmatrix}, \quad \boldsymbol{B} = \begin{bmatrix} b_1 \\ b_2 \\ 0 \end{bmatrix}, \quad \boldsymbol{C} = \begin{bmatrix} 0 & 1 & 0 \\ 0 & 0 & 1 \end{bmatrix}. \tag{16.72}$$

Let $\boldsymbol{y}_{\rm d} = [0,\ \psi_{\rm d}]^{\top} =$ constant and $\boldsymbol{e} = \boldsymbol{C}(\boldsymbol{x} - \boldsymbol{x}_{\rm d})$. The steady-state optimal solution minimizing the quadratic performance index

$$J = \min_{\boldsymbol{u}} \left\{ \frac{1}{2} \int_0^T (\boldsymbol{e}^{\top}\boldsymbol{Q}\boldsymbol{e} + \boldsymbol{u}^{\top}\boldsymbol{R}\boldsymbol{u}) {\rm d}\tau \right\} \tag{16.73}$$

where $\boldsymbol{Q} = {\rm diag}\{q_{11},\ q_{22}\} \geq 0$ and $\boldsymbol{R} = r_{11} > 0$ are the weights is (see Section 16.1.3)

$$\boldsymbol{u} = \boldsymbol{G}_1\boldsymbol{x} + \boldsymbol{G}_2\boldsymbol{y}_{\rm d} \tag{16.74}$$

where

$$\boldsymbol{G}_1 = -\boldsymbol{R}^{-1}\boldsymbol{B}^{\top}\boldsymbol{P}_{\infty} \tag{16.75}$$

$$\boldsymbol{G}_2 = -\boldsymbol{R}^{-1}\boldsymbol{B}^{\top}(\boldsymbol{A} + \boldsymbol{B}\boldsymbol{G}_1)^{-\top}\boldsymbol{C}^{\top}\boldsymbol{Q} \tag{16.76}$$

and $\boldsymbol{P}_{\infty}$ is the solution of the ARE

$$\boldsymbol{P}_{\infty}\boldsymbol{A} + \boldsymbol{A}^{\top}\boldsymbol{P}_{\infty} - \boldsymbol{P}_{\infty}\boldsymbol{B}\boldsymbol{R}^{-1}\boldsymbol{B}^{\top}\boldsymbol{P}_{\infty} + \boldsymbol{C}^{\top}\boldsymbol{Q}\boldsymbol{C} = \boldsymbol{0}. \tag{16.77}$$

The robustness of optimal autopilots for course-keeping control with a state estimator is analyzed in Holzhüter (1992).

16.1.6 Case Study: Optimal DP System for Surface Vessels

In Section 15.3.6 a nonlinear PID controller was designed for DP and the equilibrium point was rendered asymptotically stable under the assumption of full-state feedback. Output feedback in terms of a passive observer was also discussed. An alternative to the nonlinear PID controller is to formulate the problem as a linear-quadratic optimal control problem. The LQ controller will be designed under the assumption that all states can be measured. This assumption can, however, be relaxed by combining the LQ controller with a Kalman filter for optimal state estimation; see Section 13.4.6. The resulting control law is known as the LQG optimal controller, and convergence and stability of the interconnected system can be proven using a linear separation principle (Gelb *et al.* 1988).

DP Control Model

Recall from Section 6.7.3 that

$$\dot{\boldsymbol{\eta}} = \boldsymbol{R}(t)\boldsymbol{\nu} \tag{16.78}$$

$$\boldsymbol{M}\dot{\boldsymbol{\nu}} + \boldsymbol{D}\boldsymbol{\nu} = \boldsymbol{R}^{\top}(t)\boldsymbol{b} + \boldsymbol{\tau} \tag{16.79}$$

$$\dot{\boldsymbol{b}} = \boldsymbol{0}. \tag{16.80}$$

In order to incorporate the limitations of the propellers, the model is augmented by actuator dynamics. The simplest way of doing this is to define three time constants in *surge, sway* and *yaw* such that

$$\dot{\boldsymbol{\tau}} = \boldsymbol{A}_{\text{thr}}(\boldsymbol{\tau} - \boldsymbol{\tau}_{\text{com}}) \tag{16.81}$$

where $\boldsymbol{\tau}_{\text{com}}$ is the commanded thrust in surge, sway and yaw while $\boldsymbol{A}_{\text{thr}} = -\text{diag}\{1/T_{\text{surge}}, 1/T_{\text{sway}}, 1/T_{\text{yaw}}\}$ is a diagonal matrix containing the time constants. The resulting LTV state-space model becomes

$$\dot{\boldsymbol{x}}_{\text{c}} = \boldsymbol{A}_{\text{c}}(t)\boldsymbol{x}_{\text{c}} + \boldsymbol{B}_{\text{c}}\boldsymbol{\tau}_{\text{com}} \tag{16.82}$$

where the controller states are $\boldsymbol{x}_{\text{c}} := [\boldsymbol{\eta}^{\top}, \boldsymbol{\nu}^{\top}, \boldsymbol{\tau}^{\top}]^{\top}$ and

$$\boldsymbol{A}_{\text{c}}(t) = \begin{bmatrix} \boldsymbol{0}_{3\times3} & \boldsymbol{R}(t) & \boldsymbol{0}_{3\times3} \\ \boldsymbol{0}_{3\times3} & -\boldsymbol{M}^{-1}\boldsymbol{D} & \boldsymbol{M}^{-1} \\ \boldsymbol{0}_{3\times3} & \boldsymbol{0}_{3\times3} & \boldsymbol{A}_{\text{thr}} \end{bmatrix}, \quad \boldsymbol{B}_{\text{c}} = \begin{bmatrix} \boldsymbol{0}_{3\times3} \\ \boldsymbol{0}_{3\times3} \\ -\boldsymbol{A}_{\text{thr}} \end{bmatrix}. \tag{16.83}$$

This model is the basis for LQ control design.

Model-based State Estimator for DP

The Kalman filter can be designed using only position and heading angle measurements. For this purpose the filter states are chosen as $\boldsymbol{x}_{\text{f}} := [\boldsymbol{\eta}^{\top}, \boldsymbol{b}^{\top}, \boldsymbol{\nu}^{\top}]^{\top}$. The WF model is omitted for simplicity but in an industrial system six more states should be added following the approach in Section 13.4.6. The filter model takes the following form

$$\dot{\boldsymbol{x}}_{\text{f}} = \boldsymbol{A}_{\text{f}}(t)\boldsymbol{x}_{\text{f}} + \boldsymbol{B}_{\text{f}}\boldsymbol{\tau} + \boldsymbol{E}_{\text{f}}\boldsymbol{w} \tag{16.84}$$

$$\boldsymbol{y}_{\text{f}} = \boldsymbol{C}_{\text{f}}\boldsymbol{x}_{\text{f}} + \boldsymbol{\varepsilon} \tag{16.85}$$

where

$$\boldsymbol{A}_{\text{f}}(t) = \begin{bmatrix} \boldsymbol{0}_{3\times3} & \boldsymbol{0}_{3\times3} & \boldsymbol{R}(t) \\ \boldsymbol{0}_{3\times3} & \boldsymbol{0}_{3\times3} & \boldsymbol{0}_{3\times3} \\ \boldsymbol{0}_{3\times3} & \boldsymbol{M}^{-1}\boldsymbol{R}^{\top}(t) & -\boldsymbol{M}^{-1}\boldsymbol{D} \end{bmatrix}, \quad \boldsymbol{B}_{\text{f}} = \begin{bmatrix} \boldsymbol{0}_{3\times3} \\ \boldsymbol{0}_{3\times3} \\ \boldsymbol{M}^{-1} \end{bmatrix}, \quad \boldsymbol{C}_{\text{f}} = [\boldsymbol{I}_3, \ \boldsymbol{0}_{3\times6}]. \tag{16.86}$$

MATLAB

A DP model of a supply vessel length 76.2 m in surge, sway and yaw is included in the MSS toolbox as `supply.m`. The non-dimensional system matrices are (Fossen *et al.* 1996)

$$M'' = \begin{bmatrix} 1.1274 & 0 & 0 \\ 0 & 1.8902 & -0.0744 \\ 0 & -0.0744 & 0.1278 \end{bmatrix}, \quad D'' = \begin{bmatrix} 0.0358 & 0 & 0 \\ 0 & 0.1183 & -0.0124 \\ 0 & -0.0041 & 0.0308 \end{bmatrix}. \tag{16.87}$$

These values are defined in accordance to the bis system (see Appendix D.1) such that

$$M = mT^{-2}(TM''T^{-1}), \quad D = m\sqrt{g/L}\,T^{-2}(TD''T^{-1}) \tag{16.88}$$

where $T = \mathrm{diag}\{1,\ 1,\ L\}$.

Optimal Feedback Control

The control law is chosen as

$$\tau_{\mathrm{com}} = \tau_{\mathrm{LQ}} - \hat{\tau}_{\mathrm{wind}} \tag{16.89}$$

where $\hat{\tau}_{\mathrm{wind}}$ is an optional wind feedforward term based on (15.188) and τ_{LQ} is the optimal control law solving the LQ control objective

$$J = \min_{\tau_{\mathrm{LQ}}} \left\{ \frac{1}{2} \int_0^T (x_c^\top Q x_c + \tau_{\mathrm{LQ}}^\top R \tau_{\mathrm{LQ}})\, \mathrm{d}\tau \right\} \tag{16.90}$$

where $R = R^\top > 0$ and $Q = Q^\top \geq 0$ are two cost matrices to be specified by the user. The Q matrix is defined as $Q := \mathrm{diag}\{Q_1,\ Q_2,\ Q_3\}$ where the weights Q_1, Q_2 and Q_3 put penalty on position and heading η, velocity v and actuator dynamics τ, respectively. The optimal control law minimizing (16.90) is (see Section 16.1.1)

$$\tau_{\mathrm{LQ}} = \underbrace{-R^{-1} B_c^\top P}_{G} x_c \tag{16.91}$$

where P is the solution of the continuous-time Riccati differential equation

$$PA_c + A_c^\top P - PB_cR^{-1}B_c^\top P + Q = -\dot{P}. \tag{16.92}$$

Integral Action

In order to obtain zero steady-state errors in *surge*, *sway* and *yaw*, integral action must be included in the control law. Integral action can be obtained by using state augmentation. Since we want the three outputs $(x^n,\ y^n,\ \psi)$ to be regulated to zero, no more than three integral states can be augmented to the system. Define a new state variable

$$z_c := \int_0^t y_c(\tau)\mathrm{d}\tau \quad \Rightarrow \quad \dot{z}_c = y_c. \tag{16.93}$$

Here $\boldsymbol{y}_c$ is a subspace of $\boldsymbol{x}_c$ given by

$$\boldsymbol{y}_c = \boldsymbol{C}_c\boldsymbol{x}_c, \qquad \boldsymbol{C}_c = [\boldsymbol{I}_3,\ \boldsymbol{0}_{3\times6}]. \tag{16.94}$$

Next consider an augmented model with state vector $\boldsymbol{x}_a := [\boldsymbol{z}_c^\top,\ \boldsymbol{x}_c^\top]^\top$ such that

$$\dot{\boldsymbol{x}}_a = \boldsymbol{A}_a(t)\boldsymbol{x}_a + \boldsymbol{B}_a\boldsymbol{\tau}_{com} \tag{16.95}$$

where

$$\boldsymbol{A}_a(t) = \begin{bmatrix} \boldsymbol{0}_{3\times3} & \boldsymbol{C}_c \\ \boldsymbol{0}_{9\times3} & \boldsymbol{A}_c(t) \end{bmatrix}, \quad \boldsymbol{B}_a = \begin{bmatrix} \boldsymbol{0}_{3\times3} \\ \boldsymbol{B}_c \end{bmatrix}. \tag{16.96}$$

The performance index for the integral controller is

$$J = \min_{\tau_{LQ}} \left\{ \frac{1}{2}\int_0^T (\boldsymbol{x}_a^\top \boldsymbol{Q}_a \boldsymbol{x}_a + \boldsymbol{\tau}_{LQ}^\top \boldsymbol{R}\, \boldsymbol{\tau}_{LQ})\, \mathrm{d}\tau \right\} \tag{16.97}$$

where $\boldsymbol{R} = \boldsymbol{R}^\top > 0$ and

$$\boldsymbol{Q}_a = \begin{bmatrix} \boldsymbol{Q}_I & \boldsymbol{0} \\ \boldsymbol{0} & \boldsymbol{Q} \end{bmatrix} \geq 0. \tag{16.98}$$

The matrix $\boldsymbol{Q}_I = \boldsymbol{Q}_I^\top > 0$ is used to specify the integral times in *surge*, *sway* and *yaw*. The optimal PID controller is (see Section 16.1.1)

$$\boldsymbol{\tau}_{LQ} = \boldsymbol{G}_a\boldsymbol{x}_a = \boldsymbol{G}\boldsymbol{x}_c + \boldsymbol{G}_I \underbrace{\int_0^t \boldsymbol{y}_c(\tau)\,\mathrm{d}\tau}_{z_c} \tag{16.99}$$

where $\boldsymbol{G}_a = [\boldsymbol{G}_I,\ \boldsymbol{G}]$ and

$$\boldsymbol{G}_a = -\boldsymbol{R}^{-1}\boldsymbol{B}_a^\top \boldsymbol{P} \tag{16.100}$$

$$\boldsymbol{P}\boldsymbol{A}_a + \boldsymbol{A}_a^\top\boldsymbol{P} - \boldsymbol{P}\boldsymbol{B}_a\boldsymbol{R}^{-1}\boldsymbol{B}_a^\top\boldsymbol{P} + \boldsymbol{Q}_a = -\dot{\boldsymbol{P}}. \tag{16.101}$$

LQG Control – Linear Separation Principle

In practice only some of the states are measured. A minimum requirement is that the position and heading of the craft is measured such that velocities and bias terms can be estimated by a state estimator. This is usually done under the assumption that the states $\boldsymbol{x}$ can be replaced with the estimated states $\hat{\boldsymbol{x}}$ such that the optimal integral controller (16.99) can be modified as

$$\boldsymbol{\tau}_{\mathrm{LQ}} = \boldsymbol{G}\hat{\boldsymbol{x}}_{\mathrm{c}} + \boldsymbol{G}_{\mathrm{I}}\boldsymbol{C} \int_0^t \hat{\boldsymbol{x}}_{\mathrm{c}}(\tau)\, \mathrm{d}\tau \tag{16.102}$$

where the state estimate $\hat{\boldsymbol{x}}_{\mathrm{c}}$ can be computed using the following methods

- Kalman filter (Section 13.4.6)
- Passive observer (Section 13.5.1).

For the Kalman filter in cascade with the LQ controller there exists a *linear separation principle* guaranteeing that the estimated states converge to the true states (Athans and Falb 1966). This is called LQG control and it was first applied to design DP systems by Balchen et al. (1976, 1980a, 1980b) and Grimble et al. (1980a, 1980b). Optimal DP systems are used to maintain the position of offshore drilling and supply vessels (see Figure 16.6).

16.1.7 Case Study: Optimal Rudder-roll Damping Systems for Ships

The roll motion of ships and underwater vehicles can be damped by using fins alone or in combination with rudders. The main motivation for using roll stabilizing systems on merchant ships is to prevent cargo damage and to increase the effectiveness of the crew by avoiding or reducing seasickness. This is also important from a safety point of view. For naval ships critical marine operations include landing a helicopter, formation control, underway replenishment, or the effectiveness of the crew during combat.

Several *passive* and *active* (feedback control) systems have been proposed to accomplish roll reduction (Burger and Corbet 1960, Lewis 1967, Bhattacharyya 1978). Design methods for rudder-roll damping and fin stabilization systems are found in Perez (2005). Some passive solutions are:

Bilge keels: Bilge keels are fins in planes approximately perpendicular to the hull or near the turn of the bilge. The longitudinal extent varies from about 25–50% of the length of the ship. Bilge keels are widely used, are inexpensive but increase the hull resistance.

Figure 16.6 Oil production using a dynamically positioned semi-submersible. Source: Illustration by B. Stenberg.

In addition to this, they are effective mainly around the natural roll frequency of the ship. This effect significantly decreases with the speed of the ship. Bilge keels were first demonstrated in 1870.

Hull modifications: The shape and size of the ship hull can be optimized for minimum rolling using hydrostatic and hydrodynamic criteria. This must, however, be done before the ship is built.

Anti-rolling tanks: The most common anti-rolling tanks in use are free-surface tanks, U-tube tanks and diversified tanks (Holden and Fossen 2012). These systems provide damping of the roll motion even at small speeds. The disadvantages are the reduction in metacenter height due to free water surface effects and that a large amount of space is required. The earliest versions were installed about the year 1874.

The most widely used systems for *active* roll damping are:

Fin stabilizers: Fin stabilizers are highly useful devices for roll damping. They provide considerable damping if the speed of the ship is not too low. The disadvantage with additional fins is increased hull resistance and high costs associated with the installation, since at least two new hydraulic systems must be installed. Retractable fins are popular, since they are inside the hull when not in use (no additional drag). It should be noted that fins are not effective at low speed and that they cause underwater noise in addition to drag. Fin stabilizers were patented by John I. Thornycroft in 1889.

Rudder-roll damping (RRD): Roll damping by means of the rudder is relatively inexpensive compared to fin stabilizers, has approximately the same effectiveness and causes no drag or underwater noise if the system is turned off. However, RRD requires a relatively fast rudder to be effective; typically rudder rates of $\dot{\delta}_{max} = 5–20^\circ$ s^{-1} are needed. RRD will not be effective at low ship speeds.

Gyroscopic roll stabilizers: Gyroscopic roll stabilizers are typically used for boats and yachts under 100 ft (see Section 9.8.1). The ship gyroscopic stabilizer has a spinning rotor that generates a roll stabilizing moment that counteracts the wave-induced roll motions. Unlike stabilizing fins, the ship gyroscopic stabilizer can only produce a limited roll stabilizing moment and effective systems require approximately 3–5% of the craft's displacement.

For a history of ship stabilization, the interested reader is advised to consult Bennett (1991), while a detailed evaluation of different ship roll stabilization systems can be found in Sellars and Martin (1992).

Rudder-roll damping (RRD) was first suggested in the late 1970s; see Cowley and Lambert (1972, 1975), Carley (1975), Lloyd (1975) and Baitis (1980). Research in the early 1980 showed that it was indeed feasible to control the heading of a ship with at least one rudder while simultaneously using the rudder for roll damping. If only one rudder is used, this is an *underactuated control* problem. In the linear case this can be solved by *frequency separation* of the steering and roll modes since heading control can be assumed to be a low-frequency trajectory-tracking control problem while roll damping can be achieved at higher frequencies.

Before designing an RRD system the applicability of the control system in terms of effectiveness should be determined (Roberts 1993). For a large number of ships it is in fact impossible to obtain a significant roll damping effect due to limitations of the rudder servo and the relatively large inertia of the ship.

Motivated by the results in the 1970s, RRD was tested by the US Navy by Baitis et al. (1983, 1989), in Sweden by Källström (1987), Källström *et al.* (1988), Källström and Schultz (1990) and Källström and Theoren (1994), and in the Netherlands by Van Amerongen and coauthors. Van Amerongen *et al.* (1987), Van Amerongen and Van Nauta Lempke (1987) and Van der Klugt (1987) introduced LQG theory in RRD systems. A similar approach was proposed by Katebi *et al.* (1987), while adaptive RRD is discussed in Zhou (1990).

Blanke and co-workers have developed an RRD autopilot (Blanke *et al.* 1989) that was implemented by the Danish Navy on 14 ships (Munk and Blanke 1987). Sea trials show that some of the ships had less efficient RRD systems than others. In Blanke and Christensen (1993) it was shown that the cross-couplings between steering and roll were highly sensitive to parametric variations, which again resulted in robustness problems. Different loading conditions and varying rudder shapes have been identified as reasons for this (Blanke 1996). In Stoustrup *et al.* (1995) it has been shown that a robust RRD controller can be designed by separating the roll and steering specifications and then optimizing the two controllers independently. The coupling effects between the roll and yaw modes have also been measured in model scale and compared with full-scale trial results (Blanke and Jensen 1997), while a new approach to identification of steering-roll models has been presented by Blanke and Tiano (1997).

More recently H_∞ control has been used to deal with model uncertainties in RRD control systems. This allows the designer to specify frequency-dependent weights for frequency separation between the steering and roll modes; see Yang and Blanke (1997, 1998). Qualitative feedback theory (QFT) has also been applied to solve the combined RRD heading control problem under model uncertainty; see Hearns and Blanke (1998). Results from sea trials are reported in Blanke *et al.* (2000a).

Simulation and full-scale experimental results of RRD systems using a multivariate autoregressive model and the minimum AIC estimate procedure were reported by Oda et al. (1996, 1997). Experimental results with various control strategies were also reported by Sharif *et al.* (1996). A nonlinear RRD control system using sliding mode control for compensation of modeling errors was reported in Lauvdal and Fossen (1997).

A gain-scheduling algorithm for input rate and magnitude saturations in RRD damping systems has been developed by Lauvdal and Fossen (1998). This method is motivated by the automatic gain controller (AGC) by Van der Klugt (1987) and a technique developed for stabilization of integrator chains with input rate saturation.

In this section the focus will be on linear-quadratic optimal RRD. The interested reader is recommended to consult the references above and Perez (2005) for other design techniques.

Linear-quadratic Optimal RRD Control System

Consider the 4-DOF maneuvering model (6.150) in Section 6.6:

$$\dot{\boldsymbol{x}} = \boldsymbol{A}\boldsymbol{x} + \boldsymbol{B}u \tag{16.103}$$

where $\boldsymbol{x} = [v,\ p,\ r,\ \phi,\ \psi]^\top$ and

$$\phi = \boldsymbol{c}_{\text{roll}}^\top\, \boldsymbol{x}, \qquad \psi = \boldsymbol{c}_{\text{yaw}}^\top\, \boldsymbol{x}. \tag{16.104}$$

The transfer functions corresponding to (16.103) and (16.104) are

$$\frac{\phi}{\delta}(s) = \frac{b_2 s^2 + b_1 s + b_0}{s^4 + a_3 s^3 + a_2 s^2 + a_1 s + a_0} \approx \frac{K_{\text{roll}}\, \omega_{\text{roll}}^2\, (T_5 s + 1)}{(T_4 s + 1)(s^2 + 2\zeta\omega_{\text{roll}} s + \omega_{\text{roll}}^2)} \tag{16.105}$$

$$\frac{\psi}{\delta}(s) = \frac{c_3 s^3 + c_2 s^2 + c_1 s + c_0}{s(s^4 + a_3 s^3 + a_2 s^2 + a_1 s + a_0)} \approx \frac{K_{\text{yaw}}\, (T_3 s + 1)}{s(T_1 s + 1)(T_2 s + 1)}. \tag{16.106}$$

The control objective is a simultaneous heading control $\psi = \psi_{\text{d}} =$ constant and RRD ($p_{\text{d}} = \phi_{\text{d}} = 0$) using one control input. There will be a trade-off between accurate heading control (minimizing $\tilde{\psi} = \psi - \psi_{\text{d}}$) and control action needed to increase the natural frequency ω_{roll} and damping ratio ζ_{roll}. Also notice that it is impossible to regulate ϕ to a nonzero value while simultaneously controlling the heading angle to a nonzero value by means of a single rudder. This can easily be seen by performing a steady-state analysis of the closed-loop system. This suggests that the output of the controller should be specified as

$$\boldsymbol{y} = [p,\ r,\ \phi,\ \psi]^\top, \qquad \boldsymbol{y}_{\text{d}} = [0,\ 0,\ 0,\ \psi_{\text{d}}]^\top. \tag{16.107}$$

Choosing $\boldsymbol{y} = \boldsymbol{C}\boldsymbol{x}$ implies that

$$\boldsymbol{C} = \begin{bmatrix} 0 & 1 & 0 & 0 & 0 \\ 0 & 0 & 1 & 0 & 0 \\ 0 & 0 & 0 & 1 & 0 \\ 0 & 0 & 0 & 0 & 1 \end{bmatrix}. \tag{16.108}$$

Application of optimal control theory implies that the control objective should be specified as an optimization problem for course keeping, roll damping and minimum fuel consumption. The trade-off between these quantities can be expressed as

$$J = \min_{\boldsymbol{u}} \left\{ \frac{1}{2} \int_0^T (\tilde{\boldsymbol{y}}^\top \boldsymbol{Q} \tilde{\boldsymbol{y}} + \boldsymbol{u}^\top \boldsymbol{R} \boldsymbol{u}) \, \mathrm{d}\tau \right\} \tag{16.109}$$

where $\tilde{\boldsymbol{y}} = \boldsymbol{C}\tilde{\boldsymbol{x}} = \boldsymbol{x} - \boldsymbol{x}_\mathrm{d}$ and $\boldsymbol{x}_\mathrm{d} = [0,\ 0,\ 0,\ 0,\ \psi_\mathrm{d}]^\top$. Accurate steering is weighted against roll damping by specifying the cost matrix $\boldsymbol{Q} = \mathrm{diag}\{Q_\mathrm{p},\ Q_\mathrm{r},\ Q_\phi,\ Q_\psi\} \geq 0$, while $\boldsymbol{R} = \mathrm{diag}\{R_1,\ R_{21},\ \ldots,\ R_\mathrm{r}\} > 0$ weights the use of the different rudder servos.

The solution to the LQ trajectory-tracking problem is (see Section 16.1.3)

$$\boldsymbol{u} = \boldsymbol{G}_1 \boldsymbol{x} + \boldsymbol{G}_2 \boldsymbol{y}_\mathrm{d} \tag{16.110}$$

where

$$\boldsymbol{G}_1 = -\boldsymbol{R}^{-1} \boldsymbol{B}^\top \boldsymbol{P}_\infty \tag{16.111}$$

$$\boldsymbol{G}_2 = -\boldsymbol{R}^{-1} \boldsymbol{B}^\top (\boldsymbol{A} + \boldsymbol{B}\boldsymbol{G}_1)^{-\top} \boldsymbol{C}^\top \boldsymbol{Q} \tag{16.112}$$

with $\boldsymbol{P}_\infty = \boldsymbol{P}_\infty^\top > 0$ given by

$$\boldsymbol{P}_\infty \boldsymbol{A} + \boldsymbol{A}^\top \boldsymbol{P}_\infty - \boldsymbol{P}_\infty \boldsymbol{B} \boldsymbol{R}^{-1} \boldsymbol{B}^\top \boldsymbol{P}_\infty + \boldsymbol{C}^\top \boldsymbol{Q} \boldsymbol{C} = \boldsymbol{0}. \tag{16.113}$$

Frequency Separation and Bandwidth Limitations

Since $(\boldsymbol{A},\ \boldsymbol{B})$ is controllable and full-state feedback is applied, it is possible to move all the five poles of the system. The closed-loop system becomes

$$\begin{aligned} \dot{\boldsymbol{x}} &= \boldsymbol{A}\boldsymbol{x} + \boldsymbol{B}\boldsymbol{u} \\ &= \underbrace{(\boldsymbol{A} + \boldsymbol{B}\boldsymbol{G}_1^\top)}_{\boldsymbol{A}_\mathrm{c}} \boldsymbol{x} + \boldsymbol{B}\boldsymbol{G}_2 \underbrace{\boldsymbol{h}\psi_\mathrm{d}}_{\boldsymbol{y}_\mathrm{d}} \end{aligned} \tag{16.114}$$

where $\boldsymbol{h} = [0,\ 0,\ 0,\ 1]^{\top}$. The closed-loop transfer function in yaw is

$$\psi(s) = \boldsymbol{c}_{\text{yaw}}^{\top}(s\boldsymbol{I}_5 - \boldsymbol{A}_{\text{c}})^{-1}\boldsymbol{B}\boldsymbol{G}_2\boldsymbol{h}\psi_{\text{d}}(s) \tag{16.115}$$

which clearly satisfy $\lim_{t\to\infty}\psi(t) = \psi_{\text{d}}$. Note that integral action in yaw is needed in a practical implementation of the controller. Similarly, the closed-loop roll dynamics becomes

$$\phi(s) = \boldsymbol{c}_{\text{roll}}^{\top}(s\boldsymbol{I}_5 - \boldsymbol{A}_{\text{c}})^{-1}\boldsymbol{B}\boldsymbol{G}_2\boldsymbol{h}\psi_{\text{d}}(s). \tag{16.116}$$

If one rudder is used to control both ϕ and ψ, frequency separation is necessary to achieve this. Assume that the steering dynamics is slower than the frequency $1/T_{\text{l}}$ and that the natural frequency in roll is higher than $1/T_{\text{h}}$. Hence, the vertical reference unit (VRU) and compass measurements can be low- and high-pass filtered according to

$$\frac{\phi}{\phi_{\text{vru}}}(s) = h_{\text{h}}(s) = \frac{T_{\text{h}}s}{T_{\text{h}}s+1} \tag{16.117}$$

$$\frac{\psi}{\psi_{\text{compass}}}(s) = h_{\text{l}}(s) = \frac{1}{T_{\text{l}}s+1}. \tag{16.118}$$

It is also necessary to filter the roll and yaw rate measurements p and r. These signals can also be computed by numerical differentiation of ϕ_{vru} and ψ_{vru} using a state estimator. This suggests that the bandwidth of the yaw angle control system must satisfy $\omega_{\text{b}} \ll \omega_{\text{roll}}$ (frequency separation). This again implies that the low- and high-pass filters must satisfy the inequality

$$\underbrace{\omega_{\text{yaw}}}_{\substack{\text{cross-over}\\ \text{frequency}}} < \underbrace{\omega_{\text{b}}}_{\substack{\text{bandwidth}\\ \text{in yaw}}} < \underbrace{1/T_{\text{l}}}_{\substack{\text{low-pass filter}\\ \text{frequency}}} < \underbrace{1/T_{\text{h}}}_{\substack{\text{high-pass filter}\\ \text{frequency}}} < \underbrace{\omega_{\text{roll}}}_{\substack{\text{natural}\\ \text{frequency}}}$$

which clearly puts a restriction on the ships that can be stabilized. For many ships this requirement is impossible to satisfy due to limitations of the rudder servos and control forces.

Example 16.3 (RRD Control System Using One Rudder)
Let $\boldsymbol{G}_1 = [g_{11},\ g_{12},\ g_{13},\ g_{14},\ g_{15}]$ *and* $\boldsymbol{G}_2 = [0,\ 0,\ 0,\ g_{24}]$ *such that the solution (16.110) of the SISO LQ trajectory-tracking problem can be written*

$$\delta = [g_{11},\ g_{12},\ g_{13},\ g_{14},\ g_{15}]\ \boldsymbol{x} + g_{24}\psi_{\text{d}}. \tag{16.119}$$

Expanding (16.119) gives

$$\delta = \underbrace{-K_{\text{v}}v}_{\textit{sway feedback}}\ \underbrace{-K_{\text{p}}(\psi - \psi_{\text{d}}) - K_{\text{d}}r}_{\textit{PD heading controller}}\ \underbrace{-K_{\text{p}_1}p - K_{\text{p}_2}\phi}_{\textit{roll damper}} \tag{16.120}$$

where $K_v = -g_{11}$, $K_p = -g_{15} = g_{24}$, $K_d = -g_{13}$, $K_{p_1} = -g_{12}$ *and* $K_{p_2} = -g_{14}$. *Frequency separation suggests that*

$$\delta = h_l(s)\delta_{course} + h_h(s)\delta_{roll} \tag{16.121}$$

where the course and roll control systems are

$$\delta_{course} = -K_v v - K_p(\psi - \psi_d) - K_d r \tag{16.122}$$

$$\delta_{roll} = -K_{p_1} p - K_{p_2}\phi. \tag{16.123}$$

Alternatively, the gains can be computed by using pole placement. The two subsystems (6.155) and (6.156) with heading autopilot and RRD become (neglecting the interactions between the systems)

$$\begin{bmatrix} \dot{p} \\ \dot{\phi} \end{bmatrix} = \begin{bmatrix} \underbrace{a_{22} - b_{21}K_{p_1}}_{-2\zeta_{roll}\ \omega_{roll}} & \underbrace{(a_{24} - b_{21}K_{p_2})}_{-\omega_{roll}^2} \\ 1 & 0 \end{bmatrix} \begin{bmatrix} p \\ \phi \end{bmatrix} = 0 \tag{16.124}$$

$$\begin{bmatrix} \dot{v} \\ \dot{r} \\ \dot{\psi} \end{bmatrix} = \begin{bmatrix} a_{11} - b_{11}K_v & a_{13} - b_{11}K_d & -b_{11}K_p \\ a_{31} - b_{31}K_v & a_{33} - b_{31}K_d & -b_{31}K_p \\ 0 & 1 & 0 \end{bmatrix} \begin{bmatrix} v \\ r \\ \psi - \psi_d \end{bmatrix} = 0. \tag{16.125}$$

For roll we define

$$-\omega_{roll}^2 := a_{24} - b_{21}K_{r_2}, -2\zeta_{roll}\ \omega_{roll} := a_{22} - b_{21}K_{r_1}. \tag{16.126}$$

Solving for the gains gives

$$K_{r_1} = \frac{a_{22} + 2\zeta_{roll}\ \omega_{roll}}{b_{21}}, \qquad K_{r_2} = \frac{a_{24} + \omega_{roll}^2}{b_{21}} \tag{16.127}$$

where ζ_{roll} *and* ω_{roll} *are pole-placement design parameters that can be used instead of eigenvalues.*

MATLAB

The model of Son and Nomoto (see `ExRRD2.m` *in the MSS toolbox) has been used to demonstrate how an LQ optimal RRD control system can be designed. The linear state-space model for the container ship is*

```
A = [ -0.0406 -1.9614   0.2137   0.1336 0
       0.0011 -0.1326 -0.1246 -0.0331 0
      -0.0010  0.0147 -0.1163 -0.0006 0
       0        1        0        0       0
       0        0        1        0       0];

B = [ -0.0600 0.0035 0.0026 0 0]';
```

The controller gains were computed using

```
Q = diag([10000 1000 10 1])
R = 0.5
[G1,G2] = lqtracker(A,B,C,Q,R)

G1 =
     0.1631   -16.1193    -6.7655    -1.1644    -0.4472
G2 =
    -0.0000    -0.0000     0.0000     0.4472
```

Note that $g_{15} = -g_{24}$. *The open- and closed-loop poles are computed in Matlab by using the commands* `damp(A)` *and* `damp(A+B*G1)`; *see Table 16.2.*

It is seen that the natural frequency and relative damping ratio in roll are $\omega_{roll} = 0.193$ rad s^{-1} *and* $\zeta_{roll} = 0.36$, *respectively. This is improved to* $\omega_{roll} = 0.197$ rad s^{-1} *and* $\zeta_{roll} = 0.52$ *by roll feedback. It is difficult to increase the relative damping ratio further due to limitations of the steering machine* ($\dot{\delta}_{\max} = 20°$ s^{-1} *and* $\delta_{\max} = 20°$). *These values can, however, be changed in* `RRDcontainer.m`

Since the roll frequency ω_{roll} *is* 0.193 rad s^{-1} *and the cross-over frequency in yaw* ω_{yaw} *is* 0.03 rad s^{-1}, *see Figure 6.8 in Example 6.5, it is approximately one decade between the frequencies* ω_{yaw} *and* ω_{roll}. *Therefore, frequency separation can be obtained by choosing the low-pass and high-pass filter frequencies as* $1/T_l = 0.1$ rad s^{-1} *in yaw and* $1/T_h = 0.05$ rad s^{-1} *in roll, respectively. It is seen that the heading controller moves the poles to* $-0.061, -0.026$ *and* -0.131, *resulting in satisfactory course-changing capabilities (see Figure 16.7). It is also seen that the course-keeping performance is degraded during RRD. The additional yawing motion, typically 1–2° in amplitude, is the price paid for adding roll feedback to an autopilot system. Also notice that the right half-plane zero in the transfer function* ϕ/δ *given by (16.105) is unchanged since feedback only moves the poles.*

Performance Criterion for RRD

The percentage roll reduction of RRD system can be computed by using the following criterion of Oda *et al.* (1992)

$$\text{Roll reduction} = \frac{\sigma_{AP} - \sigma_{RRD}}{\sigma_{AP}} \times 100\% \tag{16.128}$$

where

σ_{AP} = standard deviation of roll rate during course-keeping (RRD off)

σ_{RRD} = standard deviation of roll rate during course-keeping (RRD on).

Table 16.2 Eigenvalues, damping ratios and frequencies for the RRD control system.

Eigenvalues		Damping		Frequencies (rad/s)	
Open loop	Closed loop	Open loop	Closed loop	Open loop	Closed loop
0	−0.061	–	1.00	–	0.016
−0.027	−0.026	1.00	1.00	0.027	0.026
−0.071 + 0.183*i*	−0.100 + 0.165*i*	0.36	0.52	0.197	0.193
−0.071 − 0.183*i*	−0.100 − 0.165*i*	0.36	0.52	0.197	0.193
−0.121	−0.131	1.00	1.00	0.121	0.131

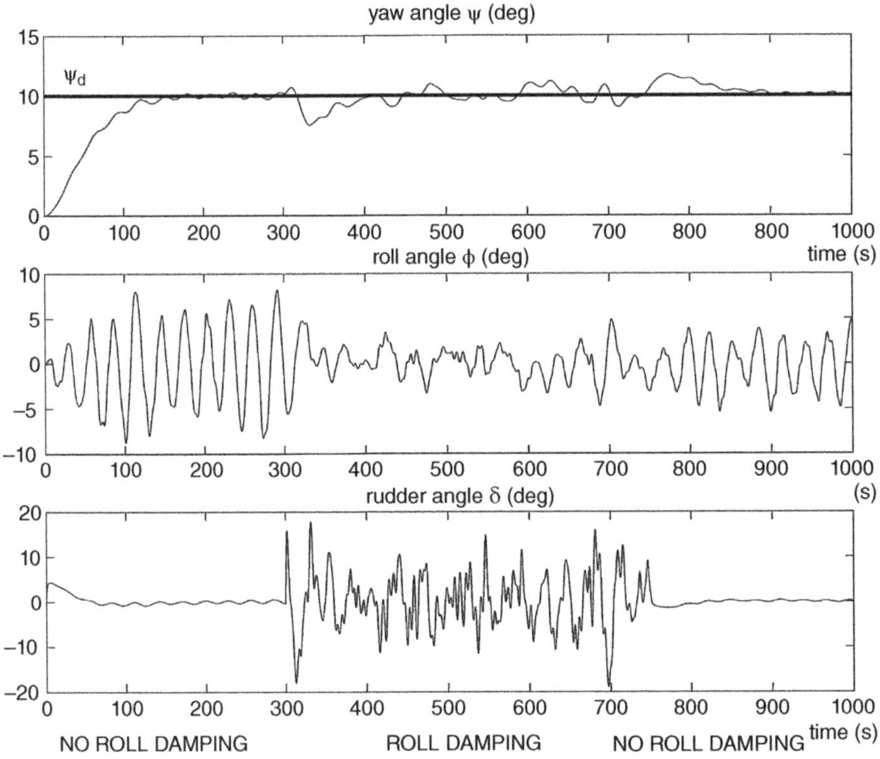

Figure 16.7 Performance of RRD control system during course-keeping and a 10° course-changing maneuver. The RRD system is active between $t = 300$–700 s.

For the case study in Example 16.3, $\sigma_{AP} = 0.0105$ and $\sigma_{RRD} = 0.0068$. This resulted in a roll reduction of approximately 35% during course-keeping. For small high-speed vessels a roll reduction as high as 50–75% can be obtained. This of course depends on the shape of the hull (hydrodynamic effects) and the capacity of the steering machine. In particular the maximum rudder rate $\dot{\delta}_{max}$ should be in the magnitude of 15–20% to obtain good results.

16.1.8 Case Study: Optimal Fin and RRD Systems for Ships

The most effective roll damping systems are those that combine stabilizing fins and rudders (Källström 1981, Roberts and Braham 1990, Perez 2005). Warship stabilization using integrated rudder and fins are discussed by Roberts (1992). Robust fin stabilizer controller design using the QFT and H_∞ design techniques have been presented by Hearns *et al.* (2000), while the performance of classical PID, optimized PID (Hickey *et al.* 2000) and H_∞ controllers are compared in Katebi *et al.* (2000). Sea trials with the *MV Barfleur* using PID and H_∞ controllers are presented in Hickey *et al.* (1997) and experimental results with a fin and RRD control system on board a frigate-size Royal Naval warship are reported in Sharif et al. (1995, 1996).

Reduction of vertical accelerations of fast ferries using fins and a T-foil is discussed by Esteban *et al.* (2000) and Giron-Sierra *et al.* (2001) while the modeling and identification results are reported in de-la-Cruz *et al.* (1998) and Aranda *et al.* (2000).

Fin stabilizers are useful for roll reduction since they are highly effective, work on a large number of ships and are more easier to control than RRD systems, even for varying load conditions and actuator configurations. Fin stabilizers are effective at high speed, but at the price of additional drag and added noise. The most economical systems are retractable fins, where additional drag is avoided during normal operation, since fin stabilizers are not needed in moderate weather. Another advantageous feature of fin stabilizing systems is that they can be used to control ϕ to a non-zero value (heel control). This is impossible with an RRD control system where the accurate control of ψ has priority.

Note that a stand-alone fin stabilization system can be constructed by simply removing the rudder inputs from the input matrix. When designing an LQ optimal fin and RRD system the following model representations can be used

$$\boldsymbol{M}\dot{\boldsymbol{v}} + \boldsymbol{D}\boldsymbol{v} = \boldsymbol{T}\boldsymbol{f}, \qquad \boldsymbol{f} = \boldsymbol{K}\boldsymbol{u}. \tag{16.129}$$

In this representation, $\boldsymbol{K}$ is the diagonal matrix of force coefficients and $\boldsymbol{T}$ is the actuator configuration matrix (see Section 11.2). We can premultiply (16.129) with $\boldsymbol{M}^{-1}$ to obtain

$$\dot{\boldsymbol{v}} = \underbrace{-\boldsymbol{M}^{-1}\boldsymbol{D}}_{\boldsymbol{A}}\boldsymbol{v} + \underbrace{\boldsymbol{M}^{-1}\boldsymbol{T}}_{\boldsymbol{B}}\boldsymbol{f}. \tag{16.130}$$

State-space models for combined steering and roll damping are found in Section 6.6.

Energy Optimal Criterion for Combined Fin and RRD

We will consider a ship equipped with r_1 rudders and r_2 fins. The total number of actuators is $r = r_1 + r_2$, implying that $\boldsymbol{u} \in \mathbb{R}^r$. The DOFs considered are *surge, sway, roll* and *yaw* implying that $\boldsymbol{v} = [u,\ v,\ p,\ r]^\top$; see Section 6.6. It is also assumed that the ship is fully actuated such that $r \geq n$. An energy optimal criterion weighting force $\boldsymbol{f}$ against accurate tracking and roll damping is

$$J = \min_{f} \left\{ \frac{1}{2} \int_0^T (e^{\top} Qe + f^{\top} R_{\mathrm{f}} f) \, \mathrm{d}t \right\}$$

$$= \min_{\tau} \left\{ \frac{1}{2} \int_0^T (e^{\top} Qe + \tau^{\top} \underbrace{(T_{\mathrm{w}}^{\dagger})^{\top} R_{\mathrm{f}} T_{\mathrm{w}}^{\dagger}}_{R_\tau} \tau) \, \mathrm{d}t \right\} \tag{16.131}$$

where $e = y - y_{\mathrm{d}}$ and $\tau = Tf$. The elements in $Q = \mathrm{diag}\{Q_p, Q_r, Q_\phi, Q_\psi\} \geq 0$ are used to weight accurate steering against roll damping. The rudder and fin servos are weighted against each other by specifying the elements in $R_{\mathrm{f}} = \mathrm{diag}\{R_{\delta_1}, R_{\delta_2}, \ldots, R_{\delta_{r_1}}, R_{f_1}, R_{f_2}, \ldots, R_{f_{r_2}}\} > 0$. If $r_1 = 0$ and $R_{\delta_1} = R_{\delta_2} = \ldots = R_{\delta_r} = 0$ only fin stabilization is obtained (no rudders).

The solution to the LQ problem (16.131) with generalized force $\tau = TKu$ as the control variable is (see Section 16.1.3)

$$\tau = G_1 x + G_2 y_{\mathrm{d}} \tag{16.132}$$

$$G_1 = -((T_{\mathrm{w}}^{\dagger})^{\top} R_{\mathrm{f}} T_{\mathrm{w}}^{\dagger})^{-1} B^{\top} P_\infty \tag{16.133}$$

$$G_2 = -((T_{\mathrm{w}}^{\dagger})^{\top} R_{\mathrm{f}} T_{\mathrm{w}}^{\dagger})^{-1} B^{\top} (A + BG_1)^{-\top} C^{\top} Q \tag{16.134}$$

where Q and R_{f} are design matrices while $P_\infty = P_\infty^{\top} > 0$ is given by

$$P_\infty A + A^{\top} P_\infty - P_\infty B[(T_{\mathrm{w}}^{\dagger})^{\top} R_{\mathrm{f}} T_{\mathrm{w}}^{\dagger}]^{-1} B^{\top} P_\infty + C^{\top} QC = \mathbf{0}. \tag{16.135}$$

16.2 State Feedback Linearization

The basic idea with feedback linearization is to transform the nonlinear system dynamics into a linear system (Freund 1973). Feedback linearization is discussed in more detail by Isidori (1989) and Slotine and Li (1991). Conventional control techniques such as pole-placement and linear-quadratic optimal control theory can then be applied to the linear system. In robotics, this technique is commonly referred to as *computed torque* control (Sciavicco and Siciliano 1996).

Feedback linearization is easily applicable to ships and underwater vehicles since these models basically are nonlinear *mass–damper–spring* systems, which can be transformed into a linear system by using a nonlinear mapping. Transformations that can be used for applications both in BODY and NED coordinates will be presented. Trajectory-tracking control in the BODY frame is used for velocity control while NED frame applications are recognized as position and attitude control. Combined position and velocity control systems will also be discussed.

16.2.1 Decoupling in the BODY Frame (Velocity Control)

The control objective is to transform the marine craft dynamics into a linear system

$$\dot{\boldsymbol{\nu}} = \boldsymbol{a}^b \tag{16.136}$$

where $\boldsymbol{a}^b$ can be interpreted as a body-fixed *commanded acceleration* vector. The body-fixed vector representation should be used to control the linear and angular velocities. Consider the nonlinear marine craft dynamics in the form

$$\boldsymbol{M}\dot{\boldsymbol{\nu}} + \boldsymbol{n}(\boldsymbol{\nu}, \boldsymbol{\eta}) = \boldsymbol{\tau} \tag{16.137}$$

where $\boldsymbol{\eta}$ and $\boldsymbol{\nu}$ are assumed to be measured and $\boldsymbol{n}$ is the nonlinear vector

$$\boldsymbol{n}(\boldsymbol{\nu}, \boldsymbol{\eta}) = \boldsymbol{C}(\boldsymbol{\nu})\boldsymbol{\nu} + \boldsymbol{D}(\boldsymbol{\nu})\boldsymbol{\nu} + \boldsymbol{g}(\boldsymbol{\eta}). \tag{16.138}$$

The nonlinearities can be canceled out by simply selecting the control law as (see Figure 16.8)

$$\boldsymbol{\tau} = \boldsymbol{M}\boldsymbol{a}^b + \boldsymbol{n}(\boldsymbol{\nu}, \boldsymbol{\eta}) \tag{16.139}$$

where the commanded acceleration vector $\boldsymbol{a}^b$ expressed in $\{b\}$ can be chosen by, for instance, pole placement or linear-quadratic optimal control theory. However, note that to investigate optimality of the original system, the optimal control and cost function must be transformed back through the nonlinear mapping.

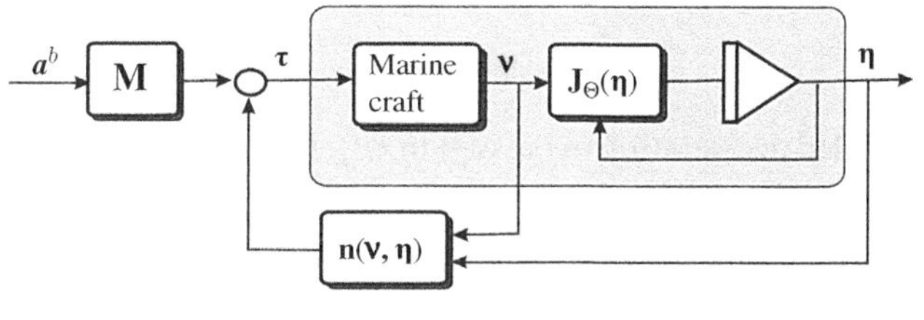

Figure 16.8 Nonlinear decoupling in the BODY frame.

Pole Placement

Let $\boldsymbol{\Lambda} > 0$ be a diagonal design matrix

$$\boldsymbol{\Lambda} = \mathrm{diag}\{\lambda_1, \ \lambda_2, \ \dots \ , \ \lambda_n\} \tag{16.140}$$

used to specify the desired control bandwidth, $\boldsymbol{v}_d$ the desired linear and angular velocity vector and $\tilde{\boldsymbol{v}} = \boldsymbol{v} - \boldsymbol{v}_d$ the velocity tracking error. Then the commanded acceleration vector can be chosen as a PI controller with acceleration feedforward

$$\boldsymbol{a}^b = \dot{\boldsymbol{v}}_d - \boldsymbol{K}_p\tilde{\boldsymbol{v}} - \boldsymbol{K}_i \int_0^t \tilde{\boldsymbol{v}}(\tau)\,d\tau. \tag{16.141}$$

Choosing the gains as

$$\boldsymbol{K}_p = 2\boldsymbol{\Lambda}, \qquad \boldsymbol{K}_i = \boldsymbol{\Lambda}^2 \tag{16.142}$$

yields the second-order error dynamics

$$\boldsymbol{M}(\dot{\tilde{\boldsymbol{v}}} - \boldsymbol{a}^b) = \boldsymbol{M}\left(\dot{\tilde{\boldsymbol{v}}} + 2\boldsymbol{\Lambda}\tilde{\boldsymbol{v}} + \boldsymbol{\Lambda}^2 \int_0^t \tilde{\boldsymbol{v}}(\tau)\,d\tau\right) = \boldsymbol{0}. \tag{16.143}$$

This implies that the poles are given by

$$(s + \lambda_i)^2 \int_0^t \tilde{v}(\tau)\,d\tau = 0 \quad (i = 1, \ \dots \ , n). \tag{16.144}$$

The reference model of Section12.1.1 can be used to generate a smooth velocity trajectory $\boldsymbol{v}_d$ for trajectory-tracking control.

16.2.2 *Decoupling in the NED Frame (Position and Attitude Control)*

For position and attitude control the dynamics are decoupled in the NED reference frame. Consider

$$\ddot{\boldsymbol{\eta}} = \boldsymbol{a}^n \tag{16.145}$$

where $\boldsymbol{a}^n$ is commanded acceleration expressed in $\{n\}$. Consider the kinematic and kinetic equations in the form

$$\dot{\boldsymbol{\eta}} = \boldsymbol{J}_\Theta(\boldsymbol{\eta})\boldsymbol{v} \tag{16.146}$$

$$\boldsymbol{M}\dot{\boldsymbol{v}} + \boldsymbol{n}(\boldsymbol{v}, \boldsymbol{\eta}) = \boldsymbol{\tau} \tag{16.147}$$

where both $\boldsymbol{\eta}$ and $\boldsymbol{\nu}$ are assumed measured. Differentiation of the kinematic equation (16.146) with respect to time yields

$$\dot{\boldsymbol{\nu}} = \boldsymbol{J}_{\Theta}^{-1}(\boldsymbol{\eta})[\ddot{\boldsymbol{\eta}} - \dot{\boldsymbol{J}}_{\Theta}(\boldsymbol{\eta})\boldsymbol{\nu}]. \tag{16.148}$$

The nonlinear control law

$$\boldsymbol{\tau} = \boldsymbol{M}\boldsymbol{a}^b + \boldsymbol{n}(\boldsymbol{\nu}, \boldsymbol{\eta}) \tag{16.149}$$

applied to (16.147) yields

$$\boldsymbol{M}(\dot{\boldsymbol{\nu}} - \boldsymbol{a}^b) = \boldsymbol{M}\boldsymbol{J}_{\Theta}^{-1}(\boldsymbol{\eta})[\ddot{\boldsymbol{\eta}} - \dot{\boldsymbol{J}}_{\Theta}(\boldsymbol{\eta})\boldsymbol{\nu} - \boldsymbol{J}_{\Theta}(\boldsymbol{\eta})\boldsymbol{a}^b] = \boldsymbol{0}. \tag{16.150}$$

Choosing

$$\boldsymbol{a}^n = \dot{\boldsymbol{J}}_{\Theta}(\boldsymbol{\eta})\boldsymbol{\nu} + \boldsymbol{J}_{\Theta}(\boldsymbol{\eta})\boldsymbol{a}^b \tag{16.151}$$

yields the linear decoupled system

$$\boldsymbol{M}^*(\ddot{\boldsymbol{\eta}} - \boldsymbol{a}^n) = \boldsymbol{0} \tag{16.152}$$

where $\boldsymbol{M}^* = \boldsymbol{J}_{\Theta}^{-\top}(\boldsymbol{\eta})\boldsymbol{M}\boldsymbol{J}_{\Theta}^{-1}(\boldsymbol{\eta}) > 0$. From (16.151) it is seen that

$$\boldsymbol{a}^b = \boldsymbol{J}_{\Theta}^{-1}(\boldsymbol{\eta})[\boldsymbol{a}^n - \dot{\boldsymbol{J}}_{\Theta}(\boldsymbol{\eta})\boldsymbol{\nu}] \tag{16.153}$$

where the commanded acceleration $\boldsymbol{a}^n$ can be chosen as a PID control law with acceleration feedforward

$$\boldsymbol{a}^n = \ddot{\boldsymbol{\eta}}_{\mathrm{d}} - \boldsymbol{K}_{\mathrm{d}}\dot{\tilde{\boldsymbol{\eta}}} - \boldsymbol{K}_{\mathrm{p}}\tilde{\boldsymbol{\eta}} - \boldsymbol{K}_{\mathrm{i}}\int_0^t \tilde{\boldsymbol{\eta}}(\tau)\,\mathrm{d}\tau \tag{16.154}$$

where $\boldsymbol{K}_{\mathrm{p}}$, $\boldsymbol{K}_{\mathrm{d}}$ and $\boldsymbol{K}_{\mathrm{i}}$ are positive definite matrices chosen such that the error dynamics

$$\ddot{\tilde{\boldsymbol{\eta}}} + \boldsymbol{K}_{\mathrm{d}}\dot{\tilde{\boldsymbol{\eta}}} + \boldsymbol{K}_{\mathrm{p}}\tilde{\boldsymbol{\eta}} + \boldsymbol{K}_{\mathrm{i}}\int_0^t \tilde{\boldsymbol{\eta}}(\tau)\,\mathrm{d}\tau = \boldsymbol{0} \tag{16.155}$$

is exponentially stable. One obvious pole-placement algorithm for PID control is

$$(s + \lambda_i)^3 \int_0^t \tilde{\eta}(\tau)\,\mathrm{d}\tau = \boldsymbol{0} \quad (i = 1, \ldots, n) \tag{16.156}$$

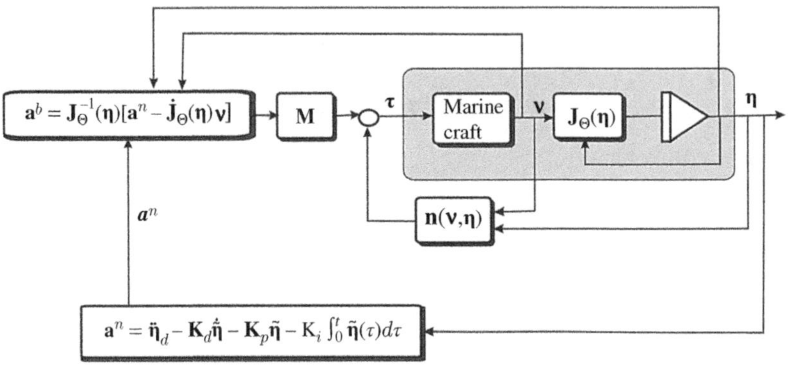

Figure 16.9 Nonlinear decoupling in the NED frame with transformation to the BODY frame.

which yields

$$\boldsymbol{K}_{\mathrm{d}} = 3\boldsymbol{\Lambda} = \mathrm{diag}\{3\lambda_1, 3\lambda_2, \ldots, 3\lambda_{\mathrm{n}}\}$$
$$\boldsymbol{K}_{\mathrm{p}} = 3\boldsymbol{\Lambda}^2 = \mathrm{diag}\{3\lambda_1^2, 3\lambda_2^2, \ldots, 3\lambda_{\mathrm{n}}^2\}$$
$$\boldsymbol{K}_{\mathrm{i}} = \boldsymbol{\Lambda}^3 = \mathrm{diag}\{\lambda_1^3, \lambda_2^3, \ldots, \lambda_{\mathrm{n}}^3\}.$$

This is shown in Figure 16.9. When implementing the trajectory-tracking controller a third-order reference model can be used to compute smooth position and attitude trajectories $\boldsymbol{\eta}_{\mathrm{d}}$ (see Section 12.1.1).

16.2.3 Case Study: Speed Control Based on Feedback Linearization

Consider the following decoupled model of a marine craft in surge

$$m\dot{u} + d_1 u + d_2|u|u = \tau. \tag{16.157}$$

From Section 16.2.1 it follows that the commanded acceleration can be calculated as

$$a^b = \dot{u}_{\mathrm{d}} - K_{\mathrm{p}}(u - u_{\mathrm{d}}) - K_i \int_0^t (u - u_{\mathrm{d}})\,\mathrm{d}\tau. \tag{16.158}$$

Hence, the speed controller takes the following form

$$\tau = m[\dot{u}_{\mathrm{d}} - K_{\mathrm{p}}(u - u_{\mathrm{d}}) - K_i \int_0^t (u - u_{\mathrm{d}})\,\mathrm{d}\tau] + d_1 u + d_2|u|u. \tag{16.159}$$

Let $\tilde{u} = u - u_{\mathrm{d}}$ such that

$$\dot{\tilde{u}} + K_{\mathrm{p}}\tilde{u} + K_i \int_0^t \tilde{u}(\tau)\,\mathrm{d}\tau = 0. \tag{16.160}$$

The equilibrium point $\tilde{u} = 0$ is globally exponentially stable if the gains are chosen as

$$K_p = 2\lambda, \qquad K_i = \lambda^2 \tag{16.161}$$

with $\lambda > 0$. In order to implement the speed controller, the following reference model can be used (see Section 12.1.1)

$$\ddot{u}_d + 2\zeta\omega_n\dot{u}_d + \omega_n^2 u_d = \omega_n^2 r^b \tag{16.162}$$

where $\zeta > 0$ and $\omega_n > 0$ are the reference model damping ratio and natural frequency while r^b is the setpoint specifying the desired surge velocity.

16.2.4 Case Study: Autopilot Based on Feedback Linearization

Consider the nonlinear model (see section 7.2.3)

$$\dot{\psi} = r \tag{16.163}$$

$$m\dot{r} + d_1 r + d_2|r|r = \tau \tag{16.164}$$

where ψ is the yaw angle. Hence, the commanded acceleration can be calculated as (Fossen and Paulsen 1992)

$$a^n = \dot{r}_d - K_d(r - r_d) - K_p(\psi - \psi_d) - K_i \int_0^t (\psi - \psi_d)\, d\tau \tag{16.165}$$

where r_d is the desired yaw rate and ψ_d is the desired heading angle. For this particular example, (16.153) implies that $a^n = a^b$. Choosing the decoupling control law as

$$\tau = m\left[\dot{r}_d - K_d(r - r_d) - K_p(\psi - \psi_d) - K_i \int_0^t (\psi - \psi_d)\, d\tau\right] + d_1 r + d_2|r|r \tag{16.166}$$

gives the closed-loop system

$$\dot{\tilde{\psi}} = \tilde{r} \tag{16.167}$$

$$\dot{\tilde{r}} + K_d\tilde{r} + K_p\tilde{\psi} = 0 \tag{16.168}$$

where $\tilde{\psi} = \psi - \psi_d$ and $\tilde{r} = r - r_d$. The reference model can be chosen as (see Section 12.1.1)

$$\psi_d^{(3)} + (2\zeta + 1)\omega_n\ddot{\psi}_d + (2\zeta + 1)\omega_n^2\dot{\psi}_d + \omega_n^3\psi_d = \omega_n^3 r^n. \tag{16.169}$$

Note that (16.166) depends on the uncertain parameters d_1 and d_2 while m is quite easy to estimate using hydrodynamic programs. Hence, care must be taken when implementing (16.166). For most craft, the control law (16.166) works very well even with $d_1 = d_2 = 0$ so the need for choosing non-zero damping parameters should be seen as a trade-off between robustness and performance.

16.3 Integrator Backstepping

Backstepping is a design methodology for construction of a feedback control law through a *recursive* construction of a control Lyapunov function. Nonlinear backstepping designs are strongly related to feedback linearization. However, while feedback linearization methods cancel all nonlinearities in the system it will be shown that when applying the backstepping design methodology more design flexibility is obtained. In particular, the designer is given the possibility to exploit "good" nonlinearities while "bad" nonlinearities can be dominated by adding nonlinear damping, for instance. Hence, additional robustness is obtained, which is important in industrial control systems since cancellation of all nonlinearities requires precise models that are difficult to obtain in practice.

16.3.1 A Brief History of Backstepping

The idea of integrator backstepping seems to have appeared simultaneously, often implicit, in the works of Koditschek (1987), Sonntag and Sussmann (1988), Tsinias (1989) and Byrnes and Isidori (1989). Stabilization through an integrator (Kokotovic and Sussmann 1989) can be viewed as a special case of stabilization through an SPR transfer function, which is a frequently used technique in the early adaptive designs (Parks 1966, Landau 1979, Narendra and Annaswamy 1989). Extensions to nonlinear cascades by using passivity arguments have been done by Ortega (1991) and Byrnes *et al.* (1991). Integrator backstepping appeared as a recursive design technique in Saberi *et al.* (1990) and was further developed by Kanellakopoulos *et al.* (1992). The relationship between backstepping and passivity has been established by Lozano *et al.* (1992). For the interested reader, a tutorial overview of backstepping is given in Kokotovic (1991).

Adaptive and nonlinear backstepping designs are described in detail by Krstic *et al.* (1995). This includes methods for parameter adaptation, tuning functions and modular designs for both full-state feedback and output feedback (observer backstepping). Sepulchre *et al.* (1997) make extensions to forwarding, passivity and cascaded designs. Also discussions on stability margins and optimality are included. The concept of vectorial backstepping was first introduced by Fossen and Berge (1997). Vectorial backstepping exploits the structural properties of nonlinear MIMO systems and this simplifies design and analysis significantly. Krstic and Deng (1998) present stochastic systems with a focus on stochastic stability and regulation.

The focus of this section is practical designs with implementation considerations for mechanical systems. This is done by exploiting the nonlinear system properties of mechanical systems such as dissipativeness (good damping), symmetry of the inertia matrix and the skew-symmetric property of the Coriolis and centripetal matrix. In addition, emphasis is placed on control design with integral action. Two techniques for integral action in nonlinear systems using backstepping designs are discussed (Loria *et al.* 1999, Fossen *et al.* 2001).

16.3.2 The Main Idea of Integrator Backstepping

Integrator backstepping is a *recursive* design technique using *control Lyapunov functions (CLF)*. The CLF concept is a generalization of Lyapunov design results (Jacobson 1977, Jurdjevic and Quinn 1978).

Definition 16.2 (Control Lyapunov Function)
A smooth positive definite and radially unbounded function $V : \mathbb{R}^n \rightarrow \mathbb{R}_+$ is called a control Lyapunov function for (Arstein 1983, Sontag 1983)

$$\dot{\boldsymbol{x}} = \boldsymbol{f}(\boldsymbol{x}, \boldsymbol{u}) \tag{16.170}$$

where $\boldsymbol{x} \in \mathbb{R}^n$ and $\boldsymbol{u} \in \mathbb{R}^r$ if

$$\inf_{\boldsymbol{u}\in\mathbb{R}^r} \left\{ \frac{\partial V}{\partial \boldsymbol{x}}(\boldsymbol{x})\boldsymbol{f}(\boldsymbol{x}, \boldsymbol{u}) \right\} < 0, \quad \forall \boldsymbol{x} \neq \boldsymbol{0}. \tag{16.171}$$

The main idea of integrator backstepping can be demonstrated by considering a nonlinear scalar system

$$\dot{x}_1 = f(x_1) + x_2 \tag{16.172}$$

$$\dot{x}_2 = u \tag{16.173}$$

$$y = x_1 \tag{16.174}$$

where $x_1 \in \mathbb{R}$, $x_2 \in \mathbb{R}$, $y \in \mathbb{R}$ and $u \in \mathbb{R}$. The second equation represents a pure integrator (see Figure 16.10).

Let the design objective be regulation of $y \rightarrow 0$ as $t \rightarrow \infty$. The only equilibrium point with $y = 0$ is $(x_1, x_2) = (0, -f(0))$ corresponding to $\dot{x}_1 = f(0) + x_2 = 0$. The design objective is to render the equilibrium point GAS or GES. Since the nonlinear system (16.172) and (16.173) consists of two states x_1 and x_2, this will be a recursive design in two steps. Equations (16.172) and (16.173) are therefore treated as two cascaded systems, each with a single input and output. The recursive design starts with the system x_1 and continues with x_2. A change of coordinates

$$\boldsymbol{z} = \boldsymbol{\phi}(\boldsymbol{x}), \qquad \boldsymbol{x} = \boldsymbol{\phi}^{-1}(\boldsymbol{z}) \tag{16.175}$$

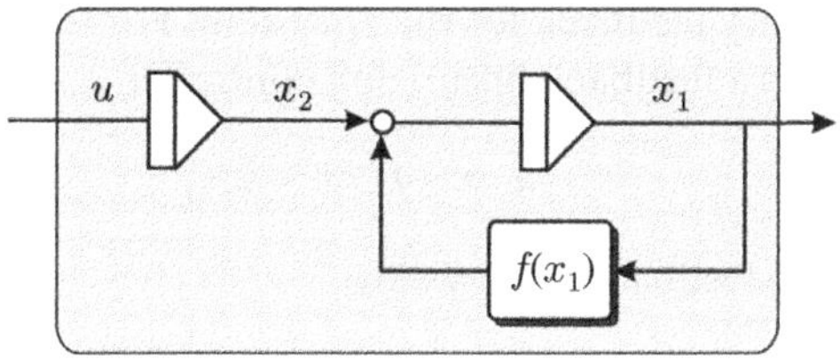

Figure 16.10 Second-order nonlinear system with one single nonlinearity $f(x_1)$ and a pure integrator at the input.

is introduced during the recursive design process where z is a new state vector and $\phi(x)$: $\mathbb{R}^n \rightarrow \mathbb{R}^n$ is a transformation to be interpreted later. The backstepping transformation is a *global diffeomorphism*, that is a mapping with smooth functions $\phi(x)$ and $\phi^{-1}(x)$.

Step 1: For the first system (16.172) the state x_2 is chosen as a *virtual control* input while it is recalled that our design objective is to regulate the output $y = x_1$ to zero. Hence, the first backstepping variable is chosen as

$$z_1 = x_1. \tag{16.176}$$

The virtual control is defined as

$$x_2 := \alpha_1 + z_2 \tag{16.177}$$

where

α_1 = stabilizing function
z_2 = new state variable.

Hence, the z_1 system can be written

$$\dot{z}_1 = f(z_1) + \alpha_1 + z_2. \tag{16.178}$$

The new state variable z_2 will not be used in the first step, but its presence is important since z_2 is needed to couple the z_1 system to the next system, that is the z_2 system to be considered in the next step. Moreover, integrator backstepping implies that the coordinates during the recursive design are changed from $x = [x_1, \ x_2]^\top$ to $z = [z_1, \ z_2]^\top$. A CLF for the z_1 system is

$$V_1 = \frac{1}{2}z_1^2 \tag{16.179}$$

$$\begin{aligned}\dot{V}_1 &= z_1\dot{z}_1 \\ &= z_1(f(z_1) + \alpha_1) + z_1 z_2.\end{aligned} \tag{16.180}$$

We now turn our attention to the design of the stabilizing function α_1 which will provide the necessary feedback for the z_1 system. For instance, choosing the stabilizing function as a feedback linearizing controller

$$\alpha_1 = -f(z_1) - k_1 z_1 \tag{16.181}$$

where $k_1 > 0$ is the feedback gain, yields

$$\dot{V}_1 = -k_1 z_1^2 + z_1 z_2 \tag{16.182}$$

and

$$\dot{z}_1 = -k_1 z_1 + z_2. \tag{16.183}$$

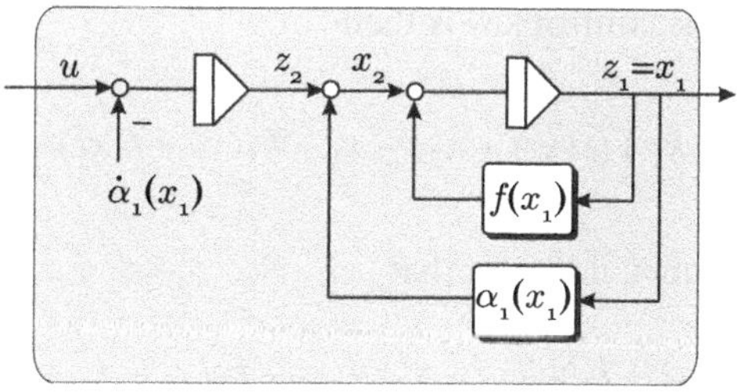

Figure 16.11 Stabilization of the x_1 system by means of the stabilizing function $\alpha_1(x_1)$. Note that $-\dot{\alpha}_1(x_1)$ when integrated cancels out the feedback term $\alpha_1(x_1)$.

A block diagram showing the stabilizing function and the new state variable is shown in Figure 16.11. Hence, if $z_2 = 0$ then the z_1 system is stabilized. We now turn our attention to the z_2 system.

Step 2: The z_2 dynamics is computed by time differentiation of (16.177)

$$\begin{aligned}\dot{z}_2 &= \dot{x}_2 - \dot{\alpha}_1 \\ &= u - \dot{\alpha}_1.\end{aligned} \tag{16.184}$$

A CLF for the z_2 system is

$$V_2 = V_1 + \frac{1}{2}z_2^2 \tag{16.185}$$

$$\begin{aligned}\dot{V}_2 &= \dot{V}_1 + \dot{z}_2 z_2 \\ &= (-k_1 z_1^2 + z_1 z_2) + \dot{z}_2 z_2 \\ &= -k_1 z_1^2 + z_2(z_1 + \dot{z}_2) \\ &= -k_1 z_1^2 + z_2(u - \dot{\alpha}_1 + z_1).\end{aligned} \tag{16.186}$$

Since our system has relative degree two, the control input u appears in the second step. Hence, choosing the control law as

$$u = \dot{\alpha}_1 - z_1 - k_2 z_2 \tag{16.187}$$

with $k_2 > 0$ yields

$$\dot{V}_2 = -k_1 z_1^2 - k_2 z_2^2 < 0, \quad \forall z_1 \neq 0, z_2 \neq 0. \tag{16.188}$$

Implementation Aspects

When implementing the control law (16.187) it is important to avoid expressions involving the time derivatives of the states. For this simple system only $\dot{\alpha}_1$ must be evaluated. This can be done by time differentiation of $\alpha_1(x_1)$ along the trajectory of x_1. Hence, $\dot{\alpha}_1$ can be computed without using the state derivatives

$$\begin{aligned}\dot{\alpha}_1 &= -\frac{\partial f(x_1)}{\partial x_1}\dot{x}_1 - k_1\dot{x}_1 \\ &= -\left(\frac{\partial f(x_1)}{\partial x_1} + k_1\right)(f(x_1) + x_2).\end{aligned} \tag{16.189}$$

The final expression for the control law is then

$$u = -\left(\frac{\partial f(x_1)}{\partial x_1} + k_1\right)(f(x_1) + x_2) - x_1 - k_2(x_2 + f(x_1) + k_1 x_1). \tag{16.190}$$

If $f(x_1) = -x_1$ (linear theory), it is seen that

$$\begin{aligned} u &= -(-1 + k_1)(-x_1 + x_2) - x_1 - k_2(x_2 - x_1 + k_1 x_1) \\ &= -\underbrace{(2 + k_1 k_2 - k_1 - k_2)}_{K_p} x_1 - \underbrace{(k_1 + k_2 - 1)}_{K_d} x_2 \end{aligned} \tag{16.191}$$

which is a standard PD control law. In general, the expression for u is a nonlinear feedback control law depending on the nonlinear function $f(x_1)$.

Backstepping Coordinate Transformation

The backstepping coordinate transformation $\boldsymbol{z} = \boldsymbol{\phi}(\boldsymbol{x})$ takes the form

$$\begin{bmatrix} z_1 \\ z_2 \end{bmatrix} = \begin{bmatrix} x_1 \\ x_2 + f(x_1) + k_1 x_1 \end{bmatrix} \tag{16.192}$$

while the inverse transformation $\boldsymbol{x} = \boldsymbol{\phi}^{-1}(\boldsymbol{z})$ is

$$\begin{bmatrix} x_1 \\ x_2 \end{bmatrix} = \begin{bmatrix} z_1 \\ z_2 - f(z_1) - k_1 z_1 \end{bmatrix}. \tag{16.193}$$

If you have performed the backstepping design procedure correctly the dynamics of the closed-loop system in (z_1, z_2) coordinates can always be written as the sum of a diagonal and skew-symmetric matrix times the state vector. This can be seen by writing the resulting dynamics in the form

$$\begin{bmatrix} \dot{z}_1 \\ \dot{z}_2 \end{bmatrix} = -\underbrace{\begin{bmatrix} k_1 & 0 \\ 0 & k_2 \end{bmatrix}}_{\text{diagonal matrix}} \begin{bmatrix} z_1 \\ z_2 \end{bmatrix} + \underbrace{\begin{bmatrix} 0 & 1 \\ -1 & 0 \end{bmatrix}}_{\text{skew-symmetrical matrix}} \begin{bmatrix} z_1 \\ z_2 \end{bmatrix} \tag{16.194}$$

or equivalently

$$\dot{\boldsymbol{z}} = -\boldsymbol{K}\boldsymbol{z} + \boldsymbol{S}\boldsymbol{z} \tag{16.195}$$

where $z = [z_1, \; z_2]^\top$, $K = \text{diag}\{k_1, \; k_2\} > 0$ and

$$S = -S^\top = \begin{bmatrix} 0 & 1 \\ -1 & 0 \end{bmatrix} \tag{16.196}$$

where S satisfies $z^\top S z = 0, \forall z$. In some cases the diagonal matrix will be a function of the state; that is $K(z) > 0$. This is the case when nonlinear damping is added or when some of the nonlinearities not are canceled by the controller.

Investigation of Stability

It is also seen that

$$V_2 = \frac{1}{2} z^\top z \tag{16.197}$$

$$\begin{aligned} \dot{V}_2 &= z^\top(-Kz + Sz) \\ &= -z^\top K z. \end{aligned} \tag{16.198}$$

Hence, Lyapunov's direct method for autonomous systems ensures that the equilibrium point $(x_1, x_2) = (0, -f(0))$ is GAS. In fact, this system will also be GES since it can be shown that the state vector x decays exponentially to zero by using Theorem A.3; that is

$$\|z(t)\|_2 \leq e^{-\beta(t-t_0)} \|z(t_0)\|_2 \tag{16.199}$$

where $\beta = \lambda_{\min}(K) > 0$ is the convergence rate.

A generalization to SISO mass–damper–spring systems is done in Section 16.3.3 while extensions to MIMO control are made in Section 16.3.6.

Backstepping Versus Feedback Linearization

The backstepping control law of the previous section is in fact equal to a feedback linearizing controller since the nonlinear function $f(x_1)$ is perfectly compensated for by choosing the stabilizing function as

$$\alpha_1 = -f(x_1) - k_1 z_1. \tag{16.200}$$

The disadvantage with this approach is that a perfect model is required. This is impossible in practice. Consequently, an approach of canceling all the nonlinearities may be sensitive for modeling errors.

One of the nice features of backstepping is that the stabilizing functions can be modified to exploit so-called "good" nonlinearities. For instance, assume that

$$f(x_1) = -a_0 x_1 - a_1 x_1^2 - a_2 |x_1| x_1 \tag{16.201}$$

where a_0, a_1 and a_2 are assumed to be *unknown* positive constants. Since both $a_0 x_1$ and $a_2 |x_1| x_1$ tend to damp out the motion these two expressions should be exploited in the control design and therefore not canceled out. On the contrary, the destabilizing term $a_1 x_1^2$ must be perfectly compensated for or dominated by adding a nonlinear damping term proportional to x_1^3 (remember that $z_1 = x_1$).

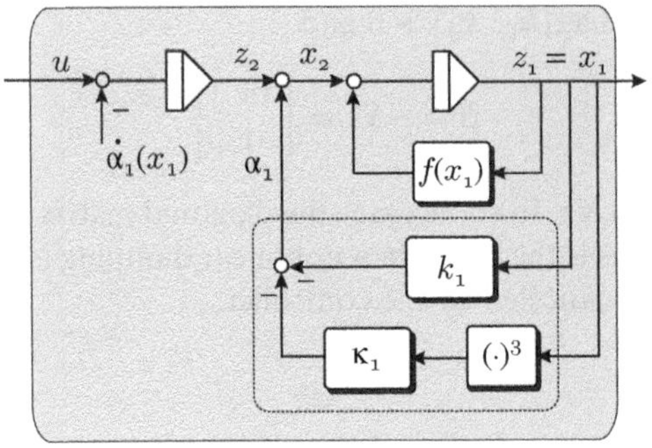

Figure 16.12 Domination of destabilizing terms by adding nonlinear damping.

Nonlinear damping suggests the following candidate for the stabilizing function

$$\alpha_1 = \underbrace{-k_1 z_1}_{\text{linear damping}} \underbrace{-\kappa_1 z_1^3}_{\text{nonlinear damping}} \tag{16.202}$$

where $k_1 > 0$ and $\kappa_1 > 0$ (see Figure 16.12). Hence,

$$\begin{aligned} \dot{z}_1 &= f(z_1) + (\alpha_1 + z_2) \\ &= -a_0 z_1 - a_1 z_1^2 - a_2 |z_1| z_1 - (k_1 + \kappa_1 z_1^2) z_1 + z_2 \\ &= -\underbrace{(a_0 + a_2|z_1| + k_1)}_{\text{good damping}} z_1 - \underbrace{a_1 z_1^2}_{\text{bad damping}} - \kappa_1 z_1^3 + z_2. \end{aligned} \tag{16.203}$$

Consider the CLF

$$V_1 = \frac{1}{2} z_1^2 \tag{16.204}$$

$$\dot{V}_1 = -(a_0 + a_2|z_1| + k_1) z_1^2 - a_1 z_1^3 - \kappa_1 z_1^4 + z_1 z_2. \tag{16.205}$$

In the next step it is seen that

$$V_2 = V_1 + \frac{1}{2} z_2^2 \tag{16.206}$$

$$\dot{V}_2 = -(\underbrace{a_0 + a_2|z_1|}_{\text{energy dissipation}} + k_1) z_1^2 \underbrace{-a_1 z_1^3}_{\text{energy dissipation/ generation}} - \kappa_1 z_1^4 + z_2(z_1 + u - \dot{\alpha}_1). \tag{16.207}$$

From this expression it can be concluded that the good damping terms contribute to the energy dissipation. The bad damping term, however, must be dominated by the nonlinear damping term. Choosing

$$u = \dot{\alpha}_1 - k_2 z_2 - z_1 \tag{16.208}$$

finally yields

$$\dot{V}_2 = -(a_0 + a_2|z_1| + k_1)z_1^2 - a_1 z_1^3 - \kappa_1 z_1^4 - k_2 z_2^2. \tag{16.209}$$

This expression can be rewritten by *completing the squares*. Consider the expression

$$\left(\frac{1}{2\sqrt{\kappa_1}}x + \sqrt{\kappa_1}y\right)^2 = \frac{1}{4\kappa_1}x^2 + xy + \kappa_1 y^2 \geq 0 \tag{16.210}$$

$$\Updownarrow$$

$$-xy - \kappa_1 y^2 = -\left(\frac{1}{2\sqrt{\kappa_1}}x + \sqrt{\kappa_1}y\right)^2 + \frac{1}{4\kappa_1}x^2. \tag{16.211}$$

Equation (16.209) with $x = a_1 z_1$ and $y = z_1^2$ yields

$$\dot{V}_2 = -\left(\frac{a_1}{2\sqrt{\kappa_1}}z_1 + \sqrt{\kappa_1}z_1^2\right)^2 + \frac{a_1^2}{4\kappa_1}z_1^2 - (a_0 + a_2|z_1| + k_1)z_1^2 - k_2 z_2^2. \tag{16.212}$$

Since

$$-\left(\frac{a_1}{2\sqrt{\kappa_1}}z_1 + \sqrt{\kappa_1}z_1^2\right)^2 \leq 0$$
$$-a_2|z_1| \leq 0 \tag{16.213}$$

it then follows that

$$\dot{V}_2 \leq -\left(a_0 + k_1 - \frac{a_1^2}{4\kappa_1}\right)z_1^2 - k_2 z_2^2. \tag{16.214}$$

Hence, by choosing the controller gains according to

$$\kappa_1 > 0, \qquad k_1 > \frac{a_1^2}{4\kappa_1} - a_0, \qquad k_2 > 0 \tag{16.215}$$

our design goal to render $\dot{V}_2 < 0$ is satisfied. Note that the controller (16.208) with (16.202) is implemented without using the unknown parameters a_0, a_1 and a_2. Hence, a robust nonlinear controller is derived by using backstepping. This result differs from feedback linearization, which is based on model cancellation.

16.3.3 Backstepping of SISO Mass–Damper–Spring Systems

The results of Section 16.3.2 can be generalized to the following class of SISO mechanical systems

$$\dot{x} = \upsilon \tag{16.216}$$

$$m\dot{\upsilon} + d(\upsilon)\upsilon + k(x)x = \tau \tag{16.217}$$

$$y = x \tag{16.218}$$

where x is the position and υ is the velocity. The mass m is constant while $d(\upsilon)$ and $k(x)$ are nonlinear damping and spring coefficients, respectively. Backstepping of the mass–damper–spring can be performed by choosing the output $e = y - y_d$ where e is the tracking error and $y_d(t) \in C^r$ is an r times differentiable (smooth) and bounded reference trajectory (see Section 12.1.1). Regulation of $y = x$ to zero is obtained by choosing $\dot{y}_d = y_d = 0$. Time differentiation of e yields the following model

$$\dot{e} = \upsilon - \dot{y}_d \tag{16.219}$$

$$m\dot{\upsilon} = \tau - d(\upsilon)\upsilon - k(x)x. \tag{16.220}$$

The backstepping control law solving this problem is derived in two recursive steps similar to the integrator backstepping example in Section 16.3.2.

Step 1: Let $z_1 = e = y - y_d$, such that

$$\dot{z}_1 = \upsilon - \dot{y}_d. \tag{16.221}$$

Taking υ as *virtual control*,

$$\upsilon = \alpha_1 + z_2 \tag{16.222}$$

where z_2 is a new state variable to be interpreted later, yields

$$\dot{z}_1 = \alpha_1 + z_2 - \dot{y}_d. \tag{16.223}$$

Next, the stabilizing function α_1 is chosen as

$$\alpha_1 = \dot{y}_d - [k_1 + n_1(z_1)]z_1 \tag{16.224}$$

where $k_1 > 0$ is a feedback gain and $n_1(z_1) \geq 0$ is a nonlinear damping term, for instance a nonlinear nondecreasing function $n_1(z_1) = \kappa_1 |z_1|^{n_1}$ with $n_1 > 0$ and $\kappa_1 \geq 0$. This yields

$$\dot{z}_1 = -[k_1 + n_1(z_1)]z_1 + z_2. \tag{16.225}$$

A CLF for z_1 is

$$V_1 = \frac{1}{2}z_1^2 \tag{16.226}$$

$$\begin{aligned}\dot{V}_1 &= z_1\dot{z}_1\\ &= -[k_1 + n_1(z_1)]z_1^2 + z_1z_2.\end{aligned} \tag{16.227}$$

Step 2: The second step stabilizes the z_2 dynamics. Moreover, from (16.222) it is seen that

$$\begin{aligned}m\dot{z}_2 &= m\dot{v} - m\dot{\alpha}_1\\ &= \tau - d(v)v - k(x)x - m\dot{\alpha}_1.\end{aligned} \tag{16.228}$$

Let V_2 be the second CLF, which is chosen to reflect the kinetic energy $\frac{1}{2}mv^2$ of the system. However, it makes sense to replace the velocity v with z_2 in order to solve the trajectory-tracking control problem. This is usually referred to as "pseudo-kinetic energy". Consider

$$V_2 = V_1 + \frac{1}{2}mz_2^2 \tag{16.229}$$

$$\begin{aligned}\dot{V}_2 &= \dot{V}_1 + mz_2\dot{z}_2\\ &= -[k_1 + n_1(z_1)]z_1^2 + z_1z_2 + z_2[\tau - d(v)v - k(x)x - m\dot{\alpha}_1].\end{aligned} \tag{16.230}$$

Since the input τ appears in $\dot{V}_2$, a value for τ can be prescribed such that $\dot{V}_2$ becomes negative definite. For instance

$$\tau = m\dot{\alpha}_1 + d(v)v + k(x)x - z_1 - k_2z_2 - n_2(z_2)z_2 \tag{16.231}$$

where $k_2 > 0$ and $n_2(z_2) = \kappa_2|z_2|^{n_2} \geq 0$ with $n_2 > 0$ can be specified by the designer. This yields

$$\dot{V}_2 = -[k_1 + n_1(z_1)]z_1^2 - [k_2 + n_2(z_2)]z_2^2. \tag{16.232}$$

When implementing the control law, $\dot{\alpha}_1$ is computed by taking the time derivative of α_1 along the trajectories of y_d and z_1, see (16.224), to obtain

$$\dot{\alpha}_1 = \frac{\partial\alpha_1}{\partial\dot{y}_d}\ddot{y}_d + \frac{\partial\alpha_1}{\partial z_1}\dot{z}_1 = \ddot{y}_d + \frac{\partial\alpha_1}{\partial z_1}(v - \dot{y}_d). \tag{16.233}$$

Hence, the state derivatives are avoided in the control law. Note that the desired state y_d is assumed to be smooth such that $\dot{y}_d$ and $\ddot{y}_d$ exist.

The resulting error dynamics is

$$\begin{bmatrix}1 & 0\\ 0 & m\end{bmatrix}\begin{bmatrix}\dot{z}_1\\ \dot{z}_2\end{bmatrix} = -\begin{bmatrix}k_1 + n_1(z_1) & 0\\ 0 & k_2 + n_2(z_2)\end{bmatrix}\begin{bmatrix}z_1\\ z_2\end{bmatrix} + \begin{bmatrix}0 & 1\\ -1 & 0\end{bmatrix}\begin{bmatrix}z_1\\ z_2\end{bmatrix}$$

$$\Updownarrow$$

$$\boldsymbol{M}\dot{\boldsymbol{z}} = -\boldsymbol{K}(\boldsymbol{z})\boldsymbol{z} + \boldsymbol{S}\boldsymbol{z}. \tag{16.234}$$

Hence, the equilibrium point $(z_1, z_2) = (0, 0)$ is GES. This can be seen from the CLF $V_2(\boldsymbol{z}) = (1/2)\boldsymbol{z}^\top \boldsymbol{M}\boldsymbol{z}$, which after time differentiation yields $\dot{V}_2(\boldsymbol{z}) = -\boldsymbol{z}^\top \boldsymbol{K}\boldsymbol{z}$ since $\boldsymbol{z}^\top \boldsymbol{S}\boldsymbol{z} = \boldsymbol{0}, \forall \boldsymbol{z}$. Note that kinetic energy has been applied in the Lyapunov analysis to achieve this.

Setpoint Regulation

Setpoint regulation is obtained by choosing $\dot{y}_d = y_d = 0$. For simplicity let $n_1(z_1) = n_2(z_2) = 0$ such that

$$z_1 = x$$

$$\alpha_1 = -k_1 z_1$$

and

$$\tau = m\dot{\alpha}_1 + d(v)v + k(x)x - z_1 - k_2 z_2. \tag{16.235}$$

Nonlinear PD Control

The backstepping control law (16.235) can also be viewed as a nonlinear PD control law

$$u = -K_p(x)x - K_d(v)v \tag{16.236}$$

by writing (16.235) as

$$\begin{aligned} u &= [d(v) - mk_1]v + [k(x) - 1]x - k_2(v + k_1 x) \\ &= [d(v) - mk_1 - k_2]v + [k(x) - 1 - k_1 k_2]x. \end{aligned} \tag{16.237}$$

Hence,

$$K_p(x) = k_1 k_2 + 1 - k(x) \tag{16.238}$$

$$K_d(v) = mk_1 + k_2 - d(v). \tag{16.239}$$

Nonlinear PID Control

The nonlinear PD controller (16.236) can be extended to include integral action by using *constant parameter adaptation* or by *augmenting an additional integrator* to the plant. More specifically:

1. *Constant parameter adaptation*: An unknown constant (or slowly varying) disturbance is added to the dynamic model. This constant or bias is estimated online by using adaptive control. The resulting system with parameter estimator can be shown to be UGAS for the case of regulation and trajectory-tracking control (Fossen *et al.* 2001).
2. *Integrator augmentation*: An additional integrator is augmented on the right-hand side of the integrator chain in order to obtain zero steady-state errors. The resulting system is proven to be GES.

The methods are presented in Sections 16.3.4 and 16.3.5. A comparative study of integral backstepping techniques is found in Skjetne and Fossen (2004).

16.3.4 Integral Action by Constant Parameter Adaptation

The constant parameter adaptation technique is based on Fossen *et al.* (2001). For simplicity a mass–damper–spring system is considered. Hence, adaptive backstepping results in a control law of PID type. Consider the system

$$\dot{x} = v \tag{16.240}$$

$$m\dot{v} + d(v)v + k(x)x = \tau + w \tag{16.241}$$

$$\dot{w} = 0. \tag{16.242}$$

The trajectory-tracking control law can be designed by considering the tracking error $z_1 = x - x_d$ with

$$\begin{aligned}\dot{z}_1 &= \dot{x} - \dot{x}_d \\ &= v - \dot{x}_d \\ &= (\alpha_1 + z_2) - v_d\end{aligned} \tag{16.243}$$

where z_2 is a new state variable and $v := \alpha_1 + z_2$ is the virtual control for z_1. Choosing the stabilizing function

$$\alpha_1 = \dot{x}_d - k_1 z_1 \tag{16.244}$$

yields

$$\dot{z}_1 = -k_1 z_1 + z_2. \tag{16.245}$$

The definition $z_2 := v - \alpha_1$ implies that

$$\dot{z}_2 = \dot{v} - \ddot{x}_d + k_1(v - \dot{x}_d) \tag{16.246}$$

$$m\dot{z}_2 = \tau - d(v)v - k(x)x + w - m\ddot{x}_d + mk_1(v - \dot{x}_d). \tag{16.247}$$

Consider the CLF

$$V_1 = \frac{1}{2}z_1^2 + \frac{1}{2p}\tilde{w}^2, \quad p > 0 \tag{16.248}$$

$$\begin{aligned}\dot{V}_1 &= z_1\dot{z}_1 + \frac{1}{p}\tilde{w}\dot{\tilde{w}} \\ &= z_1 z_2 - k_1 z_1^2 + \frac{1}{p}\tilde{w}\dot{\hat{w}}\end{aligned} \tag{16.249}$$

where $\tilde{w} = \hat{w} - w$ is the parameter estimation error. Next, consider the CLF

$$V_2 = V_1 + \frac{1}{2}mz_2^2 \tag{16.250}$$

$$
\begin{aligned}
\dot{V}_2 &= \dot{V}_1 + z_2(m\dot{z}_2) \\
&= z_1 z_2 - k_1 z_1^2 + \frac{1}{p}\tilde{w}\dot{\tilde{w}} \\
&\quad + z_2[\tau - d(v)v - k(x)x + w - m\ddot{x}_d + mk_1(v - \dot{x}_d)]
\end{aligned}
\tag{16.251}
$$

where it is noticed that $\dot{\tilde{w}} = \dot{\hat{w}}$. Choosing the control law as

$$
\tau = d(v)\alpha_1 + k(x)x - \hat{w} + m\ddot{x}_d - mk_1(v - \dot{x}_d) - z_1 - k_2 z_2 \tag{16.252}
$$

where $\alpha_1 = v - z_2$, yields

$$
\dot{V}_2 = -k_1 z_1^2 - [k_2 + d(v)]z_2^2 + \tilde{w}\left(\frac{1}{p}\dot{\hat{w}} - z_2\right). \tag{16.253}
$$

Choosing the update law as

$$
\dot{\hat{w}} = pz_2 \tag{16.254}
$$

finally yields

$$
\dot{V}_2 = -k_1 z_1^2 - [k_2 + d(v)]z_2^2. \tag{16.255}
$$

The error dynamics takes the form

$$
\begin{bmatrix} \dot{z}_1 \\ \dot{z}_2 \end{bmatrix} = \begin{bmatrix} -k_1 & 1 \\ -1 & -k_2 - d(v) \end{bmatrix} \begin{bmatrix} z_1 \\ z_2 \end{bmatrix} + \begin{bmatrix} 0 \\ -1 \end{bmatrix} \tilde{w} \tag{16.256}
$$

$$
\dot{\tilde{w}} = -p \begin{bmatrix} 0 & -1 \end{bmatrix} \begin{bmatrix} z_1 \\ z_2 \end{bmatrix} \tag{16.257}
$$

$$\Updownarrow$$

$$
\dot{\boldsymbol{z}} = \boldsymbol{h}(\boldsymbol{z}, t) + \boldsymbol{b}\tilde{w} \tag{16.258}
$$

$$
\dot{\tilde{w}} = -p\boldsymbol{b}^{\top}\left(\frac{\partial W(\boldsymbol{z}, t)}{\partial \boldsymbol{z}}\right)^{\top}. \tag{16.259}
$$

Note that the dissipative term $d(v) = d(z_2 + \alpha_1) = d(z_2 - k_1 z_1 + \dot{x}_d(t)) > 0,\ \forall v$ has not been "canceled out" in order to exploit this as good damping in the error dynamics. The price for exploiting the so-called *good nonlinearities* in the design is that the error dynamics becomes *non-autonomous*.Since the feedback gains are assumed to be positive, that is $k_1 > 0$ and $k_2 > 0, p > 0, \boldsymbol{b} = [0, -1]^{\top}$ and $\boldsymbol{b}^{\top}\boldsymbol{b} = 1 > 0$, Theorem A.5 with $W(\boldsymbol{z}) = (1/2)\boldsymbol{z}^{\top}\boldsymbol{z}$ guarantees that the non-autonomous systems (16.256)–(16.257) is UGAS.

Note that if a feedback linearizing controller is applied instead of (16.252), replacing the damping term $d(v)\alpha_1$ with $d(v)v$, the control input becomes

$$\tau = d(v)v + k(x)x - \hat{w} + m\ddot{x}_d - mk_1(v - \dot{x}_d) - z_1 - k_2 z_2. \tag{16.260}$$

The error dynamics

$$\begin{bmatrix} \dot{z}_1 \\ \dot{z}_2 \end{bmatrix} = \begin{bmatrix} -k_1 & 1 \\ -1 & -k_2 \end{bmatrix} \begin{bmatrix} z_1 \\ z_2 \end{bmatrix} + \begin{bmatrix} 0 \\ -1 \end{bmatrix} \tilde{w} \tag{16.261}$$

is *autonomous*. In this case, *Krasovskii–LaSalle's invariant set theorem* (Theorem A.2) can be used to prove GAS.

16.3.5 Integrator Augmentation Technique

Consider the second-order mass–damper–spring system

$$\dot{x} = v \tag{16.262}$$

$$m\dot{v} + d(v)v + k(x)x = \tau + w \tag{16.263}$$

$$y = x \tag{16.264}$$

where w is a constant *unknown* disturbance. Let $e = y - y_d$ denote the tracking error where y_d is the desired output. Hence,

$$\dot{e} = v - \dot{y}_d \tag{16.265}$$

$$m\dot{v} + d(v)v + k(x)x = \tau + w. \tag{16.266}$$

Nonlinear PID Control

If $w = 0$, backstepping results in a nonlinear control law of PD type similar to the result in Section 16.3.3. However, by augmenting the plant with an additional integrator at the right end of the integrator chain, as illustrated in Figure 16.13, nonlinear PID control can be obtained. Augmentation of an additional integrator $\dot{e}_I = e$ to the second-order plant (16.265) and (16.266) yields

$$\dot{e}_I = e \tag{16.267}$$

$$\dot{e} = v - \dot{y}_d \tag{16.268}$$

$$m\dot{v} + d(v)v + k(x)x = \tau + w. \tag{16.269}$$

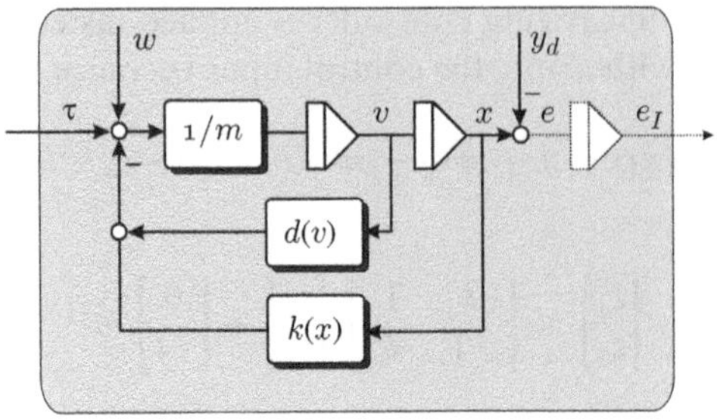

Figure 16.13 Augmentation of an additional integrator.

For simplicity let us first assume that $w = 0$. Hence, backstepping with $z_1 = e_\mathrm{I}$ results in three steps:

Step 1:

$$\begin{aligned}\dot{z}_1 &= e \\ &= \alpha_1 + z_2.\end{aligned} \tag{16.270}$$

Choosing the stabilizing function $\alpha_1 = -k_1 z_1$ yields

$$\dot{z}_1 = -k_1 z_1 + z_2. \tag{16.271}$$

Hence,

$$V_1 = \frac{1}{2} z_1^2 \tag{16.272}$$

$$\begin{aligned}\dot{V}_1 &= z_1 \dot{z}_1 \\ &= -k_1 z_1^2 + z_1 z_2.\end{aligned} \tag{16.273}$$

Step 2:

$$\begin{aligned}\dot{z}_2 &= \dot{e} - \dot{\alpha}_1 \\ &= v - \dot{y}_\mathrm{d} - \dot{\alpha}_1 \\ &= (\alpha_2 + z_3) - \dot{y}_\mathrm{d} - \dot{\alpha}_1.\end{aligned} \tag{16.274}$$

Hence,

$$V_2 = V_1 + \frac{1}{2} z_2^2 \tag{16.275}$$

$$\begin{aligned}\dot{V}_2 &= -k_1 z_1^2 + z_1 z_2 + z_2 \dot{z}_2 \\ &= -k_1 z_1^2 + z_2 (z_1 + \alpha_2 + z_3 - \dot{y}_\mathrm{d} - \dot{\alpha}_1).\end{aligned} \tag{16.276}$$

Choosing the stabilizing function $\alpha_2 = \dot{\alpha}_1 + \dot{y}_\mathrm{d} - k_2 z_2 - z_1$ yields

$$\dot{z}_2 = -z_1 - k_2 z_2 + z_3 \tag{16.277}$$

$$\dot{V}_2 = -k_1 z_1^2 - k_2 z_2^2 + z_2 z_3. \tag{16.278}$$

Step 3:

$$
\begin{aligned}
m\dot{z}_3 &= m\dot{v} - m\dot{\alpha}_2 \\
&= \tau + w - d(v)v - k(x)x - m\dot{\alpha}_2 \\
&= \tau - d(v)\alpha_2 - d(v)z_3 - k(x)x - m\dot{\alpha}_2.
\end{aligned} \tag{16.279}
$$

Let

$$
V_3 = V_2 + \frac{1}{2}mz_3^2 \tag{16.280}
$$

$$
\begin{aligned}
\dot{V}_3 &= -k_1z_1^2 - k_2z_2^2 + z_3(z_2 + m\dot{z}_3) \\
&= -k_1z_1^2 - k_2z_2^2 + z_3(z_2 + \tau - d(v)\alpha_2 - d(v)z_3 - k(x)x - m\dot{\alpha}_2).
\end{aligned} \tag{16.281}
$$

Choosing the control law as

$$
\tau = m\dot{\alpha}_2 + d(v)\alpha_2 + k(x)x - z_2 - k_3z_3 \tag{16.282}
$$

finally yields

$$
\dot{V}_3 = -k_1z_1^2 - k_2z_2^2 - (d(v) + k_3)z_3^2 < 0, \ \forall z_1 \neq 0, z_2 \neq 0, z_3 \neq 0 \tag{16.283}
$$

and

$$
m\dot{z}_3 = -[d(v) + k_3]z_3 - z_2. \tag{16.284}
$$

For the undisturbed case $w = 0$, the error dynamics takes the form

$$
\begin{bmatrix} 1 & 0 & 0 \\ 0 & 1 & 0 \\ 0 & 0 & m \end{bmatrix}
\begin{bmatrix} \dot{z}_1 \\ \dot{z}_2 \\ \dot{z}_3 \end{bmatrix}
= -\begin{bmatrix} k_1 & 0 & 0 \\ 0 & k_2 & 0 \\ 0 & 0 & d(v) + k_3 \end{bmatrix}
\begin{bmatrix} z_1 \\ z_2 \\ z_3 \end{bmatrix}
+ \begin{bmatrix} 0 & 1 & 0 \\ -1 & 0 & 1 \\ 0 & -1 & 0 \end{bmatrix}
\begin{bmatrix} z_1 \\ z_2 \\ z_3 \end{bmatrix}. \tag{16.285}
$$

Hence, the equilibrium point $(z_1, z_2, z_3) = (0, 0, 0)$ is GES and therefore the tracking error e converges to zero. If $w =$ constant, the error dynamics takes the form

$$
\begin{bmatrix} 1 & 0 & 0 \\ 0 & 1 & 0 \\ 0 & 0 & m \end{bmatrix}
\begin{bmatrix} \dot{z}_1 \\ \dot{z}_2 \\ \dot{z}_3 \end{bmatrix}
= -\begin{bmatrix} k_1 & 0 & 0 \\ 0 & k_2 & 0 \\ 0 & 0 & d(v) + k_3 \end{bmatrix}
\begin{bmatrix} z_1 \\ z_2 \\ z_3 \end{bmatrix}
+ \begin{bmatrix} 0 & 1 & 0 \\ -1 & 0 & 1 \\ 0 & -1 & 0 \end{bmatrix}
\begin{bmatrix} z_1 \\ z_2 \\ z_3 \end{bmatrix}
+ \begin{bmatrix} 0 \\ 0 \\ 1 \end{bmatrix} w.
$$

Hence, in the steady state ($\dot{\mathbf{z}} = \mathbf{0}$ and $d(v) = 0$)

$$
z_2 = k_1z_1 = e - \alpha_1 = e + k_1z_1 \Rightarrow e = 0. \tag{16.286}
$$

The equilibrium point for $w = $ constant is

$$\begin{bmatrix} z_1 \\ z_2 \\ z_3 \end{bmatrix} = \begin{bmatrix} k_1 & -1 & 0 \\ 1 & k_2 & -1 \\ 0 & 1 & k_3 \end{bmatrix}^{-1} \begin{bmatrix} 0 \\ 0 \\ 1 \end{bmatrix} w = \frac{1}{k_1 k_2 k_3 + k_1 + k_3} \begin{bmatrix} 1 \\ k_1 \\ 1 + k_1 k_2 \end{bmatrix} w. \tag{16.287}$$

Therefore it can be concluded that for the case $w = $ constant the equilibrium point (z_1, z_2, z_3) is GES but (z_1, z_2, z_3) will converge to the constant non-zero values given by (16.287), even though $e = 0$. This shows that augmentation of an additional integrator when performing backstepping leads to zero steady-state errors in the case of regulation under the assumption of a constant disturbance w.

16.3.6 Case Study: Backstepping Design for Mass–Damper–Spring

The concept of vectorial backstepping was first introduced by Fossen and Berge (1997) and Fossen and Grøvlen (1998). Consider a MIMO nonlinear mass–damper–spring system in the form

$$\dot{\boldsymbol{x}} = \boldsymbol{v} \tag{16.288}$$

$$\boldsymbol{M}\dot{\boldsymbol{v}} + \boldsymbol{D}(\boldsymbol{v})\boldsymbol{v} + \boldsymbol{K}(\boldsymbol{x})\boldsymbol{x} = \boldsymbol{B}\boldsymbol{u} \tag{16.289}$$

where $\boldsymbol{x} \in \mathbb{R}^n$ is the position vector, $\boldsymbol{v} \in \mathbb{R}^n$ is the velocity vector, $\boldsymbol{u} \in \mathbb{R}^r$ $(r \geq n)$ is the control input vector, $\boldsymbol{D}(\boldsymbol{v}) \in \mathbb{R}^{n \times n}$ represents a matrix of damping coefficients, $\boldsymbol{K}(\boldsymbol{x}) \in \mathbb{R}^{n \times n}$ is a matrix of spring coefficients, $\boldsymbol{M} \in \mathbb{R}^{n \times n}$ is the inertia matrix and $\boldsymbol{B} \in \mathbb{R}^{n \times r}$ is the input matrix. Hence, backstepping can be performed in *two vectorial steps*.

Step 1: For the first system (16.288) consider $\boldsymbol{v}$ as the control and let

$$\boldsymbol{v} = \boldsymbol{\sigma} + \boldsymbol{\alpha}_1 \tag{16.290}$$

where

$\boldsymbol{\sigma} = \tilde{\boldsymbol{v}} + \boldsymbol{\Lambda}\tilde{\boldsymbol{x}}$ New state vector used for tracking control

$\boldsymbol{\alpha}_1$ Stabilizing vector field to be defined later.

Here $\tilde{\boldsymbol{v}} = \boldsymbol{v} - \boldsymbol{v}_{\mathrm{d}}$ and $\tilde{\boldsymbol{x}} = \boldsymbol{x} - \boldsymbol{x}_{\mathrm{d}}$ are the velocity and position tracking errors, respectively, and $\boldsymbol{\Lambda} > 0$ is a diagonal matrix of positive elements. The definition of the $\boldsymbol{\sigma}$ vector is motivated by Slotine and Li (1987), who introduced $\boldsymbol{\sigma}$ as a *sliding variable* (see Section 16.4) when designing their adaptive robot controller. It turns out that this transformation has the nice property of transforming the nonlinear state-space model (16.288) and (16.289) to the form

$$\begin{aligned} \boldsymbol{M}\dot{\boldsymbol{\sigma}} + \boldsymbol{D}(\boldsymbol{v})\boldsymbol{\sigma} &= \boldsymbol{M}\dot{\boldsymbol{v}} + \boldsymbol{D}(\boldsymbol{v})\boldsymbol{v} - \boldsymbol{M}\dot{\boldsymbol{v}}_{\mathrm{r}} - \boldsymbol{D}(\boldsymbol{v})\boldsymbol{v}_{\mathrm{r}} \\ &= \boldsymbol{B}\boldsymbol{u} - \boldsymbol{M}\dot{\boldsymbol{v}}_{\mathrm{r}} - \boldsymbol{D}(\boldsymbol{v})\boldsymbol{v}_{\mathrm{r}} - \boldsymbol{K}(\boldsymbol{x})\boldsymbol{x} \end{aligned} \tag{16.291}$$

where $\boldsymbol{v}_{\mathrm{r}}$ can be interpreted as a *virtual reference trajectory*

$$\begin{aligned}\boldsymbol{v}_{\mathrm{r}} &= \boldsymbol{v} - \boldsymbol{\sigma} \\ &= \boldsymbol{v}_{\mathrm{d}} - \boldsymbol{\Lambda}\tilde{\boldsymbol{x}}.\end{aligned} \tag{16.292}$$

The position error dynamics of Step 1 can therefore be written

$$\begin{aligned}\dot{\tilde{\boldsymbol{x}}} &= \boldsymbol{v} - \boldsymbol{v}_{\mathrm{d}} \\ &= \boldsymbol{\sigma} + \boldsymbol{\alpha}_1 - \boldsymbol{v}_{\mathrm{d}} \; (\boldsymbol{\alpha}_1 = \boldsymbol{v}_{\mathrm{r}} = \boldsymbol{v} - \boldsymbol{\sigma}) \\ &= -\boldsymbol{\Lambda}\tilde{\boldsymbol{x}} + \boldsymbol{\sigma}.\end{aligned} \tag{16.293}$$

Hence,

$$V_1 = \frac{1}{2}\tilde{\boldsymbol{x}}^{\top}\boldsymbol{K}_{\mathrm{p}}\tilde{\boldsymbol{x}}, \qquad \boldsymbol{K}_{\mathrm{p}} = \boldsymbol{K}_{\mathrm{p}}^{\top} > 0 \tag{16.294}$$

and

$$\begin{aligned}\dot{V}_1 &= \tilde{\boldsymbol{x}}^{\top}\boldsymbol{K}_{\mathrm{p}}\dot{\tilde{\boldsymbol{x}}} \\ &= \tilde{\boldsymbol{x}}^{\top}\boldsymbol{K}_{\mathrm{p}}(-\boldsymbol{\Lambda}\tilde{\boldsymbol{x}} + \boldsymbol{\sigma}) \\ &= -\tilde{\boldsymbol{x}}^{\top}\boldsymbol{K}_{\mathrm{p}}\boldsymbol{\Lambda}\tilde{\boldsymbol{x}} + \boldsymbol{\sigma}^{\top}\boldsymbol{K}_{\mathrm{p}}\tilde{\boldsymbol{x}}.\end{aligned} \tag{16.295}$$

Step 2: In the second step, a CLF motivated by *pseudo-kinetic energy* is chosen according to

$$V_2 = \frac{1}{2}\boldsymbol{\sigma}^{\top}\boldsymbol{M}\boldsymbol{\sigma} + V_1, \quad \boldsymbol{M} = \boldsymbol{M}^{\top} > 0 \tag{16.296}$$

$$\begin{aligned}\dot{V}_2 &= \boldsymbol{\sigma}^{\top}\boldsymbol{M}\dot{\boldsymbol{\sigma}} + \dot{V}_1 \\ &= \boldsymbol{\sigma}^{\top}(\boldsymbol{B}\boldsymbol{u} - \boldsymbol{M}\dot{\boldsymbol{v}}_{\mathrm{r}} - \boldsymbol{D}(\boldsymbol{v})\boldsymbol{v}_{\mathrm{r}} - \boldsymbol{K}(\boldsymbol{x})\boldsymbol{x} - \boldsymbol{D}(\boldsymbol{v})\boldsymbol{\sigma}) - \tilde{\boldsymbol{x}}^{\top}\boldsymbol{K}_{\mathrm{p}}\boldsymbol{\Lambda}\tilde{\boldsymbol{x}} + \boldsymbol{\sigma}^{\top}\boldsymbol{K}_{\mathrm{p}}\tilde{\boldsymbol{x}} \\ &= \boldsymbol{\sigma}^{\top}(\boldsymbol{B}\boldsymbol{u} - \boldsymbol{M}\dot{\boldsymbol{v}}_{\mathrm{r}} - \boldsymbol{D}(\boldsymbol{v})\boldsymbol{v}_{\mathrm{r}} - \boldsymbol{K}(\boldsymbol{x})\boldsymbol{x} - \boldsymbol{D}(\boldsymbol{v})\boldsymbol{\sigma} + \boldsymbol{K}_{\mathrm{p}}\tilde{\boldsymbol{x}}) - \tilde{\boldsymbol{x}}^{\top}\boldsymbol{K}_{\mathrm{p}}\boldsymbol{\Lambda}\tilde{\boldsymbol{x}}.\end{aligned} \tag{16.297}$$

This suggests that the control law is chosen as

$$\boldsymbol{B}\boldsymbol{u} = \boldsymbol{M}\dot{\boldsymbol{v}}_{\mathrm{r}} + \boldsymbol{D}(\boldsymbol{v})\boldsymbol{v}_{\mathrm{r}} + \boldsymbol{K}(\boldsymbol{x})\boldsymbol{x} - \boldsymbol{K}_{\mathrm{p}}\tilde{\boldsymbol{x}} - \boldsymbol{K}_{\mathrm{d}}\,\boldsymbol{\sigma}, \quad \boldsymbol{K}_{\mathrm{d}} > 0 \tag{16.298}$$

which results in

$$\dot{V}_2 = -\boldsymbol{\sigma}^{\top}(\boldsymbol{D}(\boldsymbol{v}) + \boldsymbol{K}_{\mathrm{d}})\boldsymbol{\sigma} - \tilde{\boldsymbol{x}}^{\top}\boldsymbol{K}_{\mathrm{p}}\boldsymbol{\Lambda}\tilde{\boldsymbol{x}}. \tag{16.299}$$

Since V_2 is positive definite and $\dot{V}_2$ is negative definite it follows from Theorem A.3 that the equilibrium point $(\tilde{\boldsymbol{x}}, \boldsymbol{\sigma}) = (\boldsymbol{0}, \boldsymbol{0})$ is GES. Moreover, convergence of $\boldsymbol{\sigma} \rightarrow \boldsymbol{0}$ and $\tilde{\boldsymbol{x}} \rightarrow \boldsymbol{0}$ implies that $\tilde{\boldsymbol{v}} \rightarrow \boldsymbol{0}$. When implementing the control law (16.298) it is assumed that $\boldsymbol{B}$ has a pseudoinverse

$$\boldsymbol{B}^{\dagger} = \boldsymbol{B}^{\top}(\boldsymbol{B}\boldsymbol{B}^{\top})^{-1} \tag{16.300}$$

or simply $\boldsymbol{B}^{-1}$ for the square case $r = n$.

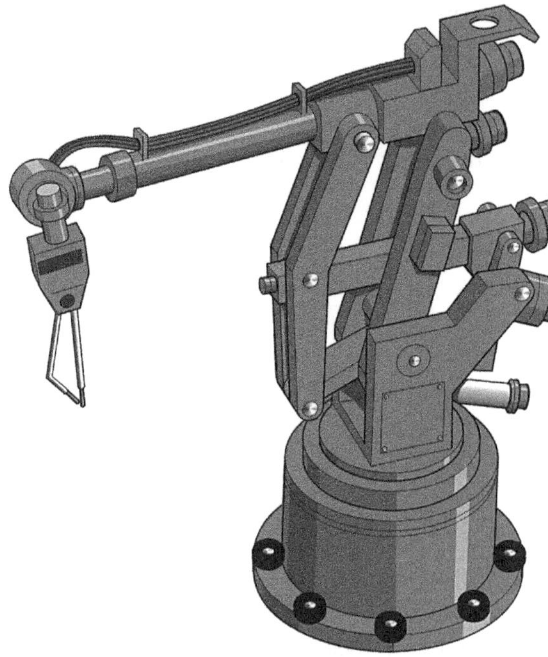

Figure 16.14 Robot manipulator.

16.3.7 Case Study: Backstepping Design for Robot Manipulators

This example is based on the results of Fossen and Berge (1997). Consider the nonlinear robot model (Sciavicco and Siciliano 1996)

$$\dot{\boldsymbol{q}} = \boldsymbol{v} \tag{16.301}$$

$$\boldsymbol{M}(\boldsymbol{q})\dot{\boldsymbol{v}} + \boldsymbol{C}(\boldsymbol{q}, \boldsymbol{v})\boldsymbol{v} + \boldsymbol{g}(\boldsymbol{q}) = \boldsymbol{\tau} \tag{16.302}$$

where $\boldsymbol{M}(\boldsymbol{q}) = \boldsymbol{M}^{\top}(\boldsymbol{q}) > 0$ is the inertia matrix, $\boldsymbol{C}(\boldsymbol{q}, \boldsymbol{v})$ is a matrix of Coriolis and centripetal forces defined in terms of the *Christoffel symbols* and $\boldsymbol{g}(\boldsymbol{q})$ is a vector of gravitational forces and moments. $\boldsymbol{q} \in \mathbb{R}^n$ is a vector of joint angles, $\boldsymbol{v} \in \mathbb{R}^n$ is a vector of joint angular rates and $\boldsymbol{\tau} \in \mathbb{R}^n$ is a vector of control torques. Vectorial backstepping of a robot manipulator (see Figure 16.14) can be done in two steps:

Step 1: Define the *virtual control vector*

$$\dot{\boldsymbol{q}} = \boldsymbol{v} := \boldsymbol{\sigma} + \boldsymbol{\alpha}_1 \tag{16.303}$$

where $\boldsymbol{\sigma}$ is a new state variable and $\boldsymbol{\alpha}_1$ is stabilizing vector field, which can be chosen as

$$\boldsymbol{\alpha}_1 = \boldsymbol{\nu}_\text{r}, \qquad \boldsymbol{\nu}_\text{r} = \boldsymbol{\nu}_\text{d} - \boldsymbol{\Lambda}\tilde{\boldsymbol{q}} \tag{16.304}$$

where $\boldsymbol{\Lambda} > 0$ is a diagonal design matrix and $\tilde{\boldsymbol{q}} = \boldsymbol{q} - \boldsymbol{q}_\text{d}$ is the tracking error. Combining (16.303) and (16.304) yields

$$\tilde{\boldsymbol{\nu}} = -\boldsymbol{\Lambda}\tilde{\boldsymbol{q}} + \boldsymbol{\sigma} \tag{16.305}$$

where $\dot{\tilde{\boldsymbol{q}}} = \tilde{\boldsymbol{\nu}}$.

Step 2: Consider the CLF

$$V = \frac{1}{2}(\boldsymbol{\sigma}^\top \boldsymbol{M}(\boldsymbol{q})\boldsymbol{\sigma} + \tilde{\boldsymbol{q}}^\top \boldsymbol{K}_\text{q}\tilde{\boldsymbol{q}}) > 0, \qquad \forall \boldsymbol{\sigma} \neq \boldsymbol{0}, \tilde{\boldsymbol{q}} \neq \boldsymbol{0} \tag{16.306}$$

$$\begin{aligned}\dot{V} &= \boldsymbol{\sigma}^\top \boldsymbol{M}(\boldsymbol{q})\dot{\boldsymbol{\sigma}} + \frac{1}{2}\boldsymbol{\sigma}^\top \dot{\boldsymbol{M}}(\boldsymbol{q})\boldsymbol{\sigma} + \tilde{\boldsymbol{q}}^\top \boldsymbol{K}_\text{q}\tilde{\boldsymbol{\nu}} \\ &= \boldsymbol{\sigma}^\top \boldsymbol{M}(\boldsymbol{q})\dot{\boldsymbol{\sigma}} + \frac{1}{2}\boldsymbol{\sigma}^\top \dot{\boldsymbol{M}}(\boldsymbol{q})\boldsymbol{\sigma} - \tilde{\boldsymbol{q}}^\top \boldsymbol{K}_\text{q}\boldsymbol{\Lambda}\tilde{\boldsymbol{q}} + \tilde{\boldsymbol{q}}^\top \boldsymbol{K}_\text{q}\boldsymbol{\sigma}.\end{aligned} \tag{16.307}$$

Equations (16.303) and (16.304) can be combined to give

$$\begin{aligned}\boldsymbol{M}(\boldsymbol{q})\dot{\boldsymbol{\sigma}} &= \boldsymbol{M}(\boldsymbol{q})\dot{\boldsymbol{\nu}} - \boldsymbol{M}(\boldsymbol{q})\dot{\boldsymbol{\alpha}} \\ &= \boldsymbol{\tau} - \boldsymbol{M}(\boldsymbol{q})\dot{\boldsymbol{\nu}}_\text{r} - \boldsymbol{C}(\boldsymbol{q},\boldsymbol{\nu})\boldsymbol{\nu}_\text{r} - \boldsymbol{g}(\boldsymbol{q}) - \boldsymbol{C}(\boldsymbol{q},\boldsymbol{\nu})\boldsymbol{\sigma}.\end{aligned} \tag{16.308}$$

Substituting (16.308) into (16.307) yields

$$\begin{aligned}\dot{V} &= \boldsymbol{\sigma}^\top(\boldsymbol{\tau} - \boldsymbol{M}(\boldsymbol{q})\dot{\boldsymbol{\nu}}_\text{r} - \boldsymbol{C}(\boldsymbol{q},\boldsymbol{\nu})\boldsymbol{\nu}_\text{r} - \boldsymbol{g}(\boldsymbol{q}) + \boldsymbol{K}_\text{q}\tilde{\boldsymbol{q}}) \\ &\quad + \boldsymbol{\sigma}^\top\left(\frac{1}{2}\dot{\boldsymbol{M}}(\boldsymbol{q}) - \boldsymbol{C}(\boldsymbol{q},\boldsymbol{\nu})\right)\boldsymbol{\sigma} - \tilde{\boldsymbol{q}}^\top \boldsymbol{K}_\text{q}\boldsymbol{\Lambda}\tilde{\boldsymbol{q}} \\ &= \boldsymbol{\sigma}^\top(\boldsymbol{\tau} - \boldsymbol{M}(\boldsymbol{q})\dot{\boldsymbol{\nu}}_\text{r} - \boldsymbol{C}(\boldsymbol{q},\boldsymbol{\nu})\boldsymbol{\nu}_\text{r} - \boldsymbol{g}(\boldsymbol{q}) + \boldsymbol{K}_\text{q}\tilde{\boldsymbol{q}}) - \tilde{\boldsymbol{q}}^\top \boldsymbol{K}_\text{q}\boldsymbol{\Lambda}\tilde{\boldsymbol{q}}.\end{aligned} \tag{16.309}$$

Here the skew-symmetric property $\boldsymbol{\sigma}^\top(1/2\dot{\boldsymbol{M}}(\boldsymbol{q}) - \boldsymbol{C}(\boldsymbol{q},\boldsymbol{\nu}))\boldsymbol{\sigma} = \boldsymbol{0}, \forall \boldsymbol{\sigma}$ has been applied. The backstepping control law is chosen as

$$\boldsymbol{\tau} = \boldsymbol{M}(\boldsymbol{q})\dot{\boldsymbol{\nu}}_\text{r} + \boldsymbol{C}(\boldsymbol{q},\boldsymbol{\nu})\boldsymbol{\nu}_\text{r} + \boldsymbol{g}(\boldsymbol{q}) - \boldsymbol{K}_\text{d}\boldsymbol{\sigma} - \boldsymbol{K}_\text{q}\tilde{\boldsymbol{q}} \tag{16.310}$$

where $\boldsymbol{K}_\text{d} = \boldsymbol{K}_\text{d}^\top > 0$ and $\boldsymbol{K}_q = \boldsymbol{K}_q^\top > 0$ are design matrices. This finally yields

$$\dot{V} = -\boldsymbol{\sigma}^\top \boldsymbol{K}_\text{d}\boldsymbol{\sigma} - \tilde{\boldsymbol{q}}^\top \boldsymbol{K}_\text{q}\boldsymbol{\Lambda}\tilde{\boldsymbol{q}} < 0, \quad \forall \boldsymbol{\sigma} \neq \boldsymbol{0}, \tilde{\boldsymbol{q}} \neq \boldsymbol{0} \tag{16.311}$$

and GES follows. The control law (16.310) is equivalent to the control law of Slotine and Li (1987) with perfectly known parameters (non-adaptive case) except for the additional feedback term $\boldsymbol{K}_\text{q}\tilde{\boldsymbol{q}}$ which is necessary to obtain exponential stability.

16.3.8 Case Study: Backstepping Design for Surface Craft

A MIMO nonlinear backstepping technique for marine craft where the nonlinear system properties are exploited is presented below (Fossen and Strand 1998). Consider a marine craft described by the following equations of motion

$$\dot{\boldsymbol{\eta}} = \boldsymbol{J}_{\Theta}(\boldsymbol{\eta})\boldsymbol{\nu} \tag{16.312}$$

$$\boldsymbol{M}\dot{\boldsymbol{\nu}} + \boldsymbol{C}(\boldsymbol{\nu})\boldsymbol{\nu} + \boldsymbol{D}(\boldsymbol{\nu})\boldsymbol{\nu} + \boldsymbol{g}(\boldsymbol{\eta}) = \boldsymbol{\tau} \tag{16.313}$$

$$\boldsymbol{\tau} = \boldsymbol{B}\boldsymbol{u}. \tag{16.314}$$

This model describes the motion of a craft in 6 DOFs. It is assumed that the craft is fully actuated such that $\boldsymbol{B}\boldsymbol{B}^{\top}$ is invertible. The system (16.312)–(16.314) satisfies the following properties:

(i) $\boldsymbol{M} = \boldsymbol{M}^{\top} > 0,\ \dot{\boldsymbol{M}} = \boldsymbol{0}$
(ii) $\boldsymbol{C}(\boldsymbol{\nu}) = -\boldsymbol{C}^{\top}(\boldsymbol{\nu})$
(iii) $\boldsymbol{D}(\boldsymbol{\nu}) > 0$
(iv) $\boldsymbol{B}\boldsymbol{B}^{\top}$ is non-singular.

Recall that $\boldsymbol{J}_{\Theta}(\boldsymbol{\eta})$ is the Euler angle transformation matrix (not defined for $\theta = \pm 90^{\circ}$) and assume that the reference trajectories given by $\boldsymbol{\eta}_{\mathrm{d}}^{(3)}, \ddot{\boldsymbol{\eta}}_{\mathrm{d}}, \dot{\boldsymbol{\eta}}_{\mathrm{d}}$ and $\boldsymbol{\eta}_{\mathrm{d}}$ are smooth and bounded. The virtual reference trajectories are defined as

$$\dot{\boldsymbol{\eta}}_{\mathrm{r}} := \dot{\boldsymbol{\eta}}_{\mathrm{d}} - \boldsymbol{\Lambda}\tilde{\boldsymbol{\eta}} \tag{16.315}$$

$$\boldsymbol{\nu}_{\mathrm{r}} := \boldsymbol{J}_{\Theta}^{-1}(\boldsymbol{\eta})\dot{\boldsymbol{\eta}}_{\mathrm{r}}, \qquad \theta \neq \pm 90^{\circ} \tag{16.316}$$

where $\tilde{\boldsymbol{\eta}} = \boldsymbol{\eta} - \boldsymbol{\eta}_{\mathrm{d}}$ is the tracking error and $\boldsymbol{\Lambda} > 0$ is a diagonal design matrix. Furthermore, let

$$\boldsymbol{\sigma} = \dot{\boldsymbol{\eta}} - \dot{\boldsymbol{\eta}}_{\mathrm{r}} = \dot{\tilde{\boldsymbol{\eta}}} + \boldsymbol{\Lambda}\tilde{\boldsymbol{\eta}}. \tag{16.317}$$

The marine craft dynamics (16.312) and (16.313) can be written

$$\boldsymbol{M}^{*}(\boldsymbol{\eta})\ddot{\boldsymbol{\eta}} + \boldsymbol{C}^{*}(\boldsymbol{\nu},\boldsymbol{\eta})\dot{\boldsymbol{\eta}} + \boldsymbol{D}^{*}(\boldsymbol{\nu},\boldsymbol{\eta})\dot{\boldsymbol{\eta}} + \boldsymbol{g}^{*}(\boldsymbol{\eta}) = \boldsymbol{J}_{\Theta}^{-\top}(\boldsymbol{\eta})\boldsymbol{\tau} \tag{16.318}$$

where

$$\begin{aligned}
\boldsymbol{M}^{*}(\boldsymbol{\eta}) &= \boldsymbol{J}_{\Theta}^{-\top}(\boldsymbol{\eta})\boldsymbol{M}\boldsymbol{J}_{\Theta}^{-1}(\boldsymbol{\eta})\\
\boldsymbol{C}^{*}(\boldsymbol{\nu},\boldsymbol{\eta}) &= \boldsymbol{J}_{\Theta}^{-\top}(\boldsymbol{\eta})[\boldsymbol{C}(\boldsymbol{\nu}) - \boldsymbol{M}\boldsymbol{J}_{\Theta}^{-1}(\boldsymbol{\eta})\dot{\boldsymbol{J}}_{\Theta}(\boldsymbol{\eta})]\boldsymbol{J}_{\Theta}^{-1}(\boldsymbol{\eta})\\
\boldsymbol{D}^{*}(\boldsymbol{\nu},\boldsymbol{\eta}) &= \boldsymbol{J}_{\Theta}^{-\top}(\boldsymbol{\eta})\boldsymbol{D}(\boldsymbol{\nu})\boldsymbol{J}_{\Theta}^{-1}(\boldsymbol{\eta})\\
\boldsymbol{g}^{*}(\boldsymbol{\eta}) &= \boldsymbol{J}_{\Theta}^{-\top}(\boldsymbol{\eta})\boldsymbol{g}(\boldsymbol{\eta}).
\end{aligned}$$

Hence,

$$\begin{aligned}M^*(\eta)\dot{\sigma} &= -C^*(\nu,\eta)\sigma - D^*(\nu,\eta)\sigma + J_\Theta^{-\top}(\eta)Bu \\ &\quad - M^*(\eta)\ddot{\eta}_r - C^*(\nu,\eta)\dot{\eta}_r - D^*(\nu,\eta)\dot{\eta}_r - g^*(\eta)\end{aligned} \tag{16.319}$$

or equivalently

$$\begin{aligned}M^*(\eta)\dot{\sigma} &= -C^*(\nu,\eta)\sigma - D^*(\nu,\eta)\sigma \\ &\quad + J_\Theta^{-\top}(\eta)[Bu - M\dot{\nu}_r - C(\nu)\nu_r - D(\nu)\nu_r - g(\eta)].\end{aligned} \tag{16.320}$$

Step 1: Consider the error dynamics

$$\dot{\eta} - \dot{\eta}_d = J_\Theta(\eta)(\nu - \nu_d). \tag{16.321}$$

Let ν be the virtual control vector defined by

$$J_\Theta(\eta)\nu := \sigma + \alpha_1. \tag{16.322}$$

The position error dynamics can therefore be written

$$\begin{aligned}\dot{\tilde{\eta}} &= J_\Theta(\eta)(\nu - \nu_d) \\ &= \sigma + \alpha_1 - J_\Theta(\eta)\nu_d \qquad \{\alpha_1 = \dot{\eta}_r = \dot{\eta}_d - \Lambda\tilde{\eta}, \quad \dot{\eta}_d = J_\Theta(\eta)\nu_d\} \\ &= -\Lambda\tilde{\eta} + \sigma.\end{aligned} \tag{16.323}$$

Hence, a CLF is

$$V_1 = \frac{1}{2}\tilde{\eta}^\top K_p\tilde{\eta}, \qquad K_p = K_p^\top > 0 \tag{16.324}$$

$$\begin{aligned}\dot{V}_1 &= \tilde{\eta}^\top K_p\dot{\tilde{\eta}} \\ &= -\tilde{\eta}^\top K_p\Lambda\tilde{\eta} + \sigma^\top K_p\tilde{\eta}.\end{aligned} \tag{16.325}$$

Step 2: In the second step a CLF motivated by the *pseudo-kinetic energy* is chosen

$$V_2 = \frac{1}{2}\sigma^\top M^*(\eta)\sigma + V_1, \qquad M^* = (M^*)^\top > 0 \tag{16.326}$$

$$\begin{aligned}\dot{V}_2 &= \sigma^\top M^*(\eta)\dot{\sigma} + \frac{1}{2}\sigma^\top \dot{M}^*(\eta)\sigma + \dot{V}_1 \\ &= -\sigma^\top[C^*(\nu,\eta) + D^*(\nu,\eta)]\sigma \\ &\quad + \sigma^\top J_\Theta^{-\top}(\eta)[Bu - M\dot{\nu}_r - C(\nu)\nu_r - D(\nu)\nu_r - g(\eta)] \\ &\quad + \frac{1}{2}\sigma^\top \dot{M}^*(\eta)\sigma - \tilde{\eta}^\top K_p\Lambda\tilde{\eta} + \sigma^\top K_p\tilde{\eta}.\end{aligned} \tag{16.327}$$

Using the skew-symmetric property

$$\sigma^\top(\dot{M}^*(\eta) - 2C^*(\nu,\eta))\sigma = 0, \quad \forall \nu,\eta,\sigma \tag{16.328}$$

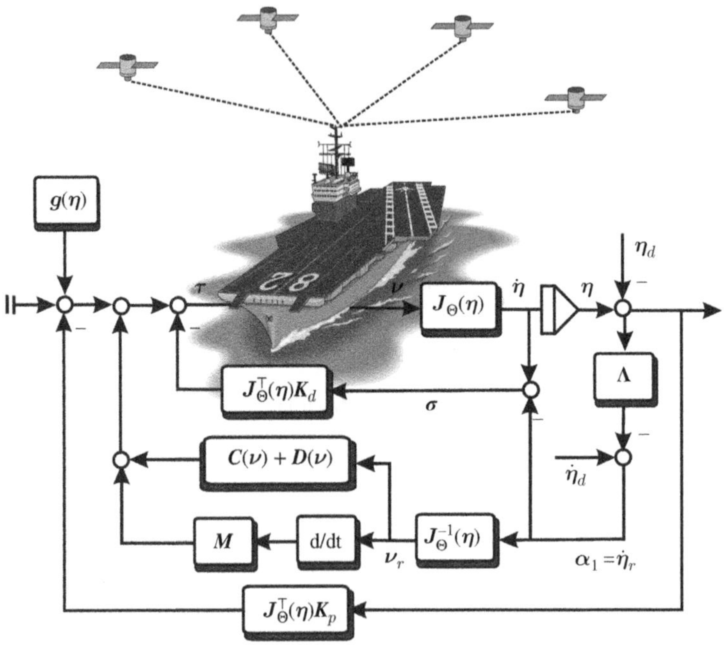

Figure 16.15 Nonlinear MIMO backstepping controller for 6-DOF trajectory-tracking control.

yields

$$\dot{V}_2 = \boldsymbol{\sigma}^\top \boldsymbol{J}_\Theta^{-\top}(\boldsymbol{\eta})[\boldsymbol{B}\boldsymbol{u} - \boldsymbol{M}\dot{\boldsymbol{v}}_\mathrm{r} - \boldsymbol{C}(\boldsymbol{v})\boldsymbol{v}_\mathrm{r} - \boldsymbol{D}(\boldsymbol{v})\boldsymbol{v}_\mathrm{r} - \boldsymbol{g}(\boldsymbol{\eta}) \\ + \boldsymbol{J}_\Theta^\top(\boldsymbol{\eta})\boldsymbol{K}_p\tilde{\boldsymbol{\eta}}] - \boldsymbol{\sigma}^\top \boldsymbol{D}^*(\boldsymbol{v},\boldsymbol{\eta})\boldsymbol{\sigma} - \tilde{\boldsymbol{\eta}}^\top \boldsymbol{K}_p \boldsymbol{\Lambda}\tilde{\boldsymbol{\eta}}. \quad (16.329)$$

Hence, the control law can be chosen as (see Figure 16.15)

$$\boldsymbol{\tau} = \boldsymbol{M}\dot{\boldsymbol{v}}_\mathrm{r} + \boldsymbol{C}(\boldsymbol{v})\boldsymbol{v}_\mathrm{r} + \boldsymbol{D}(\boldsymbol{v})\boldsymbol{v}_\mathrm{r} + \boldsymbol{g}(\boldsymbol{\eta}) - \boldsymbol{J}_\Theta^\top(\boldsymbol{\eta})\boldsymbol{K}_p\tilde{\boldsymbol{\eta}} - \boldsymbol{J}_\Theta^\top(\boldsymbol{\eta})\boldsymbol{K}_\mathrm{d}\,\boldsymbol{\sigma} \\ \boldsymbol{u} = \boldsymbol{B}^\dagger \boldsymbol{\tau} \quad (16.330)$$

where $\boldsymbol{K}_\mathrm{d} > 0$. This results in

$$\dot{V}_2 = -\boldsymbol{\sigma}^\top(\boldsymbol{D}^*(\boldsymbol{v},\boldsymbol{\eta}) + \boldsymbol{K}_\mathrm{d})\boldsymbol{\sigma} - \tilde{\boldsymbol{\eta}}^\top \boldsymbol{K}_\mathrm{p}\boldsymbol{\Lambda}\tilde{\boldsymbol{\eta}}. \quad (16.331)$$

Since V_2 is positive definite and $\dot{V}_2$ is negative definite it follows from Theorem A.3 that the equilibrium point $(\tilde{\boldsymbol{\eta}}, \boldsymbol{\sigma}) = (\boldsymbol{0}, \boldsymbol{0})$ is exponentially stable. In addition, $\boldsymbol{\sigma} \to \boldsymbol{0}$ and (16.317) imply that $\dot{\tilde{\boldsymbol{\eta}}} \to \boldsymbol{0}$.

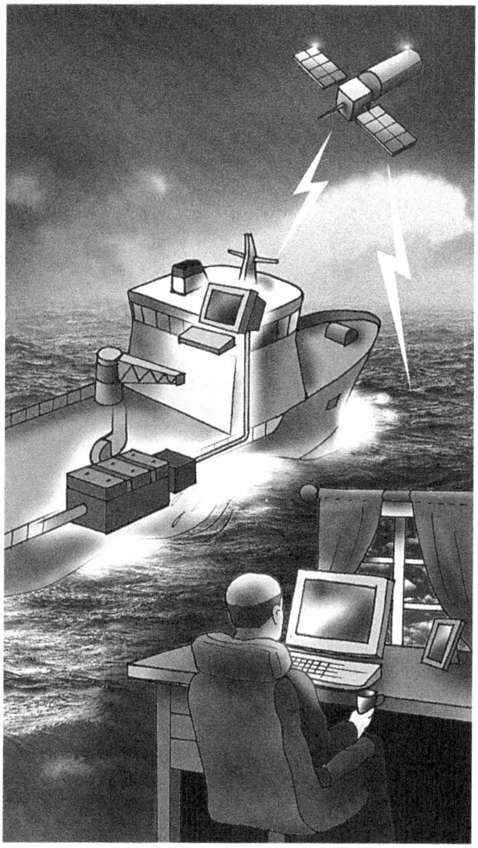

Figure 16.16 Dynamic positioning of a supply vessel using measurements from a global navigation satellite system. Source: Illustration by B. Stenberg.

Example 16.4 (Backstepping Design for 3-DOF Surface Vessels)

Stationkeeping and maneuvering of ships and semi-submersibles are well described by 3-DOF (surge, sway and yaw) maneuvering models (see Figure 16.16). In this case the Euler angle transformation matrix reduces to $\boldsymbol{J}_{\Theta}(\boldsymbol{\eta}) \equiv \boldsymbol{R}(\psi)$ which is the rotation matrix in yaw. Consider the 3-DOF low-speed maneuvering model (see Section 6.5.1)

$$\dot{\boldsymbol{\eta}} = \boldsymbol{R}(\psi)\boldsymbol{\nu} \tag{16.332}$$

$$\boldsymbol{M}\dot{\boldsymbol{\nu}} + \boldsymbol{C}(\boldsymbol{\nu})\boldsymbol{\nu} + \boldsymbol{D}(\boldsymbol{\nu})\boldsymbol{\nu} = \boldsymbol{\tau}. \tag{16.333}$$

The PD control law is obtained from (16.330), which yields

$$\boldsymbol{\tau} = \boldsymbol{M}\dot{\boldsymbol{\nu}}_{\mathrm{r}} + \boldsymbol{C}(\boldsymbol{\nu})\boldsymbol{\nu}_{\mathrm{r}} + \boldsymbol{D}(\boldsymbol{\nu})\boldsymbol{\nu}_{\mathrm{r}} - \boldsymbol{R}^{\top}(\psi)\boldsymbol{K}_{p}\tilde{\boldsymbol{\eta}} - \boldsymbol{R}^{\top}(\psi)\boldsymbol{K}_{\mathrm{d}}\,\boldsymbol{\sigma}. \tag{16.334}$$

Integral action can easily be included by using adaptive backstepping or integral augmentation techniques as explained in Sections 16.3.4 and 16.3.5.

16.3.9 Case Study: Autopilot Based on Backstepping

A nonlinear backstepping controller can be designed by writing the autopilot model (7.29) in SISO strict feedback form:

$$\dot{\psi} = r \tag{16.335}$$

$$m\dot{r} + d(r)r = \delta \tag{16.336}$$

where $m = T/K$ and $d(r) = H_N(r)/K$. The only nonlinearity in this model is due to the maneuvering characteristic $H_N(r)$.

In Section 16.3.3 it was shown that the backstepping controller for this system is

$$\delta = m\dot{\alpha}_1 + d(r)r - z_1 - k_2 z_2 - n_2(z_2)z_2 \tag{16.337}$$

$$\alpha_1 = r_d - [k_1 + n_1(z_1)]z_1 \tag{16.338}$$

where $k_1 > 0$ and $k_2 > 0$ are two feedback gains and $n_i(z_i) \geq 0$ $(i = 1, 2)$ are two optional nonlinear damping terms, for instance chosen as non-decreasing functions $n_i(z_i) = \kappa_i |z_i|^{n_i}$ with $n_i \geq 1$ and $\kappa_i \geq 0$ $(i = 1, 2)$ as design parameters. The following change of coordinates is needed to implement the controller

$$z_1 = \psi - \psi_d \tag{16.339}$$

$$z_2 = r - \alpha_1. \tag{16.340}$$

The backstepping controller includes a PD term as well as reference feedforward. In addition the nonlinear damping terms $n_i(z_i)$ $(i = 1, 2)$ can be used to improve the performance and stability of the closed-loop system. In a practical implementation of the control (16.337)–(16.338) it is necessary to map z_1 to the smallest signed angle (see Section 13.4.3).

When using feedback linearization all the nonlinearities in $H_N(r)$ are compensated for. This requires that the dissipative terms are known with good accuracy, which is not true in many cases. The backstepping controller gives more design flexibility with respect to the damping terms. In fact, it is possible to exploit good damping terms such as $n_3 r^3$ and $n_1 r$ in $H_N(r)$ instead of canceling them. This is straightforward in setpoint regulation; see Krstic *et al.* (1995), for instance. In trajectory-tracking control, however, it is not clear how good damping with respect to a time-varying reference trajectory should be defined. A discussion on backstepping versus feedback linearization is found in Section 16.3.2.

Extensions to integral action can be done by using the method of Loria *et al.* (1999) and Fossen *et al.* (2001), which is referred to as *backstepping with integral action*. Alternatively, an *integrator augmentation technique* can be applied. Both these methods are described in detail in Sections 16.3.4 and 16.3.5.

16.3.10 Case Study: Path-following Controller for Underactuated Marine Craft

For floating rigs, semi-submersibles and supply vessels, trajectory-tracking control in *surge, sway* and *yaw* is easily achieved since independent control forces and moments are simultaneously available in all degrees of freedom. For slow speed, this is referred to as DP and the craft is controlled by means of tunnel thrusters, azimuths and main propellers. Conventional craft, on the other hand, are usually equipped with one or two main propellers for forward speed control and rudders for turning control. The minimum configuration for waypoint tracking control is one main propeller and a single rudder. This means that only two controls are available, thus rendering the ship underactuated for the task of 3-DOF trajectory-tracking control.

Conventional waypoint guidance systems are usually designed by reducing the output space from 3-DOF position and heading to 2-DOF heading and surge (Healey and Marco 1992). In its simplest form this involves the use of a classical autopilot system where the commanded yaw angle ψ_d is generated such that the *cross-track error* is minimized. A path-following control system is usually designed such that the ship moves forward with reference speed u_d at the same time as the cross-track error to the path is minimized. As a result, ψ_d and u_d are tracked using only two controls.

This section is based on Fossen *et al.* (2003) and presents a maneuvering controller involving an LOS guidance system and a nonlinear feedback trajectory-tracking controller. The desired output is reduced from (x^n_d, y^n_d, ψ_d) to ψ_d and u_d using an LOS projection algorithm. The tracking task $\psi(t) \rightarrow \psi_d(t)$ is then achieved using only one control (normally the rudder), while tracking of the speed assignment u_d is performed by the remaining control (the main propeller). Since we are dealing with segments of straight lines, the LOS projection algorithm will guarantee that the task of path-following is satisfied.

First, an LOS guidance procedure is derived. This includes a projection algorithm and a waypoint switching algorithm. To avoid large bumps in ψ_d when switching, and to provide the necessary derivatives of ψ_d to the controller, the commanded LOS heading is fed through a reference model. Second, a nonlinear 2-DOF tracking controller is derived using the backstepping technique. Three stabilizing functions $\boldsymbol{\alpha} := [\alpha_1, \alpha_2, \alpha_3]^\top$ are defined where α_1 and α_3 are specified to satisfy the tracking objectives in the controlled surge and yaw modes. The stabilizing function α_2 in the uncontrolled sway mode is left as a free design variable. By assigning dynamics to α_2, the resulting controller becomes a dynamic feedback controller so that $\alpha_2(t) \rightarrow v(t)$ during path following. This is an appealing idea that adds to the extensive theory of backstepping. The presented design technique results in a robust controller for underactuated ships since integral action can be implemented for both path-following and speed control.

Problem Statement

The problem statement is stated as a maneuvering problem with the following two objectives (Skjetne *et al.* 2004):

LOS geometric task: Force the marine craft position $\boldsymbol{p} = [x^n,\ y^n]^\top$ to converge to a desired path by forcing the course angle χ to converge to (see Section 12.4)

$$\chi_d = \text{atan2}(y^n_{\text{los}} - y^n,\ x^n_{\text{los}} - x^n) \tag{16.341}$$

where the LOS position $\boldsymbol{p}_{\text{los}} = [x^n_{\text{los}},\ y^n_{\text{los}}]^\top$ is the point along the path to which the craft should be pointed.

Dynamic task: Force the speed u to converge to a desired speed assignment u_d according to

$$\lim_{t\to\infty} [u(t) - u_d(t)] = 0 \tag{16.342}$$

where u_d is the desired speed composed along the body-fixed x_b axis.

A conventional trajectory-tracking control system for 3 DOFs is usually implemented using a standard PID autopilot in series with an LOS algorithm. Hence, a state-of-the-art autopilot system can be modified to take the LOS reference angle as input (see Figure 15.21). This adds flexibility since the default commercial autopilot system can be used together with the LOS guidance system. The speed can be adjusted manually by the captain or automatically using the path speed profile.

Consider the 3-DOF nonlinear maneuvering model in the following form

$$\dot{\boldsymbol{\eta}} = \boldsymbol{R}(\psi)\boldsymbol{v} \tag{16.343}$$

$$\boldsymbol{M}\dot{\boldsymbol{v}} + \boldsymbol{N}(\boldsymbol{v})\boldsymbol{v} = \begin{bmatrix} (1-t)T \\ Y_\delta \delta \\ N_\delta \delta \end{bmatrix} := \begin{bmatrix} \tau_1 \\ Y_\delta \delta \\ \tau_3 \end{bmatrix} \tag{16.344}$$

where $\boldsymbol{\eta} = [x^n,\ y^n,\ \psi]^\top$, $\boldsymbol{v} = [u,\ v,\ r]^\top$ and

$$\boldsymbol{R}(\psi) = \begin{bmatrix} \cos(\psi) & -\sin(\psi) & 0 \\ \sin(\psi) & \cos(\psi) & 0 \\ 0 & 0 & 1 \end{bmatrix}. \tag{16.345}$$

The matrices $\boldsymbol{M}$ and $\boldsymbol{N}$ are

$$\boldsymbol{M} = \begin{bmatrix} m_{11} & 0 & 0 \\ 0 & m_{22} & m_{23} \\ 0 & m_{32} & m_{33} \end{bmatrix} = \begin{bmatrix} m - X_{\dot{u}} & 0 & 0 \\ 0 & m - Y_{\dot{v}} & mx_g - Y_{\dot{r}} \\ 0 & mx_g - N_{\dot{v}} & I_z - N_{\dot{r}} \end{bmatrix}$$

$$\boldsymbol{N}(\boldsymbol{v}) = \begin{bmatrix} n_{11} & 0 & 0 \\ 0 & n_{22} & n_{23} \\ 0 & n_{32} & n_{33} \end{bmatrix} = \begin{bmatrix} -X_u & 0 & 0 \\ 0 & -Y_v & mu - Y_r \\ 0 & -N_v & mx_g u - N_r \end{bmatrix}.$$

The control force and moment in surge and yaw are denoted τ_1 and τ_3, respectively, while sway is left uncontrolled. Note that the rudder angle δ affects the sway equation but it will not be used to actively control sway. The controller computes τ_1 and τ_3 which can be allocated to thrust T and rudder angle δ using

$$\tau_1 = (1-t)T \tag{16.346}$$

$$\tau_3 = N_\delta \delta \tag{16.347}$$

where t is the thrust deduction number. This gives

$$T = \frac{1}{1-t}\tau_1 \tag{16.348}$$

$$\delta = \frac{1}{N_\delta}\tau_3. \tag{16.349}$$

Backstepping Design

The design is based on the model (16.343) and (16.344) where $\boldsymbol{M} = \boldsymbol{M}^\top > 0$. Define the error signals z_1 and $\boldsymbol{z}_2$ according to

$$z_1 := \chi - \chi_\mathrm{d} = \psi - \psi_\mathrm{d} \tag{16.350}$$

$$\boldsymbol{z}_2 := [z_{2,1},\ z_{2,2},\ z_{2,3}]^\top = \boldsymbol{\nu} - \boldsymbol{\alpha} \tag{16.351}$$

where χ_d is the desired LOS angle, u_d is the desired surge velocity and $\boldsymbol{\alpha} := [\alpha_1,\ \alpha_2,\ \alpha_3]^\top$ is a vector of stabilizing functions to be specified later. Next, let

$$\boldsymbol{h} = [0,\ 0,\ 1]^\top \tag{16.352}$$

such that

$$\begin{aligned}\dot{z}_1 &= r - r_\mathrm{d} = \boldsymbol{h}^\top \boldsymbol{\nu} - r_\mathrm{d} \\ &= \alpha_3 + \boldsymbol{h}^\top \boldsymbol{z}_2 - r_\mathrm{d}\end{aligned} \tag{16.353}$$

where $r_\mathrm{d} = \dot{\psi}_\mathrm{d}$ and

$$\begin{aligned}\boldsymbol{M}\dot{\boldsymbol{z}}_2 &= \boldsymbol{M}\dot{\boldsymbol{\nu}} - \boldsymbol{M}\dot{\boldsymbol{\alpha}} \\ &= \boldsymbol{\tau} - \boldsymbol{N}(\boldsymbol{\nu})\boldsymbol{\nu} - \boldsymbol{M}\dot{\boldsymbol{\alpha}}.\end{aligned} \tag{16.354}$$

Consider the CLF

$$V = \frac{1}{2}z_1^2 + \frac{1}{2}\boldsymbol{z}_2^\top \boldsymbol{M}\boldsymbol{z}_2, \qquad \boldsymbol{M} = \boldsymbol{M}^\top > 0 \tag{16.355}$$

$$\dot{V} = z_1\dot{z}_1 + \boldsymbol{z}_2^\top \boldsymbol{M}\dot{\boldsymbol{z}}_2 \tag{16.356}$$

$$= z_1(\alpha_3 + \boldsymbol{h}^\top \boldsymbol{z}_2 - r_\mathrm{d}) + \boldsymbol{z}_2^\top(\boldsymbol{\tau} - \boldsymbol{N}(\boldsymbol{\nu})\boldsymbol{\nu} - \boldsymbol{M}\dot{\boldsymbol{\alpha}}). \tag{16.357}$$

Choosing the virtual control α_3 as

$$\alpha_3 = -cz_1 + r_\mathrm{d} \tag{16.358}$$

while α_1 and α_2 are yet to be defined gives

$$\begin{aligned}\dot{V} &= -cz_1^2 + z_1\boldsymbol{h}^\top \boldsymbol{z}_2 + \boldsymbol{z}_2^\top(\boldsymbol{\tau} - \boldsymbol{N}(\boldsymbol{\nu})\boldsymbol{\nu} - \boldsymbol{M}\dot{\boldsymbol{\alpha}}) \\ &= -cz_1^2 + \boldsymbol{z}_2^\top(\boldsymbol{h}z_1 + \boldsymbol{\tau} - \boldsymbol{N}(\boldsymbol{\nu})\boldsymbol{\nu} - \boldsymbol{M}\dot{\boldsymbol{\alpha}}).\end{aligned} \tag{16.359}$$

Suppose we can assign

$$\boldsymbol{\tau} = \begin{bmatrix} \tau_1 \\ Y_\delta \delta \\ \tau_3 \end{bmatrix} = \boldsymbol{M}\dot{\boldsymbol{\alpha}} + \boldsymbol{N}(\boldsymbol{v})\boldsymbol{v} - \boldsymbol{K}\boldsymbol{z}_2 - \boldsymbol{h}z_1 \tag{16.360}$$

where $\boldsymbol{K} = \text{diag}\{k_1,\ k_2,\ k_3\} > 0$. This results in

$$\dot{V} = -cz_1^2 - \boldsymbol{z}_2^\top \boldsymbol{K}\boldsymbol{z}_2 < 0, \quad \forall z_1 \neq 0, \boldsymbol{z}_2 \neq \boldsymbol{0} \tag{16.361}$$

and by standard Lyapunov arguments, this guarantees that $(z_1,\ \boldsymbol{z}_2)$ is bounded and converges to zero. However, note from (16.360) that it is only possible to prescribe values for τ_1 and τ_3; that is

$$\tau_1 = m_{11}\dot{\alpha}_1 + n_{11}u - k_1(u - \alpha_1) \tag{16.362}$$

$$\tau_3 = m_{32}\dot{\alpha}_2 + m_{33}\dot{\alpha}_3 + n_{32}v + n_{33}r - k_3(r - \alpha_3) - z_1. \tag{16.363}$$

Choosing $\alpha_1 = u_\text{d}$ clearly solves the dynamic task since the closed-loop surge dynamics becomes

$$m_{11}(\dot{u} - \dot{u}_\text{d}) + k_1(u - u_\text{d}) = 0. \tag{16.364}$$

The second equation in (16.360) results in a dynamic equality constraint

$$m_{22}\dot{\alpha}_2 + m_{23}\dot{\alpha}_3 + n_{22}v + n_{23}r - k_2(v - \alpha_2) = \frac{Y_\delta}{N_\delta}\tau_3 \tag{16.365}$$

affected by the control input τ_3. Substituting (16.363) into this expression yields

$$\left(m_{22} - \frac{Y_\delta}{N_\delta}m_{32}\right)\dot{\alpha}_2 + \left(m_{23} - \frac{Y_\delta}{N_\delta}m_{33}\right)\dot{\alpha}_3 + \left(n_{22} - \frac{Y_\delta}{N_\delta}n_{32}\right)v$$
$$+ \left(n_{23} - \frac{Y_\delta}{N_\delta}n_{33}\right)r - k_2(v - \alpha_2) + \frac{Y_\delta}{N_\delta}(k_3(r - \alpha_3) + z_1) = 0.$$

Application of $\dot{\alpha}_3 = c^2 z_1 - cz_{2,3} + \dot{r}_\text{d}$, $\alpha_3 = -cz_1 + r_\text{d}$, $v = \alpha_2 + z_{2,2}$ and $r = \alpha_3 + z_{2,3}$ then gives

$$\left(m_{22} - \frac{Y_\delta}{N_\delta}m_{32}\right)\dot{\alpha}_2 = -\left(n_{22} - \frac{Y_\delta}{N_\delta}n_{32}\right)\alpha_2 + \gamma(z_1, \boldsymbol{z}_2, r_\text{d}, \dot{r}_\text{d}) \tag{16.366}$$

where

$$\begin{aligned}\gamma(z_1, \boldsymbol{z}_2, r_\text{d}, \dot{r}_\text{d}) = &-\left(m_{23} - \frac{Y_\delta}{N_\delta}m_{33}\right)(c^2 z_1 - cz_{2,3} + \dot{r}_\text{d}) - \left(n_{22} - \frac{Y_\delta}{N_\delta}n_{32}\right)z_{2,2} \\ &-\left(n_{23} - \frac{Y_\delta}{N_\delta}n_{33}\right)(-cz_1 + r_\text{d} + z_{2,3}) + k_2 z_{2,2} - \frac{Y_\delta}{N_\delta}(k_3 z_{2,3} + z_1).\end{aligned} \tag{16.367}$$

The variable α_2 becomes a dynamic state of the controller according to (16.366). Furthermore, $m_{22} > (Y_\delta/N_\delta)m_{32}$ and $n_{22} > (Y_\delta/N_\delta)n_{32}$ imply that (16.366) is a stable differential equation driven by the converging error signals $(z_1, \boldsymbol{z}_2)$ and the bounded reference signals $(r_\text{d}, \dot{r}_\text{d})$ within the expression of $\gamma(\cdot)$. Since $z_{2,2}(t) \to 0$, it follows that $|\alpha_2(t) - v(t)| \to 0$ as

$t \to \infty$. The main result is summarized by Theorem 16.1, which is a modification of Fossen *et al.* (2003).

Theorem 16.1 ***(LOS Backstepping Controller for Underactuated Marine Craft)***
The LOS maneuvering problem for the 3-DOF underactuated marine craft (16.343) and (16.344) is solved using the control laws

$$\tau_1 = m_{11}\dot{u}_d + n_{11}u - k_1(u - u_d) \tag{16.368}$$

$$\tau_3 = m_{32}\dot{\alpha}_2 + m_{33}\dot{\alpha}_3 + n_{32}v + n_{33}r - k_3(r - \alpha_3) - \text{ssa}(z_1) \tag{16.369}$$

where ssa: $\mathbb{R} \to [-\pi, \pi)$ *maps the unconstrained angle* $z_1 \in \mathbb{R}$ *to the smallest difference between the angles (see Section 13.4.3),* $k_1 > 0$, $k_3 > 0$, $z_1 := \psi - \psi_d$, $z_2 := [u - u_d, v - \alpha_2, r - \alpha_3]^\top$ *and*

$$\alpha_3 = -cz_1 + r_d, \qquad c > 0 \tag{16.370}$$

$$\dot{\alpha}_3 = -c(r - r_d) + \dot{r}_d. \tag{16.371}$$

The reference signals u_d, $\dot{u}_d$, ψ_d, r_d *and* $\dot{r}_d$ *are provided by the LOS guidance system, while* α_2 *is given by*

$$\left(m_{22} - \frac{Y_\delta}{N_\delta}m_{32}\right)\dot{\alpha}_2 = -\left(n_{22} - \frac{Y_\delta}{N_\delta}n_{32}\right)\alpha_2 + \gamma(z_1, z_2, r_d, \dot{r}_d).$$

This results in an asymptotically stable equilibrium point $(z_1, z_2) = (0, \mathbf{0})$ *while the stabilizing function* α_2 *satisfies*

$$\lim_{t\to\infty} |\alpha_2(t) - v(t)| = 0. \tag{16.372}$$

Note that the smooth reference signal ψ_d *must be differentiated twice to produce* r_d *and* $\dot{r}_d$ *while* u_d *must be differentiated once to give* $\dot{u}_d$*. This is most easily achieved by using reference models represented by low-pass filters (see Section 12.1.1).*

Proof. *The closed-loop system is*

$$\begin{bmatrix}\dot{z}_1\\ \dot{z}_2\end{bmatrix} = \begin{bmatrix}-c & \boldsymbol{h}^\top\\ -\boldsymbol{M}^{-1}\boldsymbol{h} & -\boldsymbol{M}^{-1}\boldsymbol{K}\end{bmatrix}\begin{bmatrix}z_1\\ z_2\end{bmatrix} \tag{16.373}$$

$$\bar{m}\dot{\alpha}_2 = -\bar{n}\alpha_2 + \gamma(z_1, z_2, r_d, \dot{r}_d) \tag{16.374}$$

where

$$\bar{m} = \left(m_{22} - \frac{Y_\delta}{N_\delta}m_{32}\right), \qquad \bar{n} = \left(n_{22} - \frac{Y_\delta}{N_\delta}n_{32}\right). \tag{16.375}$$

From the Lyapunov arguments (16.355) and (16.361), the equilibrium $(z_1, z_2) = (0, \mathbf{0})$ *of the* z *subsystem is asymptotically stable. The unforced* α_2 *subsystem (*$\gamma = 0$*) is clearly exponentially stable. Since* $(z_1, z_2) \in \mathcal{L}_\infty$ *and* $(r_d, \dot{r}_d) \in \mathcal{L}_\infty$*, then* $\gamma \in \mathcal{L}_\infty$*. This implies that the*

α_2 subsystem is input-to-state stable from γ to α_2. This is seen by applying, for instance, $V_2 = (1/2)\bar{m}\alpha_2^2$ which, differentiated along the solutions of α_2, gives $\dot{V}_2 \leq -(1/2)\bar{n}\alpha_2^2$ for all $|\alpha_2| \geq \frac{2}{\bar{n}}|\gamma(z_1, z_2, r_d, \dot{r}_d)|$. By standard comparison functions, it is then possible to show that for all $|\alpha_2| \geq \frac{2}{\bar{n}}|\gamma(z_1, z_2, r_d, \dot{r}_d)|$ then

$$|\alpha_2(t)| \leq |\alpha_2(0)|e^{-\frac{\bar{n}}{2}t}. \tag{16.376}$$

Hence, α_2 converges to the bounded set $\{\alpha_2 : |\alpha_2| \leq \frac{2}{\bar{n}}|\gamma(z_1, z_2, r_d, \dot{r}_d)|\}$ since $z_{2,2}(t) \to 0$ as $t \to \infty$. Since z_1 is mapped to the smallest signed angle by using $ssa(z_1)$ only local asymptotic stability can be guaranteed (see Section 13.4.3).

16.3.11 Case Study: Weather Optimal Position Control

Conventional DP systems for ships and free-floating rigs are usually designed for station-keeping by specifying a desired constant position (x_d^n, y_d^n) and a desired constant heading angle ψ_d. In order to minimize the ship fuel consumption, the desired heading ψ_d should in many operations be chosen such that the yaw moment is zero. For vessels with port/starboard symmetry, this means that the mean environmental forces due to wind, waves and ocean currents act through the centerline of the vessel. Then the ship must be rotated until the yaw moment is zero.

Unfortunately, it is impossible to measure or compute the direction of the mean environmental force with sufficient accuracy. Hence, the desired heading ψ_d is usually taken to be the measurement of the mean wind direction, which can be easily measured. In practice, however, this can result in large offsets from the true mean direction of the total environmental force. The main reason for this is the unmeasured ocean current force component and waves that do not coincide with the wind direction. Hence, the DP system can be operated under highly non-optimal conditions if fuel saving is the issue. A small offset in the optimal heading angle will result in a large use of thrust.

One popular method for computing the weather optimal heading ψ_d is to monitor the resulting thruster forces in the x_b and y_b directions. Hence, the bow of the ship can be turned in one direction until the thruster force in the y_b direction approaches zero. This method is appealing but the main catch in doing this is that the total resulting thruster forces in the x_b and y_b directions have to be computed since there are no sensors doing this job directly. The sensors only measure the angular speed and pitch angle of the propellers. Hence, the thrust for each propeller must be computed by using a model of the thruster characteristic, resulting in a fairly rough estimate of the total thruster force in each direction.

Another principle, proposed by Pinkster (1971) and Pinkster and Nienhuis (1986), is to control the (x^n, y^n) positions using a PID feedback controller, in addition to feedback from the yaw velocity, such that the vessel tends toward the optimal heading. This principle, however, requires that the rotation point of the vessel is located a certain distance forward of the center of gravity, or even fore of the bow, and it also puts restrictions on the thruster configuration and the number of thrusters installed.

This section describes the weather optimal position controller (WOPC) by Fossen and Strand (2001). The control objective is that the vessel heading should adjust automatically to the mean environmental forces (wind, waves and ocean currents) such that a minimum amount of energy is used in order to save fuel and reduce NO_x/CO_x emissions without using any environmental sensors. This is particularly useful for shuttle tankers and FPSOs, which can be located at the same position for a long time. Also DP-operated supply vessels that must keep their position for days in loading/off-loading operations have a great WOPC fuel-saving potential.

The ship can be exponentially stabilized on a circle arc with constant radius by letting the bow of the ship point toward the origin of the circle. In order to maintain a fixed position at the same time, a translatory circle center control law is designed. The circle center is translated such that the Cartesian position is constant, while the bow of the ship is automatically turned up against the mean environmental force to obtain weathervaning. This approach is motivated by a pendulum in the gravity field where gravity is the unmeasured quantity. The circular motion of the controlled ship, where the mean environmental force can be interpreted as an unknown force field, copies the dynamics of a pendulum in the gravity field (see Figure 16.17).

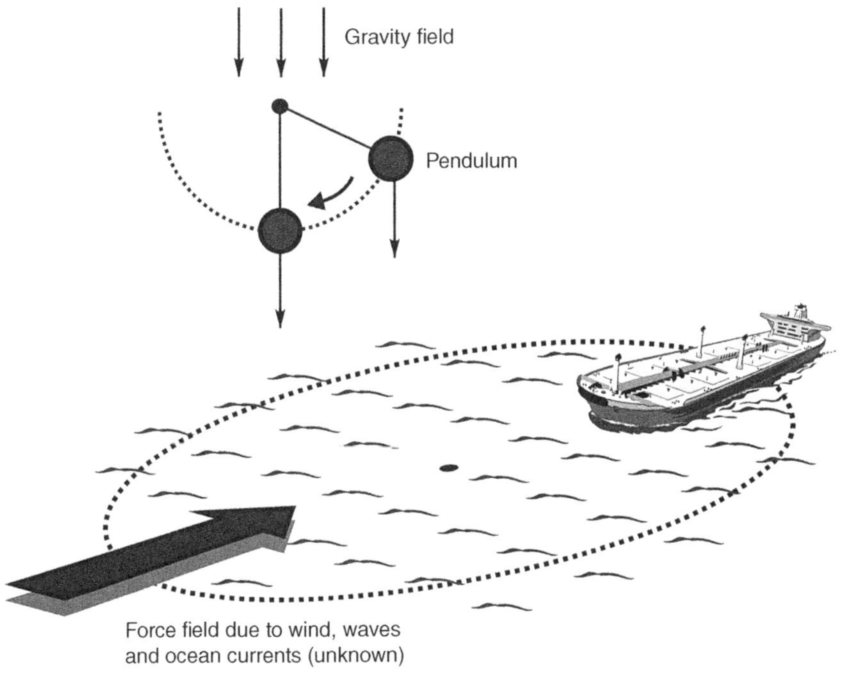

Figure 16.17 The principle of WOPC using the equivalence to a pendulum in the gravity field where gravity is the unmeasured quantity.

3-DOF Equations of Motion using Polar Coordinates

Consider a marine craft in 3 DOFs

$$\dot{\boldsymbol{\eta}} = \boldsymbol{R}(\psi)\boldsymbol{\nu} \tag{16.377}$$

$$\boldsymbol{M}\dot{\boldsymbol{\nu}} + \boldsymbol{C}(\boldsymbol{\nu})\boldsymbol{\nu} + \boldsymbol{D}(\boldsymbol{\nu})\boldsymbol{\nu} = \boldsymbol{\tau} + \boldsymbol{w} \tag{16.378}$$

where the north-east positions and heading are represented by $\boldsymbol{\eta} = [x^n, \ y^n, \ \psi]^\top$ and the body-fixed velocities are represented by $\boldsymbol{\nu} = [u, \ v, \ r]^\top$. It is assumed that $\boldsymbol{M} = \boldsymbol{M}^\top > 0$, $\dot{\boldsymbol{M}} = \boldsymbol{0}$ and $\boldsymbol{D}(\boldsymbol{\nu}) > 0$. Unmodeled external forces and moment due to wind, ocean currents and waves are lumped together into a body-fixed disturbance vector $\boldsymbol{w} \in \mathbb{R}^3$ to be interpreted later.

The Cartesian coordinates $(x^n, \ y^n)$ are related to the *polar coordinates* $(\rho, \ \gamma)$ by

$$x^n = x_0^n + \rho\cos(\gamma), \qquad y^n = y_0^n + \rho\sin(\gamma) \tag{16.379}$$

where $(x_0^n, \ y_0^n)$ is the origin of a circle with radius ρ and polar angle γ given by

$$\rho = \sqrt{(x^n - x_0^n)^2 + (y^n - y_0^n)^2} \tag{16.380}$$

$$\gamma = \text{atan2}((y^n - y_0^n), \ (x^n - x_0^n)). \tag{16.381}$$

Time differentiation of (16.379) yields

$$\dot{x}^n = \dot{x}_0^n + \dot{\rho}\cos(\gamma) - \rho\sin(\gamma)\dot{\gamma} \tag{16.382}$$

$$\dot{y}^n = \dot{y}_0^n + \dot{\rho}\sin(\gamma) + \rho\cos(\gamma)\dot{\gamma}. \tag{16.383}$$

Define the state vectors

$$\boldsymbol{p}_0 := [x_0^n, \ y_0^n]^\top, \qquad \boldsymbol{x} := [\rho, \ \gamma, \ \psi]^\top. \tag{16.384}$$

From (16.382) and (16.383) a new kinematic relationship can be derived in terms of the vectors $\boldsymbol{p}_0$ and $\boldsymbol{x}$ as

$$\dot{\boldsymbol{\eta}} = \boldsymbol{R}(\gamma)\boldsymbol{H}(\rho)\dot{\boldsymbol{x}} + \boldsymbol{L}\dot{\boldsymbol{p}}_0 \tag{16.385}$$

$$\boldsymbol{H}(\rho) = \begin{bmatrix} 1 & 0 & 0 \\ 0 & \rho & 0 \\ 0 & 0 & 1 \end{bmatrix}, \qquad \boldsymbol{L} = \begin{bmatrix} 1 & 0 \\ 0 & 1 \\ 0 & 0 \end{bmatrix}. \tag{16.386}$$

From (16.385) the Cartesian kinematics (16.377) can be replaced by a differential equation for the polar coordinates

$$\dot{x} = T(x)\nu - T(x)R^{\top}(\psi)L\dot{p}_0 \tag{16.387}$$

$$T(x) = H^{-1}(\rho)\underbrace{R^{\top}(\gamma)R(\psi)}_{R^{\top}(\gamma-\psi)}. \tag{16.388}$$

Note that the conversion between Cartesian and polar coordinates is only a local diffeomorphism, since the radius must be kept larger than a minimum value, that is $\rho > \rho_{\min} > 0$, in order to avoid the singular point $\rho = 0$.

Marine Craft Model Transformation

The marine craft model (16.378) can be expressed in polar coordinates by using (16.387) and substituting

$$\nu = T^{-1}(x)\dot{x} + R^{\top}(\psi)L\dot{p}_0 \tag{16.389}$$

$$\dot{\nu} = T^{-1}(x)\ddot{x} + \dot{T}^{-1}(x)\dot{x} + R^{\top}(\psi)L\ddot{p}_0 + \dot{R}^{\top}(\psi)L\dot{p}_0 \tag{16.390}$$

such that

$$M\dot{\nu} + C(\nu)\nu + D(\nu)\nu = \tau + w$$

$$\Updownarrow \ \rho > 0$$

$$M_x(x)\ddot{x} + C_x(\nu, x)\dot{x} + D_x(\nu, x)\dot{x} = T^{-\top}(x)[q(\nu, x, \dot{p}_0, \ddot{p}_0) + \tau + w] \tag{16.391}$$

where

$$M_x(x) = T^{-\top}(x)MT^{-1}(x)$$

$$C_x(\nu, x) = T^{-\top}(x)(C(\nu) - MT^{-1}(x)\dot{T}(x))T^{-1}(x)$$

$$D_x(\nu, x) = T^{-\top}(x)D(\nu)T^{-1}(x)$$

$$q(\nu, x, \dot{p}_0, \ddot{p}_0) = -MR^{\top}(\psi)L\ddot{p}_0 - M\dot{R}^{\top}(\psi)L\dot{p}_0 - [C(\nu) + D(\nu)]R^{\top}(\psi)L\dot{p}_0.$$

Here $M_x(x)$, $C_x(\nu, x)$ and $D_x(\nu, x)$ can be shown to satisfy

$$M_x(x) = M_x^{\top}(x) > 0, \quad D_x(\nu, x) > 0, \quad \forall x.$$

The marine craft dynamics also satisfy the skew-symmetric property

$$z^{\top}(\dot{M}_x - 2C_x)z = 0, \quad \forall z, x. \tag{16.392}$$

Weather Optimal Control Objectives

The steady-state LF motion of the craft and also the craft's equilibrium position depend on the *unknown* environmental forces acting on the hull. Let the environmental forces due to *wind, waves* and *ocean currents* be represented by

- A slowly-varying mean force F_e that attacks the craft at a point (l_x, l_y) expressed in $\{b\}$.
- A slowly-varying mean direction β_e relative to the Earth-fixed frame (see Figure 16.18).

The WF motion is assumed to be filtered out of the measurements by using a wave filter (see Chapter 13). Since there are no sensors that can be used to measure (F_e, β_e) and (l_x, l_y) with sufficient accuracy, it is impossible to use feedforward from the environmental forces. This leads to the following assumptions:

A1: *The unknown mean environmental force F_e and its direction β_e are assumed to be constant or at least slowly varying* for a given sea state.
A2: *The unknown attack point (l_x, l_y) is constant for each constant F_e.*

These are good assumptions since the control system is only supposed to counteract the slowly varying motion components of the environmental forces.

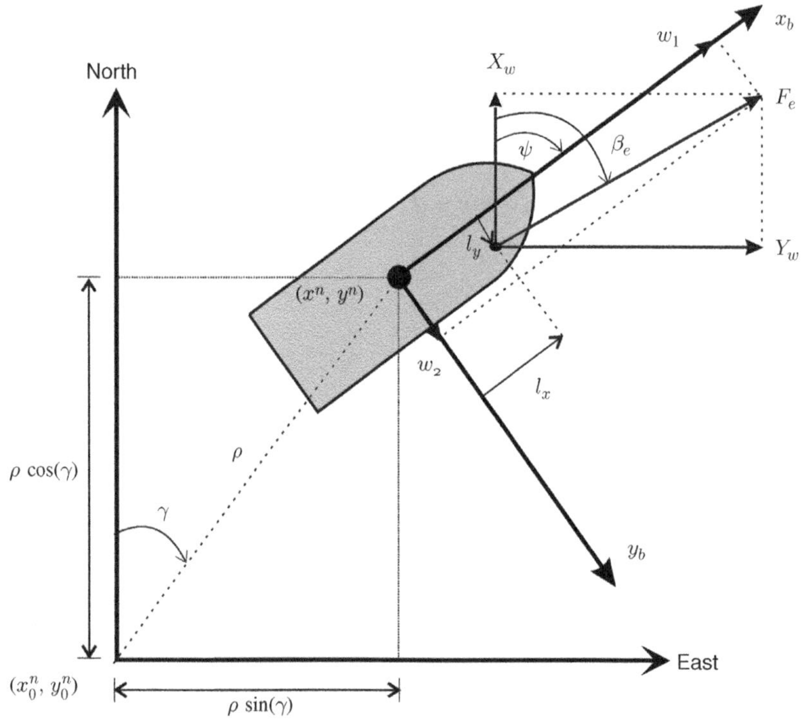

Figure 16.18 Environmental force F_e decomposed into components w_1 and w_2.

From Figure 16.18 the environmental force vector $\boldsymbol{w} \in \mathbb{R}^3$ can be expressed as

$$\boldsymbol{w} = \begin{bmatrix} w_1 \\ w_2 \\ w_3 \end{bmatrix} = \begin{bmatrix} F_e \cos(\beta_e - \psi) \\ F_e \sin(\beta_e - \psi) \\ l_x F_e \sin(\beta_e - \psi) - l_y F_e \cos(\beta_e - \psi) \end{bmatrix}. \tag{16.393}$$

Note that the environmental forces vary with the heading angle ψ of the craft. Consequently,

$$F_e = \sqrt{w_1^2 + w_2^2}, \qquad \beta_e = \psi + \tan^{-1}(w_2/w_1). \tag{16.394}$$

The environmental force components X_w and Y_w with attack point (l_x, l_y) are shown in Figure 16.18. The weather optimal control objectives can be satisfied by using the following definitions (Fossen and Strand 2001).

Definition 16.3 (Weather Optimal Heading)
The weather optimal heading angle ψ_{opt} is given by the equilibrium state where the yaw moment $w_3 = 0$ at the same time as the bow of the craft is turned up against weather (mean environmental forces); that is $w_2 = 0$. This implies that $\psi_{opt} = \beta_e$, $l_x = \text{constant}$ and $l_y = 0$ such that

$$\boldsymbol{w} = [-F_e, \ 0, \ 0]^\top. \tag{16.395}$$

Hence, the mean environmental force attacks the craft in the bow, which has the minimum drag coefficient for water and wind loads.

Definition 16.4 (Weather Optimal Positioning)
Weather optimal positioning (stationkeeping) is defined as the equilibrium state (16.395) with the additional requirement that the position $(x^n, y^n) = (x^n_d, y^n_d)$ is kept constant.

Weather optimal heading control (WOHC) is obtained by restricting the craft's movement to a circle with constant radius $\rho = \rho_d$ and at the same time force the craft's bow to point towards the center of the circle until the weather optimal heading angle $\psi = \psi_{opt}$ is reached (see Figure 16.19). An analogy to this is a pendulum in a gravity field (see Figure 16.17). The position $(x^n, y^n) = (x^n_0 + \rho\cos(\gamma), \ y^n_0 + \rho\sin(\gamma))$ will vary until the weather optimal heading angle is reached. This is obtained by specifying the desired polar coordinates according to

$$\rho_d = \text{constant}, \qquad \dot{\gamma}_d = 0, \qquad \psi_d = \pi + \gamma. \tag{16.396}$$

The requirement $\rho_d = \text{constant}$ implies that the craft follows a circular arc with a constant radius. The second requirement $\dot{\gamma}_d = 0$ implies that the tangential speed $\rho\dot{\gamma}$ is kept small while the last requirement $\psi_d = \pi + \gamma$ ensures that the craft's bow points toward the center of the circle.

Weather optimal positioning control is used to maintain a fixed Earth-fixed position $(x^n, y^n) = (x^n_d, y^n_d)$ by moving the circle center $\boldsymbol{p}_0 = [x^n_0, y^n_0]^\top$ and at the same time keep $\psi = \psi_{opt}$. This is referred to as *translatory circle center control.*

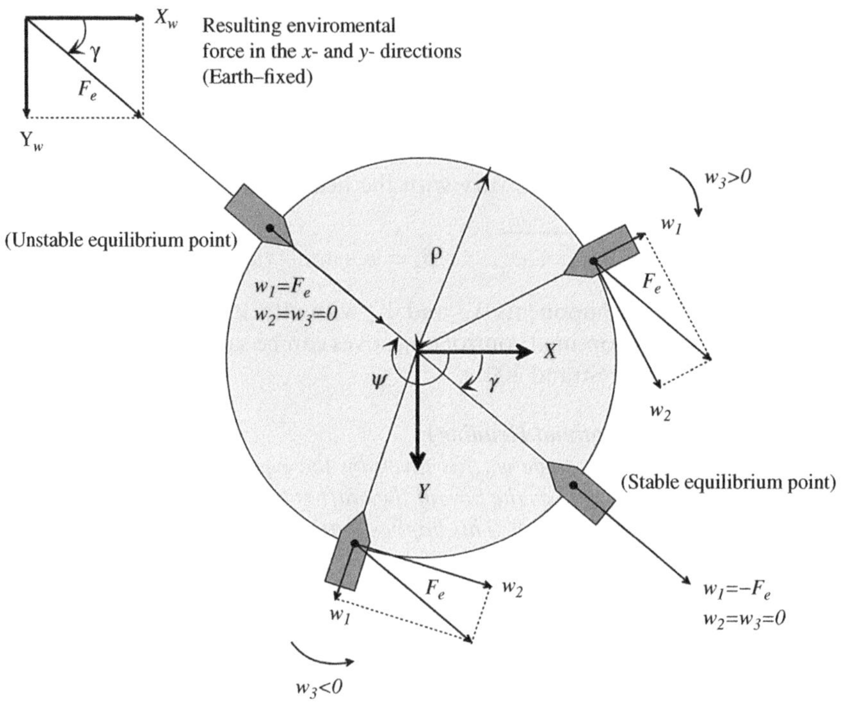

Figure 16.19 Stable and unstable equilibrium points for WOPC.

Nonlinear and Adaptive Control Design

The WOPC positioning controller is derived by using the polar coordinate representation. The backstepping design methodology with extension to integral control (Fossen *et al.* 2001) is used to derive the feedback controller (see Section 16.3). Note that conventional PID control can be used as well.

The WOPC controller will be derived in three successive steps:

1. *Nonlinear backstepping (PD control)*: The ship is forced to move along a circular arc with desired radius ρ_d, with minimum tangential velocity $\rho\dot{\gamma}$ and desired heading ψ_d.
2. *Adaptive backstepping (PID control)*: Integral action is included to compensate for the unknown environmental force F_e.
3. *Translational control of the circle center*: The circle center (x_0^n, y_0^n) is translated such that the ship maintains a constant position $(x_\text{d}^n, y_\text{d}^n)$ even though it is moving along a virtual circular arc. Hence, the captain of the ship will only notice that the ship is rotating a yaw angle ψ about a constant position $(x_\text{d}^n, y_\text{d}^n)$ until the weather optimal heading ψ_opt is reached.

Nonlinear Backstepping (PD Control)

The control objective (16.396) is specified in polar coordinates. Since the transformed system (16.391) is of order two, backstepping is performed in two *vectorial steps*, resulting in a nonlinear PD control law. First, a *virtual reference trajectory* is defined as

$$\dot{\boldsymbol{x}}_{\rm r} := \dot{\boldsymbol{x}}_{\rm d} - \boldsymbol{\Lambda} \boldsymbol{z}_1 \tag{16.397}$$

where $\boldsymbol{z}_1 = \boldsymbol{x} - \boldsymbol{x}_{\rm d}$ is the tracking error and $\boldsymbol{\Lambda} > 0$ is a diagonal design matrix. Furthermore, let $\boldsymbol{z}_2$ denote a measure of tracking defined according to

$$\boldsymbol{z}_2 := \dot{\boldsymbol{x}} - \dot{\boldsymbol{x}}_{\rm r} = \dot{\boldsymbol{z}}_1 + \boldsymbol{\Lambda} \boldsymbol{z}_1. \tag{16.398}$$

From (16.398), the following expressions are obtained

$$\dot{\boldsymbol{x}} = \boldsymbol{z}_2 + \dot{\boldsymbol{x}}_{\rm r}, \qquad \ddot{\boldsymbol{x}} = \dot{\boldsymbol{z}}_2 + \ddot{\boldsymbol{x}}_{\rm r}. \tag{16.399}$$

This implies that the marine craft model (16.391) can be expressed in terms of $\boldsymbol{z}_2$, $\dot{\boldsymbol{x}}_{\rm r}$ and $\ddot{\boldsymbol{x}}_{\rm r}$ as

$$\boldsymbol{M}_x \dot{\boldsymbol{z}}_2 + \boldsymbol{C}_x \boldsymbol{z}_2 + \boldsymbol{D}_x \boldsymbol{z}_2 = \boldsymbol{T}^{-\top} \boldsymbol{\tau} + \boldsymbol{T}^{-\top} \boldsymbol{q}(\cdot) - \boldsymbol{M}_x \ddot{\boldsymbol{x}}_{\rm r} - \boldsymbol{C}_x \dot{\boldsymbol{x}}_{\rm r} - \boldsymbol{D}_x \dot{\boldsymbol{x}}_{\rm r} + \boldsymbol{T}^{-\top} \boldsymbol{w}. \tag{16.400}$$

Step 1: Let $\boldsymbol{z}_1$ be the first error variable, which from (16.398) has the dynamics

$$\dot{\boldsymbol{z}}_1 = -\boldsymbol{\Lambda} \boldsymbol{z}_1 + \boldsymbol{z}_2. \tag{16.401}$$

A CLF for the first step is

$$V_1 = \frac{1}{2} \boldsymbol{z}_1^\top \boldsymbol{K}_{\rm p} \boldsymbol{z}_1 \tag{16.402}$$

$$\dot{V}_1 = -\boldsymbol{z}_1^\top \boldsymbol{K}_{\rm p} \boldsymbol{\Lambda} \boldsymbol{z}_1 + \boldsymbol{z}_1^\top \boldsymbol{K}_{\rm p} \boldsymbol{z}_2 \tag{16.403}$$

where $\boldsymbol{K}_{\rm p} = \boldsymbol{K}_{\rm p}^\top > 0$ is a constant design matrix.

Step 2: In the second step the CLF is motivated by the "pseudo-kinetic energy"

$$V_2 = V_1 + \frac{1}{2} \boldsymbol{z}_2^\top \boldsymbol{M}_x \boldsymbol{z}_2, \qquad \boldsymbol{M}_x = \boldsymbol{M}_x^\top > 0 \tag{16.404}$$

$$\dot{V}_2 = \dot{V}_1 + \boldsymbol{z}_2^\top \boldsymbol{M}_x \dot{\boldsymbol{z}}_2 + \frac{1}{2} \boldsymbol{z}_2^\top \dot{\boldsymbol{M}}_x \boldsymbol{z}_2 \tag{16.405}$$

which by substitution of (16.403) and (16.400) gives

$$\begin{aligned} \dot{V}_2 = & -\boldsymbol{z}_1^\top \boldsymbol{K}_{\rm p} \boldsymbol{\Lambda} \boldsymbol{z}_1 + \frac{1}{2} \boldsymbol{z}_2^\top (\dot{\boldsymbol{M}}_x - 2\boldsymbol{C}_x) \boldsymbol{z}_2 - \boldsymbol{z}_2^\top \boldsymbol{D}_x \boldsymbol{z}_2 + \boldsymbol{z}_2^\top \boldsymbol{T}^{-\top} \boldsymbol{w} \\ & + \boldsymbol{z}_2^\top (\boldsymbol{K}_{\rm p} \boldsymbol{z}_1 + \boldsymbol{T}^{-\top} \boldsymbol{\tau} + \boldsymbol{T}^{-\top} \boldsymbol{q}(\cdot) - \boldsymbol{M}_x \ddot{\boldsymbol{x}}_{\rm r} - \boldsymbol{C}_x \dot{\boldsymbol{x}}_{\rm r} - \boldsymbol{D}_x \dot{\boldsymbol{x}}_{\rm r}). \end{aligned} \tag{16.406}$$

By using the property (16.392) and choosing the nonlinear PD control law as

$$\boldsymbol{\tau} = \boldsymbol{T}^\top (\boldsymbol{M}_x \ddot{\boldsymbol{x}}_{\rm r} + \boldsymbol{C}_x \dot{\boldsymbol{x}}_{\rm r} + \boldsymbol{D}_x \dot{\boldsymbol{x}}_{\rm r} - \boldsymbol{K}_{\rm p} \boldsymbol{z}_1 - \boldsymbol{K}_{\rm d} \boldsymbol{z}_2) - \boldsymbol{q}(\cdot) \tag{16.407}$$

where $\boldsymbol{K}_\mathrm{d} > 0$ is a strictly positive design matrix, it is seen that

$$\dot{V}_2 = -\boldsymbol{z}_1^\top \boldsymbol{K}_\mathrm{p}\boldsymbol{\Lambda}\boldsymbol{z}_1 - \boldsymbol{z}_2^\top (\boldsymbol{K}_\mathrm{d} + \boldsymbol{D}_x)\boldsymbol{z}_2 + \boldsymbol{z}^\top \boldsymbol{T}^{-\top}\boldsymbol{w}. \tag{16.408}$$

Note that the dissipative term $\boldsymbol{z}_2^\top \boldsymbol{D}_x \boldsymbol{z}_2 > 0, \forall \boldsymbol{z}_2 \neq \boldsymbol{0}$ is exploited in the design as it appears in the expression for $\dot{V}_2$. With the control law (16.407) the closed-loop dynamics becomes

$$\boldsymbol{M}_x\dot{\boldsymbol{z}}_2 + (\boldsymbol{C}_x + \boldsymbol{D}_x + \boldsymbol{K}_\mathrm{d})\boldsymbol{z}_2 + \boldsymbol{K}_\mathrm{p}\boldsymbol{z}_1 = \boldsymbol{T}^{-\top}\boldsymbol{w}. \tag{16.409}$$

The error dynamics of the resulting system becomes *non-autonomous* since

$$\begin{bmatrix} \boldsymbol{K}_\mathrm{p} & \boldsymbol{0}_{3\times3} \\ \boldsymbol{0}_{3\times3} & \boldsymbol{M}_x \end{bmatrix} \begin{bmatrix} \dot{\boldsymbol{z}}_1 \\ \dot{\boldsymbol{z}}_2 \end{bmatrix} = - \begin{bmatrix} \boldsymbol{K}_\mathrm{p}\boldsymbol{\Lambda} & \boldsymbol{0}_{3\times3} \\ \boldsymbol{0}_{3\times3} & \boldsymbol{C}_x + \boldsymbol{D}_x + \boldsymbol{K}_\mathrm{d} \end{bmatrix} \begin{bmatrix} \boldsymbol{z}_1 \\ \boldsymbol{z}_2 \end{bmatrix}$$

$$+ \begin{bmatrix} \boldsymbol{0}_{3\times3} & \boldsymbol{K}_\mathrm{p} \\ -\boldsymbol{K}_\mathrm{p} & \boldsymbol{0}_{3\times3} \end{bmatrix} \begin{bmatrix} \boldsymbol{z}_1 \\ \boldsymbol{z}_2 \end{bmatrix} + \begin{bmatrix} \boldsymbol{0}_{3\times1} \\ \boldsymbol{T}^{-\top} \end{bmatrix} \boldsymbol{w}$$

$$\Updownarrow$$

$$\mathcal{M}(\boldsymbol{x})\dot{\boldsymbol{z}} = -\mathcal{K}(\boldsymbol{x}, \boldsymbol{\nu})\boldsymbol{z} + \mathcal{S}\boldsymbol{z} + \bar{\mathcal{B}}(\boldsymbol{x})\boldsymbol{w} \tag{16.410}$$

where the different matrices are defined as

$$\mathcal{M}(\boldsymbol{x}) = \mathcal{M}^\top(\boldsymbol{x}) = \begin{bmatrix} \boldsymbol{K}_\mathrm{p} & \boldsymbol{0}_{3\times3} \\ \boldsymbol{0}_{3\times3} & \boldsymbol{M}_x(\boldsymbol{x}) \end{bmatrix}$$

$$\mathcal{K}(\boldsymbol{x}, \boldsymbol{\nu}) = \begin{bmatrix} \boldsymbol{K}_\mathrm{p}\boldsymbol{\Lambda} & \boldsymbol{0}_{3\times3} \\ \boldsymbol{0}_{3\times3} & \boldsymbol{C}_x(\boldsymbol{x}, \boldsymbol{\nu}) + \boldsymbol{D}_x(\boldsymbol{x}, \boldsymbol{\nu}) + \boldsymbol{K}_\mathrm{d} \end{bmatrix} > 0$$

$$\mathcal{S} = -\mathcal{S}^\top = \begin{bmatrix} \boldsymbol{0}_{3\times3} & \boldsymbol{K}_\mathrm{p} \\ -\boldsymbol{K}_\mathrm{p} & \boldsymbol{0}_{3\times3} \end{bmatrix}, \quad \bar{\mathcal{B}}(\boldsymbol{x}) = \begin{bmatrix} \boldsymbol{0}_{3\times1} \\ \boldsymbol{T}^{-\top}(\boldsymbol{x}) \end{bmatrix}.$$

In the absence of disturbances, $\boldsymbol{w} \equiv \boldsymbol{0}$, the origin $\boldsymbol{z} = \boldsymbol{0}$ is uniformly locally exponentially stable (ULES) according to Lyapunov. Global results cannot be achieved due to the local diffeomorphism between the Cartesian and polar coordinates; that is the transformation matrix $\boldsymbol{T}(\boldsymbol{x})$ is singular for $\rho = 0$. In addition the yaw angle is defined on $[0, 2\pi)$ and not $\mathbb{R}$.

With disturbances $\boldsymbol{w} \neq \boldsymbol{0}$, the closed-loop system is input-to-state stable (ISS). We will now show how adaptive backstepping can be used to compensate a non-zero disturbance vector $\boldsymbol{w} \neq \boldsymbol{0}$.

PID Control by Adaptive Backstepping

Since the mean disturbance $\boldsymbol{w}$ is non-zero this will result in a steady-state offset when using a PD controller. The craft is, however, restricted to move along a circular arc with $\boldsymbol{w}$ as a force field. Therefore there will be a stable and an unstable equilibrium point on the circle arc (similar to a pendulum in the gravity field); see Figure 16.17. The stable equilibrium point is given by

$$\boldsymbol{w} = \boldsymbol{\phi}F_\mathrm{e} = [-1,\ 0,\ 0]^\top F_\mathrm{e}. \tag{16.411}$$

Since the disturbance F_e is assumed to be slowly varying, adaptive backstepping can be applied to obtain an integral effect in the system. Thus, in the analysis it will be assumed that $\dot{F}_e = 0$. Let the estimate of F_e be denoted as $\hat{F}_e$ and $\tilde{F}_e = \hat{F}_e - F_e$. An additional step in the derivation of the backstepping control law must be performed in order to obtain an adaptive update law for $\hat{F}_e$.

Step 3: The adaptive update law is found by adding the square parameter estimation error to V_2. Consequently,

$$V_3 = V_2 + \frac{1}{2\mu}\tilde{F}_e^2, \quad \mu > 0 \tag{16.412}$$

$$\dot{V}_3 = \dot{V}_2 + \frac{1}{\mu}\dot{\tilde{F}}_e\tilde{F}_e. \tag{16.413}$$

The nonlinear control law (16.407) is modified to

$$\boldsymbol{\tau} = \boldsymbol{T}^\top(\boldsymbol{M}_x\ddot{\boldsymbol{x}}_r + \boldsymbol{C}_x\dot{\boldsymbol{x}}_r + \boldsymbol{D}_x\dot{\boldsymbol{x}}_r - \boldsymbol{K}_p\boldsymbol{z}_1 - \boldsymbol{K}_d\boldsymbol{z}_2) - \boldsymbol{q}(\cdot) - \boldsymbol{\phi}\hat{F}_e \tag{16.414}$$

where the last term $\boldsymbol{\phi}\hat{F}_e$ provides integral action. Hence, the $\boldsymbol{z}_2$ dynamics becomes

$$\boldsymbol{M}_x\dot{\boldsymbol{z}}_2 + (\boldsymbol{C}_x + \boldsymbol{D}_x + \boldsymbol{K}_d)\boldsymbol{z}_2 + \boldsymbol{K}_p\boldsymbol{z}_1 = -\boldsymbol{T}^{-\top}\boldsymbol{\phi}\tilde{F}_e. \tag{16.415}$$

This implies that

$$\begin{aligned}\dot{V}_3 &= -\boldsymbol{z}_1^\top \boldsymbol{K}_p\boldsymbol{\Lambda}\boldsymbol{z}_1 - \boldsymbol{z}_2^\top(\boldsymbol{K}_d + \boldsymbol{D}_x)\boldsymbol{z}_2 - \boldsymbol{z}_2^\top\boldsymbol{T}^{-\top}\boldsymbol{\phi}\tilde{F}_e + \frac{1}{\mu}\dot{\tilde{F}}_e\tilde{F}_e\\ &= -\boldsymbol{z}_1^\top \boldsymbol{K}_p\boldsymbol{\Lambda}\boldsymbol{z}_1 - \boldsymbol{z}_2^\top(\boldsymbol{K}_d + \boldsymbol{D}_x)\boldsymbol{z}_2 + \tilde{F}_e(-\boldsymbol{\phi}^\top\boldsymbol{T}^{-1}\boldsymbol{z}_2 + \frac{1}{\mu}\dot{\tilde{F}}_e).\end{aligned} \tag{16.416}$$

The adaptive law $\dot{\hat{F}}_e = \dot{\tilde{F}}_e$ is chosen as

$$\dot{\hat{F}}_e = \mu\boldsymbol{\phi}^\top\boldsymbol{T}^{-1}\boldsymbol{z}_2, \quad \mu > 0 \tag{16.417}$$

such that

$$\dot{V}_3 = -\boldsymbol{z}_1^\top\boldsymbol{K}_p\boldsymbol{\Lambda}\boldsymbol{z}_1 - \boldsymbol{z}_2^\top(\boldsymbol{K}_d + \boldsymbol{D}_x)\boldsymbol{z}_2 \leq 0. \tag{16.418}$$

The resulting error dynamics for the adaptive backstepping controller is a non-autonomous system

$$\mathcal{M}(\boldsymbol{x})\dot{\boldsymbol{z}} = [-\mathcal{K}(\boldsymbol{x}, \boldsymbol{v}) + \mathcal{S}]\boldsymbol{z} + \mathcal{B}(\boldsymbol{x})\tilde{F}_e \tag{16.419}$$

$$\dot{\tilde{F}}_e = -\mu\mathcal{B}^\top(\boldsymbol{x})\boldsymbol{z} \tag{16.420}$$

where

$$\mathcal{B}(\boldsymbol{x}) = \begin{bmatrix} \boldsymbol{0}_{3\times1} \\ -\boldsymbol{T}^{-\top}(\boldsymbol{x})\boldsymbol{\phi} \end{bmatrix}. \tag{16.421}$$

In order to satisfy the control objective, the controller gains must be chosen according to

$$\boldsymbol{K}_{\mathrm{p}} = \begin{bmatrix} k_{\mathrm{p}_1} & 0 & 0 \\ 0 & 0 & 0 \\ 0 & 0 & k_{\mathrm{p}_3} \end{bmatrix}, \quad \boldsymbol{K}_{\mathrm{d}} = \begin{bmatrix} k_{\mathrm{d}_1} & 0 & 0 \\ 0 & k_{\mathrm{d}_2} & 0 \\ 0 & 0 & k_{\mathrm{d}_3} \end{bmatrix}, \quad \boldsymbol{\Lambda} = \begin{bmatrix} \lambda_1 & 0 & 0 \\ 0 & 0 & 0 \\ 0 & 0 & \lambda_3 \end{bmatrix}. \tag{16.422}$$

Note that $k_{\mathrm{p}_2} = 0$ and $\lambda_2 = 0$. This implies that the craft is free to move along the circular arc with tangential velocity $\rho\dot{\gamma}$. The gain $k_{\mathrm{d}_2} > 0$ is used to increase the tangential damping (D control) while the radius ρ and heading ψ are stabilized by using PID control.

Since the controller gains k_{p_2} and λ_2 are chosen to be zero, the matrices

$$\boldsymbol{K}_{\mathrm{p}} \geq 0, \qquad \boldsymbol{\Lambda} \geq 0 \tag{16.423}$$

are only positive semi-definite. Hence, V_3 is positive semi-definite. Uniform local asymptotic stability (ULAS) of the equilibrium $(\boldsymbol{z}, \tilde{\boldsymbol{F}}_{\mathrm{e}}) = (\boldsymbol{0}, 0)$ can, however, be proven since the error dynamics $(\boldsymbol{z}_1, \boldsymbol{z}_2)$ is ISS. Consider the reduced-order system $(\boldsymbol{z}_{1r}, \boldsymbol{z}_2)$ given by

$$\boldsymbol{z}_{1r} = \boldsymbol{E}\boldsymbol{z}_1, \quad \boldsymbol{E} = \begin{bmatrix} 1 & 0 & 0 \\ 0 & 0 & 1 \end{bmatrix}. \tag{16.424}$$

This implies that

$$\begin{aligned} \dot{\boldsymbol{z}}_{1r} &= -\boldsymbol{E}\boldsymbol{\Lambda}\boldsymbol{z}_1 + \boldsymbol{E}\boldsymbol{z}_2 \\ &= -(\boldsymbol{E}\boldsymbol{\Lambda}\boldsymbol{E}^{\top})\boldsymbol{z}_{1r} + \boldsymbol{E}\boldsymbol{z}_2. \end{aligned} \tag{16.425}$$

Note that the last step is possible since the diagonal matrices $\boldsymbol{\Lambda} = \mathrm{diag}\{\lambda_1, 0, \lambda_3\}$ satisfy

$$\boldsymbol{\Lambda}\boldsymbol{E}^{\top}\boldsymbol{z}_{1r} = \boldsymbol{\Lambda}\boldsymbol{z}_1. \tag{16.426}$$

Hence, the error dynamics (16.419) and (16.420) can be transformed to

$$\mathcal{M}_{\mathrm{r}}(\boldsymbol{x})\dot{\boldsymbol{z}}_{\mathrm{r}} = [-\mathcal{K}_{\mathrm{r}}(\boldsymbol{x}, \boldsymbol{v}) + \mathcal{S}_{\mathrm{r}}]\boldsymbol{z}_{\mathrm{r}} + \mathcal{B}_{\mathrm{r}}(\boldsymbol{x})\tilde{F}_{\mathrm{e}} \tag{16.427}$$

$$\dot{\hat{F}}_{\mathrm{e}} = -\mu\mathcal{B}_{\mathrm{r}}^{\top}(\boldsymbol{x})\boldsymbol{z}_{\mathrm{r}} \tag{16.428}$$

where $\boldsymbol{z}_{\mathrm{r}} = [\boldsymbol{z}_{1r}^{\top}, \boldsymbol{z}_2^{\top}]^{\top}$ and

$$\mathcal{M}_{\mathrm{r}}(\boldsymbol{x}) = \mathcal{M}_{\mathrm{r}}^{\top}(\boldsymbol{x}) = \begin{bmatrix} \boldsymbol{E}\boldsymbol{K}_{\mathrm{p}}\boldsymbol{E}^{\top} & \boldsymbol{0}_{2\times3} \\ \boldsymbol{0}_{3\times2} & \boldsymbol{M}_x(\boldsymbol{x}) \end{bmatrix}$$

$$\mathcal{K}_{\mathrm{r}}(\boldsymbol{x}, \boldsymbol{v}) = \begin{bmatrix} (\boldsymbol{E}\boldsymbol{K}_{\mathrm{p}}\boldsymbol{E}^{\top})(\boldsymbol{E}\boldsymbol{\Lambda}\boldsymbol{E}^{\top}) & \boldsymbol{0}_{2\times3} \\ \boldsymbol{0}_{3\times2} & \boldsymbol{C}_x(\boldsymbol{x}, \boldsymbol{v}) + \boldsymbol{D}_x(\boldsymbol{x}, \boldsymbol{v}) + \boldsymbol{K}_{\mathrm{d}} \end{bmatrix} > 0$$

$$\mathcal{S}_{\mathrm{r}} = -\mathcal{S}_{\mathrm{r}}^{\top} = \begin{bmatrix} \boldsymbol{0}_{2\times2} & \boldsymbol{E}\boldsymbol{K}_{\mathrm{p}} \\ -\boldsymbol{K}_{\mathrm{p}}\boldsymbol{E}^{\top} & \boldsymbol{0}_{3\times3} \end{bmatrix}, \quad \mathcal{B}_{\mathrm{r}}(\boldsymbol{x}) = \begin{bmatrix} \boldsymbol{0}_{2\times1} \\ \boldsymbol{T}^{-\top}(\boldsymbol{x})\boldsymbol{\phi} \end{bmatrix}$$

where the fact that $\boldsymbol{K}_{\mathrm{p}}\boldsymbol{E}^{\top}\boldsymbol{z}_{1r} = \boldsymbol{K}_{\mathrm{p}}\boldsymbol{z}_1$ for $\boldsymbol{K}_{\mathrm{p}} = \mathrm{diag}\{k_{\mathrm{p}_1}, 0, k_{\mathrm{p}_3}\}$ has been applied.

Non-autonomous Lyapunov Analysis

Even though the Lyapunov function V_3 corresponding to the states $(z_1,\ z_2)$ is only positive semi-definite (since $\boldsymbol{K}_{\rm p}$ is positive semi-definite) the Lyapunov function V_{3r} corresponding to the new output $(z_{1r},\ z_2)$ is positive definite. Using the fact that the closed-loop system governed by $(z_1,\ z_2)$ is ISS, asymptotic tracking is guaranteed by

$$V_{3r} = \frac{1}{2}\left[z_{1r}^{\top}(\boldsymbol{E}\boldsymbol{K}_{\rm p}\boldsymbol{E}^{\top})z_{1r} + z_2^{\top}\boldsymbol{M}_x z_2 + \frac{1}{\mu}\tilde{F}_{\rm e}^2\right] > 0 \tag{16.429}$$

$$\dot{V}_{3r} = -z_{1r}^{\top}(\boldsymbol{E}\boldsymbol{K}_{\rm p}\boldsymbol{E}^{\top})(\boldsymbol{E}\boldsymbol{\Lambda}\boldsymbol{E}^{\top})z_{1r} - z_2^{\top}(\boldsymbol{K}_{\rm d} + \boldsymbol{D}_x)z_2 \leq 0 \tag{16.430}$$

where $\boldsymbol{E}\boldsymbol{K}_{\rm p}\boldsymbol{E}^{\top} > 0$ and $\boldsymbol{E}\boldsymbol{\Lambda}\boldsymbol{E}^{\top} > 0$. Hence, $z_{1r},\ z_2, \tilde{F}_{\rm e} \in \mathcal{L}_{\infty}$. Note that $\dot{V}_3$ is only negative semi-definite since a negative term proportional to $-\tilde{F}_{\rm e}^2$ is missing in the expression for $\dot{V}_3$. ULES of the equilibrium point $(z_{1r},\ z_2, \tilde{F}_{\rm e}) = (\mathbf{0},\ \mathbf{0},\ 0)$ follows by using the stability theorem of Fossen *et al.* (2001) for nonlinear *non-autonomous* systems (see Appendix A.2.4). Since, the closed-loop system $(z_1,\ z_2)$ is ISS it is sufficient to consider the reduced order system $(z_{1r},\ z_2)$ with output $z_{1r} = \boldsymbol{E}z_1$ in the stability analysis. According to Appendix A.2.4, we can choose $\boldsymbol{x}_1 = [z_{1r}^{\top}, z_2^{\top}]^{\top}$, $\boldsymbol{x}_2 = \tilde{F}_{\rm e}, \boldsymbol{P} = \mu$ and $W(\boldsymbol{x}_1, t) = (1/2)\boldsymbol{x}_1^{\top}\boldsymbol{x}_1$. Then the equilibrium point $(z_{1r},\ z_2, \tilde{F}_{\rm e}) = (\mathbf{0},\ \mathbf{0},\ 0)$ of the nonlinear error system (16.419) and (16.420) is ULES since

$$\text{rank}\{(\mathcal{M}_{\rm r}^{-1}(\boldsymbol{x})\mathcal{B}_{\rm r}(\boldsymbol{x}))^{\top}\ (\mathcal{M}_{\rm r}^{-1}(\boldsymbol{x})\mathcal{B}_{\rm r}(\boldsymbol{x}))\} = 1, \quad \forall \boldsymbol{x}$$

and

$$\begin{aligned}
\max\{\|\boldsymbol{h}(\boldsymbol{x}_1, t)\|, \|\boldsymbol{x}_1\|\} &= \max\{\|\mathcal{M}_{\rm r}^{-1}(\boldsymbol{x})[-\mathcal{K}_{\rm r}(\boldsymbol{x}, \boldsymbol{v}) + \mathcal{S}_{\rm r}]\boldsymbol{x}_1\|, \|\boldsymbol{x}_1\|\} \\
&\leq \rho_1(\|\boldsymbol{x}_1\|)\|\boldsymbol{x}_1\| \\
\|\boldsymbol{B}(\boldsymbol{x}, t)\| &= \|\mathcal{M}_{\rm r}^{-1}(\boldsymbol{x})\mathcal{B}_{\rm r}(\boldsymbol{x})\| \leq \rho_2(\|\boldsymbol{x}_1\|) \\
\max\left\{\left\|\frac{\partial \boldsymbol{B}(\boldsymbol{x}, t)}{\partial t}\right\|, \left\|\frac{\partial \boldsymbol{B}(\boldsymbol{x}, t)}{\partial \boldsymbol{x}_i}\right\|\right\} &= \max\left\{\left\|\frac{\partial \mathcal{M}_{\rm r}^{-1}(\boldsymbol{x})\mathcal{B}_{\rm r}(\boldsymbol{x})}{\partial \boldsymbol{x}_i}\right\|\right\} \leq \rho_3(\|\boldsymbol{x}_1\|).
\end{aligned}$$

Translational Control of the Circle Center

Weather optimal position control can be achieved by moving the circle center coordinates $\boldsymbol{p}_0 = [x_0^n,\ y_0^n]^{\top}$ dynamically such that the craft maintains a constant position. In order to meet the fixed position control objective, an update law for the circle center $\boldsymbol{p}_0$ must be derived. The Cartesian Earth-fixed position of the craft is given by

$$\boldsymbol{p} = \boldsymbol{L}^{\top}\boldsymbol{\eta} \tag{16.431}$$

where $\boldsymbol{L}$ is defined in (16.386). Let $\tilde{\boldsymbol{p}} = \boldsymbol{p} - \boldsymbol{p}_{\rm d}$ denote the corresponding deviation from the desired position vector $\boldsymbol{p}_{\rm d} = [x_{\rm d}^n,\ y_{\rm d}^n]^{\top}$. The desired position can either be constant (regulation) or a smooth time-varying reference trajectory. The control law for translation of the circle center is derived by considering the following CLF

$$V_p = \frac{1}{2}\tilde{\boldsymbol{p}}^\top \tilde{\boldsymbol{p}} \tag{16.432}$$

$$\dot{V}_p = \tilde{\boldsymbol{p}}^\top(\dot{\boldsymbol{p}} - \dot{\boldsymbol{p}}_\mathrm{d}) = \tilde{\boldsymbol{p}}^\top(\boldsymbol{L}^\top \dot{\boldsymbol{\eta}} - \dot{\boldsymbol{p}}_\mathrm{d}). \tag{16.433}$$

By using (16.385), $\boldsymbol{L}^\top \boldsymbol{L} = \boldsymbol{I}_2$ and $\dot{\boldsymbol{x}} = \boldsymbol{z}_2 + \dot{\boldsymbol{x}}_\mathrm{r}$ it is seen that

$$\begin{aligned} \dot{V}_\mathrm{p} &= \tilde{\boldsymbol{p}}^\top \left(\boldsymbol{L}^\top(\boldsymbol{R}(\gamma)\boldsymbol{H}(\rho)\dot{\boldsymbol{x}} + \boldsymbol{L}\dot{\boldsymbol{p}}_0) - \dot{\boldsymbol{p}}_\mathrm{d}\right) \\ &= \tilde{\boldsymbol{p}}^\top \left(\dot{\boldsymbol{p}}_0 - \dot{\boldsymbol{p}}_\mathrm{d} + \boldsymbol{L}^\top \boldsymbol{R}(\gamma)\boldsymbol{H}(\rho)\dot{\boldsymbol{x}}_\mathrm{r}\right) + \tilde{\boldsymbol{p}}^\top \boldsymbol{L}^\top \boldsymbol{R}(\gamma)\boldsymbol{H}(\rho)\boldsymbol{z}_2. \end{aligned} \tag{16.434}$$

Now, by choosing the circle center update law as

$$\dot{\boldsymbol{p}}_0 = \dot{\boldsymbol{p}}_\mathrm{d} - \boldsymbol{L}^\top \boldsymbol{R}(\gamma)\boldsymbol{H}(\rho)\dot{\boldsymbol{x}}_\mathrm{r} - k_0\, \tilde{\boldsymbol{p}} \tag{16.435}$$

where $k_0 > 0$, it is seen that

$$\dot{V}_\mathrm{p} = -k_0 \tilde{\boldsymbol{p}}^\top \tilde{\boldsymbol{p}} + \tilde{\boldsymbol{p}}^\top\, \boldsymbol{L}^\top \boldsymbol{R}(\gamma)\boldsymbol{H}(\rho)\boldsymbol{z}_2. \tag{16.436}$$

In (16.436) a cross-term in $\tilde{\boldsymbol{p}}$ and $\boldsymbol{z}_2$ is noted. In order to guarantee that the time derivative of the total system $V_\mathrm{wopc} = V_{3r} + V_\mathrm{p}$ is negative semi-definite, the weather optimal controller (16.414) must be modified such that the cross-term in (16.436) is canceled.

Weather Optimal Position Control (WOPC)

The cross-terms involving $\tilde{\boldsymbol{p}}$ and $\boldsymbol{z}_2$ in $\dot{V}_\mathrm{p}$ can be removed by modifying the nonlinear controller (16.414) to

$$\begin{aligned} \boldsymbol{\tau} = \boldsymbol{T}^\top(&\boldsymbol{M}_x \ddot{\boldsymbol{x}}_\mathrm{r} + \boldsymbol{C}_x \dot{\boldsymbol{x}}_\mathrm{r} + \boldsymbol{D}_x \dot{\boldsymbol{x}}_\mathrm{r} \boldsymbol{K}_p \boldsymbol{z}_1 - \boldsymbol{K}_\mathrm{d} \boldsymbol{z}_2) \\ &- \boldsymbol{q}(\cdot) - \boldsymbol{\phi} \hat{\boldsymbol{F}}_\mathrm{e} - \boldsymbol{T}^\top \boldsymbol{H}^\top(\rho)\boldsymbol{R}^\top(\gamma)\boldsymbol{L}\tilde{\boldsymbol{p}}. \end{aligned} \tag{16.437}$$

The last term in $\boldsymbol{\tau}$ implies that

$$\dot{V}_{3r} = -\boldsymbol{z}_{1r}^\top(\boldsymbol{E}\boldsymbol{K}_p\boldsymbol{E}^\top)(\boldsymbol{E}\boldsymbol{\Lambda}\boldsymbol{E}^\top)\boldsymbol{z}_{1r} - \boldsymbol{z}_2^\top(\boldsymbol{K}_\mathrm{d} + \boldsymbol{D}_x)\boldsymbol{z}_2 - \tilde{\boldsymbol{p}}^\top \boldsymbol{L}^\top \boldsymbol{R}(\gamma)\boldsymbol{H}(\rho)\boldsymbol{z}_2. \tag{16.438}$$

Consider

$$V_{\text{wopc}} = V_{3r} + V_{\text{p}} \tag{16.439}$$

$$\dot{V}_{\text{wopc}} = -z_{1r}^{\top}(\boldsymbol{E}\boldsymbol{K}_{\text{p}}\boldsymbol{E}^{\top})(\boldsymbol{E}\boldsymbol{\Lambda}\boldsymbol{E}^{\top})z_{1r} - z_2^{\top}(\boldsymbol{K}_{\text{d}} + \boldsymbol{D}_x)z_2 - k_0\tilde{\boldsymbol{p}}^{\top}\tilde{\boldsymbol{p}} \tag{16.440}$$

and therefore the equilibrium point $(z_{1r},\ z_2,\ \tilde{F}_{\text{e}}, \tilde{\boldsymbol{p}}) = (\mathbf{0},\ \mathbf{0},\ 0, \mathbf{0})$ is ULES. The term $\ddot{\boldsymbol{p}}_0$ is needed in the expression for $\boldsymbol{q}(\cdot)$. This term is computed from (16.435) as

$$\begin{aligned}\ddot{\boldsymbol{p}}_0 &= \ddot{\boldsymbol{p}}_{\text{d}} - k_0(\dot{\boldsymbol{p}} - \dot{\boldsymbol{p}}_{\text{d}}) \\ &\quad - \boldsymbol{L}^{\top}\boldsymbol{R}(\gamma)\boldsymbol{H}(\rho)\ddot{\boldsymbol{x}}_{\text{r}} - \boldsymbol{L}^{\top}\dot{\boldsymbol{R}}(\gamma)\boldsymbol{H}(\rho)\dot{\boldsymbol{x}}_{\text{r}} - \boldsymbol{L}^{\top}\boldsymbol{R}(\gamma)\dot{\boldsymbol{H}}(\rho)\dot{\boldsymbol{x}}_{\text{r}}.\end{aligned} \tag{16.441}$$

Experiment 1: Weather Optimal Heading Control (WOHC)

The proposed WOHC system has been implemented and tested experimentally using a model ship of scale 1:70. A ducted fan was used to generate wind forces. The length of

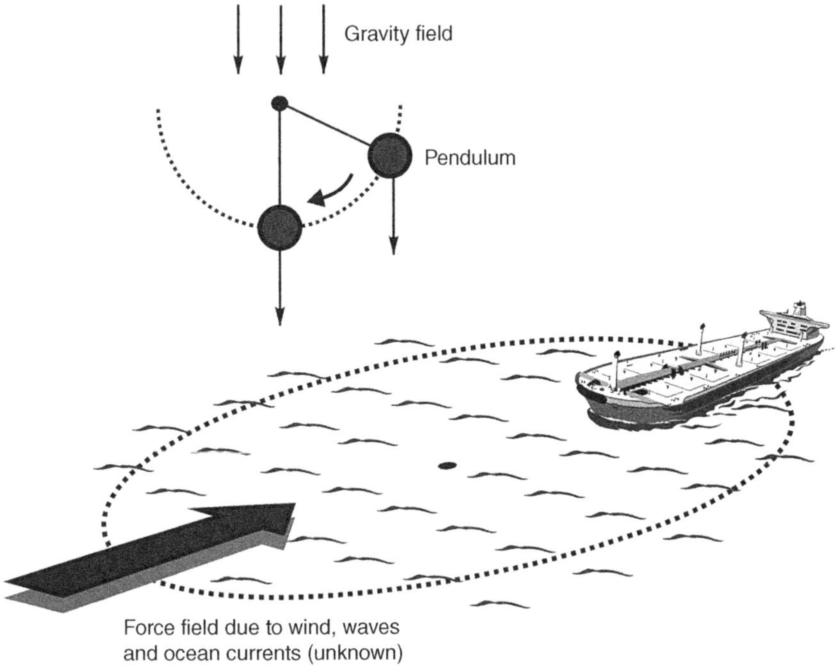

Figure 16.20 WOHC experiment showing the circular motion of the ship when the circle center controller is turned off (WOHC).

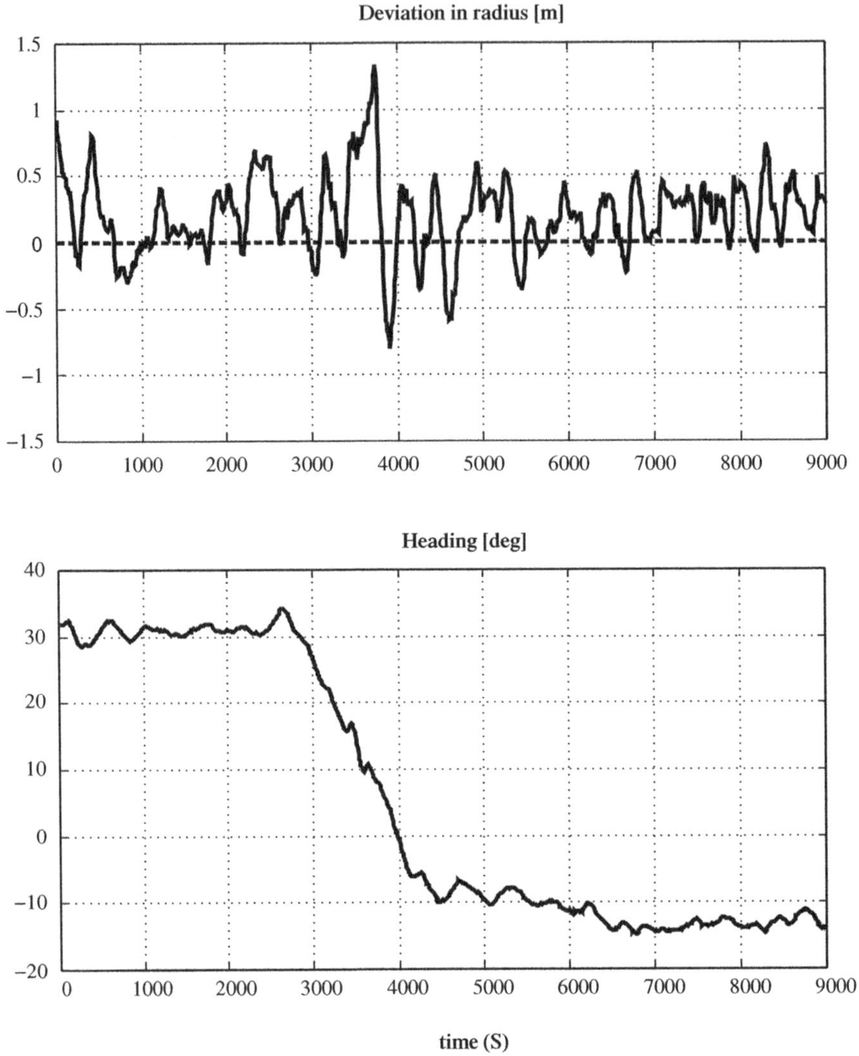

Figure 16.21 WOHC experiment showing the performance of the radius regulator (upper plot) and weather optimal heading (lower plot) versus time (s).

the model ship is $L_{\mathrm{m}} = 1.19$ m and the mass is $m_{\mathrm{m}} = 17.6$ kg. The experimental results are scaled to full scale by considering a supply vessel with mass $m_{\mathrm{s}} = 4500$ tons using the bis system (see Appendix D.1).

In the first experiment the ship was allowed to move on the circle arc and the circle center controller (16.435) was turned off; that is $x_0^n =$ constant and $y_0^n =$ constant. This is referred to as WOHC. The fixed origin and circle arc are shown in Figure 16.20.

Note that the initial heading is approximately 30 degrees (see Figure 16.21), while the position $(x^n,\ y^n) \approx (13, -43)$. These values are those obtained when the fan was initially directed at 210° in the opposite direction of the ship heading.

After 3000 s the fan was slowly rotated to 165°, corresponding to a weather optimal heading of −15° (see Figure 16.21). During this process, the ship starts to move on the circle arc with heading towards the circle center until it is stabilized to its new heading at −15°. The new position on the circle arc is $(x^n,\ y^n) \approx (3, 20)$. This clearly demonstrates that the ship heading converges to the optimal value by copying the dynamics of a pendulum in the gravity field. This is done without using any external wind sensor.

In the next experiment, the circle center is translated online in order to obtain a constant position $(x^n,\ y^n)$.

Experiment 2: Weather Optimal Position Control (WOPC)

In the second experiment the ship should maintain its position by activating the circle center controller (16.435). The performance during stationkeeping and translation of the circle is shown in Figures 16.22–16.24. The position controller works within an accuracy of ±1 m, which is the accuracy of the GNSS system.

Again the weather optimal heading is changed from approximately 23° to 2° but this time without changing the position $(x^n,\ y^n)$ of the ship. The position deviations and the weather optimal heading are shown in Figure16.23. These values are obtained by moving the fan from an initial angle of 203° to 182°.

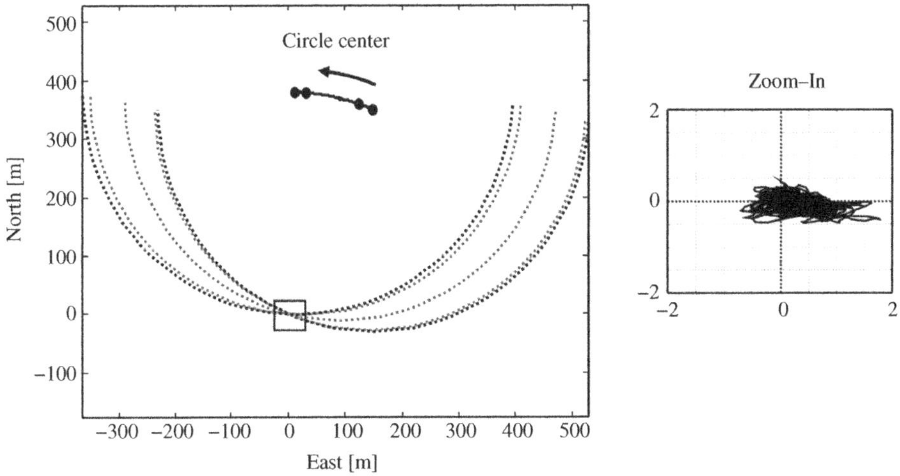

Figure 16.22 WOPC experiment showing how the circle center is moved to obtain stationkeeping to $(x^n_d,\ y^n_d) = (0,\ 0)$.

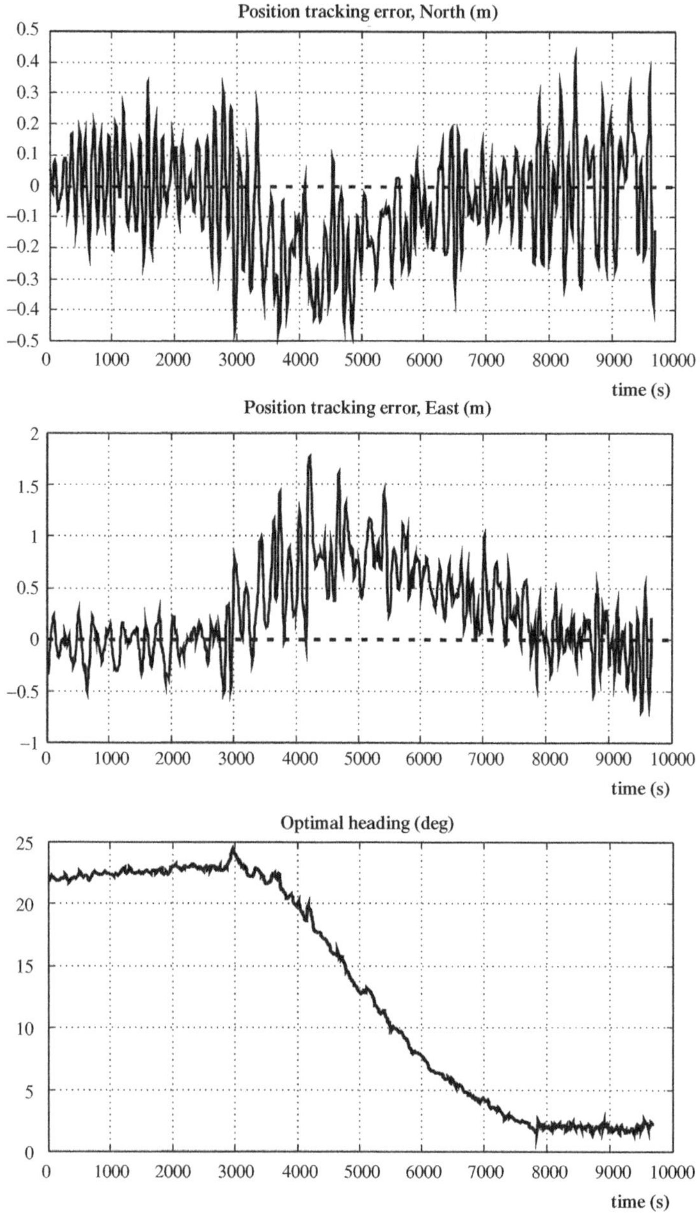

Figure 16.23 WOPC experiment showing the north and east position accuracies (upper plots) and weather optimal heading (lower plot) versus time (seconds). The position accuracy is within ±1 m while the heading changes from 23° to 2° as the fan is rotated.

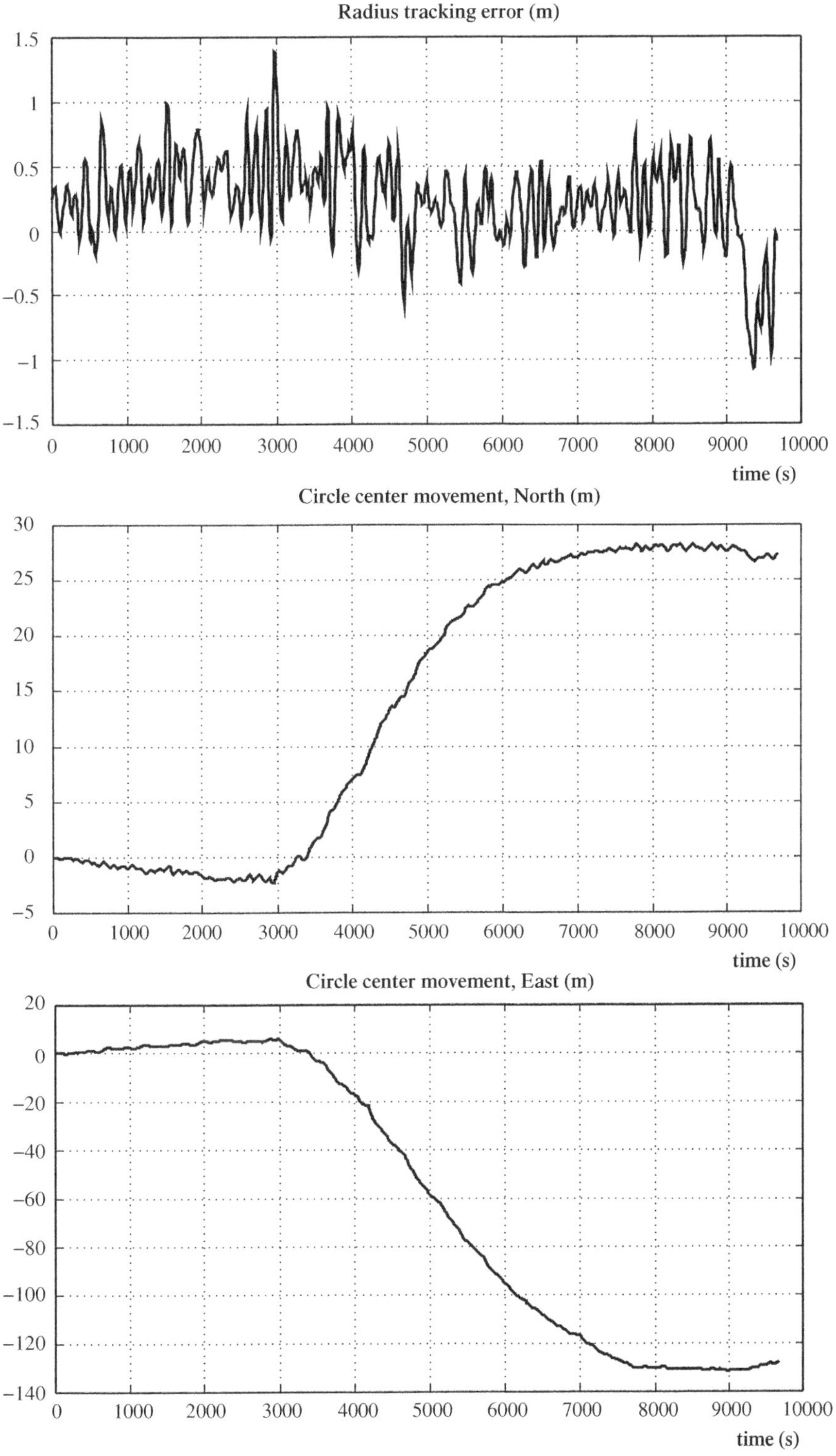

Figure 16.24 WOPC experiment showing the deviation for the radius regulator (upper plot) and the translation of the circle center (x_0^n, y_0^n) (lower plots) versus time in seconds. The radius deviation is within ± 1 m during the rotation of the fan.

16.4 Sliding Mode Control

A robust nonlinear control method is *sliding mode control* (Utkin 1977), which incorporates techniques to handle

- Structured (parametric) uncertainties
- Unstructured uncertainties (unmodeled dynamics).

Robustness is achieved by altering the dynamics of the nonlinear system by application of a discontinuous control signal. The state-feedback control law is not a continuous function of time. Instead, it can switch from one continuous structure to another based on the current position in the state space. Hence, sliding mode control is a variable structure control method.

Sliding mode techniques are discussed by Slotine and Li (1991), Utkin (1992) and Shtessel *et al.* (2014) while applications to marine craft are found in Yoerger and Slotine (1985), Healey and Lienard (1993) and McGookin et al. (2000a, 2000b), for instance. This section considers the following sliding mode methods and their application to marine craft:

- Conventional integral sliding mode control (SMC)
- Adaptive-gain super-twisting algorithm (STA).

16.4.1 Conventional Integral SMC for Second-order Systems

Conventional sliding mode control (Sections 16.4.1 and 16.4.2), or to be more precise first-order sliding mode control (Utkin 1992; Shtessel *et al.* 2014), can be used to design robust autopilots for heading, course, speed and depth control. Section 16.4.3 extends these results to second-order sliding modes known as super-twisting sliding mode control. Higher-order sliding mode algorithms have the advantage of driving the sliding variable σ and its consecutive derivatives to zero in the presence of disturbances and uncertainties.

Consider the second-order nonlinear system

$$\ddot{x} = f(\dot{x}, x) + bu + d \tag{16.442}$$

where $x \in \mathbb{R}$ is the state, $u \in \mathbb{R}$ is the control input and $|d| \leq d^{\max}$ is an unknown bounded disturbance. The *sliding surface* S is defined as

$$S := \{x : \sigma(x) = 0\} \tag{16.443}$$

and *sliding mode* takes place if the states x evolve with time such that $\sigma(x(t_r)) = 0$ for some finite $t_r \in \mathbb{R}^+$ and $\sigma(x(t)) = 0$ for all $t > t_r$.

Integral SMC is based on the *sliding variable*

$$\sigma := \dot{\tilde{x}} + 2\lambda\tilde{x} + \lambda^2 \int_0^t \tilde{x}(\tau)\,\mathrm{d}\tau \tag{16.444}$$

where $\tilde{x} = x - x_{\text{d}}$ is the tracking error and $\lambda > 0$ is a design parameter specifying the bandwidth of the controller. For $\sigma = 0$ this expression describes a sliding surface (manifold) with exponentially stable dynamics. To see this, let us define a second sliding variable

$$\sigma_0 := \tilde{x} + \lambda \int_0^t \tilde{x}(\tau)\, \mathrm{d}\tau \tag{16.445}$$

such that the manifold $\sigma = 0$ corresponding to (16.444) can be rewritten as

$$\sigma = \dot{\sigma}_0 + \lambda \sigma_0 = 0. \tag{16.446}$$

Hence, both σ_0 and $\tilde{x}$ converge exponentially to zero since the linear system

$$\begin{bmatrix} \dot{\tilde{x}} \\ \dot{\sigma}_0 \end{bmatrix} = \begin{bmatrix} -\lambda & 1 \\ 0 & -\lambda \end{bmatrix} \begin{bmatrix} \tilde{x} \\ \sigma_0 \end{bmatrix} \tag{16.447}$$

has two real eigenvalues at $-\lambda$. This ensures that the tracking error $\tilde{x} \to 0$ on the manifold $\sigma = 0$. Consequently, the *control objective* is reduced to finding a nonlinear control law which ensures that

$$\lim_{t\to\infty} \sigma = 0. \tag{16.448}$$

A graphical interpretation of the *sliding variable* is given in Figure 16.25. It is seen that a trajectory starting at $\sigma > 0$ will move toward the error-state sliding surface $S = \{\tilde{x} : \sigma(\tilde{x}) = 0\}$. When $\sigma = 0$ is reached the trajectory will continue moving on the straight line corresponding to $\sigma = 0$ toward the equilibrium point $\sigma_0 = 0$. Similar behavior is observed when starting with a negative value of σ.

For the system (16.442), the control objective (16.448) can be satisfied by considering the CLF

$$V = \frac{1}{2}\sigma^2 \tag{16.449}$$

$$\begin{aligned} \dot{V} &= \sigma\dot{\sigma} \\ &= \sigma(f(\dot{x}, x) + bu + d - \ddot{x}_{\text{d}} + 2\lambda\dot{\tilde{x}} + \lambda^2\tilde{x}). \end{aligned} \tag{16.450}$$

The control law is chosen as

$$u = \frac{1}{b}(\ddot{x}_{\text{d}} - 2\lambda\dot{\tilde{x}} - \lambda^2\tilde{x} - f(\dot{x}, x) - K_\sigma \text{sgn}(\sigma)) \tag{16.451}$$

where $K_\sigma > d^{\max}$. The signum function is defined as

$$\text{sgn}(\sigma) := \begin{cases} 1 & \text{if } \sigma > 0 \\ 0 & \text{if } \sigma = 0 \\ -1 & \text{otherwise} \end{cases}. \tag{16.452}$$

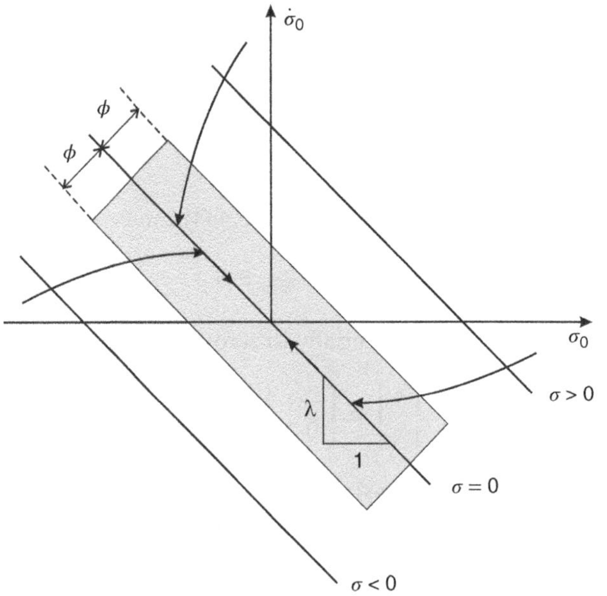

Figure 16.25 Graphical interpretation of the sliding surface $S = \{\tilde{x} : \sigma(\tilde{x}) = 0\}$, the sliding variable $\sigma = \dot{\sigma}_0 + \lambda\sigma_0$ and boundary layer $\phi > 0$.

Since $|\sigma| = \text{sgn}(\sigma)\sigma$ and $K_\sigma > d^{\max}$, Equation (16.450) is equivalent to

$$\begin{aligned} \dot{V} &= \sigma d - K_\sigma |\sigma| \\ &< 0, \quad \forall \sigma \neq 0. \end{aligned} \tag{16.453}$$

Consequently, the equilibrium point $\sigma = 0$ is GAS and σ converge to zero in finite time. In view of (16.447) the tracking error $\tilde{x}$ also converges to zero.

It is well known that the switching term $K_\sigma \text{sgn}(\sigma)$ can lead to chattering for large values of K_σ. Chattering in the controller can, however, be eliminated by replacing the signum function with a saturating function. Slotine and Li (1991) suggest smoothing out the control discontinuity inside a boundary layer according to

$$\text{sat}(\sigma) = \begin{cases} \text{sgn}(\sigma) & \text{if } |\sigma/\phi| > 1 \\ \sigma/\phi & \text{otherwise} \end{cases} \tag{16.454}$$

where $\phi > 0$ can be interpreted as the boundary layer thickness. This substitution will assign a low-pass filter structure to the dynamics inside the boundary layer (see Figure 16.25). Another possibility is to replace $K_\sigma \text{sgn}(\sigma)$ with $K_\sigma \tanh(\sigma/\phi)$, where $\phi > 0$ is a design parameter used to shape the slope of $\tanh(\cdot)$ close to the origin.

The control law (16.451) is in fact a saturated PID controller so constant disturbances will be removed in steady state. The case studies in Sections 16.4.4 and 16.4.5 show how integral SMCs can be further improved to compensate for structured and unstructured uncertainties by increasing the gain K_σ.

16.4.2 Conventional Integral SMC for Third-order Systems

Consider the third-order nonlinear system

$$x^{(3)} = f(\ddot{x}, \dot{x}, x) + bu + d \tag{16.455}$$

where $x \in \mathbb{R}$ is the state, $u \in \mathbb{R}$ is the control input and $|d| \leq d^{\max}$ is an unknown bounded disturbance. Define the sliding variable

$$\sigma = \ddot{\tilde{x}} + 3\lambda\dot{\tilde{x}} + 3\lambda^2\tilde{x} + \lambda^3 \int_0^t \tilde{x}(\tau)\, \mathrm{d}\tau \tag{16.456}$$

where $\tilde{x} = x - x_\mathrm{d}$ and $\lambda > 0$. Consider the Lyapunov function candidate

$$V = \frac{1}{2}\sigma^2 \tag{16.457}$$

$$\begin{aligned}\dot{V} &= \sigma\dot{\sigma}\\ &= \sigma(f(\ddot{x}, \dot{x}, x) + bu + d - x_\mathrm{d}^{(3)} + 3\lambda\ddot{\tilde{x}} + 3\lambda^2\dot{\tilde{x}} + \lambda^3\tilde{x}).\end{aligned} \tag{16.458}$$

Choosing the control law as

$$u = \frac{1}{b}(x_\mathrm{d}^{(3)} - 3\lambda\ddot{\tilde{x}} - 3\lambda^2\dot{\tilde{x}} - \lambda^3\tilde{x} - f(\ddot{x}, \dot{x}, x) - K_\sigma \mathrm{sgn}(\sigma)) \tag{16.459}$$

with $K_\sigma > d^{\max}$ gives

$$\begin{aligned}\dot{V} &= \sigma d - K_\sigma|\sigma|\\ &< 0, \quad \forall \sigma \neq 0.\end{aligned} \tag{16.460}$$

Consequently, the equilibrium point $\sigma = 0$ is GAS and the sliding variable σ and thus $\tilde{x}$ both converge to zero in finite time.

16.4.3 Super-twisting Adaptive Sliding Mode Control

One of the most powerful second-order continuous sliding mode control algorithms is the super-twisting algorithm (STA) that handles a relative degree equal to one (Levant 2003, 2005; Shtessel *et al.* 2007, 2010). The STA generates a continuous control signal that drives the sliding variable σ and its derivative $\dot{\sigma}$ to zero in finite time in the presence of matched disturbances, which are assumed to be upper bounded. The STA also attenuates the chattering due to the signum function as opposed to conventional SMC. The main reason for this is that the discontinuous signum function is integrated by the STA.

Consider the nonlinear system

$$x^{(n)} = f(\boldsymbol{x}, t) + b(\boldsymbol{x}, t)u + d \tag{16.461}$$

where $\boldsymbol{x} = [x,\ \dot{x}, ..., x^{(n-1)}]^{\top} \in \mathbb{R}^{n}$, $u \in \mathbb{R}$ is the control input and $d \in \mathbb{R}$ is a bounded disturbance with unknown bound.

The control objective is to drive the sliding variable σ and its derivative $\dot{\sigma}$ to zero in finite time in the presence of the bounded perturbation d with unknown boundary using continuous control. The conventional SMC (Sections 16.4.1 and 16.4.2) can robustly handle such a problem if the boundary of the perturbation is known. The main disadvantages of the conventional SMC is control chattering due to the signum function and the need to know the upper bound on the disturbance. On the contrary, the adaptive-gain STA does not need to know the upper bound on the disturbance and it has also chattering attenuation.

For marine craft autopilot design the model (16.461) with $n = 2$ (heading autopilot systems) and $n = 3$ (depth autopilot systems) are of particular interest. For these cases, the sliding variable σ is chosen as

$$(n = 2) \qquad \sigma = \dot{\tilde{x}} + \lambda\tilde{x} \tag{16.462}$$

$$(n = 3) \qquad \sigma = \ddot{\tilde{x}} + 2\lambda\dot{\tilde{x}} + \lambda^2\tilde{x}. \tag{16.463}$$

Consequently, we define

$$\sigma := x^{(n-1)} - x_r^{(n-1)} \tag{16.464}$$

where

$$(n = 2) \qquad \dot{x}_{\rm r} = \dot{x}_{\rm d} - \lambda\tilde{x} \tag{16.465}$$

$$(n = 3) \qquad \ddot{x}_{\rm r} = \ddot{x}_{\rm d} - 2\lambda\dot{\tilde{x}} - \lambda^2\tilde{x}. \tag{16.466}$$

Time differentiation of (16.464) gives

$$\dot{\sigma} = \underbrace{f(\boldsymbol{x}, t) + d - x_{\rm r}^{(n)}}_{\phi(\boldsymbol{x},t)} + b(\boldsymbol{x}, t)u \tag{16.467}$$

where it is assumed that

$$|\phi(\boldsymbol{x}, t)| \leq \delta|\sigma|^{1/2} \tag{16.468}$$

for some $\delta > 0$ and $b(\boldsymbol{x}, t) \neq 0$ for all $\boldsymbol{x}$ and $t \in [0, \infty)$. Note that the boundary δ must exist but it can be unknown. From this it follows that

$$\begin{aligned}\dot{\sigma} &= \phi(\boldsymbol{x}, t) + b(\boldsymbol{x}, t)u \\ &= \phi(\boldsymbol{x}, t) + \omega, \qquad u = b^{-1}(\boldsymbol{x}, t)\omega.\end{aligned} \tag{16.469}$$

We are now ready to formulate the main result.

Theorem 16.2 (Adaptive Super-twisting Control Law (Shtessel* et al. *2010))
Consider the system (16.469) and suppose that the nonlinear vector field ϕ satisfies (16.468) for some unknown constant $\delta > 0$. Then for any initial conditions $\boldsymbol{x}(0)$ the sliding surface $\sigma = 0$ will be reached in finite time for the adaptive-gain STA

$$\omega = -\alpha|\sigma|^{1/2}\,\mathrm{sgn}(\sigma) + v \tag{16.470}$$

$$\dot{v} = -\beta\,\mathrm{sgn}(\sigma) \tag{16.471}$$

where the adaptive gains are

$$\dot{\alpha} = \begin{cases} \omega^* \sqrt{\frac{\gamma^*}{2}} & \text{if } \sigma \neq 0 \\ 0 & \text{otherwise} \end{cases} \tag{16.472}$$

$$\beta = 2\varepsilon\alpha + \lambda^* + 4\varepsilon^2 \tag{16.473}$$

and $\omega^* > 0$, $\gamma^* > 0$, $\lambda^* > 0$, and $\varepsilon > 0$ are four constants. The stability proof guarantees that $\sigma \to 0$ and $\dot{\sigma} \to 0$ in finite time.

Proof: *See Shtessel* et al. *(2010).*

16.4.4 Case Study: Heading Autopilot Based on Conventional Integral SMC

A marine craft is exposed to wind, waves and ocean currents. Consequently, it is necessary to include integral action to compensate for the drift forces. Consider the sliding variable

$$\sigma := \tilde{r} + 2\lambda\tilde{\psi} + \lambda^2 \int_0^t \tilde{\psi}(\tau)\,\mathrm{d}\tau \tag{16.474}$$

where $\tilde{\psi} = \psi - \psi_d$ and $\tilde{r} = r - r_d$ are the tracking errors, and $\lambda > 0$. Feedback from σ is a combination of three terms similar to those used in a PID controller. Consider the nonlinear model (7.30) of the yaw dynamics (Norrbin 1963)

$$T\dot{r} + n_3 r^3 + n_1 r = K\delta + d_r + \tau_{\text{wind}} \tag{16.475}$$

where $|d_r| < d_r^{\max}$ is an unknown yaw rate disturbance due to unmodeled dynamics, ocean currents, waves and parametric uncertainty, τ_{wind} is the wind moment and δ is the rudder angle. Note that $n_1 = 1$ for a course stable ship and $n_1 = -1$ for a course-unstable ship. Hence, the parameter n_1 is perfectly known while n_3, T and K are uncertain. Next, we write $\sigma = r - r_r$ where r_r is defined as

$$r_r := r_d - 2\lambda\tilde{\psi} - \lambda^2 \int_0^t \tilde{\psi}(\tau)\,\mathrm{d}\tau \tag{16.476}$$

where $\tilde{\psi} = \psi - \psi_d$. From this it follows that

$$\begin{aligned} T\dot{\sigma} &= T\dot{r} - T\dot{r}_r \\ &= K\delta + d_r + \tau_{\text{wind}} - (n_3 r^2 + n_1)r - T\dot{r}_r \\ &= K\delta + d_r + \tau_{\text{wind}} - (n_3 r^2 + n_1)(\sigma + r_r) - T\dot{r}_r. \end{aligned} \tag{16.477}$$

Consider the Lyapunov function candidate $V = (1/2)T\sigma^2$ and

$$\begin{aligned} \dot{V} &= \sigma T\dot{\sigma} \\ &= \sigma[K\delta + d_r + \tau_{\text{wind}} - (n_3 r^2 + n_1)(\sigma + r_r) - T\dot{r}_r] \\ &= -(n_3 r^2 + n_1)\sigma^2 + \sigma[K\delta + d_r + \tau_{\text{wind}} - (n_3 r^2 + n_1)r_r - T\dot{r}_r]. \end{aligned} \tag{16.478}$$

Let the control law be chosen as

$$\delta = \frac{1}{\hat{K}}(\hat{T}\dot{r}_r + (\hat{n}_3 r^2 + n_1)r_r - \hat{\tau}_{\text{wind}} - K_d\sigma - K_\sigma \text{sgn}(\sigma)) \tag{16.479}$$

where $K_\sigma > 0$ and $K_d \geq 0$ (optional gain for course-unstable ships) while $\hat{T}, \hat{K}$ and $\hat{n}_3$ are the estimates of T, K and n_3, respectively. Similar, $\hat{\tau}_{\text{wind}}$ is an estimate of the wind yaw moment based on measured wind speed and direction. Hence,

$$\begin{aligned} \dot{V} = &-\left(n_3 r^2 + n_1 + \frac{K}{\hat{K}}K_d\right)\sigma^2 - \frac{K}{\hat{K}}K_\sigma|\sigma| \\ &+ \left[d_r + \left(\tau_{\text{wind}} - \frac{K}{\hat{K}}\hat{\tau}_{\text{wind}}\right) - \left(T - \frac{K}{\hat{K}}\hat{T}\right)\dot{r}_r \right. \\ &\left. - \left(n_3 - \frac{K}{\hat{K}}\hat{n}_3\right)r^2 r_r - \left(1 - \frac{K}{\hat{K}}\right)n_1 r_r\right]\sigma. \end{aligned} \tag{16.480}$$

For course unstable ships ($n_1 = -1$) we choose $K_d > \hat{K}/K$ while for course stable ships ($n_1 = 1$) no additional feedback is needed and thus $K_d = 0$. This guarantees that

$$-\left(n_3 r^2 + n_1 + \frac{K}{\hat{K}} K_d\right)\sigma^2 < 0, \qquad \forall \sigma \neq 0. \tag{16.481}$$

In order to guarantee that $\dot{V} < 0$, the switching gain K_σ must be chosen large enough so that the remainder of the terms in (16.480) are dominated. Consequently,

$$\begin{aligned} K_\sigma \geq \frac{\hat{K}}{K} d_r^{\max} + \left|\left(\frac{\hat{K}}{K}\tau_{\text{wind}} - \hat{\tau}_{\text{wind}}\right)\right| + \left|\left(\frac{\hat{K}}{K}T - \hat{T}\right)\dot{r}_r\right| \\ + \left|\left(\frac{\hat{K}}{K}n_3 - \hat{n}_3\right)r^2 r_r\right| + \left|\left(\frac{\hat{K}}{K} - 1\right)n_1 r_r\right|. \end{aligned} \tag{16.482}$$

The switching term $K_\sigma \text{sgn}(\sigma)$ can lead to chattering for large values of K_σ. Hence, K_σ should be treated as a design parameter with (16.482) as a guideline. Recall that Lyapunov stability analysis results in conservative requirements for all gains. One way to find an estimate of K_σ is to assume, for instance, 20% uncertainty in all elements.

The sliding variable should be modified in a practical implementation using

$$\sigma := \tilde{r} + 2\lambda\, \text{ssa}(\tilde{\psi}) + \lambda^2 \int_0^t \text{ssa}(\tilde{\psi}(\tau))\, \text{d}\tau \tag{16.483}$$

where ssa: $\mathbb{R} \rightarrow [-\pi, \pi)$ maps the unconstrained angle $\tilde{\psi} \in \mathbb{R}$ to the smallest difference between the angles (see Section 13.4.3).

A modern autopilot must have both course-keeping and turning capabilities. This can be obtained in one design by using a reference model to compute the desired states ψ_d, r_d and $\dot{r}_d$ needed for turning. A third-order filter for this purpose was derived in Section 12.1.1, for instance

$$\frac{\psi_d}{\psi_{\text{ref}}}(s) = \frac{\omega_n^3}{(s+\omega_n)(s^2 + 2\zeta\omega_n s + \omega_n^2)} \tag{16.484}$$

where the setpoint ψ_{ref} is the operator input, ζ is the relative damping ratio and ω_n is the natural frequency. Note that $\lim_{t\rightarrow\infty} \psi_d(t) = \psi_{\text{ref}}$ and that $\dot{\psi}_d$ and $\ddot{\psi}_d$ are smooth and bounded for steps in ψ_{ref}. This is the main motivation for choosing a third-order model since a second-order model will result in steps in $\ddot{\psi}_d$ for steps in ψ_{ref}.

MATLAB

The conventional integral SMC has been implemented in the MSS toolbox script `ExSMC.m` where a nonlinear model

$$\dot{\psi} = r \tag{16.485}$$

$$T\dot{r} + n_3 r^3 + n_1 r = K\delta + d_r \tag{16.486}$$

of the ROV Sefakkel has been used to generate the measurements. The ROV Sefakkel is 45 m long training ship (Van Amerongen 1982). The unknown disturbance term d_r is chosen as $d_r = K\delta_0$ where $\delta_0 = 1^\circ$ is a rudder offset caused by environmental disturbances. The numerical values for T and K for varying speeds $U \in [1, ...7]$m s^{-1} are computed using interpolation and extrapolation of the ROV Sefakkel data set

```
data = [...                          % data = [ U K T ]
         2       0.15    33
         2.6     0.19    33
         3.6     0.29    33
         4       0.37    33
         5       0.50    31
         6.2     0.83    43 ];
```

For the ROV Sefakkel $n_3 = 0.4$ and $n = 1$ (course stable). The rudder dynamics is modeled using (see Section 9.5.2)

```
if abs(delta_c) >= delta_max                  % saturated rudder angle
   delta_c = sign(delta_c) * delta_max;
end
delta_dot = delta_c - delta;                  % saturated rudder rate
if abs(delta_dot) >= Ddelta_max
   delta_dot = sign(delta_dot) * Ddelta_max;
end
```

The ship model including rudder dynamics is implemented in the MSS toolbox function

```
[psi_dot,r_dot,delta_dot] = ROVzefakkel(r,U,delta,delta_c,d_r)
```

Figures 16.26 and 16.27 show the performance of the integral SMC for yaw angle commands ψ_{ref} which are filtered using a third-order reference model for generation of ψ_d, r_d and $\dot{r}_d$ as described in Section 12.1.1. Figure 16.26 illustrates the chattering problem when using the sgn(σ) function in the feedback controller. This problem is avoided by using tanh(σ/ϕ) instead as shown in Figure 16.27 where the control signal is smooth. A similar behavior is obtained for sat(σ). All these options can be tested by specifying the `flag` parameter equal to `1,2,3` in the script.

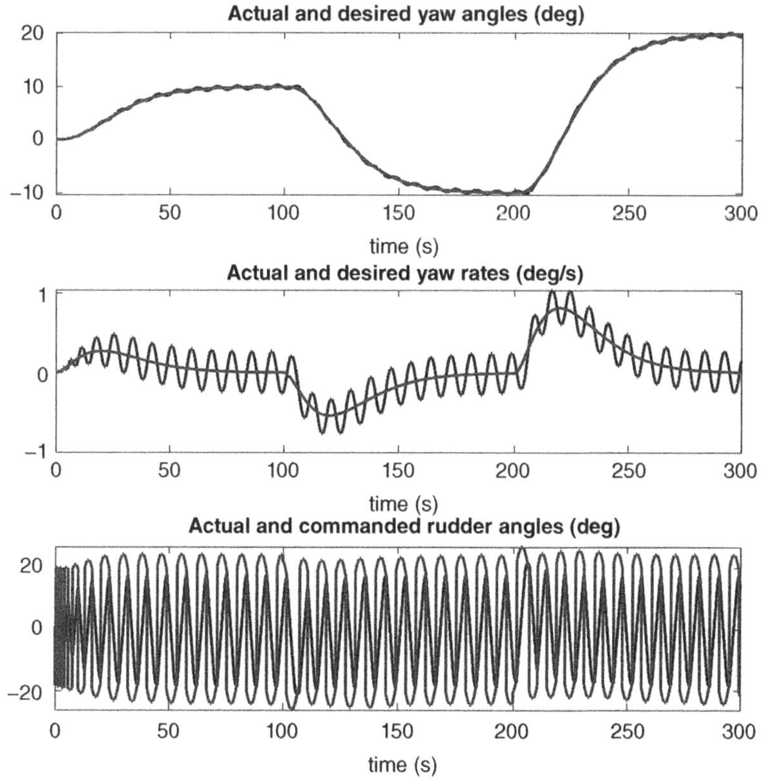

Figure 16.26 Conventional integral sliding mode controller for heading control. The chattering is due to the $\mathrm{sgn}(\sigma)$ term in the controller.

16.4.5 Case Study: Depth Autopilot for Diving Based on Conventional Integral SMC

Pitch and depth control of underwater vehicles is usually done by using control surfaces, thrusters and ballast systems. For a neutrally buoyant vehicle, stern rudders are effective for diving and depth changing maneuvers, since they require relatively little control energy compared to thrusters. Consider the longitudinal model in Section 8.2.1 where it was shown that the decoupled pitch equation is

$$(I_y - M_{\dot{q}})\ddot{\theta} - M_q\dot{\theta} + \mathrm{BG}_z W\theta = M_{\delta_S}\delta_S + d_\theta \tag{16.487}$$

The control input δ_S is the dive plane deflection while $|d_\theta| \leq d_\theta^{\max}$ is bounded disturbance due to unmodeled dynamics, ocean currents and coupling terms. The dive planes are located in the stern of the vehicle as shown in Section 9.7. The model (16.487) is equivalent to the second-order system

$$\ddot{\theta} + a_1\dot{\theta} + a_0\theta = b_0\delta_S + d_\theta \tag{16.488}$$

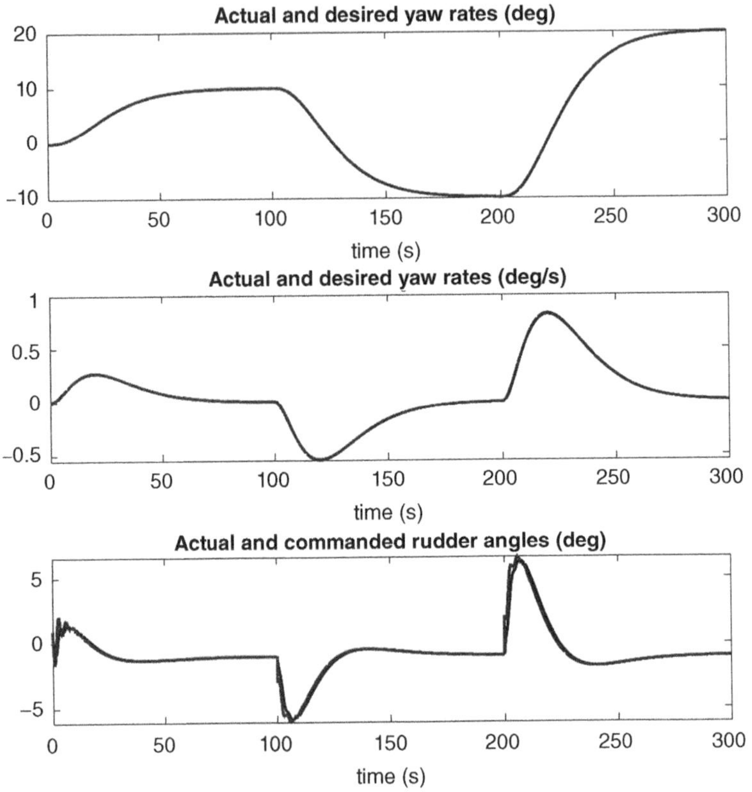

Figure 16.27 Conventional integral sliding mode controller for heading control. Chattering is avoided by replacing $\text{sgn}(\sigma)$ with $\tanh(\sigma/\phi)$ in the controller.

where $a_0 = \text{BG}_z W/(I_y - M_{\dot{q}})$, $a_1 = -M_q/(I_y - M_{\dot{q}})$ and $b_0 = M_{\delta_S}/(I_y - M_{\dot{q}})$. From (2.36) it follows that

$$\begin{aligned} \dot{z}^n &= -u\sin(\theta) + v\cos(\theta)\sin(\phi) + w\cos(\theta)\cos(\phi) \\ &\approx -U\theta + d_z \end{aligned} \tag{16.489}$$

where d_z is an unknown disturbance term due to kinematic inaccuracies. This is based on the assumption that θ, ϕ, v and w are small, and $u \approx U = \text{constant}$. Hence,

$$(z^n)^{(3)} = -U\ddot{\theta} + \ddot{d}_z. \tag{16.490}$$

This gives the third-order system

$$(z^n)^{(3)} = \underbrace{a_1 U\dot{\theta} + a_0 U\theta}_{f(\cdot)} - b_0 U\delta_S + \underbrace{(\ddot{d}_z - Ud_\theta)}_{d}. \tag{16.491}$$

The resulting disturbance term d is assumed to be upper bounded by $|d| \leq d^{\max}$. The next step is to design an integral SMC for depth control, which is robust for the uncertain dynamics given by d. Following the approach of Section (16.4.2), we define

$$\sigma = \ddot{\tilde{z}}^n + 3\lambda\dot{\tilde{z}}^n + 3\lambda^2\tilde{z}^n + \lambda^3 \int_0^t \tilde{z}^n(\tau)\,\mathrm{d}\tau \tag{16.492}$$

where $\tilde{z}^n = z^n - z^n_{\mathrm{d}}$ and $\lambda > 0$. The control law is chosen as

$$\delta_{\mathrm{S}} = -\frac{1}{b_0 U}((z^n_{\mathrm{d}})^{(3)} - 3\lambda\ddot{\tilde{z}}^n - 3\lambda^2\dot{\tilde{z}}^n - \lambda^3\tilde{z}^n - a_1 U\dot{\theta} - a_0 U\theta - K_\sigma \mathrm{sgn}(\sigma)). \tag{16.493}$$

Consequently, $\sigma \to 0$ and thus $\tilde{z}^n \to 0$. When implementing the control law, higher-order derivatives in z^n can be avoided by using the measured pitch angle θ and pitch rate q to compute

$$\dot{z}^n \approx -U\theta \tag{16.494}$$

$$\ddot{z}^n \approx -Uq. \tag{16.495}$$

Finally, the reference signal for depth-changing maneuvers are computed using a linear reference model

$$\frac{z^n_{\mathrm{d}}}{z^n_{\mathrm{ref}}}(s) = \frac{\omega_{\mathrm{n}}^3}{(s+\omega_{\mathrm{n}})(s^2 + 2\zeta\omega_{\mathrm{n}} s + \omega_{\mathrm{n}}^2)} \tag{16.496}$$

where z^n_{ref} is the operator input. The corresponding state-space model is

$$\dot{\boldsymbol{x}}_{\mathrm{d}} = \boldsymbol{A}_{\mathrm{d}}\boldsymbol{x}_{\mathrm{d}} + \boldsymbol{b}_{\mathrm{d}}\, z^n_{\mathrm{ref}} \tag{16.497}$$

where $\boldsymbol{x}_{\mathrm{d}} = [z^n_{\mathrm{d}}, \dot{z}^n_{\mathrm{d}}, \ddot{z}^n_{\mathrm{d}}]^\top$ and

$$\boldsymbol{A}_{\mathrm{d}} = \begin{bmatrix} 0 & 1 & 0 \\ 0 & 0 & 1 \\ -\omega_n^3 & -(2\zeta+1)\omega_{\mathrm{n}}^2 & -(2\zeta+1)\omega_{\mathrm{n}} \end{bmatrix}, \quad \boldsymbol{b}_{\mathrm{d}} = \begin{bmatrix} 0 \\ 0 \\ \omega_{\mathrm{n}}^3 \end{bmatrix}. \tag{16.498}$$

Note that a step in the command z^n_{ref} will give a step in $(z^n_{\mathrm{d}})^{(3)}$ while $\ddot{z}^n_{\mathrm{d}}$, $\dot{z}^n_{\mathrm{d}}$ and z^n_{d} will be low-pass filtered and therefore smooth signals in the sliding mode controller. We also notice that the steady-state position for a constant reference signal z^n_{ref} is

$$\lim_{t\to\infty} z^n_{\mathrm{d}} = z^n_{\mathrm{ref}}. \tag{16.499}$$

16.4.6 Case Study: Heading Autopilot Based on the Adaptive-gain Super Twisting Algorithm

We will revisit the nonlinear yaw model (7.30)

$$\dot{\psi} = r \tag{16.500}$$

$$T\dot{r} + n_3 r^3 + n_1 r = K\delta + d_r + \tau_{\text{wind}} \tag{16.501}$$

where d_r is an unknown disturbance due to unmodeled dynamics, ocean currents, waves and parametric uncertainty. The sliding variable is chosen as

$$\sigma := \tilde{r} + \lambda\tilde{\psi} \tag{16.502}$$

where $\tilde{\psi} = \psi - \psi_d$ and $\tilde{r} = r - r_d$ are the tracking errors, and $\lambda > 0$. Since $\sigma = r - r_r$ and

$$r_r = r_d - \lambda\tilde{\psi} \tag{16.503}$$

we obtain

$$\dot{\sigma} = \underbrace{\frac{1}{T}(d_r + \tau_{\text{wind}} - n_3 r^3 - n_1 r) - \dot{r}_r +}_{\phi(\cdot)} \underbrace{\frac{K}{T}}_{\text{b}} \delta. \tag{16.504}$$

Hence, ω given by (16.469) relates to the rudder angle according to

$$\omega = \frac{K}{T}\delta \quad \Rightarrow \quad \delta = \frac{T}{K}\omega. \tag{16.505}$$

Finally, the control signal ω is computed using the adaptive-gain STA given by Theorem 16.2. This is mathematically equivalent to

$$\omega = -\alpha|\sigma|^{1/2}\,\text{sgn}(\sigma) + \upsilon \tag{16.506}$$

$$\dot{\upsilon} = -\beta\,\text{sgn}(\sigma) \tag{16.507}$$

where the gains are chosen as

$$\dot{\alpha} = \begin{cases} \alpha_0 & \text{if } \sigma \neq 0 \\ 0 & \text{otherwise} \end{cases} \tag{16.508}$$

$$\beta = \beta_0 \tag{16.509}$$

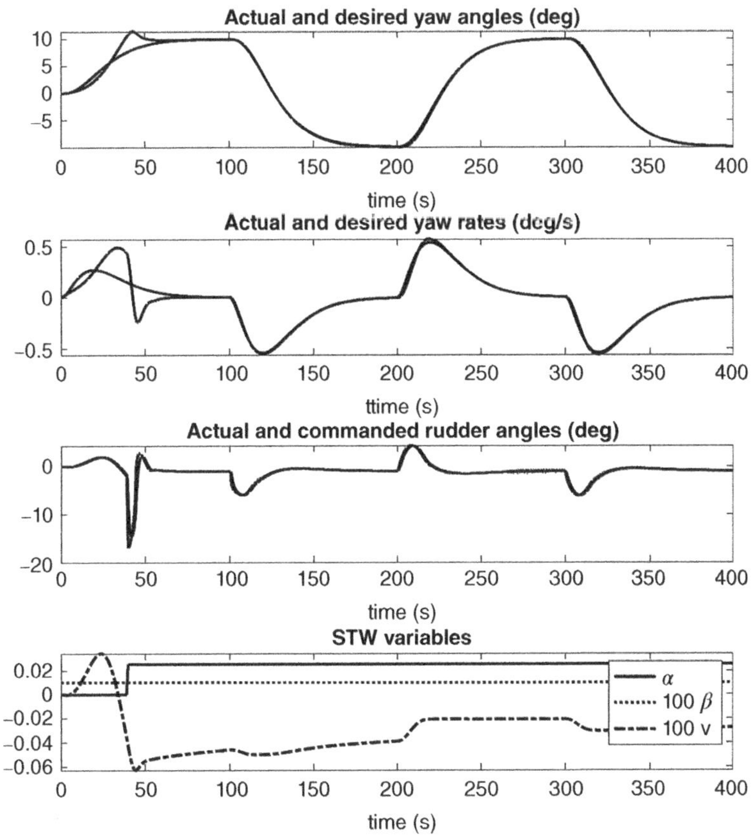

Figure 16.28 Super twisting adaptive sliding mode controller for heading control.

where $\alpha_0 > 0$ and $\beta_0 > 0$ are two tunable parameters. Note that the STA does not require knowledge of the parameters bounds for the unknown quantities in $\phi(\cdot)$.

It is also possible to compensate for the wind yaw moment by modifying the sliding variable and control input as

$$\dot{\sigma} = \underbrace{\frac{1}{T}(d_\mathrm{r} + (\tau_\mathrm{wind} - \hat{\tau}_\mathrm{wind}) - n_3 r^3 - n_1 r) - \dot{r}_\mathrm{r}}_{\phi(\cdot)} + \underbrace{\frac{K}{T}\delta + \frac{1}{T}\hat{\tau}_\mathrm{wind}}_{\omega} \tag{16.510}$$

implying that

$$\delta = \frac{T}{K}\omega - \frac{1}{K}\hat{\tau}_\mathrm{wind}. \tag{16.511}$$

MATLAB

The MSS toolbox function

```
[psi_dot,r_dot,delta_dot] = ROVzefakkel(r,U,delta,delta_c,d_r)
```

is used to generate measurements for the adaptive-gain STA, which is implemented in the MSS toolbox example script `ExSTA.m`. The ROV Sefakkel is 45 m long training ship (Van Amerongen 1982). The unknown disturbance term d_r is chosen as $d_r = K\delta_0$ where $\delta_0 = 1^\circ$ is a rudder offset caused by environmental disturbances. The numerical values for T and K in the model depend on the cruise speed $U \in [1, ...7]\text{m s}^{-1}$ while $n_3 = 0.4$ and $n = 1$ (course stable).

Figure 16.28 show the performance of the adaptive-gain STA for yaw angle commands ψ_{ref} which are filtered using a third-order reference model for generation of ψ_d, r_d and $\dot{r}_d$ as described in Section 12.1.1. The STA is implemented using (16.506)–(16.509) with

$$\alpha_0 = 0.03, \qquad \beta_0 = 0.0001.$$

In addition, the sigmoid function $\text{sgn}(\sigma)$ in (16.507) has been replaced by $\tanh(\sigma/\phi)$ to further improve performance and avoid chattering. It is seen from Figure 16.28 that the controller adapts quite well by adjusting the states α and v. The control input δ is free from chattering. Also note that the control signal ω does not depend on the ship model parameters but when mapped to rudder angle commands δ_c accurate estimates of K and T are needed to implement (16.511) successfully.

Part Three

Appendices

A

Nonlinear Stability Theory

This appendix briefly reviews some useful results from nonlinear stability theory. The methods are classified according to

- Lyapunov stability of nonlinear *autonomous* systems $\dot{\boldsymbol{x}} = \boldsymbol{f}(\boldsymbol{x})$, that is systems where $\boldsymbol{f}(\boldsymbol{x})$ does not explicitly depend on the time t.
- Lyapunov stability of nonlinear *non-autonomous* systems $\dot{\boldsymbol{x}} = \boldsymbol{f}(\boldsymbol{x}, t)$, that is systems where $\boldsymbol{f}(\boldsymbol{x}, t)$ does depend on t explicitly.

A.1 Lyapunov Stability for Autonomous Systems

Before stating the main Lyapunov theorems for *autonomous* systems, the concepts of stability and convergence are briefly reviewed (Khalil 2002).

A.1.1 Stability and Convergence

Consider the nonlinear time-invariant system

$$\dot{\boldsymbol{x}} = \boldsymbol{f}(\boldsymbol{x}), \quad \boldsymbol{x}(0) = \boldsymbol{x}_0 \tag{A.1}$$

where $\boldsymbol{x} \in \mathbb{R}^n$ and $\boldsymbol{f} : \mathbb{R}^n \rightarrow \mathbb{R}^n$ is assumed to be *locally Lipschitz* in $\boldsymbol{x}$; that is for each point $\boldsymbol{x} \in D \subset \mathbb{R}^n$ there exists a neighborhood $D_0 \in D$ such that

$$\|\boldsymbol{f}(\boldsymbol{x}) - \boldsymbol{f}(\boldsymbol{y})\| \leq L\|\boldsymbol{x} - \boldsymbol{y}\|, \quad \forall \boldsymbol{x}, \boldsymbol{y} \in D_0 \tag{A.2}$$

where L is called the Lipschitz constant on D_0.

Let $\boldsymbol{x}_e$ denote the equilibrium point of (A.1) given by

$$\boldsymbol{f}(\boldsymbol{x}_e) = \boldsymbol{0}. \tag{A.3}$$

Handbook of Marine Craft Hydrodynamics and Motion Control, Second Edition. Thor I. Fossen.

The solutions $\boldsymbol{x}(t)$ of (A.1) are

- *Bounded*, if there exists a non-negative function $0 < \gamma(\boldsymbol{x}(0)) < \infty$ such that

$$\|\boldsymbol{x}(t)\| \leq \gamma(\boldsymbol{x}(0)), \quad \forall t \geq 0. \tag{A.4}$$

In addition, the equilibrium point $\boldsymbol{x}_e$ of (A.1) is

- *Stable* if, for each $\epsilon > 0$, there exists a $\delta(\epsilon) > 0$ such that

$$\|\boldsymbol{x}(0)\| < \delta(\epsilon) \Rightarrow \|\boldsymbol{x}(t)\| < \epsilon, \quad \forall t \geq 0. \tag{A.5}$$

- *Unstable*, if it is not stable.
- *Attractive* if, for each $r > 0, \epsilon > 0$, there exists a $T(r, \epsilon) > 0$ such that

$$\|\boldsymbol{x}(0)\| \leq r \Rightarrow \|\boldsymbol{x}(t)\| \leq \epsilon, \quad \forall t \geq T(r, \epsilon). \tag{A.6}$$

  Attractivity implies convergence, that is $\lim_{t\to\infty} \|\boldsymbol{x}(t)\| = \mathbf{0}$.
- *(Locally) asymptotically stable (AS)*, if the equilibrium point $\boldsymbol{x}_e$ is stable and attractive.
- *Globally stable (GS)*, if the equilibrium point $\boldsymbol{x}_e$ is stable and $\delta(\epsilon)$ can be chosen to satisfy $\lim_{\epsilon\to\infty}\delta(\epsilon) = \infty$.
- *Global asymptotically stable (GAS)*, if the equilibrium point $\boldsymbol{x}_e$ is stable for all $\boldsymbol{x}(0)$ (region of attraction $\mathbb{R}^n$).
- *(Locally) exponentially stable (ES)*, if there exist positive constants α, λ and r such that

$$\|\boldsymbol{x}(0)\| < r \Rightarrow \|\boldsymbol{x}(t)\| < \alpha \mathrm{e}^{-\lambda t}\|\boldsymbol{x}(0)\|, \ \forall t \geq 0. \tag{A.7}$$

- *Globally exponentially stable (GES)*, if there exist positive constants α, λ and r such that for all $\boldsymbol{x}(0)$ (region of attraction $\mathbb{R}^n$):

$$\|\boldsymbol{x}(t)\| < \alpha e^{-\lambda t}\|\boldsymbol{x}(0)\|, \ \forall t \geq 0. \tag{A.8}$$

Different theorems for investigation of stability and convergence will now be presented. A guideline for which theorem that should be applied is given in Table A.1 whereas the different theorems are listed in the forthcoming sections.

Notice that for non-autonomous systems, GAS is replaced by *uniform global asymptotic stability* (UGAS) since uniformity is a necessary requirement in the case of time-varying nonlinear systems.

Table A.1 Classification of theorems for stability and convergence.

Autonomous systems	$V > 0, \dot{V} < 0$	$V > 0, \dot{V} \leq 0$	Lyapunov's direct method	GAS/GES
			Krasovskii–LaSalle's theorem	GAS
Non-autonomous systems	$V > 0, \dot{V} < 0$	$V \geq 0, \dot{V} \leq 0$	LaSalle–Yoshizawa's theorem	UGAS
			Barbalat's lemma	Convergence

A.1.2 Lyapunov's Direct Method

Theorem A.1 (Lyapunov's Direct Method)
Let $\boldsymbol{x}_e = \mathbf{0}$ be the equilibrium point of (A.1) and assume that $\boldsymbol{f}(\boldsymbol{x})$ is locally Lipschitz in $\boldsymbol{x}$. Let $V : \mathbb{R}^n \rightarrow \mathbb{R}_+$ be a continuously differentiable function $V(\boldsymbol{x})$ satisfying

$$(i)\ V(\mathbf{x}) > 0 \text{ (positive definite) and } V(0) = 0 \tag{A.9}$$

$$(ii)\ \dot{V}(\mathbf{x}) = \frac{\partial V(\mathbf{x})}{\partial \mathbf{x}} \mathbf{f}(\mathbf{x}) \leq -W(x) \leq 0 \tag{A.10}$$

$$(iii)\ V(\mathbf{x}) \rightarrow \infty \text{ as } \|\mathbf{x}\| \rightarrow \infty \text{ (radially unbounded).} \tag{A.11}$$

Then the equilibrium point $\mathbf{x}_e$ is GS if $W(x) \geq 0$ (positive semi-definite) and GAS if $W(\mathbf{x}) > 0$ (positive definite) for all $\mathbf{x} \neq \mathbf{0}$.

Proof. *Khalil (2002).*

The requirement that $W(\boldsymbol{x}) > 0$ such that $\dot{V}(\boldsymbol{x}) < 0$ is in many cases difficult to satisfy. This is illustrated in the following example.

Example A.1 (Stability of a Mass–Damper–Spring System)
Consider the nonlinear mass–damper–spring system

$$\dot{x} = v \tag{A.12}$$

$$m\dot{v} + d(v)v + kx^2 = 0 \tag{A.13}$$

where $m > 0, d(v) > 0, \forall v$ and $k > 0$ (see Figure A.1). Let us choose $V(\boldsymbol{x})$ as the sum of kinetic energy $(1/2)mv^2$ and potential energy $(1/2)kx^2$ such that

$$V(\boldsymbol{x}) = \frac{1}{2}(mv^2 + kx^2) = \frac{1}{2}\boldsymbol{x}^\top \begin{bmatrix} m & 0 \\ 0 & k \end{bmatrix} \boldsymbol{x} \tag{A.14}$$

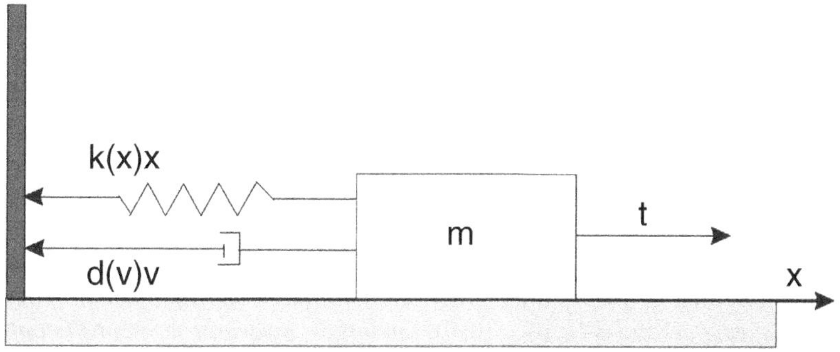

Figure A.1 Mass–damper–spring system.

where $\boldsymbol{x} = [v, \ x]^{\top}$ *results in*

$$\begin{aligned}\dot{V}(\boldsymbol{x}) &= mv\dot{v} + kx\dot{x} \\ &= v(m\dot{v} + kx) \\ &= -d(v)v^2 \\ &= -\boldsymbol{x}^{\top}\begin{bmatrix} d(v) & 0 \\ 0 & 0 \end{bmatrix}\boldsymbol{x}. \end{aligned} \tag{A.15}$$

Hence, only stability can be concluded from Theorem A.1, since $\dot{V}(\boldsymbol{x}) = 0$ *for all* $v = 0$*. However, GAS can in many cases also be proven for systems with a negative semi-definite* $\dot{V}(\boldsymbol{x})$ *thanks to the invariant set theorem of* Krasovskii–LaSalle; *see LaSalle and Lefschetz (1961) and LaSalle (1966).*

A.1.3 Krasovskii–LaSalle's Theorem

The theorem of Krasovskii–LaSalle can be used to check a nonlinear *autonomous* system for GAS in the case of a negative semi-definite $\dot{V}(\boldsymbol{x})$.

Theorem A.2 (Krazovskii–LaSalle's Theorem)
Let $V : \mathbb{R}^n \rightarrow \mathbb{R}_+$ *be a continuously differentiable positive definite function such that*

$$V(\boldsymbol{x}) \rightarrow \infty \text{ as } \|\boldsymbol{x}\| \rightarrow \infty \tag{A.16}$$

$$\dot{V}(\boldsymbol{x}) \leq 0, \quad \forall \boldsymbol{x}. \tag{A.17}$$

Let Ω *be the set of all points where* $\dot{V}(\boldsymbol{x}) = 0$*, that is*

$$\Omega = \{\boldsymbol{x} \in \mathbb{R}^n | \dot{V}(\boldsymbol{x}) = 0\} \tag{A.18}$$

and M *be the largest invariant set in* Ω*. Then all solutions* $\boldsymbol{x}(t)$ *converge to* M*. If* $M = \{\boldsymbol{x}_e\}$ *then the equilibrium point* $\boldsymbol{x}_e$ *of (A.1) is GAS.*

Proof. *LaSalle (1966).*

Example A.2 (Example A.1: Stability of a Mass–Damper–Spring System, Continued)
Again consider the mass–damper–spring system of Example A.1. The set Ω *is found by requiring that*

$$\dot{V}(\boldsymbol{x}) = -d(v)v^2 \equiv 0 \tag{A.19}$$

which is true for $v = 0$*. Therefore,*

$$\Omega = \{(x \in \mathbb{R}, v = 0)\}. \tag{A.20}$$

Now, $v = 0$ *implies that* $m\dot{v} = -kx$*, which is non-zero when* $x \neq 0$*. Hence, the system cannot get "stuck" at a point other than* $x = 0$*. Since the equilibrium point of the mass–damper–spring system is* $(x, v) = (0, 0)$*, the largest invariant set* M *in* Ω *contains only one point, namely* $(x, v) = (0, 0)$*. Hence, the equilibrium point of (A.1) is GAS according to Theorem A.2.*

A.1.4 Global Exponential Stability

The following theorem is useful to guarantee global exponential stability.

Theorem A.3 (Global Exponential Stability)
Let $\boldsymbol{x}_\mathrm{e} = \boldsymbol{0}$ be the equilibrium point of (A.1) and assume that $\boldsymbol{f}(\boldsymbol{x})$ is locally Lipschitz in $\boldsymbol{x}$. Let $V : \mathbb{R}^n \to \mathbb{R}_+$ be a continuously differentiable and radially unbounded function satisfying

$$V(\boldsymbol{x}) = \boldsymbol{x}^\top \boldsymbol{P}\boldsymbol{x} > 0, \quad \forall \boldsymbol{x} \neq \boldsymbol{0} \tag{A.21}$$

$$\dot{V}(\boldsymbol{x}) \leq -\boldsymbol{x}^\top \boldsymbol{Q}\boldsymbol{x} < 0, \quad \forall \boldsymbol{x} \neq \boldsymbol{0} \tag{A.22}$$

with constant matrices $\boldsymbol{P} = \boldsymbol{P}^\top > 0$ and $\boldsymbol{Q} = \boldsymbol{Q}^\top > 0$. Then the equilibrium point $\boldsymbol{x}_\mathrm{e}$ is GES and the state vector satisfies

$$\|\boldsymbol{x}(t)\|_2 \leq \sqrt{\frac{\lambda_{\max}(\boldsymbol{P})}{\lambda_{\min}(\boldsymbol{P})}} e^{-\alpha t} \|\boldsymbol{x}(0)\|_2 \tag{A.23}$$

where

$$\alpha = \frac{\lambda_{\min}(\boldsymbol{Q})}{2\lambda_{\max}(\boldsymbol{P})} > 0 \tag{A.24}$$

is a bound on the convergence rate.

Proof. *Since $V(\boldsymbol{x})$ is bounded by*

$$0 < \lambda_{\min}(\boldsymbol{P})\|\boldsymbol{x}(t)\|_2^2 \leq V(\boldsymbol{x}) \leq \lambda_{\max}(\boldsymbol{P})\|\boldsymbol{x}(t)\|_2^2, \quad \forall \boldsymbol{x} \neq \boldsymbol{0} \tag{A.25}$$

it is seen that

$$-\|\boldsymbol{x}(t)\|_2^2 \leq -\frac{1}{\lambda_{\max}(\boldsymbol{P})} V(\boldsymbol{x}). \tag{A.26}$$

Hence, it follows from (A.22) that

$$\begin{aligned}
\dot{V}(\boldsymbol{x}) &\leq -\boldsymbol{x}^\top \boldsymbol{Q}\boldsymbol{x} \\
&\leq -\lambda_{\min}(\boldsymbol{Q})\|\boldsymbol{x}(t)\|_2^2 \\
&\leq -\underbrace{\frac{\lambda_{\min}(\boldsymbol{Q})}{\lambda_{\max}(\boldsymbol{P})}}_{2\alpha} V(\boldsymbol{x}).
\end{aligned} \tag{A.27}$$

Integration of $\dot{V}(\boldsymbol{x}(t))$ yields

$$V(\boldsymbol{x}(t)) \leq \mathrm{e}^{-2\alpha t} V(\boldsymbol{x}(0)). \tag{A.28}$$

Finally, (A.25) implies

$$\lambda_{\min}(\boldsymbol{P})\|\boldsymbol{x}(t)\|_2^2 \leq e^{-2\alpha t} \lambda_{\max}(\boldsymbol{P})\|\boldsymbol{x}(0)\|_2^2 \tag{A.29}$$

$$\|\boldsymbol{x}(t)\|_2 \leq \sqrt{\frac{\lambda_{\max}(\boldsymbol{P})}{\lambda_{\min}(\boldsymbol{P})}} \mathrm{e}^{-\alpha t} \|\boldsymbol{x}(0)\|_2. \tag{A.30}$$

This shows that $\|x(t)\|_2$ will converge exponentially to zero with convergence rate α.

A.2 Lyapunov Stability of Non-autonomous Systems

In this section several useful theorems for convergence and stability of time-varying nonlinear systems

$$\dot{x} = f(x, t), \qquad x(0) = x_0 \tag{A.31}$$

where $x \in \mathbb{R}^n, t \in \mathbb{R}_+$ and $f : \mathbb{R}^n \times \mathbb{R}_+ \to \mathbb{R}^n$ is assumed to be *locally Lipschitz* in x and uniformly in t, are briefly reviewed.

A.2.1 Barbălat's Lemma

Lemma A.1 (Barbălat's Lemma)
Let $\phi : \mathbb{R}_+ \to \mathbb{R}$ be a uniformly continuous function and suppose that $\lim_{t\to\infty} \int_0^t \phi(\tau)\mathrm{d}\tau$ exists and is finite; then

$$\lim_{t\to\infty} \phi(t) = 0. \tag{A.32}$$

Proof. *Barbălat (1959).*

Note that *Barbălat's lemma* only guarantees *global convergence*. This result is particularly useful if there exists a uniformly continuous function $V : \mathbb{R}^n \times \mathbb{R}_+ \to \mathbb{R}$ satisfying

(i) $V(x, t) \geq 0$

(ii) $\dot{V}(x, t) \leq 0$

(iii) $\dot{V}(x, t)$ is uniformly continuous.

Hence, according to Barbălat's lemma, $\lim_{t\to\infty} \dot{V}(x, t) = 0$. The requirement that $\dot{V}$ should be uniformly continuous can easily be checked by using

$$\ddot{V}(x, t) \text{ is bounded} \Rightarrow \dot{V}(x, t) \text{ is uniformly continuous.}$$

A.2.2 LaSalle–Yoshizawa's Theorem

For non-autonomous systems the following theorem of LaSalle (1966) and Yoshizawa (1968) is quite useful.

Theorem A.4 (LaSalle–Yoshizawa's Theorem)
Let $x_e = \mathbf{0}$ be the equilibrium point of (A.31) and assume that $f(x, t)$ is locally Lipschitz in x. Let $V : \mathbb{R}^n \times \mathbb{R}_+ \to \mathbb{R}_+$ be a continuously differentiable function $V(x, t)$ satisfying

$$(i)\ V(x, t) > 0 \ (\textit{positive definite}) \textit{ and } V(0) = 0 \tag{A.33}$$

$$(ii)\ \dot{V}(\boldsymbol{x},t) = \frac{\partial V(\boldsymbol{x},t)}{\partial t} + \frac{\partial V(\boldsymbol{x},t)}{\partial \boldsymbol{x}}\boldsymbol{f}(x,t) \leq -W(\boldsymbol{x}) \leq 0 \tag{A.34}$$

$$(iii)\ V(\boldsymbol{x},t) \rightarrow \infty \text{ as } \|\boldsymbol{x}\| \rightarrow \infty \text{ (radially unbounded)} \tag{A.35}$$

where $W(\boldsymbol{x})$ is a continuous function. Then all solutions $\boldsymbol{x}(t)$ of (A.31) are uniformly globally bounded and

$$\lim_{t\rightarrow\infty} W(\boldsymbol{x}(t)) = 0. \tag{A.36}$$

In addition, if $W(\boldsymbol{x}) > 0$ (positive definite), then the equilibrium point $\boldsymbol{x}_e = \boldsymbol{0}$ of (A.31) is UGAS.

Proof. *LaSalle (1966) and Yoshizawa (1968).*

A.2.3 On USGES of Proportional Line-of-sight Guidance Laws

Several of the LOS guidance laws in Chapter 12 are described by the non-autonomous system

$$\dot{y} = -\frac{U}{\sqrt{\Delta^2 + y^2}}\,y \tag{A.37}$$

where both U and Δ can be time varying but bounded by $0 < \Delta_{min} \leq \Delta \leq \Delta_{max}$ and $0 < U_{min} \leq U \leq U_{max}$. Fossen and Pettersen (2014, theorem 1) have shown that the equilibrium point $y = 0$ of (A.37) is uniformly semi-globally exponentially stable (USGES). USGES is slightly weaker than GES but GES cannot be achieved for this type of system due to the kinematic representation, which introduces saturation through the trigonometric functions.

The USGES stability property is important for marine craft that are exposed to environmental disturbances, in particular to quantify robustness. In particular, it is seen from lemma 9.2 in Khalil (2002) that the USGES property implies that we always can choose a region of attraction in which we have exponential convergence sufficiently large. Hence, we can always satisfy the condition for which the solution of the perturbed system will be uniformly bounded irrespective of the size of the perturbation. USGES thus provides stronger robustness properties than UGAS.

The stability proof (Fossen and Pettersen 2014) is based on the Lyapunov function candidate $V(t,y) = (1/2)y^2$. Hence,

$$\dot{V}(t,y) = -\frac{U}{\sqrt{\Delta^2 + y^2}}\,y^2 \leq 0. \tag{A.38}$$

Since $V(t,y) > 0$ and $\dot{V}(t,y) \leq 0$ it follows that

$$|y(t)| \leq |y(t_0)|, \quad \forall t \geq t_0 \tag{A.39}$$

and by Khalil (2002, theorem 4.8) the origin $y = 0$ is uniformly stable. Next, we define

$$\phi(t,y) := \frac{U}{\sqrt{\Delta^2 + y^2}}. \tag{A.40}$$

For each $r > 0$ and all $|y(t)| \leq r$, we have

$$\phi(t, y_e) \geq \frac{U_{\min}}{\sqrt{\Delta_{\max}^2 + r^2}} := c(r) \tag{A.41}$$

Consequently,

$$\begin{aligned} \dot{V}(t, y) &= -2\phi(t, y)V(t, y) \\ &\leq -2c(r)V(t, y), \qquad \forall |y(t)| \leq r. \end{aligned} \tag{A.42}$$

In view of (A.39), the above holds for all trajectories generated by the initial conditions $y(t_0)$. Consequently, we can invoke the comparison lemma (Khalil 2002, lemma 3.4) by noticing that the linear system $\dot{z} = -2c(r)z$ has the solution $z(t) = \mathrm{e}^{-2c(r)(t-t_0)}z(t_0)$, which implies that $\dot{v}(t) \leq \mathrm{e}^{-2c(r)(t-t_0)}v(t_0)$ for $v(t) = V(t, y(t))$. Therefore,

$$y(t) \leq \mathrm{e}^{-c(r)(t-t_0)}y(t_0) \tag{A.43}$$

for all $t \geq t_0$, $|y(t_0)| \leq r$ and any $r > 0$. Hence, we can conclude that the equilibrium point $y = 0$ is USGES (Loria and Panteley 2004, definition 2.7).

A.2.4 UGAS when Backstepping with Integral Action

When designing industrial control systems it is important to include integral action in the control law in order to compensate for slowly varying and constant disturbances. This is necessary to avoid steady-state errors both in regulation and tracking. The integral part of the controller can be provided by using *adaptive backstepping* (Krstic *et al.* 1995) under the assumption of constant disturbances (see Section 16.3.4). Unfortunately, the resulting error dynamics in this case often becomes non-autonomous, which again implies that *Krasovskii–LaSalle's theorem* cannot be used. An alternative theorem for this case will be stated by considering the nonlinear system

$$\dot{\boldsymbol{x}} = \boldsymbol{f}(\boldsymbol{x}, \boldsymbol{u}, \boldsymbol{\theta}, t) \tag{A.44}$$

where $\boldsymbol{x} \in \mathbb{R}^n, \boldsymbol{u} \in \mathbb{R}^n$ and $\boldsymbol{\theta} \in \mathbb{R}^p$ $(p \leq n)$ is a constant *unknown* parameter vector. Furthermore, assume that there exists an adaptive control law

$$\boldsymbol{u} = \boldsymbol{u}(\boldsymbol{x}, \boldsymbol{x}_{\mathrm{d}}, \hat{\boldsymbol{\theta}}) \tag{A.45}$$

$$\dot{\hat{\boldsymbol{\theta}}} = \boldsymbol{\phi}(\boldsymbol{x}, \boldsymbol{x}_{\mathrm{d}}) \tag{A.46}$$

where $\boldsymbol{x}_{\mathrm{d}} \in C^r$ and $\hat{\boldsymbol{\theta}} \in \mathbb{R}^p$, such that the error dynamics can be written

$$\dot{\boldsymbol{z}} = \boldsymbol{h}(\boldsymbol{z}, t) + \boldsymbol{B}(t)\tilde{\boldsymbol{\theta}} \tag{A.47}$$

$$\dot{\tilde{\boldsymbol{\theta}}} = -\boldsymbol{P}\boldsymbol{B}(t)^{\top}\left(\frac{\partial W(\boldsymbol{z}, t)}{\partial \boldsymbol{z}}\right)^{\top}, \quad \boldsymbol{P} = \boldsymbol{P}^{\top} > 0 \tag{A.48}$$

where $W(z, t)$ is a suitable C^1 function and $\tilde{\boldsymbol{\theta}} = \hat{\boldsymbol{\theta}} - \boldsymbol{\theta}$ is the parameter estimation error. The parameter estimate $\hat{\boldsymbol{\theta}}$ can be used to compensate for a constant disturbance, that is integral action. Hence, the conditions in the following theorem can be used to establish UGAS when backstepping with integral action. The conditions are based on Loria *et al.* (1999) and Fossen *et al.* (2001).

Theorem A.5 (UGAS/ULES when Backstepping with Integral Action)
The origin of the system (A.47) and (A.48) is UGAS if $\boldsymbol{B}^{\top}(t)\boldsymbol{B}(t)$ is invertible for all t, $\boldsymbol{P} = \boldsymbol{P}^{\top} > 0$, there exists a continuous, non-decreasing function $\rho : \mathbb{R}_+ \rightarrow \mathbb{R}_+$ such that

$$\max\left\{ \|\boldsymbol{h}(z,t)\|, \left\| \frac{\partial W(z,t)}{\partial z} \right\| \right\} \leq \rho(\|z\|)\|z\| \tag{A.49}$$

and there exist class-$\mathcal{K}_\infty$ functions α_1 and α_2 and a strictly positive real number $c > 0$ such that $W(z, t)$ satisfy

$$\alpha_1(\|z\|) \leq W(z,t) \leq \alpha_2(\|z\|) \tag{A.50}$$

$$\frac{\partial W(z,t)}{\partial t} + \frac{\partial W(z,t)}{\partial z}\boldsymbol{h}(z,t) \leq -c\|z\|^2. \tag{A.51}$$

If, in addition, $\alpha_2(s) \propto s^2$ for sufficiently small s then the origin is ULES.

Proof. *Fossen* et al. *(2001).*

Theorem A.5 implies that both $z \rightarrow \boldsymbol{0}$ and $\tilde{\boldsymbol{\theta}} \rightarrow \boldsymbol{0}$ when $t \rightarrow \infty$. The following example illustrates how a UGAS integral controller can be derived.

Example A.3 (UGAS Integral Controller)
Consider the non-autonomous system

$$\dot{x} = -a(t)x + \theta + u \tag{A.52}$$

$$u = -K_{\mathrm{p}}x - \hat{\theta} \tag{A.53}$$

$$\dot{\hat{\theta}} = px \tag{A.54}$$

where $0 < a(t) \leq a_{\max}$, $\theta =$ constant, $K_{\mathrm{p}} > 0$ and $p > 0$. This is a PI controller since

$$u = -K_{\mathrm{p}}x - p\int_0^t x(\tau)\mathrm{d}\tau \tag{A.55}$$

Choosing $z = x$, the error dynamics can be written

$$\dot{z} = -(a(t) + K_{\mathrm{p}})z - \tilde{\theta} \tag{A.56}$$

$$\dot{\tilde{\theta}} = pz \tag{A.57}$$

which is in the form (A.47) and (A.48) with $W(z) = (1/2)z^2$ *and* $\boldsymbol{B} = 1$. *Since* $\boldsymbol{B}^\top \boldsymbol{B} = 1 > 0$ *and*

$$\max\{|a(t)z + K_p z|, |z|\} \leq \rho |z| \tag{A.58}$$

with $\rho = a_{\max} + K_p$, *the equilibrium point* $z = 0$ *is UGAS according to Theorem A.5. Note that the LaSalle–Yoshizawa theorem fails for this case since*

$$V(z,t) = W(z) + \frac{1}{2p}\tilde{\theta}^2 \tag{A.59}$$

$$\begin{aligned} \dot{V}(z,t) &= z\dot{z} + \frac{1}{p}\tilde{\theta}\dot{\tilde{\theta}} \\ &= -[a(t) + K_p]z^2 \\ &\leq 0 \end{aligned} \tag{A.60}$$

which by LaSalle–Yoshizawa only shows UGS and $z(t) \to 0$, *but not* $\tilde{\theta} \to 0$.

B

Numerical Methods

From a physical point of view, marine craft kinematics and kinetics are most naturally derived in the continuous-time domain using *Newtonian* or *Lagrangian* dynamics. In the implementation of a control law, it is desirable to represent the nonlinear dynamics in discrete time. This appendix discusses methods for discretization of linear and nonlinear systems, numerical integration and differentiation.

B.1 Discretization of Continuous-time Systems

This section discusses discretization of linear state-space models with extensions to nonlinear systems. For notational simplicity, let $t_k = kt$ such that $\boldsymbol{x}[k] = \boldsymbol{x}(t_k)$ and $\boldsymbol{x}[k+1] = \boldsymbol{x}(t_k + h)$ where h is the sampling time and k is the sampling index. The *forward shift operator* z is defined by

$$\boldsymbol{x}[k+1] := z\boldsymbol{x}[k]. \tag{B.1}$$

B.1.1 State-space Models

Consider the linear continuous-time model

$$\dot{\boldsymbol{x}} = \boldsymbol{A}\boldsymbol{x} + \boldsymbol{B}\boldsymbol{u}. \tag{B.2}$$

Assume that $\boldsymbol{u}$ is piecewise constant over the sampling interval h and equal to $\boldsymbol{u}[k]$. Hence, the solution of (B.2) is

$$\boldsymbol{x}[k+1] = \mathrm{e}^{\boldsymbol{A}h}\boldsymbol{x}[k] + \left(\int_{kh}^{(k+1)h} \mathrm{e}^{\boldsymbol{A}[(k+1)h-\tau]}\boldsymbol{B}\mathrm{d}\tau \right) \boldsymbol{u}[k] \tag{B.3}$$

Handbook of Marine Craft Hydrodynamics and Motion Control, Second Edition. Thor I. Fossen.

where we have moved the constant term $\boldsymbol{u}[k]$ outside the integral. The integral can be solved by changing the variables

$$\lambda = (k+1)h - \tau, \qquad \lambda = \begin{cases} 0, & \tau = (k+1)h \\ h, & \tau = kh \end{cases} . \tag{B.4}$$

Consequently,

$$\int_{kh}^{(k+1)h} \mathrm{e}^{A[(k+1)h-\tau]}\boldsymbol{B}\mathrm{d}\tau = \int_{h}^{0} \mathrm{e}^{A\lambda}\boldsymbol{B}(-\mathrm{d}\lambda) = \int_{0}^{h} \mathrm{e}^{A\lambda}\boldsymbol{B}\mathrm{d}\lambda. \tag{B.5}$$

Finally, the discrete-time state-space model becomes

$$\boldsymbol{x}[k+1] = \boldsymbol{A}_{\mathrm{d}}\boldsymbol{x}[k] + \boldsymbol{B}_{\mathrm{d}}\boldsymbol{u}[k] \tag{B.6}$$

where

$$\boldsymbol{A}_{\mathrm{d}} = e^{Ah} \tag{B.7}$$

$$\boldsymbol{B}_{\mathrm{d}} = \int_{0}^{h} e^{A\tau}\boldsymbol{B}\mathrm{d}\tau. \tag{B.8}$$

MATLAB

The matrices $\boldsymbol{A}_{\mathrm{d}}$ and $\boldsymbol{B}_{\mathrm{d}}$ can be computed in Matlab as

```
[Ad,Bd] = c2d(A, B, h)
```

The function `c2d` discretizes the continuous-time dynamic system model using zero-order hold on the inputs.

Explicit Solution

Consider the state-transition matrix

$$\boldsymbol{\Phi} = \mathrm{e}^{Ah}. \tag{B.9}$$

If the inverse matrix $\boldsymbol{A}^{-1}$ for the system (B.3) exists, an explicit solution of (B.8) is found by

$$\boldsymbol{A}_{\mathrm{d}} = \boldsymbol{\Phi} \tag{B.10}$$

$$\boldsymbol{B}_{\mathrm{d}} = \boldsymbol{A}^{-1}(\boldsymbol{\Phi} - \boldsymbol{I}_n)\boldsymbol{B}. \tag{B.11}$$

Example B.1 (Discretization of a First-order Linear System)
Consider the SISO linear system

$$\dot{x} = ax + bu \tag{B.12}$$

$$y = cx + du. \tag{B.13}$$

Application of (B.10) and (B.11) yields

$$x[k+1] = e^{ah}x[k] + \frac{b}{a}(e^{ah} - 1)u[k] \tag{B.14}$$

$$y[k] = cx[k] + du[k]. \tag{B.15}$$

B.1.2 Computation of the Transition Matrix

The transition matrix $\boldsymbol{\Phi}$ can be computed numerically as

$$\boldsymbol{\Phi} = e^{\boldsymbol{A}h} = \boldsymbol{I}_n + \boldsymbol{A}h + \frac{1}{2!}\boldsymbol{A}^2h^2 + \cdots + \frac{1}{m!}\boldsymbol{A}^m h^m + \cdots. \tag{B.16}$$

MATLAB

The transition matrix can be computed in Matlab as

```
PHI = expm(A * h)
```

If the inverse matrix $\boldsymbol{A}^{-1}$ exists

$$\boldsymbol{A}_{\mathrm{d}} = \boldsymbol{\Phi} = \boldsymbol{I}_n + \boldsymbol{A}h + \frac{1}{2!}\boldsymbol{A}^2h^2 + \cdots + \frac{1}{m!}\boldsymbol{A}^m h^m + \cdots \tag{B.17}$$

$$\boldsymbol{B}_{\mathrm{d}} = \boldsymbol{A}^{-1}(\boldsymbol{\Phi} - \boldsymbol{I}_n)\boldsymbol{B} = \left(h + \frac{1}{2!}\boldsymbol{A}h^2 + \cdots + \frac{1}{m!}\boldsymbol{A}^{m-1}h^m + \cdots\right)\boldsymbol{B}. \tag{B.18}$$

A first-order approximation (Euler discretization) is obtained from (B.17) and (B.18) by setting $m = 1$,

$$\boldsymbol{A}_{\mathrm{d}} \approx \boldsymbol{I}_n + h\boldsymbol{A} \tag{B.19}$$

$$\boldsymbol{B}_{\mathrm{d}} \approx h\boldsymbol{B}. \tag{B.20}$$

B.2 Numerical Integration Methods

In this section numerical solutions to the nonlinear time-varying system

$$\dot{\boldsymbol{x}} = \boldsymbol{f}(\boldsymbol{x}, \boldsymbol{u}, t) \tag{B.21}$$

where the control input $\boldsymbol{u}$ is assumed to be constant over the sampling interval h (zero-order hold), are discussed. Four different methods will be presented.

B.2.1 Euler's Method

A frequently used method for numerical integration is forward Euler

$$\boldsymbol{x}[k+1] = \boldsymbol{x}[k] + h\boldsymbol{f}(\boldsymbol{x}[k], \boldsymbol{u}[k], t_k). \tag{B.22}$$

The global truncation error for Euler's method is of order $O(h)$.

Applying Euler's method to a second-order system

$$\dot{x} = v \tag{B.23}$$

$$m\dot{v} + dv + kx = \tau \tag{B.24}$$

yields

$$v[k+1] = v[k] + h\left(\frac{1}{m}\tau[k] - \frac{d}{m}v[k] - \frac{k}{m}x[k]\right) \tag{B.25}$$

$$x[k+1] = x[k] + hv[k]. \tag{B.26}$$

It should be noted that Euler's method should only be applied to a well-damped second-order system and not an undamped oscillator. In fact an undamped oscillator will yield an unstable solution, as seen from Figure B.1, where the circle in the upper left-hand plot represents the stable region. An undamped oscillator will have eigenvalues on the imaginary axis, which clearly lie outside the circle.

Forward and Backward Euler Integration

A stable method for the undamped second-order system can be obtained by combining the *forward* and *backward* methods of Euler (dotted line in the upper left-hand plot in Figure B.1) according to

$$\text{Forward Euler:} \quad v[k+1] = v[k] + h\left(\frac{1}{m}\tau[k] - \frac{d}{m}v[k] - \frac{k}{m}x[k]\right) \tag{B.27}$$

$$\text{Backward Euler:} \quad x[k+1] = x[k] + hv[k+1]. \tag{B.28}$$

Extension to Nonlinear Systems

The combined forward and backward methods can be extended to the more general nonlinear system

$$\dot{\boldsymbol{\eta}} = \boldsymbol{J}_\Theta(\boldsymbol{\eta})\boldsymbol{\nu} \tag{B.29}$$

$$\dot{\boldsymbol{\nu}} = \boldsymbol{M}^{-1}[\boldsymbol{Bu} - \boldsymbol{C}(\boldsymbol{\nu})\boldsymbol{\nu} - \boldsymbol{D}(\boldsymbol{\nu})\boldsymbol{\nu} - \boldsymbol{g}(\boldsymbol{\eta})]. \tag{B.30}$$

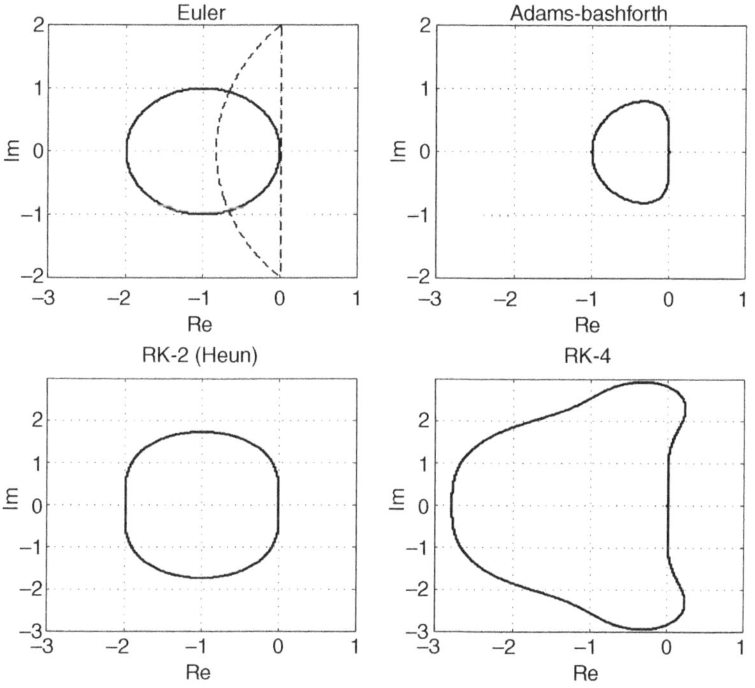

Figure B.1 Stability regions for the Euler, Adams–Bashford, RK-2 and RK-4 methods.

The discrete-time state-space model is

$$\boldsymbol{v}[k+1] = \boldsymbol{v}[k] + h\boldsymbol{M}^{-1}[\boldsymbol{B}\boldsymbol{u}[k] - \boldsymbol{C}(\boldsymbol{v}[k])\boldsymbol{v}[k] - \boldsymbol{D}(\boldsymbol{v}[k])\boldsymbol{v}[k] - \boldsymbol{g}(\boldsymbol{\eta}[k])] \tag{B.31}$$

$$\boldsymbol{\eta}[k+1] = \boldsymbol{\eta}[k] + h\boldsymbol{J}_{\Theta}(\boldsymbol{\eta}[k])\boldsymbol{v}[k+1]. \tag{B.32}$$

B.2.2 Adams–Bashford's Second-order Method

Adams–Bashford integration is more computationally intensive than the schemes of Euler. For instance, the two-step Adams–Bashford integration

$$\boldsymbol{x}[k+1] = \boldsymbol{x}[k] + h\left[\frac{3}{2}\boldsymbol{f}(\boldsymbol{x}[k], \boldsymbol{u}[k], t_k) - \frac{1}{2}\boldsymbol{f}(\boldsymbol{x}[k-1], \boldsymbol{u}[k-1], t_{k-1})\right] \tag{B.33}$$

implies that the old value

$$\dot{\boldsymbol{x}}[k-1] = \boldsymbol{f}(\boldsymbol{x}[k-1], \boldsymbol{u}[k-1], t_{k-1}) \tag{B.34}$$

must be stored. The global truncation error for this method is of order $O(h^2)$. The advantage with this method compared to Euler integration is seen from Figure B.1.

B.2.3 Runge–Kutta Second-order Method

Heun's integration method or Runge–Kutta's second-order method (RK-2) is implemented as

$$\boldsymbol{k}_1 = \boldsymbol{f}(\boldsymbol{x}[k], \boldsymbol{u}[k], t_k) \tag{B.35}$$

$$\boldsymbol{k}_2 = \boldsymbol{f}(\boldsymbol{x}[k] + h\boldsymbol{k}_1, \boldsymbol{u}[k], t_k + h) \tag{B.36}$$

$$\boldsymbol{x}[k+1] = \boldsymbol{x}[k] + \frac{h}{2}(\boldsymbol{k}_1 + \boldsymbol{k}_2). \tag{B.37}$$

The global truncation error for Heun's method is of order $O(h^2)$.

B.2.4 Runge–Kutta Fourth-order Method

An extension of Heun's integration method to the fourth order (RK-4) is

$$\boldsymbol{k}_1 = h\boldsymbol{f}(\boldsymbol{x}[k], \boldsymbol{u}[k], t_k) \tag{B.38}$$

$$\boldsymbol{k}_2 = h\boldsymbol{f}(\boldsymbol{x}[k] + \boldsymbol{k}_1/2, \boldsymbol{u}[k], t_k + h/2) \tag{B.39}$$

$$\boldsymbol{k}_3 = h\boldsymbol{f}(\boldsymbol{x}[k] + \boldsymbol{k}_2/2, \boldsymbol{u}[k], t_k + h/2) \tag{B.40}$$

$$\boldsymbol{k}_4 = h\boldsymbol{f}(\boldsymbol{x}[k] + \boldsymbol{k}_3/2, \boldsymbol{u}[k], t_k + h) \tag{B.41}$$

$$\boldsymbol{x}[k+1] = \boldsymbol{x}[k] + \frac{1}{6}(\boldsymbol{k}_1 + 2\boldsymbol{k}_2 + 2\boldsymbol{k}_3 + \boldsymbol{k}_4). \tag{B.42}$$

The global truncation error for the RK-4 method is of order $O(h^4)$.

B.3 Numerical Differentiation

Numerical differentiation is usually sensitive to noisy measurements. Nevertheless, a reasonable estimate $\dot{\eta}_{\rm f}$ of the time derivative $\dot{\eta}$ of a signal η can be obtained by using a *filtered differentiation*. The simplest filter is obtained by the first-order high-pass filter

$$\dot{\eta}_{\rm f}(s) = \frac{Ts}{Ts+1}\,\eta(s) \tag{B.43}$$

corresponding to the continuous-time system

$$\dot{x} = ax + bu \tag{B.44}$$

$$y = cx + du \tag{B.45}$$

with $u = \eta$, $y = \dot{\eta}_f$, $a = -1/T$, $b = 1/T$, $c = -1$ and $d = 1$. Using the results from Example B.1, the following discrete-time filter equations can be used to differentiate a time-varying signal

$$x[k+1] = e^{-h/T} x[k] - (e^{-h/T} - 1)u[k] \tag{B.46}$$

$$y[k] = -x[k] + u[k]. \tag{B.47}$$

C

Model Transformations

When deriving the equations of motion it is convenient to express inertia, Coriolis-centripetal, damping, gravitational and buoyancy forces in different reference points in the body-fixed frame $\{b\}$, such as the CO, CB, CF and CG to exploit structural properties of the model. Rigid-body, hydrostatic and hydrodynamic data are unusually computed in different reference frames and care must be taken when transforming the data to the CO, which is the reference point used to design the guidance, navigation and motion control systems.

The main tool for 6-DOF model transformations is the *system transformation matrix*, which will be derived below. The system transformation matrix can be used to express the equations of motion about an arbitrarily point but also to transform the data of the model matrices to different reference points.

C.1 Transforming the Equations of Motion to an Arbitrarily Point

The rigid-body translational and rotational parts of the system inertia matrix is decoupled if the coordinate system is located in the CG. Hence, it is convenient to express the rigid-body kinetics in the CG. However, it is common to compute hydrodynamic added mass and damping in the CF or the CO (see Section 2.1.1). Since the model matrices and vectors are computed in several coordinate origins such as the CG, CF and CO it is necessary to transform the data between the different origins such that the equations of motion can be expressed about the CO. This creates a need for a common framework for systematic transformation of model matrices and vectors to a common coordinate origin.

C.1.1 System Transformation Matrix

The system transformation matrix is used to transform the generalized velocities, accelerations and forces between two points in the BODY frame. It will be assumed that NED

Handbook of Marine Craft Hydrodynamics and Motion Control, Second Edition. Thor I. Fossen.

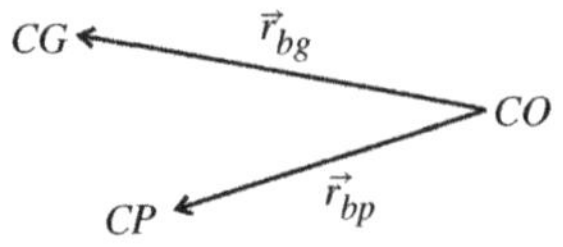

Figure C.1 Definitions of vectors and reference points.

is an approximative inertial frame when deriving the transformation. The system transformation matrix is derived from the distance vector

$$\vec{r}_{\mathrm{np}} = \vec{r}_{\mathrm{nb}} + \vec{r}_{\mathrm{bp}}. \tag{C.1}$$

Time differentiation of $\vec{r}_{\mathrm{bp}}$ in a moving reference frame $\{b\}$ using (3.8) gives

$$\vec{v}_{\mathrm{np}} = \vec{v}_{\mathrm{nb}} + \left(\frac{{}^{\mathrm{b}}\mathrm{d}}{\mathrm{d}t}\vec{r}_{\mathrm{bp}} + \vec{\omega}_{\mathrm{nb}} \times \vec{r}_{\mathrm{bp}} \right). \tag{C.2}$$

For a rigid body

$$\frac{{}^{\mathrm{b}}\mathrm{d}}{\mathrm{d}t}\vec{r}_{\mathrm{bp}} = \vec{0} \tag{C.3}$$

such that

$$\vec{v}_{\mathrm{np}} = \vec{v}_{\mathrm{nb}} + \vec{\omega}_{\mathrm{nb}} \times \vec{r}_{\mathrm{bp}}. \tag{C.4}$$

Consequently,

$$\begin{aligned} \boldsymbol{v}^b_{np} &= \boldsymbol{v}^b_{\mathrm{nb}} + \boldsymbol{\omega}^b_{\mathrm{nb}} \times \boldsymbol{r}^b_{\mathrm{bp}} \\ &= \boldsymbol{v}^b_{\mathrm{nb}} - \boldsymbol{S}(\boldsymbol{r}^b_{\mathrm{bp}})\boldsymbol{\omega}^b_{\mathrm{nb}} \\ &= \boldsymbol{v}^b_{\mathrm{nb}} + \boldsymbol{S}^{\top}(\boldsymbol{r}^b_{\mathrm{bp}})\boldsymbol{\omega}^b_{\mathrm{nb}} \end{aligned} \tag{C.5}$$

where $\boldsymbol{r}^b_{\mathrm{bp}} = [x_{\mathrm{p}}, y_{\mathrm{p}}, z_{\mathrm{p}}]^{\top}$ is the vector from the CO to the CP expressed in $\{b\}$ and $\boldsymbol{\omega}^b_{np} = \boldsymbol{\omega}^b_{\mathrm{nb}}$ is the angular velocity vector.

Definition C.1 (System Transformation Matrix)
The system transformation matrix

$$\boldsymbol{H}(\boldsymbol{r}^b_{\mathrm{bp}}) = \begin{bmatrix} \boldsymbol{I}_3 & \boldsymbol{S}^{\top}(\boldsymbol{r}^b_{\mathrm{bp}}) \\ \boldsymbol{0}_{3\times3} & \boldsymbol{I}_3 \end{bmatrix}, \quad \boldsymbol{H}^{-1}(\boldsymbol{r}^b_{\mathrm{bp}}) = \begin{bmatrix} \boldsymbol{I}_3 & \boldsymbol{S}(\boldsymbol{r}^b_{\mathrm{bp}}) \\ \boldsymbol{0}_{3\times3} & \boldsymbol{I}_3 \end{bmatrix} \tag{C.6}$$

transforms the linear and angular velocity vectors between the two points CO and CP in the $\{b\}$ *frame according to*

$$\begin{bmatrix} \boldsymbol{v}^b_{np} \\ \boldsymbol{\omega}^b_{np} \end{bmatrix} = \boldsymbol{H}(\boldsymbol{r}^b_{\mathrm{bp}}) \begin{bmatrix} \boldsymbol{v}^b_{\mathrm{nb}} \\ \boldsymbol{\omega}^b_{\mathrm{nb}} \end{bmatrix} \tag{C.7}$$

$$\Updownarrow$$

$$\boldsymbol{v}_{\mathrm{np}} = \boldsymbol{H}(\boldsymbol{r}^b_{\mathrm{bp}})\boldsymbol{v} \tag{C.8}$$

where $\boldsymbol{v} := \boldsymbol{v}_{\mathrm{nb}}$. *Hence,*

$$\boldsymbol{v} = \boldsymbol{H}^{-1}(\boldsymbol{r}^b_{\mathrm{bp}})\boldsymbol{v}_{\mathrm{np}}. \tag{C.9}$$

The generalized force vector $\boldsymbol{\tau}$ *can be transformed from the CO to an arbitrary point CP by*

$$\begin{bmatrix} \boldsymbol{f}^b_b \\ \boldsymbol{m}^b_b \end{bmatrix} = \begin{bmatrix} \boldsymbol{f}^b_p \\ \boldsymbol{r}^b_{\mathrm{bp}} \times \boldsymbol{f}^b_p + \boldsymbol{m}^b_p \end{bmatrix} = \begin{bmatrix} \boldsymbol{I}_3 & \boldsymbol{0}_{3\times 3} \\ \boldsymbol{S}(\boldsymbol{r}^b_{\mathrm{bp}}) & \boldsymbol{I}_3 \end{bmatrix} \begin{bmatrix} \boldsymbol{f}^b_p \\ \boldsymbol{m}^b_p \end{bmatrix} \tag{C.10}$$

$$\Updownarrow$$

$$\boldsymbol{\tau} = \boldsymbol{H}^{\top}(\boldsymbol{r}^b_{\mathrm{bp}})\boldsymbol{\tau}^p. \tag{C.11}$$

MATLAB

The system transformation matrix is implemented in the MSS toolbox as

```
function H = Hmtrx(r)
% H = Hmtrx(r)  6x6 system transformation matrix

S = Smtrx(r);
H = [ eye(3)       S'
      zeros(3,3) eye(3) ];
```

C.1.2 Equations of Motion About an Arbitrarily Point

Definition C.1 implies that the equations of motion can be represented at an arbitrary defined point CP by using the transformation matrix $\boldsymbol{H}(\boldsymbol{r}^b_{\mathrm{bp}})$. Consider the nonlinear equations of motion expressed in $\{b\}$ with coordinate origin CO

$$\boldsymbol{M}\dot{\boldsymbol{v}} + \boldsymbol{C}(\boldsymbol{v})\boldsymbol{v} + \boldsymbol{D}(\boldsymbol{v})\boldsymbol{v} + \boldsymbol{g}(\boldsymbol{\eta}) = \boldsymbol{\tau}. \tag{C.12}$$

This expression can be transformed to a point CP in $\{b\}$ by substituting (C.9) into (C.12) and premultiplying both sides of (C.12) with $\boldsymbol{H}^{-\top}(\boldsymbol{r}^b_{\text{bp}})$ to obtain

$$\underbrace{\boldsymbol{H}^{-\top}(\boldsymbol{r}^b_{\text{bp}})\boldsymbol{M}\boldsymbol{H}^{-1}(\boldsymbol{r}^b_{\text{bp}})}_{\boldsymbol{M}^{\text{p}}}\dot{\boldsymbol{v}}_{\text{np}} + \underbrace{\boldsymbol{H}^{-\top}(\boldsymbol{r}^b_{\text{bp}})\boldsymbol{C}(\boldsymbol{v})\boldsymbol{H}^{-1}(\boldsymbol{r}^b_{\text{bp}})}_{\boldsymbol{C}^{\text{p}}(\boldsymbol{v})}\boldsymbol{v}_{\text{np}} + \underbrace{\boldsymbol{H}^{-\top}(\boldsymbol{r}^b_{\text{bp}})\boldsymbol{D}(\boldsymbol{v})\boldsymbol{H}^{-1}(\boldsymbol{r}^b_{\text{bp}})}_{\boldsymbol{D}^{\text{p}}(\boldsymbol{v})}\boldsymbol{v}_{\text{np}} + \underbrace{\boldsymbol{H}^{-\top}(\boldsymbol{r}^b_{\text{bp}})\boldsymbol{g}(\boldsymbol{\eta})}_{\boldsymbol{g}^{\text{p}}(\boldsymbol{\eta})} = \underbrace{\boldsymbol{H}^{-\top}(\boldsymbol{r}^b_{\text{bp}})\boldsymbol{\tau}}_{\boldsymbol{\tau}^{\text{p}}}. \tag{C.13}$$

C.2 Matrix and Vector Transformations

The result (C.13) implies that a matrix $\boldsymbol{A}$ and a vector $\boldsymbol{\tau}$ can be transformed from the CO to an arbitrarily coordinate origin CP in $\{b\}$ by

$$\boldsymbol{A}^{\text{p}} = \boldsymbol{H}^{-\top}(\boldsymbol{r}^b_{\text{bp}})\boldsymbol{A}\boldsymbol{H}^{-1}(\boldsymbol{r}^b_{\text{bp}}) \tag{C.14}$$

$$\boldsymbol{\tau}^{\text{p}} = \boldsymbol{H}^{-\top}(\boldsymbol{r}^b_{\text{bp}})\boldsymbol{\tau}. \tag{C.15}$$

The superscript p denotes quantities that have been transformed to the coordinate origin CP, while quantiles in the default coordinate origin CO are written without the superscript b to simplify the notation. Consequently, the inverse transformation from the CP to the CO becomes

$$\boldsymbol{A} = \boldsymbol{H}^{\top}(\boldsymbol{r}^b_{\text{bp}})\boldsymbol{A}^{\text{p}}\boldsymbol{H}(\boldsymbol{r}^b_{\text{bp}}) \tag{C.16}$$

$$\boldsymbol{\tau} = \boldsymbol{H}^{\top}(\boldsymbol{r}^b_{\text{bp}})\boldsymbol{\tau}^{\text{p}}. \tag{C.17}$$

The matrix-vector transformations for mass, Coriolis-centripetal, damping and gravitational forces are summarized in Table C.1.

Table C.1 Matrix-vector transformations between the CO and the CP.

Location of the CP

$$\boldsymbol{r}^b_{\mathrm{bp}} = [x_{\mathrm{p}},\ y_{\mathrm{p}},\ z_{\mathrm{p}}]^\top,\ \text{vector from the CO to the CP expressed in } \{b\}$$

System transformation matrix

$$\boldsymbol{H}(\boldsymbol{r}^b_{\mathrm{bp}}) = \begin{bmatrix} \boldsymbol{I}_3 & \boldsymbol{S}^\top(\boldsymbol{r}^b_{\mathrm{bp}}) \\ \boldsymbol{0}_{3\times3} & \boldsymbol{I}_3 \end{bmatrix}, \quad \boldsymbol{H}^{-1}(\boldsymbol{r}^b_{\mathrm{bp}}) = \begin{bmatrix} \boldsymbol{I}_3 & \boldsymbol{S}(\boldsymbol{r}^b_{\mathrm{bp}}) \\ \boldsymbol{0}_{3\times3} & \boldsymbol{I}_3 \end{bmatrix}$$

Transformation: CO → CP

$$\boldsymbol{M}^{\mathrm{p}} = \boldsymbol{H}^{-\top}(\boldsymbol{r}^b_{\mathrm{bp}})\boldsymbol{M}\boldsymbol{H}^{-1}(\boldsymbol{r}^b_{\mathrm{bp}}) \tag{C.18}$$

$$\boldsymbol{C}^{\mathrm{p}}(\boldsymbol{\nu}) = \boldsymbol{H}^{-\top}(\boldsymbol{r}^b_{\mathrm{bp}})\boldsymbol{C}(\boldsymbol{\nu})\boldsymbol{H}^{-1}(\boldsymbol{r}^b_{\mathrm{bp}}) \tag{C.19}$$

$$\boldsymbol{D}^{\mathrm{p}}(\boldsymbol{\nu}) = \boldsymbol{H}^{-\top}(\boldsymbol{r}^b_{\mathrm{bp}})\boldsymbol{D}(\boldsymbol{\nu})\boldsymbol{H}^{-1}(\boldsymbol{r}^b_{\mathrm{bp}}) \tag{C.20}$$

$$\boldsymbol{g}^{\mathrm{p}}(\boldsymbol{\eta}) = \boldsymbol{H}^{-\top}(\boldsymbol{r}^b_{\mathrm{bp}})\boldsymbol{g}(\boldsymbol{\eta}) \tag{C.21}$$

Transformation: CP → CO

$$\boldsymbol{M} = \boldsymbol{H}^\top(\boldsymbol{r}^b_{\mathrm{bp}})\boldsymbol{M}^{\mathrm{p}}\boldsymbol{H}(\boldsymbol{r}^b_{\mathrm{bp}}) \tag{C.22}$$

$$\boldsymbol{C}(\boldsymbol{\nu}) = \boldsymbol{H}^\top(\boldsymbol{r}^b_{\mathrm{bp}})\boldsymbol{C}^{\mathrm{p}}(\boldsymbol{\nu})\boldsymbol{H}(\boldsymbol{r}^b_{\mathrm{bp}}) \tag{C.23}$$

$$\boldsymbol{D}(\boldsymbol{\nu}) = \boldsymbol{H}^\top(\boldsymbol{r}^b_{\mathrm{bp}})\boldsymbol{D}^{\mathrm{p}}(\boldsymbol{\nu})\boldsymbol{H}(\boldsymbol{r}^b_{\mathrm{bp}}) \tag{C.24}$$

$$\boldsymbol{g}(\boldsymbol{\eta}) = \boldsymbol{H}^\top(\boldsymbol{r}^b_{\mathrm{bp}})\boldsymbol{g}^{\mathrm{p}}(\boldsymbol{\eta}) \tag{C.25}$$

D

Non-dimensional Equations of Motion

When designing ship control systems it is often convenient to make the equations of motion non-dimensional such that the model parameters can be treated as constants with respect to the instantaneous speed U defined by

$$U = \sqrt{u^2 + v^2} = \sqrt{(u_0 + \Delta u)^2 + \Delta v^2} \tag{D.1}$$

where u_0 is the *service speed* and Δu and Δv are small perturbations in the surge and sway velocities, respectively. Hence, when moving on a straight path

$$U \approx u_0. \tag{D.2}$$

During course-changing maneuvers the instantaneous speed will decrease due to increased resistance during the turn.

D.1 Non-dimensionalization

Non-dimensionalization is the partial or full removal of units from an equation involving physical quantities by a suitable substitution of variables. The most commonly used methods for marine craft are the *Prime system* of SNAME (1950) and the *Bis system* of Norrbin (1970).

Prime system: This system uses the craft's speed U, the length $L = L_{pp}$ (the length between the fore and aft perpendiculars), the time unit L/U and the mass unit $1/2\rho L^3$ or $1/2\rho L^2 T$ as normalization variables. The latter is inspired by wing theory, where the reference area $A = LT$ is used instead of $A = L^2$. The prime system cannot be used for low-speed applications such as dynamic ship positioning, since normalization of the velocities u, v and w implies dividing by the speed U, which can be zero for a dynamically positioned ship. As a consequence, the prime system is mostly used in ship maneuvering.

Handbook of Marine Craft Hydrodynamics and Motion Control, Second Edition. Thor I. Fossen.

Bis System: This system can be used for zero speed as well as high-speed applications since division of speed U is avoided. The bis system is based on the use of the length $L = L_{pp}$, with the time unit $\sqrt{L/g}$ such that speed becomes $\sqrt{Lg} > 0$. In addition, the body mass density ratio $\mu = m/\rho\nabla$, where m is the mass unit and ∇ is the hull contour displacement, is applied. The density ratio μ takes the following values:

$\mu < 1$ Underwater vehicles (ROVs, AUVs and submarines)
$\mu = 1$ Floating ships/rigs and neutrally buoyant underwater vehicles
$\mu > 1$ Heavy torpedoes(typically$\mu = 1.3$–1.5).

The prime and bis system variables used for non-dimensionalization are given in Table D.1. The non-dimensional quantities will be distinguished from those with dimension by applying the notation $(\cdot)'$ for the Prime system and $(\cdot)''$ for the Bis system.

D.1.1 Non-dimensional Hydrodynamic Coefficients

The procedure of making a hydrodynamic coefficient dimensionless is easiest illustrated by an example. For instance, the hydrodynamic coefficient Y_r can be made nondimensional by using the prime and bis systems. First, let us determine the dimension of Y_r. Consider

$$\underbrace{Y}_{[\mathrm{N}=\mathrm{kg\ m\ s^{-2}}]} = \underbrace{Y_r}_{[\mathrm{unknown}]}\underbrace{r}_{[\mathrm{rad\ s^{-1}}]}.$$

Table D.1 Prime and Bis system variables used for non-dimensionalization.

Unit	Prime system I	Prime system II	Bis system
Length	L	L	L
Mass	$\frac{1}{2}\rho L^3$	$\frac{1}{2}\rho L^2 T$	$\mu\rho\nabla$
Inertia moment	$\frac{1}{2}\rho L^5$	$\frac{1}{2}\rho L^4 T$	$\mu\rho\nabla L^2$
Time	$\frac{L}{U}$	$\frac{L}{U}$	$\sqrt{L/g}$
Reference area	L^2	LT	$\mu\,\frac{2\nabla}{L}$
Position	L	L	L
Angle	1	1	1
Linear velocity	U	U	$\sqrt{Lg}$
Angular velocity	$\frac{U}{L}$	$\frac{U}{L}$	$\sqrt{g/L}$
Linear acceleration	$\frac{U^2}{L}$	$\frac{U^2}{L}$	g
Angular acceleration	$\frac{U^2}{L^2}$	$\frac{U^2}{L^2}$	$\frac{g}{L}$
Force	$\frac{1}{2}\rho U^2 L^2$	$\frac{1}{2}\rho U^2 LT$	$\mu\rho g\nabla$
Moment	$\frac{1}{2}\rho U^2 L^3$	$\frac{1}{2}\rho U^2 L^2 T$	$\mu\rho g\nabla L$

Hence, the unknown dimension must be [kg m s^{-1}] since [rad] is a non-dimensional unit. The non-dimensional values Y'_{r} and Y''_{r} are found by using the *mass*, *length* and *time* entries from Table D.1. Consequently,

$$Y'_{\mathrm{r}} = \frac{Y_{\mathrm{r}}}{\frac{\left[\frac{1}{2}\rho L^3\right][L]}{[L/U]}} = \frac{1}{\frac{1}{2}\rho L^3 U} Y_{\mathrm{r}} \tag{D.3}$$

$$Y''_{\mathrm{r}} = \frac{Y_{\mathrm{r}}}{\frac{[\mu\rho\nabla][L]}{[\sqrt{L/g}]}} = \frac{1}{\mu\rho\nabla\sqrt{Lg}} Y_{\mathrm{r}}. \tag{D.4}$$

For a floating ship, Y''_{r} can be further simplified since $\mu = 1$ and $m = \rho\nabla$. Hence,

$$Y''_{\mathrm{r}} = \frac{1}{m\sqrt{Lg}} Y_{\mathrm{r}}. \tag{D.5}$$

D.1.2 Non-dimensional Nomoto Models

The gain and time constants in Nomoto's first- and second-order models can be made invariant with respect to U and L by using

$$K' = (L/U)\,K \tag{D.6}$$

$$T' = (U/L)\,T. \tag{D.7}$$

This suggests that the first-order Nomoto model

$$T\dot{r} + r = K\delta \tag{D.8}$$

can be expressed as

$$(L/U)\,T'\,\dot{r} + r = (U/L)\,K'\,\delta \tag{D.9}$$

or

$$\dot{r} = -\left(\frac{U}{L}\right)\frac{1}{T'}r + \left(\frac{U}{L}\right)^2 \frac{K'}{T'}\delta. \tag{D.10}$$

This representation is quite useful since the non-dimensional gain and time constants will typically be in the range $0.5 < K' < 2$ and $0.5 < T' < 2$ for most ships. An extension to Nomoto's second-order model is given by

$$(L/U)^2\,T'_1\,T'_2\,\psi^{(3)} + (L/U)\,(T'_1 + T'_2)\,\ddot{\psi} + \dot{\psi} = (U/L)\,K'\,\delta + K'\,T'_3\,\dot{\delta} \tag{D.11}$$

where the non-dimensional time constants T_i' are defined as $T_i' = T_i\ (U/L)$ for $i = 1, 2, 3$ and the non-dimensional gain constant is $K' = (L/U)\ K$.

D.1.3 Non-dimensional Maneuvering Models

Consider the linear maneuvering model (6.138) in non-dimensional form

$$\boldsymbol{M}'\dot{\boldsymbol{\nu}}' + \boldsymbol{N}'\boldsymbol{\nu}' = \boldsymbol{b}'\delta'. \tag{D.12}$$

Transforming the states $\boldsymbol{\nu}'$ and control input δ' to dimensional quantities yields

$$(\boldsymbol{T}\boldsymbol{M}'\boldsymbol{T}^{-1})\dot{\boldsymbol{\nu}} + \frac{U}{L}(\boldsymbol{T}\boldsymbol{N}'\boldsymbol{T}^{-1})\boldsymbol{\nu} = \frac{U^2}{L}\boldsymbol{T}\boldsymbol{b}'\delta \tag{D.13}$$

where $\delta = \delta'$ and

$$\boldsymbol{T} = \text{diag}\{1,\ 1/L\}. \tag{D.14}$$

Expanding (D.13) yields

$$\begin{bmatrix} m_{22}' & Lm_{26}' \\ \frac{1}{L}m_{62}' & m_{66}' \end{bmatrix} \begin{bmatrix} \dot{v} \\ \dot{r} \end{bmatrix} + \frac{U}{L} \begin{bmatrix} n_{22}' & Ln_{26}' \\ \frac{1}{L}n_{62}' & n_{66}' \end{bmatrix} \begin{bmatrix} v \\ r \end{bmatrix} = \frac{U^2}{L} \begin{bmatrix} b_1' \\ \frac{1}{L}\ b_2' \end{bmatrix} \delta \tag{D.15}$$

where m_{ij}', d_{ij}' and b_i' are defined according to prime systems I or II in Table D.1.

D.2 6-DOF Procedure for Non-dimensionalization

A systematic procedure for non-dimensionalization of 6-DOF models is easily derived by defining the transformation matrix

$$\boldsymbol{T} = \text{diag}\left\{\ 1,\ 1,\ 1,\ \frac{1}{L},\ \frac{1}{L},\ \frac{1}{L}\ \right\} \tag{D.16}$$

$$\boldsymbol{T}^{-1} = \text{diag}\{\ 1,\ 1,\ 1,\ L,\ L, L\ \}. \tag{D.17}$$

Consider the non-dimensional model

$$\boldsymbol{M}'\dot{\boldsymbol{\nu}}' + \boldsymbol{N}'\boldsymbol{\nu}' + \boldsymbol{G}'\boldsymbol{\eta}' = \boldsymbol{\tau}'. \tag{D.18}$$

When designing marine craft simulators and gain-scheduled controllers it is convenient to perform the numerical integration in real time using dimensional time. Consequently, it is standard to use the non-dimensional hydrodynamic coefficients as input to the simulator or controller, while the states $\boldsymbol{\nu}, \boldsymbol{\eta}$ and input $\boldsymbol{\tau}$ should have their physical dimensions. For the *Prime system*this is obtained by applying the following transformation to (D.18). This is mathematically equivalent to

$$\boldsymbol{M}'\left(\frac{L}{U^2}\boldsymbol{T}^{-1}\dot{\boldsymbol{\nu}}\right)+\boldsymbol{N}'\left(\frac{1}{U}\boldsymbol{T}^{-1}\boldsymbol{\nu}\right)+\boldsymbol{G}'\left(\frac{1}{L}\boldsymbol{T}^{-1}\boldsymbol{\eta}\right)=\frac{1}{\frac{1}{2}\rho U^2L^2}\boldsymbol{T}\boldsymbol{\tau} \tag{D.19}$$

such that

$$(\boldsymbol{T}\boldsymbol{M}'\boldsymbol{T}^{-1})\dot{\boldsymbol{\nu}}+\left(\frac{U}{L}\right)(\boldsymbol{T}\boldsymbol{N}'\boldsymbol{T}^{-1})\boldsymbol{\nu}+\left(\frac{U}{L}\right)^2(\boldsymbol{T}\boldsymbol{G}'\boldsymbol{T}^{-1})\boldsymbol{\eta}=\frac{1}{\frac{1}{2}\rho L^3}\boldsymbol{T}^2\boldsymbol{\tau}. \tag{D.20}$$

Consequently,

$$\boldsymbol{M}=\frac{\rho L^3}{2}\boldsymbol{T}^{-2}(\boldsymbol{T}\boldsymbol{M}'\boldsymbol{T}^{-1}) \tag{D.21}$$

$$\boldsymbol{N}=\frac{\rho L^2U}{2}\boldsymbol{T}^{-2}(\boldsymbol{T}\boldsymbol{N}'\boldsymbol{T}^{-1}) \tag{D.22}$$

$$\boldsymbol{G}=\frac{\rho LU^2}{2}\boldsymbol{T}^{-2}(\boldsymbol{T}\boldsymbol{G}'\boldsymbol{T}^{-1}). \tag{D.23}$$

Notice that $\boldsymbol{\nu}, \boldsymbol{\eta}$ and the input vector $\boldsymbol{\tau}$ now have physical dimensions while $\boldsymbol{M}', \boldsymbol{N}'$ and $\boldsymbol{G}'$ are non-dimensional. Similarly, *Bis system* scaling with $\mu = 1$ gives

$$(\boldsymbol{T}\boldsymbol{M}''\boldsymbol{T}^{-1})\dot{\boldsymbol{\nu}}+\sqrt{\frac{g}{L}}(\boldsymbol{T}\boldsymbol{N}''\boldsymbol{T}^{-1})\boldsymbol{\nu}+\frac{g}{L}(\boldsymbol{T}\boldsymbol{G}''\boldsymbol{T}^{-1})\boldsymbol{\eta}=\frac{1}{m}\boldsymbol{T}^2\boldsymbol{\tau}. \tag{D.24}$$

Consequently,

$$\boldsymbol{M}=m\boldsymbol{T}^{-2}(\boldsymbol{T}\boldsymbol{M}''\boldsymbol{T}^{-1}) \tag{D.25}$$

$$\boldsymbol{N}=m\sqrt{\frac{g}{L}}\boldsymbol{T}^{-2}(\boldsymbol{T}\boldsymbol{N}''\boldsymbol{T}^{-1}) \tag{D.26}$$

$$\boldsymbol{G}=m\frac{g}{L}\boldsymbol{T}^{-2}(\boldsymbol{T}\boldsymbol{G}''\boldsymbol{T}^{-1}). \tag{D.27}$$

The 6-DOF non-dimensionalization procedure is summarized in Table D.2.

Table D.2 Variables used for non-dimensionalization of 6-DOF models.

	Prime system	Bis system
Acceleration	$\dot{\boldsymbol{\nu}} = \frac{U^2}{L}\boldsymbol{T}\dot{\boldsymbol{\nu}}'$	$\dot{\boldsymbol{\nu}} = g\ \boldsymbol{T}\dot{\boldsymbol{\nu}}''$
Velocity	$\boldsymbol{\nu} = U\boldsymbol{T}\boldsymbol{\nu}'$	$\boldsymbol{\nu} = \sqrt{Lg}\boldsymbol{T}\boldsymbol{\nu}''$
Position/attitude	$\boldsymbol{\eta} = L\ \boldsymbol{T}\boldsymbol{\eta}'$	$\boldsymbol{\eta} = L\ \boldsymbol{T}\boldsymbol{\eta}''$
Control forces/moments	$\boldsymbol{\tau} = \frac{1}{2}\rho U^2 L^2 \boldsymbol{T}^{-1}\boldsymbol{\tau}'$	$\boldsymbol{\tau} = \mu\rho g\nabla \boldsymbol{T}^{-1}\boldsymbol{\tau}''$

References

2nd ISSC (1964). Report of the Seakeeping Committee. In: *Proceedings of the 2nd International Ship and Offshore Structures Congress*. Delft, August 1964, p. 19.

12th ITTC (1969). Report of the Seakeeping Committee. In: *Proceedings of the 12th International Towing Tank Conference*. Rome, September 1969, pp. 775–779.

14th ITTC (1975). Discussion and Recommendations for an ITTC 1975 Maneuvering Trial Code. In: *Proceedings of the 14th International Towing Tank Conference*. Ottawa, September 1975, pp. 348–365.

15th ITTC (1978). Report of the Seakeeping Committee. In: *Proceedings of the 15th International Towing Tank Conference*. The Hague, Netherlands, September 1978, pp. 55–70.

17th ITTC (1984). Report of the Seakeeping Committee. In: *Proceedings of the 17th International Towing Tank Conference*. The Hague, Netherlands, September 1984, pp. 531–534.

Aarset, M. F., J. P. Strand and T. I. Fosen (1998). Nonlinear Vectorial Observer Backstepping With Integral Action and Wave Filtering for Ships. In: *Proceedings of the IFAC Conference on Control Applications in Marine Systems (CAMS'98)*, pp. 83–89.

Abkowitz, M. A. (1964). Lectures on Ship Hydrodynamics – Steering and Maneuverability. Technical Report Hy-5. Hydro- and Aerodynamics Laboratory. Denmark Technical University, Lyngby, Denmark.

Acheson, D. J. (1990). *Elementary Fluid Dynamics*. Oxford Applied Mathematics and Computing Science Series. Claredon Press, Oxford.

Allensworth, T. (1999). A Short History of Sperry Marine. Internet. https://www.sperrymarine.com/corporate-history/sperry-marine [Accessed January 28, 2020].

Aranovskiy, S.V., A. A. Bobtsov, A. S. Kremlev and G. V. Luk'yanova (2007). A Robust Algorithm for Identification of the Frequency of a Sinusoidal Signal. *Journal of Computer and Systems Sciences International* 46(3), 371–376.

Aranda, J., J. M. de-la-Cruz, J. M. Diaz, B. de-Andres, P. Ruiperez, S. Esteban and J. M. Giron (2000). Modelling of a High Speed Craft by a Nonlinear Least Squares Method with Constraints. In: *Proceedings of the 5th IFAC Conference on Maneuvering and Control of Marine Craft (MCMC'2000)*. Aalborg, Denmark, pp. 227–232.

Arimoto, S. and F. Miyazaki (1984). Stability and Robustness of PID Feedback Control for Robot Manipulators of Sensory Capability. In: *Proceedings of the 1st International Symposium on Robotics Research* (M. Brady and R. Paul, Eds), pp. 783–799. MIT Press.

Arstein, Z. (1983). Stabilization with Relaxed Controls. *Nonlinear Analysis* 7, 1163–1173.

Asare, H. and D. Wilson (1986). Design of Computed Torque Model Reference Adaptive Control for Space-Based Robotic Manipulators. ASME, Winter Annual Meeting (WAM), pp. 195–204.

Athans, M. and Falb, P. L. (Eds) (1966). *Optimal Control*. McGraw-Hill Book Company. New York, NY.

Handbook of Marine Craft Hydrodynamics and Motion Control, Second Edition. Thor I. Fossen.

Bailey, P. A., W. G. Price and P. Temarel (1998). A Unified Mathematical Model Describing the Maneuvering of a Ship Travelling in a Seaway. *Transactions of the Royal Institute of Naval Architects* Transactions RINA-140, 131–149.

Baitis, A. E. (1980). The Development and Evaluation of a Rudder Roll Stabilization System for the WHEC Hamiltonian Class. Technical Report DTNSRDC. Naval Ship Research and Development Center. Bethesda, MD.

Baitis, E., D. A. Woolaver and T. A. Beck (1983). Rudder Roll Stabilization for Coast Guards Cutters and Frigates. *Naval Engineers Journal*, pp. 267–282.

Baitis, E., D. A. Woolaver and T. A. Beck (1989). Ship Roll Stabilization in the U.S. Navy. *Naval Engineers Journal*, pp. 43–53.

Balchen, J. G., N. A. Jenssen and S. Sælid (1976). Dynamic Positioning Using Kalman Filtering and Optimal Control Theory. In: *Proceedings of the IFAC/IFIP Symposium on Automation in Offshore Oil Field Operation*. Bergen, Norway, pp. 183–186.

Balchen, J. G., N. A. Jenssen and S. Sælid (1980a). Dynamic Positioning of Floating Vessels Based on Kalman Filtering and Optimal Control. In: *Proceedings of the 19th IEEE Conference on Decision and Control*. New York, NY, pp. 852–864.

Balchen, J. G., N. A. Jenssen, E. Mathisen and S. Sælid (1980b). Dynamic Positioning System Based on Kalman Filtering and Optimal Control. *Modeling, Identification and Control* 1(3), 135–163.

Batista, P, C. Silvestre, and P. J. Oliveira (2011a). GES Attitude Observers–Part I: Multiple General Vector Observations. In: *Proceedings of the IFAC World Congress*, pp. 2985–2990, Milan, Italy.

Batista, P, C. Silvestre, and P. J. Oliveira (2011b). GES Attitude Observers–Part II: Single Vector Observations. In: *Proceedings of the IFAC World Congress*, pp. 2991–2996, Milan, Italy.

Barbălat (1959). Systèmes d'Équations Différentielles d'Oscillations Non Linéaires. *Revue de Mathématiques Pures et Appliquées*.4(2), 267–270. Académie de la République Populaire Roumaine (in French).

Barbier, C., P. Sen and M. Downie (1994). Parallel Dynamic Programming and Ship Voyage Management. *Concurrency: Practice and Experience* 6(8), 673–696.

Barbour, N. and G. Schmidt (1998). Inertial Sensor Technology Trends. In: *Proceedings of the Workshop on Autonomous Underwater Vehicles*, pp. 52–55.

Barnitsas, M. M., D. Ray and P. Kinley (1981). *KT, KQ and Efficiency Curves for the Wageningen B-Series Propellers*. <Open Access: http://deepblue.lib.umich.edu/handle/2027.42/3557 >.

Beard, R. W. and T. W. McLain (2012). *Small Unmanned Aircraft. Theory and Practice*. Princeton University Press.

Bech, M. I. (1968). The Reversed Spiral Test as Applied to Large Ships. In: *Shipping World and Shipbuilder*, pp. 1753–1754.

Bech, M. I. and L. Wagner Smith (1969). Analogue Simulation of Ship Maneuvers. Technical Report Hy-14. Hydro- and Aerodynamics Laboratory. Lyngby, Denmark.

Belleter, D. J., R. Galeazzi and T. I. Fossen (2015). Experimenal Verification of a Globally Exponentially Stable Nonlinear Wave Encounter Frequency Estimator. *Ocean Engineering* 97(15), 48–56.

Bennet, S. (1979). *A History of Control Engineering 1800–1930*. Peter Peregrinus. London.

Bennett, S. (1991). Ship Stabilization: History. In: *Concise Encyclopedia of Traffic and Transportation Systems* (Markos Papageorgiou, Ed.), pp. 454–459. Pergamon Press.

Berge, S. P. and T. I. Fossen (1997). Robust Control Allocation of Overactuated Ships: Experiments With a Model Ship. In: *Proceedings of the 4th IFAC Conference on Manoeuvring and Control of Marine Craft*. Brijuni, Croatia, pp. 166–171.

Berge, S. P. and T. I. Fossen (2000). On the Properties of the Nonlinear Ship Equations of Motion. *Journal of Mathematical and Computer Modelling of Dynamical Systems* 6(4), 365–381.

Bertram, V. (2004). *Practical Ship Hydrodynamics*. Butterworth Heinemann.

Bhat, S. P. and D. S. Bernstein (2000). A Topological Obstruction to Continuous Global Stabilization of Rotational Motion and the Unwinding Phenomenon. *Systems and Control Letters* 39(1), 63–70.

Bhattacharyya, R. (1978). *Dynamics of Marine Vehicles*. John Wiley & Sons, Inc. New York, NY.

Blanke, M. (1981). Ship Propulsion Losses Related to Automated Steering and Prime Mover Control. PhD thesis. The Technical University of Denmark, Lyngby.

Blanke, M. (1994). Optimal Speed Control for Cruising. In: *Proceedings of the 3rd Conference on Marine Craft Maneuvering and Control*, Southampton, UK.

Blanke, M. (1996). Uncertainty Models for Rudder-Roll Damping Control. In: *Proceedings of the IFAC World Congress*. Vol. Q. Elsevier. San Francisco, CA, pp. 285–290.

Blanke, M. and A. Christensen (1993). Rudder-Roll Damping Autopilot Robustness due to Sway-Yaw-Roll Couplings. In: *Proceedings of the 10th International Ship Control Systems Symposium (SCSS'93)*. Ottawa, Canada, pp. A.93–A.119.

Blanke, M. and A. G. Jensen (1997). Dynamic Properties of Container Vessel with Low Metacentric Height. *Transactions of the Institute of Measurement and Control* 19(2), 78–93.

Blanke, M. and A. Tiano (1997). Multivariable Identification of Ship Steering and Rol Motions. *Transactions of the Institute of Measurement and Control* 19(2), 62–77.

Blanke, M., P. Haals and K. K. Andreasen (1989). Rudder Roll Damping Experience in Denmark. In: *Proceedings of the IFAC Workshop on Expert Systems and Signal Processing in Marine Automation*. Lyngby, Denmark, pp. 149–160.

Blanke, M., J. Adrian, K.-E. Larsen and J. Bentsen (2000a). Rudder Roll Damping in Coastal Region Sea Conditions. In: *Proceedings 5th IFAC Conference on Manoeuvring and Control of Marine Craft (MCMC'00)*. Aalborg, Denmark, 23–25 August, pp. 39–44.

Blanke, M., K. P. Lindegaard and T. I. Fossen (2000b). Dynamic Model for Thrust Generation of Marine Propellers. In: *Proceedings of the IFAC Conference of Manoeuvreing of Marine Craft (MCMC'00)*, Aalborg, Denmark, 23–25 August.

Blendermann, W. (1986). Die Windkräfte am Schiff. Technical Report Bericht Nr. 467. Institut für Schiffbau der Universität Hamburg (in German).

Blendermann, W. (1994). Parameter Identification of Wind Loads on Ships. *Journal of Wind Engineering and Industrial Aerodynamics* 51, pp. 339–351.

Bordignon, K. A. and W. C. Durham (1995). Closed-Form Solutions to Constrained Control Allocation Problem. *Journal of Guidance, Control and Dynamics* 18(5), pp. 1000–1007.

Breivik, M. (2010). Topics in Guided Motion Control. PhD thesis. Department of Engineering Cybernetics, Norwegian University of Science and Technology, Trondheim, Norway.

Breivik, M. and T. I. Fossen (2004a). Path following for marine surface vessels. In: *Proceedings of the OTO'04*. Kobe, Japan.

Breivik, M. and T. I. Fossen (2004b). Path Following of Straight Lines and Circles for Marine Surface Vessels. In: *Proceedings of the IFAC CAMS'04*. Ancona, Italy.

Breivik, M. and T. I. Fossen (2005a). A Unified Concept for Controlling a Marine Surface Vessel Through the Entire Speed Envelope. In: *Proceedings of the IEEE MED'05*. Cyprus.

Breivik, M. and T. I. Fossen (2005b). Principles of guidance-based path following in 2D and 3D. In: *Proceedings of the CDC-ECC'05*. Seville, Spain.

Breivik, M. and T.I. Fossen (2009). Guidance Laws for Autonomous Underwater Vehicles. In: *Intelligent Underwater Vehicles* (A. V. Inzartsev, Ed.). Chap. 4. I-Tech Education and Publishing. <Open Access: http://books.i-techonline.com>. ISBN 978-953-7619-49-7.

Breivik, M. and G. Sand (2009). Jens Glad Balchen: A Norwegian Pioneer in Engineering Cybernetics. *Modeling, Identification and Control* 30(3), pp. 101–125.

Breivik, M., J. P. Strand and T. I. Fossen (2006). Guided Dynamic Positioning for Fully Actuated Marine Surface Vessels. In: *Proceedings of the 7th IFAC MCMC'06*. Lisbon, Portugal.

Breivik, M., V. E. Hovstein and T. I. Fossen (2008). Ship formation control: A guided leader-follower approach. In: *Proceedings of the 17th IFAC World Congress*. Seoul, Korea.

Bretschneider, C. L. (1959). Wave Variability and Wave Spectra for Wind Generated Gravity Waves. Technical Report. Beach Erosion Board, Corps. of Engineers. 118 (Technical Memo).

Bretschneider, C. L. (1969). *Wave and Wind Loads. Section 12 of Handbook of Ocean and Underwater Engineering*. McGraw-Hill. New York, NY.

Brian, Adrian (2003). *Ship Hydrostatics and Stability*. Butterworth-Heinemann. Oxford, UK.

Brian, D., O. Andersen and J. B. Moore (1989). *Optimal Control: Linear Quadratic Methods*. Prentice Hall. London.

Britting, K. R. (1971). *Inertial Navigation Systems Analysis*. Wiley Interscience.

Brown, R. G. and Y. C. Hwang (2012). *Introduction to Random Signals and Applied Kalman Filtering*. John Wiley & Sons, Inc. New York, NY.

Burger, W. and A. G. Corbet (1960). *Ship Stabilizers. Their Design and Operation in Correcting the Rolling of Ships. A Handbook for Merchant Navy Officers*. Pergamon Press Ltd. London.

Byrnes, C. I. and A. Isidori (1989). New Results and Examples in Nonlinear Feedback Stabilization. *Systems and Control Letters*. 12, 437–442.

Byrnes, C. I., A. Isidori and J. C. Willems (1991). Passivity, Feedback Equivalence, and the Global Stabilization of Minimum Phases Nonlinear Systems. *IEEE Transactions on Automatic Control* 36, 1228–1240.

Børhaug, E. and K. Y. Pettersen (2005). Cross-Track Control for Underactuated Autonomous Vehicles. In: *Proc. of the IEEE CDC'12*, Seville, Spain, pp. 602–608.

Børhaug, E. and K. Y. Pettersen (2006). LOS Path Following for Underwater Underactuated Vehicles. In: *Proceedings of the IFAC MCMC'07*. Lisbon, Portugal.

Børhaug, E., A. Pavlov and K. Y. Pettersen (2008), Integral LOS control for Path Following of Underactuated Marine Surface Vessels in the presence of Constant Ocean Currents, *Proc. of the 47th IEEE Conference on Decision and Control*, pp. 4984–4991, Cancun, Mexico.

Caharija, W., K.Y. Pettersen, M. Bibuli, P. Calado, E. Zereik, J. Braga, J. T. Gravdahl, A. J. Sørensen, M. Milovanovic and G. Bruzzone (2016). Integral Line-of-Sight Guidance and Control of Underactuated Marine Vehicles: Theory, Simulations and Experiments, *IEEE Transactions on Control Systems Technology* 24(5), 1623–1642.

Calvert, S. (1989). Optimal Weather Routing Procedures for Vessels on Oceanic Voyages. PhD thesis. Institute of Marine Studies, Polytechnic South West, UK.

Carley, J. B. (1975). Feasibility Study of Steering and Stabilizing by Rudder. In: *Proceedings of the 4th International Ship Control Systems Symposium (SCSS'75)*. The Hague, Netherlands.

Carlton, J. S. (1994). *Marine Propellers and Propulsion*. Oxford, U.K., Butterworth-Heinemann Ltd.

Chislett, M. S. and J. Strøm-Tejsen (1965a). Planar Motion Mechanism Tests and Full-Scale Steering and Maneuvering Predictions for a Mariner Class Vessel. Technical Report Hy-5. Hydro- and Aerodynamics Laboratory. Lyngby, Denmark.

Chislett, M. S. and J. Strøm-Tejsen (1965b). Planar Motion Mechanism Tests and Full-Scale Steering and Maneuvering Predictions for a Mariner Class Vessel. Technical Report Hy-6. Hydro- and Aerodynamics Laboratory. Lyngby, Denmark.

Chalmers, T. W. (1931). The Automatic Stabilisation of Ships. Chapman and Hall, London.

Chen, C.-T. (2012). *Linear System Theory and Design*. Oxford University Press.

Chiaverini, S. (1993). *Estimate of the two Smallest Singular Values of the Jacobian Matrix: Application to Damped Least-Squares Inverse Kinematics, Journal of Robotic Systems* 10(8), 991–1008.

Chiaverini, S., B. Siciliano and O. Egeland (1994). *Review of the Damped Least-Squares Inverse Kinematics with Experiments on an Industrial Robot Manipulator*. 2(2), 123–134.

Choi, H. T., A. Hanai, S. K. Choi and J. Yuh (2003). Development of an Underwater Robot, ODIN-III. Proc. of the 2003 IEEURSJ International Conference on Intelligent Robots and Systems Las Vegas, Nevada. October 2003.

Chou, J. C. K. (1992). Quaternion Kinematic and Dynamic Differential Equations. *IEEE Transactions on Robotics and Automation* 8(1), 53–64.

Clarke, D. (2003). The foundations of steering and manoeuvring. In: *Proceedings of the IFAC Conference on Maneuvering and Control of Marine Craft (MCMC'03)*. Girona, Spain.

Clarke, D. and J. R. Horn (1997). Estimation of hydrodynamic derivatives. In: *Proceedings of the 11th Ship Control Systems Symposium, Southampton, Volume 2, 14th-19th April 1997*, pp. 275–289.

Coates, E. M., T. I. Fossen and A. Loria (2021). On Ship Heading Control Using the Smallest Signed Angle. Submitted to the *International Journal of Control*.

Cowley, W. E. and T. H. Lambert (1972). The Use of Rudder as a Roll Stabilizer. In: *Proceedings of the 3rd International Ship Control Systems Symposium (SCSS'72)*. Bath, UK.

Cowley, W. E. and T. H. Lambert (1975). Sea Trials on a Roll Stabilizer Using the Ship's Rudder. In: *Proceedings of the 4th International Ship Control Systems Symposium (SCSS'75)*. The Hague, Netherlands.

Craig, J. J. (1989). *Introduction to Robotics*. Addison-Wesley. Reading, MA.

Crassidis, J. L., F. L. Markley and Y. Cheng (2007). Survey of Nonlinear Attitude Estimation Methods. *Journal of Guidance, Control and Dynamics* 30(1), 12–28.

Cristi, R., F. A. Papoulias and A. J. Healey (1990). Adaptive Sliding Mode Control of Autonomous Underwater Vehicles in the Dive Plane. *IEEE Journal of Oceanic Engineering* 15(3), 152–160.

Cummins, W. E. (1962). The Impulse Response Function and Ship Motions. Technical Report 1661. David Taylor Model Basin. Hydrodynamics Laboratory, USA.

Davidson, K. S. M. and L. I. Schiff (1946). Turning and Course Keeping Qualities. *Transactions of SNAME*.

Defant, A. (1961). *Physical Oceanography*. Pergamon Press. London.

De Kat, J. O. and J. E. W. Wichers (1991). Behavior of a Moored Ship in Unsteady Current, Wind and Waves. *Marine Technology* 28(5), 251–264.

de-la-Cruz, J. M., J. Aranda, P. Ruiperez, J. M. Diaz and A. Maron (1998). Identification of the Vertical Plane Motion Model of a High Speed Craft by Model Testing in Irregular Waves. In: *Proceedings of the IFAC Conference on Control Applications in Marine Systems (CAMS'98)*. Fukuoaka, Japan, pp. 257–262.

Dieudonné, J. (1953). Collected French Papers on the Stability of Route of Ships at Sea, 1949–1950 (Translated by H. E. Saunders and E. N. Labouvie) Technical Report DTMB-246. Naval Ship Research and Development Center. Washington D.C.

DnV (1990). *Rules for Classification of Steel Ships: Dynamic Positioning Systems, Part 6, Chapter 7*. Det norske Vertitas, Veritasveien 1, N-1322 Høvik, Norway.

Donha, D. C., D. S. Desanj, M. R. Katebi and M. J. Grimble (1998). H_∞ adaptive Controller for Autopilot Applications. *Spesial Issue on Marine Systems of the International Journal of Adaptive Control and Signal Processing* 12(8), 623–648.

Draper, C. S. (1971). Guidance is Forever. *Navigation* 18(1), 26–50.

Dubins, L. (1957). On Curves of Minimal Length with a Constraint on Average Curvature and with Prescribed Initial and Terminal Positions and Tangents. *American Journal of Mathematics* 79, 497–516.

Durham, W. C. (1993). Constrained Control Allocation. *Journal of Guidance, Control and Dynamics* 16(4), 717–725.

Durham, W. C. (1994a). Attainable Moments for the Constrained Control Allocation Problem. *Journal of Guidance, Control and Dynamics* 17(6), 1371–1373.

Durham, W. C. (1994b). Constrained Control Allocation: Three Moment Problem. *Journal of Guidance, Control and Dynamics* 17(2), 330–336.

Durham, W. C. (1999). Efficient, Near-Optimal Control Allocation. *Journal of Guidance, Control and Dynamics* 22(2), 369–372.

Dyne, G. and P. Trägårdh (1975). Simuleringsmodell för 350 000 tdw tanker i fullast- och ballastkonditioner på djupt vatten. Technical Report 2075-1. Swedish State Shipbuilding Experimental Tank (SSPA). Gothenburg, Sweden.

Egeland, O. and J. T. Gravdahl (2002). *Modeling and Simulation for Automatic Control*. Marine Cybernetics. Trondheim, Norway.

Encarnacao, P. and A. Pascoal (2001). Combined Trajectory Tracking and Path Following for Marine Craft. In: *Proceedings of the 9th Mediterranean Conference on Control and Automation*. Dubbrovnik, Croatia.

Encarnacao, P., A. Pascoal and M. Arcak (2000). Path Following for Autonomous Marine Craft. In: *Proceedings of the 5th MCMC'00*. Aalborg, Denmark, pp. 117–122.

Enshaei, H. and R. Brimingham (2012). Monitoring Dynamic Stability via Ship's Motion Responses. In: *Proceeding of the 11th International Conference on the Stability of Ships and Ocean Vehicles*, Athens, Greece, pp. 707–717.

Esteban, S., J. M. de-la-Cruz, J. M. Giron-Sierra, B. de-Andres, J. M. Diaz and J. Aranda (2000). Fast Ferry Vertical Accelerations Reduction with Active Flaps and T-foil. In: *Proceedings of the 5th Conference on Maneuvering and Control of Marine Craft (MCMC'2000)*. Aalborg, Denmark, pp. 233–238.

Euler, L. (1776). Novi Commentarii Academiae Scientairum Imperialis Petropolitane, Vol. XX.

Faltinsen, O. M. (1990). *Sea Loads on Ships and Offshore Structures*. Cambridge University Press.

Faltinsen, O. M. (2005). *Hydrodynamic of High-Speed Marine Vehicles*. Cambridge University Press.

Faltinsen, O. M. and B. Sortland (1987). Slow Drift Eddy Making Damping of a Ship. *Applied Ocean Research* 9(1), 37–46.

Farrell, J. A. (2008). *Aided Navigation: GPS with High Rate Sensors*. McGraw-Hill. New York, NY.

Farrell, J. A. and M. Barth (1998). *The Global Positioning System and Inertial Navigation*. McGraw-Hill. New York, NY.

Fathi, D. (2004). *ShipX Vessel Responses (VERES)*. Marintek AS, Trondheim, Norway. http://www.sintef.no/.

Fedyaevsky, K. K. and G. V. Sobolev (1963). *Control and Stability in Ship Design*. State Union Shipbuilding House.

Feldman, J. (1979). DTMSRDC Revised Standard Submarine Equations of Motion. Technical Report DTNSRDC-SPD-0393-09. Naval Ship Research and Development Center. Washington D.C.

Fjellstad, O. E. and T. I. Fossen (1994). Quaternion Feedback Regulation of Underwater Vehicles. In: *Proceedings of the 3rd IEEE Conference on Control Applications (CCA'94)*. Glasgow, pp. 857–862.

Fjellstad, O. E., T. I. Fossen and O. Egeland (1992). Adaptive Control of ROVs with Actuator Dynamics and Saturation. In: *Proceedings of the 2nd International Offshore and Polar Engineering Conference (ISOPE)*. San Francisco, CA.

Forssell, B. (1991). *Radio Navigation Systems*. Prentice Hall. Englewood Cliffs, NJ.

Fossen, T. I. (1991). Nonlinear Modeling and Control of Underwater Vehicles. PhD thesis. Department of Engineering Cybernetics, Norwegian University of Science and Technology. Trondheim, Norway.

Fossen, T. I. (1993). Comments on "Hamiltonian Adaptive Control of Spacecraft". *IEEE Transactions on Automatic Control* 38(4), 671–672.

Fossen, T. I. (1994). *Guidance and Control of Ocean Vehicles*. John Wiley & Sons. Ltd. ISBN 0-471-94113-1.

Fossen, T. I. (2000a). A Survey on Nonlinear Ship Control: From Theory to Practice. In: *Proceedings of the IFAC Conference on Maneuvering and Control of Marine Craft* (G. Roberts, Ed.). Elsevier Science. Netherlands, pp. 1–16. Plenary Talk.

Fossen, T. I. (2000b). Recent Developments in Ship Control Systems Design. In *World Superyacht Review*. Sterling Publications Limited, London, pp. 115–116.

Fossen, T. I. (2005). Nonlinear Unified State-Space Model for Ship Maneuvering and Control in a Seaway. *International Journal of Bifurcation and Chaos* IJBC-15(9), 2717–2746. Also in the Proceedings of the ENOC'05 (Plenary Talk), Eindhoven, Netherlands, August 2005, pp. 43–70.

Fossen, T. I. (2012). How to Incorporate Wind, Waves and Ocean Currents in the Marine Craft Equations of Motion. In: *Proceedings of the IFAC Conference on Maneuvering and Control of Marine Craft (MCMC'12)*, 19-21 September, Arenzano, Italy 2012.

Fossen, T. I. and M. Blanke (2000). Nonlinear Output Feedback Control of Underwater Vehicle Propellers using Feedback from Estimated Axial Flow Velocity. *IEEE Journal of Oceanic Engineering* 25(2), 241–255.

Fossen, T. I. and S. P. Berge (1997). Nonlinear Vectorial Backstepping Design for Global Exponential Tracking of Marine Vessels in the Presence of Actuator Dynamics. In: *Proceedings of the IEEE Conference on Decision and Control (CDC'97)*. San Diego, CA, pp. 4237–4242.

Fossen, T. I. and O. E. Fjellstad (1995). Nonlinear Modelling of Marine Vehicles in 6 Degrees of Freedom. *International Journal of Mathematical Modelling of Systems* 1(1), 17–28.

Fossen, T. I. and Å. Grøvlen (1998). Nonlinear Output Feedback Control of Dynamically Positioned Ships Using Vectorial Observer Backstepping. *IEEE Transactions on Control Systems Technology* 6(1), 121–128.

Fossen, T. I. and T. A. Johansen (2006). A Survey of Control Allocation Methods for Ships and Underwater Vehicles. In: *Proceedings of the 14th IEEE Mediterrania Conference on Control and Automation*. Ancona, Italy.

Fossen, T. I. and T. Lauvdal (1994). Nonlinear Stability Analysis of Ship Autopilots in Sway, Roll and Yaw. *Proceedings of the Conference on Marine Craft Maneuvering and Control (MCMC'94)*. Southampton, UK.

Fossen, T. I. and M. Paulsen (1992). Adaptive Feedback Linearization Applied to Steering of Ships. In: *Proceedings of the 1st IEEE Conference on Control Applications*. Dayton, Ohio, pp. 1088–1093.

Fossen, T. I. and T. Perez (2004). Marine Systems Simulator (MSS). <https://github.com/cybergalactic/MSS>.

Fossen, T. I. and T. Perez (2009). Kalman Filtering for Positioning and Heading Control of Ships and Offshore Rigs. *IEEE Control Systems Magazine* 29(6), 32–46.

Fossen, T. I. and S. I. Sagatun (1991). Adaptive Control of Nonlinear Systems: A Case Study of Underwater Robotic Systems. *Journal of Robotic Systems* 8(3), 393–412.

Fossen, T. I. and Ø. N. Smogeli (2004). Nonlinear Time-Domain Strip Theory Formulation for Low-Speed Maneuvering and Station-Keeping. *Modelling, Identification and Control* 25(4), 201–221.

Fossen, T. I. and J. P. Strand (1998). Nonlinear Ship Control (Tutorial Paper). In: *Proceedings of the IFAC CAMS'98*. Fukuoaka, Japan, pp. 1–75.

Fossen, T. I. and J. P. Strand (1999a). A Tutorial on Nonlinear Backstepping: Applications to Ship Control. *Modelling, Identification and Control* 20(2), 83–135.

Fossen, T. I. and J. P. Strand (1999b). Passive Nonlinear Observer Design for Ships Using Lyapunov Methods: Experimental Results with a Supply Vessel. *Automatica* 35(1), 3–16.

Fossen, T. I. and J. P. Strand (2000). Position and Velocity Observer Design. In: *The Ocean Engineering Handbook* (F. El-Hawary, Ed.). Chap. 3, pp. 189–206. CRC Press.

Fossen, T. I. and J. P. Strand (2001). Nonlinear Passive Weather Optimal Positioning Control (WOPC) System for Ships and Rigs: Experimental Results. *Automatica* 37(5), 701–715.

Fossen, T. I., S. I. Sagatun and A. J. Sørensen (1996). Identification of Dynamically Positioned Ships. *Journal of Control Engineering Practice* 4(3), 369–376.

Fossen, T. I., A. Loria and A. Teel (2001). A Theorem for UGAS and ULES of (Passive) Nonautonomous Systems: Robust Control of Mechanical Systems and Ships. *International Journal of Robust and Nonlinear Control* 11, 95–108.

Fossen, T. I., M. Breivik and R. Skjetne (2003). Line-of-Sight Path Following of Underactuated Marine Craft. In: *Proceedings of the IFAC MCMC'03*. Girona, Spain.

Fossen, T. I., T. A. Johansen and T. Perez (2009). A Survey of Control Allocation Methods for Underwater Vehicles. In: *Underwater Vehicles* (A. V. Inzartsev, Ed.). Chap. 7, pp. 109–128. In-Tech Education and Publishing.

Fossen, T. I. and K. Y. Pettersen (2014). On Uniform Semiglobal Exponential Stability (USGES) of Proportional Line-of-Sight Guidance Laws. *Automatica* 50, 2912–2917.

Fotakis, J., M. J. Grimble and B. Kouvaritakis (1982). A Comparison of Characteristic Locus and Optimal Designs for Dynamic Ship Positioning Systems. *IEEE Transactions on Automatic Control* 27(6), 1143–1157.

Fredriksen, E. and K. Y. Pettersen (2006). *Global κ-exponential Waypoint Maneuvering of Ships: Theory and Experiments, Automatica* 42(4), 677–687.

Frenet, F. (1847). Sur les Courbes à Double Courbure, Thése, Toulouse. *Journal de Mathématiques*, (in French).

Freund, E. (1973). Decoupling and Pole Assignment in Nonlinear Systems. *Electronics Letter*.

Fung, P. T-K. and M. J. Grimble (1981). Self-Tuning Control of Ship Positioning Systems. In: *Self-Tuning and Adaptive Control: Theory and Applications* (C. J. Harris and S. A. Billings, Eds.), pp. 308–331. Peter Peregrinus Ltd. on behalf of IEE.

Fung, P. T-K. and M. J. Grimble (1983). Dynamic Ship Positioning Using a Self Tuning Kalman Filter. *IEEE Transactions on Automatic Control* 28(3), 339–349.

Fujii, H. (1960). Experimental Researches on Rudder Performance (1). *J. Zosen Kiokai*, 107, 105–111, (in Japanese).

Fujii, H. and T. Tsuda (1961). Experimental Researches on Rudder Performance (2). *J. Zosen Kiokai* 110, 31–42, (in Japanese).

Fujii, H. and T. Tsuda (1962). Experimental Researches on Rudder Performance (3). *J. Zosen Kiokai* 111, 51–58, (in Japanese).

Gade, K. (2016). The Seven Ways to Find Heading. *The Royal Institute of Navigation* 69, 955–970.

Gelb, A. and W. E. Vander Velde (1968). *Multiple-Input Describing Functions and Nonlinear System Design*. McGraw-Hill Book Company.

Gelb, A., J. F. Kasper, Jr., R. A. Nash, Jr., C. F. Price and A. A. Sutherland, Jr. (1988). *Applied Optimal Estimation*. MIT Press. Boston, MA.

Gertler, M. and S. C. Hagen (1960). Handling Criteria for Surface Ships. Technical Report DTMB-1461. Naval Ship Research and Development Center. Washington D.C.

Gertler, M. and G. R. Hagen (1967). Standard Equations of Motion for Submarine Simulation. Technical Report DTMB-2510. Naval Ship Research and Development Center. Washington D.C.

Giron-Sierra, J. M., S. Esteban, B. De Andres, J. M. Diaz and J. M. Riola (2001). Experimental Study of Controlled Flaps and T-foil for Comfort Improvements of a Fast Ferry. In: *Proceedings of the IFAC Conference on Control Applications in Marine Systems (CAMS'01)*. Glasgow, UK.

Goldstein, H. (1980). *Classical Mechanics*. Addison-Wesley. Reading, MA.

Graver, J. G. (2006). *Underwater Gliders: Dynamics, Control and Design*. PhD thesis. Princeton University, New Jersey, USA.

Grewal, M. S., L. R. Weill and A. P. Andrews (2001). *Global Positioing Systems, Inertial Navigation and Integration*. John Wiley & Sons, Inc. New York, NY.

Grimble, M. J. (1978). Relationship Between Kalman and Notch Filters Used in Dynamic Ship Positioning Systems. *Electronics Letters* EL-14(13), 399–400.

Grimble, M. J. and M. A. Johnson (1989). *Optimal Control and Stochastic Estimation. Theory and Applications*. John Wiley & Sons, Ltd, Chichester, UK.

Grimble, M. J., R. J. Patton and D. A. Wise (1979). The Design of Dynamic Positioning Systems using Extended Kalman Filtering Techniques. In: *Proceedings of OCEANS'79*, pp. 488–497.

Grimble, M. J., R. J. Patton and D. A. Wise (1980). The Design of Dynamic Positioning Control Systems Using Stochastic Optimal Control Theory. *Optimal Control Applications and Methods* 1, 167–202.

Grimble, M. J., R. J. Patton and D. A. Wise (1980b). Use of Kalman Filtering Techniques in Dynamic Ship Positioning Systems. *IEE Proceedings* 127D(3), 93–102.

Grip, H. F., T. I. Fossen, T. A. Johansen and A. Saberi (2011). Attitude Estimation Based on Time-Varying Reference Vectors with Biased Gyro and Vector Measurements. In: *Proceedings of the 2011 IFAC World Congress*. Milan, Italy.

Grip, H. F., T. I. Fossen, T. A. Johansen and A. Saberi (2012). Attitude Estimation Using Biased Gyro and Vector Measurements with Time-Varying Reference Vector. *IEEE Transactions on Automatic Control* 57(5), 1332–1338.

Grip, H. F., T. I. Fossen, T. A. Johansen and A. Saberi (2016). Nonlinear Observer for Attitude, Position and Velocity: Theory and Experiments. Chapter 17 In *Multisensor Attitude Estimation* (H. Fourati and D. C. Belkhia, Eds.), CRC Press (Taylor & Francis Group), pp. 291–314.

Grøvlen, Å. and T. I. Fossen (1996). Nonlinear Control of Dynamic Positioned Ships Using Only Position Feedback: An Observer Backstepping Approach. In: *Proceedings of the 35th IEEE Conference on Decision and Control (CDC'96)*, pp. 3388–3393.

Hagiwara, H. (1989). Weather Routing of Sail-Assisted Motor Vessels. PhD thesis. University of Delft, Netherlands.

Hamel, T. and R. Mahony (2006). Attitude estimation on SO (3) based on direct inertial measurements. In: *Proceedings of the IEEE International Conference Robotics Automation*, pp. 2170–2175, Orlando, FL.

Harvey, A. C. (1989). *Forecasting Structural Time Series Models and the Kalman Filter*. Cambridge University Press.

Hasselmann, K. *et al.* (1973). Measurements of Wind-Wave Growth and Swell Decay during the Joint North Sea Wave Project (JONSWAP). *Deutschen Hydrografischen Zeitschrift*.

Hauser, J. and R. Hindmann (1995). Maneuvering Regulation from Trajectory Tracking: Feedback Linearizable Systems. In: *Proceedings IFAC Symposium on Nonlinear Control Systems Design*. Lake Tahoe, CA, pp. 595–600.

Healey, A. J. and D. Lienard (1993). Multivariable Sliding Mode Control for Autonomous Diving and Steering of Unmanned Underwater Vehicles. *IEEE Journal of Ocean Engineering*, OE-18(3), 327–339.

Healey, A. J. and D. B. Marco (1992). Slow Speed Flight Control of Autonomous Underwater Vehicles: Experimental Results with the NPS AUV II. In: *Proceedings of the 2nd International Offshore and Polar Engineering Conference (ISOPE)*. San Francisco, CA, pp. 523–532.

Healey, A. J., S. M Rock, S. Cody, D. Miles and J. P. Brown (1995). Toward an Improved Understanding of Thruster Dynamics for Underwater Vehicles. *IEEE Journal of Oceanic Engineering* 29(4), 354–361.

Hearns, G. and M. Blanke (1998). Quantitative Analysis and Design of a Rudder Roll Damping Controller. In: *Proceedings IFAC Conference on Control Applications in Marine Systems (CAMS'98)*, pp. 115–120. Fukuoka, Japan.

Hearns, E., R. Katebi and M. Grimble (2000). Robust Fin Roll Stabilizer Controller Design. In: *Proceedings of the 5th IFAC Conference on Maneuvering and Control of Marine Craft*. Aalborg, Denmark, pp. 81–86.

Hegrenæs, Øyvind (2010). Autonomous Navigation for Underwater Vehicles. PhD thesis. Department of Engineering Cybernetics, Norwegian University of Science and Technology, Trondheim, Norway.

Heiskanen, W. A. and H. Moritz (1967). *Physical Geodesy*. Freeman. London.

Hickey, N. A., M. J. Grimble, M. A. Johnson, M. R. Katebi and R. Melville (1997). Robust Fin Roll Stabilization of Surface Ships. In: *Proceedings of the IEEE Conference on Decision and Control (CDC'97)*. Vol. 5, pp. 4225–4230.

Hickey, N. A., M. A. Johnson, M. R. Katebi and M. J. Grimble (2000). PID Controller Optimization for Fin Roll Stabilization. In: *Proceedings of the IEEE Conference of Control Applications (CCA'2000)*. Hawaii, HI, pp. 1785–1790.

Hoerner, S. F. (1965). *Fluid Dynamic Drag*. Hartford House.

Hofmann-Wellenhof, B., H. Lichtenegger and J. Collins (1994). *Global Positioning System: Theory and Practice*. 3rd ed. Springer Verlag. New York, NY.

Holden, C. and T. I. Fossen (2012). A Nonlinear 7-DOF Model for U-Tanks of Arbitrary Shape. *Ocean Engineering* 45, 22–37.

Holzhüter, T. (1992). On the Robustness of Course Keeping Autopilots. In: *Proceedings of IFAC Workshop on Control Applications in Marine Systems (CAMS'92)*. Genova, Italy, pp. 235–244.

Holzhüter, T. (1997). LQG Approach for the High-Precision Track Control of Ships. *IEE Proceedings on Control Theory and Applications* 144(2), 121–127.

Holzhüter, T. and R. Schultze (1996). On the Experience with a High-Precision Track Controller for Commercial Ships. *Control Engineering Practice* 4(3), 343–350.

Holzhüter, T. and H. Strauch (1987). A Commercial Adaptive Autopilot for Ships: Design and Experimental Experience. In: *Proceedings of the 10th IFAC World Congress*. July 27-31, Munich, Germany, pp. 226–230.
Hooft, J. P. (1994). The cross-flow drag on manoeuvring ship. *Oceanic Engineering* 21(3), 329–342.
Hua, M.-D (2010). Attitude Estimation for Accelerated Vehicles using GPS/INS Measurements. *Control Engineering Practice* 18, 723–732.
Hua, M. D., G. Ducard, T. Hamel, R. Mahony and K. Rudin (2014). Implementation of a Nonlinear Attitude Estimator for Aerial Robotic Vehicles. *IEEE Transactions on Control Systems Technology* 22(1), 201–213.
Hughes, P. C. (1986). *Spacecraft Attitude Dynamics*. John Wiley & Sons, Inc. New York, NY.
Humphreys, D. E. and K. W. Watkinson (1978). Prediction of the Acceleration Hydrodynamic Coefficients for Underwater Vehicles from Geometric Parameters. Technical Report NCSL-TR-327-78. Naval Coastal System Center. Panama City, Florida.
Hydroid Inc. (2019). Marine Robots (AUVs) <www.hydroid.com>. Retrieved 12 June 2019.
Ikeda, Y., K. Komatsu, Y. Himeno and N. Tanaka (1976). On Roll Damping Force of Ship: Effects of Friction of Hull and Normal Force of Bilge Keels. *Journal of the Kansai Society of Naval Architects* 142, 54–66.
Imlay, F. H. (1961). The Complete Expressions for Added Mass of a Rigid Body Moving in an Ideal Fluid. Technical Report DTMB 1528. David Taylor Model Basin. Washington D.C.
Ingram, M. J., R. C. Tyce and R. G. Allen (1996). Dynamic Testing of State of the Art Vertical Reference Units. In: *Proceedings of the Oceans 96 MTS/IEEE*. IEEE, pp. 1533–1538.
Isherwood, R. M. (1972). Wind Resistance of Merchant Ships. *RINA Transcripts* 115, 327–338.
Isidori, A. (1989). *Nonlinear Control Systems*. Springer-Verlag. Berlin.
ISO 2631-1 (1997). Mechanical Vibration and Shock. Evaluation of Human Exposure to Whole-Body Vibration – Part 1: General Requirements.
ISO 2631-3 (1985). Evaluation of Human Exposure to Whole-Body Vibration – Part 3: Evaluation of Whole Body z-axis Vertical Vibration in the Frequency Range 0.1 to 0.63 Hz.
Jacobson, D. H. (1977). *Extensions to Linear-Quadratic Control, Optimization and Matrix Theory*. Academic Press. New York, NY.
Jensen, J. J., A. E. Mansour and A. S. Olsen (2004). Estimation of Ship Motions using Closed-Form Expressions. *Ocean Engineering* 31, 61–85.
Johansen, T. A. and T. I. Fossen (2013). Control Allocation - A Survey. Automatica 49, 1087-1103.
Johansen, T. A., T. I. Fossen and S. P. Berge (2004). Constraint Nonlinear Control Allocation with Singularity Avoidance using Sequential Quadratic Programming. *IEEE Transactions on Control Systems Technology* 12, 211–216.
Johansen, T. A., T. I. Fossen and P. Tøndel (2005). Efficient Optimal Constrained Control Allocation via Multi-Parametric Programming. *AIAA Journal of Guidance, Control and Dynamics* 28, 506–515.
Jouffroy, J. and T. I. Fossen (2010). A Tutorial on Incremental Stability Analysis using Contraction Theory. *Modelling, Identification and Control* 31(3), 93–106.
Joung, T.-H., K. Sammut, F. He and S.-K. Lee (2012). Shape Optimization of an Autonomous Underwater Vehicle with a Ducted Propeller using Computational Fluid Dynamics Analysis. *International Journal of Naval Architecture and Ocean Engineering* 4, 44–56.
Journée, J.M.J. and W.W. Massie (2001). *Offshore Hydrodynamics*. Delft University of Technology.
Jurdjevic, V. and J. P. Quinn (1978). Controllability and Stability. *Journal of Differential Equations* 28, 381–389.
Källström, C. G. (1981). Control of Yaw and Roll by Rudder/Fin Stabilization System. In: *Proceedings of the 6th International Ship Control Systems Symposium (SCSS'81)*, Vol. 2. Paper F2 3-1. Ottawa, Canada.
Källström, C. G. (1987). Improved Operational Effectiveness of Naval Ships by Rudder Roll Stabilization. In: *NAVAL'87, Asian Pacific Naval Exhibition and Conference*. Singapore.
Källström, C. G. and W. L. Schultz (1990). An Integrated Rudder Control System for Roll Damping and Maintenance. In: *Proceedings of the 9th International Ship Control Systems Symposium (SCSS'90)*. Bethesda, MD, pp. 3.278–3.296.
Källström, C. G. and K. Theoren (1994). Rudder-Roll Stabilization an Improved Control Law. In: *Proceedings of the IEEE Conference on Control Applications*. New York, NY, pp. 1099–1105.
Källström, C. G., P. Wessel and S. Sjölander (1988). Roll Reduction by Rudder Control. In: *Proceedings of the SNAME Spring Meeting/STAR Symposium*. Pittsburg, PE, pp. 67–76.
Kalman, R. E. (1960). A New Approach to Linear Filtering and Prediction Problems. *ASME Transactions, Series D: Journal of Basic Engineering* 82, 35–42.

Kalman, R. E. and R. S. Bucy (1961). New Results in Linear Filtering and Prediction Theory. *ASME Transactions, Series D: Journal of Basic Engineering* 83, 95–108.

Kane, T. R., P. W. Likins and D. A. Levinson (1983). *Spacecraft Dynamics*. McGraw-Hill. New York, NY.

Kanellakopoulos, I., P. V. Kokotovic and A. S. Morse (1992). A Toolkit for Nonlinear Feedback Design. *Systems and Control Letters* 18, 83–92.

Katebi, M. R., D. K. K. Wong and M. J. Grimble (1987). LQG Autopilot and Rudder Roll Stabilization Control System Design. In: *Proceedings of the 8th International Ship Control Systems Symposium (SCSS'87)*. The Hague, Netherlands, pp. 3.69–3.84.

Katebi, M. R., M. J. Grimble and Y. Zhang (1997a). H_∞ Robust Control Design for Dynamic Ship Positioning. In: *IEE Proceedings on Control Theory and Applications*, Vol. 144 2, pp. 110–120.

Katebi, M. R., Y. Zhang and M. J. Grimble (1997b). Nonlinear Dynamic Ship Positioning. In: *Proceedings of the 13th IFAC World Cogress*, Vol. Q, pp. 303–308.

Katebi, M. R., N. A. Hickey and M. J. Grimble (2000). Evaluation of Fin Roll Stabilizer Controller Design. In: *Proccedings of the 5th IFAC Conference on Maneuvering and Control of Marine Craft*. Aalborg, Denmark, pp. 69–74.

Kayton, M. and W. L. Fried (1997). *Avionics Navigation Systems*. John Wiley & Sons, Inc. New York, NY.

Kelasidi, E., P. Liljebäck, K. Y. Pettersen and J. T. Gravdahl (2017). Integral Line-of-Sight Guidance for Path Following Control of Underwater Snake Robots: Theory and Experiments, *IEEE Transactions on Robotics* 33(3), 610–628.

Kempf, G. (1932). Measurements of the Propulsive and Structural Characteristic of Ships. In: *Transactions of SNAME* 40, 42–57.

Kempf, G. (1944). Manöveriernorm für Schiffe. In: *Hansa, Deutsche Schiffahrs-zeitschrift*, Heft 27/28, pp. 372–276 (in German).

Khalil, H. K. (2002). *Nonlinear Systems*. Macmillan. New York, NY.

Kijima, K., T. Katsuno, Y. Nakiri and Y. Furukawa (1990). On the Manoeuvring Performance of a Ship with the Parameter of Loading Condition. *J. Society of Naval Architects of Japan* 168, 141–148.

Kirchhoff, G. (1869). *Über die Bewegung eines Rotationskorpers in einer Flussigkeit. Crelle's Journal*, No. 71, pp. 237–273 (in German).

Kitamura, F., H. Sato, K. Shimada and T. Mikami (1997). Estimation of Wind Force Acting on Huge Floating Ocean Structures. In: *Proceedings of the Oceans '97. MTS/IEEE Conference*, Vol. 1. pp. 197–202.

Koditschek, D. E. (1987). Adaptive Techniques for Mechanical Systems. In: *Proceedings of the 5th Yale Workshop on Adaptive Systems*. New Haven, CT, pp. 259–265.

Kokotovic, P. V. (1991). The Joy of Feedback: Nonlinear and Adaptive. *IEEE Control Systems Magazine* 12, 7–17. Bode Price Lecture.

Kokotovic, P. V. and H. J. Sussmann (1989). A Positive Real Condition for Global Stabilization of Nonlinear Systems. *Systems and Control Letters* 13, 125–133.

Korotkin, A. I. *Added Masses of Ship Structures*. (2009). Springer Science.

Kostov, V. and E. Degtiariova-Kostova (1993). Suboptimal Paths in the Problem of a Planar Motion with Bounded Derivative of the Curvature. Technical Report. Research Report 2051, INRIA, France.

Koyama, T. (1967). *On the Optimum Automatic Steering System of Ships at Sea*. J.S.N.A., Vol. 122.

Kristiansen, E. and O. Egeland (2003). Frequency Dependent Added Mass in Models for Controller Design for Wave Motion Damping. In: *Proceedings of the IFAC Conference on Maneuvering and Control of Marine Craft (MCMC'03)*. Girona, Spain.

Kristiansen, E., A. Hjuslstad and O. Egeland (2005). State-Space Representation of Radiation Forces in Time-Domain Vessel Models. *Oceanic Engineering* 32, 2195–2216.

Krstic, M. and H. Deng (1998). *Stabilization of Nonlinear Uncertain Systems*. Springer-Verlag. Berlin.

Krstic, M., I. Kanellakopoulos and P. V. Kokotovic (1995). *Nonlinear and Adaptive Control Design*. John Wiley & Sons, Inc. New York, NY.

Kurokawa, H. (1997). Constrained Steering Law of Pyramid-Type Control Moment Gyros and Ground Tests. *AIAA Journal of Guidance, Control and Dynamics* 20(3), 445–449.

Kurokawa, H. (1997). Survey of Theory and Steering Laws of Single-Gimbal Control Moment Gyros. *AIAA Journal of Guidance, Control and Dynamics* 30(5), 1331–1340.

Kvasnica, M., P. Grieder and M. Baotic (2004). Multi-Parametric Toolbox (MPT). Viewed December 26, 2009, <www.control.ee.ethz.ch/~mpt>.

Lamb, H. (1932). *Hydrodynamics*. Cambridge University Press. London.

Landau, Y. D. (1979). *Adaptive Control – The Model Reference Approach*. Marcel Dekker Inc. New York, NY.
Lapierre, L. and D. Soetanto (2007). Nonlinear Path-Following Control of an AUV. *Ocean Engineering* 34, 1734–1744.
LaSalle, J. P. (1966). Stability Theory for Ordinary Differential Equations. *Journal of Differential Equations* 4, 57–65.
LaSalle, J. and S. Lefschetz (1961). *Stability by Lyapunov's Direct Method*. Academic Press. Baltimore, MD.
Lauvdal, T. and T. I. Fossen (1997). Nonlinear Rudder-Roll Damping of Non-Minimum Phase Ships Using Sliding Mode Control. In: *Proceedings of the European Control Conference*. Brussels, Belgium.
Lauvdal, T. and T. I. Fossen (1998). Rudder Roll Stabilization of Ships Subject to Input Rate Saturations Using a Gain Scheduled Control Law. In: *Proceedings of the IFAC Conference on Control Applications in Marine Systems (CAMS'98)*, pp. 121–127. Fukuoka, Japan.
Lawrence, D. A., E. W. Frew and W. J. Pisano (2008). Lyapunov Vector Fields for Autonomous Unmanned Aircraft Flight Control. *Journal of Guidance, Control and Dynamics* 31(5), 1220–1229.
Leick, A. (1995). *GPS: Satellite Surveying*. Chap. Appendix G, pp. 534–537. John Wiley & Sons, Ltd.
Leite, A .J. P., J. A. P. Aranha, C. Umeda and M. B. de Conti (1998). Current Forces in Tankers and Bifurcation of Equilibrium of Turret Systems: Hydrodynamic Model and Experiments. *Applied Ocean Research* 20, 145–156.
Lekkas, A. M. and T. I. Fossen (2012). A Time-Varying Lookahead Distance Guidance Law for Path Following. *Proc. of the IFAC MCMC'12*, 19-21 September, Arenzano, Italy.
Lekkas, A. M. and T. I. Fossen (2013). Line-of-Sight Guidance for Path Following of Marine Vehicles. Chapter 5, In *"Advanced in Marine Robotics"*, LAP LAMBERT Academic Publishing (O. Gal, Ed.), pp. 63–92. ISBN 978-3-659-41689-7.
Lekkas, A. M., A. R. Dahl, M. Breivik and T. I. Fossen (2013). Continuous-Curvature Path Generation using Fermat's Spiral. *Modeling, Identification and Control* 34(4), 183–198.
Lekkas, A. M. and T. I. Fossen (2014). Integral LOS Path Following for Curved Paths Based on a Monotone Cubic Hermite Spline Parametrization. *IEEE Transactions on Control Systems Technology* 22(6), 2287–2301.
Levant, A. (2003). Higher-Order Sliding Modes, Differentiation and Output Feedback Control *International Journal of Control* 76(9/10), 924–941.
Levant, A. (2005). Homogeneity Approach to High-Order Sliding Mode Design *Automatica* 41(5), 823–830.
Lewandowski, E. M. (2004). *The Dynamics of Marine Craft: Maneuvering and Seakeeping*. World Scientific. Washington D. C.
Lewis, F. M. (1967). The Motion of Ships in Waves. In: *Principles of Naval Architecture* (J. P. Comstock, Ed.). pp. 669–692. Society of Naval Architects and Marine Engineers, 601 Pavonia Avenue, Jersey City, NJ 07306.
Lewis, E. V. (Ed.) (1989). *Principles of Naval Architecture*. 2nd ed. Society of Naval Architects and Marine Engineers (SNAME).
Li, B. (2016). *Dynamics and Control of Autonomous Underwater Vehicles with Internal Actuators*. PhD thesis. Faculty of College of Engineering and Computer Science, Florida Atlantic University, Boca Raton, FL, USA.
Li, B. and T.-C. Su (2016). Nonlinear Heading Control of an Autonomous Underwater Vehicle with Internal Actuators. *Ocean Engineering* 125, 103–112.
Li, J.-W. and C. Shao (2008). Tracking Control of Autonomous Underwater Vehicles with Internal Moving Mass. *Acta Automatica Sinica* 34(10), 1319–1323.
Lin, Ching-Fang (1992). *Modern Navigation, Guidance and Control Processing*. Prentice-Hall Inc. Englewood Cliffs, New Jersey 07632.
Lindegaard, K.-P. (2003). *Acceleration Feedback in Dynamic Positioning Systems*. PhD thesis. Department of Engineering Cybernetics, Norwegian University of Science and Technology. Trondheim, Norway.
Lindegaard, K. P. and T. I . Fossen (2001a). A Model Based Wave Filter for Surface Vessels using Position, Velocity and Partial Acceleration Feedback. In: *Proceedings of the IEEE Conference on Decision and Control (CDC'2001)*. IEEE. Orlando, FL, pp. 946–951.
Lindegaard, K. P. and T. I. Fossen (2001b). On Global Model Based Observer Designs for Surface Vessels. In: *5th IFAC Conference on Control Applications in Marine Systems (CAMS'2001)*. Elsevier Science.
Lindegaard, K.-P. and T. I. Fossen (2003). Fuel Efficient Control Allocation for Surface Vessels with Active Rudder Usage: Experiments with a Model Ship. *IEEE Transactions on Control Systems Technology* 11, 850–862.
Lindfors, I. (1993). Thrust Allocation Method for the Dynamic Positioning System. In: *10th International Ship Control Systems Symposium (SCSS'93)*. Ottawa, Canada, pp. 3.93–3.106.

Liu, J. and R. Hekkenberg (2017). Sixty Years of Research on Ship Rudders: Effects of Design Choices on Rudder Performance. *Ships and Offshore Structures* 12(4), 495–512.

Ljung, L. (1999). *System Identification: Theory for the User*. Prentice Hall. Englewood Cliffs, New Jersey.

Lloyd, A. E. J. M. (1975). Roll Stabilization by Rudder. In: *Proceedings of the 4th International Ship Control Systems Symposium (SCSS'75)*. The Hague, Netherlands.

Lloyd, A. R. J. M. (1989). *Seakeeping; Ship Behavior in Rough Water*. Ellis Horwood Ltd.

Lo, H. (1991). Dynamic Ship Routing through Stochastic, Spatially Dependent Ocean Currents. PhD thesis. The Ohio State University, Columbus, OH.

Lo, H. K. and M. R. McCord (1995). Routing through Dynamic Ocean Currents: General Heuristics and Empirical Results in the Gulf Stream Region. *Transportation Research Part B: Methodological* 29(2), 109–124.

Lo, H. K. and M. R. McCord (1998). Adaptive Ship Routing through Stochastic Ocean Currents: General Formulations and Empirical Results. *Transportation Research Part A: Policy and Practice* 32(7), 547–561.

Loria, A., T. I. Fossen and A. Teel (1999). UGAS and ULES of Non-Autonomous Systems: Applications to Integral Action Control of Ships and Manipulators. In: *Proceedings of the 5th European Control Conference (ECC'99)*. Karlsruhe, Germany.

Loria, A., T. I. Fossen and E. Panteley (2000). A Separation Principle for Dynamic Positioning of Ships: Theoretical and Experimental Results. *IEEE Transactions on Control Systems Technology* 8(2), 332–343.

Loria, A. and E. Panteley (2004). Cascaded Nonlinear Time-Varying Systems: Analysis and Design. Ch. 2. In: *Advanced Topics in Control Systems Theory*, Springer-Verlag London (F. Lamnabhi-Lagarrigue, A. Loria and E. Panteley Eds.), pp. 23–64.

Lozano, R., B. Brogliato and I. D. Landau (1992). Passivity and Global Stabilization of Cascaded Nonlinear Systems. *IEEE Transactions on Automatic Control* 37, 1386–1388.

Lozano, R., B. Brogliato, O. Egeland and B. Maschke (2000). Dissipative Systems Analysis and Control. Theory and Applications. Springer Verlag.

Lyapunov, M. A. (1907). Probleme Général de la Stabilité de Mouvement. *Ann. Fac. Sci. Toulouse* 9, 203–474. (Translation of a paper published in Comm. Soc. Math. Kharkow 1893, reprinted in Annual Math Studies, Vol. 17, Princeton, 1949).

Lyshevski, S. E. (2001). Autopilot Design for Highly Maneuverable Multipurpose Underwater Vehicles. In: *Proceedings of the American Control Conference*. Arlington, VA, pp. 131–136.

MacKunis, W., K. Dupree, N. Fitz-Coy and W. E. Dixon (2008). *Adaptive Satellite Attitude Control in the Presence of Inertia and CMG Gimbal Friction Uncertainties. Journal of the Astronautical Sciences* 56(1), 121–134.

Madureira, L., A. Sousa, J. Braga, P. Calado, P. Dias, R. Martins, J. Pinto and J. Sousa (2013). The Light Autonomous Underwater Vehicle: Evolutions and Networking. In: *Proc. of the 2013 MTS/IEEE OCEANS*. Bergen, Norway, 10–14 June 2013.

Matsumoto, K. and K. Suemitsu (1980). The Prediction of Manoeuvring Performances by Captive Model Tests. *Journal of the Kansai Society of Naval Architects*, No. 176.

McCord, M. and S. Smith (1995). Beneficial Voyage Characteristics for Routing through Dynamic Currents. *Transportation Research Record* 1511, 19–25.

McGee, T. G., S. Spry and J. K. Hedrick (2006). Optimal Path Planning in a Constant Wind with a Bounded Turning Rate. In: *Proceedings AIAA Conference on Guidance, Navigation and Control*. Ketstone, CO. pp. 4261–4266.

McGookin, E. W., D. J. Murray-Smith, Y. Lin and T. I. Fossen (2000a). Experimental Results from Supply Ship Autopilot Optimisation Using Genetic Algorithms. *Transactions of the Institute of Measurement and Control* 22(2), 141–178.

McGookin, E. W., D. J. Murray-Smith, Y. Lin and T. I. Fossen (2000b). Ship Steering Control System Optimisation Using Genetic Algorithms. *Journal of Control Engineering Practise* 8, 429–443.

Mahony, R., T. Hamel and J.M. Pflimlin (2008). Nonlinear Complementary Filters on the Special Orthogonal Group. *IEEE Transactions on Automatic Control* 53(5), 1203–1218.

Markley, F. L. and J. L. Crassidis (2014). *Fundamentals of Spacecraft Attitude Determination and Control*, Volume 33 of *Space Technology Library*. Springer-Verlag New York, 1 edition.

Matrosov, V. M. (1962). On the Stability of Motion. *Prikl. Mat. Meh.* 26, 885–895.

Maybeck, P. S. (1979). *Stochastic Models, Estimation and Control*. Academic Press. New York, NY.

Meirovitch, L. (1990). *Dynamics and Control of Structures*. Wiley Interscience. New York, NY.

Meirovitch, L. and M. K. Kwak (1989). State Equations for a Spacecraft with Flexible Appendages in Terms of Quasi-Coordinates. *Applied Mechanics Reviews* 42(11), 161–170.

Micaelli, A. and C. Samson (1993). Trajectory Tracking for Unicycle-Type and Two-Steering-Wheels Mobile Robots. Internal Report 2097. INRIA. Sophia-Antipolis, France.

Minorsky, N. (1922). Directional Stability of Automatic Steered Bodies. *Journal of the American Society of Naval Engineers* 34(2), 280–309.

Morgan, J. M. (Ed.) (1978). *Dynamic Positioning of Offshore Vessels*. Petroleum, Tulsa, OK.

Munk, T. and M. Blanke (1987). Simple Command of a Complex Ship. In: *Proceedings of the 8th Ship Control Systems Symposium*. The Hague, Netherlands.

Myring, D. F. (1976). A Theoretical Study of Body Drag in Subcritical Axisymmetric Flow. *Aeronautical Quarterly* 27(3), 186–94.

Nakamura, Y. and H. Hanafusa (1986). Inverse Kinematic Solutions with Singularity Robustness for Robot Manipulator Control. *Journal of Dynamic Systems, Measurements and Control* 108(3), 163–171.

Narendra, K. S. and A. M. Annaswamy (1989). *Stable Adaptive Systems*. Prentice Hall Inc. Boston, MA.

NASA (2013). Mission Planning and Analysis Division. Euler Angles, Quaternions, and Transformation Matrices. NASA. Retrieved 12 January 2013.

Nelson, R. C. (1998). *Flight Stability and Automatic Control*. McGraw-Hill Int.

Nelson, D. R, D. B. Barber, T. W. McLain and R. W. Beard (2007). Vector Field Path Following for Miniature Air Vehicles *IEEE Transactions on Robotics* 33(3), 519–529.

Neumann, G. (1952). *On Wind-Generated Ocean Waves with Special Reference to the Problem of Wave Forecasting*. New York University, College of Engineering Research Division, Department of Meteorology and Oceanography. Prepared for the Naval Res.

Newman, J. N. (1977). *Marine Hydrodynamics*. MIT Press. Cambridge, MA.

Nocedal, J. and S. J. Wright (1999). *Numerical Optimization*. Springer-Verlag, New York, NY.

Nomoto, K., T. Taguchi, K. Honda and S. Hirano (1957). On the Steering Qualities of Ships. Technical Report. International Shipbuilding Progress, Vol. 4.

Norrbin, N. H. (1963). On the Design and Analyses of the Zig-Zag Test on Base of Quasi Linear Frequency Response. Technical Report B 104-3. The Swedish State Shipbuilding Experimental Tank (SSPA). Gothenburg, Sweden.

Norrbin, N. H. (1965). Zig-Zag Provets Teknik och Analys. Technical Report 12. The Swedish State Shipbuilding Experimental Tank (SSPA). Gothenburg, Sweden (in Swedish).

Norrbin, N. H. (1970). Theory and Observation on the Use of a Mathematical Model for Ship Maneuvering in Deep and Confined Waters. In: *Proceedings of the 8th Symposium on Naval Hydrodynamics*. Pasadena, CA.

Norrbin, N. H. (1972). On the Added Resistance due to Steering on a Straight Course. In: *Proceedings of the 13th ITTC*. Berlin, Hamburg, Germany.

Norsok Standard (1999). Actions and Actions Effects Rev. 1.0. N-003.

OCIMF (1977). Prediction of Wind and Current Loads on VLCCs. Oil Companies International Marine Forum, London, pp. 1–77.

Oda, H. *et al.* (1992). Rudder Roll Stabilization Control System through Multivariable Auto Regressive Model. In: *Proceedings of IFAC Workshop on Control Applications in Marine Systems (CAMS'92)*. Genova, Italy, pp. 113–127.

Oda, H., K. Ohtsu and T. Hotta (1996). Statistical Analysis and Design of a Rudder-Roll Stabilization System. *Control Engineering Practice* 4(3), 351–358.

Oda, H., K. Igarashi and K. Ohtsu (1997). Simulation Study and Full-Scale Experiment of Rudder-Roll Stabilization Systems. In: *Proceedings of the 11th Ship Control Systems Symposium*. Southampton, UK, pp. 299–313.

Ogilvie, T. F. (1964). Recent Progress Towards the Understanding and Prediction of Ship Motions. In: *Proceedings of the 5th Symposium on Naval Hydrodynamics*, pp. 3–79.

O'Hanlon, J. F. and M. E. McCauley (1974). Motion Sickness Incidence as a Function of Vertical Sinusoidal Motion. *Aerospace Medicine* 45(4), 366–369.

Ortega, R. (1991). Passivity Properties for Stabilization of Cascaded Nonlinear Systems. *Automatica* 27, 423–424.

Ortega, R., A. Loria, P. J. Nicklasson and H. Sira-Ramirez (1998). *Passivity-Based Control of Euler–Lagrange Systems: Mechanical, Electrical and Electromechanical Applications*. Springer Verlag.

Padadakis, N. A. and A. N. Perakis (1990). Deterministic Minimal Time Vessel Routing. *Operations Research* 38(3), 416–438.
Panteley, E. and A. Loria (1998). On Global Uniform Asymtotic Stability of Nonlinear Time-Varying Noautonomous Systems in Cascade. *System Control Letters* 33(2), 131–138.
Papoulias, F. A. (1991). Bifurcation analysis of line of sight vehicle guidance using sliding modes. *International Journal of Bifurcation and Chaos* 1(4), 849–865.
Parkinson, B. W. and Spilker, J. J. (Eds) (1995). *Global Positioning System: Theory and Applications*. American Institute of Aeronautics and Astronautics Inc. Washington D. C.
Parks, P. C. (1966). Lyapunov Redesign of Model Reference Adaptive Control Systems. *IEEE Transactions on Automatic Control* 11, 362–367.
Perez, T. (2005). *Ship Motion Control: Course Keeping and Roll Reduction using Rudder and Fins*. Advances in Industrial Control Series. Springer-Verlag, London, UK.
Perez, T. and T. I. Fossen (2007). Kinematic Models for Seakeeping and Manoeuvring of Marine Vessels. *Modeling, Identification and Control* 28(1), 19–30.
Perez, T. and T. I. Fossen (2008). Time- vs. Frequency-domain Identification of Parametric Radiation Force Models for Marine Structures at Zero Speed. *Modeling, Identification and Control* 29(1), 1–19.
Perez, T. and T. I. Fossen (2009). A Matlab Toolbox for Parametric Identification of Radiation-Force Models of Ships and Offshore Structures. *Modeling, Identification and Control* 30(1), 1–15.
Perez, T. and T. I. Fossen (2011). Practical Aspects of Frequency-Domain Identification of Dynamic Models of Marine Structures from Hydrodynamic Data. *IEEE Journal of Oceanic Engineering*.
Perez, T. and P. D. Steinmann (2007). Modelling and Performance of an Active Heave Compensator for Offshore Operations. In: *Proceedings of the IFAC Conference on Control Applications in Marine Systems CAMS'09*. Bol, Croatia, pp. 30–36.
Perez, T. and P. D. Steinmann (2009). Analysis of Ship Roll Gyrostabiliser Control. In: *Proceedings of the 8th IFAC International Conference on Manoeuvring and Control of Marine Craft MCMC'09*. Guaruja (SP), Brazil. pp. 30–36.
Pettersen, K. Y. and E. Lefeber (2001). Waypoint Tracking Control of Ships. In: *Proc. of the IEEE Conf. on Decision and Control*, Orlando, Florida, pp. 940–945.
Phillips, O. M. (1958). The Equilibrium Range in the Spectrum of Wind Generated Waves. *Journal of Fluid Mechanics* 4(4), 426–434.
Pierson, W. J. and L. Moskowitz (1963). A Proposed Spectral Form for Fully Developed Wind Seas Based on the Similarity Theory of S. A. Kitaigorodsku. U.S. Naval Oceanograhic Office Contract 62306-1042.
Pinkster, J. (1971). Dynamic Positioning of Vessels at Sea. International Centre for Mechanical Structures, Udine, Italy.
Pinkster, J. A. and U. Nienhuis (1986). Dynamic Positioning of Large Tankers at Sea. In: *Proceedings of the Offshore Technology Conference (OTC'86)*. Houston, TX.
Price, W. G. and R. E. D. Bishop (1974). *Probalistic Theory of Ship Dynamics*. Chapman and Hall. London.
Reid, R. E., A. K. Tugcu and B. C. Mears (1984). The Use of Wave Filter Design in Kalman Filter State Estimation of the Automatic Steering Problem of a Tanker in a Seaway. *IEEE Transactions on Automatic Control* 29(7), 577–584.
REMUS (2019). Woods Hole Oceanographic Institution <http://www.whoi.edu/>. Retrieved 12 June 2019.
Rios-Neto, A. and J. J Da Cruz (1985). A Stochastic Rudder Control Law for Ship Path-Following Autopilots, *Automatica* 21(4), 371–384.
Roberts, G. N. (1992). Ship Roll Damping Using Rudder and Stabilizing Fins. In: *Proceedings of the IFAC Workshop on Control Applications in Marine Systems (CAMS'92)*. Genova, Italy, pp. 129–138.
Roberts, G. N. (1993). A Method to Determine the Applicability of Rudder Roll Stabilization for Ships. In: *Proceedings of the 12th IFAC World Congress*, pp. 1009–1012.
Roberts, G. N. and S. W. Braham (1990). Warship Roll Stabilization Using Integrated Rudder and Fins. In: *9th International Ship Control Systems Symposium (SCSS'90)*. Bethesda, MD, pp. 1.234–1.248.
Robertson, A. and R. Johansson (1998). Comments on "Nonlinear Output Feedback Control of Dynamically Positioned Ships Using Vectorial Observer Backstepping". *IEEE Transactions on Control Systems Technology* 6(3), 439–441.
Ross, A. (2008). Nonlinear Manoeuvring Models for Ships: a Lagrangian Approach. PhD thesis. Department of Engineering Cybernetics, Norwegian University of Science and Technology.

Ross, A. and T. I. Fossen (2005). Identification of Frequency-Dependent Viscous Damping. In: *Proceedings of OCEANS'05 MTS/IEEE*. Washington D.C.

Ross, A., T. Perez and T. I. Fossen (2006). Clarification of the Low-Frequency Modelling Concept for Marine Craft. In: *Proceedings of MCMC'06*. Lisbon.

Routh, E. J. (1877). *A Treatise on The Stability on Motion*. Macmillan. London, UK.

Saberi, A., P. V. Kokotovic and H. J. Sussmann (1990). Global Stabilization of Partially Linear Composite Systems. *SIAM J. Control Opt.* 28, 1491–1503.

Sælid, S. and N. A. Jenssen (1983). Adaptive Ship Autopilot with Wave Filter. *Modeling, Identification and Control* 4(1), 33–46.

Sælid,S., N. A. Jenssen and J. G. Balchen (1983). Design and Analysis of a Dynamic Positioning System Based on Kalman Filtering and Optimal Control. *IEEE Transactions on Automatic Control* 28(3), 331–339.

Sagatun, S. I. and T.I. Fossen (1991). Lagrangian Formulation of Underwater Vehicles' Dynamics. In: *Proceedings of the IEEE International Conference on Systems, Man and Cybernetics*. Charlottesville, VA, pp. 1029–1034.

Sagatun, S. I., T. I. Fossen and K.-P. Lindegaard (2001). Inertance Control of Underwater Installations. In: *Proceedings of the 5th IFAC Conference on Control Applications in Marine Systems CAMS'2001*. Glasgow, UK.

Salcudean, S. (1991). A Globally Convergent Angular Velocity Observer for Rigid Body Motion. *IEEE Transaction on Automatic Control* 36(12), 1493–1497.

Shuster, M. D. and S. D. Oh (1981). Three-Axis Attitude Determination from Vector Observations. *Journal of Guidance, Control and Dynamics* 4(1), 70–77.

Samson, C. (1992). Path Following and Time-Varying Feedback Stabilization of a Wheeled Mobile Robot. In: *Proceedings of ICARV*, pp. RO–13.1.

Sargent, J. S. and P. N. Cowgill (1976). Design Considerations for Dynamically Positioned Utility Vessels. In: *Proceedings of the 8th Offshore Technology Conference*. Dallas, TX.

Sarpkaya, T. (1981). *Mechanics of Wave Forces on Offshore Structures*. Van Nostrand Reinhold Company. New York, NY.

Scheuer, A. and C. Laugier (1998). Planning Sub-Optimal and Continuous-Curvature Paths for Car-Like Robots. In: *Proceedings IEEE-RSJ International Conference on Intelligent Robots and Systems*.

Schlick, O. (1904). Gyroscopic Effects of Flying Weels on Board Ships. Transactions of The Institution of Naval Architects INA, 1904.

Sciavicco, L. and B. Siciliano (1996). *Modeling and Control of Robot Manipulators*. McGraw-Hill Companies Inc.

Sellars, F. H. and J. P. Martin (1992). Selection and Evaluation of Ship Roll Stabilization Systems. *Marine Technology* 29(2), 84–101.

Sepulchre, R., M. Jankovic and P. Kokotovic (1997). *Constructive Nonlinear Control*. Springer-Verlag. Berlin.

Serret, J. A. (1851). Sur Quelques Formules Relatives à la Théorie des Courbes à Double Courbure. *Journal de Mathématiques Pures et Appliquées*, 16., (in French).

Sharif, M. T., G. N. Roberts and R. Sutton (1995). Sea-Trial Experimental Results of Fin/Rudder Roll Stabilization. *Control Engineering Practice* 3(5), 703–708.

Sharif, M. T., G. N. Roberts and R. Sutton (1996). Final Experimental Results of Full Scale Fin/Rudder Roll Stabilization Sea Trials. *Control Engineering Practice* 4(3), 377–384.

Shepperd, S. W. (1978). Quaternion from Rotation Matrix. *Journal of Guidance and Control* 1(3), 223–224.

Shtessel, Y., I. Shkolnikov and A. Levant (2007). Smooth Second Order Sliding Modes: Missile Guidance Application *Automatica* 43(8), 1470–1476.

Shtessel, Y., J. A. Moreno, F. Plestan, L. M. Fridman and A. S. Poznyak (2010). Super-twisting Adaptive Sliding Mode Control: A Lyapunov Design. In: *Proceedings of the 49th IEEE Conference on Decision and Control*, December 15–17, Atlanta, GA, USA.

Shtessel, Y., C. Edwards, L. Fridman and A. Levant (2014). *Sliding Mode Control and Observation*. Birkhäuser, New York, NY.

Shneydor, N. A. (1998). *Missile Guidance and Pursuit: Kinematics, Dynamics and Control*. Horwood Publishing Ltd.

Siouris, G. M. (2004). *Missile Guidance and Control Systems*. Springer-Verlag New York, NY.

Skejic, R., M. Breivik, T. I. Fossen and O. M. Faltinsen (2009). Modeling and Control of Underway Replenishment Operations in Calm Water. In: *Proceedings of the IFAC MCMC'09*. Sao Paulo, Brazil.

Skjetne, R. and T. I. Fossen (2004). On Integral Control in Backstepping: Analysis of Different Techniques. In: *Proceedings of the 2004 American Control Conference*. Boston, MA, pp. 1899–1904.

Skjetne, R., T. I. Fossen and P. V. Kokotovic (2002). Output Maneuvering for a Class of Nonlinear Systems. In: *Proceedings of the IFAC World Congress*. Barcelona, Spain.

Skjetne, R., T. I. Fossen and P. V. Kokotovic (2004). Robust Output Maneuvering for a Class of Nonlinear Systems. *Automatica*.

Slotine, J. J. E. and M. D. Di Benedetto (1990). Hamiltonian Adaptive Control of Spacecraft. *IEEE Transactions on Automatic Control* 35(7), 848–852.

Slotine, J. J. E. and W. Li (1987). Adaptive Manipulator Control. A Case Study. In: *Proceedings of the 1987 IEEE Conference on Robotics and Automation*. Raleigh, North Carolina, pp. 1392–1400.

Slotine, J. J. E. and W. Li (1991). *Applied Nonlinear Control*. Prentice-Hall International Englewood Cliffs, New Jersey 07632.

Smith, J. E. (1977). *Mathematical Modeling and Digital Simulation for Engineers and Scientists*. John Wiley & Sons, Inc. New York, NY.

SNAME (1950). The Society of Naval Architects and Marine Engineers. Nomenclature for Treating the Motion of a Submerged Body Through a Fluid. In: *Technical and Research Bulletin No. 1–5*.

SNAME (1989). The Society of Naval Architects and Marine Engineers. Guide For Sea Trials. In: *Technical and Research Bulletin No. 3–47*.

Solà, J. (2016). Quaternion Kinematics for the Error-State Kalman Filter. Technical Report, IRI-TR-16-02, Institut de Robòtica i Informàtica Industrial, CSIC-UPC.

Son, K. H. and K. Nomoto (1981). On the Coupled Motion of Steering and Rolling of a High Speed Container Ship. J.S.N.A., Japan, Vol. 150, pp. 232–244 (in Japanese).

Son, K. H. and K. Nomoto (1982). On the Coupled Motion of Steering and Rolling of a High Speed Container Ship. *Naval Architect of Ocean Engineering* 20, 73–83. From J.S.N.A., Japan, Vol. 150, 1981.

Sonntag, E. D. and H. J. Sussmann (1988). Further Comments on the Stabilizability of the Angular Velocity of a Rigid Body. *Systems and Control Letters* 12, 437–442.

Sontag, E. D. (1983). A Lyapunov-like Characterization of Asymptotic Controllability. *SIAM Journal of Control and Optimization* 21, 462–471.

Sørdalen, O. J. (1997a). Full Scale Sea Trials with Optimal Thrust Allocation. In: *Proceedings of the 4th IFAC Conference on Manoeuvring and Control of Marine Craft (MCMC'97)*. Brijuni, Croatia, pp. 150–155.

Sørdalen, O. J. (1997b). Optimal Thrust Allocation for Marine Vessels. *Control Engineering Practice* 5(9), 1223–1231.

Sørensen, A. J. and J. P. Strand (1998). Positioning of Semi-submersible with Roll and Pitch Damping. In: *Proceedings of the IFAC Conference on Control Applications in Marine Systems (CAMS'98)*. Fukuoka, Japan, pp. 67–73.

Sørensen, A. J. and J. P. Strand (2000). Positioning of Small-Waterplane-Area Marine Constructions with Roll and Pitch Damping. *Journal of Control Engineering in Practice* 8(2), 205–213.

Sørensen, A. J., S. I. Sagatun and T. I. Fossen (1995). The Design of a Dynamic Positioning System Using Model Based Control. In: *Proceedings of the IFAC Workshop on Control Applications in Marine Systems (CAMS'95)*. Trondheim, Norway, pp. 16–26.

Sørensen, A. J., S. I. Sagatun and T. I. Fossen (1996). Design of a Dynamic Positioning System Using Model Based Control. *Journal of Control Engineering Practice* 4(3), 359–368.

Sørensen, A. J., T. I. Fossen and J. P. Strand (2000). Design of Controllers for Positioning of Marine Vessels. In: *The Ocean Engineering Handbook* (F. El-Hawary, Ed.). Chap. 3, pp. 207–218. CRC Press, USA.

Sørensen, A. J., B. Leira, J. P. Strand and C. M. Larsen (2001). Optimal Setpoint Chasing in Dynamic Positioning of Deep-Water Drilling and Intervention Vessels. *International Journal of Robust and Nonlinear Control* 11, 1187–1205.

Spjøtvold, J. (2008). Parametric Programming in Control Theory. PhD thesis. Department of Engineering Cybernetics, Norwegian University of Science and Technology, Trondheim.

Stevens, B. L. and F. Lewis (1992). *Aircraft Control and Simulation*. John Wiley & Sons, Ltd.

Stoustrup, J., H. H. Niemann and M. Blanke (1995). A Multi-Objective H_∞ Solution to the Rudder Roll Damping Problem. In: *Proceedings of the IFAC Workshop on Control Applications in Marine Systems (CAMS'95)*, pp. 238–246.

Strand, J. P. (1999). Nonlinear Position Control Systems Design for Marine Vessels. PhD thesis. Department of Engineering Cybernetics, Norwegian University of Science and Technology. Trondheim, Norway.

Strand, J. P. and T. I. Fossen (1999). Nonlinear Passive Observer for Ships with Adaptive Wave Filtering. In: *New Directions in Nonlinear Observer Design* (H. Nijmeijer and T. I. Fossen, Eds.). Chap. I-7, pp. 113–134. Springer-Verlag. London.

Strand, J. P. and A. J. Sørensen (2000). Marine Positioning Systems. In: *Ocean Engineering Handbook* (F. El-Hawary, Ed.). Chap. 3, pp. 163–176. CRC Press, USA.

Strand, J. P., A. J. Sørensen and T. I. Fossen (1998). Design of Automatic Thruster Assisted Position Mooring Systems for Ships. *Modeling, Identification and Control* 19(2), 61–75.

Sun, Z., L. Zhang, G. Jin and X. Yang (2010). Analysis of Inertia Dyadic Uncertainty for Small Agile Satellite with Control Moment Gyros. In: *Proceedings of the 2010 IEEE International Conference on Mechatronics and Automation*, August 4-7, Xi'an, China.

Söding, H. (1999). Limits of Potential Theory in Rudder Flow Predictions In: *Twenty-Second Symposium on Naval Hydrodynamics* Chapter: 10 Weinblum Lecture, pp. 622–637. The National Academies Press, USA.

Techy, L. and C. A. Woolsey (2009). Minimum-Time Path Planning for Unmanned Aerial Vehicles in Steady Uniform Winds. *AIAA Journal of Guidance, Control, and Dynamics* 32(6), 1736–1746.

Techy, L. and C. A. Woolsey (2010). Planar Path Planning for Flight Vehicles in Wind with Turn Rate and Acceleration Bounds. In: *Proceedings AIAA Conference on Guidance, Navigation and Control*. Anchorage, AK.

Thienel, J. and R. M. Sanner (2003). A Coupled Nonlinear Spacecraft Attitude Controller and Observer with an Unknown Constant Gyro Bias and Gyro Noise. *IEEE Transactions on Automatic Control* 48(11), 2011–2015.

Tinker, S. J. (1982). Identification of Submarine Dynamics from Free-Model Test. In: *Proceedings of the DRG Seminar*. Netherlands.

Titterton, D. H. and J. L. Weston (1997). *Strapdown Inertial Navigation Technology*. IEE. London, UK.

Thornton, B., T. Ura, Y. Nose and S. Turnock (2005). Internal Actuation of Underwater Robots Using Control Moment Gyros. In: *Proceedings Oceans 2005-Europe*, pp. 591–598.

Thornton, B., T. Ura, Y. Nose and S. Turnock (2007). Zero-G Class Underwater Robots: Unrestricted Attitude Control Using Control Moment Gyros. *IEEE Journal of Oceanic Engineering* 32(3), 565–583.

Thornton, B., T. Ura and Y. Nose (2008). Combined Energy Storage and Three-Axis Attitude Control of a Gyroscopically Actuated AUV. In: *Proceedings Oceans*. Quebec City, QC, Canada.

Tøndel, P., T. A. Johansen and A. Bemporad (2003a). An algorithm for multi-parametric quadratic programming and explicit MPC solutions. *Automatica* 39, 489–497.

Tøndel, P., T. A. Johansen and A. Bemporad (2003b). Evaluation of Piecewise Affine Control via Binary Search Tree. *Automatica* 39, 743–749.

Torsethaugen, K. (1996). Model for a Doubly Peaked Wave Spectrum. Technical Report STF22 A96204. SINTEF, Trondheim, Norway. Prepared for Norsk Hydro.

Triantafyllou, M. S. and A. M. Amzallag (1984). A New Generation of Underwater Unmanned Tethered Vehicles Carrying Heavy Equipment at Large Depths. Technical Report MITSG 85-30TN. MIT Sea Grant. Boston, MA.

Triantafyllou, M. S. and F. S. Hover (2002). Maneuvering and Control of Marine Vehicles. Internet. <http://ocw.mit.edu/13/13.49/f00/lecture-notes/>[Accessed November 15, 2002].

Triantafyllou, M. S., M. Bodson and M. Athans (1983). Real Time Estimation of Ship Motions Using Kalman Filtering Techniques. *IEEE Journal of Ocean Engineering* 8(1), 9–20.

Tsinias, J. (1989). Sufficient Lyapunov-Like Conditions for Stabilization. *Mathematics of Control, Signals and Systems* 2, 343–357.

Tsourdos, A., B. White and M. Shanmugavel (2010). Cooperative Path Planning of Unmanned Aerial Vehicles. John Wiley & Sons, Ltd.

Tzeng, C. Y. (1998a). Analysis of the Pivot Point for a Turning Ship. *Journal of Marine Science and Technology* 6(1), 39–44.

Tzeng, C. Y. (1998b). Optimal Control of a Ship for a Course-Changing Maneuver. *Journal of Optimization Theory and Applications* JOTA-97(2), 281–297.

Utkin, V. I. (1977). Variable Structure Systems with Sliding Modes. *IEEE Transactions on Automatic Control* 22(2), 212–222.

Utkin, V. I. (1992). *Sliding Modes in Control and Optimization*. Springer-Verlag, Berlin.

Van Amerongen, J. (1982). Adaptive Steering of Ships – A Model Reference Approach to Improved Maneuvering and Economical Course Keeping. PhD thesis. Delft University of Technology, Netherlands.

Van Amerongen, J. (1984). Adaptive Steering of Ships – A Model Reference Approach. *Automatica* 20(1), 3–14.

Van Amerongen, J. and H. R. Van Nauta Lemke (1978). Optimum Steering of Ships with an Adaptive Autopilot. In: *Proceedings of the 5th Ship Control Systems Symposium*. Annapolis, MD.
Van Amerongen, J. and H. R. Van Nauta Lemke (1980). Criteria for Optimum Steering of Ships. In: *Proceedings of Symposium on Ship Steering Automatic Control*. Genova, Italy.
Van Amerongen, J. and H. R. Van Nauta Lempke (1987). Adaptive Control Aspects of a Rudder Roll Stabilization System. In: *Proceedings of the 10th IFAC World Congress*. Munich, Germany, pp. 215–219.
Van Amerongen, J., P. G. M. Van der Klugt and J. B. M. Pieffers (1987). Rudder Roll Stabilization – Controller Design and Experimental Results. In: *Proceedings of the 8th International Ship Control Systems Symposium (SCSS'87)*. The Hague, Netherlands, pp. 1.120–1.142.
Van Berlekom, W. B., P. Trägårdh and A. Dellhag (1974). Large Tankers – Wind Coefficients and Speed Loss Due to Wind and Sea. In: *Meeting at the Royal Institution of Naval Architects, April 25, 1974*. London, pp. 41–58.
Van der Klugt, P. G. M. (1987). Rudder Roll Stabilization. PhD thesis. Delft University of Technology, Delft, Netherlands.
Vik, B. (2000). Nonlinear Design and Analysis of Integrated GPS and Inertial Navigation Systems. PhD thesis. Department of Engineering Cybernetics, Norwegian University of Science and Technology. Trondheim, Norway.
Vik, B. and T. I. Fossen (2001). Nonlinear Observer Design for Integration of GPS and Inertial Navigation Systems. In: *Proceedings of the Conference on Decision and Control (CDC'2001)*. Orlando, FL.
Vik, B., A. Shiriaev and T. I. Fossen (1999). Nonlinear Observer Design for Integration of DGPS and INS. In: *New Directions in Nonlinear Observer Design* (H. Nijmeijer and T. I. Fossen, Eds). Chap. I-8, pp. 135–160. Springer-Verlag. London.
Wagner, Von B. (1967). Windkräfte an Überwasserschiffen. In: *Schiff und Hafen, Heft 12, 19. Jahrgang*, pp. 894–900. (in German).
WAMIT (2010). WAMIT User Manual. <www.wamit.com>.
Webster, W. C. and J. Sousa (1999). Optimum Allocation for Multiple Thrusters. In: *Proceedings of the International Society of Offshore and Polar Engineers Conference (ISOPE'99)*. Brest, France.
Wendel, K. (1956). Hydrodynamic Masses and Hydrodynamic Moments of Inertia. Technical Report. TMB Translation 260.
Wie, B. (1998). Space Vehicle Dynamics and Control, *AIAA Education Series, American Institute of Aeronautics and Astronautics*
Wilson, P. A. (2018). *Basic Naval Architecture. Ship Stability*. Springer Nature Switzerland AG.
Whitcomb, L. L. and D. Yoerger (1999). Development, Comparison, and Preliminary Experimental Validation of Nonlinear Dynamic Thruster Models. *IEEE Journal of Oceanic Engineering* 24(4), 481–494.
Woolsey, C. A. and N. E. Leonard (2002a).Stabilizing Underwater Vehicle Motion using Internal Rotors. *Automatica* 38, 2053–2062.
Woolsey, C. A. and N. E. Leonard (2002b). Moving Mass Control for Underwater Vehicles In: *Proceedings of the American Control Conference*, Anchorage, AK, USA, May 8-10.
World Geodetic System (1984). Its Definition and Relationships with Local Geoedtic Systems. DMA TR 8350.2, 2nd ed., Defense Mapping Agency, Fairfax, VA.
World Magnetic Model (2020). Internet. <https://www.ngdc.noaa.gov/geomag/WMM/> [Accessed February 20, 2020].
Yakubovich, V. A. (1973). A Frequency Theorem in Control Theory. *Siberian Mathematical Journal* 14(2), 384–420.
Yang, C. and M. Blanke (1997). A Robust Roll Damping Controller. In: *Proceedings of the IFAC Conference on Manoeuvring and Control (MCMC'97)*. Croatia.
Yang, C. and M. Blanke (1998). Rudder-Roll Damping Controller Design Using Mu Synthesis. In: *Proceedings IFAC Conference on Control Applications in Marine Systems (CAMS'98)*, pp. 127–132. Fukuoaka, Japan.
Yanushevsky, R. (2008). *Modern Missile Guidance*. CRC Press.
Yasukawa, H. and Y. Yoshimura (2015). Introduction of MMG Standard Method for Ship Maneuvering Predictions. *Journal of Marine Science Technology*, 20:37–52.
Yoerger, D. R. and J. J. E. Slotine (1985). Robust Trajectory Control of Underwater Vehicles. *IEEE Journal of Oceanic Engineering* 10(4), 462–470.

Yoerger, D. R., J. G. Cooke and J.- J. E. Slotine (1991). The Influence of Thruster Dynamics on Underwater Vehicle Behavior and their Incorporation into Control System Design, *IEEE Journal of Oceanic Engineering* 15(3), 167–178.

Yoshizawa, T. (1968). *Stability Theory by Lyapunov's Second Method.* The Mathematical Society of Japan.

Yu, Z. and J. Falnes (1995). Spate-Space Modelling of a Vertical Cylinder in Heave. *Applied Ocean Research* 17, 265–275.

Yu, Z. and J. Falnes (1998). Spate-Space Modelling of Dynamic Systems in Ocean Engineering. *Journal of Hydrodynamics, China Ocean Press*, pp. 1–17.

Zhou, W. W. (1990). A New Approach for Adaptive Rudder Roll Stabilization Control. In: *9th International Ship Control Systems Symposium (SCSS'90)*. Bethesda, MD, pp. 1.115–1.125.

Zhu, J. (1993). Exact Conversion of Earth-Centered Earth-Fixed Coordinates to Geodetic Coordinates. *AIAA Journal of Guidance, Control and Dynamics* 162(2), 389–391.

Xu, R., G. Tanga, L. Hana and D. Xiea (2019). Trajectory Tracking Control for a CMG-Based Underwater Vehicle with Input Saturation in 3D Space. *Ocean Engineering* 173, 587–598.

Index

Printed and bound by CPI Group (UK) Ltd, Croydon, CR0 4YY

27/10/2024

14580374-0003